W9-APO-401

Philip Daber...
1/18/94

DESIGN OF MACHINERY

AN INTRODUCTION TO THE SYNTHESIS AND ANALYSIS OF MECHANISMS AND MACHINES

McGraw-Hill Series in Mechanical Engineering

Jack P. Holman, *Southern Methodist University*
John R. Lloyd, *Michigan State University*
Consulting Editors

Anderson: *Modern Compressible Flow: With Historical Perspective*
Arora: *Introduction to Optimum Design*
Bray and Stanley: *Nondestructive Evaluation: A Tool for Design, Manufacturing and Service*
Culp: *Principles of Energy Conversion*
Dally: *Packaging of Electronic Systems: A Mechanical Engineering Approach*
Dieter: *Engineering Design: A Materials and Processing Approach*
Eckert and Drake: *Analysis of Heat and Mass Transfer*
Edwards and McKee: *Fundamentals of Mechanical Component Design*
Heywood: *Internal Combustion Engine Fundamentals*
Hinze: *Turbulence*
Hutton: *Applied Mechanical Vibrations*
Juvinall: *Engineering Considerations of Stress, Strain, and Strength*
Kays and Crawford: *Convective Heat and Mass Transfer*
Kane and Levinson: *Dynamics: Theory and Applications*
Kimbrell: *Kinematics Analysis and Synthesis*
Martin: *Kinematics and Dynamics of Machines*
Norton: *Design of Machinery: An Introduction to the Synthesis and Analysis of Mechanisms and Machines*
Phelan: *Fundamentals of Mechanical Design*
Raven: *Automatic Control Engineering*
Rosenberg and Karnopp: *Introduction to Physics*
Schlichting: *Boundary-Layer Theory*
Shames: *Mechanics of Fluids*
Sherman: *Viscous Flow*
Shigley: *Kinematic Analysis of Mechanisms*
Shigley and Uicker: *Theory of Machines and Mechanisms*
Shigley and Mischke: *Mechanical Engineering Design*
Stoeker and Jones: *Refrigeration and Air Conditioning*
Vanderplaats: *Numerical Optimization: Techniques for Engineering Design with Applications*
White: *Viscous Flow*
Zeid: *CAD/CAM Theory and Practice*

DESIGN OF MACHINERY

AN INTRODUCTION TO THE SYNTHESIS AND ANALYSIS OF MECHANISMS AND MACHINES

Robert L. Norton

Worcester Polytechnic Institute

Worcester, Massachusetts

McGraw-Hill, Inc.

New York   St. Louis   San Francisco   Auckland   Bogotá
Caracas   Lisbon   London   Madrid   Mexico   Milan
Montreal   New Delhi   Paris   San Juan   Singapore
Sydney   Tokyo   Toronto

DESIGN OF MACHINERY

An Introduction to the Synthesis and Analysis of Mechanisms and Machines

Copyright © 1992 by McGraw-Hill Inc. All rights reserved. Printed in the United States of America. Except as permitted under the United States Copyright Act of 1976, no part of this publication may be reproduced or distributed in any form or by any means, or stored in a data base or retrieval system, without the prior written permission of the publisher.

THIRD PRINTING, 1993

4 5 6 7 8 9 0 DOC DOC 9 0 9 8 7 6 5 4 3

P/N 047799-X
PART OF
ISBN 0-07-909702-2

The editor was John J. Corrigan;
the production supervisor was Friederich W. Schulte;
the book design was done by Wanda Siedlecka.
The body text was set in Palatino, and headings set in Avant Garde.
Comp Associates, Worcester, Mass., made the color separations.
R. R. Donnelley & Sons was printer and binder.

All text, drawings, and equations in this book were prepared and typeset electronically, by the author, on a *Macintosh*® IIx computer using Microsoft *Word*®, Adobe *Illustrator*®, *PowerDraw*® by Engineered Software, *MathType*® by Design Science, and Aldus *Pagemaker*® desktop publishing software. Printer's film color separations were made on a *Varityper* 4300 laser typesetter directly from the author's disks using Aldus *PrePrint*. All *clip art* illustrations are courtesy of *Dubl-Click Software Inc.*, 9316 Deering Ave, Chatsworth CA 91311, reprinted from their *Industrial Revolution* and *Old Earth Almanac* series with their permission (and with the author's thanks).

Library of Congress Cataloging-in-Publication Data

Norton, Robert L.
    Design of machinery: an introduction to the synthesis and analysis of mechanisms and machines / Robert L. Norton.
        p. cm. —(McGraw-Hill series in mechanical engineering)
    Includes bibliographical references and index.
    ISBN 0-07-909702-2 (set)
        1. Machinery—Design.   2. Machinery, Kinematics.   3. Machinery, Dynamics of.   I. Title.   II. Series.
    TJ230.N63   1992   91-7510
    621.8'15—dc20

ABOUT THE AUTHOR

Robert L. Norton earned undergraduate degrees in both mechanical engineering and industrial technology at Northeastern University, and an MS in engineering design at Tufts University. He is a registered professional engineer in Massachusetts and New Hampshire. He has extensive industrial experience in engineering design and manufacturing, and many years experience teaching mechanical engineering, engineering design, computer science, and related subjects at Northeastern University, Tufts University, and Worcester Polytechnic Institute. At Polaroid Corporation for ten years, he designed cameras, related mechanisms, and high-speed automated machinery. He spent three years at Jet Spray Cooler Inc., Waltham, Mass., designing food-handling machinery and products. For five years he helped develop artificial-heart and noninvasive assisted-circulation (counterpulsation) devices at the Tufts New England Medical Center and Boston City Hospital. Since leaving industry to join academia, he has continued as an independent consultant on engineering projects ranging from disposable medical products to high-speed production machinery. He holds thirteen U.S. patents.

Norton has been on the faculty of Worcester Polytechnic Institute since 1981, and is currently Professor of Mechanical Engineering and head of the design group in that department. He teaches undergraduate and graduate courses in mechanical engineering with emphasis on design, kinematics, and dynamics of machinery. He also directs the operation of several general-purpose and computer-aided-design laboratories. He is the author of numerous technical papers and journal articles covering kinematics, dynamics of machinery, cam design and manufacturing, computers in education, and engineering education. He is a member of the American Society of Mechanical Engineers and the American Society of Engineering Education. Rumors about the transplantation of a 68040 microprocessor into his brain are decidedly untrue (though he could use some additional RAM). As for the unobtainium* ring, well, that's another story.

* See index.

This book is dedicated to the memory of my father,

Harry J. Norton, Sr.

who sparked a young boy's interest in engineering;

to the memory of my mother,

Kathryn Warren Norton

who made it all possible;

to my wife,

Nancy Norton

who provides unflagging patience and support;

and to my children,

Robert, Mary, and Thomas,

who make it all worthwhile.

CONTENTS

PREFACE

When I hear, I forget
When I see, I remember
When I do, I understand
Ancient Chinese Proverb

This text is intended for the kinematics and dynamics of machinery topics which are often given as a single course, or two-course sequence, in the junior year of most mechanical engineering programs. The usual prerequisites are first courses in statics, dynamics and calculus. Usually, the first semester, or portion, is devoted to kinematics, and the second to dynamics of machinery. These courses are ideal vehicles for introducing the mechanical engineering student to the process of design, since mechanisms tend to be intuitive for the typical mechanical engineering student to visualize and create. While this text attempts to be thorough and complete on the topics of analysis, it also emphasizes the synthesis and design aspects of the subject to a greater degree than most texts in print on these subjects. Also, it emphasizes the use of computer-aided engineering as an approach to the design and analysis of this class of problems by providing software that can enhance student understanding. While the mathematical level of this text is aimed at second- or third-year university students, it is presented *de novo* and should be understandable to the technical school student as well.

Part I of this text is suitable for a one-semester or one-term course in kinematics. Part II is suitable for a one-semester or one-term course in dynamics of machinery. Alternatively, both topic areas can be covered in one semester with less emphasis on some of the topics covered in the text.

The writing and style of presentation in the text is designed to be clear, informal, and easy to read. Many example problems and solution techniques are presented and spelled out in detail, both verbally and graphically. All the illustrations are done with computer-drawing or drafting programs. Some scanned photographic images are also included. The entire text, including equations and artwork, is printed directly from computer disk by laser typesetting for maximum clarity and quality. Many suggested readings are provided in the bibliography. Short problems, and where appropriate, many longer, unstructured design project assignments are provided at the ends of chapters. These projects provide an opportunity for the students *to do and understand*.

The author's approach to these courses and this text is based on over 30 years' experience in mechanical engineering design, both in industry and as a consultant. He has taught these subjects since 1967, both in evening school to practicing engineers and in day school to younger students. His approach to the course has evolved a great deal in that time, from a traditional approach, emphasizing graphical analysis of many structured problems, through emphasis on algebraic methods as computers became available, through requiring students to write their own computer programs, to the current state described above.

The one constant throughout has been the attempt to convey the art of the design process to the students in order to prepare them to cope with *real* engineering problems in practice. Thus, the author has always promoted design within these courses. Only recently, however, has technology provided a means to more effectively accomplish this goal, in the form of the graphics microcomputer. This text attempts to be an improvement over those currently available by providing up-to-date methods and techniques for analysis and synthesis which take full advantage of the graphics microcomputer, and by emphasizing design as well as analysis. The text also provides a more complete, modern, and thorough treatment of cam design than existing texts in print on the subject.

The author has written several menu driven, interactive, student-friendly computer programs for the design and analysis of mechanisms and machines. These programs are designed to enhance the student's understanding of the basic concepts in these courses while simultaneously allowing more comprehensive and realistic problem and project assignments to be done in the limited time available, than could ever be done with manual solution techniques, whether graphical or algebraic. Unstructured, realistic design problems which have many valid solutions are assigned. Synthesis and analysis are equally emphasized. The analysis methods presented are up to date, using vector equations and matrix techniques wherever applicable. Manual graphical analysis methods are de-emphasized. The graphics output from the computer programs allows the student to see the results of variation of parameters rapidly and accurately and reinforces learning.

These computer programs are distributed, on a floppy disk, with this book which also contains instructions for their use on any IBM compatible, MS DOS computer. Programs FOURBAR, FIVEBAR and SIXBAR analyze the kinematics of those types of linkages. Program DYNAFOUR does a complete dynamic analysis of the fourbar linkage in addition to its kinematics. Program DYNACAM allows the design and dynamic analysis of cam-follower systems. Program ENGINE analyzes the slider-crank linkage as used in the internal combustion engine and provides a complete dynamic analysis of single and multicylinder engine configurations, allowing the mechanical dynamic design of engines to be done. Program MATRIX is a general purpose linear equation system solver. All these programs, except MATRIX, provide dynamic, graphical animation of the designed devices. The reader is strongly urged to make use of these programs in order to investigate the results of variation of parameters in these kinematic devices. The programs are designed to enhance and augment the text rather than be a substitute for it. The converse is also true. Many solutions to the book's examples and to the problem sets are provided on the program disk as files to be read into these programs. Many of these solutions can be animated on the computer screen for a better demonstration of the concept than is possible on the printed page. The instructor and students are both encouraged to take advantage of the computer programs provided.

The author's intention is that synthesis topics be introduced first to allow the students to work on some simple design tasks early in the term while still mastering the analysis topics. Though this is not the "traditional" approach to the teaching of this material, the author believes that it is a superior method to that of initial concentration on detailed analysis of mechanisms for which the student has no concept of origin or purpose. Chapters 1 and 2 are introductory. Those instructors wishing to pursue the traditional approach of teaching analysis before synthesis can leave Chapters 3 and 5 on linkage synthesis for later consumption. Chapters 4, 6, and 7 on position, velocity, and acceleration analysis are sequential and build upon each other. In fact, some of the problem sets are common among these three chapters so that students can use their position solutions to find velocities and then later use both to find the accelerations in the same linkages. Chapter 9 on cams is more extensive and complete than that of other kinematics texts and takes a design approach. Chapter 10 on gear trains is introductory. The dynamic force treatment in Part II uses matrix methods for the solution of the system simultaneous equations. Graphical force analysis is not emphasized. Balancing of rotating machinery and linkages is covered in Chapter 13. Chapters 14 and 15 use the internal combustion engine as an example to pull together many dynamic concepts in a design context. Chapter 17 presents an introduction to dynamic systems modelling and uses the cam-follower system as the example. Chapters 3, 9, 12, 14, and 15 provide open ended project problems as well as structured problem sets. The assignment and execution of unstructured project problems can greatly enhance the student's understanding of the concepts as described by the proverb in the epigraph to this preface.

ACKNOWLEDGMENTS: The sources of photographs and other nonoriginal art used in the text are acknowledged in the captions and opposite the title page, but the author would also like to express his thanks for the cooperation of all those individuals and companies who generously made these items available. The author would also like to thank those who have reviewed various sections of the text and who made many useful suggestions for improvement. Mr. John Titus of the University of Minnesota reviewed Chapter 5 on analytical synthesis and Mr. Dennis Klipp of Keyes Fiber Corp., Waterville, Maine, reviewed Chapter 9 on cam design. Professor William J. Crochetiere and Mr. Homer Eckhardt of Tufts University, Medford, Mass., reviewed Chapter 17. Mr. Eckhardt and Professor Crochetiere of Tufts, and Professor Charles Warren of the University of Alabama taught from and reviewed Part I. Professor Holly K. Ault of Worcester Polytechnic Institute thoroughly reviewed the entire text while teaching from the pre-publication, class-test versions of the complete book. Professor Michael Keefe of the University of Delaware provided many helpful comments. Special thanks go to Professor John Sullivan of Worcester Polytechnic Institute for the use of an Apple scanner. Sincere thanks also go to the large number of undergraduate students and graduate teaching assistants who caught many typos and errors in the text and in the programs while using the pre-publication versions. The author takes full responsibility for any errors that may remain and invites from all readers their criticisms, suggestions for improvement, and identification of errors in the text or programs, so that both can be improved in future versions.

Robert L. Norton

Take to Kinematics. It will repay you.
It is more fecund than geometry;
it adds a fourth dimension to space.

CHEBYSCHEV TO SYLVESTER,
 1873

KINEMATICS OF
MECHANISMS

Chapter 1

INTRODUCTION

Inspiration most often strikes
those who are hard at work
ANONYMOUS

1.0 PURPOSE

In this text we will explore the topics of **kinematics** and **dynamics of machinery** in respect to the **synthesis of mechanisms** in order to accomplish desired motions or tasks, and also the **analysis of mechanisms** in order to determine their rigid-body dynamic behavior. These topics are fundamental to the broader subject of **machine design**. On the premise that we cannot analyze anything until it has been synthesized into existence, we will first explore the topic of **synthesis of mechanisms**. Then we will investigate techniques of **analysis of mechanisms**. All this will be directed toward developing your ability to design viable mechanism solutions to real, unstructured engineering problems by using a **design process**. We will begin with careful definitions of the terms used in these topics.

1.1 KINEMATICS AND KINETICS

KINEMATICS   *The study of motion without regard to forces*

KINETICS   *The study of forces on systems in motion*

These two concepts are really *not* physically separable. We arbitrarily separate them for instructional reasons in engineering education. It is also valid in engineering design practice to first consider the desired kinematic motions and their consequences, and then subsequently investigate the kinetic forces associated with those motions. The student should realize that the division between **kinematics** and **kinetics** is quite arbitrary and is done largely for convenience. One cannot design most dynamic mechanical systems without taking both topics into thorough consideration. It is quite logical to consider them in the order listed since, from Newton's second law, $\mathbf{F} = m\mathbf{a}$, one typically needs to know the **accelerations** (a) in order to compute the dynamic **forces** ($\mathbf{F}$) due to the motion of the system's

mass (*m*). There are also many situations in which the applied forces are known and the resultant accelerations are to be found.

One principle aim of **kinematics** is to create (design) the desired motions of the subject mechanical parts and then mathematically compute the positions, velocities, and accelerations which those motions will create on the parts. Since, for most earthbound mechanical systems, the mass remains essentially constant with time, defining the accelerations as a function of time then also defines the dynamic forces as a function of time. **Stresses**, in turn, will be a function of both applied and inertial (*m***a**) forces. Since engineering design is charged with creating systems which will not fail during their expected service life, the goal is to keep stresses within acceptable limits for the materials chosen and the environmental conditions encountered. This obviously requires that all system forces be defined and kept within desired limits. In machinery which moves (the only interesting kind), the largest forces encountered are often those due to the dynamics of the machine itself. These dynamic forces are proportional to acceleration, which brings us back to kinematics, the foundation of mechanical design. Very basic and early decisions in the design process involving kinematic principles can be crucial to the success of any mechanical design. A design which has poor kinematics will prove troublesome and perform badly.

1.2   MECHANISMS AND MACHINES

A **mechanism** is a device which transforms motion to some desirable pattern and typically develops very low forces and transmits little power. A **machine** typically contains mechanisms which are designed to provide significant forces and transmit significant power. Examples of both are shown in Figure 1-1. Some examples of common mechanisms are a pencil sharpener, a camera shutter, an analog clock, a folding chair, an adjustable desk lamp, and an umbrella. Some examples of machines which possess motions similar to the mechanisms listed above are a food blender, a bank vault door, an automobile transmission, a bulldozer, a robot, and an amusement park ride. There is no clear-cut dividing line between mechanisms and machines. They differ in degree rather than in kind. If the forces or energy levels within the device are significant, it is considered a machine; if not, it is considered a mechanism. A useful working **definition of a mechanism** is *A system of elements arranged to transmit **motion** in a predetermined fashion.* This can be converted to a definition of a **machine** by adding the words ***and energy*** after ***motion***.

Mechanisms, if lightly loaded and run at slow speeds, can sometimes be treated strictly as kinematic devices; that is, they can be analyzed kinematically without regard to forces. Machines (and mechanisms running at higher speeds), on the other hand, must first be treated as mechanisms, a kinematic analysis of their velocities and accelerations must be done, and then they must be subsequently analyzed as dynamic systems in which their static and dynamic forces due to those accelerations are analyzed using the principles of kinetics. **Part I** of this text deals with *Kinematics of Mechanisms*, and **Part II** with *Dynamics of Machinery*. The techniques of mechanism synthesis presented in Part I are applicable to the design of both mechanisms and machines, since in each case some collection of moveable members must be created to provide and control the desired motions and geometry.

(*a*) A mechanism

(*b*) A machine

FIGURE 1-1

Mechanisms and machines

1.3 A BRIEF HISTORY OF KINEMATICS

Machines and mechanisms have been devised by people since the dawn of history. The ancient Egyptians devised primitive machines to accomplish the building of the pyramids and other monuments. Though the wheel and pulley (on an axle) were not known to the Old Kingdom Egyptians, they made use of the lever, the inclined plane (or wedge), and probably the log roller. The origin of the wheel and axle is not definitively known. Its first appearance seems to have been in Mesopotamia about 3000 to 4000 B.C.

A great deal of design effort was spent from early times on the problem of timekeeping as more sophisticated clockworks were devised. Much early machine design was directed toward military applications (catapults, wall scaling apparatus, etc.). The term **civil engineering** was later coined to differentiate civilian from military applications of technology. **Mechanical engineering** had its beginnings in machine design as the inventions of the industrial revolution required more complicated and sophisticated solutions to motion control problems. **James Watt** (1736-1819) probably deserves the title of first kinematician for his synthesis of a straight-line linkage (see Figure 3-24a) to guide the very long stroke pistons in the then new steam engines. Since the planer was yet to be invented (in 1817), no means then existed to machine a long, straight guide to serve as a crosshead in the steam engine. Watt was certainly the first on record to recognize the value of the motions of the coupler link in the fourbar linkage. **Oliver Evans** (1755-1819) an early American inventor, also designed a straight-line linkage for a steam engine. **Euler** (1707-1783) was a contemporary of Watt, though they apparently never met. Euler presented an analytical treatment of mechanism in his *Mechanica sive Motus Scienta Analytice Exposita* (1736-1742), which included the concept that planar motion is composed of two independent components, namely, translation of a point and rotation of the body about that point. Euler also suggested the separation of the problem of dynamic analysis into the "geometrical" and the "mechanical" in order to simplify the determination of the system's dynamics. Two of his contemporaries, **d'Alembert** and **Kant**, also proposed similar ideas. This is the origin of our division of the topic into kinematics and kinetics as described above.

In the early 1800's, L'Ecole Polytechnic in Paris, France, was the repository of engineering expertise. **Lagrange** and **Fourier** were among its faculty. One of its founders was **Gaspard Monge** (1746-1818), inventor of descriptive geometry (which incidentally was kept as a military secret by the French government for 30 years because of its value in planning fortifications). Monge created a course in elements of machines and set about the task of classifying all mechanisms and machines known to mankind! His colleague, **Hachette**, completed the work in 1806 and published it as what was probably the first mechanism text in 1811. **Andre Marie Ampere** (1775-1836), also a professor at L'Ecole Polytechnic, set about the formidable task of classifying "all human knowledge." In his *Essai sur la Philosophie des Sciences*, he was the first to use the term "**cinematique**," from the Greek word for motion, to describe the study of motion without regard to forces, and suggested that "this science ought to include all that can be said with respect to motion in its different kinds, independently of the forces by which it is produced." His term was later anglicized to *kinematics* and germanized to *kinematik*.

Robert Willis (1800-1875) wrote the text *Principles of Mechanism* in 1841 while a professor of natural philosophy at the University of Cambridge, England. He attempted to systematize the task of mechanism synthesis. He counted five ways of obtaining relative motion between input and output links: rolling contact, sliding contact, linkages, wrapping connectors (belts, chains), and tackle (rope or chain hoists). **Franz Reuleaux** (1829-1905), published *Theoretische Kinematik* in 1875. Many of his ideas are still current and useful. **Alexander Kennedy** (1847-1928) translated Reuleaux into English in 1876. This text became the foundation of modern kinematics and is still in print! (See bibliography at end of chapter.) He provided us with the concept of a kinematic pair (joint), whose shape and interaction define the type of motion transmitted between elements in the mechanism. Reuleaux defined six basic mechanical components: the link, the wheel, the cam, the screw, the ratchet, and the belt. He also defined "higher" and "lower" pairs, higher having line or point contact (as in a ball bearing) and lower having surface contact (as in pin joints). Reuleaux is generally considered the father of modern kinematics and is responsible for the symbolic notation of skeletal, generic linkages used in all modern kinematics texts.

In this century, prior to World War II, most theoretical work in kinematics was done in Europe, especially in Germany. Few research results were available in English. In the United States, kinematics was largely ignored until the 1940's, when **A.E.R. DeJonge** wrote *What Is Wrong with "Kinematics" and "Mechanisms,"*[1] which called upon the U.S. mechanical engineering education establishment to pay attention to the European accomplishments in this field. Since then, much new work has been done, especially in kinematic synthesis, by American and European engineers and researchers such as **J. Denavit**, **A. Erdman**, **F. Freudenstein**, **A.S. Hall**, **R. Hartenberg**, **R. Kaufman**, **B. Roth**, **G. Sandor** and **A. Soni**, (all of the U.S.) and **K. Hain** (of Germany). Many of these U.S. researchers have applied the computer to the solution of previously intractable problems, both of analysis and synthesis, making practical use of many of the theories of their predecessors. This text will make much use of the availability of computers to allow more efficient analysis and synthesis of solutions to machine design problems. Several computer programs are included with this book for your use.

1.4　THE DESIGN PROCESS

Design, Invention, Creativity

These are all familiar terms but may mean different things to different people. These terms can encompass a wide range of activities from styling the newest look in clothing, to creating impressive architecture, to engineering a machine for the manufacture of facial tissues. **Engineering design**, which we are concerned with here, embodies all three of these activities as well as many others. The word **design** is derived from the Latin **designare**, which means *"to designate, or mark out."* Webster's gives several definitions, the most applicable being *"to outline, plot, or plan, as action or work. . . to conceive, invent – contrive."* **Engineering design** has been defined as *". . . the process of applying the various techniques and scientific principles for the purpose of defining a device, a process or a system in sufficient detail to permit its* *realization . . . Design may be simple or enormously complex, easy or difficult, mathematical or nonmathematical; it may involve a trivial problem or one of great importance."* **Design** is a universal constituent of engineering practice. But the complexity of engineering subjects usually requires that the student be served with a collection of **structured, set-piece problems** designed to elucidate a particular concept or concepts related to the particular topic. These textbook problems typically take the form of *"given A, B, C, and D, find E."* Unfortunately, real-life engineering problems are almost never so structured. Real design problems more often take the form of: *"What we need is a framus to stuff this widget into that hole within the time allocated to the transfer of this other gizmo."* The new engineering graduate will search in vain among his or her textbooks for much guidance to solve such a problem. This **unstructured problem** statement usually leads to what is commonly called **"blank paper syndrome."** Engineers often find themselves staring at a blank sheet of paper pondering how to begin solving such an ill-defined problem.

Blank Paper
Syndrome

Much of engineering education deals with topics of **analysis** which means *to decompose, to take apart, to resolve into its constituent parts.* This is quite necessary. The engineer must know how to analyze systems of various types, mechanical, electrical, thermal or fluid. Analysis requires a thorough understanding of both the appropriate mathematical techniques and the fundamental physics of the system's function. But, before any system can be analyzed, it must exist, and a blank sheet of paper provides little substance for analysis. Thus the first step in any engineering design exercise is that of **synthesis**, which means *putting together.*

The design engineer, in practice, regardless of discipline, continuously faces the challenge of *structuring the unstructured problem*. Inevitably, the problem as posed to the engineer is ill-defined and incomplete. Before any attempt can be made to *analyze the situation* he or she must first carefully define the problem, using an engineering approach, to ensure that any proposed solution will solve the right problem. Many examples exist of excellent engineering solutions which were ultimately rejected because they solved the wrong problem, i.e., a different one than the client really had.

Much research has been devoted to the definition of various "design processes" intended to provide means to structure the unstructured problem and lead to a viable solution. Some of these processes present dozens of steps, others only a few. The one presented in Table 1-1 contains ten steps and has, in the author's experience, proven successful in 30 years of practice in engineering design.

ITERATION Before discussing each of these steps in detail it is necessary to point out that this is not a process in which one proceeds from step one through ten in a linear fashion. Rather it is, by its nature, an iterative process in which progress is made haltingly, two steps forward and one step back. It is inherently *circular*. To **iterate** means *to repeat, to return to a previous state.* If, for example, your apparently great idea, upon analysis, turns out to violate the second law of thermodynamics, you can return to the ideation step and get a better idea! Or, if necessary, you can return to an earlier step in the process, perhaps the background research, and learn more about the problem. With the understanding that the actual execution of the process involves iteration, for simplicity, we will now discuss each step in the order listed in Table 1-1

TABLE 1-1   A Design Process

1	Identification of Need
2	Background Research
3	Goal Statement
4	Task Specifications
5	Ideation and Invention
6	Analysis
7	Selection
8	Detailed Design
9	Prototyping and Testing
10	Production

Identifying the need

Reinventing the
wheel

Grass shorteners

Identification of Need

This first step is often done for you by someone, boss or client, saying "What we need is . . . " Typically this statement will be brief and lacking in detail. It will fall far short of providing you with a structured problem statement. For example, the problem statement might be "We need a better lawn mower."

Background Research

This is the most important phase in the process, and is unfortunately often the most neglected. The term research, used in this context, should *not* conjure up visions of white-coated scientists mixing concoctions in test tubes. Rather this is research of a more mundane sort, gathering background information on the relevant physics, chemistry, or other aspects of the problem. Also it is desirable to find out if this, or a similar problem, has been solved before. There is no point in reinventing the wheel. If you are lucky enough to find a ready-made solution on the market, it will no doubt be more economical to purchase it than to build your own. Most likely this will not be the case, but you may learn a great deal about the problem to be solved by investigating the existing "art" associated with similar technologies and products. The patent literature and technical publications in the subject area are obvious sources of information. Clearly, if you find that the solution exists and is covered by a patent still in force, you have only a few ethical choices: buy the patentee's existing solution, design something which does not conflict with the patent, or drop the project. It is very important that sufficient energy and time be expended on this research and preparation phase of the process in order to avoid the embarrassment of concocting a great solution to the wrong problem. Most inexperienced (and some experienced) engineers give too little attention to this phase and jump too quickly into the ideation and invention stage of the process. *This must be avoided!* You must discipline yourself to *not* try to solve the problem before thoroughly preparing yourself to do so.

Goal Statement

Once the background of the problem area as originally stated is fully understood, you will be ready to recast that problem into a more coherent goal statement. This new problem statement should have three characteristics. It should be concise, be general, and be uncolored by any terms which predict a solution. It should be couched in terms of **functional visualization**, meaning to visualize its function, rather than any particular embodiment. For example, if the original statement of need was *"Design a Better Lawn Mower,"* after research into the myriad of ways to cut grass that have been devised over the ages, the wise designer might restate the goal as **"Design a Means to Shorten Grass."** The original problem statement has a built-in trap in the form of the *colored* words "lawn mower." For most people, this phrase will conjure up a vision of something with whirring blades and a noisy engine. For the **ideation** phase to be most successful, it is necessary to avoid such images and to state the problem generally, clearly, and concisely. As an exercise, list ten ways to shorten grass. Most of them would not occur to you had you been asked for ten better lawn mower designs. You should use **functional visualization** to avoid unnecessarily limiting your creativity!

TABLE 1-2   Task Specifications

1	Device to have self-contained power supply.
2	Device to be corrosion resistant.
3	Device to cost less than $100.00.
4	Device to emit < 80 dB sound intensity at 50 feet.
5	Device to shorten 1/4 acre of grass per hour.
6	etc . . . etc.

Task Specifications [1]

When the background is understood, and the goal clearly stated, you are ready to formulate a set of **task specifications**. These should be *performance specifications,* ***not*** *design specifications.* The difference is that **performance specifications** define ***what*** *the system must do*, while **design specifications** define ***how*** *it must do it.* At this stage of the design process it is unwise to attempt to specify *how* the goal is to be accomplished. That is left for the **ideation** phase. The purpose of the **task specifications** is to carefully define and constrain the problem so that it both *can be solved*, and *can be shown to have been solved* after the fact. A sample set of task specifications for our "Grass Shortener" is shown in Table 1-2.

Note that these specifications constrain the design without overly restricting the engineer's design freedom. It would be inappropriate to require a gasoline engine for specification 1, since other possibilities exist which will provide the desired mobility. Likewise, to demand stainless steel for all components in specification 2 would be unwise, since corrosion resistance can be obtained by many other, less-expensive means. In short, the task specifications serve to define the problem in as complete and as general a manner as possible, and they serve as a contractual definition of what is to be accomplished. The finished design can be measured for compliance to these specifications.

Task Specs

Lorem
Ipsum
Dolor amet
Euismod
Volutpat
Laoreet
Adipiscing

Ideation and Invention

This step is full of both fun and **frustration**. This phase is potentially the most satisfying to most designers, but it is also the most difficult. A great deal of research has been done to explore the phenomenon of "**creativity**." It is, most agree, a common human trait. It is certainly exhibited to a very high degree by all young children. The rate and degree of development that occurs in the human from birth through the first few years of life certainly requires some innate creativity. Some have claimed that our methods of Western education tend to stifle children's natural creativity by encouraging conformity and restricting individuality. From "coloring within the lines" in kindergarten to imitating the textbook's writing patterns in later grades, individuality is suppressed in favor of a socializing conformity. This is perhaps necessary to avoid anarchy but probably does have the effect of reducing the individual's ability to think creatively. Some claim that creativity can be

---

[1]   Orson Welles, famous author and filmmaker, once said, *"The enemy of art is the absence of limitations."* We can paraphrase that as *The enemy of design is the absence of specifications.*

taught, some that it is only inherited. No hard evidence exists for either theory. It is probably true that one's lost or suppressed creativity can be rekindled. Other studies suggest that most everyone underutilizes his or her potential creative abilities. You can enhance your creativity through various techniques.

CREATIVE PROCESS Many techniques have been developed to enhance or inspire creative problem solving. In fact, just as design processes have been defined, so has the *creative process* shown in Table 1-3. This creative process can be thought of as a subset of the design process and to exist within it. The ideation and invention step can thus be broken down into these four substeps.

IDEA GENERATION is the most difficult of these steps. Even very creative people have difficulty in inventing "on demand." Many techniques have been suggested to improve the yield of ideas. The most important technique is that of *deferred judgment*, which means that your criticality should be temporarily suspended. Do not try to judge the quality of your ideas at this stage. That will be taken care of later, in the **analysis** phase. The goal here is to obtain as large a *quantity* of potential designs as possible. Even superficially ridiculous suggestions should be welcomed, as they may trigger new insights and suggest other more realistic and practical solutions.

BRAINSTORMING is a technique for which some claim great success in generating creative solutions. This technique requires a group, preferably 6 to 15 people, and attempts to circumvent the largest barrier to creativity, which is *fear of ridicule*. Most people, when in a group, will not suggest their real thoughts on a subject, for fear of being laughed at. Brainstorming's rules require that no one is allowed to make fun of or criticize anyone's suggestions, no matter how ridiculous. One participant acts as "scribe" and is duty bound to record all suggestions, no matter how apparently silly. When done properly, this technique can be fun and can sometimes result in a "feeding frenzy" of ideas which build upon each other. Large quantities of ideas can be generated in a short time. Judgment on their quality is deferred to a later time.

When working alone, other techniques are necessary. **Analogies** and **inversion** are often useful. Attempt to draw analogies between the problem at hand and other physical contexts. If it is a mechanical problem, convert it by analogy to a fluid or electrical one. Inversion turns the problem inside out. For example, consider what you want moved to be stationary and vice versa. Insights often follow. Another useful aid to creativity is the use of **synonyms**. Define the action verb in the problem statement, and then list as many synonyms for that verb as possible. For example:

PROBLEM STATEMENT: Move this object from point A to point B.

The action verb is "move." Some synonyms are push, pull, slip, slide, shove, throw, eject, jump, spill, etc.

By whatever means, the aim in this **ideation** step is to generate a large number of ideas without particular regard to quality. But, at some point, your "mental well" will go dry. You will have then reached the step in the creative process called **frustration** . It is time to leave the problem and do something else for a time. While your conscious mind is occupied with other concerns, your subconscious mind will still be hard at work on the problem. This is the step called

TABLE 1-3 The Creative Process

5a Idea Generation

5b Frustration

5c Incubation

5d Eureka!

Brainstorming

incubation. Suddenly, at a quite unexpected time and place, an idea will pop into your consciousness, and it will seem to be the obvious and "right" solution to the problem . . . **Eureka!** Most likely, later analysis will discover some flaw in this solution. If so, back up and **iterate!** More ideation, perhaps more research, and possibly even a redefinition of the problem may be necessary.

Analysis

Once you are at this stage, you have structured the problem, at least temporarily, and can now apply more sophisticated analysis techniques to examine the performance of the design in the **Analysis phase** of the design process. (These analysis methods will be discussed in detail in the following chapters.) Further iteration will be required as problems are discovered from the analysis. Repetition of as many earlier steps in the design process as necessary must be done to ensure the success of the design.

Frustration

Selection

When the technical analysis indicates that you have some potentially viable designs, the optimum or best available one must be **selected** for **detailed design, prototyping,** and **testing.** The selection process usually involves a comparative analysis of the available design solutions. A **decision matrix** sometimes helps to identify the best solution by forcing you to consider a variety of factors in a systematic way. A decision matrix for our better grass shortener is shown in Figure 1-2. Each design occupies a row in the matrix. The columns are assigned categories in which the designs are to be judged, such as cost, ease of use, efficiency, performance, reliability, and any others you deem appropriate to the particular problem. Each category is then assigned a **weight factor**, which measures its relative importance. For example, reliability may be a more important criterion to the user than cost,

	Cost	Safety	Performance	Reliability	RANK
Weight Factor	.35	.30	.15	.20	1.0
Design 1	3 1.05	6 1.80	4 .60	9 1.80	5.3
Design 2	4 1.40	2 .60	7 1.05	2 .40	3.5
Design 3	1 .35	9 2.70	4 .60	5 1.00	4.7
Design 4	9 3.15	1 .30	6 .90	7 1.40	5.8
Design 5	7 2.45	4 1.20	2 .30	6 1.20	5.2

Eureka!

FIGURE 1-2

A decision matrix

or vice versa. You as the design engineer, have to exercise your judgment as to the selection and weighting of these categories. The body of the matrix is then filled with numbers which rank each design on a convenient scale, such as 1 to 10, in each of the categories. Note that this is ultimately a *subjective ranking* on your part. You must examine the designs and decide on a score for each. The scores are then multiplied by the weight factors (which are usually chosen so as to sum to a convenient number such as 1) and the products summed for each design. The weighted scores then give a ranking of designs. Be cautious in applying these results. Remember the source and subjectivity of your scores and the weight factors! There is a temptation to put more faith in these results than is justified. After all, they look impressive! They can even be taken out to several decimal places! (But they shouldn't be.) The real value of a decision matrix is that it breaks the problem into more tractable pieces and forces you to think about the relative value of each design in many categories. You can then make a more informed decision as to the "best" design.

Detailed Design

This step usually includes the creation of a complete set of assembly and detail drawings or **Computer-Aided Design** (CAD) part files, for *each and every part* used in the design. Each detail drawing must specify all the dimensions and the material specifications necessary to make that part. From these drawings (or CAD files) a prototype test model (or models) must be constructed for physical testing. Most likely the tests will discover more flaws, requiring further **iteration**.

Prototyping and Testing

MODELS: Ultimately, one cannot be sure of the correctness or viability of any design until it is built and tested. This usually involves the construction of a prototype physical model. A mathematical model, while very useful, can never be as complete and accurate a representation of the actual physical system as a physical model, due to the need to make simplifying assumptions. Prototypes are often very expensive but may be the most economical way to prove a design, short of building the actual, full-scale device. Prototypes can take many forms, from working scale models to full-size, but simplified, representations of the concept. Scale models introduce their own complications in regard to proper scaling of the physical parameters. For example, volume of material varies as the cube of linear dimensions, but surface area varies as the square. Heat transfer to the environment may be proportional to surface area, while heat generation may be proportional to volume. So linear scaling of a system, either up or down, may lead to behavior different from that of the full-scale system. One must exercise caution in scaling physical models. You will find as you begin to design linkage mechanisms that a **simple cardboard model** of your chosen link lengths, coupled together with thumbtacks for pivots, will tell you a great deal about the quality and character of the mechanism's motions. You should get into the habit of making such simple articulated models for all your linkage designs.

TESTING of the model or prototype may range from simply actuating it and observing its function to attaching extensive instrumentation to accurately measure displacements, velocities, accelerations, forces, temperatures, and other parameters.

Tests may need to be done under controlled environmental conditions such as high or low temperature or humidity. The microcomputer has made it possible to measure many phenomena more accurately and inexpensively than could be done before.

Production

Finally, with enough time, money and perseverance, the design will be ready for production. This might consist of the manufacture of a single final version of the design, but more likely will mean making thousands or even millions of your widget. The danger, expense, and embarrassment of finding flaws in your design after making large quantities of defective devices should inspire you to use the greatest care in the earlier steps of the design process to ensure that it is properly engineered.

The **design process** is widely used in engineering. Engineering is usually defined in terms of what an engineer does, but engineering can also be defined in terms of *how* the engineer does what he or she does. **Engineering** is *as much a method, an approach, a process, a state of mind for problem solving, as it is an activity.* The engineering approach is that of thoroughness, attention to detail, and consideration of all the possibilities. While it may seem a contradiction in terms to emphasize "attention to detail" while extolling the virtues of open-minded, free-wheeling, creative thinking, it is not. The two activities are not only compatible; they are symbiotic. It ultimately does no good to have creative, original ideas if you do not, or cannot, carry out the execution of those ideas and "reduce them to practice." To do this you must discipline yourself to suffer the nitty-gritty, nettlesome, tiresome details which are so necessary to the completion of any one phase of the creative design process. For example, to do a creditable job in the design of anything, you must *completely* define the problem. If you leave out some detail of the problem definition, you will end up solving the wrong problem. Likewise, you must *thoroughly* research the background information relevant to the problem. You must *exhaustively* pursue conceptual potential solutions to your problem. You must then *extensively* analyze these concepts for validity. And, finally, you must *detail* your chosen design down to the last nut and bolt to be confident it will work. If you wish to be a good designer and engineer, you must discipline yourself to do things thoroughly and in a logical, orderly manner, even while thinking great creative thoughts and iterating to a solution. Both attributes, creativity and attention to detail, are necessary for success in engineering design.

1.5   MULTIPLE SOLUTIONS

Note that by the nature of the design process, there is **not** any **one** correct answer or solution to any design problem. Unlike the structured "engineering textbook" problems, which most students are used to, there is no right answer "in the back of the book" for any real design problem. There are as many potential solutions as there are designers willing to attempt them. Some solutions will be better than others, but many will work. Some will not! There is no "one right answer" in design engineering, which is what makes it interesting. The only way to determine the relative merits of various potential design solutions is by thorough analysis, which usually will include physical testing of constructed prototypes. Because this is a very expensive process, it is desirable to do as much analysis on paper, or in the computer, as possible before actually building the device. Where feasible, mathematical models of the design, or parts of the design, should be created. These may take many forms, depending on the type of physical system involved. In the design of mechanisms and machines it is usually possible to write the equations for the rigid-body dynamics of the system, and solve them in "closed form" with (or without) a computer. Accounting for the elastic deformations of the members of the mechanism or machine usually requires more complicated approaches using **Finite Difference** techniques or the **Finite Element Method** (FEM).

1.6   HUMAN FACTORS ENGINEERING

With few exceptions, all machines are designed to be used by humans. Even robots must be programmed by a human. **Human factors engineering** is the study of the human-machine interaction and is defined as *an applied science that coordinates the design of devices, systems, and physical working conditions with the capacities and requirements of the worker.* The machine designer must be aware of this subject and design devices to "fit the man" rather than expect the man to adapt to fit the machine. The term **ergonomics** is synonymous with *human factors engineering.* We often see reference to the good or bad ergonomics of an automobile interior or a household appliance. A machine designed with poor ergonomics will be uncomfortable and tiring to use and may even be dangerous.

Make the machine
fit the man

There is a wealth of human factors data available in the literature. Some references are noted in the bibliography. The type of information which might be needed for a machine design problem ranges from dimensions of the human body and their distribution among the population by age and gender, to the ability of the human to withstand accelerations in various directions on the body, to typical strengths and force generating ability in various positions. Obviously, if you are designing a device that will be controlled by a human (a grass shortener, perhaps), you need to know how much force the user can exert with hands held in various positions, what the user's reach is, and how much noise the ears can stand without damage. If your device will carry the user on it, you need data on the limits of acceleration which the body can tolerate. Data on all these topics exist. Much of it was developed by the government which regularly tests the ability of military personnel to withstand extreme environmental conditions. Part of the background research of any machine design problem should include some investigation of human factors.

1.7   THE ENGINEERING REPORT

Communication of your ideas and results is a very important aspect of engineering. Many engineering students picture themselves in professional practice spending most of their time doing calculations of a nature similar to those they have done as students. Fortunately, this is seldom the case, as it would be very boring. Actually, engineers spend the largest percentage of their time communicating with others, either orally or in writing. Engineers write proposals and technical reports, give presentations, and interact with support personnel. When your design is done, it is usually necessary to present the results to your client, peers, or employer. The usual form of presentation is a formal engineering report. Thus, it is very important for the engineering student to develop his or her communication skills. *You may be the cleverest person in the world, but no one will know that if you cannot communicate your ideas clearly and concisely.* In fact, if you cannot explain what you have done, you probably don't understand it yourself. To give you some experience in this important skill, the design project assignments in later chapters are intended to be written up in formal engineering reports. Information on the writing of engineering reports can be found in the suggested readings in the bibliography at the end of this chapter.

1.8   UNITS

There are several systems of units used in engineering. The most common in the United States are the **U.S. foot-pound-second system (fps)**, the **U.S. inch-pound-second system (ips)**, and the **System International (SI)**. All systems are created from the choice of three of the quantities in the general expression of Newton's second law

$$F = \frac{mL}{t^2} \tag{1.1a}$$

where F is force, m is mass, L is length, and t is time. The units for any three of these variables can be chosen and the other is then derived in terms of the chosen units. The three chosen units are called *base units*, and the remaining one is then a *derived unit*.

Most of the confusion that surrounds the conversion of computations between either one of the U.S. systems and the SI system is due to the fact that the SI system uses a different set of base units than the U.S. systems. Both U.S. systems choose *force, length,* and *time* as the base units. Mass is then a derived unit in the U.S. systems, and they are referred to as *gravitational systems* because the value of mass is dependent on the local gravitational constant. The SI system chooses **mass,** *length,* and *time* as the base units and force is the derived unit. SI is then referred to as an *absolute system* since the mass is a base unit whose value is not dependent on local gravity.

The **U.S. foot-pound-second (fps)** system requires that all lengths be measured in feet (ft), forces in pounds (lb), and time in seconds (sec). Mass is then derived from Newton's law as

$$m = \frac{Ft^2}{L} \tag{1.1b}$$

and the units are:

Pounds seconds squared per **foot** (lb-sec^2/ft) = **slugs**

The **U.S. inch-pound-second (ips)** system requires that all lengths be measured in inches (in), forces in pounds (lb) and time in seconds (sec). Mass is still derived from Newton's law, equation 1.1b, but the units are now:

Pounds seconds squared per **inch** (lb-sec^2/in) = **blobs**

This mass unit is not slugs! It is worth twelve slugs or one blob. [2]

----
[2] See next page

Weight is defined as the force exerted on an object by gravity. Probably the most common units error that students make is to mix up these two unit systems (**fps** and **ips**) when converting weight units (which are pounds force) to mass units. Note that the gravitational acceleration constant (g) on earth at sea level is approximately 32.2 **feet** per second squared which is equivalent to 386.4 **inches** per second squared. The relationship between mass and weight is:

Mass = weight / gravitational acceleration

$$m = \frac{W}{g} \tag{1.2}$$

It should be obvious that, if you measure all your lengths in **inches** and then use $g = 32.2$ **feet**/sec^2 to compute mass, you will have an error of a *factor of 12* in your results. This is a serious error, large enough to crash the airplane you designed. Even worse off is the student who neglects to convert weight to mass *at all* in his calculations. He will have an error of either 32.2 or 386.4 in his results. This is enough to sink the ship!

To even further add to the student's confusion about units is the common use of the unit of **pounds mass** (lb$_m$). This unit is often used in fluid dynamics and thermodynamics and comes about through the use of a slightly different form of Newton's equation:

$$F = \frac{ma}{g_c} \tag{1.3}$$

where m = mass in lb$_m$, a = acceleration and g_c = the gravitational constant.

The value of the **mass** of an object measured in **pounds mass** (lb$_m$) is *numerically equal* to its **weight** in **pounds force** (lb$_f$). However the student *must remember to divide* the value of m in lb$_m$ by g_c when substituting into this form of Newton's equation. Thus the lb$_m$ will be divided either by 32.2 or by 386.4 when calculating the dynamic force. The result will be the same as when the mass is expressed in either slugs or blobs in the $F = ma$ form of the equation. Remember that in round numbers at sea level on earth:

1 lb$_m$ = 1 lb$_f$ 1 slug = 32.2 lb$_f$ 1 blob = 386.4 lb$_f$

The **SI** system requires that lengths be measured in meters (m), mass in kilograms (kg), and time in seconds (sec). This is sometimes also referred to as the **mks** system. Force is derived from Newton's law, equation 1.1b and the units are:

kilogram-meters per second^2 (kg-m/sec^2) = newtons

Thus in the SI system there are distinct names for mass and force which helps alleviate confusion. When converting between SI and U.S. systems, be alert to the fact that mass converts from kilograms (kg) to either slugs (sl) or blobs (bl), and force converts from newtons (N) to pounds (lb). The gravitational constant (g) in the SI system is approximately 9.81 m/sec^2.

---

[2] It is unfortunate that the mass unit in the **ips** system has never officially been given a name such as the term *slug* used for mass in the **fps** system. The author boldly suggests (with tongue only slightly in cheek) that this unit of mass in the **ips** system be called a *blob* (bl) to distinguish it more clearly

TABLE 1-4 Variables and Units

Base Units in Boldface - Abbreviations in ( )

Variable	Symbol	ips unit	fps unit	SI unit
Force	F	**pounds (lb)**	**pounds (lb)**	newtons (N)
Length	l	**inches (in)**	**feet (ft)**	**meters (m)**
Time	t	**seconds (sec)**	**seconds (sec)**	**seconds (sec)**
Mass	m	lb-sec^2/in (bl)	lb-sec^2/ft (sl)	**kilograms (kg)**
Weight	W	pounds (lb)	pounds (lb)	newtons (N)
Velocity	v	in/sec	ft/sec	m/sec
Acceleration	a	in/sec^2	ft/sec^2	m/sec^2
Jerk	j	in/sec^3	ft/sec^3	m/sec^3
Angle	θ	degrees (deg)	degrees (deg)	degrees (deg)
Angle	θ	radians (rad)	radians (rad)	radians (rad)
Angular velocity	ω	radians/sec	radians/sec	radians/sec
Angular acceleration	α	radians/sec^2	radians/sec^2	radians/sec^2
Angular jerk	φ	radians/sec^3	radians/sec^3	radians/sec^3
Torque	T	lb-in	lb-ft	N-m
Mass moment of inertia	I	lb-in-sec^2	lb-ft-sec^2	N-m-sec^2
Energy	E	in-lb	ft-lb	joules
Power	P	in-lb/sec	ft-lb/sec	watts
Volume	V	in^3	ft^3	m^3
Weight density	v	lb/in^3	lb/ft^3	N/m^3
Mass density	ρ	bl/in^3	sl/ft^3	kg/m^3

from the *slug* (sl), and to help the student avoid some of the common units errors listed above. **Twelve *slugs* = one *blob*.** *Blob* does not sound any sillier than slug, is easy to remember, implies mass, and has a convenient abbreviation (bl) which is an anagram for the abbreviation for pound (lb). Besides, if you have ever seen a garden slug, you know it looks just like a *"little blob."*

The principal system of units used in this textbook will be the U.S. **ips** system. Most machine design in the United States is still done in this system. Table 1-4 shows some of the variables used in this text and their units. The student is cautioned to always check the units in any equation written for a problem solution, whether in school or in professional practice after graduation. If properly written, an equation should cancel all units across the equal sign. If it does not, then you can be *absolutely sure it is **incorrect.*** Unfortunately, a unit balance in an equation does not guarantee that it is correct, as many other errors are possible. Always double-check your results. You might save a life.

1.9   WHAT'S TO COME

In this text we will explore the topic of **machine design** in respect to the **synthesis of mechanisms** in order to accomplish desired motions or tasks, and also the Analysis of these mechanisms in order to determine their rigid-body dynamic behavior. On the premise that we cannot analyze anything until it has been synthesized into existence, we will first explore the topic of synthesis of mechanisms. Then we will investigate the analysis of those and other mechanisms for their kinematic behavior. Finally, we will (in Part II) deal with the dynamic analysis of the forces and torques generated by these moving machines.

1.10   REFERENCES

1. **de Jonge, A. E. R.**, "What is Wrong with 'Kinematics' and 'Mechanisms'? ," *Mech. Eng.*, vol. 64, April, 1942, pp. 273-278.

1.11   BIBLIOGRAPHY

For additional information on the history of kinematics, the following are recommended:

Brown, H. T., *Five Hundred and Seven Mechanical Movements* , Brown, Coombs & Co., New York, 1869, republished by USM Corporation, Beverly, Mass., 1970.

de Jonge, A. E. R., "A Brief Account of Modern Kinematics," *Trans. ASME,* vol. 65, August, 1943, pp. 663-683.

Ferguson, E. S., *Kinematics of Mechanisms from the Time of Watt.*, U.S. Natl. Museum Bull. 228, Paper 27, 1962, pp. 185-230. (Obtainable from the Superintendent of Documents, U.S. Government Printing Office, Washington D.C.)

Freudenstein, F., "Trends in the Kinematics of Mechanisms," *Appl. Mech. Revs,* vol. 12, September, 1959, pp. 587-590.

Hartenberg, R. and **Denavit, J.**, *Kinematic Synthesis of Linkages,* Chap. 1, McGraw-Hill Inc., New York, 1964, pp. 1-27.

Reuleaux, F., *The Kinematics of Machinery,* Dover Publications, New York, 1963.

Strandh, S., *A History of the Machine,* A&W Publishers, New York, 1979.

For additional information on **creativity and the design process**, *the following are recommended:*

Alger, J. R. M. and **Hays, C. V.**, *Creative Synthesis in Design,* , Prentice-Hall, Englewood Cliffs, NJ, 1964.

Allen, M. S., *Morphological Creativity*, Prentice-Hall, Englewood Cliffs, NJ, 1962.

Buhl, H. R., *Creative Engineering Design.*, Iowa State University Press, Ames, IA, 1960.

Gordon, W. J. J., *Synectics*, Harper & Row, New York, 1962.

Haefele, J. W., *Creativity and Innovation*, Van Nostrand Reinhold, New York, 1962.

Harrisberger, L., *Engineersmanship*, 2nd ed., Brooks/Cole, Monterey, CA, 1982.

Osborn, A. F., *Applied Imagination*, Scribner's, 1963.

Pleuthner, W., "Brainstorming." *Machine Design*, Jan. 12, 1956.

Taylor, C. W., *Widening Horizons in Creativity*, John Wiley & Sons, New York, 1964.

Von Fange, E.K., *Professional Creativity*, Prentice-Hall, Englewood Cliffs, NJ, 1959.

For additional information on **Human Factors**, *the following are recommended:*

Bailey, R. W., *Human Performance Engineering: A Guide for System Designers*, Prentice-Hall, New York, 1982.

Burgess, R. W., *Designing for Humans: The Human Factor in Engineering*, Petrocelli Books Inc., 1986.

Clark, T. S., and Corlett, E. N., *The Ergonomics of Workspaces and Machines*, Taylor and Francis Inc., 1984.

Huchinson, R. D., *New Horizons for Human Factors in Design*, McGraw-Hill, New York, 1981.

McCormick, E. J., *Human Factors Engineering*, McGraw-Hill Inc., New York, 1964.

Osborne, D. J., *Ergonomics at Work*, John Wiley & Sons, New York, 1987.

Pheasant, S., *Bodyspace: Anthropometry, Ergonomics & Design*, Taylor and Francis Inc., 1986.

Salvendy, G., *Handbook of Human Factors*, John Wiley & Sons, New York, 1987.

Sanders, M. S., *Human Factors in Engineering and Design*, McGraw-Hill Inc., New York, 1987.

Woodsen, W. E., *Human Factors Design Handbook*, McGraw-Hill Inc., New York, 1981.

For additional information on **writing engineering reports**, *the following are recommended:*

Barrass, R., *Scientists Must Write*, Chapman and Hall, John Wiley & Sons, New York, 1978.

Crouch, W. G., and **Zetler, R. L.**, *A Guide to Technical Writing*, 3rd ed., The Ronald Press Co., New York, 1964.

Davis, D. S., *Elements of Engineering Reports*, Chemical Publishing Co., New York, 1963.

Gray, D. E., *So, You Have to Write a Technical Report*, Information Resources Press, Washington D.C., 1970.

Michaelson, H. B., *How to Write and Publish Engineering Papers and Reports*, ISI Press, Phila. Pa., 1982.

Nelson, J. R., *Writing the Technical Report*, 3ed., McGraw-Hill Inc., New York, 1952

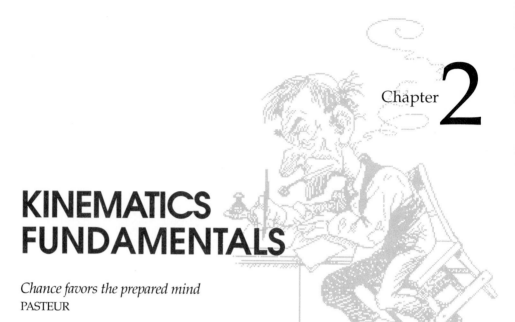

Chapter 2

KINEMATICS FUNDAMENTALS

Chance favors the prepared mind
PASTEUR

2.0   INTRODUCTION

This chapter will present definitions of a number of terms and concepts fundamental to the synthesis and analysis of mechanisms. It will also present some very simple but powerful analysis tools which are useful in the synthesis of mechanisms.

2.1   DEGREES OF FREEDOM (*DOF*)

Any mechanical system can be classified according to the number of **degrees of freedom** (*DOF*) which it possesses. The system's ***DOF*** is equal to *the number of independent parameters (measurements) which are needed to uniquely define its position in space at any instant of time.* Note that *DOF* is defined with respect to a selected frame of reference. Figure 2-1 shows a pencil laying on a flat piece of paper with an xy coordinate system added. If we constrain this pencil to always remain in the plane of the paper, three parameters (*DOF*) are required to completely define the position of the pencil on the paper, two linear coordinates (x, y) to define the position of any one point on the pencil, and one angular coordinate (θ) to define the angle of the pencil with respect to the axes. The minimum number of measurements needed to define its position are shown in the figure as x, y, and θ. This system of the pencil in a plane then has **three *DOF*.** Note that the particular parameters chosen to define its position are not unique. Any alternate set of three parameters could be used. There is an infinity of sets of parameters possible, but in this case there must be three parameters per set, **such as two lengths and an angle,** to define the system's position because *a rigid body in plane motion has three DOF.*

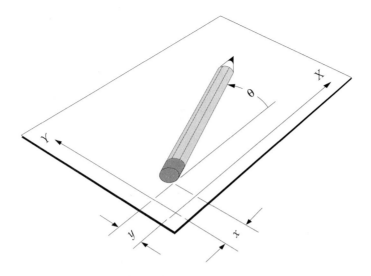

FIGURE 2-1

A rigid body in the plane has three *DOF*

  Now allow the pencil to exist in a three-dimensional world. Hold it above your desktop and move it about. You now will need six parameters to define its **six DOF.** One possible set of parameters which could be used are **three lengths,** (x, y, z), plus **three angles** (θ, ϕ, ρ). *Any rigid body in three-space has six degrees of freedom.* Try to identify these six *DOF* by moving your pencil or pen with respect to your desktop.

  The pencil in these examples represents a **rigid body**, or **link**, which for purposes of kinematic analysis we will assume to be incapable of deformation. This is merely a convenient fiction to allow us to more easily define the gross motions of the body. We can later superpose any deformations due to external or inertial loads onto our kinematic motions to obtain a more complete and accurate picture of the body's behavior. But remember, we are typically facing a *blank sheet of paper* at the beginning stage of the design process. We cannot determine deformations of a body until we define its size, shape, material properties, and loadings. Thus, at this stage we will assume, for purposes of initial kinematic synthesis and analysis, that *our kinematic bodies are rigid and massless.*

2.2 TYPES OF MOTION

A rigid body free to move within a reference frame will, in the general case, have **complex motion**, which is a simultaneous combination of **rotation** and **translation**. In three-dimensional space, there may be rotation about any axis (any skew axis or one of the three principal axes) and also simultaneous translation which can be resolved into components along three axes. In a plane, or two-dimensional space, complex motion becomes a combination of simultaneous rotation about one axis (perpendicular to the plane) and also translation resolved into components along two axes in the plane. For simplicity, we will limit our present discussions to the case of **planar (2-D) kinematic systems**. We will define these terms as follows for our purposes, in planar motion:

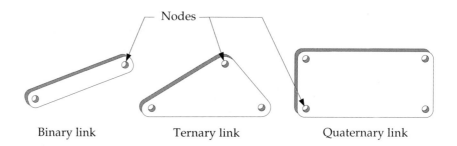

Nodes

Binary link　　　　　　　Ternary link　　　　　　　Quaternary link

FIGURE 2-2

Links of different order

Pure rotation -

the body possesses one point (center of rotation) which has no motion with respect to the "stationary" frame of reference. All other points on the body describe arcs about that center. A reference line drawn on the body through the center changes only its angular orientation.

Pure translation -

all points on the body describe parallel (curvilinear or rectilinear) paths. A reference line drawn on the body changes its linear position but does not change its angular orientation.

Complex motion -

a simultaneous combination of rotation and translation. Any reference line drawn on the body will change both its linear position and its angular orientation. Points on the body will travel non-parallel paths, and there will be, at every instant, a center of rotation, which will continuously change location.

Translation and **rotation** represent independent motions of the body. Each can exist without the other. If we define a 2-D coordinate system as shown in Figure 2-1, the x and y terms represent the translation components of motion, and the θ term represents the rotation component.

2.3　　LINKS, JOINTS, AND KINEMATIC CHAINS

We will begin our exploration of the kinematics of mechanisms with an investigation of the subject of **linkage design**. Linkages are the basic building blocks of all mechanisms. We will show in later chapters that all common forms of mechanisms (cams, gears, belts, chains) are in fact variations on a common theme of linkages. Linkages are made up of links and joints.

A **link**, as shown in Figure 2-2, is an (assumed) rigid body which possesses at least two **nodes** which are *points for attachment to other links.*

Binary link　　　　　　　*- one with two nodes.*

Ternary link　　　　　　　*- one with three nodes.*

Quaternary link　　　　　*- one with four nodes.*

etc.

A **joint** is *a connection between two or more links (at their nodes), which connection allows some motion, or potential motion, between the connected links.* **Joints** (also called **kinematic pairs**) can be classified in several ways:

1 By the number of degrees of freedom allowed at the joint.

2 By the type of contact between the elements, line, point, or surface.

3 By the type of physical closure of the joint: either **force** or **form** closed.

4 By the number of links joined (order of the joint).

Figure 2-3 shows examples of both one and two-freedom joints commonly found in planar mechanisms. Figure 2-3a shows two forms of a planar, **one-freedom** joint (or pair), namely, a rotating pin joint and a translating slider joint. These are both typically referred to as **full joints** or as **lower pairs**. The pin joint allows one rotational *DOF*, and the slider joint allows one translational *DOF* between the joined links. These are both special cases of another common, one-freedom joint, the screw and nut (not shown). Motion of either the nut or the screw with respect to the other results in helical motion. If the helix angle is made zero, the nut rotates without advancing and it becomes the pin joint. If the helix angle is made 90 degrees, the nut will translate along the axis of the screw, and it becomes the slider joint.

The term **lower pair** was coined by Reuleaux[1] to describe joints with surface contact, as with a pin surrounded by a hole. He used the term **higher pair** to describe joints with point or line contact. But if there is any clearance between pin and hole (as there must be for motion), so-called surface contact in the pin joint actually becomes line contact, as the pin contacts only one "side" of the hole. Likewise, at a microscopic level, a block sliding on a flat surface actually has contact only at discrete points, which are the tops of the surfaces' asperities. The main practical advantage of lower pairs over higher pairs is their better ability to trap lubricant between their enveloping surfaces. This is especially true for the rotating pin joint. The lubricant is more easily squeezed out of a higher pair, non-enveloping joint. As a result, the pin joint is preferred for low wear and long life, even over its lower pair cousin, the slider joint.

Figure 2-3b shows examples of two-freedom joints which simultaneously allow two independent, relative motions, namely translation and rotation, between the joined links. Paradoxically, this **two-freedom joint** is usually referred to as a "**half joint**," with its two freedoms placed in the denominator. The **half joint** is sometimes also called a **roll-slide joint** because it allows both rolling and sliding. A joystick, or ball-and-socket joint (Figure 2-3c), is an example of a three-freedom joint, which allows three independent angular motions between the two links joined. This *ball joint* would typically be used in a three-dimensional mechanism, one example being the ball joints in an automotive suspension system.

A joint with more than one freedom is usually also a **higher pair** (the ball joint being an exception). Full joints (lower pairs) and half joints (higher pairs) are both used in planar (2-D), and also in spatial (3-D) mechanisms. Note that if you do not allow the two links in Figure 2-3b connected by a half joint to slide, perhaps by providing a high friction coefficient between them, you can "lock out" the translating (Δx) freedom and make it behave as a full joint. This is then called a

Rotating full pin joint (form closed)

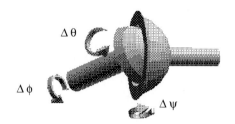

(c) Ball and socket joint - three *DOF*

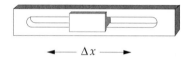

Translating full slider joint (form closed)

(a) Full joints - one *DOF*

First order pin joint - one *DOF*
(two links joined)

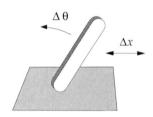

Link against plane (force closed)

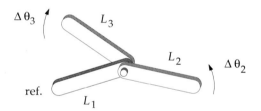

Second order pin joint - two *DOF*
(three links joined)

(d) Order of joints

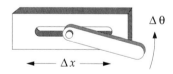

Pin in slot (form closed)

(b) Half joints - two *DOF*

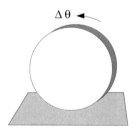

May roll, slide, or roll- slide,
depending on friction

(e) Rolling joint - one or two *DOF*

FIGURE 2-3

Joints of various types

pure rolling joint and has rotational freedom ($\Delta\theta$) only. A common example of this type of joint is your automobile tire rolling against the road, as shown in Figure 2-3e. In normal use there is pure rolling and no sliding at this joint, unless, of course, you encounter an icy road or become too enthusiastic about accelerating or cornering. If you lock your brakes on ice, this joint converts to a pure sliding one like the slider block in Figure 2-3a. Friction determines the actual number of freedoms at this kind of joint. It can be **pure roll**, **pure slide**, or **roll-slide**.

Note that to visualize the degree of freedom of a joint in a mechanism, it is helpful to "mentally disconnect" the two links which create the joint from the rest of the mechanism. You can then more easily see how many freedoms the two joined links have with respect to one another.

Figure 2-3b also shows examples of both **form closed** and **force closed** joints. A **form closed** joint is kept together or *closed by its geometry*. A pin in a hole or a slider in a two-sided slot are form closed. In contrast, a **force closed** joint, such as a pin in a half-bearing or a slider on a surface, *requires some external force to keep it together or closed*. This force could be supplied by gravity, by a spring, or any external means. There can be substantial differences in the behavior of a mechanism due to the choice of force or form closure, as we shall see. The choice should be carefully considered. In linkages, form closure is usually preferred, and it is easy to accomplish. But for cam-follower systems, force closure is often preferred. This topic will be explored further in later chapters.

Figure 2-3d shows examples of joints of various orders, where **order** is defined as *the number of links joined minus one*. It takes two links to make a single joint; thus the simplest joint combination of two links has order one. As additional links are placed on the same joint, the order is increased on a one for one basis. Joint order has significance in the proper determination of overall degree of freedom for the assembly. We gave definitions for a **mechanism** and a **machine** in Chapter 1. With the kinematic elements of links and joints now defined, we can define those devices more carefully based on Reuleaux's classifications of the kinematic chain, mechanism, and machine.[1]

A kinematic chain is defined as:

An assemblage of links and joints, interconnected in a way to provide a controlled output motion in response to a supplied input motion.

A mechanism is defined as:

A kinematic chain in which at least one link has been "grounded," or attached, to the frame of reference (which itself may be in motion).

A machine is defined as:

A combination of resistant bodies arranged to compel the mechanical forces of nature to do work accompanied by determinate motions.

By Reuleaux's definition[1] a machine is *a collection of mechanisms arranged to transmit forces and do work.* He viewed all energy or force transmitting devices as machines which utilize mechanisms as their building blocks to provide the necessary motion constraints.

We will now define a **crank** as *a link which makes a complete revolution and is pivoted to ground*, a **rocker** as *a link which has oscillatory (back and forth) rotation and is pivoted to ground*, and a **coupler** (or connecting rod) *which has complex motion and is not pivoted to ground*. **Ground** is defined as *any link or links that are fixed* (non-moving) with respect to the reference frame. Note that the reference frame may in fact itself be in motion.

2.4   DETERMINING DEGREE OF FREEDOM

The concept of **degree of freedom** (*DOF*) is fundamental to both the synthesis and analysis of mechanisms. We need to be able to quickly determine the *DOF* of any collection of links and joints which may be suggested as a solution to a problem. **Degree of freedom** of a system can be defined as:

DOF

the number of inputs which need to be provided in order to create a predictable output;

also:

the number of independent coordinates required to define its position.

At the outset of the design process, some general definition of the desired output motion is usually available. The number of inputs needed to obtain that output may or may not be specified. Cost is the principal constraint here. Each required input will need some type of actuator, either a human operator or a "slave" in the form of a motor, solenoid, air cylinder, or other energy conversion device. (These devices are discussed in Section 2.15.) These multiple input devices will have to have their actions coordinated by a "controller," which must have some intelligence. This control is now often provided by a computer but can also be mechanically programmed into the mechanism design. There is no require-ment that a mechanism have only one *DOF*, although that is often desirable for simplicity. Some machines have many *DOF*. For example, picture the number of control levers on a bulldozer or crane.

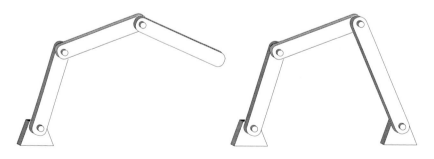

(*a*) Open mechanism chain   (*b*) Closed mechanism chain

FIGURE 2-4

Mechanism chains

Kinematic chains or mechanisms may be either **open** or **closed**. Figure 2-4 shows both open and closed mechanisms. A closed mechanism will have no open attachment points or **nodes** and may have one or more degrees of freedom. An open mechanism of more than one link will always have more than one degree of freedom, thus requiring as many actuators (motors) as it has *DOF*. A common example of an open mechanism is an industrial robot. *An open kinematic chain of two binary links and one joint* is called a **dyad**. The sets of links shown in Figures 2-3a and 2-3b are **dyads**.

Reuleaux limited his definitions to closed kinematic chains and to mechanisms having only one *DOF*, which he called *constrained*.[1] The somewhat broader definitions above are perhaps better suited to current-day applications. A multi-*DOF* mechanism, such as a robot, will be constrained in its motions as long as the necessary number of inputs are supplied to control all its *DOF*.

To determine the overall *DOF* of any mechanism, we must account for the number of links and joints, and for the interactions among them. The *DOF* of any assembly of links can be predicted from an investigation of the **Gruebler condition**.[2]. Any link in a plane has three *DOF*. Therefore, a system of L unconnected links in the same plane will have $3L$ *DOF*, as shown in Figure 2-5a in which two unconnected links have a total of six *DOF*. When these two links are connected by a **full joint** in Figure 2-5b, D_{y1} and D_{y2} are combined as D_y, and D_{x1} and D_{x2} are combined as D_x. This removes two *DOF*, leaving four *DOF*. In Figure 2-5c the half joint removes only one *DOF* from the system (because it has two *DOF*), leaving the system of two links connected by a half joint with a total of five *DOF*. In addition, when any link is grounded or attached to the reference frame, all three of its *DOF* will be removed. This reasoning leads to **Gruebler's Equation**:

$$DOF = 3L - 2J - 3G \qquad (2.1a)$$

where:

L = *number of links*
J = *number of joints*
G = *number of grounded links*

Note that in any real mechanism, even if more than one link is grounded, the net effect will be to create one larger, higher-order ground link, as there is only one ground plane. Thus G is always one, and Gruebler's equation becomes:

$$DOF = 3(L-1) - 2J \qquad (2.1b)$$

The value of J in equations 2.1a and 2.1b must reflect the value of all joints in the mechanism. That is, half joints count as 1/2 because they only remove one *DOF*. It is less confusing if we use **Kutzbach's** modification of Gruebler's equation in this form:

$$DOF = 3(L-1) - 2J_1 - J_2 \qquad (2.1c)$$

where:

L = *number of links*
J_1 = *number of full joints*
J_2 = *number of half joints*

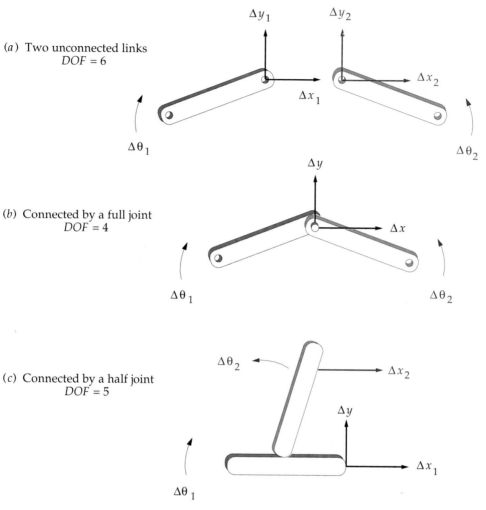

(a) Two unconnected links
      DOF = 6

(b) Connected by a full joint
      DOF = 4

(c) Connected by a half joint
      DOF = 5

FIGURE 2-5

Joints remove *DOF*

The value of J_1 and J_2 in these equations must still be carefully determined to account for all full, half, and multiple joints in any linkage. Multiple joints count as one less than the number of links joined at that joint and add to the "full" (J_1) category. The *DOF* of any proposed mechanism can be quickly ascertained from this expression before investing any time in more detailed design. It is interesting to note that this equation has no information in it about link sizes or shapes, only their quantity. Figure 2-6a shows a mechanism with *DOF* = 1 and only full joints in it.

Figure 2-6b shows a structure with *DOF* = 0, which happens to contain both half and multiple joints. Note the schematic notation used to show the ground link. The ground link need not be drawn in outline as long as all the grounded joints are shown. Note also the joints labelled "**multiple**" and "**half**" in Figures 2-6a and 2-6b. As an exercise, compute the *DOF* of these examples with **Kutzbach's** equation.

(*a*) Full and multiple
jointed linkage

Note:
There are no
half joints in
this linkage

$L = 8, \quad J = 10$

$DOF = 1$

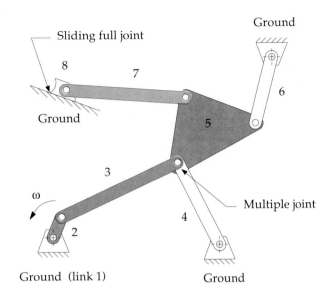

(*b*) Linkage with full, half
and multiple joints

$L = 6, \quad J = 7.5$

$DOF = 0$

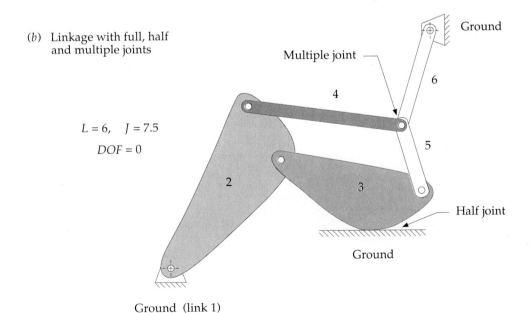

FIGURE 2-6

Linkages containing joints of various types

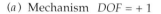

(*a*) Mechanism *DOF* = + 1   (*b*) Structure *DOF* = 0   (*c*) Preloaded structure *DOF* = –1

FIGURE 2-7

Mechanisms, structures and preloaded structures

2.5   MECHANISMS AND STRUCTURES

The degree of freedom of an assembly of links completely predicts its character. There are only three possibilities. *If the DOF is positive, it will be a **mechanism**, and the links will have relative motion. If the DOF is exactly zero, then it will be a **structure**, and no motion is possible. If the DOF is negative, then it is a **preloaded structure**, which means that no motion is possible and some stresses may also be present at the time of assembly. Figure 2-7 shows examples of these three cases. One link is grounded in each case.

Figure 2-7a shows four links joined by four full joints which, from the Gruebler equation, gives one *DOF*. It will move, and only one input is needed to give predictable results.

Figure 2-7b shows three links joined by three full joints. It has zero *DOF* and is thus a **structure**. Note that all three pins can be inserted into their respective pairs of link holes (nodes) without straining the structure, as a position can always be found to allow assembly.

Figure 2-7c shows two links joined by two full joints. It has a *DOF* of minus one, making it a **preloaded structure**. In order to insert the two pins without straining the links, the center distances of the holes in both links must be exactly the same. Practically speaking, it is impossible to make two parts exactly the same. There will always be some manufacturing error, even if very small. Thus you may have to force the second pin into place, creating some stress in the links. The structure will then be preloaded. You have probably met a similar situation in a course in Applied Mechanics in the form of an indeterminate beam, one in which there were too many supports or constraints for the equations available. An indeterminate beam also has negative *DOF*, while a *simply supported* beam has zero *DOF*.

Both structures and preloaded structures are commonly encountered in engineering. In fact the true structure of zero *DOF* is rare in engineering practice. Most buildings, bridges, and machine frames are preloaded structures, due to the use of welded and riveted joints rather than pin joints. Even simple structures like the chair you are sitting in are often preloaded. Since our concern here is with mechanisms, we will concentrate on devices with positive *DOF* only.

2.6   NUMBER SYNTHESIS

The term **number synthesis** has been coined to mean *the determination of the number and order of links and joints necessary to produce motion of a particular DOF.* **Order** in this context refers to the number of nodes per link, i.e., **binary, ternary, quaternary,** etc. The value of number synthesis is to allow the exhaustive determination of all possible combinations of links which will yield any chosen *DOF*. This then equips the designer with a definitive catalog of potential linkages to solve a variety of motion control problems.

As an example we will now derive all the possible link combinations for one *DOF*, including sets of up to eight links, and link orders up to and including hexagonal links. For simplicity we will assume that the links will be connected with only full rotating joints. We can later introduce half joints, multiple joints, and sliding joints through linkage transformation. First let's look at some interesting attributes of linkages as defined by the above assumption regarding full joints.

Hypothesis: If all joints are full joints, an odd number of *DOF* requires an even number of links and vice versa.

Proof: **Given:** All even integers can be denoted by $2m$ or by $2n$, and all odd integers can be denoted by $2m - 1$ or by $2n - 1$, where n and m are any positive integers. The number of joints must be a positive integer.

Let : L = number of links, J = number of joints, and $DOF = 2m$ (i.e., all even numbers)

Then: rewriting Gruebler's equation (Eq 2-1b) to solve for J,

$$J = \frac{3}{2}(L-1) - \frac{DOF}{2}$$ (2.1d)

Try: Substituting $DOF = 2m$, and $L = 2n$ (i.e., both any even numbers):

$$J = 3n - m - \frac{3}{2}$$ (2.1e)

This cannot result in J being a positive integer as required.

Try: $DOF = 2m - 1$ and $L = 2n - 1$ (i.e., both any odd numbers):

$$J = 3n - m - \frac{7}{2}$$ (2.1f)

This also cannot result in J being a positive integer as required.

Try: $DOF = 2m - 1$, and $L = 2n$ (i.e., odd-even):

$$J = 3n - m - 2$$ (2.1g)

This is a positive integer for $m \geq 1$ and $n \geq 2$.

Try: $DOF = 2m$ and $L = 2n - 1$ (i.e., even-odd):

$$J = 3n - m - 3 \qquad (2.1h)$$

This is a positive integer for $m \geq 1$ and $n \geq 2$.

So, for our example of one *DOF* mechanisms, we can only consider combinations of 2, 4, 6, 8 . . . links. Letting the order of the links be represented by:

$$
\begin{aligned}
B &= \textit{number of binary links} \\
T &= \textit{number of ternary links} \\
Q &= \textit{number of quaternaries} \\
P &= \textit{number of pentagonals} \\
H &= \textit{number of hexagonals}
\end{aligned}
$$

the total number of links in any mechanism will be:

$$L = B + T + Q + P + H + \cdots \qquad (2.2a)$$

Since *two link nodes* are needed to make *one joint*:

$$J = \frac{Nodes}{2} \qquad (2.2b)$$

and

$$Nodes = Order\ of\ Link \times No.\ of\ Links\ of\ that\ Order \qquad (2.2c)$$

then

$$J = \frac{(2B + 3T + 4Q + 5P + 6H + \cdots)}{2} \qquad (2.3)$$

Substitute Equations 2.2a and 2.3 into Gruebler's equation (2.1b)

$$DOF = 3(B + T + Q + P + H - 1) - 2\left(\frac{(2B + 3T + 4Q + 5P + 6H)}{2}\right)$$

$$(2.4)$$

$$DOF = B - Q - 2P - 3H - 3$$

Note what is missing from this equation! The ternary links have dropped out. The *DOF* is independent of the number of ternary links in the mechanism. But because each ternary link has three nodes, it can only create or remove 3/2 joints. So we must add or subtract ternary links in pairs to maintain an integer number of joints. *The addition or subtraction of ternary links in pairs will not affect the DOF of the mechanism.*

In order to determine all possible combinations of links for a particular *DOF*, we must combine equations 2.1b and 2.3:

$$\frac{3}{2}(L - 1) - \frac{DOF}{2} = \frac{(2B + 3T + 4Q + 5P + 6H)}{2}$$

$$(2.5)$$

$$3L - 3 - DOF = 2B + 3T + 4Q + 5P + 6H$$

Now combine Equation 2.5 with Equation 2.2a to eliminate *B*:

$$L - 3 - DOF = T + 2Q + 3P + 4H \qquad (2.6)$$

We will now solve equations 2.2a and 2.6 simultaneously (by progressive substitution) to determine all compatible combinations of links for $DOF = 1$, up to eight links. The strategy will be to start with the smallest number of links, and the highest-order link possible with that number, eliminating impossible combinations.

(Note: *L must be even for odd DOF.*)

CASE 1. $L = 2$

$$L - 4 = T + 2Q + 3P + 4H = -2 \qquad (2.7a)$$

This requires a negative number of links, so $L = 2$ is impossible.

CASE 2. $L = 4$

$$L - 4 = T + 2Q + 3P + 4H = 0; \qquad \text{so: } T = Q = P = H = 0$$
$$(2.7b)$$
$$L = B + 0 = 4; \qquad\qquad B = 4$$

The simplest one *DOF* linkage is four binary links - the ***fourbar linkage***.

CASE 3. $L = 6$

$$L - 4 = T + 2Q + 3P + 4H = 2; \qquad \text{so: } \qquad P = H = 0 \qquad (2.7c)$$

T may only be 0, 1, or 2, $\qquad\qquad Q$ may only be 0, or 1

If $Q = 0$ then T must be 2 and:

$$L = B + 2T + 0Q = 6; \qquad B = 4, \qquad T = 2 \qquad (2.7d)$$

If $Q = 1$ then T must be 0 and:

$$L = B + 0T + 1Q = 6; \qquad B = 5, \qquad Q = 1 \qquad (2.7e)$$

There are then two possibilities for $L = 6$. Note that one of them is in fact the simpler fourbar with two ternaries added as was predicted above.

TABLE 2-1   One *DOF* Mechanisms

Links	B	T	Q	P	H
4	4	–	–	–	–
6	4	2	–	–	–
6	5	–	1	–	–
8	7	–	–	–	1
8	4	4	–	–	–
8	5	2	1	–	–
8	6	–	2	–	–
8	6	1	–	1	–

CASE 4. $L = 8$

A tabular approach is needed with this many links:

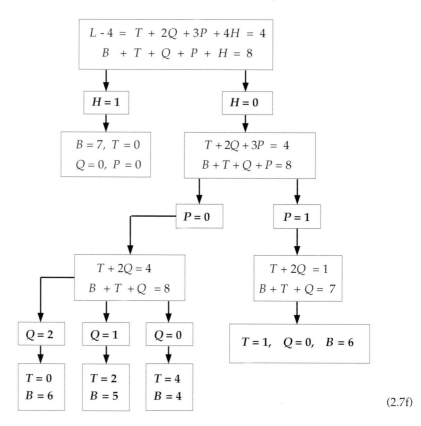

$$(2.7f)$$

From this analysis we can see that, for one *DOF,* there is only one possible four link configuration, two six link configurations and five possibilities for eight links using binary through hexagonal links. Table 2-1 (on previous page) shows a summary of all these possible linkages for the one degree of freedom case.

2.7   PARADOXES

Because the Gruebler criterion pays no attention to link sizes or shapes, it *can give misleading results* in the face of unique geometric configurations. For example, Figure 2-8a shows a structure ($DOF = 0$) with the ternary links of arbitrary shape. This link arrangement is sometimes called the "**E-Quintet**," because of its resemblance to a capital **E** and the fact that it has five links, including the ground. It is the next simplest **structural** building block to the "**delta triplet**."

Figure 2-8b shows the same E-Quintet with the ternary links straight and parallel and with equispaced nodes. The three binaries are also equal in length. With this very unique geometry, you can see that it will move despite Gruebler's prediction to the contrary.

(*a*) The E - Quintet with $DOF = 0$
agrees with Gruebler equation

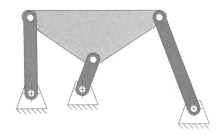

(*b*) The E - Quintet with $DOF = 1$
Gruebler still predicts $DOF = 0$

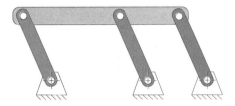

Full joint -
pure rolling
no slip

(*c*) Rolling cylinders - $DOF = 1$
Gruebler predicts zero

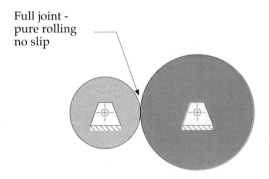

FIGURE 2-8

Gruebler paradoxes

Figure 2-8c shows a very common mechanism which also disobeys Gruebler's criterion. The joint between the two wheels can be postulated to allow no slip, provided that sufficient friction is available. If no slip occurs, then this is a one-freedom, or full, joint which allows only relative angular motion ($\Delta\theta$) between the wheels. With that assumption, there are 3 links and 3 full joints, from which Gruebler's equation predicts zero DOF. However, this linkage does move (actual $DOF = 1$), because the center distance, or length of link 1, is exactly equal to the sum of the radii of the two wheels.

There are other examples of paradoxes which disobey the Gruebler criterion due to their unique geometry. The designer needs to be alert to these possible inconsistencies.

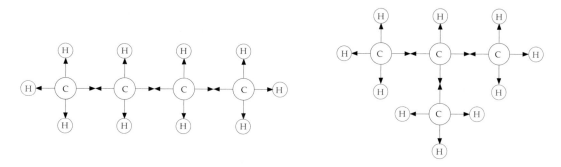

(*a*) Hydrocarbon isomers n-butane and isobutane

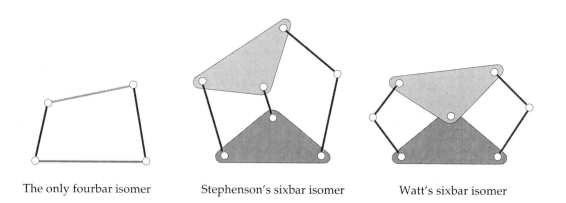

The only fourbar isomer   Stephenson's sixbar isomer   Watt's sixbar isomer

(*b*) All valid isomers of the fourbar and sixbar linkages

Structural subchain
reduces three links
to a zero *DOF*
"delta triplet" truss

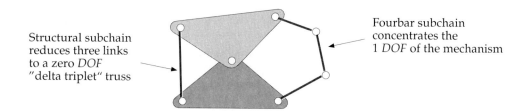

Fourbar subchain
concentrates the
1 *DOF* of the mechanism

(*c*) An invalid sixbar isomer which reduces to the simpler fourbar

FIGURE 2-9

Isomers

2.8   ISOMERS

The word **isomer** is from the Greek and means *having equal parts.* Isomers in chemistry are compounds which have the same number and type of atoms but which are interconnected differently and thus have different physical properties. Figure 2-9a shows two hydrocarbon isomers, n-butane and isobutane. Note that each has the same number of carbon and hydrogen atoms (C_4H_{10}), but they are differently interconnected and have different properties.

Linkage isomers are analogous to these chemical compounds in that the **links** (like atoms) have various **nodes** (electrons) available to connect to other links' nodes. The assembled linkage is analogous to the chemical compound. Depending on the particular connections of available links, the assembly will have different motion properties. The number of isomers possible from a given collection of links (such as in Table 2-1) is far from obvious. In fact the problem of mathematically predicting the number of isomers of all link combinations has yet to be solved. Brute force techniques have been applied to the determination of isomers for various cases involving relatively few links.

Figure 2-9b shows all the isomers for the simple cases of $DOF = 1$ with 4 and 6 links. Note that there is only one isomer for the case of 4 links. An isomer is only unique if the interconnections between its types of links are different. That is, all binary links are considered equal, just as all hydrogen atoms are equal in the chemical analogue. Link lengths and shapes do not figure into the Gruebler criterion or the condition of isomerism. The six link case of 4 binaries and 2 ternaries has only two valid isomers. These are known as the **Watt's chain** and the **Stephenson's chain** after their discoverers. Note the different interconnections of the ternaries to the binaries in these two examples. The Watt's chain has the two ternaries directly connected, but the Stephenson's chain does not.

There is also a third potential isomer for this case of six links, as shown in Figure 2-9c, but it fails the test of **distribution of degree of freedom**, which requires that the overall DOF (here $DOF = 1$) be uniformly distributed throughout the linkage and not concentrated in a subchain. Note that this arrangement (Figure 2-9c) has a **structural subchain** of $DOF = 0$ in the triangular formation of the two ternaries and in the single binary connecting them. This creates a truss, called the **delta triplet**. The remaining three binaries in series form a fourbar chain ($DOF = 1$) with the structural subchain of the two ternaries and the single binary effectively reduced to a structure which acts like a single link. Thus this arrangement has been reduced to the simpler case of the fourbar linkage despite its six bars. This is an **invalid isomer** and is rejected. It is left as an exercise for the reader to find all the valid isomers of the 8 bar, $DOF = 1$ cases.

2.9   LINKAGE TRANSFORMATION

The synthesis techniques described above give the designer a toolkit of basic linkages of particular DOF. If we now relax the arbitrary constraint which restricted us to only full, rotating joints, we can transform these basic linkages to a wider variety of mechanisms with even greater usefulness. There are several transformation techniques which we can apply.

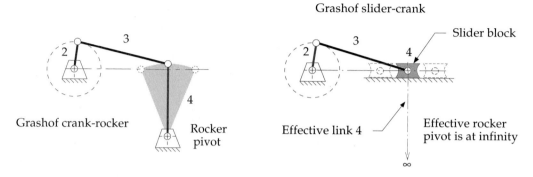

Grashof crank-rocker

Grashof slider-crank

(*a*) Transforming a fourbar crank-rocker to a slider-crank

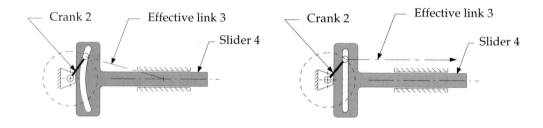

(*b*) Transforming the slider-crank to the Scotch yoke

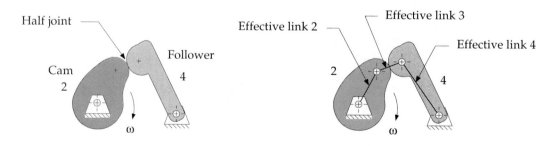

(*c*) Cam-follower has an effective fourbar equivalent

FIGURE 2-10

Linkage transformation

1   Any full rotating joint can be replaced by a sliding full joint with no change in *DOF* of the mechanism.

2   Any full joint can be replaced by a half joint, but this will increase the *DOF* by one.

3   Removal of a link will reduce the *DOF* by one.

4   The combination of items 2 and 3 above will keep the original *DOF* unchanged.

5   Any ternary or higher-order link can be partially "shrunk" to a lower-order link by coalescing nodes. This will create a multiple joint but will not change the *DOF* of the mechanism.

6   Complete shrinkage of a higher-order link is equivalent to its removal. A multiple joint will be created, and the *DOF* will be reduced.

Figure 2-10a shows a fourbar crank-rocker linkage transformed into the fourbar slider-crank by the application of rule #1. It is still a fourbar linkage. Link 4 has become a sliding block. The Gruebler's equation is unchanged at one *DOF* because the slider block provides a full joint against link 1, as did the pin joint it replaces. Note that this transformation from a rocking output link to a slider output link is equivalent to increasing the length (radius) of rocker link 4 until its arc motion at the joint between links 3 and 4 becomes a straight line. Thus the slider block is equivalent to an infinitely long rocker link 4, which is pivoted at infinity along a line perpendicular to the slider axis as shown in Figure 2-10a.

Figure 2-10b shows a fourbar slider-crank transformed via rule #4 by the substitution of a half joint for the coupler. The first version shown retains the same motion of the slider as the original linkage by use of a curved slot in link 4. The effective coupler is always perpendicular to the tangent of the slot and falls on the line of the original coupler. The second version shown has the slot made straight and perpendicular to the slider axis. The effective coupler now is "pivoted" at infinity. This is called a **Scotch yoke** and gives exact *simple harmonic motion* of the slider in response to a constant speed input to the crank.

Figure 2-10c shows a fourbar linkage transformed into a **cam-follower** linkage by the application of rule #4. Link 3 has been removed and a half joint substituted for a full joint between links 2 and 4. This still has one *DOF*, and the cam-follower is in fact a fourbar linkage in another disguise, in which the coupler (link 3) has become an effective link of *variable length*. We will investigate the fourbar linkage and these variants of it in greater detail in later chapters.

Figure 2-11a shows the **Stephenson's sixbar chain** from Figure 2-9b transformed by *partial shrinkage* of a ternary link (rule #5) to create a multiple joint. It is still a one-*DOF* Stephenson's sixbar. Figure 2-11b shows the **Watt's sixbar chain** from Figure 2-9b with one ternary link *completely shrunk* to create a multiple joint. This is now a structure with *DOF* = 0. The two triangular sub-chains are obvious. Just as the fourbar chain is the basic building block of one-*DOF* mechanisms, this threebar triangle **delta triplet** is the *basic building block* of zero-*DOF* structures (trusses).

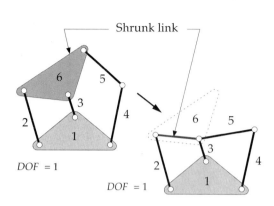

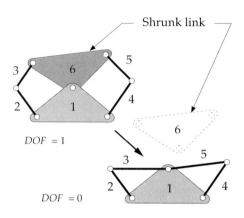

(*a*) Partial shrinkage of higher link
retains original *DOF*

(*b*) Complete shrinkage of higher link
reduces *DOF* by one

FIGURE 2-11

Link shrinkage

2.10   INTERMITTENT MOTION

Intermittent motion is *a sequence of motions and dwells.* A **dwell** is *a period in which the output link remains stationary while the input link continues to move.* There are many applications in machinery which require intermittent motion. The **cam-follower** variation on the fourbar linkage as shown in Figure 2-10c is often used in these situations. The design of that device for both intermittent and continuous output will be addressed in detail in Chapter 9. Other pure linkage **dwell mechanisms** are discussed in the next chapter.

GENEVA MECHANISM A common form of intermittent motion device is the **Geneva mechanism** shown in Figure 2-12a. This is also a transformed fourbar linkage in which the coupler has been replaced by a half joint. The input crank (link 2) is typically motor driven at a constant speed. The **Geneva wheel** is fitted with at least three equispaced, radial slots. The crank has a **pin** which enters a radial slot and causes the Geneva wheel to turn through a portion of a revolution. When the pin leaves that slot, the Geneva wheel remains stationary until the pin enters the next slot. The result is intermittent rotation of the Geneva wheel.

The crank is also fitted with an arc segment, which fits into a matching cutout on the periphery of the Geneva wheel when the pin is out of the slot. This keeps the Geneva wheel stationary and in the proper location for the next entry of the pin. The number of slots determines the number of "stops" of the mechanism, where *stop* is synonymous with *dwell*. A Geneva wheel needs a minimum of three stops to work. The maximum number of stops is limited only by the size of the wheel.

RATCHET AND PAWL Figure 2-12b shows a ratchet and pawl mechanism. The **arm** pivots about the center of the toothed **ratchet wheel** and is moved back and forth to index the wheel. The **driving pawl** rotates the ratchet wheel (or **ratchet**) in

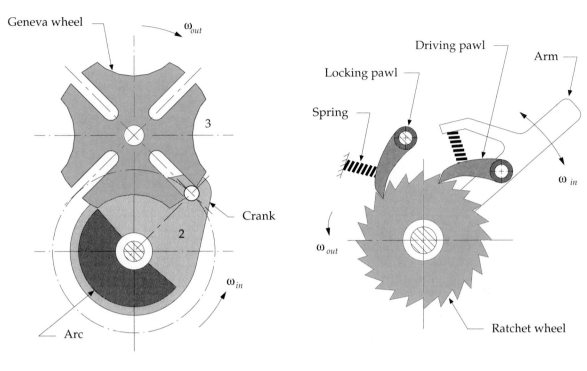

(a) Four-stop Geneva mechanism

(b) Ratchet and pawl mechanism

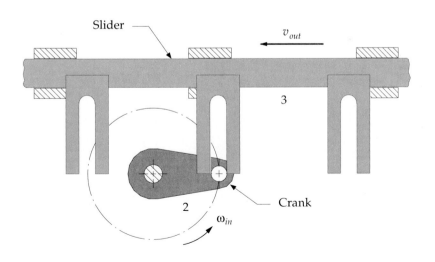

(c) Linear intermittent motion "Geneva" mechanism

FIGURE 2-12

Rotary and linear intermittent motion mechanisms

the counterclockwise direction and does no work on the return (clockwise) trip. The **locking pawl** prevents the ratchet from reversing direction while the driving pawl returns. Both pawls are usually spring-loaded against the ratchet. This mechanism is widely used in devices such as "ratchet" wrenches, winches, etc.

LINEAR GENEVA MECHANISM   There is also a variation of the Geneva mechanism which has linear translational output, as shown in Figure 2-12c. This mechanism is analogous to an open Scotch yoke device with multiple yokes. It can be used as an intermittent conveyor drive with the slots arranged along the conveyor chain or belt. It is also sometimes used with a reversing motor to get linear, reversing oscillation of a single slotted output slider.

2.11   INVERSION

It should now be apparent that there are many possible linkages for any situation. Even with the limitations imposed in the example (one *DOF*, eight links, up to hexagonal order), there are eight number synthesis linkage combinations shown in Table 2-1, and these together yield many valid isomers. In addition, we can introduce another factor, namely mechanism inversion. An **inversion** is *created by grounding a different link in the kinematic chain.* Thus there are as many inversions of a given linkage as it has links.

The motions resulting from each inversion can be quite different, but some inversions of a linkage may yield motions similar to other inversions of the same linkage. In these cases only some of the inversions may have distinctly different

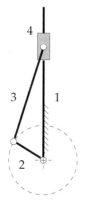  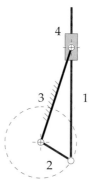 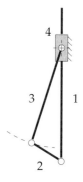

(*a*) Inversion # 1   (*b*) Inversion # 2   (*c*) Inversion # 3   (*d*) Inversion # 4
   slider block   slider block has   slider block   slider block
   translates   complex motion   rotates   is stationary

FIGURE 2-13

Four distinct inversions of the fourbar slider-crank mechanism motions. We will denote the *inversions which have distinctly different motions* as **distinct inversions.**

Figure 2-13 shows the four inversions of the fourbar slider-crank linkage, all of which have distinct motions. Inversion #1, with link 1 as ground and its slider block in pure translation, is the most commonly seen and is used for **piston engines** and **piston pumps.** Inversion #2 is obtained by grounding link 2 and gives the **Whitworth** or **crank-shaper** quick-return mechanism, in which the slider block has complex motion. (Quick-return mechanisms will be investigated further in the next chapter.) Inversion #3 is obtained by grounding link 3 and gives the slider block pure rotation. Inversion #4 is obtained by grounding the slider link 4 and is used in hand operated, **well pump** mechanisms, in which the handle is link 2 (extended) and link 1 passes down the well pipe to mount a piston on its bottom. (It is shown upside-down in the figure.)

The **Watt's sixbar chain** has two distinct inversions, and the **Stephenson's sixbar** has three distinct inversions, as shown in Figure 2-14. The pin-jointed fourbar has three distinct inversions: the crank-rocker, double-crank and double-rocker which are shown in Figures 2-15 and 2-16.

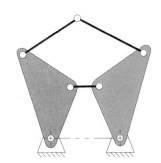 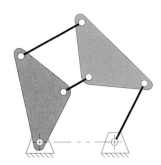 

Stephenson's sixbar inversion I   Stephenson's sixbar inversion II   Stephenson's sixbar inversion III

 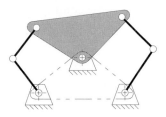

Watt's sixbar inversion I   Watt's sixbar inversion II

FIGURE 2-14

All distinct inversions of the sixbar linkage

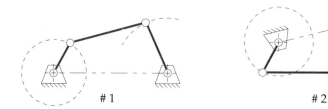

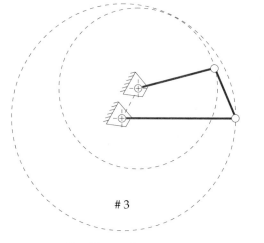

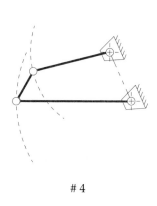

#3

Double-crank inversion
(drag link)

#4

Double-rocker inversion
(coupler rotates)

FIGURE 2-15

All inversions of the Grashof fourbar linkage

2.12   THE GRASHOF CONDITION

The **fourbar linkage** has been shown above to be the *simplest possible pin-jointed mechanism* for single degree of freedom controlled motion. It also appears in various disguises such as the **slider-crank** and the **cam-follower**. It is in fact the most common and ubiquitous device used in machinery. It is also extremely versatile in terms of the types of motion which it can be generate.

Simplicity is one mark of good design. The fewest parts that can do the job will usually give the least expensive and most reliable solution. Thus the **four-bar linkage** should be among the first solutions to motion control problems to be investigated. The **Grashof condition**[3] is a very simple relationship which predicts the behavior of a fourbar linkage's inversions based only on the link lengths.

Let: S =   *length of shortest link*
        L =   *length of longest link*
        P =   *length of one remaining link*
        Q =   *length of other remaining link*

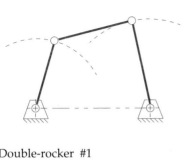

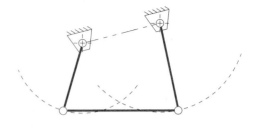

(a) Double-rocker #1

(b) Double-rocker # 2

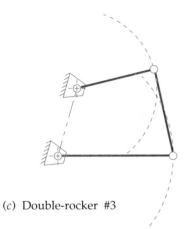

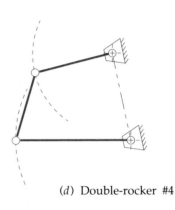

(c) Double-rocker #3

(d) Double-rocker #4

FIGURE 2-16

All inversions of the non-Grashof fourbar

Then if:

$$S + L \leq P + Q \qquad (2.8)$$

the linkage is **Grashof** and at least one link will be capable of making a full revolution with respect to the ground plane. If that inequality is not true, then the linkage is **non-Grashof** and *no* link will be capable of a complete revolution relative to the ground plane.

Note that the above statements apply regardless of the order of assembly of the links. That is, the determination of the Grashof condition can be made on a set of unassembled links. Whether they are later assembled in S, L, P, Q order or S, P, L, Q or any other order, will *not* change the Grashof condition.

The motions possible from a fourbar linkage will depend on both the Grashof condition and the **inversion** chosen. The inversions will be defined with respect to the shortest link. The motions are:

For the case, $S + L < P + Q$:

Ground either link adjacent to the shortest and you get a **crank-rocker**, in which the shortest will fully rotate and the other link pivoted to ground will oscillate.

Ground the shortest link and you will get a **double-crank**, in which both links pivoted to ground make complete revolutions as does the coupler.

Ground the link opposite the shortest and you will get a **Grashof double-rocker**, in which both links pivoted to ground oscillate and only the coupler makes a full revolution.

For the case, $S + L > P + Q$:

All inversions will be **double-rockers** in which no link can fully rotate.

For the case, $S + L = P + Q$:

Referred to as **special case Grashof**, all inversions will be either **double-cranks** or **crank-rockers** but will have "**change points**" twice per revolution of the input crank when the links all become colinear. At these change points the output behavior will become indeterminate. The linkage behavior is then unpredictable as it may assume either of two configurations. Its motion must be limited to avoid reaching the change points or an additional, out-of-phase link provided to guarantee a "carry through" of the change points. (See Figure 2-17c.)

Figure 2-15 (p. 44) shows the four possible inversions of the **Grashof case**: two crank-rockers, a double-crank (also called a drag link), and a double-rocker with rotating coupler. The two crank rockers give similar motions and so are not distinct from one another. Figure 2-16 (p. 45) shows four non-distinct inversions, all double-rockers, of a **non-Grashof linkage**.

Figure 2-17a and 2-17b show the **parallelogram** and **anti-parallelogram** configurations of the **special case Grashof** linkage. The parallelogram linkage is quite useful as it exactly duplicates the rotary motion of the driver crank at the driven crank. One common use is to couple the two windshield wiper output rockers across the width of the windshield on an automobile. The coupler of the parallelogram linkage is in curvilinear translation, remaining at the same angle while all points on it describe identical circular paths. It is often used for this parallel motion, as in truck tailgate lifts and industrial robots.

The anti-parallelogram linkage is also a double-crank, but the output crank has an angular velocity different from the input crank. Note that the change points allow the linkage to switch unpredictably between the parallelogram and anti-parallelogram forms every 180 degrees unless some additional links are provided to carry it through those positions. This can be achieved by adding an out-of-phase companion linkage coupled to the same crank, as shown in Figure 2-17c. A common application of this double parallelogram linkage was on steam locomotives, used to connect the drive wheels together. The change points were handled by providing the duplicate linkage, ninety degrees out of phase, on the other side of the locomotive. When one side was at a change point, the other side would drive it through.

The **double parallelogram** arrangement shown in Figure 2-17c is quite useful as it gives a translating coupler which remains horizontal in all positions.

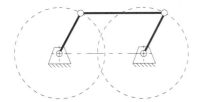

(*a*) Parallelogram form

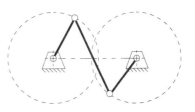

(*b*) Antiparallelogram form

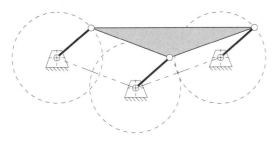

(*c*) Double-parallelogram linkage gives parallel
motion (pure curvilinear translation) to coupler
and also carries through the change points.

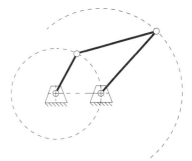

(*d*) Deltoid form

FIGURE 2-17

Some forms of the special case Grashof linkage

The two parallelogram stages of the linkage are out of phase so each carries the other through its change points. Figure 2-17d shows the **deltoid** configuration which is a crank-rocker.

There is nothing either bad or good about the Grashof condition. Linkages of all three persuasions are equally useful in their place. If, for example, your need is for a motor driven windshield wiper linkage, you may want a non-special case Grashof linkage in order to have a rotating link for the motor's input, plus a special case parallelogram stage to couple the two sides together as described above. If your need is to control the wheel motions of a car over bumps, you may want a non-Grashof linkage for short stroke oscillatory motion. If you want to exactly duplicate some input motion at a remote location, you may want a special case Grashof parallelogram linkage, as used in a drafting machine. In any case, this simply determined condition tells volumes about the behavior to be expected from a proposed fourbar linkage design prior to any construction of models or prototypes.

2.13   LINKAGES OF MORE THAN FOUR BARS

We have seen that the simplest one degree of freedom linkage is the fourbar. It is an extremely versatile and useful device. Many quite complex motion control problems can be solved with just four links and four pins. Thus in the interest of simplicity,

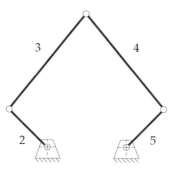 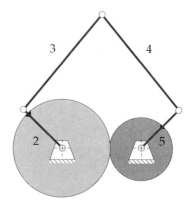

  (*a*)   Fivebar linkage – 2 *DOF*   (*b*)   Geared fivebar linkage – 1 *DOF*

FIGURE 2-18

Two forms of the fivebar linkage

designers should always first try to solve their problems with this device. However, there will be cases when a more complicated solution is necessary. Adding one link and one joint to form a fivebar (Figure 2-18a) will increase the *DOF* by one, to two. By adding a pair of gears to tie two links together, the *DOF* is reduced again to one, and the **geared fivebar mechanism (GFBM)** shown in Figure 2-18b is created.

     This mechanism will provide more complex motions than the fourbar at the expense of the added link and gearset. The student may observe the dynamic behavior of the linkage in Figure 2-18b by running the program FIVEBAR provided with this text and reading the data file named FIG2-18B into that program. Look ahead to Chapter 8 for help in running the programs. Simply accept all the default responses, and animate the linkage.

     We have already met the Watt's and Stephenson's sixbar mechanisms in Figure 2-9b in Section 2.8. The **Watt's sixbar** can be thought of as *two fourbar linkages connected in series* and sharing two links in common. The **Stephenson's sixbar** can be thought of as *two fourbar linkages connected in parallel* and sharing two links in common. Many linkages can be designed by the technique of combining multiple fourbar chains as *basic building blocks* into more complex assemblages. Many real design problems will require solutions consisting of more than four bars. Some Watt's and Stephenson's linkages are provided as built-in examples to the program SIXBAR supplied with this text. You may run that program to observe these linkages dynamically. Select any example from the menu, accept all default responses and animate the linkages.

2.14   SPRINGS AS LINKS

We have so far been dealing only with rigid links. In many mechanisms and machines, it is necessary to counterbalance the static loads applied to the device. A

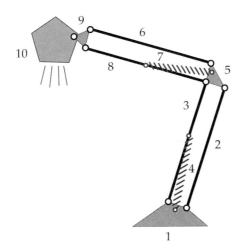

FIGURE 2-19

A spring balanced linkage

common example is the hood hinge mechanism on your automobile. Unless you have the (cheap) model with the strut that you place in a hole to hold up the hood, it will have a fourbar linkage connecting the hood to the body on each side. The hood is the coupler of a non-Grashof linkage whose two rockers are pivoted to the body. A spring is also placed between two of the links to provide a force to hold the hood in the open position. The spring in this case is a fifth link of variable length. As long as it can provide the right amount of force, it acts to reduce the *DOF* of the mechanism to zero, and it holds the system in static equilibrium. However, you can force it to again be a one *DOF* system by overcoming the spring force when you pull the hood shut.

Another example, which may now be right next to you, is the ubiquitous adjustable arm desk lamp, as shown in Figure 2-19. This device, if well designed and made, will remain stable over a fairly wide range of positions despite variation in the overturning moment due to the lamp head's changing moment arm. This is accomplished by careful design of the geometry of the spring-link relationships so that, as the spring force changes with increasing length, its moment arm also changes in a way that continually balances the changing moment of the lamp head. A linear spring can be characterized by its spring constant, $k = F/x$, where F is force and x is spring displacement. Doubling its deflection will double the force. Most coil springs of the type used in these examples are linear.

2.15   PRACTICAL CONSIDERATIONS

PIN JOINTS VERSUS SLIDERS AND HALF JOINTS   Proper material selection and good lubrication are the key to long life in any situation, such as a joint, where two materials rub together. Such an interface is called a **bearing**. Assuming the proper materials have been chosen, the choice of joint type can have a significant effect on the ability to provide good, clean lubrication over the lifetime of the machine.

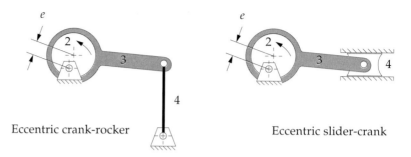

Eccentric crank-rocker   Eccentric slider-crank

FIGURE 2-20

Eccentric cranks

The simple **pin joint** (Figure 2-3a) is the clear winner here for several reasons. Its geometry of a pin in a hole is conducive to the trapping of a lubricant film within the annular interface. Seals can easily be provided at the ends of the hole, wrapped around the pin to prevent loss of lubricant. Replacement lubricant can be introduced through radial holes into the bearing interface, either continuously or periodically, without disassembly. Relatively inexpensive ball and roller bearings are also commercially available in a large variety of sizes for pin type joints. These can be obtained prelubricated and with end seals. The rolling elements in these bearings provide low-friction operation and good dimensional control. Thus it is relatively easy and inexpensive to design and build a good quality pin joint.

SLIDERS require a carefully machined and straight slot or rod (Figure 2-3a). The bearings often must be custom made, though linear ball bearings are commercially available to run over hardened shafts. Lubrication is difficult to maintain on a sliding joint. The lubricant is not geometrically captured, and it must be resupplied either by running the joint in an oil bath or by periodic manual regreasing. An open slot or shaft tends to accumulate airborne dirt particles which can act as a grinding compound when trapped in the lubricant. This will accelerate wear.

HALF JOINTS, such as a round pin in a slot (Figure 2-3b) or a cam-follower joint, suffer even more acutely from the slider's lubrication problems, because they typically have two oppositely curved surfaces in line contact, which tend to squeeze any lubricant out of the joint. This type of joint needs to be run in an oil bath for long life. This requires that the assembly be housed in an expensive, oil-tight box with seals on all protruding shafts.

So it should be obvious that the "pure" pin-jointed linkage with good bearings at the joints is a potentially superior design, all else equal, and it should be the first possibility to be explored in any machine design problem. However, there will be many problems in which the need for a straight, sliding motion or the exact dwells of a cam-follower are required. Then their practical limitations will have to be dealt with accordingly. All three joint types are used extensively in all kinds of machinery with great success. As long as the proper *attention to engineering detail* is paid, the design can be successful.

Some common examples of all three joint types can be found in an automobile. The windshield wiper mechanism is a pure pin-jointed linkage. The pistons in

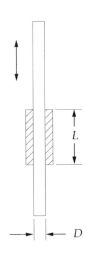

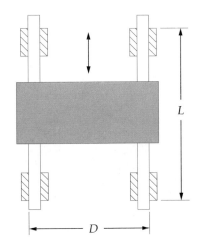

(*a*) Single rod in bushing   (*b*) Platform on two rods

FIGURE 2-21

Bearing ratio

the engine cylinders are true sliders and are bathed in engine oil. The valves in the engine are opened and closed by cam-follower joints which are drowned in engine oil. You probably change your engine oil fairly frequently. When was the last time you lubricated your windshield wiper linkage? Has this linkage (not the motor) ever failed?

SHORT LINKS It sometimes happens that the required length of a crank is so short that it is not possible to provide suitably sized pins at each end. The solution is to design the link as an **eccentric crank**, as shown in Figure 2-20. One pivot pin is enlarged to the point that it, in effect, contains the link. The outside diameter of the circular crank is the journal for the moving pivot. The fixed pivot is placed a distance *e* from the center of this circle equal to the required crank length. The distance *e* is the crank's eccentricity. This arrangement has the advantage of a large surface area within the bearing to reduce wear, though keeping the large diameter journal lubricated can be difficult.

BEARING RATIO The need for straight line motion in machinery requires extensive use of linear translating slider joints. There is a very basic geometrical relationship called *bearing ratio*, which if ignored or violated will invariably lead to problems.

The **bearing ratio** (BR) is defined as *the effective length of the slider over the effective diameter of the bearing*: $BR = L/D$. For smooth operation **this ratio should be greater than 1.5, and never less than 1**. The larger it is, the better. **Effective length** is defined as *the distance over which the moving slider contacts the stationary guide*. There need not be continuous contact over that distance. That is, two short collars, spaced far apart, are effectively as long as their overall separation plus their own lengths and are kinematically equivalent to a long tube. **Effective diameter** is *the largest distance across the stationary guides*, in any plane perpendicular to the sliding motion.

If the slider joint is simply a rod in a bushing, as shown in Figure 2-21a, the effective diameter and length are identical to the actual dimensions of the rod diameter and bushing length. If the slider is a platform riding on two rods and multiple bushings, as shown in Figure 2-21b, then the effective diameter and length are the overall width and length, respectively, of the platform assembly. It is this case that often leads to poor bearing ratios.

A common example of a device with a poor bearing ratio is a drawer in an inexpensive piece of furniture. If the only guides for the drawer's sliding motion are its sides running against the frame, it will have a bearing ratio less than 1, since it is wider than it is deep. You have probably experienced the sticking and jamming that occurs with such a drawer. A better quality chest of drawers will have a center guide with a large L/D ratio under the bottom of the drawer and will slide smoothly.

LINKAGES VERSUS CAMS   The pin-jointed linkage has all the advantages listed above. The cam-follower mechanism (Figure 2-10c) has all the problems associated with the half joint listed above. However, both are widely used in machine design, often in the same machine and in combination (cams driving linkages). So why choose one over the other?

Linkages have the disadvantage of relatively large size compared to the output displacement of the working portion, thus they are somewhat difficult to package. Cams tend to be compact in size compared to the follower displacement. Linkages are relatively difficult to synthesize, and cams are relatively easy to design (as long as a computer is available). But linkages are much easier and cheaper to manufacture than cams. Dwells are easy to get with cams, and difficult with linkages. Linkages can survive very hostile environments, with poor lubrication, whereas cams cannot, unless sealed from the environmental contaminants.

So the answer is far from clear cut. It is another *design tradeoff situation* in which you must weigh all the factors and make the best compromise. The potential advantages of the pure linkage make it important to consider it before choosing an easier design task but an ultimately more expensive solution.

MOTORS AND DRIVERS   All of the mechanisms synthesized here will require some type of driver to provide the input motion and energy. There are many possibilities. If the design requires a continuous rotary input motion, such as for a Grashof linkage, a slider-crank, or a cam-follower, then a motor is the logical choice.

Motors come in a wide variety of types. The most common energy source for a motor is electricity, but compressed air and pressurized hydraulic fluid are also used to power air and hydraulic motors. Gasoline or diesel engines are another possibility. **Electric motors** are made in several designs, among which are **AC**, **DC**, **servo**, and **stepping**.

AC and **DC** refer to *alternating current* and *direct current* respectively. AC is typically supplied by the power companies and, in the U.S., will be alternating at 60 hertz, at about ±110 volts or ±220 volts. DC current is typically supplied from battery sources and is found in any portable vehicle, whether boat, automobile, or aircraft. Battery voltages are in multiples of 1.5 volts, with 12 volts being the most common.

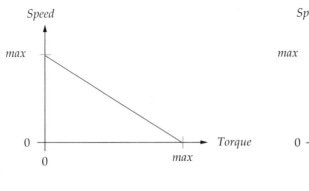

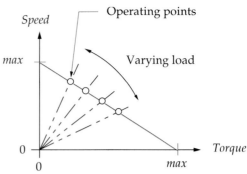

(a) *Speed-Torque* characteristic
of a PM electric motor

(b) Loadlines superposed
on *Speed-Torque* curve

FIGURE 2-22

DC permanent magnet (PM) motor *Speed-Torque* characteristic

DC Motors are made in different electrical configurations, which provide differing *torque-speed* characteristics. The *torque-speed* curve of a motor describes how it will respond to an applied load.

Figure 2-22a shows such a curve for a **permanent magnet** (PM) DC motor. Note that the torque varies greatly with speed, ranging from maximum torque at zero speed to zero torque at maximum speed. This relationship comes from the fact that *Power = Torque* x *Angular Velocity*. Since the power available from the motor is limited, an increase in torque requires a decrease in angular velocity and vice versa.

Figure 2-22b shows a family of load lines superposed on the *torque-speed* curve of the motor. These load lines represent a changing load applied to the linkage, which the motor must supply. The problem comes from the fact that *as the required load torque increases, the motor must reduce speed to supply it*. Thus the input speed will vary in response to load variations. If constant speed is desired, this is unacceptable.

One possible solution is to use a **speed-controlled DC motor** which contains circuitry that increases and decreases the current to the motor in the face of changing load to try to maintain constant speed. These DC motors will run from an AC source since the controller converts AC to DC. The cost of this solution is high, however. Another solution is to provide a **flywheel** on the input shaft, which will store kinetic energy and help smooth out the speed variations introduced by the load variations. Flywheels will be investigated in Chapter 12.

AC Motors are the most inexpensive solution to continuous rotary motion, and they can be had with a variety of *torque-speed* curves to suit various load applications. They are limited to a few speeds, which are a function of the 60 Hz AC line frequency. The most common AC motor *no load speeds* are 1725 and 3450 revolutions per minute (RPM), which represent some slippage from the more expensive synchronous AC motor speeds of 1800 and 3600 rpm. If other output speeds are needed, a gearbox speed reducer is attached to the motor's output shaft.

These AC and DC motors are designed to provide continuous rotary output. Though they can be stalled against a load, they will not tolerate a full-current, zero-velocity stall for more than a few minutes without overheating.

GEARMOTORS can be purchased with integral gearboxes for a large variety of output speeds. The design of gearboxes is covered in Chapter 10.

SERVOMOTORS are fast response, closed-loop controlled motors capable of providing a programmed function of acceleration or velocity, as well as of holding a fixed position against a load. Closed loop means that *sensors on the output device being moved feed back information on its position, velocity, and acceleration.* Circuitry in the motor controller responds to the fed back information by reducing or increasing (or reversing) the current flow to the motor. Thus precise positioning of the output device is possible, as is control of the speed and shape of its time response to changes in load or input commands. These are very expensive devices and are commonly used in applications such as moving the flight control surfaces in aircraft and guided missiles. These are relatively small and have lower power and torque capacity compared to larger AC/DC motors.

STEPPER MOTORS are designed to position an output device. Unlike servomotors, these are open loop, meaning they *receive no feedback as to whether the output device has responded as requested.* Thus they can get out of phase with the desired program. They will, however, happily sit energized for an indefinite period, holding the output in one position. Their internal construction consists of a number of magnetic strips arranged around the circumference of both the rotor and stator. When energized, the rotor will move one step, to the next magnet, for each pulse received. Thus, these are **intermittent motion** devices and do not provide continuous rotary motion like other motors. The number of magnetic strips determines their resolution (typically a few degrees per step). They are relatively small compared to AC/DC motors and have low torque capacity. They are moderately expensive and require special controllers.

AIR AND HYDRAULIC MOTORS have more limited application than electric motors, simply because they require the availability of a compressed air or hydraulic source. Both of these devices are less energy efficient than direct electric to mechanical conversion of electric motors, because of the losses associated with the conversion of the energy first from chemical or electrical to fluid pressure and then to mechanical form. Every energy conversion involves some losses. Air motors find widest application in factories and shops, where high-pressure compressed air is available for other reasons. A common example is the air impact wrench used in automotive repair shops. Although individual air motors and air cylinders are relatively cheap, these pneumatic systems are quite expensive when the cost of all the ancillary equipment is included. Hydraulic motors are most often found within machines or systems such as construction equipment (cranes), aircraft, and ships, where high-pressure hydraulic fluid is provided for many purposes. Hydraulic systems are very expensive when the cost of all the ancillary equipment is included.

AIR AND HYDRAULIC CYLINDERS are linear actuators (piston in cylinder) which provide a limited stroke, straight-line output from a pressurized fluid flow input of either compressed air or hydraulic fluid (oil). They are the method of choice if you need a linear motion input. However, they share the same high cost, low efficiency, and complication factors as listed under their motor equivalents above.

Another problem is that of control. Most motors, left to their own devices, will tend to run at a constant speed. A linear actuator, when subjected to a constant pressure fluid source, typical of most compressors, will respond with more nearly constant acceleration, which means its velocity will increase linearly with time. This can result in severe impact loads on the driven mechanism when the actuator comes to the end of its stroke at maximum velocity. Servovalve control of the fluid flow, to slow the actuator at the end of its stroke, is possible but is quite expensive.

The most common application of fluid power cylinders is in farm and construction equipment such as tractors and bulldozers, where open loop (non-servo) hydraulic cylinders actuate the bucket or blade through linkages. The cylinder and its piston become two of the links (slider and track) in a slider-crank mechanism.

SOLENOIDS are electromechanical (AC or DC) linear actuators which share some of the limitations of air cylinders, and they possess a few more of their own. They are *energy inefficient*, are limited to very short strokes (about 1 inch), and develop a force which varies exponentially over the stroke, and they deliver high impact loads. They are, however, inexpensive, reliable, and have very rapid response times. They cannot handle much power, and they are typically used as control or switching devices rather than as devices which do large amounts of work on a system.

A common application of solenoids is in camera shutters, where a small solenoid is used to pull the latch and trip the shutter action when you push the button to take the picture. Its nearly instantaneous response is an asset here, and very little work is being done. Another application is in electric door or trunk locking systems in automobiles, where the click of their impact can be clearly heard when you turn the key to lock or unlock the mechanism.

2.16 REFERENCES

1 **Reuleaux, F.,** *The Kinematics of Machinery,* Dover Publications, NY 1963, pp. 29-55.

2 **Gruebler, M.,** *Getriebelehre,* Springer Verlag, OHG, Berlin, 1917.

3 **Grashof, F.,** *Theoretische Maschinenlehre,* vol. 2, Voss, Hamburg, 1883.

2.17 PROBLEMS

*2-1 Find three (or other number as assigned) of the following common devices. Sketch careful kinematic diagrams and find their total degrees of freedom.

    a.   An automobile hood hinge mechanism
    b.   An automobile hatchback lift mechanism
    c.   An electric can opener
    d.   A folding ironing board
    e.   A folding card table
    f.   A folding beach chair
    g.   A baby swing
    h.   A folding baby walker
    i.   A drafting machine
    j.   A fancy corkscrew
    k.   A windshield wiper mechanism

* Answers in Appendix E

       l.   A dump truck dump mechanism
       m.   A trash truck dumpster mechanism
       n.   A station wagon tailgate mechanism
       o.   An automobile jack
       p.   A collapsible auto radio antenna
       q.   A record turntable and tone arm

For questions 2-2 to 2-5 please describe clearly the physical frame of reference used, and the direction and type of motion anticipated. Be verbose!

2-2   How many *DOF* does a vehicle have with respect to the road surface?

       a.   If the vehicle is a "Go-Cart" with no chassis suspension.
       b.   If the vehicle is an automobile with a suspended body structure on the chassis.

2-3   How many *DOF* do you have in your wrist and hand combined?

2-4   How many *DOF* does an unrestrained 2-year-old passenger traveling in the moving automobile of question 2-2b have with respect to the highway? Describe them verbally. Do not just cite a number.

*2-5   How many *DOF* do the following have in their normal environment?

       a.   A submerged submarine   b. An earth-orbiting satellite
       c.   A surface ship   d. A motorcycle
       e.   The print head in a 9-pin dot matrix computer printer
       f.   The pen in an XY plotter

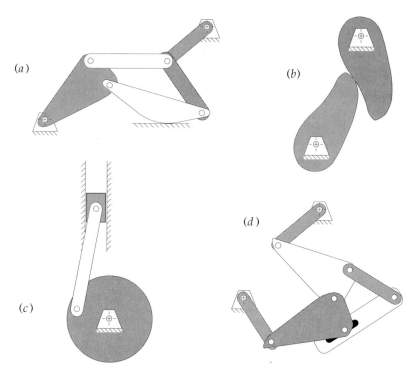

FIGURE P2-1

2-6 Calculate the *DOF* of the mechanisms shown in Figure P2-1.

*2-7 Describe the motion of the following items as pure rotation, pure translation, or complex planar motion.

a. A windmill
b. A bicycle (in the vertical plane, not turning)
c. A conventional "double-hung" window
d. The keys on a computer keyboard
e. The hand of a clock
f. A hockey puck on the ice
g. The pen in an XY plotter
h. The print head in a computer printer
i. A "casement" window

*2-8 How many *DOF* do the following joints have?

a. Your knee
b. Your ankle
c. Your shoulder
d. Your hip
e. Your knuckle

*2-9 Are the joints in problem 2-8 force closed or form closed?

*2-10 Identify the items in Figure P2-1 as mechanisms, structures, or preloaded structures.

2-11 Use number synthesis to find all the possible link combinations for two *DOF*, up to nine links, to hexagonal order, using only full revolute joints.

2-12 Find all the valid isomers of the eightbar one *DOF* linkage combination of five binaries, two ternaries, and one quaternary (see Table 2-1).

2-13 Use linkage transformation to create a mechanism with two sliding full joints from Stephenson's sixbar linkage.

2-14 Use linkage transformation to create a mechanism with a half joint from Stephenson's sixbar linkage.

*2-15 Calculate the Grashof condition of the fourbar mechanisms defined below. Build cardboard models of the linkages and describe the motions of each inversion. Link lengths are in inches (or double given numbers for centimeters).

a.   2   4.5   7   9
b.   2   3.5   7   9
c.   2   4.0   6   8

2-16 Describe the *torque-speed* relationship of an electric motor.

2-17 Describe the difference between a cam-follower (half) joint and a pin joint.

2-18 Go look at an automobile hood hinge mechanism of the type described in Section 2.14. Sketch it carefully. Calculate its *DOF* and Grashof condition. Make a cardboard model. Analyze it with a free-body diagram. Describe how it keeps the hood up. (Hint: If you don't have a car available to examine, pay a visit to your friendly local car dealer.)

2-19 Find an adjustable arm desk lamp of the type shown in Figure P2-2. Measure it and sketch it to scale. Calculate its *DOF* and Grashof condition. Make a cardboard model. Analyze it with a free-body diagram. Describe how it keeps itself stable. Are there any positions in which it loses stability? Why?

FIGURE P2-2

Desk lamp

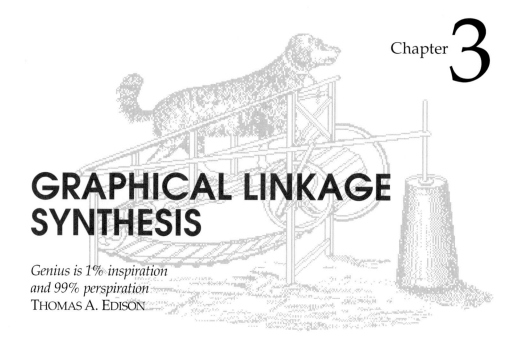

Chapter 3

GRAPHICAL LINKAGE SYNTHESIS

*Genius is 1% inspiration
and 99% perspiration*
THOMAS A. EDISON

3.0   INTRODUCTION

Most engineering design practice involves a combination of synthesis and analysis. Most engineering courses deal primarily with analysis techniques for various situations. However, one cannot analyze anything until it has been synthesized into existence. Many machine design problems require the creation of a device with particular motion characteristics. Perhaps you need to move a tool from position *A* to position *B* in a particular time interval. Perhaps you need to trace out a particular path in space to insert a part into an assembly. The possibilities are endless, but a common denominator is often the need for a linkage to generate the desired motions. So, we will now explore some simple synthesis techniques to enable you to create potential linkage design solutions for some typical kinematic applications.

3.1   SYNTHESIS

QUALITATIVE SYNTHESIS means *the creation of potential solutions in the absence of a well-defined algorithm which configures or predicts the solution*. Since most real design problems will have many more variables present than you will have equations to describe the system's behavior, you cannot simply solve the equations to get a solution. Nevertheless you must work in this fuzzy context to create a potential solution and to also judge its **quality**. You can then analyze the proposed solution to determine its viability, and iterate between synthesis and analysis, as outlined in the **design process**, until you are satisfied with the result. Several tools and techniques exist to assist you in this process. The traditional tool is the drafting board, on which you lay out, to scale, multiple orthographic views of the design, and investigate its motions by drawing arcs, showing multiple positions, and using transparent, movable overlays. *Computer-Aided Drafting (CAD)* systems can speed this process to some degree, but you will probably find that the quickest way to get a sense of the quality of your linkage design is to model it, to scale, in cardboard or drafting *Mylar®* and see the motions directly.

Other tools are available in the form of computer programs such as FOUR-BAR, FIVEBAR, SIXBAR, DYNAFOUR, DYNACAM, MATRIX and ENGINE (all included with this text), some of which do synthesis, but these are mainly analysis tools. They can analyze a trial mechanism solution so rapidly that their dynamic graphical output gives almost instantaneous visual feedback on the quality of the design. The process then becomes one of *qualitative design by successive analysis* which is really an iteration between synthesis and analysis. Very many trial solutions can be examined in a short time using these *Computer-Aided Engineering* (CAE) tools. We will develop the mathematical solutions used in these programs in subsequent chapters in order to provide the proper foundation for understanding their operation. But, if you want to try these programs to reinforce some of the concepts in these early chapters, you may do so. Chapter 8 is a manual for the use of these programs, and it can be read at any time without loss of continuity. Reference will be made to program features which are germane to topics in each chapter, as they are introduced. Datafiles for input to these computer programs are also provided on disk for example problems and figures in these chapters. The datafile names are noted near the figure or example. The student is encouraged to input these sample files to the programs in order to observe more dynamic examples than the printed page can provide. These examples can be run by merely accepting the defaults provided for all inputs.

TYPE SYNTHESIS refers to *the definition of the proper type of mechanism best suited to the problem* and is a form of qualitative synthesis. This is perhaps the most difficult task for the student as it requires some experience and knowledge of the various types of mechanisms which exist and which also may be feasible from a performance and manufacturing standpoint. As an example, assume that the task is to design a device to track the straight-line motion of a part on a conveyor belt and spray it with a chemical coating as it passes by. This has to be done at high, constant speed, with good accuracy and repeatability, and it must be reliable. Moreover, the solution must be inexpensive. Unless you have had the opportunity to see a wide variety of mechanical equipment, you might not be aware that this task could conceivably be accomplished by any of the following devices:

- *A straight-line linkage*
- *A cam and follower*
- *An air cylinder*
- *A hydraulic cylinder*
- *A robot*
- *A solenoid*

Each of these solutions, while possible, may not be optimal or even practical. More detail needs to be known about the problem to make that judgment, and that detail will come from the research phase of the design process. The straight-line linkage may prove to be too large and to have undesirable accelerations, the cam and follower will be expensive, though accurate and repeatable. The air cylinder itself is inexpensive but is noisy and unreliable. The hydraulic cylinder is more expensive, as is the robot. The solenoid, while cheap, has high impact loads and high impact velocity. So, you can see that the choice of device type can have a large effect on the quality of the design. A poor choice at the type synthesis stage can create insoluble problems later on. The design might have to be scrapped after completion, at great expense. **Design is essentially an exercise in tradeoffs.** Each proposed type of solution in this example has good and bad points. Seldom will there be a clear-cut, obvious solution to a real engineering design problem. It will be your job as a design engineer to balance these conflicting features and find a solution which gives the best tradeoff of functionality against cost, reliability, and all other factors of interest. Remember, *an engineer can do, with one dollar, what any fool can do for ten dollars.* Cost is always an important constraint in engineering design.

QUANTITATIVE SYNTHESIS, OR ANALYTICAL SYNTHESIS means the generation of one or more solutions of a particular type which you know to be suitable to the problem, and more importantly, one for which there is a synthesis algorithm defined. As the name suggests, this type of solution can be quantified, as some set of equations exists which will give a numerical answer. Whether that answer is a good or suitable one is still a matter for the judgment of the designer and requires analysis and iteration to optimize the design. Often the available equations are fewer than the number of potential variables, in which case you must assume some reasonable values for enough unknowns to reduce the remaining set to the number of available equations. Thus some qualitative judgment enters into the synthesis in this case as well. Except for very simple cases, a CAE tool is needed to do quantitative synthesis. One example of such a tool is the program LINCAGES, by A. Erdman et al, of the University of Minnesota[1] which solves the three-position and four-position linkage synthesis problems. The computer programs provided with this text also allow you to do three-position **analytical synthesis** and general linkage **design by successive analysis.** The fast computation of these programs allows one to analyze the performance of many trial mechanism designs in a short time and promotes rapid iteration to a better solution.

DIMENSIONAL SYNTHESIS of a linkage *is the determination of the proportions (lengths) of the links necessary to accomplish the desired motions* and can be a form of quantitative synthesis if an algorithm is defined for the particular problem, but can also be a form of qualitative synthesis if there are more variables than equations. The latter situation is more common for linkages. (Dimensional synthesis of cams is quantitative.) Dimensional synthesis assumes that, through *type synthesis,* you have already determined that a linkage (or a cam) is the most appropriate solution to the problem. This chapter discusses **graphical dimensional synthesis** of linkages in detail. Chapter 5 presents methods of **analytical linkage synthesis.**

3.2   FUNCTION, PATH, AND MOTION GENERATION

FUNCTION GENERATION is defined as *the correlation of an input motion with an output motion in a mechanism.* A function generator is conceptually a "black box" which delivers some predictable output in response to a known input. Historically, before the advent of electronic computers, mechanical function generators found wide application in artillery rangefinders and shipboard gun aiming systems, and many other tasks. They are, in fact, **mechanical analog computers**. The development of inexpensive digital electronic microcomputers for control systems coupled with the availability of compact servo and stepper motors has reduced the demand for these mechanical function generator linkage devices. Many such applications can now be served more economically and efficiently with electromechanical devices.[1] Moreover, the computer-controlled electromechanical function generator is programmable, allowing rapid modification of the function generated as demands change. For this reason, while presenting some simple examples in this chapter and a general, analytical design method in Chapter 5, we will not emphasize mechanical linkage function generators in this text. Note however that the cam-follower system, discussed extensively in Chapter 9, is in fact a form of mechanical function generator, and it is typically capable of higher force and power levels per dollar than electromechanical systems.

PATH GENERATION is defined as *the control of a **point** in the plane such that it follows some prescribed path*. This is typically accomplished with at least four bars, wherein a point on the coupler traces the desired path. Specific examples are presented in the section on coupler curves below. Note that no attempt is made in path generation to control the orientation of the link which contains the point of interest. However, it is common for the timing of the arrival of the point at particular locations along the path to be defined. This case is called *path generation with prescribed timing* and is analogous to function generation in that a particular output function is specified. Path and function generation will be dealt with in Chapter 5.

MOTION GENERATION is defined as *the control of a **line** in the plane such that it assumes some prescribed set of sequential positions*. Here orientation of the link containing the line is important. This is a more general problem than path generation, and in fact, path generation is a subset of motion generation. An example of a motion generation problem is the control of the "bucket" on a bulldozer. The bucket must assume a set of positions to dig, pick up, and dump the excavated earth. Conceptually, the motion of a line, painted on the side of the bucket, must be made to assume the desired positions. A linkage is the usual solution.

PLANAR MECHANISMS VERSUS SPATIAL MECHANISMS The above discussion of controlled movement has assumed that the motions desired are planar (2-D). We live in a three dimensional world, however, and our mechanisms must function in that world. **Spatial mechanisms** *are* 3-D *devices*. Their design and analysis is much more complex than that of **planar mechanisms**, which *are* 2-D *devices*. The study of spatial mechanisms is beyond the scope of this introductory text. Some references for further study are in the bibliography. However the study of planar mechanisms is not as practically limiting as it might first appear since many devices in three dimensions are constructed of multiple sets of two dimensional (planar) devices coupled together. An example is any folding chair. It will have some sort of linkage in the left side plane which allows folding. There will be an identical linkage on the right side of the chair. These two XY planar linkages will be connected by some structure along the Z direction, which ties the two planar linkages into a

---

[1] It is worth noting that the day is long past when a mechanical engineer can be content to remain ignorant of electronics and electromechanics. Virtually all modern machines are controlled by electronic devices. Mechanical engineers must understand their operation.

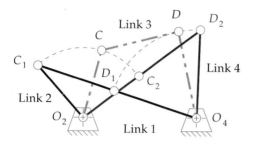

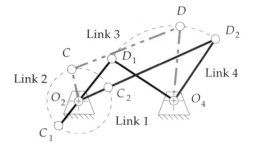

(a) Double-rocker toggle positions   (b) Crank-rocker toggle positions

FIGURE 3-1

Linkages in toggle

three dimensional assembly. Many real mechanisms are arranged in this way, as **duplicate planar linkages**, displaced in the Z direction in parallel planes and rigidly connected. When you open the hood of a car, take note of the hood hinge mechanism. It will be duplicated on each side of the car. The hood and the car body tie the two planar linkages together into a 3-D assembly. Look and you will see many other such examples of assemblies of planar linkages into 3-D configurations. So, the 2-D techniques of synthesis and analysis presented here will prove to be of practical value in designing in 3-D as well.

3.3   LIMITING CONDITIONS

The manual, graphical, dimensional synthesis techniques presented in this chapter and the computerizable, analytical synthesis techniques presented in Chapter 5 are reasonably rapid means to obtain a trial solution to a motion control problem. Once a potential solution is found, it must be evaluated for its quality. There are many criteria which may be applied. In later chapters, we will explore the analysis of these mechanisms in detail. However, one does not want to expend a great deal of time analyzing, in great detail, a design which can be shown to be inadequate by some simple and quick evaluations.

TOGGLE One important test can be applied within the synthesis procedures described below. You need to check that the linkage can in fact reach all of the specified design positions without encountering a limit or **toggle** position. Linkage synthesis procedures often only provide that the particular positions specified will be obtained. They say nothing about the linkage's behavior between those positions. Figure 3-1a shows a non-Grashof fourbar linkage in an arbitrary position CD (dashed lines), and also in its two **toggle positions**, C_1D_1, (solid black lines) and C_2D_2 (solid red lines). *The toggle positions are determined by the **colinearity** of two of the moving links.* A fourbar double-rocker mechanism (whether Grashof or non-Grashof) will have at least two of these toggle positions in which the linkage assumes a triangular configuration. When in a triangular (or

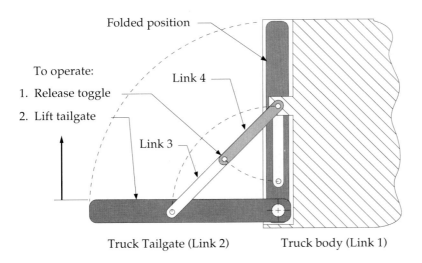

Folded position

To operate:

1. Release toggle

2. Lift tailgate

Link 4

Link 3

Truck Tailgate (Link 2)   Truck body (Link 1)

FIGURE 3-2

Deltoid toggle linkage used to control truck tailgate motion

toggle) position it will not allow further input motion in one direction from one of its rocker links (either of link 2 from position C_1D_1 or link 4 from position C_2D_2). The other rocker will then have to be driven to get the linkage out of toggle. A Grashof fourbar crank-rocker linkage will also assume two toggle positions as shown in Figure 3-1b, when the shortest link (crank O_2C) is colinear with the coupler CD (link 3), either *extended colinear* ($O_2C_2D_2$) or *overlapping colinear* ($O_2C_1D_1$). It cannot be *back driven* from the rocker O_4D (link 4) through these colinear positions, but when the crank O_2C (link 2) is driven, it will carry through both toggles because it is Grashof. Note that these toggle positions also define the limits of motion of the driven rocker (link 4), at which its angular velocity will go through zero. Use program FOURBAR to read the datafiles FIG3-1A and FIG3-1B and animate these examples.

After synthesizing a **double-rocker** solution to a multiposition (motion generation) problem, you **must** check for the presence of toggle positions *between* your design positions. *The easiest way to do this is with a cardboard model of the linkage design.* It is important to realize that a toggle condition is only undesirable if it is preventing your linkage from getting from one desired position to the other. In other circumstances the toggle is very useful. It can provide a self-locking feature when a linkage is moved slightly beyond the toggle position and against a fixed stop. Any attempt to reverse the motion of the linkage then causes it merely to jam harder against the stop. It must be manually pulled "over center," out of toggle, before the linkage will move. You have encountered many examples of this application, as in card table or ironing board leg linkages and also pickup truck or station wagon tailgate linkages. An example of such a toggle linkage is shown in Figure 3-2. It happens to be a special case Grashof linkage in the deltoid configuration (see also Figure 2-16d), which provides a locking toggle position when open, and folds on top of itself when closed, to save space. We will analyze the toggle condition in more detail in a later chapter.

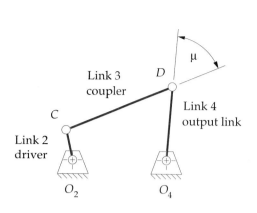

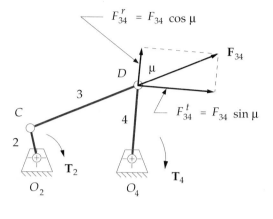

(a) Linkage transmission angle μ

(b) Static forces at a linkage joint

FIGURE 3-3

Transmission angle

TRANSMISSION ANGLE Another useful test that can be very quickly applied to a linkage design to judge its quality is the measurement of its transmission angle. This can be done analytically, graphically on the drawing board, or with the cardboard model for a rough approximation. (Extend the links beyond the pivot to measure the angle.) The **transmission angle** μ is shown in Figure 3-3a and is defined as *the angle between the output link and the coupler. It is usually taken as the absolute value of the acute angle of the pair of angles at the intersection of the two links and varies continuously from some minimum to some maximum value as the linkage goes through its range of motion.* It is a measure of the quality of force and velocity transmission at the joint. Note in Figure 3-2 that the linkage cannot be moved from the open position shown by any force applied to the tailgate, link 2, as the transmission angle is then between links 3 and 4 and is zero at that position. But a force applied to link 3 as the input link will move it. The transmission angle is now between links 4 and 1 and is 45 degrees.

Figure 3-3b shows a torque T_2 applied to link 2. Even before any motion occurs, this causes a static, colinear force F_{34} to be applied by link 3 to link 4 at point D. Its radial and tangential components F_{34}^r and F_{34}^t are resolved parallel and perpendicular to link 4, respectively. Ideally, we would like all of the force F_{34} to go into producing output torque T_4 on link 4. However, only the tangential component creates torque on link 4. The radial component F_{34}^r provides only tension or compression in link 4. This radial component only increases pivot friction and does not contribute to the output torque. Therefore, the optimum value for the **transmission angle** is 90°. When μ is less than 45° the radial component will be larger than the tangential component. Most machine designers try to keep the **minimum transmission angle above about 35°** to promote smooth running and good force transmission. However, if in your particular design there will be little or no external force or torque applied to link 4, you may be able to get away with even lower values of μ. This angle provides one rapid means to judge the quality of a newly synthesized linkage. If it is unsatisfactory, you can iterate through the synthesis procedure to improve the design. We will investigate the transmission angle in more detail in later chapters.

3.4 DIMENSIONAL SYNTHESIS

Dimensional synthesis of a linkage is *the determination of the proportions (lengths) of the links necessary to accomplish the desired motions.* This section assumes that, through *type synthesis*, you have determined that a linkage is the most appropriate solution to the problem. Many techniques exist to accomplish this task of **dimensional synthesis of a fourbar linkage**. The simplest and quickest methods are graphical. These work well for up to three design positions. Beyond that number, a numerical, analytical synthesis approach as described in Chapter 5, using a computer, is usually necessary.

Note that the principles used in these graphical synthesis techniques are simply those of **Euclidean geometry**. The rules for bisection of lines and angles, properties of parallel and perpendicular lines, and definitions of arcs, etc., are all that are needed to generate these linkages. **Compass**, **protractor**, and **rule** are the only tools needed for graphical linkage synthesis. Refer to any introductory (high school) text on geometry if your geometric theorems are rusty.

Two-Position Synthesis

Two-position synthesis subdivides into two categories: **rocker output** (pure rotation) and **coupler output** (complex motion). Rocker output is most suitable for situations in which a Grashof crank-rocker is desired and is, in fact, a trivial case of *function generation* in which the output function is defined as two discrete angular positions of the rocker. Coupler output is more general and is a simple case of *motion generation* in which two positions of a line in the plane are defined as the output. This solution will frequently lead to a double-rocker. However, the fourbar double-rocker can be motor driven by the addition of a **dyad** (twobar chain), which makes the final result a **Watt's sixbar** containing a **Grashof fourbar subchain**. We will now explore the synthesis of each of these types of solution for the two-position problem.

✒️ EXAMPLE 3-1

Rocker Output - Two Positions with Angular Displacement. (Function Generation)

Problem: Design a fourbar Grashof crank-rocker to give 45° of rocker rotation with equal time forward and back, from a constant speed motor input.

Solution: (see Figure 3-4)

1 Draw the output link $O_4 B$ in both extreme positions, B_1 and B_2 in any convenient location, such that the desired angle of motion θ_4 is subtended.

2 Draw the chord $B_1 B_2$ and extend it in any convenient direction.

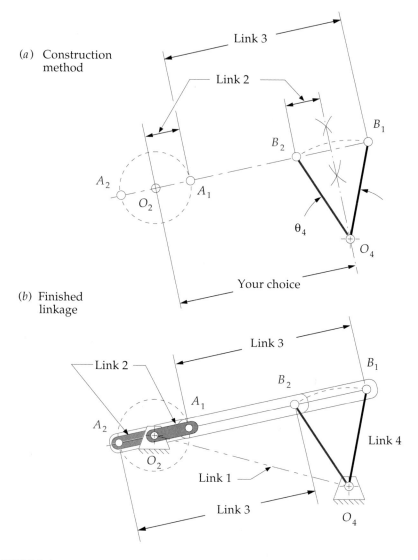

(a) Construction method

(b) Finished linkage

FIGURE 3-4

Two-position function synthesis with rocker output (non-quick-return)

3 Select a convenient point O_2 on line B_1B_2 extended.

4 Bisect line segment B_1B_2, and draw a circle of that radius about O_2.

5 Label the two intersections of the circle and B_1B_2 extended, A_1 and A_2.

6 Measure the length of the coupler as A_1 to B_1 or A_2 to B_2.

7 Measure ground length 1, crank length 2, and rocker length 4.

8 Find Grashof condition. If non-Grashof, redo steps 3 to 8 with O_2 further from O_4.

9   Make a cardboard model of the linkage and articulate it to check its function and its transmission angles.

10   You can input the file FIG3-4 to program FOURBAR to see this example come alive.

Note several things about this synthesis process. We started with the output end of the system, as it was the only aspect defined in the problem statement. We had to make many quite arbitrary decisions and assumptions to proceed because there were many more variables than we could have provided "equations" for. We are frequently forced to make "free choices" of "a convenient angle or length." These free choices are actually definitions of design parameters. A poor choice will lead to a poor design. Thus these are **qualitative synthesis** approaches and require an iterative process, even for this simple an example. The first solution you reach will probably not be satisfactory, and several attempts (iterations) should be expected to be necessary. As you gain more experience in designing kinematic solutions you will be able to make better choices for these design parameters with fewer iterations. **The value of making a simple cardboard model of your design cannot be overstressed!** You will get the *most insight* into your design's quality for the *least effort* by making, articulating, and studying the cardboard model. These general observations will hold for most of the linkage synthesis examples presented.

✎❑EXAMPLE 3-2

Rocker Output - Two Positions with Complex Displacement. (Motion Generation)

Problem: Design a fourbar linkage to move link CD from position C_1D_1 to C_2D_2.

Solution: (see Figure 3-5)

1   Draw the link CD in its two desired positions, C_1D_1 and C_2D_2 in the plane as shown.

2   Draw construction lines from point C_1 to C_2 and from point D_1 to D_2.

3   Bisect line C_1C_2 and line $D_1 D_2$ and extend their perpendicular bisectors to intersect at O_4. Their intersection is the **rotopole**

4   Select a convenient radius and draw an arc about the rotopole to intersect both lines O_4C_1 and O_4C_2. Label the intersections B_1 and B_2.

5   Do steps 2 to 8 of Example 3-1 above to complete the linkage.

6   Make a cardboard model of the linkage and articulate it to check its function and its transmission angles.

Note that Example 3-2 reduces to the method of Example 3-1 once the **rotopole** is found. Thus a link represented by a line in complex motion can be reduced to the simpler problem of pure rotation and moved to any two positions in the plane as the rocker on a fourbar linkage. The next example moves the same link through the same two positions as the coupler of a fourbar linkage.

(*a*) Finding the
   rotopole for
   Example 3-2

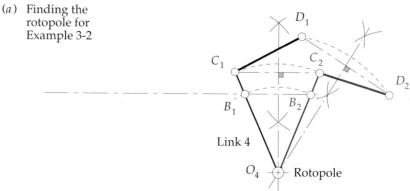

(*b*) Constructing the linkage
   by the method in Example 3-1

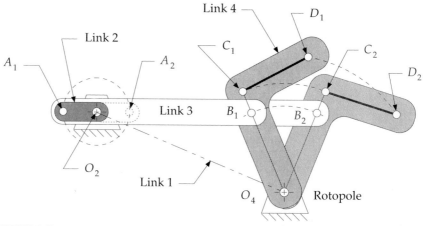

FIGURE 3-5

Two-position motion synthesis with rocker output (non quick-return)

✎EXAMPLE 3-3

Coupler Output - Two Positions with Complex Displacement (Motion Generation)

Problem: Design a fourbar linkage to move the link CD shown from position C_1D_1 to C_2D_2 (with moving pivots at C and D).

Solution: (see Figure 3-6)

1 Draw the link CD in its two desired positions, C_1D_1 and C_2D_2, in the plane as shown.

2 Draw construction lines from point C_1 to C_2 and from point D_1 to D_2.

3 Bisect line C_1C_2 and line D_1D_2 and extend the perpendicular bisectors in convenient directions. The rotopole will **not** be used in this solution.

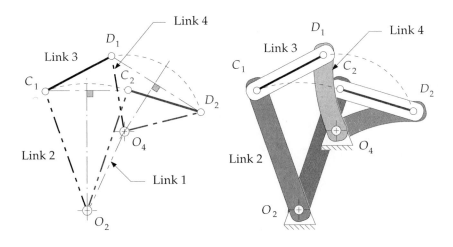

(a) Two-position synthesis   (b) Finished non-Grashof fourbar

FIGURE 3-6

Two-position motion synthesis with coupler output

4 Select any convenient point on each bisector as the fixed pivots O_2 and O_4 respectively.

5 Connect O_2 with C_1 and call it link 2. Connect O_4 with D_1 and call it link 4.

6 Line C_1D_1 is link 3. Line O_2O_4 is link 1.

7 Check the Grashof condition, and repeat steps 4 to 7 if unsatisfied. Note that any Grashof condition is potentially acceptable in this case.

8 Construct a cardboard model and check its function to be sure it can get from initial to final position without encountering any limit (toggle) positions.

9 Check transmission angles.

Input file FIG3-6 to program FOURBAR to see Example 3-3. Note that this example had nearly the same problem statement as Example 3-2, but the solution is quite different. Thus a link can also be moved to any two positions in the plane as the coupler of a fourbar linkage, rather than as the rocker. However, to limit its motions to those two coupler positions as extrema, two additional links are necessary. These additional links can be designed by the method shown in Example 3-4 and Figure 3-7.

EXAMPLE 3-4

Adding a Dyad (Twobar Chain) to Control Motion in Example 3-3.

Problem: Design a **dyad** to control and limit the extremes of motion of the previous example linkage to its two design positions.

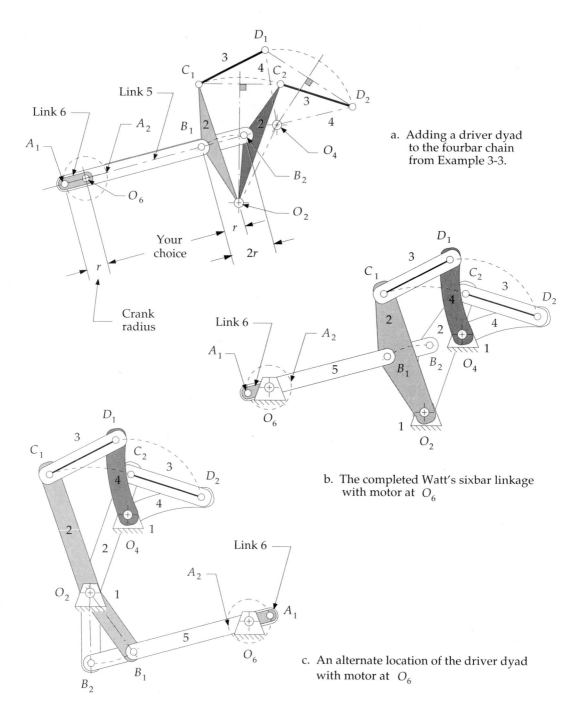

a. Adding a driver dyad to the fourbar chain from Example 3-3.

b. The completed Watt's sixbar linkage with motor at O_6

c. An alternate location of the driver dyad with motor at O_6

FIGURE 3-7

Driving a non-Grashof linkage with a dyad (non quick-return)

Solution: (see Figure 3-7a)

1   Select a convenient point on link 2 of the linkage designed in Example 3-3. Note that it need not be on the line O_2C_1. Label this point B_1.

2   Draw an arc about center O_2 through B_1 to intersect the corresponding line O_2B_2 in the second position of link 2. Label this point B_2. The chord B_1B_2 provides us with the same problem as in Example 3-1.

3   Do steps 2 to 9 of Example 3-1 above to complete the linkage, except add links 5 and 6 and center O_6 rather than links 2 and 3 and center O_2. Link 6 will be the driver crank. The fourbar subchain of links O_6, A_1, B_1, O_2 must be a Grashof crank-rocker.

Note that we have used the approach of Example 3-1 to add a **dyad** to serve as a *driver stage* for our existing fourbar. This results in a **sixbar Watt's mechanism** whose first stage is Grashof as shown in Figure 3-7b. Thus we can drive this with a motor on link 6. Note also that we can locate the motor center O_6 anywhere in the plane by judicious choice of point B_1 on link 2. If we had put B_1 below center O_2, the motor would be to the right of links 2, 3, and 4 as shown in Figure 3-7c. There is *an infinity of driver dyads* possible which will drive any double-rocker assemblage of links. Input the files FIG3-7B and FIG3-7C to program SIXBAR to see Example 3-4 in motion for these two solutions.

Three-Position Synthesis with Specified Moving Pivots

Three-position synthesis allows the definition of three positions of a line in the plane and will create a fourbar linkage configuration to move it to each of those positions. This is a **motion generation** problem. The synthesis technique is a logical extension of the method used in Example 3-3 for two-position synthesis with coupler output. The resulting linkage may be of any Grashof condition and will usually require the addition of a dyad to control and limit its motion to the positions of interest. Compass, protractor, and rule are the only tools needed in this graphical method.

EXAMPLE 3-5

Coupler Output - Three Positions with Complex Displacement (Motion Generation)

Problem: Design a fourbar linkage to move the link CD shown from position C_1D_1 to C_2D_2 and then to position C_3D_3. Moving pivots are at C and D. Find the fixed pivot locations.

Solution: (see Figure 3-8)

1   Draw link CD in its three design positions in the plane C_1D_1, C_2D_2, C_3D_3 as shown.

2   Draw construction lines from point C_1 to C_2 and from point C_2 to C_3.

3   Bisect line C_1C_2 and line C_2C_3 and extend their perpendicular bisectors until they intersect. Label their intersection O_2.

4   Repeat steps 2 and 3 for lines D_1D_2 and D_2D_3. Label the intersection O_4.

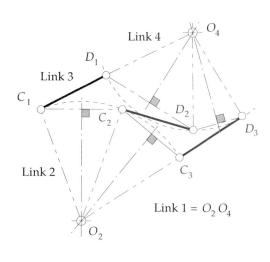

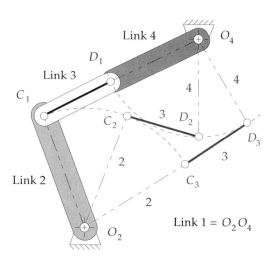

(a) Construction method   (b) Finished non-Grashof fourbar

FIGURE 3-8

Three-position motion synthesis

5   Connect O_2 with C_1 and call it link 2. Connect O_4 with D_1 and call it link 4.

6   Line C_1D_1 is link 3. Line O_2O_4 is link 1.

7   Check the Grashof condition. Note that any Grashof condition is potentially accept-able in this case.

8   Construct a cardboard model and check its function to be sure it can get from initial to final position without encountering any limit (toggle) positions.

9   Construct a driver dyad according to the method in Example 3-4 using an extension of link 3 to attach the dyad.

Note that while a solution is usually obtainable for this case, it is possible that you may not be able to move the linkage continuously from one position to the next without disassembling the links and reassembling them to get them past a limiting position. That will obviously be unsatisfactory. In the particular solution presented in Figure 3-8, note that links 3 and 4 are in toggle at position one, and links 2 and 3 are in toggle at position three. In this case we will have to drive link 3 with a driver dyad, since any attempt to drive either link 2 or link 4 will fail at the toggle positions. No amount of torque applied to link 2 at position C_1 will move link 4 away from point D_1, and driving link 4 will not move link 2 away from position C_3. Input the file FIG3-8 to program FOURBAR to see Example 3-5.

Three-Position Synthesis with Alternate Moving Pivots

Another potential problem is the possibility of an undesirable location of the fixed pivots O_2 and O_4 with respect to your packaging constraints. For example, if the fixed pivot for a windshield wiper linkage design ends up in the middle of the windshield, you may want to redesign it. Example 3-6 shows a way to obtain an alternate configuration for the same three-position motion as in the above example. And, the method shown in Example 3-8 below allows you to specify the location of the fixed pivots in advance and then find the locations of the moving pivots on link 3 compatible with those fixed pivots.

✎ EXAMPLE 3-6

Coupler Output - Three Positions with Complex Displacement - Alternate Attachment Points for Moving Pivots. (Motion Generation)

Problem: Design a fourbar linkage to move the link CD shown from position C_1D_1 to C_2D_2 and then to position C_3D_3. Use different moving pivots than CD. Find the fixed pivot locations.

Solution: (see Figure 3-9)

1 Draw the link CD in its three desired positions in the plane C_1D_1, C_2D_2, C_3D_3, as done in Example 3-5.

2 Define new attachment points E_1 and F_1 which have a fixed relationship between C_1D_1 and E_1F_1 within the link. Now use E_1F_1 to define the three positions of the link.

3 Draw construction lines from point E_1 to E_2 and from point E_2 to E_3.

4 Bisect line E_1E_2 and line E_2E_3 and extend the perpendicular bisectors until they intersect. Label the intersection O_2.

5 Repeat steps 2 and 3 for lines F_1F_2 and F_2F_3. Label the intersection O_4.

6 Connect O_2 with E_1 and call it link 2. Connect O_4 with F_1 and call it link 4.

7 Line E_1F_1 is link 3. Line O_2O_4 is link 1.

8 Check the Grashof condition. Note that any Grashof condition is potentially acceptable in this case.

9 Construct a cardboard model and check its function to be sure it can get from initial to final position without encountering any limit (toggle) positions. If not, change locations of points E and F and repeat steps 3 to 9.

10 Construct a driver dyad acting on link 2 according to the method in Example 3-4.

Note that the shift of the attachment points on link 3 from CD to EF has resulted in a shift of the locations of fixed pivots O_2 and O_4 as well. Thus they may now be in more favorable locations than they were in Example 3-5. It is important to understand that any two points on link 3, such as E and F, can serve to fully define that link as a rigid body, and that there is an infinity of such sets of points to choose from. While points C and D have some particular location in the plane which is defined by the linkage's function, points E and F can be anywhere on link 3, thus creating an infinity of solutions to this problem.

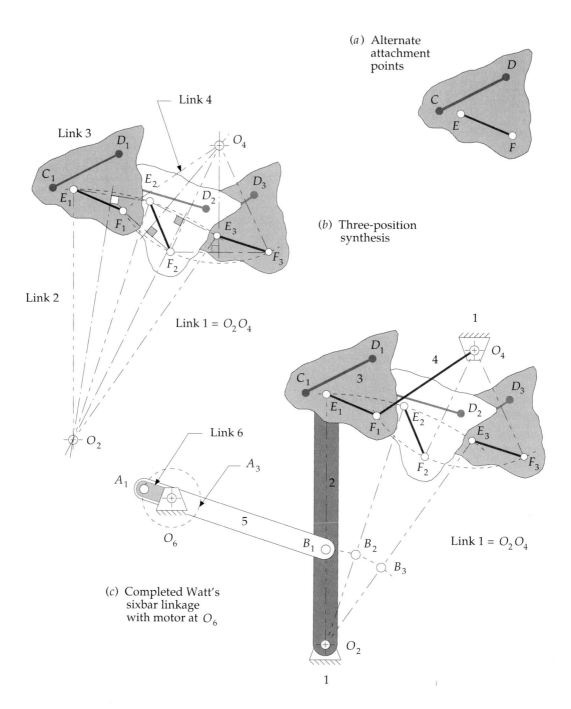

(*a*) Alternate attachment points

(*b*) Three-position synthesis

Link 4

Link 3

Link 2

Link 1 = O_2O_4

Link 6

Link 1 = O_2O_4

(*c*) Completed Watt's sixbar linkage with motor at O_6

FIGURE 3-9

Three-position synthesis with alternate moving pivots

The solution in Figure 3-9 is different from that of Figure 3-8 in several respects. It avoids the toggle positions, thus can be driven by a dyad acting on one of the rockers, as shown in Figure 3-9c, and the transmission angles are better. However, the toggle positions of Figure 3-8 might actually be of value if a self-locking feature was desired. *Recognize that both of these solutions are to the same problem*, and the solution in Figure 3-8 is just a special case of that in Figure 3-9. Both solutions may be useful. Line CD moves through the same three positions with both designs. There is an infinity of other solutions to this problem waiting to be found as well. Input the file FIG3-9C to program SIXBAR to see Example 3-6.

Three-Position Synthesis with Specified Fixed Pivots

Even though one can probably find an acceptable solution to the three-position problem by the methods described in the two preceding examples, it can be seen that the designer will have little direct control over the location of the fixed pivots since they are one of the results of the synthesis process. It is common for the designer to have some constraints on acceptable locations of the fixed pivots, since they will be limited to locations at which the ground plane of the package is accessible. It would be preferable if we could define the fixed pivot locations as well as the three positions of the moving link, and then synthesize the appropriate attachment points, E and F, to the moving link to satisfy these more realistic constraints. The principle of **inversion** can be applied to this problem. Examples 3-5 and 3-6 showed how to find the required fixed pivots for three chosen positions of the moving pivots. Inverting this problem allows specification of the fixed pivot locations and determination of the required moving pivots for those locations. The first step is to find the three positions of the ground plane which correspond to the three desired coupler positions. This is done by **inverting the linkage** [2] as shown in Figure 3-10 and Example 3-7.

✎ EXAMPLE 3-7

Three-Position Synthesis with Specified Fixed Pivots – Inverting the Three-Position Motion Synthesis Problem.

Problem: Invert a fourbar linkage which moves the link CD shown from position C_1D_1 to C_2D_2 and then to position C_3D_3. Use specified fixed pivots O_2 and O_4.

Solution: First find the inverted positions of the ground link corresponding to the three coupler positions specified. (see Figure 3-10)

1   Draw the link CD in its three desired positions in the plane C_1D_1, C_2D_2, C_3D_3, as was done in Example 3-5 and as shown in Figure 3-10a.

2   Draw the ground link O_2O_4 in its desired position in the plane with respect to the first coupler position C_1D_1 as shown in Figure 3-10a.

3   Draw construction arcs from point C_2 to O_2 and from point D_2 to O_2 whose radii define the sides of triangle $C_2O_2D_2$. This defines the relationship of the fixed pivot O_2 to the coupler line CD in the second coupler position as shown in Figure 3-10b.

4   Draw construction arcs from point C_2 to O_4 and from point D_2 to O_4 to define the

[2] This method and example were supplied by Mr. Homer Eckhardt, Consulting Engineer, Lincoln, Ma.

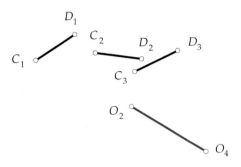

(a) Original coupler three position problem with specified pivots

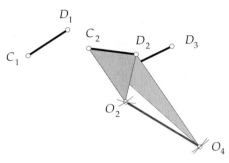

(b) Position of the ground plane relative to the second coupler position

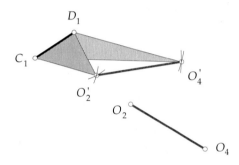

(c) Transferring second ground plane position to reference location at first position

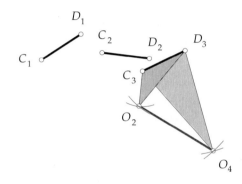

(d) Position of the ground plane relative to the third coupler position

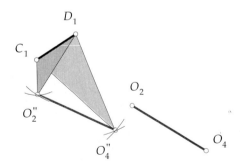

(e) Transferring third ground plane position to reference location at first position

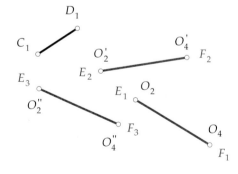

(f) The three inverted positions of the ground plane corresponding to the original coupler position

FIGURE 3-10

Inverting the three-position motion synthesis problem

triangle $C_2O_4D_2$. This defines the relationship of the fixed pivot O_4 to the coupler line CD in the second coupler position as shown in Figure 3-10b.

5 Now transfer this relationship back to the first coupler position C_1D_1 so that the ground plane position $O_2'O_4'$ bears the same relationship to C_1D_1 as O_2O_4 bore to the second coupler position C_2D_2. By doing this, we have pretended that the ground plane moved from O_2O_4 to $O_2'O_4'$ instead of the coupler moving from C_1D_1 to C_2D_2. That is, we have *inverted* the problem.

6 Repeat the process for the third coupler position as shown in Figure 3-10d and transfer the third relative ground link position to the first, or reference, position as shown in Figure 3-10e.

7 The three inverted positions of the ground plane which correspond to the three desired coupler positions are labelled O_2O_4, $O_2'O_4'$, and $O_2''O_4''$ and have also been renamed E_1F_1, E_2F_2, and E_3F_3 as shown in Figure 3-10f. These correspond to the three coupler positions shown in Figure 3-10a. Note that the original three lines C_1D_1, C_2D_2, and C_3D_3 are not now needed for the linkage synthesis.

We can use these three new lines E_1F_1, E_2F_2 and E_3F_3 to find the attachment points GH (moving pivots), on link 3 which will allow the desired fixed pivots O_2 and O_4 to be used for the three specified output positions. In effect we will now consider the ground plane O_2O_4 to be a coupler moving through the inverse of the original three positions, find the "ground pivots" GH needed for that inverted motion and put them on the real coupler instead. The inversion process done in Example 3-7 and Figure 3-10 has swapped the roles of coupler and ground plane. The remaining task is identical to that done in Example 3-5 and Figure 3-8. The result of the synthesis then must be reinverted to obtain the solution.

✒️□EXAMPLE 3-8

Finding the Moving Pivots for Three Positions and Specified Fixed Pivots

Problem: Design a fourbar linkage to move the link CD shown from position C_1D_1 to C_2D_2 and then to position C_3D_3. Use specified fixed pivots O_2 and O_4. Find the required moving pivot locations on the coupler by inversion.

Solution: Using the inverted ground link positions E_1F_1, E_2F_2, and E_3F_3 found in Example 3-7, find the fixed pivots for that inverted motion, then reinvert the resulting linkage to create the moving pivots for the three positions of coupler CD which use the selected fixed pivots O_2 and O_4 as shown in Figure 3-10a (see also Figure 3-11).

1 Start with the inverted three positions in the plane as shown in Figures 3-10f and 3-11a. Lines E_1F_1, E_2F_2, and E_3F_3 define the three positions of the inverted link to be moved.

2 Draw construction lines from point E_1 to E_2 and from point E_2 to E_3.

3 Bisect line E_1E_2 and line E_2E_3 and extend the perpendicular bisectors until they intersect. Label the intersection G.

4 Repeat steps 2 and 3 for lines F_1F_2 and F_2F_3. Label the intersection H.

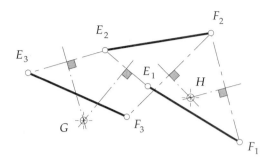

(*a*) Construction to find rotopoles *G* and *H*

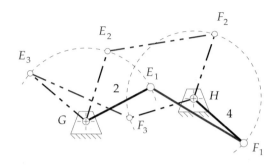

(*b*) The correct inversion of desired linkage

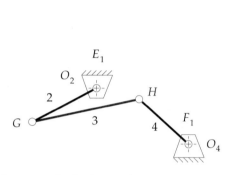

(*c*) Reinvert to obtain the result

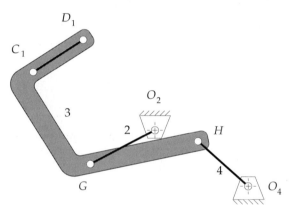

(*d*) Re-place line *CD* on link 3

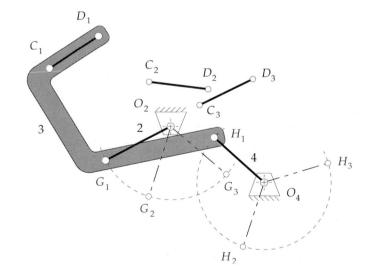

(*e*) The three positions
(link 4 driving *CCW*)

FIGURE 3-11

Constructing the linkage for three positions with specified fixed pivots by inversion

5   Connect G with E_1 and call it link 2. Connect H with F_1 and call it link 4. See Figure 3-11b.

6   In this inverted linkage, line E_1F_1 is the coupler, link 3. Line GH is the "ground" link 1.

7   We must now reinvert the linkage to return to the original arrangement. Line E_1F_1 is really the ground link O_2O_4 and GH is really the coupler. Figure 3-11c shows the reinversion of the linkage in which points G and H are now the moving pivots on the coupler and E_1F_1 has resumed its real identity as ground link O_2O_4. (See Figure 3-10e.)

8   Figure 3-11d reintroduces the original line C_1D_1 in its correct relationship to line O_2O_4 at the initial position as shown in the original problem statement in Figure 3-10a. This forms the required coupler plane and defines a minimal shape of link 3.

9   The angular motions required to reach the second and third positions of line CD shown in Figure 3-11e are the same as those defined in Figure 3-11b for the linkage inversion. The angle F_1HF_2 in Figure 3-11b is the same as angle $H_1O_4H_2$ in Figure 3-11e and F_2HF_3 is the same as angle $H_2O_4H_3$. The angular excursions of link 2 retain the same relationship between Figures 3-11b and e as well. The angular motions of links 2 and 4 are the same for both inversions as the link excursions are relative to one another.

10 Check the Grashof condition. Note that any Grashof condition is potentially acceptable in this case provided that the linkage has mobility among all three positions. This solution is a non-Grashof linkage.

11 Construct a cardboard model and check its function to be sure it can get from initial to final position without encountering any limit (toggle) positions. In this case links 3 and 4 reach a toggle position between points H_1 and H_2. This means that this linkage cannot be driven from link 2 as it will hang up at that toggle position. It must be driven from link 4.

By inverting the original problem, we have reduced it to a more tractable form which allows a direct solution by the general method of three-position synthesis from Examples 3-5 and 3-6.

Position Synthesis for More Than Three Positions

It should be obvious that the more constraints we impose on these synthesis problems, the more complicated the task becomes to find a solution. When we define more than three positions of the output link, the difficulty increases substantially.

FOUR-POSITION SYNTHESIS does not lend itself as well to manual graphical solutions, though Hall[3] does present one approach. Probably the best approach is that used by Sandor and Erdman[4] and others, which is a quantitative synthesis method and requires a computer to execute it. Briefly, a set of simultaneous vector equations is written to represent the desired four positions of the entire linkage. These are then solved after some free choices of variable values are made by the designer. The computer program LINCAGES[1] by Erdman et al, and the program KINSYN[5] by Kaufman, both provide a convenient and user-friendly computer graphics based means to make the necessary design choices to solve the four-position problem. See Chapter 5 for further discussion.

3.5   QUICK-RETURN MECHANISMS

Many machine design applications have need for a difference in average velocity between their "forward" and "return" strokes. Typically some external work is being done by the linkage on the forward stroke, and the return stroke needs to be accomplished as rapidly as possible so that a maximum of time will be available for the working stroke. Many arrangements of links will provide this feature. The only problem is to synthesize the right one!

Fourbar Quick-Return

The linkage synthesized in Example 3-1 is perhaps the simplest example of a four-bar linkage design problem (see Figure 3-4 and program FOURBAR diskfile FIG3-4). It is a crank-rocker which gives two positions of the rocker with equal time for the forward stroke and the return stroke. This is called a *non quick-return* linkage, and it is a special case of the more general **quick-return** case. The reason for its non quick-return state is the positioning of the crank center O_2 on the chord B_1B_2 extended. This results in equal crank angles of 180 degrees being swept out by the crank as it drives the rocker from one extreme (toggle position) to the other. If the crank is rotating at constant angular velocity, as it will tend to when motor driven, then each 180 degree sweep, forward and back, will take the same time interval. Try this with your cardboard model from Example 3-1 by rotating the crank at uniform velocity and observing the rocker motion and velocity.

If the crank center O_2 is located off the chord B_1B_2 extended, as shown in Figure 3-1b and Figure 3-12, then unequal angles will be swept by the crank between the toggle positions (defined as colinearity of crank and coupler). Unequal angles will give unequal time, when the crank rotates at constant velocity. These angles are labelled α and β in Figure 3-12. Their ratio α/β is called the **time ratio** (T_R) and defines the degree of quick-return of the linkage. Note that the term **quick-return** is arbitrarily used to describe this kind of linkage. If the crank is rotated in the opposite direction, it will be a **quick-forward** mechanism. Given a completed linkage, it is a trivial task to estimate the time ratio by measuring or calculating the angles α and β. It is a more difficult task to design the linkage for a chosen time ratio. Hall[6] provides a graphical method to synthesize a quick-return Grashof fourbar. To do so we need to compute the values of α and β which will give the specified time ratio. We can write two equations involving α and β and solve them simultaneously.

$$T_R = \frac{\alpha}{\beta} \qquad\qquad \alpha + \beta = 360 \qquad\qquad (3.1)$$

We also must define a construction angle,

$$\delta = |180 - \alpha| = |180 - \beta| \qquad\qquad (3.2)$$

which will be used to synthesize the linkage.

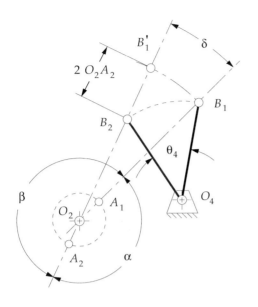

(a) Construction of a quick-return   (b) The finished linkage in its
    Grashof crank-rocker   two toggle positions

FIGURE 3-12

Quick-return Grashof fourbar crank-rocker linkage

EXAMPLE 3-9

Fourbar Crank-Rocker Quick-Return Linkage for Specified Time Ratio

Problem: Redesign Example 3-1 to provide a time ratio of 1 : 1.25 with 45° output
rocker motion.

Solution: (see Figure 3-12)

1   Draw the output link O_4B in both extreme positions, in any convenient location, such
that the desired angle of motion, θ_4, is subtended.

2   Calculate α, β, and δ using equations 3.1 and 3.2.

For this example, $\alpha = 160°$, $\beta = 200°$, $\delta = 20°$

3   Draw a construction line through point B_1 at any convenient angle.

4   Draw a construction line through point B_2 at angle δ from the first line.

5   Label the intersection of the two construction lines O_2.

6   The line O_2O_4 now defines the ground link.

7   Calculate the lengths of crank and coupler by measuring O_2B_1 and O_2B_2 and solve
simultaneously:

$$Coupler + Crank = O_2B_1$$

$$Coupler - Crank = O_2B_2$$

or you can construct the crank length by swinging an arc centered at O_2 from B_1 to cut line O_2B_2 extended. Label that intersection B_1'. The line B_2B_1' is twice the crank length. Bisect this line segment to measure crank length O_2A_1.

8 Calculate the Grashof condition. If non-Grashof, repeat steps 3 to 8 with O_2 further from O_4.

9 Make a cardboard model of the linkage and articulate it to check its function.

10 Check the transmission angles.

This method works well for time ratios down to about 1:1.5. Beyond that value the transmission angles become poor, and a more complex linkage is needed. Input the file FIG3-12 to program FOURBAR to see Example 3-9.

Sixbar Quick-Return

Larger time ratios, up to about 1:2, can be obtained by designing a sixbar linkage. The strategy here is to first design a fourbar drag link mechanism which has the desired time ratio between its driver crank and its driven or "dragged" crank, and then add a dyad (twobar) output stage, driven by the dragged crank. This dyad can be arranged to have either a rocker or a translating slider as the output link. First the drag link fourbar will be synthesized, then the dyad will be added.

✎ EXAMPLE 3-10

Sixbar Drag Link Quick-Return Linkage for Specified Time Ratio

Problem: Provide a time ratio of 1 : 1.4 with 90° rocker motion.

Solution: (see Figure 3-13)

1 Calculate α and β using equations 3.1. For this example,

$$\alpha = 150 \text{ degrees}, \quad \beta = 210 \text{ degrees}$$

2 Draw a line of centers XX at any convenient location.

3 Choose a crank pivot location O_2 on line XX and draw an axis YY perpendicular to XX through O_2.

4 Draw a circle of convenient radius O_2A about center O_2.

5 Lay out angle α with vertex at O_2, symmetrical about quadrant one.

6 Label points A_1 and A_2 at the intersections of the lines subtending angle α and the circle of radius O_2A .

7 Set compass to a convenient radius AC long enough to cut XX in two places on either side of O_2 when swung from both A_1 and A_2. Label the intersections C_1 and C_2.

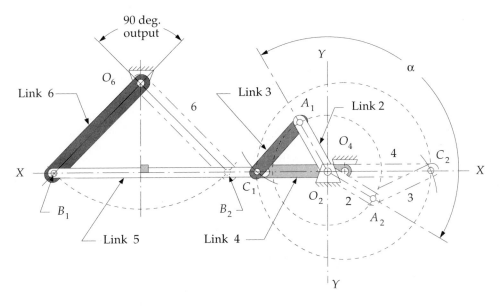

(*a*) Rocker output sixbar drag link quick-return mechanism

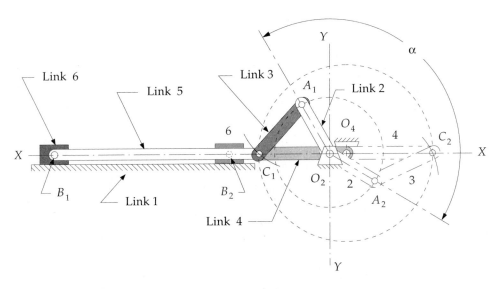

(*b*) Slider output sixbar drag link quick-return mechanism

FIGURE 3-13

Synthesizing a sixbar drag link quick-return mechanism

8　The line O_2A_1 is the driver crank, link 2, and line A_1C_1 is the coupler, link 3.

9　The distance C_1C_2 is twice the driven (dragged) crank length. Bisect it to locate the fixed pivot O_4.

10　The line O_2O_4 now defines the ground link. Line O_4C_1 is the driven crank, link 4.

11　Calculate the Grashof condition. If non-Grashof, repeat steps 7 to 11 with a shorter radius in step 7.

12　Invert the method of Example 3-1 to create the output dyad using XX as the chord and O_4C_1 as the driving crank. The points B_1 and B_2 will lie on line XX and be spaced apart a distance $2O_4C_1$. The pivot O_6 will lie on the perpendicular bisector of B_1B_2, at a distance from line XX which subtends the specified output rocker angle.

13　Check the transmission angles.

This linkage provides a quick-return when a constant-speed motor is attached to link 2. Link 2 will go through angle α while link 4 (which is dragging the output dyad along) goes through the first 180 degrees, from position C_1 to C_2. Then, while link 2 completes its cycle through β degrees, the output stage will complete another 180 degrees from C_2 to C_1. Since angle β is greater than α, the forward stroke takes longer. Note that the chordal stroke of the output dyad is twice the crank length O_4C_1. This is independent of the angular displacement of the output link which can be tailored by moving the pivot O_6 closer to or further from the line XX.

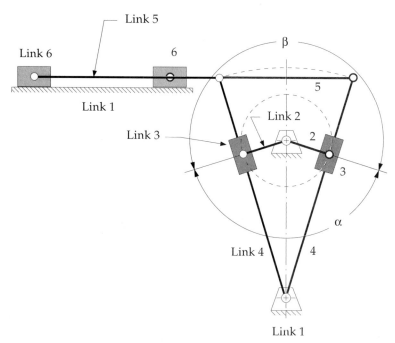

FIGURE 3-14

Crank-shaper quick-return mechanism

The transmission angle at the joint between link 5 and link 6 will be optimized if the fixed pivot O_6 is placed on the perpendicular bisector of the chord B_1B_2 as shown in Figure 3-13a. If a translating output is desired, the slider (link 6) will be located on line XX, and will oscillate between B_1 and B_2 as shown in Figure 3-13b. The arbitrarily chosen size of this or any other linkage can be scaled up or down, simply by multiplying all link lengths by the same scale factor. Thus a design made to arbitrary size can be fit to any package. Input the file FIG3-13A to program SIXBAR to see Example 3-10 in action.

CRANK-SLIDER QUICK-RETURN A commonly used mechanism, capable of large time ratios is shown in Figure 3-14. It is often used in metal shaper machines to provide a slow cutting stroke and a quick-return stroke when the tool is doing no work. It is inversion #2 of the slider-crank mechanism as was shown in Figure 2-13b. It is very easy to synthesize this linkage by simply moving the rocker pivot O_4 along the vertical centerline O_2O_4 while keeping the two extreme positions of link 4 tangent to the circle of the crank, until the desired time ratio (α/β) is achieved. Note that the angular displacement of link 4 is then defined as well. Link 2 is the input and link 6 is the output.

Depending on the relative lengths of the links this mechanism is known as a **Whitworth** or as a **crank-shaper** mechanism. If the ground link is the shortest, then it will behave as a double crank linkage, or *Whitworth mechanism*, with both pivoted links making full revolutions as shown in Figure 2-13b. If the driving crank is the shortest link, then it will behave as a crank-rocker linkage, or *crank-shaper mechanism*, as shown in Figure 3-14. They are the same inversion as the slider block is in complex motion in each case.

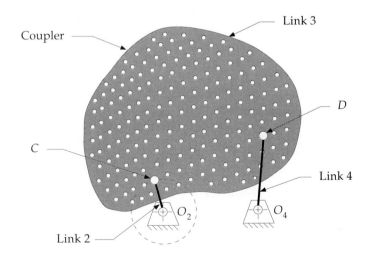

FIGURE 3-15

The fourbar coupler extended to include a large number of coupler points

(a) Flattened ellipse

(b) Kidney bean

(c) Banana

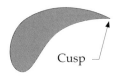

Cusp

(d) Teardrop

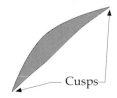

Cusps

(e) Scimitar

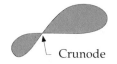

Crunode

(f) Figure eight

FIGURE 3-16

A Cursory Catalog of
coupler curve shapes

3.6   COUPLER CURVES

The **coupler** is the most interesting link in any linkage. It is in complex motion, and thus points on the coupler can have path motions of high degree. In general, the more links, the higher the degree of curve generated, where degree here means *the highest power of any term in its equation.* A curve (function) can have *up to* as many intersections (roots) with any straight line as the degree of the function. The *fourbar slidercrank* has, in general, fourth-degree coupler curves, the *pin-jointed fourbar,* up to sixth degree. The geared fivebar, the sixbar, and more complicated assemblies will have still higher degree curves. All linkages which possess one or more "floating" coupler links will generate coupler curves. It is interesting to note that coupler curves will be closed curves even for non-Grashof linkages. The coupler (or any link) can be extended infinitely in the plane. Figure 3-15 shows a fourbar linkage with its coupler extended to include a large number of points, each of which will describe a different **closed curve**. Note that these points may be anywhere on the coupler, including along line *CD.* There are, of course, an infinity of points on the coupler, each of which generates a different coupler curve. These coupler curves can be used to generate useful path motions for machine design problems. These curves are capable of *approximating straight lines* and *large circle arcs* with remote centers. Recognize that the coupler curve is a solution to the path generation problem described in Section 3.2. It is not necessarily a solution to the motion generation problem, since the attitude or orientation of a line on the coupler is not predicted by the information contained in the path. Nevertheless it is a very useful device. As we shall see, approximate straight-line motions, dwell motions, and complicated symphonies of timed motions are available from even the lowly fourbar and its infinite variety of often surprising dances.

FOURBAR COUPLER CURVES come in a variety of shapes which can be crudely categorized as shown in Figure 3-16. There is an infinite range of variation between these generalized shapes. Some features of interest are the **cusp** and the **crunode**. A **cusp** is *a sharp point on the curve which has the useful property of instantaneous zero velocity.* The simplest example of a curve with a cusp is the cycloid curve which is generated by a point on the rim of a wheel rotating on a flat surface. When the point touches the surface, it has the same (zero) velocity as all points on the stationary surface, provided there is pure rolling and no slip between the elements. Anything attached to a cusp point will come smoothly to a stop along one path and then accelerate smoothly away from that point on a different path. The cusp's feature of zero velocity has value in such applications as stamping and feeding processes. Note that the *acceleration at the cusp is not zero* and may be quite large. A **crunode** *creates a figure-eight-shaped curve which contains a double point at the crossover.* The curve at this point has two slopes (i.e., velocities) but does not go through zero velocity.

The **Hrones and Nelson** (H&N) atlas of fourbar coupler curves[7] is a useful reference which can provide the designer with a starting point for further design and analysis. It contains about 7000 coupler curves and defines the linkage geometry for each of its Grashof crank-rocker linkages. Figure 3-17a reproduces a page from this book. The H&N atlas is logically arranged, with all linkages defined by their link ratios, based on a unit length crank. The coupler is shown as a matrix of fifty coupler points for each linkage geometry, arranged ten to a page. Thus each linkage geometry occupies five pages. Each page contains a schematic "key" in the upper right corner which defines the link ratios.

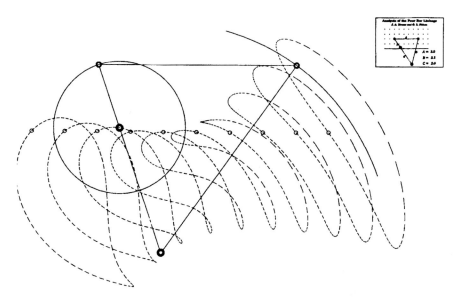

(*a*) A page from the Hrones and Nelson atlas of fourbar coupler curves

(*Hrones, J. A., and Nelson, G. L., Analysis of the Fourbar Linkage,*
MIT Technology Press, Cambridge Ma., 1951, Reprinted with permission)

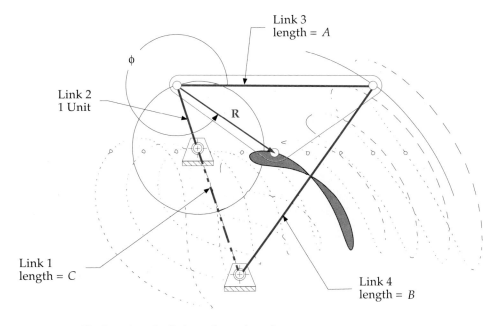

(*b*) Creating the linkage from the atlas

FIGURE 3-17

Selecting a coupler curve and constructing the linkage from the Hrones and Nelson atlas

Film

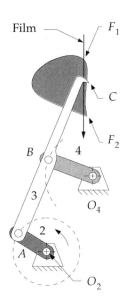

FIGURE 3-18

Movie camera film
advance mechanism

Figure 3-17b shows a "fleshed out" linkage drawn on top of the atlas page to illustrate its relationship to the atlas information. The double circles in Figure 3-17a define the fixed pivots. The crank is always of unit length. The ratios of the other link lengths to the crank are given on each page. The actual link lengths can be scaled up or down to suit your package constraints. Any one of the ten coupler points shown can be used by incorporating it into a triangular coupler link. The location of the chosen coupler point can be scaled from the atlas and is defined within the coupler by the position vector $\mathbf{R}$ whose constant angle ϕ is measured with respect to the line of centers of the coupler. The H&N coupler curves are shown as dashed lines. Each dash station represents **five degrees** of crank rotation. So, for an assumed constant crank velocity, the dash spacing is proportional to path velocity. The changes in velocity and the quick-return nature of the coupler path motion can be clearly seen from the dash spacing.

One can peruse this linkage atlas resource and find an approximate solution to virtually any path generation problem. Then one can take the tentative solution from the atlas to a CAD resource such as the FOURBAR program and further refine the design, based on the complete analysis of positions, velocities, and accelerations provided by the program. The only data needed for the FOURBAR program are the four link lengths and the location of the chosen coupler point with respect to the line of centers of the coupler link as shown in the figure. These parameters can easily be changed within the program to change and refine the design. Input the file FIG3-17B to program FOURBAR to animate the linkage in that figure.

An example of an application of a fourbar linkage to a practical problem is shown in Figure 3-18 which is a movie camera (or projector) film advance mechanism.[3] Point O_2 is the crank pivot which is motor driven at constant speed. Point O_4 is the rocker pivot, and points A and B are the moving pivots. Points A, B, and C define the coupler where C is the coupler point of interest. A movie is really a series of still pictures, each "frame" of which is projected for a fraction of a second on the screen. Between each picture, the film must be moved very quickly from one frame to the next while the shutter is closed to blank the screen. The whole cycle takes only 1/24 of a second. The human eye's response time is too slow to notice the flicker associated with this discontinuous stream of still pictures, so it appears to us to be a continuum of changing images.

The linkage shown is cleverly designed to provide the required motion. A hook is cut into the coupler of this fourbar Grashof crank-rocker at point C which generates the coupler curve shown. The hook will enter one of the sprocket holes in the film as it passes point F_1. Notice that the direction of motion of the hook at that point is nearly perpendicular to the film, so it enters the sprocket hole cleanly. It then turns abruptly downward and follows a crudely approximate straight line as it rapidly pulls the film downward to the next frame. The film is separately guided in a straight track called the "gate." The shutter (driven by another linkage from the same driveshaft at O_2) is closed during this interval of film motion, blanking the screen. At point F_2 there is a cusp on the coupler curve which causes the hook to decelerate smoothly to zero velocity in the vertical direction, and then as smoothly accelerate up and out of the sprocket hole. The abrupt transition of direction at the cusp allows the hook to back out of the hole without jarring the film, which would make the image jump on the

[3] From *Die Wissenschaftliche und Angewandte Photographie*, ed. by Kurt Michel, vol. 3, Harald Weise, *Die Kinematographische Kamera*, p. 202, Springer Verlag, OHG, Vienna, 1955.

screen as the shutter opens. The rest of the coupler curve motion is essentially "wasting time" as it proceeds up the back side, to be ready to enter the film again to repeat the process. Input the file FIG3-18 to program FOURBAR to animate this linkage.

Some advantages of using this type of device for this application are that it is very simple and inexpensive (only four links, one of which is the frame of the camera), is extremely reliable, has low friction if good bearings are used at the pivots, and can be reliably timed with the other events in the overall camera mechanism through common shafting from a single motor. There are a myriad of other examples of fourbar coupler curves used in machines and mechanisms of all kinds.

One other example of a very different application is that of the automobile suspension (Figure 3-19). Typically, the up and down motions of the car's wheels are

b. Multi-link true spatial linkage used to control rear wheel motion

(Courtesy of Mercedes-Benz of North America Inc.)

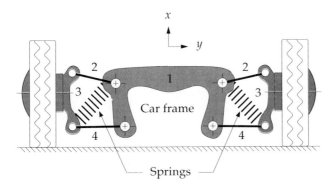

a. Fourbar planar linkages are duplicated in parallel planes,
displaced in the z direction, behind the links shown

FIGURE 3-19

Linkages used in automotive chassis suspensions

controlled by some combination of planar fourbar linkages, arranged in duplicate to provide three-dimensional control as described in Section 3.2. Only a few manufacturers currently use a true spatial linkage in which the links are not arranged in parallel planes. In all cases the wheel assembly is attached to the coupler of the linkage assembly, and its motion is along a coupler curve. The orientation of the wheel is also of concern in this case, so this is not strictly a path generation problem. By controlling the paths of multiple points on the wheel (tire contact patch, wheel center, etc.), motion generation is achieved. Figure 3-19a shows a set of parallel planar fourbar linkages suspending the wheel. The coupler curve of the wheel center is nearly a straight line over the small vertical displacement required. This is desirable as the idea is to keep the tire perpendicular to the ground for best traction under all cornering and attitude changes of the car body. This is an application in which a non-Grashof linkage is perfectly acceptable, as full rotation of the wheel in this plane might have some undesirable results and surprise the driver. Limit stops are of course provided to prevent such behavior, so even a Grashof linkage could be used. The springs support the weight of the vehicle and provide a fifth, variable length "force link" which stabilizes the mechanism as was described in Section 2.14. The function of the fourbar linkage is solely to guide and control the wheel motions. Figure 3-19b shows a true spatial linkage of seven links (including frame and wheel) and nine joints (some of which are ball/socket joints) used to control the motion of the rear wheel. These links do not move in parallel planes but rather control the three dimensional motion of the coupler which carries the wheel assembly.

GEARED FIVEBAR COUPLER CURVES (Figure 3-20) are more complex than the fourbar variety. Because there are three more independent design variables in the geared fivebar compared to the fourbar, (an additional link ratio, a gear ratio, and

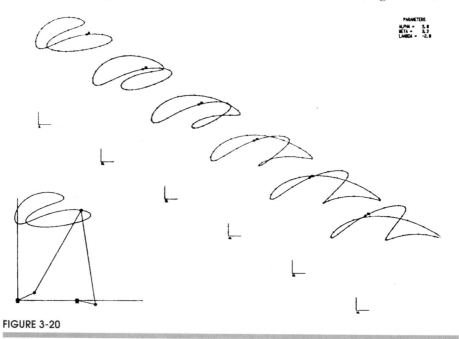

FIGURE 3-20

A page from the Zhang, Norton, Hammond atlas of geared fivebar coupler curves phase angle), the coupler curves are of higher degree than those of the fourbar. This means that the curves can be more convoluted, having more cusps and crunodes (loops). In fact, if the gear ratio used is noninteger, the input link will have to make a number of revolutions equal to the factor necessary to make the ratio an integer before the coupler curve pattern will repeat. The Zhang, Norton, Hammond (ZNH) *Atlas of Geared FiveBar Mechanisms* (GFBM)[8] shows typical coupler curves for these linkages limited to symmetrical geometry, (e.g., link 2 = link 5 and link 3 = link 4) and gear ratios of ±1 and ±2. A page from the ZNH atlas is reproduced in Figure 3-20. Additional pages are in the appendices. Each page shows the family of coupler curves obtained by variation of the phase angle for a particular set of link ratios and gear ratio. A key in the upper right corner of each page defines the ratios α = link 3/link 2, β = link 1/link 2, λ = gear 5/gear 2. Symmetry defines links 4 and 5 as noted above. The phase angle ϕ is defined on the axes drawn at each coupler curve and can be seen to have a significant effect on the resulting coupler curve shape.

This reference atlas is intended to be used as a starting point for a geared fivebar linkage design. The link ratios, gear ratio, and phase angle can be input to the program FIVEBAR, *and then varied* to observe the effects on coupler curve shape, velocities, and accelerations. Asymmetry of links can be introduced, and a coupler point location other than the pin joint between links 3 and 4 defined within the FIVEBAR program as well. Note that program FIVEBAR expects the gear ratio to be in the form gear 2/gear 5 which is the inverse of the ratio λ in the ZNH atlas.

3.7   COGNATES

It sometimes happens that a good solution to a linkage synthesis problem will be found that satisfies path generation constraints but which has the fixed pivots in inappropriate locations for attachment to the available ground plane or frame. Or the discovered linkage may be non-Grashof, when a Grashof mechanism is desired. In such cases, the use of a **cognate** to the linkage may be helpful. The term **cognate** was used by Hartenberg and Denavit[9] to describe *a linkage, of different geometry, which generates the same coupler curve.* Samuel Roberts (1875) and Chebyschev (1878) independently discovered the theorem which now bears their names:

Roberts-Chebyschev Theorem
Three different planar, pin-jointed fourbar linkages will trace identical coupler curves.

Hartenberg and **Denavit** presented extensions of this theorem to the slider-crank and the sixbar linkages:

Two different planar slider-crank linkages will trace identical coupler curves.

The coupler-point curve of a planar fourbar linkage is also described by the joint of a dyad of an appropriate sixbar linkage.

Figure 3-21a shows a fourbar linkage for which we want to find the two cognates. The first step is to release the fixed pivots O_A and O_B. While holding the coupler stationary, rotate links 2 and 4 into colinearity with the line of centers (A_1B_1) of link 3 as shown in Figure 3-21b. We can now construct lines parallel to all sides of the links in the original linkage to create the **Cayley diagram** in Figure 3-21c. This

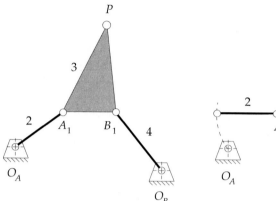

(a) Original fourbar linkage
(Cognate #1)

(b) Align links 2 and 4 with coupler

(c) Construct lines
    parallel to all
    sides of the original
    fourbar linkage
    to create cognates

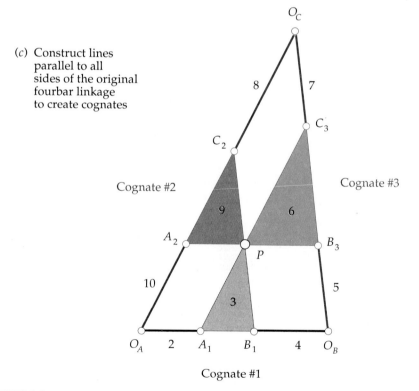

FIGURE 3-21

Cayley diagram to find cognates of a fourbar linkage

(a) Return links 2 and 4 to their fixed pivots O_A and O_B. Point O_C will assume its proper location.

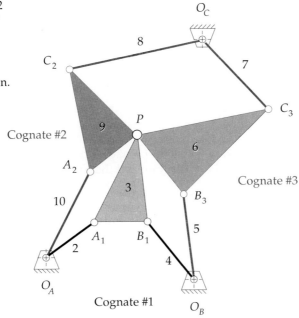

(b) Separate the three cognates. Point P has the same path motion in each cognate.

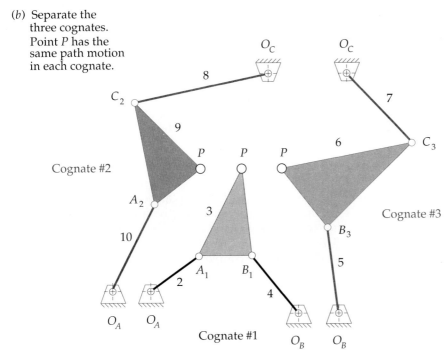

FIGURE 3-22

Roberts diagram of three fourbar cognates

schematic arrangement defines the lengths and shapes of links 5 through 10 which belong to the cognates. All three fourbars share the original coupler point P and will thus generate the same path motion on their coupler curves.

In order to find the correct location of the fixed pivot O_C, it is necessary to return the ends of links 2 and 4 to the original locations of the fixed pivots O_A and O_B as shown in Figure 3-22a. The other links will follow this motion, maintaining the parallelogram relationships between links, and fixed pivot O_C will then be in its proper location on the ground plane. This configuration is called a **Roberts diagram** of three fourbar linkage cognates which share the same coupler curve. Note that the Roberts diagram can be drawn directly from the original linkage without resort to the Cayley diagram by noting that the parallelograms which form the other cognates are also present in the Roberts diagram and the couplers form similar triangles.

The ten-link Roberts configuration (Cayley's nine plus the ground) can now be articulated up to the toggle positions of the linkages, and coupler point P will describe the original coupler path which is the same for all three cognates. The cognates can then be separated as shown in Figure 3-22b and any one of the three linkages used to generate the same coupler curve. Note that any combination of Grashof and non-Grashof linkages is possible within a set of cognates. It is important to realize that while the coupler path displacement is the same for all three cognates, its velocities and accelerations will not be the same since the input crank angular velocities are not all equal in the Roberts configuration.

Program FOURBAR will automatically calculate the two other cognates for any linkage configuration input to it. Each of the cognates can then be examined for Grashof condition and for their velocities and accelerations. The program also draws the Cayley diagram for the set of cognates. Input the datafile FIG3-21 to program FOURBAR and select 5 (plot) from main menu; then select 10 from the plot menu to display the Cayley diagram of Figure 3-21. Also input the files COGNATE1, COGNATE2, and COGNATE3 to animate and view the motion of each cognate shown in Figure 3-22. Their coupler curves (at least those portions that each cognate can reach) will be seen to be identical.

Chebyschev also discovered that any fourbar coupler curve can be duplicated with a **geared fivebar mechanism** whose *gear ratio is plus one*. The link lengths will be different from those of the fourbar, but they can be determined directly from the fourbar. Figure 3-23a shows the construction method, as described by Hall[2], to obtain the geared fivebar which will give the same coupler curve as a fourbar. The original fourbar is $O_A A_1 B_1 O_B$ (links 1, 2, 3, 4). The fivebar is $O_A A_2 P B_2 O_B$ (links 1, 5, 6, 7, 8). The two linkages share only the coupler point P and fixed pivots O_A and O_B. The fivebar is constructed by simply drawing link 6 parallel to link 2, link 7 parallel to link 4, link 5 parallel to $A_1 P$, and link 8 parallel to $B_1 P$.

A three-gear set is needed to couple links 5 and 8 with a ratio of plus one (gear 5 and gear 8 are the same diameter and have the same direction of rotation, due to the idler gear), as shown in Figure 3-23b. Link 5 is attached to gear 5 as is link 8 to gear 8. This construction technique may be applied to each of the three fourbar cognates, yielding three geared fivebars (each of which may or may not be Grashof). The three fivebar cognates can actually be seen in the Roberts diagram. Note in the example shown, a non-Grashof double-rocker fourbar yields a Grashof fivebar,

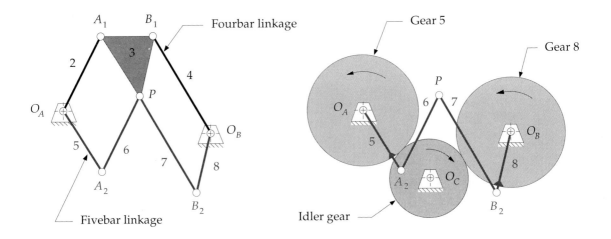

(a) Construction of equivalent fivebar linkage   (b) Resulting geared fivebar linkage

FIGURE 3-23

A geared fivebar linkage cognate of a fourbar linkage

which can be motor driven. This conversion to a GFBM linkage could be an advantage when the "right" coupler curve has been found on a non-Grashof fourbar linkage, but continuous output through the fourbar's toggle positions is needed. Thus we can see that there are at least seven linkages which will generate the same coupler curve, three fourbars, three GFBM's and a sixbar.

Program FOURBAR will calculate the equivalent geared fivebar configuration for any fourbar linkage and output its data to a disk file which can be input to program FIVEBAR for analysis. The file FIG3-23A can be input to FOURBAR to animate the linkage shown in that figure. Then also input the file FIG3-23B to program FIVEBAR to see the motion of the equivalent geared fivebar linkage. Note that the original fourbar is a double-rocker, so cannot reach all of the portions of the coupler curve. But, the geared fivebar equivalent can make a full revolution and traverse the entire coupler path. Note that this FIVEBAR diskfile was automatically created by program FOURBAR upon calculation of the fourbar linkage and storage of its data on disk.

3.8   STRAIGHT-LINE MECHANISMS

A very common application of coupler curves is for the generation of approximate straight lines. In fact the first recorded coupler curve application is that of **Watt's straight-line linkage**, shown in Figure 3-24a. This double-rocker linkage is still frequently used in automobile suspension systems to guide the rear axle up and down in a straight line. The **Chebyschev straight-line linkage**, a Grashof double-rocker, is shown in Figure 3-24b. Richard Roberts (1789-1864), (not to be confused with Samuel Roberts of the cognates) discovered the **Roberts straight-line linkage** shown in Figure 3-24c. It is also a double-rocker. The Hoekens linkage

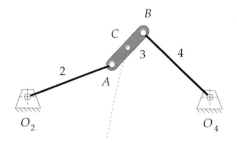

(a) Watt's straight-line linkage

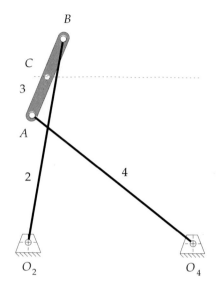

(b) Chebyshev straight-line linkage

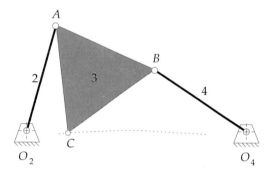

(c) Roberts straight-line linkage

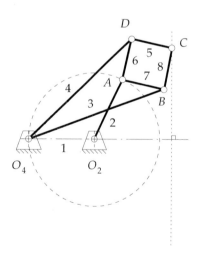

(e) Peaucellier exact straight-line linkage

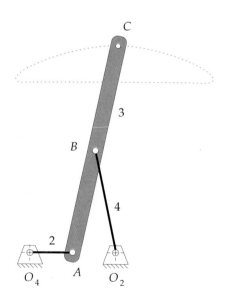

(d) Hoekens straight-line linkage

FIGURE 3-24

Some common straight-line linkages

in Figure 3-24d is a Grashof crank-rocker, which can be an advantage in some cases. In addition, the Hoekens linkage has the feature of very nearly constant velocity along the center portion of its straight-line motion.

These straight-line linkages are built-in examples in program FOURBAR. It is interesting to note that the **Hoeckens** and **Chebyschev** linkages are cognates of one another. A quick look in the Hrones and Nelson atlas of coupler curves will reveal a large number of curves with **approximate straight-line** segments. They are quite common.

To generate an **exact straight line** requires more links than four. The geared fivebar linkage is capable of exact straight line generation as can be seen by reading the file STRAIGHT into program FIVEBAR. Any symmetrical, Grashof, geared fivebar linkage with a gear ratio of –1 and a phase angle of 180° will generate an exact straight line at the joint between links 3 and 4.

Peaucellier (1864) discovered an **exact straight-line** mechanism of eight bars, shown in Figure 3-24e. Links 5, 6, 7, 8 form a rhombus. Link 3 equals link 4, and when O_2O_4 exactly equals O_2A, point C generates an *arc of infinite radius*, i.e., **a straight line**. By moving the pivot O_2 left or right from the position shown, changing only the length of link 1, this mechanism *will generate true circle arcs with radii much larger than the link lengths.*

3.9 DWELL MECHANISMS

A common requirement in machine design problems is the need for a dwell in the output motion. A **dwell** is defined as *zero output motion for some nonzero input motion.* In other words, the motor keeps going, but the output link stops moving. Many production machines perform a series of operations which involve feeding a part or tool into a workspace, and then holding it there (in a dwell) while some task is performed. Then the part must be removed from the workspace, and perhaps held in a second dwell while the rest of the machine "catches up" by indexing or performing some other tasks. Cams and followers (Chapter 9) are often used for these tasks because it is trivially easy to create a dwell with a cam. But, there is always a tradeoff in engineering design, and cams have their problems of high cost and wear as described in Section 2.15. It is also possible to obtain dwells with "pure" linkages of only links and pin joints, which have the advantage over cams of low cost and high reliability. Dwell linkages are more difficult to design than are cams with dwells. Linkages will usually yield only an approximate dwell but will be much cheaper to make and maintain than cams. Thus they may be well worth the effort.

Single-Dwell Linkages

There are two usual approaches to designing single-dwell linkages. Both result in **sixbar mechanisms,** and both require you to first find a fourbar with a suitable coupler curve. A **dyad** is then added to provide an output link with the desired dwell characteristic. The first approach to be discussed requires the design or definition of a fourbar with a coupler curve that contains an approximate circle arc portion, which "arc" occupies the desired portion of the input link (crank) cycle designated as the dwell. An atlas of coupler curves is invaluable for this part of the task.

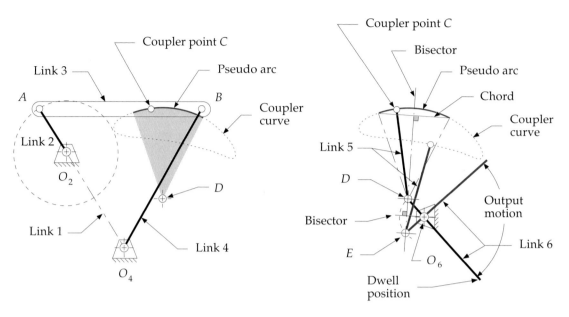

(a) Chosen fourbar crank-rocker with pseudo arc section for 60° of link 2 rotation

(b) Construction of the output dwell dyad

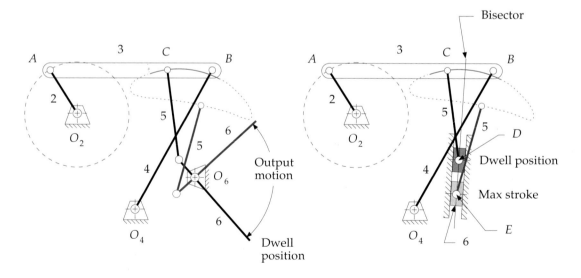

(c) Completed sixbar single-dwell linkage rocker output option

(d) Completed sixbar single-dwell linkage slider output option

FIGURE 3-25

Design of a sixbar single-dwell mechanism

✎EXAMPLE 3-11

Single-Dwell Mechanism with Only Pin Joints.

Problem: Design a sixbar linkage for 90° rocker motion over 300 crank degrees with dwell for the remaining 60°.

Solution: (see Figure 3-25)

1   Search the H&N Atlas for a fourbar linkage with a coupler curve having an approximate (pseudo) circle arc portion which occupies 60° of crank motion (12 dashes). The chosen fourbar is shown in Figure 3-25a.

2   Lay out this linkage to scale including the coupler curve and find the approximate center of the chosen coupler curve pseudo arc using graphical geometric techniques. To do so, draw the chord of the arc and construct its perpendicular bisector as shown in Figure 3-25b. The center will lie on the bisector. Label this point D.

3   Set your compass to the approximate radius of the coupler arc. This will be the length of link 5 which is to be attached at the coupler point C.

4   Trace the coupler curve with the compass point, while keeping the compass pencil lead on the perpendicular bisector, and find the extreme location along the bisector that the compass lead will reach. Label this point E.

5   The line segment DE represents the maximum displacement that a link of length CD, attached at C, will reach along the bisector.

6   Construct a perpendicular bisector of the line segment DE, and extend it in a convenient direction.

7   Locate fixed pivot O_6 on the bisector of DE such that lines O_6D and O_6E subtend the desired output angle, in this example, 90°.

8   Draw link 6 from D (or E) through O_6 and extend to any convenient length. This is the output link which will dwell for the specified portion of the crank cycle.

9   Check the transmission angles.

10   Make a cardboard model of the linkage and articulate it to check its function.

This linkage dwells because, during the time that the coupler point C is traversing the pseudo arc portion of the coupler curve, the other end of link 5, attached to C and the same length as the arc radius, is essentially stationary at its other end, which is the arc center. However the dwell at point D will have some "jitter" or oscillation, due to the fact that D is only an approximate center of the pseudo arc on the sixth-degree coupler curve. When point C leaves the arc portion, it will smoothly drive link 5 from point D to point E, which will in turn rotate the output link 6 through its arc as shown in Figure 3-25c. Note that we can have any angular displacement of link 6 we desire with the same links 2 to 5, as they alone completely define the dwell aspect. Moving pivot O_6 left and right along the bisector of line DE will change the angular displacement of link 6 but not its timing. In fact, a slider block could be substituted for link 6 as shown in Figure 3-25d, and linear translation along

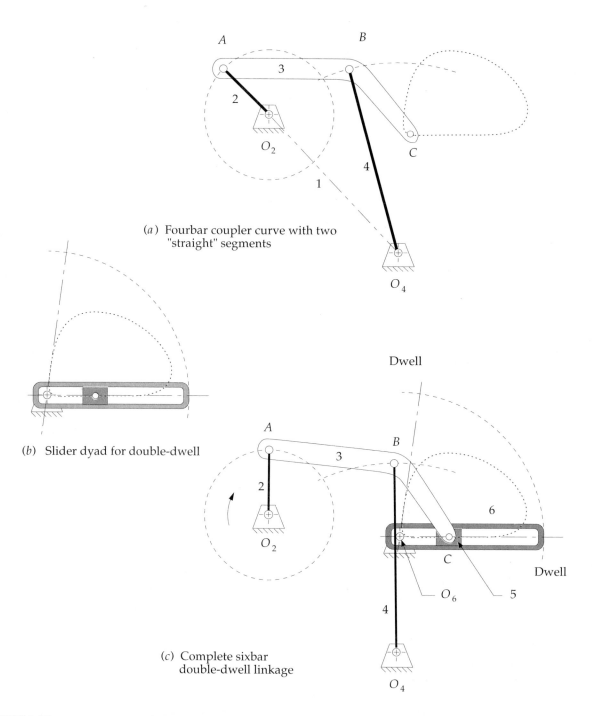

(a) Fourbar coupler curve with two "straight" segments

(b) Slider dyad for double-dwell

(c) Complete sixbar double-dwell linkage

FIGURE 3-26

Double-dwell sixbar linkage

line *DE* with the same timing and dwell at *D* will result. Input the file FIG3-25C to program SIXBAR and animate to see the linkage of Example 3-11 in motion. The dwell in the motion of link 6 can be clearly seen in the animation, including the jitter due to its approximate nature.

Double-Dwell Linkages

It is also possible, using a fourbar coupler curve, to create a double-dwell output motion. One approach is the same as that used in the single-dwell of Example 3-11. Now a coupler curve is needed which has *two* approximate circle arcs of the same radius but with different centers, both convex or both concave. A link 5 of length equal to the radius of the two arcs will be added such that it and link 6 will remain nearly stationary at each of the arcs' centers, while the coupler point traverses the circular parts of its path. Motion of the output link 6 will occur only when the coupler point is between those arc portions. Higher-order linkages, such as the geared fivebar, can be used to create multiple-dwell outputs by a similar technique since they possess coupler curves with multiple, approximate circle arcs. See the built-in example double-dwell linkage in program SIXBAR for a demonstration of this approach.

A second approach uses a coupler curve with two approximate straight-line segments of appropriate duration. If a pivoted slider block (link 5) is attached to the coupler at this point, and link 6 is allowed to slide in link 5, it only remains to choose a pivot O_6 at the intersection of the straight-line segments extended. The result is shown in Figure 3-26. While block 5 is traversing the "straight-line" segments of the curve, it will not impart any angular motion to link 6. The approximate nature of the fourbar straight line causes some jitter in these dwells also.

EXAMPLE 3-12

Double-Dwell Mechanism.

Problem: Design a sixbar linkage for 80° rocker output motion over 20 crank degrees with dwell for 160°, return motion over 140° and second dwell for 40°.

Solution: (see Figure 3-26)

1   Search the H&N atlas for a fourbar linkage with a coupler curve having two approximate straight-line portions. One should occupy 160° of crank motion (32 dashes), and the second 40° of crank motion (8 dashes). This is a wedge shaped curve as shown in Figure 3-26a.

2   Lay out this linkage to scale including the coupler curve and find the intersection of two tangent lines colinear with the straight segments. Label this point O_6.

3   Design link 6 to lay along these straight tangents, pivoted at O_6. Provide a slot in link 6 to accommodate slider block 5 as shown in Figure 3-26b.

4   Connect slider block 5 to the coupler point C on link 3 with a pin joint. The finished sixbar is shown in Figure 3-26c.

5   Check the transmission angles.

It should be apparent that these linkage dwell mechanisms have some disadvantages. Besides being difficult to synthesize, they give only approximate dwells which have some jitter on them. Also, they tend to be large for the output motions obtained, so do not package well. The acceleration of the output link can also be very high as in Figure 3-26, when block 5 is near pivot O_6. (Note the large angular displacement of link 6 resulting from a small motion of link 5.) Nevertheless they may be of value in situations where a completely stationary dwell is not required, and the low cost and high reliability of a linkage are important factors. Program SIXBAR has both single-dwell and double-dwell example linkages built in.

3.10   REFERENCES

1   **Erdman, A. G.**, and **Gustafson, J. E.**, "LINCAGES: Linkage INteractive Computer Analysis and Graphically Enhanced Synthesis," *ASME Paper No. 77-DTC-5*, 1977.

2   **Hall, A. S.**, *Kinematics and Linkage Design*, Waveland Press, P.O. Box 400, Prospect Heights, Ill., 1986, p. 134.

3   **Ibid**, p. 146.

4   **Sandor G.N.** and **Erdman, A.G.**, *Advanced Mechanism Design*, Prentice-Hall, Englewood Cliffs, N.J., 1984, pp. 177-187.

5   **Kaufman, R.E.**, "Mechanism Design by Computer," *Machine Design*, October, 1978, pp. 94-100.

6   **Hall, A. S.**, *Kinematics and Linkage Design*, Waveland Press, P.O. Box 400, Prospect Heights, Ill., 1986, pp. 33-34.

7   **Hrones, J. A.**, and **Nelson, G. L.**, *Analysis of the Fourbar Linkage*, MIT Technology Press, Cambridge, Mass., 1951.

8   **Zhang, C.**, **Norton, R. L.**, and **Hammond, T.**, "Optimization of Parameters for Specified Path Generation Using an Atlas of Coupler Curves of Geared Five-Bar Linkages," *Mechanism and Machine Theory*, vol. 19, no. 6, pp. 459-466, 1984, Pergamon Press, London.

9   **Hartenberg, R. S.**, and **Denavit, J.**, "Cognate Linkages," *Machine Design*, April 16, 1959, pp. 149-152.

3.11   BIBLIOGRAPHY

*For additional information on **type synthesis**, the following are recommended:*

Chironis, N. P., *Machine Devices and Instrumentation*. McGraw-Hill Inc., New York, 1966.

Chironis, N. P., *Mechanisms, Linkages, and Mechanical Controls*. McGraw-Hill, New York, 1965.

Jones, F., Horton, H., and **Newell, J.**, *Ingenious Mechanisms for Engineers*, vols. I to IV, Industrial Press, New York, 1930-1967.

*For additional information on **dimensional linkage synthesis**, the following are recommended:*

Dijksman, E. A., *Motion Geometry of Mechanisms*, Cambridge University Press, Cambridge, U.K., 1976.

Hall, A. S., *Kinematics and Linkage Design*, Waveland Press, P.O. Box 400, Prospect Heights, Ill., 1961.

Hain, K., *Applied Kinematics*, McGraw-Hill Inc., New York, 1967.

Hartenberg, R., and Denavit, J., *Kinematic Synthesis of Linkages*, McGraw-Hill Inc, New York, 1964.

Molian, S., *Mechanism Design*, Cambridge University Press, Cambridge, U.K., 1982.

Sandor, G., and Erdman, A., *Advanced Mechanism Design*, Prentice-Hall, Englewood Cliffs, N.J., 1984.

3.12 PROBLEMS

*3-1 Define the following examples as path, motion, or function generation cases.
   a.   A telescope aiming (star tracking) mechanism
   b.   A backhoe bucket control mechanism
   c.   A thermostat adjusting mechanism
   d.   A computer printer head moving mechanism
   e.   An XY plotter pen control mechanism

3-2 Design a fourbar Grashof crank-rocker for 90 degrees of output rocker motion with no quick-return. (See Example 3-1.) Build a cardboard model and determine the toggle positions and the minimum transmission angle.

*3-3 Design a fourbar mechanism to give the two positions shown in Figure P3-1 of output rocker motion with no quick-return. (See Example 3-2.) Build a cardboard model and determine the toggle positions and the minimum transmission angle.

3-4 Design a fourbar mechanism to give the two positions shown in Figure P3-1 of coupler motion. (See Example 3-3.) Build a cardboard model and determine the toggle positions and the minimum transmission angle. Add a driver dyad. (See Example 3-4.)

---

* Answers in Appendix E

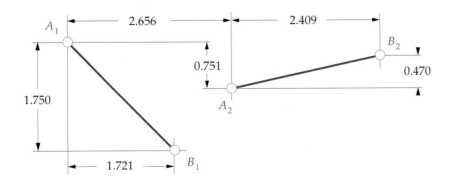

FIGURE P3-1

Problems 3-3 and 3-4

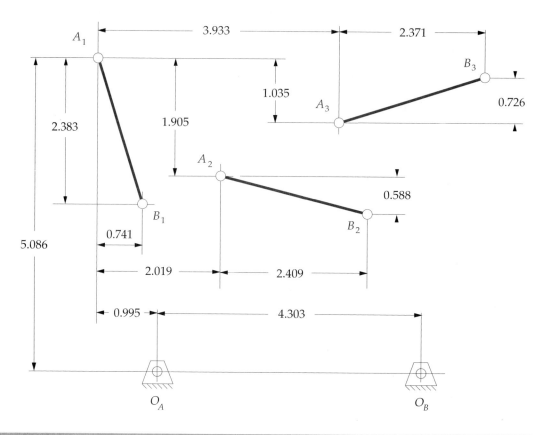

FIGURE P3-2

Problems 3-5 to 3-7

*3-5   Design a fourbar mechanism to give the three positions of coupler motion with no quick-return shown in Figure P3-2. (See also Example 3-5.) Ignore the points O_A and O_B shown. Build a cardboard model and determine the toggle positions and the minimum transmission angle. Add a driver dyad. (See Example 3-4.)

*3-6   Design a fourbar mechanism to give the three positions shown in Figure P3-2 using the fixed pivots O_A and O_B shown. Build a cardboard model and determine the toggle positions and the minimum transmission angle. Add a driver dyad. (See Example 3-4).

3-7   Repeat Problem 3-2 with a quick-return time ratio of 1 : 1.4. (See Example 3-9.)

*3-8   Design a sixbar drag link quick-return linkage for a time ratio of 1 : 2, and output rocker motion of 60 degrees. (See Example 3-10.)

3-9   Design a crank shaper quick-return mechanism for a time ratio of 1 : 3. (See Figure 3-14.)

*3-10   Select a linkage from the Hrones and Nelson atlas (or use Figure 3-17), and manually find its two cognates. Draw the Cayley and Roberts diagrams. Check your results with program FOURBAR.

*3-11 Find the three equivalent geared fivebar linkages for the three fourbar cognates in Figure 3-22a. Check your results by comparing the coupler curves with programs FOURBAR and FIVEBAR.

3-12 Design a sixbar single-dwell linkage for a dwell of 90 degrees of crank motion, with an output rocker motion of 45 degrees. (See Example 3-11.)

3-13 Design a sixbar double-dwell linkage for a dwell of 90 degrees of crank motion, with an output rocker motion of 60 degrees, followed by a second dwell of about 60 degrees of crank motion. (See Examples 3-11 and 3-12.)

3-14 Complete Example 3-5 by adding a driver dyad to link 3, and build a working cardboard model. Scale the diagram carefully for dimensions. (**Hint:** Use stiff cardboard and tacks. Dimensions must be very accurate in this example for it to work.)

3.13 PROJECTS

These larger-scale project statements deliberately lack detail and structure and are loosely defined. Thus, they are similar to the kind of "identification of need" problem statement commonly encountered in engineering practice. It is left to the student to structure the problem through background research and to create a clear goal statement and set of task specifications before attempting to design a solution. This design process is spelled out in Chapter 1 and should be followed in all of these examples. These projects can be done as an exercise in mechanism synthesis alone or can be revisited and thoroughly analyzed by the methods presented in later chapters as well. All results should be documented in a professional engineering report.

P3-1 The tennis coach needs a better tennis ball server for practice. This device must fire a sequence of standard tennis balls from one side of a standard tennis court over the net such that they land and bounce within each of the three court areas defined by the court's white lines. The order and frequency of a ball's landing in any one of the three court areas must be random. The device should operate automatically and unattended except for the refill of balls. It should be capable of firing 50 balls between reloads. The timing of ball releases should vary. For simplicity, a motor driven pin-jointed linkage design is preferred.

P3-2 A quadraplegic patient has lost all motion except that of her head. She can only move a small "mouth stick" to effect a switch closure. She was an avid reader before her injury, and would like again to be able to read standard hardcover books without the need of a person to turn pages for her. Thus, a reliable, simple, and inexpensive automatic page turner is needed. The book may be placed in the device by an assistant. It should accommodate as wide a range of book sizes as possible. Book damage is to be avoided and safety of the user is paramount.

P3-3 Grandma's off her rocker again! Junior's run down to the Bingo parlor to fetch her, but we've got to do something about her rocking chair before she gets back. She's been complaining that her arthritis makes it too painful to push the rocker. So, for her 100th birthday in 2 weeks, we're going to surprise her with a new, automated, motorized rocking chair. The only constraints placed on the problem are that the device must be safe and must provide interesting and pleasant motions, similar to those of her present *Boston* rocker, to all parts of the occupant's body. Since simplicity is the mark of good design, a linkage solution with only full pin joints is preferred.

P3-4 The local amusement park's business is suffering as a result of the proliferation of computer game parlors. They need a new and more exciting ride which will attract new customers. The only constraints are that it must be safe, provide excitement, and not subject the occupants to excessive accelerations or velocities. Also it must be as compact as possible, since space is limited. Continuous rotary input and full pin joints are preferred.

P3-5 The student section of ASME is sponsoring a spring fling on campus. They need a mechanism for their "Dunk the Professor" booth which will carry the unfortunate (untenured) volunteer into and out of the water tub. The contestants will provide the inputs to a multiple *DOF* mechanism. If they know their kinematics, they can provide a combination of inputs which will dunk the victim.

P3-6 The National House of Flapjacks wants to automate their flapjack production. They need a mechanism which will automatically flip the flapjacks "on the fly" as they travel through the griddle on a continuously moving conveyor. This mechanism must track the constant velocity of the conveyor, pick up a pancake, flip it over and place it back onto the conveyor.

P3-7 Many varieties and shapes of computer video monitors now exist. Their long-term use leads to eyestrain and body fatigue. There is a need for an adjustable stand which will hold the video monitor and the separate keyboard at any position the user deems comfortable. The computer's Central Processor Unit (CPU) can be remotely located. This device should be free standing to allow use with a comfortable chair, couch, or lounge of the user's choice. It should not require the user to assume the conventional "seated at a desk" posture to use the computer. It must be stable in all positions and safely support the equipment's weight.

P3-8 Most small boat trailers must be submerged in the water to launch or retrieve the boat. This greatly reduces the life of the trailer, especially in salt water. A need exists for a trailer that will remain on dry land while it launches or retrieves the boat. No part of the trailer should get wet. User safety is of greatest concern, as is protection of the boat from damage.

P3-9 The "Save the Skeet" foundation has requested a more humane skeet launcher be designed. While they have not yet succeeded in passing legislation to prevent the wholesale slaughter of these little devils, they are concerned about the inhumane aspects of the large accelerations imparted to the skeet as it is launched into the sky for the sportsman to shoot it down. The need is for a skeet launcher that will smoothly accelerate the clay pigeon onto its desired trajectory.

P3-10 The coin operated "kid bouncer" machines found outside supermarkets typically provide a very unimaginative rocking motion to the occupant. There is a need for a superior "bouncer" which will give more interesting motions while remaining safe for small children.

P3-11 Horseback riding is a very expensive hobby or sport. There is a need for a horseback riding simulator to train prospective riders sans the expensive horse. This device should provide similar motions to the occupant as she would feel in the saddle under various gaits such as a walk, canter, gallop etc. A more advanced version might contain jumping motions as well. User safety is most important.

P3-12   The nation is on a fitness craze. Many exercise machines have been devised. There is still room for improvement to these devices. They are typically designed for the young, strong athlete. There is also a need for an ergonomically optimum exercise machine for the older person who needs gentler exercise.

P3-13   A paraplegic patient needs a device to get himself from his wheelchair into the Jacuzzi with no assistance. He has good upper body and arm strength. Safety is paramount.

P3-14   The Army has requested a mechanical walking device to test army boots for durability. It should mimic a person's walking motion and provide forces similar to an average soldier's foot.

P3-15   NASA wants a zero-G machine for astronaut training. It must carry one person and provide a negative 1 G acceleration for as long as possible.

P3-16   The Amusement Machine Co. Inc. wants a portable "Whip" ride which will give two or four passengers a thrilling but safe ride, and which can be trailed behind a pickup truck from one location to another.

P3-17   The Air Force has requested a pilot training simulator which will give potential pilots exposure to G forces similar to those they will experience in dogfight maneuvers.

P3-18   Cheers needs a better "Mechanical Bull" simulator for their "yuppie" bar in Boston. It must give a thrilling "bucking bronco" ride but be safe.

P3-19   Despite the improvements in handicap access, many curbs block wheelchairs from public places. Design an attachment for a conventional wheelchair which will allow it to get up over a curb.

P3-20   A carpenter needs a dumping attachment to fit in her pickup truck so she can dump building materials. She can't afford to buy a dump truck.

P3-21   The same carpenter wants an inexpensive lift gate designed to fit her full-sized pickup truck, in order to lift and lower heavy cargo to the truck bed.

P3-22   This carpenter is very demanding (and lazy). She also wants a device to lift sheets of "sheet rock" into place on ceiling or walls to hold it while nailing.

P3-23   Click and Clack, the tappet brothers, need a better transmission jack for their Good News Garage. This device should position a transmission under a car (on a lift) and allow it to be maneuvered into place safely and quickly.

P3-24   A paraplegic who was an avid golfer before his injury, wants a mechanism to allow him to stand up in his wheelchair in order to once again play golf. It must not interfere with normal wheelchair use, though it could be removed from the chair when he is not golfing.

P3-25   A wheelchair lift is needed to raise chair and person 3 feet from the garage floor to the level of the first floor of the house. Safety and reliability are of major concern, as is cost.

Chapter 4

POSITION ANALYSIS

Theory is the distilled essence of practice
RANKINE

4.0   INTRODUCTION

Once a tentative mechanism design has been **synthesized**, it must then be **analyzed**. A principal goal of kinematic analysis is to determine the accelerations of all the moving parts in the assembly. **Dynamic forces** are proportional to acceleration, from Newton's second law. We need to know the dynamic forces in order to calculate the **stresses** in the components. The design engineer must ensure that the proposed mechanism or machine will not fail under its operating conditions. Thus the stresses in the materials must be kept well below allowable levels. To calculate the stresses, we need to know the static and dynamic forces on the parts. To calculate the dynamic forces, we need to know the **accelerations**. In order to calculate the accelerations, we must first find the **positions** of all the links or elements in the mechanism for each increment of input motion, and then differentiate the position equations versus time to find **velocities**, and then differentiate again to obtain the expressions for acceleration. For example, in a simple Grashof fourbar linkage, we would probably want to calculate the positions, velocities, and accelerations of the output links (coupler and rocker) for perhaps every two degrees (180 positions) of input crank position for one revolution of the crank.

This can be done by any of several methods. We could use a **graphical approach** to determine the position, velocity, and acceleration of the output links for all 180 positions of interest, or we could **derive the general equations** of motion for any position, differentiate for velocity and acceleration, and then solve these **analytical expressions** for our 180 (or more) crank locations. A computer will make this latter task much more palatable. If we choose to use the graphical approach to analysis, we will have to do an independent graphical solution for each of the positions of interest. None of the information obtained graphically for the first position will be applicable to the second position or to any others. In contrast, once the analytical solution is derived for a particular mechanism, it can be quickly solved (with a computer) for all positions. If you want information for more than 180 positions, it only means you will have to wait longer for the computer to generate those data. The derived equations are the same. So, have another cup of coffee while the computer crunches the numbers! In this chapter, we will present and derive analytical solutions to the position analysis problem for various planar mechanisms. We will also discuss graphical solutions which are useful for checking your analytical results. In Chapters 6 and 7 we will do the same for velocity and acceleration analysis of planar mechanisms.

It is interesting to note that the **graphical position analysis** of linkages is a truly trivial exercise, while the algebraic approach to position analysis is much more complicated. If you can draw the linkage to scale, you have then solved the position analysis problem graphically. It only remains to measure the link angles on the scale drawing to protractor accuracy. The converse is true for velocity and especially for acceleration analysis. The analytical solutions for these are less complicated to derive than is the analytic position solution. However, graphical velocity and acceleration analysis becomes quite complex and difficult. Moreover, the graphical vector diagrams must be redone *de novo* (meaning literally *from new*) for each of the linkage positions of interest. This is a very tedious exercise and was the only practical method available in the days *B.C.* (*Before Computer*), not so long ago. The proliferation of inexpensive microcomputers in recent years has truly revolutionized the practice of engineering. As a graduate engineer, you will never be far from a computer of sufficient power to solve this type of problem. Thus, in this text we will emphasize analytical solutions which are easily solved with a microcomputer. The computer programs provided with this text use the same analytical techniques as derived in the text.

Geez Joe, - now I wish I took that programming course!

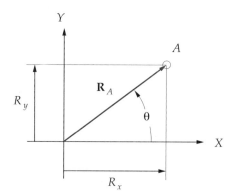

Polar form:

$$\left| R_A \right| @ \underline{\diagup \theta}$$

Cartesian form :

$$R_x , \ R_y$$

FIGURE 4-1

A position vector in the plane

4.1   COORDINATE SYSTEMS

Coordinate systems and reference frames exist for the pleasure and convenience of the engineer who defines them. In the next chapters we will freely adorn our systems with multiple coordinate systems as we see fit, to aid in understanding and solving the problem. We will denote one of these as the *global* or *absolute* coordinate system, and the others will be *local* coordinate systems within the global framework. The global system is often taken to be attached to Mother Earth, though it could as well be attached to another ground plane such as the frame of an automobile. If our goal is to analyze the motion of a windshield wiper blade, we may not care to include the gross motion of the automobile in the analysis. In that case a global coordinate system attached to the car would be useful and we could consider it to be an **absolute** coordinate system. Even if we use the earth as an absolute reference frame, we must realize that it is not stationary either, and as such is not very useful as a reference frame for a space probe. Though we will speak of absolute positions, velocities and accelerations, keep in mind that ultimately, until we discover some stationary point in the universe, all motions are really relative. The term **inertial reference frame** is used to denote *a system which itself has no acceleration.* All angles will be measured according to the *right hand rule* in this text. That is, **counterclockwise angles**, angular velocities and angular accelerations *are positive in sign.*

Local coordinate systems are typically attached to a link at some point of interest. This might be a pin joint, a center of gravity, or a line of centers of a link. These local coordinate systems may be either rotating or non-rotating as we desire. If we want to measure the angle of a link as it rotates in the global system, we probably will want to attach a non-rotating coordinate system to some point on the link (say a pin joint). This non-rotating system will move with its origin on the link but remains always parallel to the global system. If we want to measure some parameters within a link, independent of its rotation, then we will want to construct a rotating coordinate system along some line on the link. This system will both move and rotate with the link in the global system. Most often we will need to have both types of local coordinate systems on our moving links to do a complete analysis. Obviously we must define the positions and angles of these moving, local coordinate systems in the global system at all positions of interest.

4.2   POSITION AND DISPLACEMENT

Position

The **position** of a point in the plane can be defined by the use of a **position vector** as shown in Figure 4-1. The choice of **reference axes** is arbitrary and is selected to suit the observer. A two-dimensional vector has two attributes, which can be expressed in either *polar* or *cartesian* coordinates. The **polar form** provides the magnitude and the angle of the vector. The **cartesian form** provides the X and Y components of the vector. Each form is directly convertible into the other by:

the Pythagorean theorem:

$$R_A = \sqrt{R_x^2 + R_y^2}$$

and trigonometry: (4.0)

$$\theta = \arctan\left(\frac{R_y}{R_x}\right)$$

Displacement

Displacement of a point is the change in its position and can be defined as *the straight-line distance between the initial and final position of a point which has moved in the reference frame.* Note that displacement is not necessarily the same as the path length which the point may have traveled to get from its initial to final position. Figure 4-2a shows a point in two positions, A and B. The curved line depicts the path along which the point traveled. The position vector $\mathbf{R}_{BA}$ defines the displacement of the point B with respect to point A. Figure 4-2b defines this situation more rigorously and with respect to a reference frame XY. The notation $\mathbf{R}$ will be used to denote a position vector. The vectors $\mathbf{R}_A$ and $\mathbf{R}_B$ define, respectively, the absolute positions of points A and B with respect to this *global* XY reference frame. The vector $\mathbf{R}_{BA}$ denotes the difference in position, or the *displacement*, between A and B. This can be expressed as the *position difference equation:*

$$\mathbf{R}_{BA} = \mathbf{R}_B - \mathbf{R}_A \qquad (4.1a)$$

This expression is read: *The position of B with respect to A is equal to the (absolute) position of B minus the (absolute) position of A,* where *absolute* means with respect to the origin of the *global* reference frame. Note that this expression could also be written as:

$$\mathbf{R}_{BA} = \mathbf{R}_{BO} - \mathbf{R}_{AO} \qquad (4.1b)$$

with the second subscript O denoting the origin of the XY reference frame. When a position vector is rooted at the origin of the reference frame, it is customary to omit the second subscript. It is understood, in its absence, to be the origin. Also, a vector

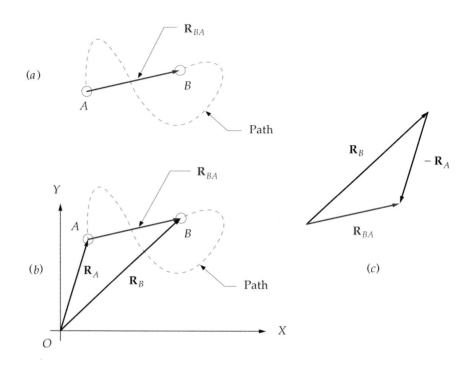

FIGURE 4-2

Position difference and relative position

referred to the origin, such as $\mathbf{R}_A$, is often called an absolute vector. This means that it is taken with respect to a reference frame which is assumed to be stationary, e.g. *the ground*. It is important to realize, however, that the ground is usually also in motion in some larger frame of reference. Figure 4-2c shows a graphical solution to equation 4.1.

In our example of Figure 4-2, we have tacitly assumed so far that this point, which is first located at A and later at B, is, in fact, the same particle, moving within the reference frame. It could be, for example, one automobile moving along the road from A to B. With that assumption, it is conventional to refer to the vector $\mathbf{R}_{BA}$ as a **position difference**. There is, however, another situation which leads to the same diagram and equation but needs a different name. Assume now that points A and B in Figure 4-2b represent not the same particle but two independent particles moving in the same reference frame, as perhaps two automobiles traveling on the same road. The vector equations 4.1 and the diagram in Figure 4-2b still are valid, but we now refer to $\mathbf{R}_{BA}$ as a **relative position**, or **apparent position**. We will use the relative position term here. A more formal way to distinguish between these two cases is as follows:

CASE 1:   *One body in two successive positions* => ***position difference***

CASE 2:   *Two bodies simultaneously in separate positions* => ***relative position***

This may seem a rather fine point to distinguish, but the distinction will prove useful, and the reasons for it more clear, when we analyze velocities and accelerations, especially when we encounter (CASE 2 type) situations in which the two bodies occupy the same position at the same time but have different motions.

4.3   TRANSLATION, ROTATION, AND COMPLEX MOTION

So far we have been dealing with a particle, or point, in plane motion. It is more interesting to consider the motion of a **rigid body**, or link. Figure 4-3a shows a link AB denoted by a position vector $\mathbf{R}_{BA}$. An axis system has been set up at the root of the vector, at point A, for convenience.

Translation

Figure 4-3b shows link AB moved to a new position $A'B'$ by translation through the displacement AA' or BB' which are equal, i.e., $\mathbf{R}_{A'A} = \mathbf{R}_{B'B}$.

A definition of translation is:

all points on the body have the same displacement.

As a result the link retains its angular orientation. Note that the translation need not be along a straight path. The curved lines from A to A' and B to B' are the **curvilinear translation** path of the link. There is no rotation of the link if these paths are parallel. If the path happens to be straight, then it will be the special case of **rectilinear translation**, and the path and the displacement will be the same.

Rotation

Figure 4-3c shows the same link AB moved from its original position at the origin by rotation through an angle. Point A remains at the origin, but B moves through the position difference vector $\mathbf{R}_{B'B} = \mathbf{R}_{B'A} - \mathbf{R}_{BA}$.

A definition of rotation is:

different points in the body undergo different displacements and thus there is a displacement difference between any two points chosen.

The link now changes its angular orientation in the reference frame, and all points have different displacements.

Complex Motion

The general case of **complex motion** is the sum of the translation and rotation components. Figure 4-3d shows the same link moved through both the translation and the rotation applied above. Note that the order in which these two components are added is immaterial. The resulting complex displacement will be the same whether you first rotate and then translate or vice versa. This is because the two factors are independent. The total complex displacement of point B is defined by the following expression:

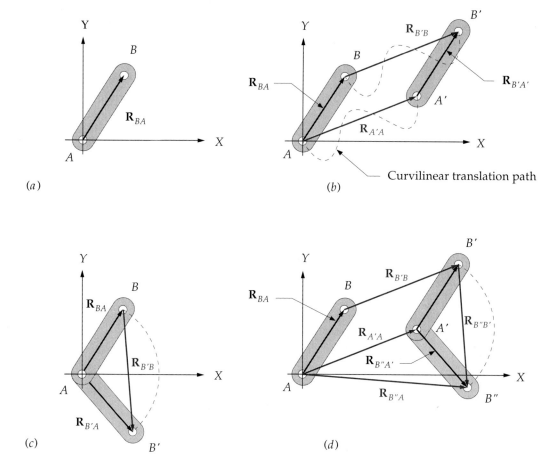

(a)

(b)

Curvilinear translation path

(c)

(d)

FIGURE 4-3

Translation, rotation and complex motion

Total Displacement = Translation Component + Rotation Component

$$\mathbf{R}_{B''B} = \mathbf{R}_{B'B} + \mathbf{R}_{B''B'} \tag{4.1c}$$

The new absolute position of point B referred to the origin at A is:

$$\mathbf{R}_{B''A} = \mathbf{R}_{A'A} + \mathbf{R}_{B''A'} \tag{4.1d}$$

Note that the above two formulas are merely applications of the position difference equation 4.1a. See also Section 2.2 for definitions and discussion of *rotation, translation,* and *complex motion.* These motion states can be expressed as the following theorems.

Theorems

Euler's theorem:

the general displacement of a rigid body with one point fixed is a rotation about some axis.

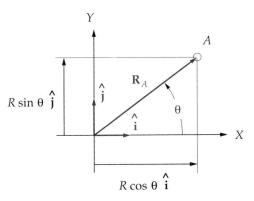

Polar form :

$$\left| \mathbf{R}_A \right| \ @ \diagup \theta$$

Cartesian form :

$$R \cos \theta \ \hat{\mathbf{i}} \ , \ R \sin \theta \ \hat{\mathbf{j}}$$

FIGURE 4-4

Unit vector notation for position vectors

This applies to pure rotation as defined in Section 2.2 and above. Chasles (1793-1880) provided a corollary to Euler's theorem now known as:

Chasles' theorem:

any displacement of a rigid body is equivalent to the sum of a translation of any one point on that body and a rotation of the body about an axis through that point.

This describes complex motion as defined in Section 2.2 and above. Note that equation 4.1c is an expression of Chasles' theorem.

4.4   COMPLEX NUMBERS AS VECTORS

There are many ways to represent vectors. They may be defined in **polar coordinates**, by their *magnitude* and *angle*, or in **cartesian coordinates** as x and y components. These forms are of course easily convertible from one to the other using equations 4.0. The position vectors in Figures 4-1 to 4-3 can be represented as any of these expressions:

Polar Form	Cartesian Form	
$R @ \angle \theta$	$r \cos \theta \hat{\mathbf{i}} + r \sin \theta \hat{\mathbf{j}}$	(4.2)
$r e^{j\theta}$	$r \cos \theta + j r \sin \theta$	(4.3)

Equation 4.2 uses **unit vectors** to represent the x and y vector component directions in the cartesian form. Figure 4-4 shows the unit vector notation for a position vector. Equation 4.3 uses **complex number notation** wherein the X direction component is called the *real portion* and the Y direction component is called the *imaginary portion*. This unfortunate term *imaginary* comes about because of the use of the notation j to represent the square root of minus one, which of course cannot be evaluated numerically. However, this *imaginary* number is used in a **complex number** as an **operator,** *not as a value.* Figure 4-5a shows the **complex plane** in which the *real* axis represents the X-directed component of the vector in the plane, and the *imaginary* axis represents the Y-directed component of the same vector. So, any term in a complex number which has no j operator is an x component, and a j indicates a y component.

Polar form: $R\,e^{\,j\theta}$

Cartesian form: $R\cos\theta + j\,R\sin\theta$

$$R = \left|\mathbf{R}_A\right|$$

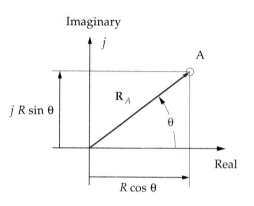

$j\,R\sin\theta$

$R\cos\theta$

(a) Complex number representation of a position vector

$\mathbf{R}_C = j^2 R = -R$

$\mathbf{R}_B = jR$

$+\theta$

$\mathbf{R}_A$

$\mathbf{R}_D = j^3 R = -jR$

(b) Vector rotations in the complex plane

FIGURE 4-5

Complex number representation of vectors in the plane

Note in Figure 4-5b that each multiplication of the vector $\mathbf{R}_A$ by the operator j results in a *counterclockwise rotation* of the vector through 90 degrees. The vector $\mathbf{R}_B = j\mathbf{R}_A$ is directed along the *positive imaginary* or j axis. The vector $\mathbf{R}_C = j^2\,\mathbf{R}_A$ is directed along the *negative real* axis because $j^2 = -1$ and thus $\mathbf{R}_C = -\mathbf{R}_A$. In similar fashion, $\mathbf{R}_D = j^3\,\mathbf{R}_A = -j\mathbf{R}_A$ and this component is directed along the *negative j axis*.

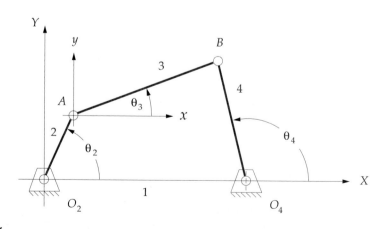

FIGURE 4-6

Measurement of angles in the fourbar linkage

One advantage of using this complex number notation to represent planar vectors comes from the **Euler identity**:

$$e^{\pm j\theta} = \cos\theta \pm j\sin\theta \qquad (4.4a)$$

Any two-dimensional vector can be represented by the compact polar notation on the left side of equation 4.4a. There is no easier function to differentiate or integrate, since it is its own derivative:

$$\frac{de^{j\theta}}{d\theta} = je^{j\theta} \qquad (4.4b)$$

We will use this **complex number notation** for vectors to develop and derive the equations for position, velocity and acceleration of linkages.

4.5   POSITION ANALYSIS OF LINKAGES

Graphical Analysis

For any one degree of freedom linkage, such as a fourbar, only one parameter is needed to completely define the positions of all the links. The parameter usually chosen is the angle of the input link. This is shown as θ_2 in Figure 4-6. We want to find θ_3 and θ_4. The link lengths are known. Note that we will consistently number the ground link as 1 and the driver link as 2 in these examples.

The graphical analysis of this problem is trivial. If we draw the linkage carefully to scale in a particular position (given θ_2), then it is only necessary to measure the angles of links 3 and 4 with a protractor. Note that all link angles are measured from a positive X axis A *local xy* axis system, parallel to the *global XY* system, has been created at point A to measure θ_3. The accuracy of this graphical solution will be limited by our care and drafting ability and by the crudity of the protractor used. Nevertheless, a very rapid approximate solution can be found. But, this solution is only valid for the particular position (value of θ_2) chosen. For each additional position analysis we must completely redraw the linkage. This can become burdensome if we need a complete analysis at every 1 or 2° increment of θ_2. In that case we will be better off to derive an analytical solution for θ_3 and θ_4 which can be solved by computer. To do so we will represent the links as **position vectors**.

The Vector Loop Equation for a Fourbar Linkage

Figure 4-7 shows the same fourbar linkage as in Figure 4-6, but the links are now drawn as position vectors which form a vector loop. Note that this loop closes on itself. Thus, the sum of the vectors in the loop must be zero. The lengths of the vectors are the link lengths and thus are known. The particular linkage position shown is defined by one input angle θ_2 because it is a one *DOF* mechanism. We want to solve for the unknown angles θ_3 and θ_4.

Note that the directions of the position vectors are chosen so as to define their angles where we desire them to be measured. By definition, *the angle of a vector is always*

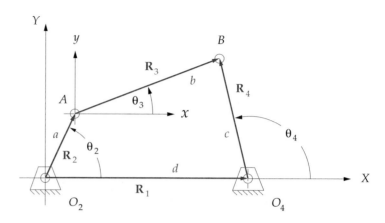

FIGURE 4-7

Position vector loop for a fourbar linkage

*measured at its root, **not at its head.*** We would like angle θ_4 to be measured at the fixed pivot O_4, so vector $\mathbf{R}_4$ is arranged to have its root at that point. We would like to measure angle θ_3 at the point where links 2 and 3 join, so vector $\mathbf{R}_3$ is rooted there. A similar logic dictates the arrangement of vectors $\mathbf{R}_1$ and $\mathbf{R}_2$. Note that the X (*real*) axis is taken for convenience along link 1 and the origin of the global coordinate system is taken at point O_2, the root of the input link vector, $\mathbf{R}_2$. These choices of vector directions and senses, as indicated by their arrowheads, lead to this vector loop equation:

$$\mathbf{R}_2 + \mathbf{R}_3 - \mathbf{R}_4 - \mathbf{R}_1 = 0 \tag{4.5a}$$

An **alternate notation** for these position vectors is to use the labels of the points at the vector **tips** and **roots** (*in that order*) as subscripts. The second subscript is conventionally omitted if it is the origin of the global coordinate system (point O_2):

$$\mathbf{R}_A + \mathbf{R}_{BA} - \mathbf{R}_{BO_4} - \mathbf{R}_{O_4} = 0 \tag{4.5b}$$

Next, we substitute the complex number notation for each position vector. To simplify the notation and minimize the use of subscripts, we will denote the scalar lengths of the four links as a, b, c, and d. These are so labelled in Figure 4-7. The equation then becomes:

$$ae^{j\theta_2} + be^{j\theta_3} - ce^{j\theta_4} - de^{j\theta_1} = 0 \tag{4.5c}$$

These are three forms of the same vector equation, and as such can be solved for two unknowns. There are four variables in this equation, namely the four link angles. The link lengths are all constant in this particular linkage. Also, the value of the angle of link 1 is fixed (at zero) since this is the ground link. **Theta2** is the *independent variable* which we will control with a motor or other driver device. That leaves the angles of links 3 and 4 to be found. We need algebraic expressions which define θ_3 and θ_4 as functions only of the constant link lengths and the one input angle, θ_2. These expressions will be of the form:

$$\theta_3 = f\{a,b,c,d,\theta_2\}$$

$$\theta_4 = g\{a,b,c,d,\theta_2\}$$

(4.5d)

To solve the polar form, vector equation 4.5c, we must substitute the *Euler equivalents* (Eq. 4.4a) for the $e^{j\theta}$ terms, and then separate the resulting cartesian form vector equation into two scalar equations which can be solved simultaneously for θ_3 and θ_4. Substituting equation 4.4a into equation 4.5c:

$$a(\cos\theta_2 + j\sin\theta_2) + b(\cos\theta_3 + j\sin\theta_3) - c(\cos\theta_4 + j\sin\theta_4) - d(\cos\theta_1 + j\sin\theta_1) = 0 \quad (4.5e)$$

This equation can now be separated into its real and imaginary parts and each set to zero.

real part (x component):

$$a\cos\theta_2 + b\cos\theta_3 - c\cos\theta_4 - d\cos\theta_1 = 0$$

but: $\theta_1 = 0$, so: (4.6a)

$$a\cos\theta_2 + b\cos\theta_3 - c\cos\theta_4 - d = 0$$

imaginary part (y component):

$$ja\sin\theta_2 + jb\sin\theta_3 - jc\sin\theta_4 - jd\sin\theta_1 = 0$$

but: $\theta_1 = 0$, and the j's divide out, so: (4.6b)

$$a\sin\theta_2 + b\sin\theta_3 - c\sin\theta_4 = 0$$

The scalar equations 4.6a and 4.6b can now be solved simultaneously for θ_3 and θ_4. To solve this set of two simultaneous trigonometric equations is straightforward but tedious. Some substitution of trigonometric identities will simplify the expressions. The first step is to rewrite equations 4.6a and 4.6b so as to isolate one of the two unknowns on the left side. We will isolate θ_3 and solve for θ_4 in this example.

$$b\cos\theta_3 = -a\cos\theta_2 + c\cos\theta_4 + d \qquad (4.6c)$$

$$b\sin\theta_3 = -a\sin\theta_2 + c\sin\theta_4 \qquad (4.6d)$$

Now square both sides of equations 4.6c and 4.6d and add them:

$$b^2(\sin^2\theta_3 + \cos^2\theta_3) = (-a\sin\theta_2 + c\sin\theta_4)^2 + (-a\cos\theta_2 + c\cos\theta_4 + d)^2 \quad (4.7a)$$

Note that the quantity in parentheses on the left side is equal to 1, eliminating θ_3 from the equation, leaving only θ_4 which can now be solved for.

$$b^2 = (-a\sin\theta_2 + c\sin\theta_4)^2 + (-a\cos\theta_2 + c\cos\theta_4 + d)^2 \qquad (4.7b)$$

The right side of this expression must now be expanded and terms collected.

$$b^2 = a^2 + c^2 + d^2 - 2ad\cos\theta_2 + 2cd\cos\theta_4 - 2ac(\sin\theta_2\sin\theta_4 + \cos\theta_2\cos\theta_4) \qquad (4.7c)$$

To further simplify this expression, the constants K_1, K_2, and K_3 are defined in terms of the constant link lengths in equation 4.7c:

$$K_1 = \frac{d}{a} \qquad K_2 = \frac{d}{c} \qquad K_3 = \frac{a^2 - b^2 + c^2 + d^2}{2ac} \qquad (4.8a)$$

and:

$$K_1 \cos \theta_4 - K_2 \cos \theta_2 + K_3 = \cos \theta_2 \cos \theta_4 + \sin \theta_2 \sin \theta_4 \qquad (4.8b)$$

If we substitute the identity $\cos(\theta_2 - \theta_4) = \cos \theta_2 \cos \theta_4 + \sin \theta_2 \sin \theta_4$, we get the form known as Freudenstein's equation.

$$K_1 \cos \theta_4 - K_2 \cos \theta_2 + K_3 = \cos(\theta_2 - \theta_4) \qquad (4.8c)$$

In order to reduce equation 4.8b to a more tractable form for solution, it will be useful to substitute the *half angle identities* which will convert the $\sin \theta_4$ and $\cos \theta_4$ terms to $\tan \theta_4$ terms:

$$\sin \theta_4 = \frac{2 \tan\left(\dfrac{\theta_4}{2}\right)}{1 + \tan^2\left(\dfrac{\theta_4}{2}\right)}; \qquad \cos \theta_4 = \frac{1 - \tan^2\left(\dfrac{\theta_4}{2}\right)}{1 + \tan^2\left(\dfrac{\theta_4}{2}\right)} \qquad (4.9)$$

This results in the following simplified form, where the link lengths and known input value (θ_2) terms have been collected as constants A, B, and C.

$$A \tan^2\left(\frac{\theta_4}{2}\right) + B \tan\left(\frac{\theta_4}{2}\right) + C = 0$$

$$(4.10a)$$

where:
$$A = \cos \theta_2 - K_1 - K_2 \cos \theta_2 + K_3$$
$$B = -2 \sin \theta_2$$
$$C = K_1 - (K_2 + 1)\cos \theta_2 + K_3$$

Note that equation 4.10a is quadratic in form, and the solution is:

$$\tan\left(\frac{\theta_4}{2}\right) = \frac{-B \pm \sqrt{B^2 - 4AC}}{2A}$$

$$(4.10b)$$

$$\theta_{4_{1,2}} = 2 \arctan\left(\frac{-B \pm \sqrt{B^2 - 4AC}}{2A}\right)$$

Equation 4-10b has two solutions, obtained from the $\pm$ conditions on the radical. These two solutions, as with any quadratic equation, may be of three types: *real and equal, real and unequal, complex conjugate*. If the discriminant under the radical is negative, then the solution is complex conjugate, which simply means that the link lengths chosen are not capable of connection for the chosen value of the input angle θ_2. This can occur either when the link lengths are completely incapable of connection in any position or, in a non-Grashof linkage, when the input angle is beyond a toggle limit position. There is then no real solution for that value of input angle θ_2. Excepting this situation, the solution will usually be real and unequal, meaning there are two values of θ_4 corresponding to any one value of θ_2. These are referred to as the **crossed** and **open** configurations of the linkage and also as the two **branches** of the linkage. In the fourbar linkage, the minus solution gives θ_4 for the open configuration and the positive solution gives θ_4 for the crossed configuration.

Figure 4-8 shows both crossed and open solutions for a Grashof crank-rocker linkage. The terms crossed and open are based on the assumption that the input link 2, for which θ_2 is defined, is placed in the first quadrant (i.e., $0 < \theta_2 < \pi/2$). A Grashof linkage is then defined as **crossed** if the two links adjacent to the shortest link cross one another, and as **open** if they do not cross one another in this position. Note that the configuration of the linkage, either crossed or open, is solely dependent upon the way that the links are assembled. You cannot predict, based on link lengths alone, which of the solutions will be the desired one. In other words, you can obtain either solution with the same linkage by simply taking apart the pin which connects links 3 and 4 in Figure 4-8, and moving those links to the only other positions at which the pin will again connect them. In so doing, you will have switched from one position solution, or **branch**, to the other.

The solution for angle θ_3 is essentially similar to that for θ_4. Returning to equations 4.6, we can rearrange them to isolate θ_4 on the left side.

$$c\cos\theta_4 = a\cos\theta_2 + b\cos\theta_3 - d \qquad (4.6e)$$
$$c\sin\theta_4 = a\sin\theta_2 + b\sin\theta_3 \qquad (4.6f)$$

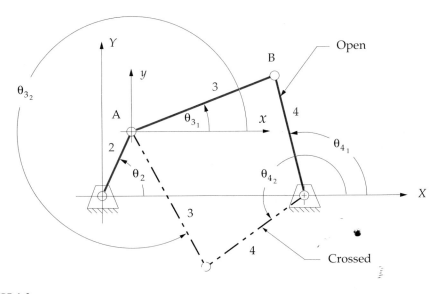

FIGURE 4-8

Open and crossed configurations of the fourbar linkage

Squaring and adding these equations will eliminate θ_4. The resulting equation can be solved for θ_3 as was done above for θ_4, yielding this expression:

$$K_1 \cos\theta_3 + K_4 \cos\theta_2 + K_5 = \cos\theta_2 \cos\theta_3 + \sin\theta_2 \sin\theta_3 \tag{4.11a}$$

The constant K_1 is the same as defined in equation 4.8b. K_4 and K_5 are:

$$K_4 = \frac{d}{b}; \qquad\qquad K_5 = \frac{c^2 - d^2 - a^2 - b^2}{2ab} \tag{4.11b}$$

This also reduces to a quadratic form:

$$D\tan^2\left(\frac{\theta_3}{2}\right) + E\tan\left(\frac{\theta_3}{2}\right) + F = 0 \tag{4.12}$$

where:
$$D = \cos\theta_2 - K_1 + K_4 \cos\theta_2 + K_5$$
$$E = -2\sin\theta_2$$
$$F = K_1 + (K_4 - 1)\cos\theta_2 + K_5$$

and the solution is:

$$\theta_{3_{1,2}} = 2\arctan\left(\frac{-E \pm \sqrt{E^2 - 4DF}}{2D}\right) \tag{4.13}$$

As with the angle θ_4, this also has two solutions, corresponding to the crossed and open branches of the linkage, as shown in Figure 4-8.

4.6   THE FOURBAR SLIDER-CRANK POSITION SOLUTION

The same vector loop approach as used above can be applied to a linkage containing sliders. Figure 4-9 shows an offset fourbar slider-crank linkage, inversion #1. The term **offset** means that *the slider axis extended does not pass through the crank pivot*. This is the general case. (The nonoffset slider-crank linkages shown in Figure 2-13 are the special cases.) This linkage could be represented by only three position vectors, $\mathbf{R}_2$, $\mathbf{R}_3$, and $\mathbf{R}_s$, but one of them ($\mathbf{R}_s$) will be a vector of varying magnitude and angle. It will be easier to use four vectors, $\mathbf{R}_1$, $\mathbf{R}_2$, $\mathbf{R}_3$, and $\mathbf{R}_4$ with $\mathbf{R}_1$ arranged parallel to the axis of sliding and $\mathbf{R}_4$ perpendicular. In effect the pair of vectors $\mathbf{R}_1$ and $\mathbf{R}_4$ are orthogonal components of the position vector $\mathbf{R}_s$ from the origin to the slider.

It simplifies the analysis to arrange one coordinate axis parallel to the axis of sliding. The variable-length, constant-direction vector $\mathbf{R}_1$ then represents the slider position with magnitude d. The vector $\mathbf{R}_4$ is orthogonal to $\mathbf{R}_1$ and defines the constant magnitude **offset** of the linkage. Note that for the special case, nonoffset version, the vector $\mathbf{R}_4$ will be zero and $\mathbf{R}_1 = \mathbf{R}_s$. The vectors $\mathbf{R}_2$ and $\mathbf{R}_3$ complete the vector loop. The coupler's position vector $\mathbf{R}_3$ is placed with its root at the slider which then defines its angle θ_3 at point B. This particular arrangement of position vectors leads to a vector loop equation similar to the pin-jointed fourbar example:

$$\mathbf{R}_2 - \mathbf{R}_3 - \mathbf{R}_4 - \mathbf{R}_1 = 0 \tag{4.14a}$$

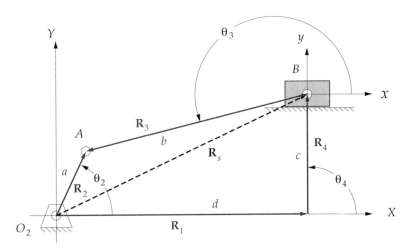

FIGURE 4-9

Position vector loop for a fourbar slider-crank linkage

Compare equation 4.14a to equation 4.5a and note that the only difference is the sign of $\mathbf{R}_3$. This is due solely to the somewhat arbitrary choice of the sense of the position vector $\mathbf{R}_3$ in each case. The angle θ_3 must always be measured at the root of vector $\mathbf{R}_3$, and in this example it will be convenient to have that angle θ_3 at the joint labelled B. Once these arbitrary choices are made it is crucial that the resulting algebraic signs be carefully observed in the equations, or the results will be completely erroneous. Letting the vector magnitudes (link lengths) be represented by a, b, c, d as shown, we can substitute the complex number equivalents for the position vectors.

$$a e^{j\theta_2} - b e^{j\theta_3} - c e^{j\theta_4} - d e^{j\theta_1} = 0 \qquad (4.14b)$$

Substitute the Euler equivalents:

$$a\left(\cos\theta_2 + j\sin\theta_2\right) - b\left(\cos\theta_3 + j\sin\theta_3\right)$$
$$-c\left(\cos\theta_4 + j\sin\theta_4\right) - d\left(\cos\theta_1 + j\sin\theta_1\right) = 0 \qquad (4.14c)$$

Separate the real and imaginary components:

real part (x component):

$$a\cos\theta_2 - b\cos\theta_3 - c\cos\theta_4 - d\cos\theta_1 = 0$$

but: $\theta_1 = 0$,   so: $\qquad\qquad\qquad\qquad\qquad\qquad\qquad\qquad (4.15a)$

$$a\cos\theta_2 - b\cos\theta_3 - c\cos\theta_4 - d = 0$$

imaginary part (y component):

$$ja\sin\theta_2 - jb\sin\theta_3 - jc\sin\theta_4 - jd\sin\theta_1 = 0$$

but: $\theta_1 = 0$, and the j's divide out, so: $\qquad\qquad\qquad\qquad (4.15b)$

$$a\sin\theta_2 - b\sin\theta_3 - c\sin\theta_4 = 0$$

We want to solve equations 4.15 simultaneously for the two unknowns, link length d and link angle θ_3. The independent variable is crank angle θ_2. Link lengths a and b, the offset c and angle θ_4 are known. But note that since we set up the coordinate system to be parallel and perpendicular to the axis of the slider block, the angle θ_1 is zero and θ_4 is 90°. Equation 4.15b can be solved for θ_3 and the result substituted into equation 4.15a to solve for d. The solution is:

$$\theta_{3_1} = \arcsin\left(\frac{a\sin\theta_2 - c}{b}\right)$$

(4.16a)

$$d = a\cos\theta_2 - b\cos\theta_3$$

(4.16b)

Note that there are again two valid solutions corresponding to the two branches of the linkage. The arcsine function is multivalued. Its evaluation will give a value between ±90° representing only one branch of the linkage. The value of d is dependent on the calculated value of θ_3. The value of θ_3 for the second branch of the linkage can be found from:

$$\theta_{3_2} = \arcsin\left(-\frac{a\sin\theta_2 - c}{b}\right) + \pi$$

(4.17)

4.7   AN INVERTED SLIDER-CRANK POSITION SOLUTION

Figure 4-10a shows inversion #3 of the common fourbar slider-crank linkage in which the sliding joint is between links 3 and 4 at point B. This is shown as an **offset** slider-crank mechanism. The slider block has pure rotation with its center offset from the slide axis. (Figure 2-13c shows the nonoffset version of this linkage in which the vector $\mathbf{R}_4$ is zero.)

The global coordinate system is again taken with its origin at input crank pivot O_2 and the positive X axis along link 1, the ground link. A local axis system has been placed at point B in order to define θ_3. Note that there is a fixed angle γ within link 4 which defines the slot angle with respect to that link.

In Figure 4-10b the links have been represented as position vectors having senses consistent with the coordinate systems that were chosen for convenience in defining the link angles. This particular arrangement of position vectors leads to the same vector loop equation as the previous slider-crank example. Equations 4.14 and 4.15 apply to this inversion as well. Note that the absolute position of point B is defined by vector $\mathbf{R}_B$ which varies in both magnitude and direction as the linkage moves. We choose to represent $\mathbf{R}_B$ as the vector difference $\mathbf{R}_2 - \mathbf{R}_3$ in order to use the actual links as the position vectors in the loop equation.

All slider linkages will have at least one link whose effective length between joints will vary as the linkage moves. In this example the length of link 3 between points A and B, designated as b, will change as it passes through the slider block on link 4. Thus the value of b will be one of the variables to be solved for in this inversion. Another variable will be θ_4, the angle of link 4. Note however, that we also have an unknown in θ_3, the angle of link 3. This is a total of three unknowns. Equations 4.15 can only be solved for two unknowns. Thus we require another equation to solve the system. There is a fixed relationship between angles θ_3 and θ_4, shown as γ in Figure 4-10, which gives the equation:

$$\theta_3 = \theta_4 + \gamma \qquad (4.18)$$

Substituting equation 4.18 into equations 4.15 yields:

$$a\cos\theta_2 - b\cos(\theta_4 + \gamma) - c\cos\theta_4 - d = 0 \qquad (4.19a)$$

$$a\sin\theta_2 - b\sin(\theta_4 + \gamma) - c\sin\theta_4 = 0 \qquad (4.19b)$$

These have only two unknowns and can be solved simultaneously for θ_4 and b. Equation 4.19b can be solved for link length b and substituted into equation 4.19a.

$$b = \frac{a\sin\theta_2 - c\sin\theta_4}{\sin(\theta_4 + \gamma)} \qquad (4.20a)$$

$$a\cos\theta_2 - \frac{a\sin\theta_2 - c\sin\theta_4}{\sin(\theta_4 + \gamma)}\cos(\theta_4 + \gamma) - c\cos\theta_4 - d = 0 \qquad (4.20b)$$

After some algebraic manipulation, equation 4.20 can be reduced to:

$$P\sin\theta_4 + Q\cos\theta_4 + R = 0$$

where:

(4.21)

$$P = a\sin\theta_2 \sin\gamma + (a\cos\theta_2 - d)\cos\gamma$$

$$Q = -a\sin\theta_2 \cos\gamma + (a\cos\theta_2 - d)\sin\gamma$$

$$R = -c\sin\gamma$$

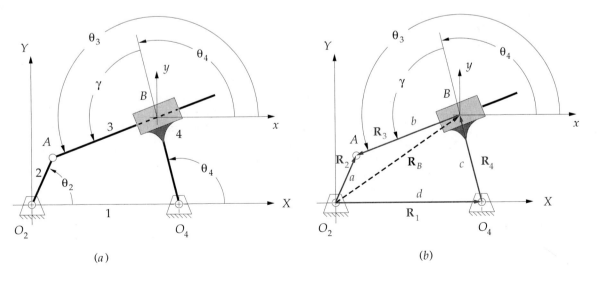

(a)   (b)

FIGURE 4-10

Inversion #3 of the slider-crank fourbar linkage

Note that the factors P, Q, R are constant for any input value of θ_2. To solve this for θ_4, it is convenient to substitute the tangent half angle identities (Eq. 4.9) for the $\sin \theta_4$ and $\cos \theta_4$ terms. This will result in a quadratic equation in $\tan(\theta_4/2)$ which can be solved for the two values of θ_4.

$$P\frac{2\tan\left(\dfrac{\theta_4}{2}\right)}{1+\tan^2\left(\dfrac{\theta_4}{2}\right)}+Q\frac{1-\tan^2\left(\dfrac{\theta_4}{2}\right)}{1+\tan^2\left(\dfrac{\theta_4}{2}\right)}+R=0 \tag{4.22a}$$

This reduces to:

$$(R-Q)\tan^2\left(\frac{\theta_4}{2}\right)+2P\tan\left(\frac{\theta_4}{2}\right)+(Q+R)=0$$

let:

$$S=R-Q;\qquad T=2P;\qquad U=Q+R$$

then:

$$S\tan^2\left(\frac{\theta_4}{2}\right)+T\tan\left(\frac{\theta_4}{2}\right)+U=0 \tag{4.22b}$$

and the solution is:

$$\theta_{4_{1,2}}=2\arctan\left(\frac{-T\pm\sqrt{T^2-4SU}}{2S}\right) \tag{4.22c}$$

As was the case with the previous examples, this also has a crossed and an open solution represented by the plus and minus signs on the radical. Note that we must also calculate the values of link length b for each θ_4 by using equation 4.20a.

4.8   LINKAGES OF MORE THAN FOUR BARS

The same approach as shown here for the fourbar linkage can be used for any number of links in a closed-loop configuration. The method is quite general. More complicated linkages may have multiple loops which will lead to more equations to be solved simultaneously.

The Geared Fivebar Linkage

Another example, which can be reduced to two equations in two unknowns, is the **geared fivebar linkage**, which was introduced in Section 2.13 and is shown in Figure 4-11a and program FIVEBAR diskfile FIG4-11. The vector loop for this linkage is shown in Figure 4-11b. It obviously has one more position vector than the fourbar. Its vector loop equation is:

$$\mathbf{R}_2+\mathbf{R}_3-\mathbf{R}_4-\mathbf{R}_5-\mathbf{R}_1=0 \tag{4.23a}$$

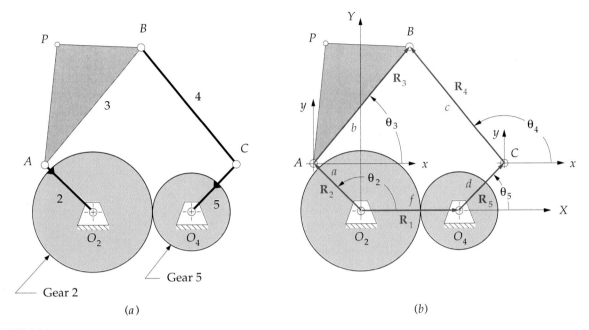

FIGURE 4-11

The geared fivebar linkage and its vector loop

Note that the vector senses are again chosen to suit the analyst's desires to have the vector angles defined at a convenient end of the respective link. Equation 4.23b substitutes the complex polar notation for the position vectors in equation 4-23a, using a, b, c, d, f to represent the scalar lengths of the links as shown the figure.

$$ae^{j\theta_2} + be^{j\theta_3} - ce^{j\theta_4} - de^{j\theta_5} - fe^{j\theta_1} = 0 \qquad (4.23b)$$

Note also that this vector loop equation has three unknown variables in it, namely the angles of links 3, 4 and 5. (The angle of link 2 is the input, or independent, variable, and link 1 is fixed with constant angle). Since a two-dimensional vector equation can only be solved for two unknowns, we will need another equation to solve this system. Because this is a geared fivebar linkage, there exists a relationship between the two geared links, here links 2 and 5. Two factors determine how link 5 behaves with respect to link 2, namely the **gear ratio** λ and the **phase angle** ϕ. The relationship is:

$$\theta_5 = \lambda\theta_2 + \phi \qquad (4.23c)$$

This allows us to express θ_5 in terms of θ_2 in equation 4.23b and reduce the unknowns to two by substituting equation 4.23c into equation 4.23b.

$$ae^{j\theta_2} + be^{j\theta_3} - ce^{j\theta_4} - de^{j(\lambda\theta_2 + \phi)} - fe^{j\theta_1} = 0 \qquad (4.24a)$$

Note that the gear ratio λ is the ratio of the diameters of the gears connecting the two links ($\lambda = dia_2 / dia_5$), and the phase angle ϕ is the *initial angle* of link 5 with respect to link 2. When link 2 is at zero degrees, link 5 is at the **phase angle** ϕ. Equation 4.23c defines the relationship between θ_2 and θ_5. Both λ and ϕ are design parameters selected by the design engineer along with the link lengths. With these parameters defined, the only unknowns left in equation 4.24 are θ_3 and θ_4.

The behavior of the geared fivebar linkage can be modified by changing the link lengths, the gear ratio, or the phase angle. The phase angle can be changed simply by lifting the gears out of engagement, rotating one gear with respect to the other, and reengaging them. Since links 2 and 5 are rigidly attached to gears 2 and 5, respectively, their relative angular rotations will be changed also. It is this fact that results in different positions of links 3 and 4 with any change in phase angle. The coupler curve's shapes will also change with variation in any of these parameters as can be seen in Figure 3-20 and in Appendix D.

The procedure for solution of this vector loop equation is the same as that used above for the fourbar linkage:

1 Substitute the Euler equivalent (Eq. 4.4a) into each term in the vector loop equation 4.24.

$$a(\cos\theta_2 + j\sin\theta_2) + b(\cos\theta_3 + j\sin\theta_3) - c(\cos\theta_4 + j\sin\theta_4)$$
$$-d\left[\cos(\lambda\theta_2 + \phi) + j\sin(\lambda\theta_2 + \phi)\right] - f(\cos\theta_1 + j\sin\theta_1) = 0 \qquad (4.24b)$$

2 Separate the real and imaginary parts of the cartesian form of the vector loop equation.

$$a\cos\theta_2 + b\cos\theta_3 - c\cos\theta_4 - d\cos(\lambda\theta_2 + \phi) - f\cos\theta_1 = 0 \qquad (4.24c)$$
$$a\sin\theta_2 + b\sin\theta_3 - c\sin\theta_4 - d\sin(\lambda\theta_2 + \phi) - f\sin\theta_1 = 0 \qquad (4.24d)$$

3 Rearrange to isolate one unknown (either θ_3 or θ_4) in each scalar equation. Note that θ_1 is zero.

$$b\cos\theta_3 = -a\cos\theta_2 + c\cos\theta_4 + d\cos(\lambda\theta_2 + \phi) + f \qquad (4.24e)$$
$$b\sin\theta_3 = -a\sin\theta_2 + c\sin\theta_4 + d\sin(\lambda\theta_2 + \phi) \qquad (4.24f)$$

4 Square both equations and add them to eliminate one unknown, say θ_3.

$$b^2 = 2c\left[d\cos(\lambda\theta_2 + \phi) - a\cos\theta_2 + f\right]\cos\theta_4$$
$$+2c\left[d\sin(\lambda\theta_2 + \phi) - a\sin\theta_2\right]\sin\theta_4$$
$$+a^2 + c^2 + d^2 + f^2 - 2af\cos\theta_2$$
$$-2d(a\cos\theta_2 - f)\cos(\lambda\theta_2 + \phi)$$
$$-2ad\sin\theta_2\sin(\lambda\theta_2 + \phi) \qquad (4.24g)$$

5 Substitute the tangent half-angle identities (Eq. 4.9) for the sine and cosine terms and manipulate the resulting equation in the same way as was done for the fourbar linkage in order to solve for θ_4.

$$A = 2c\left[d\cos(\lambda\theta_2 + \phi) - a\cos\theta_2 + f\right]$$
$$B = 2c\left[d\sin(\lambda\theta_2 + \phi) - a\sin\theta_2\right]$$
$$C = a^2 - b^2 + c^2 + d^2 + f^2 - 2af\cos\theta_2$$
$$-2d(a\cos\theta_2 - f)\cos(\lambda\theta_2 + \phi)$$
$$-2ad\sin\theta_2\sin(\lambda\theta_2 + \phi)$$
$$D = C - A; \qquad E = 2B; \qquad F = A + C$$

$$\theta_{4_{1,2}} = 2\arctan\left(\frac{-E \pm \sqrt{E^2 - 4DF}}{2D}\right) \qquad (4.24h)$$

6   Repeat steps 3 to 5 for the other unknown angle θ_3.

$$G = 2b\left[a\cos\theta_2 - d\cos(\lambda\theta_2 + \phi) - f\right]$$
$$H = 2b\left[a\sin\theta_2 - d\sin(\lambda\theta_2 + \phi)\right]$$
$$K = a^2 + b^2 - c^2 + d^2 + f^2 - 2af\cos\theta_2$$
$$-2d(a\cos\theta_2 - f)\cos(\lambda\theta_2 + \phi)$$
$$-2ad\sin\theta_2\sin(\lambda\theta_2 + \phi)$$
$$L = K - G; \qquad M = 2H; \qquad N = G + K$$

$$\theta_{3_{1,2}} = 2\arctan\left(\frac{-M \pm \sqrt{M^2 - 4LN}}{2L}\right) \qquad (4.24i)$$

Note that these derivation steps are essentially identical to those for the pin-jointed fourbar linkage once θ_2 is substituted for θ_5 using equation 4.23c.

Sixbar Linkages

WATT'S SIXBAR is essentially two fourbar linkages in series, as shown in Figure 4.12a, and can be analyzed as such. Two vector loops are drawn as shown in Figure 4.12b. These vector loop equations can be solved in succession with the results of the first loop applied as input to the second loop. Note that there is a constant angular relationship between vectors $\mathbf{R}_4$ and $\mathbf{R}_5$ within link 4. The solution for the fourbar linkage (Eqs. 4.10 and 4.13) is simply applied twice in this case. Depending on the inversion of the Watts linkage being analyzed, there may be two four-link loops or one four-link and one five-link loop. (See Figure 2-14.) In either case, if the four-link loop is analyzed first, there will not be more than two unknown link angles to be found at one time.

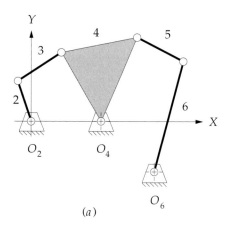

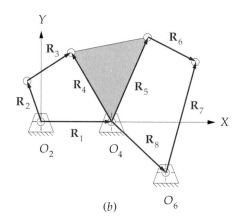

FIGURE 4-12

Watt's sixbar linkage and vector loop

STEPHENSON'S SIXBAR is a more complicated mechanism to analyze. Two vector loops can be drawn, but depending on the inversion being analyzed, either one or both loops will have five links and thus three unknown angles in it as shown in Figure 4-13a and b. However, the two loops will have at least one non-ground link in common and so a solution can be found. In the other cases an iterative solution such as a Newton-Rapheson method must be used to find the roots of the equations. Program SIXBAR is limited to the inversions which allow a closed form solution, one of which shown in Figure 4-13. Program SIXBAR does not do the iterative solution.

4.9   POSITION OF ANY POINT ON A LINKAGE

Once the angles of all the links are found it is simple and straightforward to define and calculate the position of any point on any link for any input position of

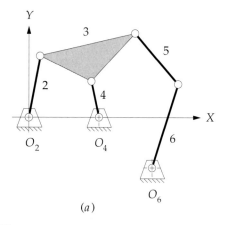

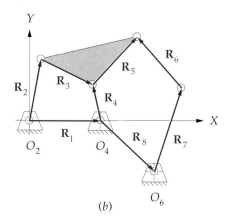

FIGURE 4-13

Stephenson's sixbar linkage and vector loop

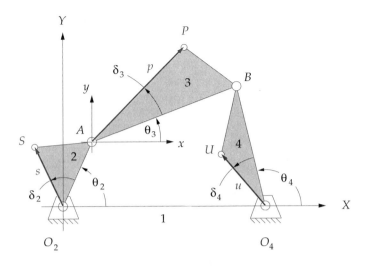

FIGURE 4-14

Positions of points on the links

the linkage. Figure 4-14 shows a fourbar linkage whose coupler, link 3, is enlarged to contain a coupler point P. The crank and rocker have also been enlarged to show points S and U which might represent the centers of gravity of those links. We want to develop algebraic expressions for the positions of these (or any) points on the links.

To find the position of point S, draw a position vector from the fixed pivot O_2 to point S. This vector $\mathbf{R}_{SO_2}$ makes an angle δ_2 with the vector $\mathbf{R}_{AO_2}$. This angle δ_2 is completely defined by the geometry of link 2 and is constant. The position vector for point S is then:

$$\mathbf{R}_{SO_2} = \mathbf{R}_S = se^{j(\theta_2 + \delta_2)} = s\left[\cos(\theta_2 + \delta_2) + j\sin(\theta_2 + \delta_2)\right] \tag{4.25}$$

The position of point U on link 4 is found in the same way, using the angle δ_4 which is a constant angular offset within the link. The expression is:

$$\mathbf{R}_{UO_4} = ue^{j(\theta_4 + \delta_4)} = u\left[\cos(\theta_4 + \delta_4) + j\sin(\theta_4 + \delta_4)\right] \tag{4.26}$$

The position of point P on link 3 can be found from the addition of two position vectors, $\mathbf{R}_A$ and $\mathbf{R}_{PA}$. $\mathbf{R}_A$ is already defined from our analysis of the link angles in equation 4.5. $\mathbf{R}_{PA}$ is the relative position of point P with respect to point A. $\mathbf{R}_{PA}$ is defined in the same way as $\mathbf{R}_S$ or $\mathbf{R}_U$, using the internal link offset angle δ_3 and the position angle of link 3, θ_3.

$$\mathbf{R}_{PA} = pe^{j(\theta_3 + \delta_3)} = p\left[\cos(\theta_3 + \delta_3) + j\sin(\theta_3 + \delta_3)\right] \tag{4.27a}$$

$$\mathbf{R}_P = \mathbf{R}_A + \mathbf{R}_{PA} \tag{4.27b}$$

Please compare equation 4.27 with equation 4.1. This is also the position difference equation.

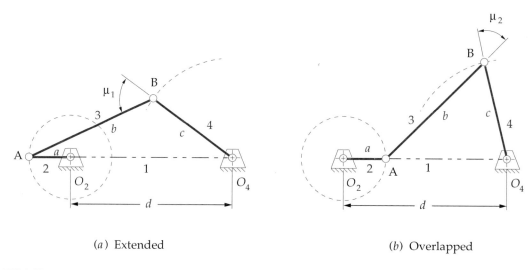

(a) Extended   (b) Overlapped

FIGURE 4-15

Extreme transmission angles in the Grashof fourbar linkage

4.10   TRANSMISSION ANGLES

The transmission angle was defined in Section 3.3 for a fourbar linkage. That definition is repeated here for your convenience.

The **transmission angle** μ is shown in Figure 3-3a and is defined as *the angle between the output link and the coupler.* It is usually taken as the absolute value of the acute angle of the pair of angles at the intersection of the two links and varies continuously from some minimum to some maximum value as the linkage goes through its range of motion. It is a measure of the quality of force transmission at the joint.

We will expand that definition here to represent the angle between any two links in a linkage, as a linkage can have many transmission angles. The angle between any output link and the coupler which drives it is a transmission angle. Now that we have developed the analytic expressions for the angles of all the links in a mechanism, it is easy to define the transmission angle algebraically. It is merely the difference between the angles of the two joined links through which we wish to pass some force or velocity. For our fourbar linkage example it will be the difference between θ_3 and θ_4. By convention we take the absolute value of the difference and force it to be an acute angle.

$$\theta_{trans} = |\theta_3 - \theta_4|$$

$$\text{if} \qquad \theta_{trans} > \frac{\pi}{2} \qquad \text{then} \quad \theta_{trans} = \pi - \theta_{trans} \qquad (4.28)$$

This computation can be done for any joint in a linkage by using the appropriate link angles.

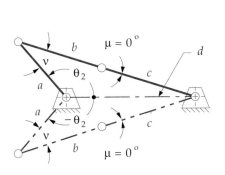

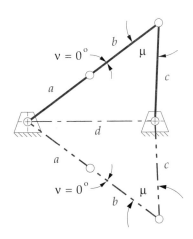

(a) Toggle positions for links b and c   (b) Toggle positions for links a and b

FIGURE 4-16

Non-Grashof double-rocker linkages in toggle

Extreme Values of the Transmission Angle

For a Grashof crank rocker fourbar linkage the extreme values of the transmission angle will occur when the crank is colinear with the ground link as shown in Figure 4-15. The values of the transmission angle in these positions are easily calculated from the law of cosines since the linkage is then in a triangular configuration. The sides of the two triangles are link 3, link 4, and either the sum or difference of links 1 and 2. One extreme value of the transmission angle occurs when links 1 and 2 are *colinear and nonoverlapping* as shown in Figure 4-15a. The other extreme transmission angle occurs when links 1 and 2 are *colinear and overlapping* as shown in Figure 4-15b. Using notation consistent with Section 4.5 and Figure 4-7 we will label the links:

a = link 2; b = link 3; c = link 4; d = link 1

For one extreme transmission angle the cosine law gives

$$\mu_1 = \arccos\left[\frac{b^2 + c^2 - (d+a)^2}{2bc}\right] \qquad (4.29a)$$

and for the other extreme transmission angle

$$\mu_2 = \arccos\left[\frac{b^2 + c^2 - (d-a)^2}{2bc}\right] \qquad (4.29b)$$

For a **Grashof double-rocker** linkage the transmission angle can vary from 0 to 90 degrees because the coupler can make a full revolution with respect to the other links. For a **non-Grashof double-rocker** linkage the transmission angle will be zero degrees in the toggle positions which occur when the output rocker c and the coupler b are colinear as shown in Figure 4-16a. In the other toggle positions when input rocker a and coupler b are colinear (Figure 4-16b), the transmission angle can be calculated from the cosine law as:

when $v = 0,$

$$\mu = \arccos\left[\frac{(a+b)^2 + c^2 - d^2}{2c(a+b)}\right] \qquad (4.30)$$

This is not the smallest value that the transmission angle μ can have in a double-rocker, as that will obviously be zero. Of course, when analyzing any linkage, the transmission angles can easily be computed and plotted for all positions using equation 4.28. Programs FOURBAR, FIVEBAR,, and SIXBAR do this. The student should investigate the variation in transmission angle for the example linkages in those programs. Diskfile FIG4-15 can be input to program FOURBAR to observe the linkage in that figure in motion.

4.11   TOGGLE POSITIONS

The input link angles which correspond to the toggle positions of the **non-Grashof double-rocker** can be calculated by the following method, using trigonometry. Figure 4-17 shows a non-Grashof fourbar linkage in a general position. A construction line h has been drawn between points A and O_4. This divides the quadrilateral loop into two triangles, O_2AO_4 and ABO_4. Equation 4.31 uses the cosine law to express the transmission angle μ in terms of link lengths and the input link angle θ_2.

$$h^2 = a^2 + d^2 - 2ad\cos\theta_2$$

also:

$$h^2 = b^2 + c^2 - 2bc\cos\mu$$

so:

$$a^2 + d^2 - 2ad\cos\theta_2 = b^2 + c^2 - 2bc\cos\mu$$

and:

$$\cos\mu = \frac{b^2 + c^2 - a^2 - d^2}{2bc} + \frac{ad}{bc}\cos\theta_2 \qquad (4.31)$$

To find the maximum and minimum values of input angle θ_2, we can differentiate equation 4.31, form the derivative of θ_2 with respect to μ and set it equal to zero.

$$\frac{d\theta_2}{d\mu} = \frac{bc}{ad}\frac{\sin\mu}{\sin\theta_2} = 0 \qquad (4.32)$$

The link lengths a, b, c, d are never zero, so this expression can only be zero when $\sin\mu$ is zero. This will be true when angle μ in Figure 4-17 is either zero or 180°. This is consistent with the definition of toggle given in Section 3.3. If μ is zero or 180°

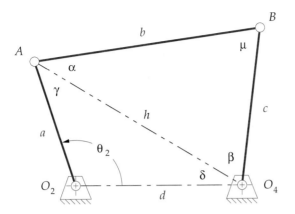

FIGURE 4-17

Finding the crank angle corresponding to the toggle positions

then $\cos\mu$ will be ± 1. Substituting these two values for $\cos\mu$ into equation 4.31 will give a solution for the value of θ_2 between zero and 180 degrees which corresponds to the toggle position of a double-rocker linkage when driven from one rocker.

$$\cos\mu = \frac{b^2 + c^2 - a^2 - d^2}{2bc} + \frac{ad}{bc}\cos\theta_2 = \pm 1$$

or:

$$\cos\theta_2 = \frac{a^2 + d^2 - b^2 - c^2}{2ad} \pm \frac{bc}{ad} \qquad (4.33)$$

and:

$$\theta_{2_{toggle}} = \arccos\left(\frac{a^2 + d^2 - b^2 - c^2}{2ad} \pm \frac{bc}{ad}\right); \quad 0 \le \theta_{2_{toggle}} \le \pi$$

Only one of these $\pm$ cases will produce an argument for the arc cosine function which lies between ± 1. The toggle angle which is in the first or second quadrant can be found from this value. The other toggle angle will then be the negative of the one found, due to the mirror symmetry of the two toggle positions about the ground link as shown in Figure 4-16. Program FOURBAR computes the values of these toggle angles for any non-Grashof linkage.

4.12 PROBLEMS

4-1 A position vector is defined as having a length equal to your height in inches (or centimeters). The tangent of its angle is defined as your weight in lbs (or kg) divided by your age in years. Calculate the data for this vector and:

a. Draw the position vector to scale on cartesian axes.
b. Write an expression for the position vector using unit vector notation.
c. Write an expression for the position vector using complex number notation, in both polar and cartesian forms.

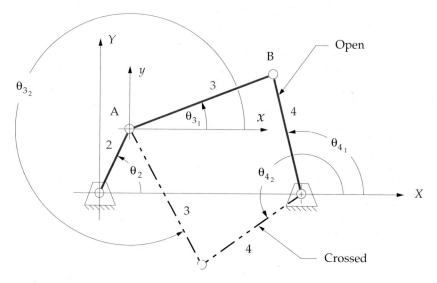

FIGURE P4-1

General configuration and terminology for the fourbar linkage

4-2 A particle is traveling along an arc of 6.5 inch radius. The arc center is at the origin of a coordinate system. When the particle is at position A, its position vector makes a 45° angle with the X axis. At position B, its vector makes a 75° angle with the X axis. Draw this system to some convenient scale and:

  a. Write an expression for the particle's position vector in position A using complex number notation, in both polar and cartesian forms.

TABLE P4-1   Data for Fourbar Analysis Problems

Row	Link 1	Link 2	Link 3	Link 4	θ_2
a	6	2	7	9	30
b	7	9	3	8	85
c	3	10	6	8	45
d	8	5	7	6	25
e	8	5	8	6	75
f	5	8	8	9	15
g	6	8	8	9	25
h	20	10	10	10	50
i	4	5	2	5	80
j	20	10	5	10	33
k	4	6	10	7	88
l	9	7	10	7	60
m	9	7	11	8	50
n	9	7	11	6	120

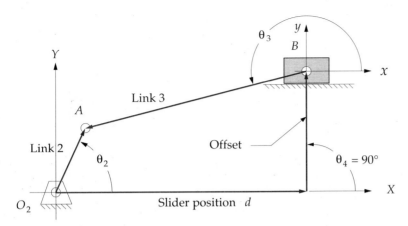

FIGURE P4-2

Open configuration and terminology for a fourbar slider-crank linkage

    b.   Write an expression for the particle's position vector in position B using complex number notation, in both polar and cartesian forms.

    c.   Write a vector equation for the position difference between points B and A. Substitute the complex number notation for the vectors in this equation and solve for the position difference numerically.

    d.   Check the result of part c with a graphical method.

4-3 Repeat problem 4-2 considering points A and B to represent separate particles, and find their relative position.

4-4 Repeat problem 4-2 with the particle's path defined as being along the line $y = -2x + 10$.

4-5 Repeat problem 4-3 with the path of the particle defined as being along the curve $y = -2x^2 - 2x + 10$.

*4-6 The link lengths and the value of θ_2 for some fourbar linkages are defined in Table P4-1. The linkage configuration and terminology are shown in Figure P4-1. For the rows assigned, draw the linkage to scale and **graphically** find all possible solutions (both open and crossed) for angles θ_3 and θ_4. Determine the Grashof condition.

---

* Answers in Appendix E

TABLE P4-2   **Data for Noninverted Slider-Crank Problems**

Row	Link 2	Link 3	Offset	θ_2
a	1.4	4	1	45
b	2	6	-3	60
c	3	8	2	-30
d	3.5	10	1	120
e	5	20	-5	225
f	3	13	0	100
g	7	25	10	330

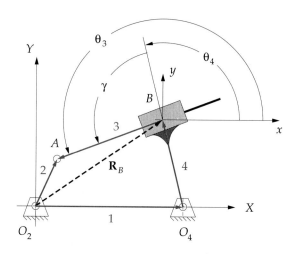

FIGURE P4-3

Configuration and terminology for inversion #3 of the fourbar slider-crank linkage

*4-7   Repeat problem 4-6 except solve by the vector loop method of Section 4.5.

4-8   Expand equation 4.7b and prove that it reduces to equation 4.7c.

*4-9   The link lengths and the value of θ_2 and offset for some fourbar slider-crank link-
       ages are defined in Table P4-2. The linkage configuration and terminology are
       shown in Figure P4-2.   For the rows assigned, draw the linkage to scale and
       graphically find all possible solutions (both open and crossed) for angle θ_3 and
       slider position d.

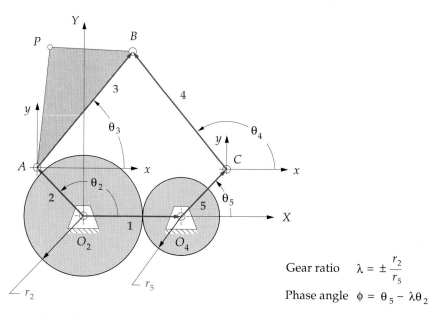

Gear ratio $\lambda = \pm \dfrac{r_2}{r_5}$

Phase angle $\phi = \theta_5 - \lambda\theta_2$

FIGURE P4-4

Open configuration and terminology for the geared fivebar linkage

TABLE P4-3   Data for Inverted Slider-Crank Problems

Row	Link 1	Link 2	Link 4	γ	θ_2
a	6	2	4	90	30
b	7	9	3	75	85
c	3	10	6	45	45
d	8	5	3	60	25
e	8	4	2	30	75
f	5	8	8	90	150

*4-10   Repeat problem 4-9 except solve by the vector loop method of Section 4.6.

*4-11   The link lengths and the value of θ_2 and γ for some inverted fourbar slider-crank linkages are defined in Table P4-3. The linkage configuration and terminology are shown in Figure P4-3. For the rows assigned, draw the linkage to scale and **graphically** find both open and crossed solutions for angles θ_3 and θ_4 and vector $\mathbf{R}_B$.

*4-12   Repeat problem 4-11 except solve by the vector loop method of Section 4.7.

*4-13   Find the transmission angles of the linkages in the assigned rows in Table P4-1.

*4-14   Find the minimum and maximum values of the transmission angle for all of the Grashof crank rocker linkages in Table P4-1.

*4-15   Find the input angles corresponding to the toggle positions of the non-Grashof linkages in Table P4-1. (For this problem, ignore the values of θ_2 given in the table.)

*4-16   The link lengths, gear ratio (λ), phase angle (ϕ), and the value of θ_2 for some geared fivebar linkages are defined in Table P4-4. The linkage configuration and terminology are shown in Figure P4-4. For the rows assigned, draw the linkage to scale and **graphically** find all possible solutions for angles θ_3 and θ_4.

*4-17   Repeat problem 4-16 except solve by the vector loop method.

TABLE P4-4   Data for Geared Fivebar Linkage Problems

Row	Link 1	Link 2	Link 3	Link 4	Link 5	λ	ϕ	θ_2
a	6	1	7	9	4	2	30	60
b	6	5	7	8	4	-2.5	60	30
c	3	5	7	8	4	-0.5	0	45
d	4	5	7	8	4	-1	120	75
e	5	9	11	8	8	3.2	-50	-39
f	10	2	7	5	3	1.5	30	120
g	15	7	9	11	4	2.5	-90	75
h	12	8	7	9	4	-2.5	60	55
i	9	7	8	9	4	-4	120	100

Chapter 5

LINKAGE SYNTHESIS

Imagination is more important than knowledge
ALBERT EINSTEIN

5.0   INTRODUCTION

With the fundamentals of position analysis established, we can now use these techniques to **synthesize linkages** for specified output positions **analytically**. The synthesis techniques presented in Chapter 3 were strictly graphical and somewhat intuitive. The **analytical synthesis** procedure is algebraic rather than graphical and is less intuitive. However, its algebraic nature makes it quite suitable for computerization. These analytical synthesis methods were originated by Sandor[1] and further developed by his students, Erdman,[2] Kaufman,[3] and Loerch et al.[4,5]

5.1   TYPES OF KINEMATIC SYNTHESIS

Erdman and Sandor[6] define three types of kinematic synthesis, **function**, **path**, and **motion generation**, which were discussed in Section 3.2. Brief definitions are repeated here for your convenience.

FUNCTION GENERATION is defined as *the correlation of an **input function** with an **output function** in a mechanism*. Typically, a double-rocker or crank-rocker is the result, with pure rotation input and pure rotation output. A slider-crank linkage can be a function generator as well, driven from either end, i.e., rotation in and translation out or vice versa.

PATH GENERATION is defined as *the control of a **point** in the plane such that it follows some prescribed path*. This is typically accomplished with a fourbar crank-rocker or double-rocker, wherein a point on the coupler traces the desired output path. No attempt is made in path generation to control the orientation of the link which contains the point of interest. The coupler curve is made to pass through a set of desired output points. However, it is common for the timing of the arrival of the coupler point at particular locations along the path to be defined. This case is called *path generation with prescribed timing* and is analogous to function generation in that a particular output function is specified.

MOTION GENERATION is defined as *the control of a **line** in the plane such that it assumes some sequential set of prescribed positions.* Here orientation of the link containing the line is important. This is typically accomplished with a fourbar crank-rocker or double-rocker, wherein a point on the coupler traces the desired output path and the linkage also controls the angular orientation of the coupler link containing the output line of interest.

5.2   PRECISION POINTS

The *points, or positions, prescribed for successive locations of the output (coupler or rocker) link in the plane* are generally referred to as **precision points** or **precision positions**. The number of precision points which can be synthesized is limited by the number of equations available for solution. The fourbar linkage can be synthesized for *up to five precision points for motion or path generation* (coupler output) and *up to seven points for function generation* (rocker output). Synthesis for two or three precision points is relatively straightforward, and each of these cases can be reduced to a system of linear simultaneous equations easily solved on a calculator. The four or more position synthesis problems involve the solution of nonlinear, simultaneous equation systems, and so are more complicated to solve, requiring a computer.

Note that these analytical synthesis procedures provide a solution which will be able to "be at" the specified precision points, but no guarantee is provided regarding the linkage's behavior between those precision points. It is possible that the resulting linkage will be incapable of moving from one precision point to another due to the presence of a toggle position or other constraint. This situation is actually no different than that of the graphical synthesis cases in Chapter 3, wherein there was also the possibility of a toggle position between design points. In fact, these analytical synthesis methods are just an alternate way to solve the same multiposition synthesis problems. One should still build a simple cardboard model of the synthesized linkage to observe its behavior and check for the presence of problems, even if the synthesis was performed by an esoteric analytical method.

5.3   TWO-POSITION MOTION GENERATION BY ANALYTICAL SYNTHESIS

Figure 5-1 shows a fourbar linkage in one position with a coupler point located at a first precision position P_1. It also indicates a second precision position (point P_2) to be achieved by the rotation of the input rocker, link 2, through an as yet unspecified angle β_2. Note also that the angle of the coupler link 3 at each of the precision positions is defined by the angles of the position vectors Z_1 and Z_2. The angle ϕ corresponds to the angle θ_3 of link 3 in its first position. This angle is unknown at the start of the synthesis and will be found. The angle α_2 represents the angular change of link 3 from position one to position two. This angle is defined in the problem statement.

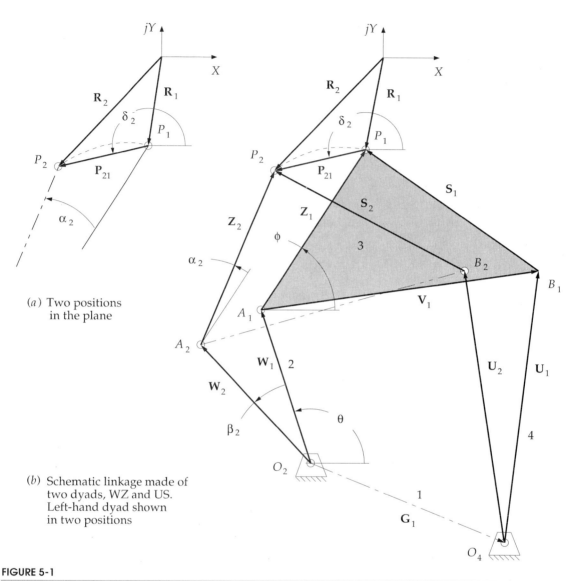

(a) Two positions in the plane

(b) Schematic linkage made of two dyads, WZ and US. Left-hand dyad shown in two positions

FIGURE 5-1

Analytical two-position synthesis

It is important to realize that the linkage as shown in the figure is schematic. Its dimensions are unknown at the outset, and are to be found by this synthesis technique. Thus, for example, the length of the position vector Z_1 as shown is not indicative of the final length of that edge of link 3, nor are the lengths (W, Z, U, V) or angles ($\theta, \phi, \sigma, \psi$) of any of the links as shown predictive of the final result.

The problem statement is:

Design a fourbar linkage which will move a line on its coupler link such that a point P on that line will be first at P_1 and later at P_2, and also rotate the line through an angle α_2 between those two precision positions. Find the lengths and angles of the four links and the coupler link dimensions A_1P_1 and B_1P_1 as shown in Figure 5-1.

The two-position analytical motion synthesis procedure is as follows:

Define the two desired precision positions in the plane with respect to an arbitrarily chosen global coordinate system XY using position vectors $\mathbf{R}_1$ and $\mathbf{R}_2$ as shown in Figure 5-1a. The change in angle α_2 of vector $\mathbf{Z}$ is the rotation required of the coupler link. Note that the position difference vector $\mathbf{P}_{21}$ defines the displacement of the output motion of point P and is defined as:

$$\mathbf{P}_{21} = \mathbf{R}_2 - \mathbf{R}_1 \tag{5.1}$$

The dyad $\mathbf{W}_1\mathbf{Z}_1$ defines the left half of the linkage. The dyad $\mathbf{U}_1\mathbf{S}_1$ defines the right half of the linkage. Note that $\mathbf{Z}_1$ and $\mathbf{S}_1$ are both embedded in the rigid coupler (link 3) and both of these vectors will undergo the same rotation through angle α_2 from position 1 to position 2. The pin-to-pin length and angle of link 3 (vector $\mathbf{V}_1$) is defined in terms of vectors $\mathbf{Z}_1$ and $\mathbf{S}_1$.

$$\mathbf{V}_1 = \mathbf{Z}_1 - \mathbf{S}_1 \tag{5.2a}$$

The ground link 1 is also definable in terms of the two dyads.

$$\mathbf{G}_1 = \mathbf{W}_1 + \mathbf{V}_1 - \mathbf{U}_1 \tag{5.2b}$$

Thus if we can define the two dyads $\mathbf{W}_1, \mathbf{Z}_1$, and $\mathbf{U}_1, \mathbf{S}_1$, we will have defined a linkage that meets the problem specifications.

We will first solve for the left-side of the linkage (vectors $\mathbf{W}_1$ and $\mathbf{Z}_1$) and later use the same procedure to solve for the right side (vectors $\mathbf{U}_1$ and $\mathbf{S}_1$). To solve for $\mathbf{W}_1$ and $\mathbf{Z}_1$ we need only write a vector loop equation around the loop which includes both positions P_1 and P_2 for the left-side dyad. We will go clockwise around the loop, starting with $\mathbf{W}_2$.

$$\mathbf{W}_2 + \mathbf{Z}_2 - \mathbf{P}_{21} - \mathbf{Z}_1 - \mathbf{W}_1 = 0 \tag{5.3}$$

Now substitute the complex number equivalents for the vectors.

$$we^{j(\theta+\beta_2)} + ze^{j(\phi+\alpha_2)} - p_{21}e^{j\delta_2} - ze^{j\phi} - we^{j\theta} = 0 \tag{5.4}$$

The sums of angles in the exponents can be rewritten as products of terms.

$$we^{j\theta}e^{j\beta_2} + ze^{j\phi}e^{j\alpha_2} - p_{21}e^{j\delta_2} - ze^{j\phi} - we^{j\theta} = 0 \tag{5.5a}$$

Simplifying and rearranging:

$$we^{j\theta}\left(e^{j\beta_2} - 1\right) + ze^{j\phi}\left(e^{j\alpha_2} - 1\right) = p_{21}e^{j\delta_2} \tag{5.5b}$$

Note that the lengths of vectors $\mathbf{W}_1$ and $\mathbf{W}_2$ are the same magnitude w because they represent the same rigid link in two different positions. The same can be said about vectors $\mathbf{Z}_1$ and $\mathbf{Z}_2$ whose common magnitude is z.

Equation 5.5 is a vector equation which contains two scalar equations and thus can be solved for two unknowns. The two scalar equations can be revealed by substituting Euler's identity (Eq. 4.4a) and separating the real and imaginary terms as was done in Section 4.5.

real part:

$$[w\cos\theta](\cos\beta_2 - 1) - [w\sin\theta]\sin\beta_2$$
$$+[z\cos\phi](\cos\alpha_2 - 1) - [z\sin\phi]\sin\alpha_2 = p_{21}\cos\delta_2 \qquad (5.6a)$$

imaginary part (with complex operator j divided out):

$$[w\sin\theta](\cos\beta_2 - 1) + [w\cos\theta]\sin\beta_2$$
$$+[z\sin\phi](\cos\alpha_2 - 1) + [z\cos\phi]\sin\alpha_2 = p_{21}\sin\delta_2 \qquad (5.6b)$$

There are eight variables in these two equations: w, θ, β_2, z, ϕ, α_2, p_{21}, and δ_2. We can only solve for two. Three of the eight are defined in the problem statement, namely α_2, p_{21}, and δ_2. Of the remaining five, w, θ, β_2, z, ϕ, we are forced to choose three as "free choices" (assumed values) in order to solve for the other two.

One strategy is to assume values for the three angles, θ, β_2, ϕ, on the premise that we may want to specify the orientation θ, ϕ of the two link vectors $\mathbf{W}_1$ and $\mathbf{Z}_1$ to suit packaging constraints, and also specify the angular excursion β_2 of link 2 to suit some driving constraint. This choice also has the advantage of leading to a set of equations which are linear in the unknowns and are thus easy to solve. For this solution, the equations can be simplified by setting the assumed and specified terms to be equal to some constants.

In equations 5.6a, let:

$$A = \cos\theta(\cos\beta_2 - 1) - \sin\theta\sin\beta_2$$
$$B = \cos\phi(\cos\alpha_2 - 1) - \sin\phi\sin\alpha_2 \qquad (5.7a)$$
$$C = p_{21}\cos\delta_2$$

and in equations 5.6b let:

$$D = \sin\theta(\cos\beta_2 - 1) + \cos\theta\sin\beta_2$$
$$E = \sin\phi(\cos\alpha_2 - 1) + \cos\phi\sin\alpha_2 \qquad (5.7b)$$
$$F = p_{21}\sin\delta_2$$

then:

$$Aw + Bz = C$$
$$Dw + Ez = F \qquad (5.7c)$$

and solving simultaneously,

$$w = \frac{CE - BF}{AE - BD}; \qquad z = \frac{AF - CD}{AE - BD} \qquad (5.7d)$$

A second strategy is to assume a length z and angle ϕ for vector $\mathbf{Z}_1$ and the angular excursion β_2 of link 2 and then solve for the vector $\mathbf{W}_1$. This is a commonly used approach. Note that the terms in square brackets in each of equations 5.6 are respectively the x and y components of the vectors $\mathbf{W}_1$ and $\mathbf{Z}_1$.

$$W_{1_x} = w\cos\theta; \qquad\qquad Z_{1_x} = z\cos\phi$$
$$W_{1_y} = w\sin\theta; \qquad\qquad Z_{1_y} = z\sin\phi \qquad\qquad (5.8a)$$

Substituting in equation 5.6,

$$W_{1_x}\left(\cos\beta_2 - 1\right) - W_{1_y}\sin\beta_2$$
$$+ Z_{1_x}\left(\cos\alpha_2 - 1\right) - Z_{1_y}\sin\alpha_2 = p_{21}\cos\delta_2$$

$$(5.8b)$$

$$W_{1_y}\left(\cos\beta_2 - 1\right) + W_{1_x}\sin\beta_2$$
$$+ Z_{1_y}\left(\cos\alpha_2 - 1\right) + Z_{1_x}\sin\alpha_2 = p_{21}\sin\delta_2$$

Z_{1x} and Z_{1y} are known from equation 5.8a with z and ϕ assumed as free choices. To further simplify the expression, combine other known terms as:

$$A = \cos\beta_2 - 1; \qquad B = \sin\beta_2; \qquad C = \cos\alpha_2 - 1$$

$$(5.8c)$$

$$D = \sin\alpha_2; \qquad E = p_{21}\cos\delta_2; \qquad F = p_{21}\sin\delta_2$$

substituting,

$$AW_{1_x} - BW_{1_y} + CZ_{1_x} - DZ_{1_y} = E$$

$$(5.8d)$$

$$AW_{1_y} + BW_{1_x} + CZ_{1_y} + DZ_{1_x} = F$$

and the solution is:

$$W_{1_x} = \frac{A\left(-CZ_{1_x} + DZ_{1_y} + E\right) + B\left(-CZ_{1_y} - DZ_{1_x} + F\right)}{-2A}$$

$$(5.8e)$$

$$W_{1_y} = \frac{A\left(-CZ_{1_y} - DZ_{1_x} + F\right) + B\left(CZ_{1_x} - DZ_{1_y} - E\right)}{-2A}$$

Either of these strategies results in the definition of a left dyad $\mathbf{W}_1\mathbf{Z}_1$ and its pivot locations which will provide the motion generation specified.

We must now repeat the process for the right-hand dyad, $\mathbf{U}_1\mathbf{S}_1$. Figure 5-2 highlights the two positions $\mathbf{U}_1\mathbf{S}_1$ and $\mathbf{U}_2\mathbf{S}_2$ of the right dyad. Vector $\mathbf{U}_1$ is initially at angle σ and moves through angle γ_2 from position 1 to 2. Vector $\mathbf{S}_1$ is initially at angle ψ. Note that the rotation of vector $\mathbf{S}$ from $\mathbf{S}_1$ to $\mathbf{S}_2$ is through the same angle α_2 as vector $\mathbf{Z}$, since they are in the same link. A vector loop equation similar to equation 5.3 can be written for this dyad.

$$\mathbf{U}_2 + \mathbf{S}_2 - \mathbf{P}_{21} - \mathbf{S}_1 - \mathbf{U}_1 = 0 \qquad (5.9a)$$

Rewrite in complex variable form and collect terms.

$$ue^{j\sigma}\left(e^{j\gamma_2} - 1\right) + se^{j\psi}\left(e^{j\alpha_2} - 1\right) = p_{21}e^{j\delta_2} \qquad (5.9b)$$

When this is expanded and the proper angles substituted, the x and y component equations become:

real part:

$$u\cos\sigma(\cos\gamma_2 - 1) - u\sin\sigma\sin\gamma_2$$
$$+s\cos\psi(\cos\alpha_2 - 1) - s\sin\psi\sin\alpha_2 = p_{21}\cos\delta_2 \qquad (5.10a)$$

imaginary part (with complex operator j divided out):

$$u\sin\sigma(\cos\gamma_2 - 1) + u\cos\sigma\sin\gamma_2$$
$$+s\sin\psi(\cos\alpha_2 - 1) + s\cos\psi\sin\alpha_2 = p_{21}\sin\delta_2 \qquad (5.10b)$$

Compare equations 5.10 to equations 5.6.

The same first strategy can be applied to equations 5.10 as was used for equations 5.6 to solve for the magnitudes of vectors $\mathbf{U}$ and $\mathbf{S}$, assuming values for angles σ, ψ, and γ_2. The quantities p_{21}, δ_2 and α_2 are defined from the problem statement as before.

In equations 5.10a let:

$$A = \cos\sigma(\cos\gamma_2 - 1) - \sin\sigma\sin\gamma_2$$
$$B = \cos\psi(\cos\alpha_2 - 1) - \sin\psi\sin\alpha_2 \qquad (5.11a)$$
$$C = p_{21}\cos\delta_2$$

and in equations 5.10b let:

$$D = \sin\sigma(\cos\gamma_2 - 1) + \cos\sigma\sin\gamma_2$$
$$E = \sin\psi(\cos\alpha_2 - 1) + \cos\psi\sin\alpha_2 \qquad (5.11b)$$
$$F = p_{21}\sin\delta_2$$

then:

$$Au + Bs = C$$
$$\qquad (5.11c)$$
$$Du + Es = F$$

and solving simultaneously,

$$u = \frac{CE - BF}{AE - BD}; \qquad\qquad s = \frac{AF - CD}{AE - BD} \qquad (5.11d)$$

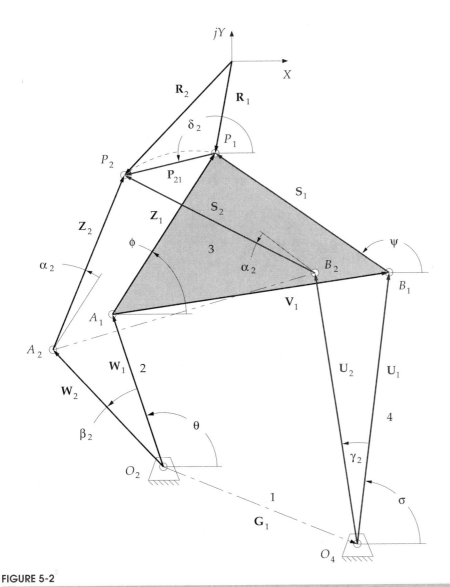

FIGURE 5-2

Right-side dyad shown in two positions

If the second strategy is used, assuming angle γ_2 and the magnitude and direction of vector $\mathbf{S}_1$ (which will define link 3), the result will be:

$$U_{1_x} = u\cos\sigma; \qquad\qquad S_{1_x} = s\cos\psi$$
$$U_{1_y} = u\sin\sigma; \qquad\qquad S_{1_y} = s\sin\psi \qquad\qquad (5.12a)$$

Substitute in equation 5.10:

$$U_{1_x}\left(\cos\gamma_2 - 1\right) - U_{1_y}\sin\gamma_2$$
$$+S_{1_x}\left(\cos\alpha_2 - 1\right) - S_{1_y}\sin\alpha_2 = p_{21}\cos\delta_2$$

$$(5.12b)$$

$$U_{1_y}\left(\cos\gamma_2 - 1\right) + U_{1_x}\sin\gamma_2$$
$$+S_{1_y}\left(\cos\alpha_2 - 1\right) + S_{1_x}\sin\alpha_2 = p_{21}\sin\delta_2$$

Let,

$$A = \cos\gamma_2 - 1; \qquad B = \sin\gamma_2; \qquad C = \cos\alpha_2 - 1$$

$$(5.12c)$$

$$D = \sin\alpha_2; \qquad E = p_{21}\cos\delta_2; \qquad F = p_{21}\sin\delta_2$$

Substitute in equation 5.12b,

$$AU_{1_x} - BU_{1_y} + CS_{1_x} - DS_{1_y} = E$$

$$AU_{1_y} + BU_{1_x} + CS_{1_y} + DS_{1_x} = F \qquad (5.12d)$$

and the solution is:

$$U_{1_x} = \frac{A\left(-CS_{1_x} + DS_{1_y} + E\right) + B\left(-CS_{1_y} - DS_{1_x} + F\right)}{-2A}$$

$$(5.12e)$$

$$U_{1_y} = \frac{A\left(-CS_{1_y} - DS_{1_x} + F\right) + B\left(CS_{1_x} - DS_{1_y} - E\right)}{-2A}$$

Note that there are infinities of possible solutions to this problem because we may choose any set of values for the three free choices of variables in this two-position case. Technically there is an infinity of solutions for each free choice. Three choices then give infinity cubed solutions! But since infinity is defined as a number larger than the largest number you can think of, infinity cubed is not any more impressively large than just plain infinity. While not strictly correct mathematically, we will, for simplicity, refer to all of these cases as having "an infinity of solutions," regardless of the power to which infinity may be raised as a result of the derivation. There are plenty of solutions to pick from, at any rate. *Unfortunately, not all will work.* Some will have toggle positions between the precision points. Others will have poor transmission angles or poor pivot locations or over-large links. Design judgment is still most important in selecting the assumed values for your free choices. Despite their name you must pay for those "free choices" later. Make a model!

5.4 COMPARISON OF ANALYTICAL AND GRAPHICAL TWO-POSITION SYNTHESIS

Note that in the **graphical solution** to this two-position synthesis problem (in Example 3-3 and Figure 3-6), we also had to make *three free choices* to solve the problem. The identical two-position synthesis problem from Figure 3-6 is reproduced in Figure 5-3. The approach taken in Example 3-3 used the two points A and B as the attachments for the moving pivots. Figure 5-3a shows the graphical construction used to find the fixed pivots O_2 and O_4. For the analytical solution we will use those points A and B as the joints of the two dyads **WZ** and **US**. These dyads meet at point P, which is the precision point. The relative position vector $\mathbf{P}_{21}$ defines the displacement of the precision point.

Note that in the graphical solution, we implicitly defined the left dyad vector **Z** by locating attachment points A and B on link 3 as shown in Figure 5-3a. This defined the two variables, z and ϕ. We also implicitly chose the value of w by selecting an arbitrary location for pivot O_2 on the perpendicular bisector. When that third choice was made, the remaining two unknowns, angles β_2 and θ, were solved for graphically at the same time, because the geometric construction was in fact a graphical "computation" for the solution of the two simultaneous equations 5.8.

The graphical and analytical methods merely represent two alternate solutions to the same problem. All of these problems can be solved both analytically and graphically. One method can provide a good check for the other.

We will now solve this problem analytically and correlate the results with the graphical solution from Chapter 3.

✎ EXAMPLE 5-1

Two-Position Analytical Motion Synthesis.

Problem: Design a fourbar linkage to move the link APB shown from position $A_1P_1B_1$ to $A_2P_2.B_2$.

Solution: (see Figure 5-3)

1   Draw the link APB in its two desired positions, $A_1P_1B_1$ and $A_2P_2B_2$, to scale in the plane as shown.

2   Measure or calculate the values of the magnitude and angle of vector $\mathbf{P}_{21}$, namely, p_{21} and δ_2. In this example they are:

$$p_{21} = 2.416; \qquad \delta_2 = 165.2°$$

3   Measure or calculate the value of the change in angle, α_2, of vector **Z** from position 1 to position 2. In this example it is:

$$\alpha_2 = 43.3°$$

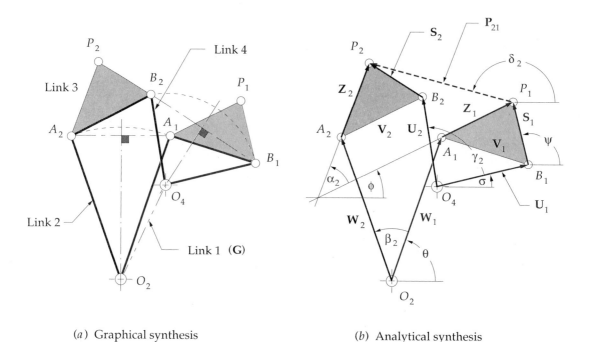

(*a*) Graphical synthesis   (*b*) Analytical synthesis

FIGURE 5-3

Two-position motion synthesis with coupler output

4   The three values in steps 2 and 3 are the only ones defined in the problem statement. We must assume three additional "free choices" to solve the problem. Method two (see equations 5.8) chooses the length z and angle ϕ of vector $\mathbf{Z}$ and β_2, the change in angle of vector $\mathbf{W}$. In order to obtain the same solution as the graphical method produced in Figure 5-3a (from the infinities of solutions available), we will choose those values consistent with the graphical solution.

$$z = 1.298; \qquad \phi = 26.5°; \qquad \beta_2 = 38.4°$$

5   Substitute these six values in equations 5.8 and obtain:

$$w = 2.467 \qquad \theta = 71.6°$$

6   Compare these to the graphical solution;

$$w = 2.48 \qquad \theta = 71°$$

which is a reasonable match given the graphical accuracy. This vector $\mathbf{W}_1$ is link 2 of the fourbar.

7   Repeat the procedure for the link 4 side of the linkage. The free choices will now be:

$$s = 1.035; \qquad \psi = 104.1°; \qquad \gamma_2 = 85.6°$$

8   Substitute these three values along with the original three values from steps 2 and 3 in equations 5.7 and obtain:

$$u = 1.486 \qquad \sigma = 15.4°$$

9   Compare these to the graphical solution:

$$u = 1.53 \qquad \sigma = 14°$$

These are a reasonable match given the graphical accuracy. This vector U_1 is link 4 of the fourbar.

10 Line A_1B_1 is link 3 and can be found from equation 5.2a. Line O_2O_4 is link 1 and can be found from equation 5.2b.

11 Check the Grashof condition, and repeat steps 4 to 7 if unsatisfied. Note that any Grashof condition is potentially acceptable in this case.

12 Construct a cardboard model and check its function to be sure it can get from initial to final position without encountering any limit (toggle) positions.

13 Check transmission angles.

Input the file FIG5-3 to program FOURBAR to see Example 5-1.

5.5   SIMULTANEOUS EQUATION SOLUTION

These methods of analytical synthesis lead to sets of linear simultaneous equations. The two-position synthesis problem results in two simultaneous equations which can be solved by direct substitution. The three-position synthesis problem will lead to a system of four simultaneous linear equations and will require a more complicated method of solution. A convenient approach to the solution of sets of linear simultaneous equations is to put them in a standard matrix form and use a numerical matrix solver to obtain the answers. Matrix solvers are built into most engineering and scientific pocket calculators. Some spreadsheet packages and equation solvers will also do a matrix solution.

As an example of this general approach, consider the following set of simultaneous equations:

$$
\begin{aligned}
-2x_1 \quad -x_2 \quad +x_3 &= -1 \\
x_1 \quad +x_2 \quad +x_3 &= 6 \\
3x_1 \quad +x_2 \quad -x_3 &= 2
\end{aligned}
\qquad (5.13a)
$$

A system this small can be solved longhand by the elimination method, but we will put it in matrix form to show the general approach which will work regardless of the number of equations. The equations 5.13a can be written as the product of two matrices set equal to a third matrix.

$$\begin{bmatrix} -2 & -1 & 1 \\ 1 & 1 & 1 \\ 3 & 1 & -1 \end{bmatrix} \times \begin{bmatrix} x_1 \\ x_2 \\ x_3 \end{bmatrix} = \begin{bmatrix} -1 \\ 6 \\ 2 \end{bmatrix} \tag{5.13b}$$

We will refer to these matrices as **A**, **B**, and **C**,

$$[\mathbf{A}] \quad \times \quad [\mathbf{B}] \quad = \quad [\mathbf{C}] \tag{5.13c}$$

where **A** is the matrix of coefficients of the unknowns, **B** is a column vector of the unknown terms and **C** is a column vector of the constant terms. When matrix **A** is multiplied by **B**, the result will be the same as the left sides of equation 5.13a. See any text on Linear Algebra such as reference 7 for a discussion of the procedure for matrix multiplication.

If equation 5.13c were a scalar equation,

$$ab = c \tag{5.14a}$$

rather than a vector (matrix) equation, it would be very easy to solve it for the unknown b when a and c are known. We would simply divide c by a to find b.

$$b = \frac{c}{a} \tag{5.14b}$$

Unfortunately, division is not defined for matrices, so another approach must be used. Note that we could also express the division in equation 5.14b as:

$$b = a^{-1}c \tag{5.14c}$$

If the equations to be solved are linearly independent, then we can find the inverse of matrix **A** and multiply it by matrix **C** to find **B**. The inverse of a matrix is defined as that matrix which when multiplied by the original matrix yields the identity matrix. The **identity matrix** is a square matrix with ones on the main diagonal and zeros everywhere else. The inverse of a matrix is denoted by adding a superscript of negative one to the symbol for the original matrix.

$$[\mathbf{A}]^{-1} \times [\mathbf{A}] = [\mathbf{I}] = \begin{bmatrix} 1 & 0 & 0 \\ 0 & 1 & 0 \\ 0 & 0 & 1 \end{bmatrix} \tag{5.15}$$

Not all matrices will possess an inverse. The determinant of the matrix must be nonzero for an inverse to exist. The class of problems dealt with here will yield matrices which have inverses provided that all data are correctly calculated for input to the matrix and represent a real physical system. The calculation of the terms of the inverse for a matrix is a complicated numerical process which requires a computer or preprogrammed pocket calculator to invert any matrix of significant size. A Gauss-Jordan-elimination numerical method is usually used to find an inverse. For our simple example in equation 5.13 the inverse of matrix **A** is found to be:

$$\begin{bmatrix} -2 & -1 & 1 \\ 1 & 1 & 1 \\ 3 & 1 & -1 \end{bmatrix}^{-1} = \begin{bmatrix} 1.0 & 0.0 & 1.0 \\ -2.0 & 0.5 & -1.5 \\ 1.0 & 0.5 & 0.5 \end{bmatrix} \qquad (5.16)$$

If the inverse of matrix **A** can be found, we can solve equation 5.13 for the unknowns **B** by multiplying both sides of the equation by the inverse of **A**. Note that unlike scalar multiplication, matrix multiplication is not commutative; i.e., **A** x **B** is not equal to **B** x **A**. We will premultiply each side of the equation by the inverse.

$$[\mathbf{A}]^{-1} \times [\mathbf{A}] \times [\mathbf{B}] = [\mathbf{A}]^{-1} \times [\mathbf{C}]$$

but:

$$[\mathbf{A}]^{-1} \times [\mathbf{A}] = [\mathbf{I}] \qquad (5.17)$$

so:

$$[\mathbf{B}] = [\mathbf{A}]^{-1} \times [\mathbf{C}]$$

The product of **A** and its inverse on the left-side of the equation is equal to the identity matrix **I**. Multiplying by the identity matrix is equivalent, in scalar terms, to multiplying by one, so has no effect on the result. Thus the unknowns can be found by premultiplying the inverse of the coefficient matrix **A** times the matrix of constant terms **C**.

This method of solution works no matter how many equations are present as long as the inverse of **A** can be found and enough computer memory and/or time is available to do the computation. Note that it is not actually necessary to find the inverse of matrix **A** to solve the set of equations. The Gauss-Jordan algorithm which finds the inverse can also be used to directly solve for the unknowns **B** by assembling the **A** and **C** matrices into an **augmented matrix** of n rows and $n + 1$ columns. The added column is the **C** vector. This approach requires fewer calculations, so it is faster and more accurate. The augmented matrix for this example is:

$$\begin{bmatrix} -2 & -1 & 1 & -1 \\ 1 & 1 & 1 & 6 \\ 3 & 1 & -1 & 2 \end{bmatrix} \qquad (5.18a)$$

The Gauss-Jordan algorithm manipulates this augmented matrix until it is in the form shown below, in which the left, square portion has been reduced to the identity matrix and the rightmost column contains the values of the column vector of unknowns. In this case the results are $x_1 = 1$, $x_2 = 2$, and $x_3 = 3$ which are the correct solution to the original equations 5.13.

$$\begin{bmatrix} 1 & 0 & 0 & 1 \\ 0 & 1 & 0 & 2 \\ 0 & 0 & 1 & 3 \end{bmatrix} \qquad (5.18b)$$

Main Menu - Program MATRIX Version X.x   mm/dd/yy

1 – INPUT   problem setup data

2 – DISPLAY   problem setup data

3 – EDIT   matrices

4 – COMPUTE   results

5 – PRINT   results

6 – HELP

<Ret> – QUIT the Program

CHOICE ?

FIGURE 5-4

Main menu for program MATRIX

The program MATRIX, supplied with this text, solves these problems with this Gauss-Jordan elimination method and operates on the augmented matrix without actually finding the inverse of **A** in explicit form. For a review of matrix algebra see reference 7.

SELECT INPUT METHOD

(Note: any selection will overwrite existing data)

1 – TYPE your DATA from KEYBOARD

2 – READ your DATA from DISKFILE

3 – Create Identity Matrix

4 – Example Matrix

5 – CLEAR all data

<Ret> – EXIT

Choice ?

FIGURE 5-5

Input menu for program MATRIX

MATRIX 1.0   Ima Student   Design # 1   mm-dd-yy   at   hh:mm

The number of equations is 3

The terms of the A matrix in row order are:

$$
\begin{array}{ccc}
-2 & 1 & 1 \\
1 & 1 & 1 \\
3 & 1 & -1
\end{array}
$$

The terms of the C vector are:

$$
\begin{array}{ccc}
-1 & 6 & 2
\end{array}
$$

FIGURE 5-6

Display data screen from program MATRIX

5.6   PROGRAM MATRIX

Program MATRIX operates similarly to the other programs supplied with this text. Hardware requirements and general operation of all programs are described in Chapter 8 which can be read at any time without loss of continuity. Program MATRIX is menu driven. Its main menu is shown in Figure 5-4. Selecting item 1 brings up the input submenu shown in Figure 5-5. You may type your matrix elements from the keyboard or read previously stored examples from diskfiles. There is a built-in example of matrix data which can be chosen from this menu to try the program. It is the same one discussed in the previous section. The **Read Data from Diskfile** choice operates identically to that of the other programs. See Chapter 8 for a detailed explanation. **Clear all Data** will reset all matrices in memory to zero or to the identity matrix and reset all labels to their default values.

Choosing **Type Data from Keyboard** prompts the user for the number of equations (up to 40) and then expects the matrix elements to be supplied in *row order*. All the elements in the first row of matrix **A** are requested in turn. Then the element from the first row of the column vector of constants, **C**, is requested. This procedure is repeated for all the rows in order. Thus the order of input of data is row by row of the **augmented matrix** as shown in equation 5.18a.

Display Data from the main menu displays the **A** and **C** matrices, as shown in Figure 5-6. These data may be sent to a diskfile from this menu choice as well. **Edit Matrices** allows any element to be changed. The **Calculate** choice causes the Gauss-Jordan reduction to be done on the augmented matrix of **A** and **C** and creates the **B** matrix of solutions. **Print Results** will then show the results

MATRIX 1.0   Ima Student   Design # 1   mm-dd-yy at hh:mm

Calculated Column Vector [B] of Results

$$\text{from} \quad [B] \;=\; [A]^{-1} \; X \; [C]$$

Row # 1	1.0000 D+00
Row # 2	2.0000 D+00
Row # 3	3.0000 D+00

FIGURE 5-7

Output data from program MATRIX

as a column vector with user supplied or default labels on each row as shown in
Figure 5-7. The results are shown in exponential notation. The D on the exponent
indicates double precision which is used on all variables for accuracy. As with all
the programs the data may be printed to the screen, to a printer, or to a diskfile
for later import to a spreadsheet or word processor. The **Help** selection brings up
part of the same submenu as do the other programs. See Chapter 8 for further
information. Note that this program does not use any units or angles in its displays,
so changing units or angle modulo is not allowed.

5.7   THREE-POSITION MOTION GENERATION BY ANALYTICAL SYNTHESIS

The same approach of defining two dyads, one at each end of the fourbar linkage, as used for two-position motion synthesis can be extended to three, four, and
five positions in the plane. The three-position motion synthesis problem will now
be addressed. Figure 5-8 shows a fourbar linkage in one general position with a
coupler point located at its first precision position P_1. Second and third precision
positions (points P_2 and P_3) are also shown. These are to be achieved by the rotation of the input rocker, link 2, through as yet unspecified angles β_2 and β_3. Note
also that the angles of the coupler link 3 at each of the precision positions are defined
by the angles of the position vectors Z_1, Z_2, and Z_3. The linkage shown in the figure
is schematic. Its dimensions are unknown at the outset and are to be found by
this synthesis technique. Thus, for example, the length of the position vector Z_1
as shown is not indicative of the final length of that edge of link 3 nor are the
lengths or angles of any of the links shown predictive of the final result.

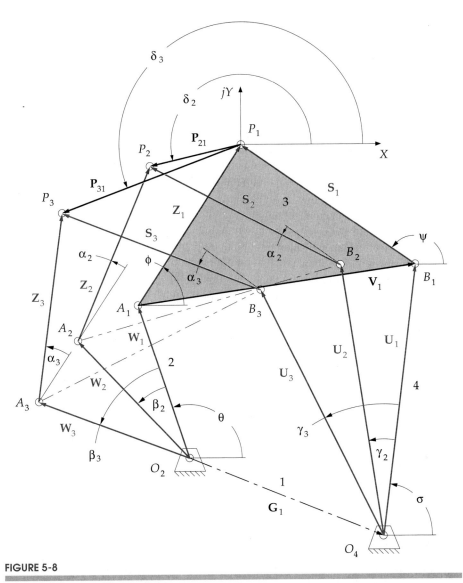

FIGURE 5-8

Analytical three-position synthesis

The problem statement is:

Design a fourbar linkage which will move a line on its coupler link such that a point P on that line will be first at P_1, later at P_2, and still later at P_3, and also rotate the line through an angle α_2 between the first two precision positions and through an angle α_3 between the first and third precision positions. Find the lengths and angles of the four links and the coupler link dimensions A_1P_1 and B_1P_1 as shown in Figure 5-8.

The three-position analytical motion synthesis procedure is as follows:

For convenience, we will place the global coordinate system XY at the first precision point P_1. We define the other two desired precision positions in the plane with respect to this global system as shown in Figure 5-8. The position difference vectors $\mathbf{P}_{21}$, drawn from P_1 to P_2 and $\mathbf{P}_{31}$, drawn from P_1 to P_3, have angles δ_2 and δ_3, respectively. The position difference vectors $\mathbf{P}_{21}$ and $\mathbf{P}_{31}$ define the displacements of the output motion of point P from point 1 to 2 and from 1 to 3, respectively.

The dyad $\mathbf{W}_1\,\mathbf{Z}_1$ defines the left half of the linkage. The dyad $\mathbf{U}_1\,\mathbf{S}_1$ defines the right half of the linkage. Vectors $\mathbf{Z}_1$ and $\mathbf{S}_1$ are both embedded in the rigid coupler (link 3), and both will undergo the same rotations, through angle α_2 from position 1 to position 2 and through angle α_3 from position 1 to position 3. The pin-to-pin length and angle of link 3 (vector $\mathbf{V}_1$) is defined in terms of vectors $\mathbf{Z}_1$ and $\mathbf{S}_1$ as in equation 5.2a. The ground link is defined by equation 5.2b as before.

As we did in the two-position case, we will first solve for the left side of the linkage (vectors $\mathbf{W}_1$ and $\mathbf{Z}_1$) and later use the same procedure to solve for the right side (vectors $\mathbf{U}_1$ and $\mathbf{S}_1$). To solve for $\mathbf{W}_1$ and $\mathbf{Z}_1$ we need to now write **two vector loop equations**, one around the loop which includes positions P_1 and P_2 and the second one around the loop which includes positions P_1 and P_3 (see Figure 5-8). We will go clockwise around the first loop for motion from position 1 to 2, starting with $\mathbf{W}_2$, and then write the second loop equation for motion from position 1 to 3 starting with $\mathbf{W}_3$.

$$\mathbf{W}_2 + \mathbf{Z}_2 - \mathbf{P}_{21} - \mathbf{Z}_1 - \mathbf{W}_1 = 0$$

(5.19)

$$\mathbf{W}_3 + \mathbf{Z}_3 - \mathbf{P}_{31} - \mathbf{Z}_1 - \mathbf{W}_1 = 0$$

Substituting the complex number equivalents for the vectors.

$$we^{j(\theta+\beta_2)} + ze^{j(\phi+\alpha_2)} - p_{21}e^{j\delta_2} - ze^{j\phi} - we^{j\theta} = 0$$

(5.20)

$$we^{j(\theta+\beta_3)} + ze^{j(\phi+\alpha_3)} - p_{31}e^{j\delta_3} - ze^{j\phi} - we^{j\theta} = 0$$

Rewriting the sums of angles in the exponents as products of terms.

$$we^{j\theta}e^{j\beta_2} + ze^{j\phi}e^{j\alpha_2} - p_{21}e^{j\delta_2} - ze^{j\phi} - we^{j\theta} = 0$$

(5.21a)

$$we^{j\theta}e^{j\beta_3} + ze^{j\phi}e^{j\alpha_3} - p_{31}e^{j\delta_3} - ze^{j\phi} - we^{j\theta} = 0$$

Simplifying and rearranging:

$$we^{j\theta}\left(e^{j\beta_2} - 1\right) + ze^{j\phi}\left(e^{j\alpha_2} - 1\right) = p_{21}e^{j\delta_2}$$

(5.21b)

$$we^{j\theta}\left(e^{j\beta_3} - 1\right) + ze^{j\phi}\left(e^{j\alpha_3} - 1\right) = p_{31}e^{j\delta_3}$$

The magnitude w of vectors $\mathbf{W}_1$, $\mathbf{W}_2$, and $\mathbf{W}_3$ is the same in all three positions because it represents the same line in a rigid link. The same can be said about vectors $\mathbf{Z}_1$, $\mathbf{Z}_2$, and $\mathbf{Z}_3$ whose common magnitude is z.

Equations 5.21 are a set of two vector equations, each of which contains two scalar equations. This set of four equations can be solved for four unknowns. The scalar equations can be revealed by substituting Euler's identity (equation 4.4a) and separating the real and imaginary terms as was done in the two-position example above.

real part:

$$w\cos\theta(\cos\beta_2 - 1) - w\sin\theta\sin\beta_2$$
$$+z\cos\phi(\cos\alpha_2 - 1) - z\sin\phi\sin\alpha_2 = p_{21}\cos\delta_2 \tag{5.22a}$$

$$w\cos\theta(\cos\beta_3 - 1) - w\sin\theta\sin\beta_3$$
$$+z\cos\phi(\cos\alpha_3 - 1) - z\sin\phi\sin\alpha_3 = p_{31}\cos\delta_3 \tag{5.22b}$$

imaginary part (with complex operator j divided out)

$$w\sin\theta(\cos\beta_2 - 1) + w\cos\theta\sin\beta_2$$
$$+z\sin\phi(\cos\alpha_2 - 1) + z\cos\phi\sin\alpha_2 = p_{21}\sin\delta_2 \tag{5.22c}$$

$$w\sin\theta(\cos\beta_3 - 1) + w\cos\theta\sin\beta_3$$
$$+z\sin\phi(\cos\alpha_3 - 1) + z\cos\phi\sin\alpha_3 = p_{31}\sin\delta_3 \tag{5.22d}$$

There are **twelve variables** in these four equations 5.22: w, θ, β_2, β_3, z, ϕ, α_2, α_3, p_{21}, p_{31}, δ_2, and δ_3. **We can solve for only four.** Six of them are defined in the problem statement, namely α_2, α_3, p_{21}, p_{31}, δ_2, and δ_3. Of the remaining six, w, θ, β_2, β_3, z, ϕ, **we must choose two as free choices** (assumed values) in order to solve for the other four. One strategy is to assume values for the two angles, β_2 and β_3, on the premise that we may want to specify the angular excursions of link 2 to suit some driving constraint. (This choice also has the benefit of leading to a set of linear equations for simultaneous solution.)

This leaves the magnitudes and angles of vectors $\mathbf{W}$ and $\mathbf{Z}$ to be found (w, θ, z, ϕ). To simplify the solution, we can substitute the following relationships to obtain the x and y components of the two unknown vectors $\mathbf{W}$ and $\mathbf{Z}$, rather than their polar coordinates.

$$W_{1_x} = w\cos\theta; \qquad\qquad Z_{1_x} = z\cos\phi$$

$$\tag{5.23}$$

$$W_{1_y} = w\sin\theta; \qquad\qquad Z_{1_y} = z\sin\phi$$

Substituting equations 5.23 into 5.22 we obtain:

$$W_{1_x}(\cos\beta_2 - 1) - W_{1_y}\sin\beta_2$$
$$+ Z_{1_x}(\cos\alpha_2 - 1) - Z_{1_y}\sin\alpha_2 = P_{21}\cos\delta_2 \qquad (5.24a)$$

$$W_{1_x}(\cos\beta_3 - 1) - W_{1_y}\sin\beta_3$$
$$+ Z_{1_x}(\cos\alpha_3 - 1) - Z_{1_y}\sin\alpha_3 = P_{31}\cos\delta_3 \qquad (5.24b)$$

$$W_{1_y}(\cos\beta_2 - 1) + W_{1_x}\sin\beta_2$$
$$+ Z_{1_y}(\cos\alpha_2 - 1) + Z_{1_x}\sin\alpha_2 = P_{21}\sin\delta_2 \qquad (5.24c)$$

$$W_{1_y}(\cos\beta_3 - 1) + W_{1_x}\sin\beta_3$$
$$+ Z_{1_y}(\cos\alpha_3 - 1) + Z_{1_x}\sin\alpha_3 = P_{31}\sin\delta_3 \qquad (5.24d)$$

These are four equations in the four unknowns W_{1_x} W_{1_y} Z_{1_x} and Z_{1_y} By setting the coefficients which contain the assumed and specified terms equal to some constants, we can simplify the notation and obtain the following solutions.

$$
\begin{array}{lll}
A = \cos\beta_2 - 1; & B = \sin\beta_2; & C = \cos\alpha_2 - 1 \\
D = \sin\alpha_2; & E = P_{21}\cos\delta_2; & F = \cos\beta_3 - 1 \\
\end{array}
$$

$$(5.25)$$

$$
\begin{array}{lll}
G = \sin\beta_3; & H = \cos\alpha_3 - 1; & K = \sin\alpha_3 \\
L = P_{31}\cos\delta_3; & M = P_{21}\sin\delta_2; & N = P_{31}\sin\delta_3 \\
\end{array}
$$

Substituting equations 5.25 in 5.24 to simplify:

$$AW_{1_x} - BW_{1_y} + CZ_{1_x} - DZ_{1_y} = E \qquad (5.26a)$$
$$FW_{1_x} - GW_{1_y} + HZ_{1_x} - KZ_{1_y} = L \qquad (5.26b)$$
$$BW_{1_x} + AW_{1_y} + DZ_{1_x} + CZ_{1_y} = M \qquad (5.26c)$$
$$GW_{1_x} + FW_{1_y} + KZ_{1_x} + HZ_{1_y} = N \qquad (5.26d)$$

This system can be put into standard matrix form:

$$
\begin{bmatrix}
A & -B & C & -D \\
F & -G & H & -K \\
B & A & D & C \\
G & F & K & H
\end{bmatrix}
\times
\begin{bmatrix}
W_{1_x} \\
W_{1_y} \\
Z_{1_x} \\
Z_{1_y}
\end{bmatrix}
=
\begin{bmatrix}
E \\
L \\
M \\
N
\end{bmatrix}
\qquad (5.27)
$$

This is of the general form of equation 5.13c. The vector of unknowns **B** can be solved for by premultiplying the inverse of the coefficient matrix **A** by the constant vector **C** or by forming the augmented matrix as in equation 5.18. For any numerical problem, the inverse of a 4 x 4 matrix can be found with many pocket calculators. The computer program MATRIX, supplied with this text, will also solve the augmented matrix equation.

Equations 5.25 and 5.26 solve the three-position synthesis problem for the left-hand side of the linkage using any pair of assumed values for β_2 and β_3. We must repeat the above process for the right-hand side of the linkage to find vectors **U** and **S**. Figure 5-8 also shows the three positions of the **US** dyad, and the angles σ, γ_2, γ_3, ψ, α_2, and α_3, which define those vector rotations for all three positions. The solution derivation for the right-side dyad, **US**, is identical to that just done for the left dyad **WZ**. The angles and vector labels are the only difference. The vector loop equations are:

$$U_2 + S_2 - P_{21} - S_1 - U_1 = 0$$

$$(5.28)$$

$$U_3 + S_3 - P_{31} - S_1 - U_1 = 0$$

Substituting, simplifying, and rearranging,

$$ue^{j\sigma}\left(e^{j\gamma_2} - 1\right) + se^{j\psi}\left(e^{j\alpha_2} - 1\right) = p_{21}e^{j\delta_2}$$

$$(5.29)$$

$$ue^{j\sigma}\left(e^{j\gamma_3} - 1\right) + se^{j\psi}\left(e^{j\alpha_3} - 1\right) = p_{31}e^{j\delta_3}$$

The solution requires that two free choices be made. We will assume values for the angles γ_2 and γ_3. Note that α_2 and α_3 are the same as for dyad **WZ**. We will, in effect, solve for angles σ and ψ by finding the x and y components of the vectors **U** and **S**. The solution is:

$A = \cos\gamma_2 - 1;$	$B = \sin\gamma_2;$	$C = \cos\alpha_2 - 1$
$D = \sin\alpha_2;$	$E = p_{21}\cos\delta_2;$	$F = \cos\gamma_3 - 1$

$$(5.30)$$

$G = \sin\gamma_3;$	$H = \cos\alpha_3 - 1;$	$K = \sin\alpha_3$
$L = p_{31}\cos\delta_3;$	$M = p_{21}\sin\delta_2;$	$N = p_{31}\sin\delta_3$

$$AU_{1_x} - BU_{1_y} + CS_{1_x} - DS_{1_y} = E \qquad (5.31a)$$

$$FU_{1_x} - GU_{1_y} + HS_{1_x} - KS_{1_y} = L \qquad (5.31b)$$

$$BU_{1_x} + AU_{1_y} + DS_{1_x} + CS_{1_y} = M \qquad (5.31c)$$

$$GU_{1_x} + FU_{1_y} + KS_{1_x} + HS_{1_y} = N \qquad (5.31d)$$

Equations 5.31 can be solved using the approach of equations 5.27 and 5.18, by changing W to U and Z to S and using the definitions of the constants given in equation 5.30 in equation 5.27.

It should be apparent that there are infinities of solutions to this three-position synthesis problem as well. An inappropriate selection of the two free choices could lead to a solution which has problems in moving among all specified positions. Thus we must check the function of the solution synthesized by this or any other method. A simple model is the quickest check.

5.8   COMPARISON OF ANALYTICAL AND GRAPHICAL THREE-POSITION SYNTHESIS

Figure 5-9 shows the same three-position synthesis problem as was done graphically in Example 3-6 in Section 3.4. Compare this figure to Figure 3-9. The labelling has been changed to be consistent with the notation in this chapter. The points P_1, P_2, and P_3 correspond to the three points labelled D in the earlier figure. Points A_1, A_2, and A_3 correspond to points E; points B_1, B_2, and B_3 correspond to points F. The old line AP becomes the present $\mathbf{Z}$ vector. Point P is the coupler point which will go through the specified precision points, P_1, P_2, and P_3. Points A and B are the attachment points for the rockers (links 2 and 4, respectively) on the coupler (link 3). We wish to solve for the coordinates of vectors $\mathbf{W}$, $\mathbf{Z}$, $\mathbf{U}$, and $\mathbf{S}$, which define not only the lengths of those links but also the locations of the fixed pivots O_2 and O_4 in the plane and the lengths of links 3 and 1. Link 1 is defined as vector $\mathbf{G}$ in Figure 5-8 and can be found from equation 5.2b. Link 3 is vector $\mathbf{V}$ found from equation 5.2a.

Four free choices must be made to constrain the problem to a particular solution out of the infinities of solutions available. In this case the values of link angles β_2, β_3, γ_2, and γ_3 have been chosen to be the same values as those which were found in the graphical solution to Example 3-6 in order to obtain the same solution as a check and comparison. Recall that in doing the graphical three-position synthesis solution to this same problem we in fact also had to make four free choices. These were the x,y coordinates of the moving pivot locations E and F in Figure 3-9 which correspond in concept to our four free choices of link angles here.

Example 3-5 also shows a graphical solution to this same problem resulting from the free choice of the x,y coordinates of points C and D on the coupler for the moving pivots (see Figure 3-8 and Example 3-5). We found some problems with toggle positions in that solution and redid it using points E and F as moving pivots in Example 3-6 and Figure 3-9. In effect the graphical three-position synthesis solution presented in Chapter 3 is directly analogous to the analytical solution presented here. For this analytical approach we choose to select the link angles β_2, β_3, γ_2, and γ_3 rather than the moving pivot locations E and F in order to force the resulting equations to be linear in the unknowns. The graphical solution done in the earlier examples is actually a solution of nonlinear equations.

 EXAMPLE 5-2

Three-Position Analytical Motion Synthesis.

Problem: Design a fourbar linkage to move the link APB shown from position $A_1P_1B_1$ to $A_2P_2B_2$ and then to position $A_3P_3B_3$.

Solution: (see Figure 5-9)

1   Draw the link APB in its three desired positions, $A_1P_1B_1$, $A_2P_2B_2$, and $A_3P_3B_3$ to scale in the plane as shown in the figure.

2   The three positions are then defined with respect to a global origin positioned at the first precision point P_1. The given data are the magnitudes and angles of the position difference vectors between precision points:

$p_{21} = 2.798$ $\delta_{21} = -31.19°$ $p_{31} = 3.919$ $\delta_{31} = -16.34°$

3   The angle changes of the coupler between precision points are:

$\alpha_2 = -45°$ $\alpha_3 = 9.3°$

4   The free choices assumed for the link angles are:

$\beta_2 = 342.3°$ $\beta_3 = 324.8°$ $\gamma_2 = 30.9°$ $\gamma_3 = 80.6°$

These defined variables and the free choices are also listed on the figure.

5   Once the free choices of link angles are made, the terms for the matrices of equation 5.27 can be defined by solving equation 5.25 for the first dyad of the linkage and equation 5.30 for the second dyad of the linkage. For this example they evaluate to:

First dyad (**WZ**):

$A = -0.0473$ $B = -0.3040$ $C = -0.2929$ $D = -0.7071$

$E = 2.3936$ $F = -0.1829$ $G = -0.5764$ $H = -0.0131$

$K = 0.1616$ $L = 3.7607$ $M = -1.4490$ $N = -1.1026$

Second dyad (**US**):

$A = -0.1419$ $B = 0.5135$ $C = -0.2929$ $D = -0.7071$

$E = 2.3936$ $F = -0.8367$ $G = 0.9866$ $H = -0.0131$

$K = 0.1616$ $L = 3.7607$ $M = -1.4490$ $N = -1.1026$

6   Program MATRIX is used to solve this matrix equation once with the values from equation 5.25 inserted to get the coordinates of vectors **W** and **Z**, and a second time with values from equation 5.31 in the matrix to get the coordinates of vectors **U** and **S**. The calculated coordinates of the link vectors from equations 5.25 to 5.31 are:

$W_x = 0.055$ $W_y = 6.832$ $Z_x = 1.179$ $Z_y = 0.940$

Link $2 = w = 6.832$

$U_x = -2.628$ $U_y = -1.825$ $S_x = -0.109$ $S_y = 1.487$

Link $4 = u = 3.2$

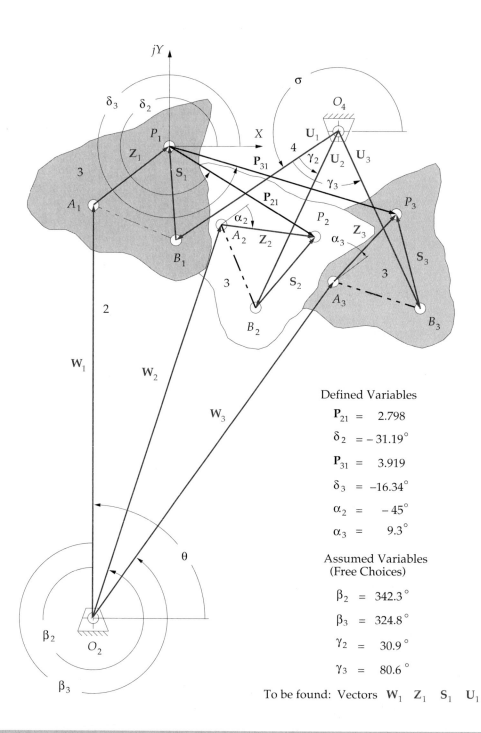

Defined Variables

$\mathbf{P}_{21}$ = 2.798

δ_2 = – 31.19°

$\mathbf{P}_{31}$ = 3.919

δ_3 = –16.34°

α_2 = – 45°

α_3 = 9.3°

Assumed Variables
(Free Choices)

β_2 = 342.3°

β_3 = 324.8°

γ_2 = 30.9°

γ_3 = 80.6°

To be found: Vectors $\mathbf{W}_1$ $\mathbf{Z}_1$ $\mathbf{S}_1$ $\mathbf{U}_1$

FIGURE 5-9

Data needed for analytical three-position synthesis

7 Equation 5.2a is used to find link 3.

$$V_x = Z_x - S_x = 1.179 - (-0.109) = 1.288$$

$$V_y = Z_y - S_y = 0.940 - 1.487 = -0.547$$

Link 3 = v = 1.399

8 The ground link is found from equation 5.2b

$$G_x = W_x + V_x - U_x = 0.055 + 1.288 - (-2.628) = 3.971$$

$$G_y = W_y + V_y - U_y = 6.832 - 0.547 - (-1.825) = 8.110$$

Link 1 = g = 9.03

9 The appropriate vector components are added together to get the locations of fixed pivots O_2 and O_4 with respect to the global origin at precision point P_1. See Figures 5-8 and 5-9.

$$O_{2x} = -Z_x - W_x = -1.179 - 0.055 = -1.234$$

$$O_{2y} = -Z_y - W_y = -0.940 - 6.832 = -7.772$$

$$O_{4x} = -S_x - U_x = -(-0.109) - (-2.628) = 2.737$$

$$O_{4y} = -S_y - U_y = -1.487 - (-1.825) = 0.338$$

Table 5-1 shows the linkage parameters as synthesized by this method. These agree with the solution found in Example 3-6 within its graphical accuracy. You may read the diskfiles EX5-2a and EX5-2b into program MATRIX to compute these results.

This problem can also be solved with program FOURBAR using the same method as derived in Section 5.7. Though the derivation was done in terms of the polar coordinates of the position difference vectors P_{21} and P_{31}, it was considered more convenient to supply the cartesian coordinates of these vectors to program FOURBAR. (It is generally more accurate to measure x,y coordinates from a sketch of the desired positions than to measure angles with a protractor.) Thus the program requests the rectangular coordinates of P_{21} and P_{31}. For this example they are:

p_{21x} = 2.394 p_{21y} = -1.449 p_{31x} = 3.761 p_{31y} = -1.103

The angles α_2 and α_3 must be measured from the diagram and supplied, in degrees. These six items constitute the set of "givens." **Note that these data are all *relative information* relating the second and third positions to the first.** No information about their absolute locations is needed. The global reference system can be taken to be anywhere in the plane. We took it to be at the first precision point position P_1 for convenience. The free choices β_2 and β_3 for the first dyad and γ_2, γ_3 for the second dyad must also be input to program FOURBAR as they also were to program MATRIX.

TABLE 5-1   Results of Analytical Synthesis for Example 5-2

Link #	Analytical Solution Length calculated (in)	Graphical Solution Length from Fig. 3-9
1	9.03	8.9
2	6.83	6.7
3	1.40	1.5
4	3.20	3.2
Coupler Pt.=	1.51 @ 61.31 deg	1.5 @ 61 deg
Open/Crossed =	CROSSED	CROSSED
Start Alpha2 =	0 rad/sec ^{2}	
Start Omega2 =	1 rad/sec	
Start Theta2 =	29 degrees	
Final Theta2 =	11 degrees	
Delta Theta2 =	–9 degrees	

Program FOURBAR then solves the matrix equation 5.27 once with the values from equation 5.25 inserted to get the coordinates of vectors **W** and **Z**, and a second time with values from equation 5.31 in the matrix to get the coordinates of vectors **U** and **S**. Equations 5.2 are then solved to find links 1 and 3, and the appropriate vector components are added together to get the locations of fixed pivots O_2 and O_4. The link lengths are returned to the main part of FOURBAR so that other linkage parameters can be calculated and the linkage animated.

Note that there are two ways to assemble any fourbar linkage, open and crossed (see Figure 4-8), and this analytical synthesis technique gives no information on which mode of assembly is necessary to get the solution desired. Thus you may have to try both modes of assembly in program FOURBAR to find the correct one after determining the proper link lengths with this method. Note also that FOURBAR always draws the linkage with the fixed link horizontal. Thus the animation of the solution is oriented differently than in Figure 5-9.

The finished linkage is the same as the one in Figure 3-8c which shows a driver dyad added to move links 2, 3, and 4 through the three precision points. You may read the diskfile FIG5-9 into program FOURBAR to see the motions of the analytically synthesized fourbar solution. The linkage will move through the three positions defined in the problem statement. The file FIG3-8c may also be read into program SIXBAR to see the full motion of the finished sixbar linkage.

5.9   SYNTHESIS FOR A SPECIFIED FIXED PIVOT LOCATION

In Example 3-8 we used graphical synthesis techniques and inversion to create a fourbar linkage for three-position motion generation with prespecified fixed pivot locations. This is a commonly encountered problem as the available locations for fixed pivots in most machines are quite limited. Loerch et al[4] show how we can use these analytical synthesis techniques to find a linkage with specified fixed pivots and three output positions for motion generation. In effect we will now take as our four free choices the x and y coordinates of the two fixed pivots instead of the angles of the links. This approach will lead to a set of nonlinear equations containing transcendental functions of the unknown angles.

Figure 5-10 shows the **WZ** dyad in three positions. Because we want to relate the fixed pivots of vectors **W** and **U** to our precision points, we will place the origin of our global axis system at precision point P_1. A position vector $\mathbf{R}_1$ can then be drawn from the root of vector $\mathbf{W}_1$ to the global origin at P_1, $\mathbf{R}_2$ to P_2, and $\mathbf{R}_3$ to P_3. The vector $-\mathbf{R}_1$ defines the location of the fixed pivot in the plane with respect to the global origin at P_1.

We will subsequently have to repeat this procedure for three positions of vector **U** at the right end of the linkage as we did with the three-position solution in Section 5.8. The procedure is presented here in detail only for the left end of the linkage (vectors **W, Z**). It is left to the reader to substitute **U** for **W** and **S** for **Z** in equations 5.32 to generate the solution for the right side.

We can write the vector loop equation for each precision position:

$$\begin{aligned}
\mathbf{W}_1 + \mathbf{Z}_1 &= \mathbf{R}_1 \\
\mathbf{W}_2 + \mathbf{Z}_2 &= \mathbf{R}_2 \\
\mathbf{W}_3 + \mathbf{Z}_3 &= \mathbf{R}_3
\end{aligned} \tag{5.32a}$$

Substitute the complex number equivalents for the vectors $\mathbf{W}_i$ and $\mathbf{Z}_i$:

$$\begin{aligned}
we^{j\theta} + ze^{j\phi} &= \mathbf{R}_1 \\
we^{j(\theta+\beta_2)} + ze^{j(\phi+\alpha_2)} &= \mathbf{R}_2 \\
we^{j(\theta+\beta_3)} + ze^{j(\phi+\alpha_3)} &= \mathbf{R}_3
\end{aligned} \tag{5.32b}$$

Expand:

$$\begin{aligned}
we^{j\theta} + ze^{j\phi} &= \mathbf{R}_1 \\
we^{j\theta}e^{j\beta_2} + ze^{j\phi}e^{j\alpha_2} &= \mathbf{R}_2 \\
we^{j\theta}e^{j\beta_3} + ze^{j\phi}e^{j\alpha_3} &= \mathbf{R}_3
\end{aligned} \tag{5.32c}$$

Note that:

$$\mathbf{W} = we^{j\theta}; \qquad\qquad \mathbf{Z} = ze^{j\phi} \tag{5.32d}$$

and:

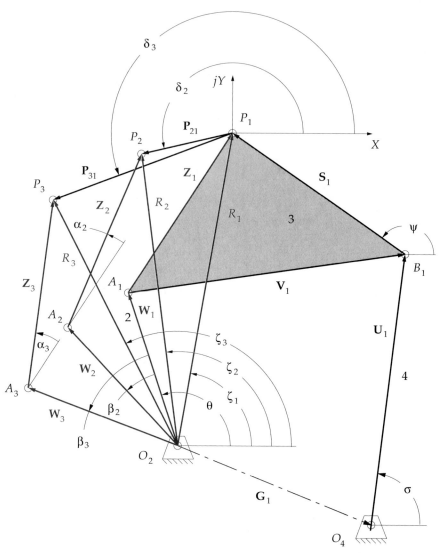

FIGURE 5-10

Three-position synthesis of a linkage with specified fixed pivot locations

$$\mathbf{W} + \mathbf{Z} = \mathbf{R}_1$$
$$\mathbf{W}e^{j\beta_2} + \mathbf{Z}e^{j\alpha_2} = \mathbf{R}_2 \tag{5.32e}$$
$$\mathbf{W}e^{j\beta_3} + \mathbf{Z}e^{j\alpha_3} = \mathbf{R}_3$$

Previously, we chose β_2 and β_3 and solved for the vectors $\mathbf{W}$ and $\mathbf{Z}$. Now we wish to, in effect, specify the x, y components of the fixed pivot O_2 ($-R_{1x}, -R_{1y}$) as our two free choices. This leaves β_2 and β_3 to be solved for. These angles are contained in transcendental expressions in the equations. Note that, if we assumed values for β_2 and β_3 as before, there could only be a solution for $\mathbf{W}$ and $\mathbf{Z}$ if the determinant of the augmented matrix of coefficients of equations 5.32e were equal to zero.

$$\begin{bmatrix} 1 & 1 & \mathbf{R}_1 \\ e^{j\beta_2} & e^{j\alpha_2} & \mathbf{R}_2 \\ e^{j\beta_3} & e^{j\alpha_3} & \mathbf{R}_3 \end{bmatrix} = 0 \qquad (5.33a)$$

Expand this determinant about the first column which contains the present unknowns β_2 and β_3:

$$\left(\mathbf{R}_3 e^{j\alpha_2} - \mathbf{R}_2 e^{j\alpha_3} \right) + e^{j\beta_2} \left(\mathbf{R}_1 e^{j\alpha_3} - \mathbf{R}_3 \right) + e^{j\beta_3} \left(\mathbf{R}_2 - \mathbf{R}_1 e^{j\alpha_2} \right) = 0 \qquad (5.33b)$$

To simplify, let:

$$A = \mathbf{R}_3 e^{j\alpha_2} - \mathbf{R}_2 e^{j\alpha_3}$$
$$B = \mathbf{R}_1 e^{j\alpha_3} - \mathbf{R}_3 \qquad (5.33c)$$
$$C = \mathbf{R}_2 - \mathbf{R}_1 e^{j\alpha_2}$$

then:

$$A + B e^{j\beta_2} + C e^{j\beta_3} = 0 \qquad (5.33d)$$

Equation 5.33d expresses the summation of vectors around a closed loop. Angles β_2 and β_3 are contained within transcendental expressions making their solution cumbersome. The procedure is similar to that used for the analysis of the fourbar linkage in Section 4.5. Substitute the complex number equivalents for all vectors in equation 5.33d. Expand using the Euler identity (Eq. 4.4a). Separate real and imaginary terms to get two simultaneous equations in the two unknowns β_2 and β_3. Square these expressions and add them to eliminate one unknown. Simplify the resulting mess and substitute the tangent half angle identities to get rid of the mixture of sines and cosines. It will ultimately reduce to a quadratic equation in the tangent of half the angle sought. The result is as follows:[1]

$$\beta_3 = 2 \arctan \left(\frac{K_2 \pm \sqrt{K_1^2 + K_2^2 - K_3^2}}{K_1 + K_3} \right)$$

and:

$$\beta_2 = \arccos \left(\frac{A_5 \sin\beta_3 + A_3 \cos\beta_3 + A_6}{A_1} \right) \qquad (5.34a)$$

or:

$$\beta_2 = \arcsin \left(\frac{A_3 \sin\beta_3 + A_2 \cos\beta_3 + A_4}{A_1} \right)$$

where:

---

[1] Note also that a two-argument arctangent function must be used to obtain the proper quadrant for angle β_3. The arccosine function returns angles in the first and second quadrants only. The arcsine function returns angles in the first and fourth quadrants only. By computing both results for β_2 one can determine which quadrant the true angle is in.

$$K_1 = A_2A_4 + A_3A_6$$
$$K_2 = A_3A_4 + A_5A_6$$
$$K_3 = \frac{A_1^2 - A_2^2 - A_3^2 - A_4^2 - A_6^2}{2}$$
(5.34b)

and:

$$A_1 = -C_3{}^2 - C_4{}^2; \qquad\qquad A_2 = C_3C_6 - C_4C_5$$
$$A_3 = -C_4C_6 - C_3C_5; \qquad\qquad A_4 = C_2C_3 + C_1C_4$$ (5.34c)
$$A_5 = C_4C_5 - C_3C_6; \qquad\qquad A_6 = C_1C_3 - C_2C_4$$

$$C_1 = R_3\cos(\alpha_2 + \zeta_3) - R_2\cos(\alpha_3 + \zeta_2)$$
$$C_2 = R_3\sin(\alpha_2 + \zeta_3) - R_2\sin(\alpha_3 + \zeta_2)$$
$$C_3 = R_1\cos(\alpha_3 + \zeta_1) - R_3\cos\zeta_3$$
$$C_4 = -R_1\sin(\alpha_3 + \zeta_1) + R_3\sin\zeta_3$$ (5.34d)
$$C_5 = R_1\cos(\alpha_2 + \zeta_1) - R_2\cos\zeta_2$$
$$C_6 = -R_1\sin(\alpha_2 + \zeta_1) + R_2\sin\zeta_2$$

The ten variables in these equations are: α_2, α_3, β_2, β_3, ζ_1, ζ_2, ζ_3, R_1, R_2, and R_3. The constants C_1 to C_6 are defined in terms of the eight known variables, R_1, R_2, R_3, ζ_1, ζ_2, and ζ_3 (which are the magnitudes and angles of position vectors $\mathbf{R}_1$, $\mathbf{R}_2$, and $\mathbf{R}_3$) and the angles α_2 and α_3 which define the change in angle of the coupler. See Figure 5-10 for depictions of these variables.

Note in equation 5.34a that there are two solutions for each angle (just as there were to the position analysis of the fourbar linkage in Section 4.5 and Figure 4-8). One solution in this case will be a trivial one wherein $\beta_2 = \alpha_2$ and $\beta_3 = \alpha_3$. The nontrivial solution is the one desired.

This procedure is then repeated, solving equations 5.34 for the right-hand end of the linkage using the desired location of fixed pivot O_4 to calculate the necessary angles γ_2 and γ_3 for link 4.

We have now reduced the problem to that of three-position synthesis without specified pivots as described in Section 5.7 and Example 5-2. In effect we have found the particular values of β_2, β_3, γ_2, and γ_3 which correspond to the solution that uses the desired fixed pivots. The remaining task is to solve for the values of W_x, W_y, Z_x, Z_y using equations 5.25 through 5.31.

✎ EXAMPLE 5-3

Three-Position Analytical Synthesis with Specified Fixed Pivots.

Problem: Design a fourbar linkage to move the line AP shown from position A_1P_1 to A_2P_2 and then to position A_3P_3 using fixed pivots O_2 and O_4 in the locations specified.

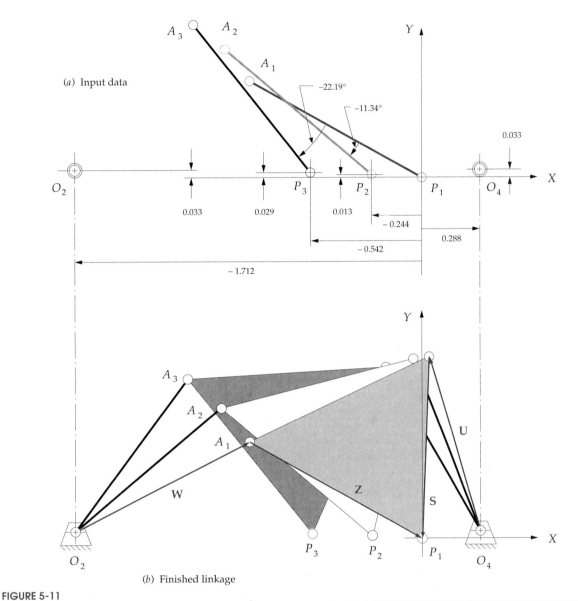

FIGURE 5-11

Three-position synthesis example for specified fixed pivots

Solution: (see Figure 5-11)

1   Draw the link AP in its three desired positions, A_1P_1, A_2P_2, and A_3P_3 to scale in the plane as shown in Figure 5-11. The three positions are defined with respect to a global origin positioned at the first precision point P_1. The given data are specified in parts 2 to 4 below.

2   The position difference vectors between precision points are:

$$P_{21x} = -0.244, \qquad P_{21y} = 0.013 \qquad\qquad P_{31x} = -0.542 \qquad P_{31y} = 0.029$$

3   The angle changes of the coupler between precision points are:

$$\alpha_2 = -11.34° \qquad\qquad \alpha_3 = -22.19°$$

4   The assumed free choices are the desired fixed pivot locations.

$$O_{2x} = -1.712 \qquad\qquad O_{2y} = 0.033 \qquad\qquad O_{4x} = 0.288 \qquad\qquad O_{4y} = 0.033$$

5   Solve equations 5.34 twice, once using the O_2 pivot location coordinates and again using the O_4 pivot location coordinates.

For pivot O_2:

$$C_1 = -0.205 \qquad C_2 = 0.3390 \qquad C_3 = 0.4028$$
$$C_4 = 0.6731 \qquad C_5 = 0.2041 \qquad C_6 = 0.3490$$
$$A_1 = -0.6152 \qquad A_2 = 0.0032 \qquad A_3 = -0.3171$$
$$A_4 = -0.0017 \qquad A_5 = -0.0032 \qquad A_6 = -0.3108$$
$$K_1 = 0.0986 \qquad K_2 = 0.0015 \qquad K_3 = 0.0907$$

The values found for the link angles to match this choice of fixed pivot location O_2 are:

$$\beta_2 = 11.96° \qquad\qquad \beta_3 = 23.96°$$

For pivot O_4:

$$C_1 = -0.3144 \qquad C_2 = -0.0231 \qquad C_3 = 0.5508$$
$$C_4 = -0.0822 \qquad C_5 = 0.2431 \qquad C_6 =- 0.0443$$
$$A_1 = -0.3102 \qquad A_2 = -0.0044 \qquad A_3 = -0.1376$$
$$A_4 = 0.0131 \qquad A_5 = 0.0044 \qquad A_6 = -0.1751$$
$$K_1 = 0.0240 \qquad K_2 = -0.0026 \qquad K_3 = 0.0232$$

The values found for the link angles to match this choice of fixed pivot location O_4 are:

$$\gamma_2 = 2.78° \qquad\qquad \gamma_3 = 9.96°$$

6   At this stage, the problem has been reduced to the same one as in the previous section; i.e., find the linkage given the free choices of the above angles β_2, β_3, γ_2, γ_3, using equations 5.25 through 5.31. The data needed for the remaining calculations are those given in steps 2, 3, and 5 of this example, namely:

for dyad 1:

$$P_{21x} \qquad P_{21y} \qquad P_{31x} \qquad P_{31y} \qquad \alpha_2 \qquad \alpha_3 \qquad \beta_2 \qquad \beta_3$$

for dyad 2:

$$P_{21x} \qquad P_{21y} \qquad P_{31x} \qquad P_{31y} \qquad \alpha_2 \qquad \alpha_3 \qquad \gamma_2 \qquad \gamma_3$$

See Example 5-2 and Section 5.7 for the procedure. A matrix solving calculator, program MATRIX, or program FOURBAR will solve this and compute the coordinates of the link vectors:

$W_x = 0.866$ $W_y = 0.500$ $Z_x = 0.846$ $Z_y = -0.533$

$U_x = -0.253$ $U_y = 0.973$ $S_x = -0.035$ $S_y = -1.006$

7   The link lengths are computed as was done in Example 5-2 and are shown in Table 5-2.

This example can be read into program FOURBAR from diskfile FIG5-11 and animated.

5.10   CENTER-POINT AND CIRCLE-POINT CIRCLES

It would be quite convenient if we could find the loci of all possible solutions to the three-position synthesis problem, as we would then have an overview of the potential locations of the ends of the vectors **W**, **Z**, **U**, and **S**. Loerch[5] shows that by holding one of the free choices (say β_2) at an arbitrary value, and then solving equations 5.25 and 5.26 while iterating the other free choice (β_3) through all possible values from 0 to 2π, a circle will be generated. This circle is the locus of all possible locations of the root of vector **W** (for the particular value of β_2 used). The root of the vector **W** is the location of the fixed pivot or *center* O_2. Thus, this circle is called a **center-point** circle. The vector **N** in Figure 5-12 defines points on the *center-point* circle with respect to the global coordinate system which is placed at precision point P_1 for convenience.

TABLE 5-2   **Fourbar Linkage Parameters for Example 5-3 Solution**

Link #	Length (in)
1	2
2	1
3	1
4	1.01
Coupler Pt.=	1 at −60.73 degrees
Open/Crossed =	OPEN
Start Alpha2 =	0 rad/sec ^{2}
Start Omega2 =	1 rad/sec
Start Theta2 =	30 degrees
Final Theta2 =	54 degrees
Delta Theta2 =	12 degrees

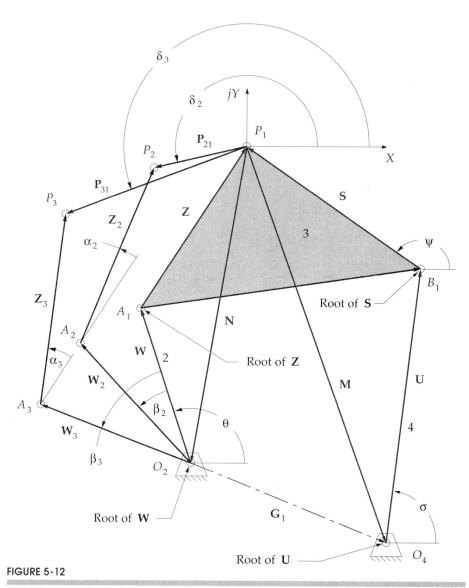

FIGURE 5-12

Definition of vectors to define center-point and circle-point circles

If the same thing is done for vector $\mathbf{Z}$, holding α_2 constant at some arbitrary value and iterating α_3 from 0 to 2π, another circle will be generated. This circle is the locus of all possible locations of the root of vector $\mathbf{Z}$ for the chosen value of α_2. Because the root of vector $\mathbf{Z}$ is joined to the tip of vector $\mathbf{W}$ and $\mathbf{W}$'s tip describes a circle about pivot O_2 in the finished linkage, this locus is called the ***circle-point* circle**. Vector $(-\mathbf{Z})$ defines points on the *circle-point* circle with respect to the global coordinate system.

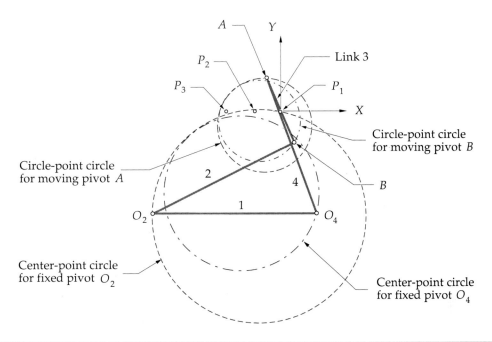

FIGURE 5-13

Circle-point and center-point circles and one possible linkage which will reach the precision points

The x,y components of vectors **W** and **Z** are defined by equations 5.25 and 5.26. Negating the x,y components of **Z** will give the coordinates of points on the circle-point circle for any assumed value of α_2 as angle α_3 is iterated from 0 to 2π. The x,y components of $\mathbf{N} = -\mathbf{Z} - \mathbf{W}$, define points on the O_2 center-point circle for any assumed value of β_2 as β_3 is iterated through 0 to 2π. Vector **W** is calculated using angles β_2 and β_3, and vector **Z** using angles α_2 and α_3, both from equations 5.25 and 5.26.

For the right-hand dyad, there will also be separate center-point circles and circle-point circles. The x,y components of $\mathbf{M} = -\mathbf{S} - \mathbf{U}$, define points on the O_4 center-point circle for any assumed value of γ_2 as γ_3 is iterated through 0 to 2π. (See Figure 5-12 and also Figure 5-8.) Negating the x,y components of **S** will give the coordinates of points on the circle-point circle for any assumed value of ψ_2 as ψ_3 is iterated through 0 to 2π. Vector **U** is calculated using angles γ_2 and γ_3, and vector **S** using angles ψ_2 and ψ_3, both from equations 5.30 and 5.31.

Note that there is still an infinity of solutions because we are choosing the value of one angle arbitrarily. Thus there will be an **infinite number of sets of center-point and circle-point circles**. A computer program which calculates and draws these circular loci can be of help in choosing a linkage design which has pivots in convenient locations. Program FOURBAR, provided with this text, will calculate the solutions to the analytical synthesis equations derived in this section, for user-selected values of all the free choices needed for three-position synthesis, both with and without specification of fixed pivot locations and will also draw the *center-point* and *circle-point* circles for the three-position case. The computer programs FOURBAR, FIVEBAR, and SIXBAR and their use are discussed in detail in Chapter 8.

Figure 5-13 shows the circle-point and center-point circles for the Chebychev example linkage in program FOURBAR for choices of $\beta_2 = 26°$, $\alpha_2 = 97.41°$, $\alpha_3 = 158.18°$ for the left dyad and $\gamma_2 = 36°$, $\psi_2 = 97.41°$, $\psi_3 = 158.18°$ for the right dyad. The program displays these circles in different colors and also increments the plot title to identify each one as it is drawn, to aid in distinguishing them. In this example the two larger circles are the center-point circles which define the loci of possible fixed pivot locations O_2 and O_4. The smaller two circles define the loci of possible moving pivot locations I_{23} and I_{34}. Note that the coordinate system has its origin at the reference precision point, in this case P_1, from which we measured all parameters used in the analysis. These circles define the pivot loci of all possible linkages which will reach the three precision points P_1, P_2 and P_3 that were specified for particular choices of angles β_2, α_2, γ_2, and ψ_2. An example linkage is drawn on the diagram to illustrate one possible solution. You may read the datafile FIG5-13 into program FOURBAR, calculate the center-point and circle-point circles from the INPUT/SYNTHE-SIZE menu choices, and then view them and the linkage motions dynamically from the PLOT menu. See Chapter 8 for more detailed information on program operation.

5.11   FOUR- AND FIVE-POSITION ANALYTICAL SYNTHESIS

The same techniques derived above for two- and three-position synthesis can be extended to four and five positions by writing more vector loop equations, one for each precision point. To facilitate this we will now put the vector loop equations in a more general form, applicable to any number of precision positions. Figure 5-8 will still serve to illustrate the notation for the general solution. The angles α_2, α_3, β_2, β_3, γ_2, and γ_3 will now be designated as α_k, β_k, and γ_k, $k = 2$ to n, where k represents the precision position and $n = 2, 3, 4$, or 5, represents the total number of positions to be solved for. The vector loop general equation set then becomes:

$$\mathbf{W}_k + \mathbf{Z}_k - \mathbf{P}_{k1} - \mathbf{Z}_1 - \mathbf{W}_1 = 0, \qquad k = 2 \text{ to } n \qquad (5.35a)$$

Which, after substituting the complex number forms and simplifying becomes:

$$we^{j\theta}\left(e^{j\beta_k} - 1\right) + ze^{j\phi}\left(e^{j\alpha_k} - 1\right) = p_{k1}e^{j\delta_k}, \qquad k = 2 \text{ to } n \qquad (5.35b)$$

This can be put in a more compact form by substituting vector notation for those terms to which it applies, let:

$$\mathbf{W} = we^{j\theta}; \qquad \mathbf{Z} = ze^{j\phi}; \qquad \mathbf{P}_{k1} = p_{k1}e^{j\delta_k} \qquad (5.35c)$$

then:

$$\mathbf{W}\left(e^{j\beta_k} - 1\right) + \mathbf{Z}\left(e^{j\alpha_k} - 1\right) = \mathbf{P}_{k1}e^{j\delta_k}, \qquad k = 2 \text{ to } n \qquad (5.35d)$$

Equation 5.35d is called the *standard form equation* by Erdman and Sandor.[6] By substituting the values of α_k, β_k, and δ_k, in equation 5.35d for all the precision positions desired, the requisite set of simultaneous equations can be written for the left dyad of the linkage. The standard form equation applies to the right-hand dyad **US** as well, with appropriate changes to variable names as required.

TABLE 5-3 Number of Variables and Free Choices for Analytical Precision Point Motion and Timed Path Synthesis.[7]

No. of Positions (n)	No. of Scalar Variables	No. of Scalar Equations	No. of Prescribed Variables	No. of Free Choices	No. of Available Solutions
2	8	2	3	3	∞^3
3	12	4	6	2	∞^2
4	16	6	9	1	∞^1
5	20	8	12	0	Finite

$$\mathbf{U}\left(e^{j\beta_k}-1\right)+\mathbf{S}\left(e^{j\alpha_k}-1\right)=\mathbf{P}_{k1}e^{j\delta_k}, \qquad k=2 \text{ to } n \qquad (5.35e)$$

The number of resulting equations, variables and free choices for each value of n is shown in Table 5-3 (after Erdman and Sandor). They provide solutions for the four- and five-position problems in reference 6. The circle-point and center-point circles of the three-position problem become higher order curves, called **Burmester curves**, in the four-position problem. Erdman's commercially available computer program LINCAGES[8] solves the **four position problem** in an interactive way, allowing the user to select center and circle pivot locations on their Burmester curve loci, which are drawn on the graphics screen of the computer.

5.12 ANALYTICAL SYNTHESIS OF A PATH GENERATOR WITH PRESCRIBED TIMING

The approach derived above for motion generation synthesis is also applicable to the case of **path generation with prescribed timing**. In path generation, the precision points are to be reached, but the angle of a line on the coupler is not of concern. Instead, the timing at which the coupler reaches the precision point is specified in terms of input rocker angle β_2. In the three-position motion generation problem we specified the angles α_2 and α_3 of vector Z in order to control the angle of the coupler. Here we instead want to specify angles β_2 and β_3 of the input rocker, to define the timing. Before, the free choices were β_2 and β_3. Now they will be α_2 and α_3. In either case, all four angles are either specified or assumed as free choices and the solution is identical. Figure 5-8 and equations 5.25, 5.26, 5.30, and 5.31 apply to this case as well. This case can be extended to as many as five precision points as shown in Table 5-3.

5.13 ANALYTICAL SYNTHESIS OF A FOURBAR FUNCTION GENERATOR

A similar process to that used for the synthesis of path generation with prescribed timing can be applied to the problem of function generation. In this case we

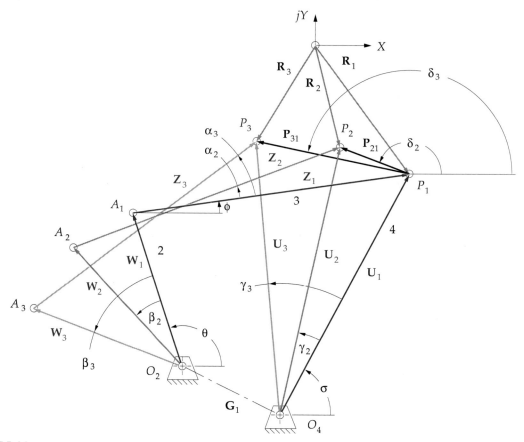

FIGURE 5-14

Fourbar function generator

do not care about motion of the coupler at all. In a fourbar function generator, the coupler exists only to **couple** the input link to the output link. Figure 5-14 shows a fourbar linkage in three positions. Note that the coupler, link 3, is merely a line from point A to point P. Point P can be thought of as a coupler point which happens to coincide with the pin joint between links 3 and 4. As such it will have simple arc motion, pivoting about O_4, rather than, for example, the higher-order path motion of the coupler point P_1 in Figure 5-12.

Our **function generator** uses *link 2 as the input link and takes the output from link 4*. The "**function**" generated is the **relationship between the angles of link 2 and link 4** for the specified three-position positions, P_1, P_2, and P_3. These are located in the plane with respect to an arbitrary global coordinate system by position vectors $\mathbf{R}_1$, $\mathbf{R}_2$, and $\mathbf{R}_3$. The function is:

$$\gamma_k = f(\beta_k), \qquad k = 1, 2, \cdots, n; \qquad n \leq 7 \qquad (5.36)$$

This is *not* a continuous function. The relationship holds only for the discrete points (k) specified.

To synthesize the lengths of the links needed to satisfy equation 5.36, we will write vector loop equations around the linkage in pairs of positions, as was done for the previous examples. However, we now wish to include both link 2 and link 4 in the loop, since link 4 is the output. See Figure 5-14.

$$\mathbf{W}_2 + \mathbf{Z}_2 - \mathbf{U}_2 + \mathbf{U}_1 - \mathbf{Z}_1 - \mathbf{W}_1 = 0$$

(5.37a)

$$\mathbf{W}_3 + \mathbf{Z}_3 - \mathbf{U}_3 + \mathbf{U}_1 - \mathbf{Z}_1 - \mathbf{W}_1 = 0$$

rearranging:

$$\mathbf{W}_2 + \mathbf{Z}_2 - \mathbf{Z}_1 - \mathbf{W}_1 = \mathbf{U}_2 - \mathbf{U}_1$$

(5.37b)

$$\mathbf{W}_3 + \mathbf{Z}_3 - \mathbf{Z}_1 - \mathbf{W}_1 = \mathbf{U}_3 - \mathbf{U}_1$$

but,

$$\mathbf{P}_{21} = \mathbf{U}_2 - \mathbf{U}_1$$
$$\mathbf{P}_{31} = \mathbf{U}_3 - \mathbf{U}_1$$

(5.37c)

substituting:

$$\mathbf{W}_2 + \mathbf{Z}_2 - \mathbf{Z}_1 - \mathbf{W}_1 = \mathbf{P}_{21}$$

(5.37d)

$$\mathbf{W}_3 + \mathbf{Z}_3 - \mathbf{Z}_1 - \mathbf{W}_1 = \mathbf{P}_{31}$$

$$we^{j(\theta+\beta_2)} + ze^{j(\phi+\alpha_2)} - ze^{j\phi} - we^{j\theta} = p_{21}e^{j\delta_2}$$

(5.37e)

$$we^{j(\theta+\beta_3)} + ze^{j(\phi+\alpha_3)} - ze^{j\phi} - we^{j\theta} = p_{31}e^{j\delta_3}$$

Note that equations 5.37d and 5.37e are identical to equations 5.19 and 5.20 derived for the three-position motion generation case and can also be put into Erdman's *standard form*[6] of equation 5.35 for the *n*-position case. The twelve variables in equation 5.37e are the same as those in equation 5.20: w, θ, β_2, β_3, z, ϕ, α_2, α_3, p_{21}, p_{31}, δ_2, and δ_3.

For the three-position function generation case the solution procedure then can be the same as that described by equations 5.20 through 5.27 for the motion synthesis problem. In other words, the solution equations are the same for **all three types** of kinematic synthesis, *function generation, motion generation,* and *path generation with prescribed timing*. This is why Erdman and Sandor called equation 5.35 the *standard form equation*. To develop the data for the function generation solution, expand equation 5.37b:

$$we^{j(\theta+\beta_2)} + ze^{j(\phi+\alpha_2)} - ze^{j\phi} - we^{j\theta} = ue^{j(\sigma+\gamma_2)} - ue^{j\sigma}$$

(5.37f)

$$we^{j(\theta+\beta_3)} + ze^{j(\phi+\alpha_3)} - ze^{j\phi} - we^{j\theta} = ue^{j(\sigma+\gamma_3)} - ue^{j\sigma}$$

There are also **twelve variables** in equation 5.37f: w, θ, z, ϕ, α_2, α_3, β_2, β_3, u, σ, γ_2, and γ_3. We can solve for any four. Four angles, β_2, β_3, γ_2, and γ_3 are specified from the function to be generated in equation 5.36. This leaves **four free choices**. In the function generation problem it is often convenient to define the length of the output rocker, u, and its initial angle σ to suit the package constraints. Thus, selecting the components u and σ, of vector U_1 can provide two convenient free choices of the four required.

With u, σ, γ_2, and γ_3 known, U_2 and U_3 can be found. Vectors P_{21} and P_{31} can then be found from equation 5.37c. Six of the unknowns in equation 5.37e are then defined, namely, β_2, β_3, p_{21}, p_{31}, δ_2, and δ_3. Of the remaining six (w, θ, z, ϕ, α_2, α_3), we must assume values for two more as free choices in order to solve for the remaining four. We will assume values (free choices) for the two angles, α_2 and α_3, (as was done for path generation with prescribed timing) and solve equations 5.37e for the components of W and Z (w, θ, z, ϕ). We have now reduced the problem to that of Section 5.7 and Example 5-2. See equations 5.20 through 5.27 for the solution.

Having chosen vector U_1 (u, σ) as a free choice in this case, we only have to solve for one dyad, WZ. Though we arbitrarily choose the length of vector U_1, the resulting function generator linkage can be scaled up or down to suit packaging constraints without affecting the input/output relation defined in equation 5.36, because it is a function of angles only. This fact is not true for the motion or path generation cases, as scaling them will change the absolute coordinates of the path or motion output precision points which were specified in the problem statement.

Table 5-4 shows the relationships between number of positions, variables, free choices and solutions for the function generation case. Note that up to seven angular output positions can be solved for.

TABLE 5-4   **Number of Variables and Free Choices for Function Generation Synthesis** [7]

No. of Positions (n)	No. of Scalar Variables	No. of Scalar Equations	No. of Defined Variables	No. of Free Choices	No. of Available Solutions
2	8	2	1	5	∞^5
3	12	4	4	4	∞^4
4	16	6	7	3	∞^3
5	20	8	10	2	∞^2
6	24	10	13	1	∞^1
7	28	12	16	0	Finite

5.14 REFERENCES

1 **Sandor**, G. N., *A General Complex Number Method for Plane Kinematic Synthesis with Applications*, Doctoral Dissertation, Columbia University, N.Y., 1959, University Microfilms, Ann Arbor, Michigan.

2 **Erdman**, A. G., "Three and Four Precision Point Kinematic Synthesis of Planar Linkages," *Mechanism and Machine Theory*, Pergamon Press, London, vol. 16, pp. 227-45.

3 **Kaufman**, R. E., "Mechanism Design by Computer," *Machine Design*, October 1978, pp. 94-100.

4 **Loerch**, R. J., A. G. Erdman, G. N. Sandor, and A. Mihda, "Synthesis of Four Bar Linkages with Specified Ground Pivots," Proc. 4th Appl. Mechanisms Conf. 1975, pp. 10.1-10.6, Oklahoma State Univ., Stillwater, OK.

5 **Loerch**, R. J., A. Erdman, and G. Sandor, "On the Existence of Circle-Point and Center-Point Circles for Three Precision Point Dyad Synthesis," J. Mech. Des., October 1979, pp. 554-562.

6 **Erdman**, A. G., and **Sandor**, G. N., *Mechanism Design: Analysis and Synthesis*, vol. 1 and *Advanced Mechanism Design, Analysis and Synthesis*, vol. 2, Prentice Hall, Englewood Cliffs, NJ, 1984.

7 **Jennings**, A., *Matrix Computation for Engineers and Scientists*, John Wiley and Sons, New York, 1977.

8 **Erdman**, A. G., and J. E. Gustafson, "LINCAGES: Linkage Interactive Computer Analysis and Graphically Enhanced Synthesis Package," ASME Paper No. 77-DTC-5, 1977.

5.15 PROBLEMS

Note that all three-position synthesis problems below may be done using a matrix solving calculator, program MATRIX, or program FOURBAR. Two-position synthesis problems can be done with a four-function calculator.

5-1 Redo problem 3-3 using the analytical methods of this chapter.

5-2 Redo problem 3-4 using the analytical methods of this chapter.

5-3 Redo problem 3-5 using the analytical methods of this chapter.

5-4 Redo problem 3-6 using the analytical methods of this chapter.

5-5 See project P3-8. Define three positions of the boat and analytically synthesize a linkage to move through them.

5-6 See project P3-20. Define three positions of the dumpster and analytically synthesize a linkage to move through them. The fixed pivots must be located on the existing truck.

5-7 See project P3-7. Define three positions of the computer monitor and analytically synthesize a linkage to move through them. The fixed pivots must be located on the floor or wall.

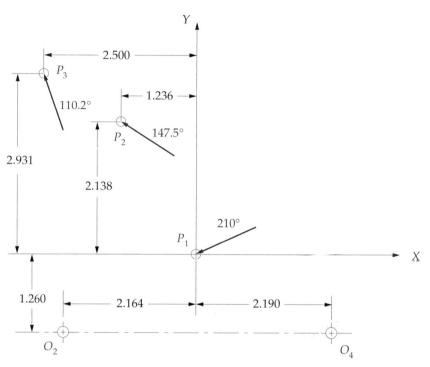

FIGURE P5-1

Data for problems 5-8 to 5-11

See Figure P5-1

*5-8 Design a linkage to carry the arrow through the two positions P_1 and P_2 at the angles shown in the figure. Use analytical synthesis without regard for the fixed pivots shown. Hint: Try the free choice values $z = 1.075$, $\phi = 204.4°$, $\beta_2 = -27°$; $s = 1.24$, $\psi = 74°$, $\gamma_2 = -40°$

5-9 Design a linkage to carry the arrow through the two positions P_2 and P_3 at the angles shown in the figure. Use analytical synthesis without regard for the fixed pivots shown. Hint: first try a rough graphical solution to create realistic values for free choices.

5-10 Design a linkage to carry the arrow through the three positions P_1, P_2 and P_3 at the angles shown in the figure. Use analytical synthesis without regard for the fixed pivots shown. Hint: Try the free choice values $\beta_2 = 30°$, $\beta_3 = 60°$; $\gamma_2 = -10°$, $\gamma_3 = 25°$.

*5-11 Design a linkage to carry the arrow through the three positions P_1, P_2 and P_3 at the angles shown in the figure. Use analytical synthesis and design it for the fixed pivots shown.

---

* Answers in Appendix E

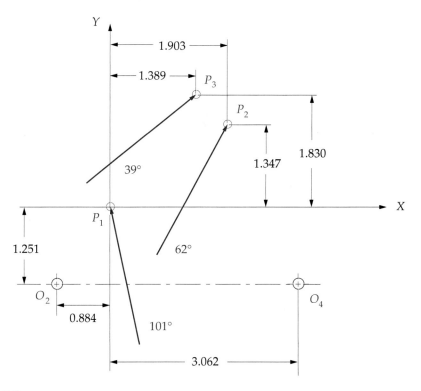

FIGURE P5-2

Data for problems 5-12 to 5-15

See Figure P5-2

5-12 Design a linkage to carry the arrow through the two positions P_1 and P_2 at the angles shown in the figure. Use analytical synthesis without regard for the fixed pivots shown. Hint: Try the free choice values $z = 2$, $\phi = 150°$, $\beta_2 = 30°$; $s = 3$, $\psi = -50°$, $\gamma_2 = 40°$.

5-13 Design a linkage to carry the arrow through the two positions P_2 and P_3 at the angles shown in the figure. Use analytical synthesis without regard for the fixed pivots shown. Hint: First try a rough graphical solution to create realistic values for free choices.

5-14 Design a linkage to carry the arrow through the three positions P_1, P_2, and P_3 at the angles shown in the figure. Use analytical synthesis without regard for the fixed pivots shown.

*5-15 Design a linkage to carry the arrow through the three positions P_1, P_2, and P_3 at the angles shown in the figure. Use analytical synthesis and design it for the fixed pivots shown.

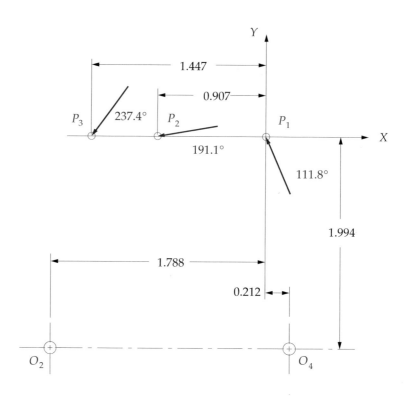

FIGURE P5-3

Data for problems 5-16 to 5-20

See Figure P5-3

5-16 Design a linkage to carry the arrow through the two positions P_1 and P_2 at the angles shown in the figure. Use analytical synthesis without regard for the fixed pivots shown.

5-17 Design a linkage to carry the arrow through the two positions P_2 and P_3 at the angles shown in the figure. Use analytical synthesis without regard for the fixed pivots shown.

5-18 Design a linkage to carry the arrow through the three positions P_1, P_2, and P_3 at the angles shown in the figure. Use analytical synthesis without regard for the fixed pivots shown.

*5-19 Design a linkage to carry the arrow through the three positions P_1, P_2, and P_3 at the angles shown in the figure. Use analytical synthesis and design it for the fixed pivots shown.

5-20 Generate the circle-point and center-point circles for the problems in the figure using program FOURBAR.

Chapter **6**

Chapter **6**

VELOCITY ANALYSIS

The faster I go, the behinder I get
ANON. PENN. DUTCH

6.0   INTRODUCTION

Once a position analysis is done, the next step is to determine the velocities of all links and points of interest in the mechanism. We need to know the velocities in our mechanism or machine, both to calculate the stored kinetic energy from $mV^2/2$, and also as a step on the way to the determination of the link's accelerations which are needed for the dynamic force calculations. Many methods and approaches exist to find velocities in mechanisms. We will examine only a few of these methods here. We will first develop manual graphical methods, which are often useful as a check on the more complete and accurate analytical solution. We will also investigate the properties of the instant center of velocity which can shed much light on a mechanism's velocity behavior with very little effort. Finally, we will derive the analytical solution for the fourbar and inverted slider-crank as examples of the general vector loop equation solution to velocity analysis problems. From these calculations we will be able to establish some indices of merit to judge our designs while they are still on the drawing board (or in the computer).

6.1   DEFINITION OF VELOCITY

Velocity is defined as *the rate of change of position with respect to time*. Position (**R**) is a vector quantity and so is velocity. Velocity can be **angular** or **linear**. **Angular velocity** will be denoted as ω and **linear velocity** as **V**.

$$\omega = \frac{d\theta}{dt}; \qquad\qquad \mathbf{V} = \frac{d\mathbf{R}}{dt} \qquad\qquad (6.1)$$

Figure 6-1 shows a link *PA* in pure rotation, pivoted at point *A* in the *xy* plane. Its position is defined by the position vector $\mathbf{R}_{PA}$. We are interested in the

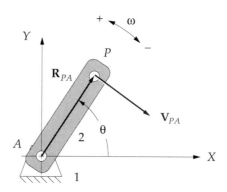

FIGURE 6-1

A link in pure rotation

velocity of point P when the link is subjected to an angular velocity ω. If we represent the position vector $\mathbf{R}_{PA}$ as a complex number in polar form,

$$\mathbf{R}_{PA} = pe^{j\theta} \tag{6.2}$$

where p is the scalar length of the vector. We can easily differentiate it to obtain:

$$\mathbf{V}_{PA} = \frac{d\mathbf{R}_{PA}}{dt} = p\, je^{j\theta}\frac{d\theta}{dt} = p\omega\, je^{j\theta} \tag{6.3}$$

Compare the right side of equation 6.3 to the right side of equation 6.2. Note that as a result of the differentiation, the velocity expression has been multiplied by the (constant) complex operator j. This causes a rotation of this velocity vector through 90 degrees with respect to the original position vector. (See also Figure 4-5b.) This 90° rotation is positive, or counterclockwise. However, the velocity expression is also multiplied by ω, which may be either positive or negative. As a result, the velocity vector will be **rotated 90 degrees** from the angle θ of the position vector **in a direction dictated by the sign of** ω. This is just mathematical verification of what you already knew, namely that *velocity is always in a direction perpendicular to the radius of rotation and is tangent to the path of motion* as shown in Figure 6-1.

Substituting the Euler identity (Eq. 4.4a) into equation 6.3 gives us the real and imaginary (or x and y) components of the velocity vector.

$$\mathbf{V}_{PA} = p\omega\, j(\cos\theta + j\sin\theta) = p\omega(-\sin\theta + j\cos\theta) \tag{6.4}$$

Note that the sine and cosine terms have swapped positions between the real and imaginary terms, due to multiplying by the j coefficient. This is evidence of the 90 degree rotation of the velocity vector versus the position vector. The former x component has become the y component, and the former y component has become a minus x component. Study Figure 4-5b to review why this is so.

The velocity $\mathbf{V}_{PA}$ in Figure 6-1 can be referred to as an **absolute velocity** since it is referenced to A, which is the origin of the global coordinate axes in that system. As such, we could have referred to it as $\mathbf{V}_{P}$, with the absence of the second subscript

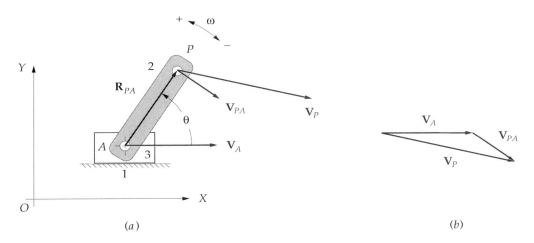

FIGURE 6-2

Velocity difference

implying reference to the global coordinate system. Figure 6-2a shows a different and slightly more complicated system in which the pivot A is no longer stationary. It has a known linear velocity $\mathbf{V}_A$ as part of the translating carriage, link 3. If ω is unchanged, the velocity of point P versus A will be the same as before, but $\mathbf{V}_{PA}$ can no longer be considered an absolute velocity. It is now a **velocity difference** and **must** carry the second subscript as $\mathbf{V}_{PA}$. The absolute velocity $\mathbf{V}_P$ must now be found from the **velocity difference equation** whose graphical solution is shown in Figure 6-2b:

$$\mathbf{V}_{PA} = \mathbf{V}_P - \mathbf{V}_A \qquad (6.5a)$$

rearranging:

$$\mathbf{V}_P = \mathbf{V}_A + \mathbf{V}_{PA} \qquad (6.5b)$$

Note the similarity of equation 6.5 to the **position difference equation** 4.1.

Figure 6-3 shows two independent bodies P and A, which could be two automobiles, moving in the same plane. If their independent velocities $\mathbf{V}_P$ and $\mathbf{V}_A$ are known, their **relative velocity** $\mathbf{V}_{PA}$ can be found from equation 6.5 arranged algebraically as:

$$\mathbf{V}_{PA} = \mathbf{V}_P - \mathbf{V}_A \qquad (6.6)$$

The graphical solution to this equation is shown in Figure 6-3b. Note that it is similar to Figure 6-2b except for a different vector being the resultant.

As we did for position analysis, we give these two cases different names despite the fact that the same equation applies. Repeating the definition from Section 4.2, modified to refer to velocity:

CASE 1:   *Two points in the same body* => *velocity difference*

CASE 2:   *Two points in different bodies* => *relative velocity*

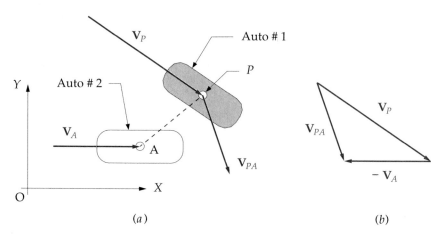

FIGURE 6-3

Relative velocity

We will find use for this semantic distinction both when we analyze linkage velocities and the velocity of slip later in this chapter.

6.2   GRAPHICAL VELOCITY ANALYSIS

Before programmable calculators and computers became universally available to engineers, graphical methods were the only practical way to solve these velocity analysis problems. With some practice and with proper tools such as a drafting machine or *CAD* package, one can fairly rapidly solve for the velocities of particular points in a mechanism for any one input position by drawing vector diagrams. However, it is a tedious process if velocities for many positions of the mechanism are to be found, because each new position requires a completely new set of vector diagrams be drawn. Very little of the work done to solve for the velocities at position 1 carries over to position 2, etc. Nevertheless, this method still has more than historical value as it can provide a quick check on the results from a computer program solution. Such a check needs only be done for a few positions to prove the validity of the program.

To solve any velocity analysis problem graphically, we need only two equations, 6.5 and 6.7 (which is merely the scalar form of Eq. 6.3):

$$|\mathbf{V}| = v = r\omega \qquad (6.7)$$

Note that the scalar equation 6.7 defines only the **magnitude** (v) of the velocity of any point on a body which is in pure rotation. In a graphical CASE 1 analysis, the **direction** of the vector due to the rotation component must be understood from equation 6.3 to be perpendicular to the radius of rotation. Thus, if the center of rotation is known, the direction of the velocity component due to that rotation is known and its sense will be consistent with the angular velocity ω of the body.

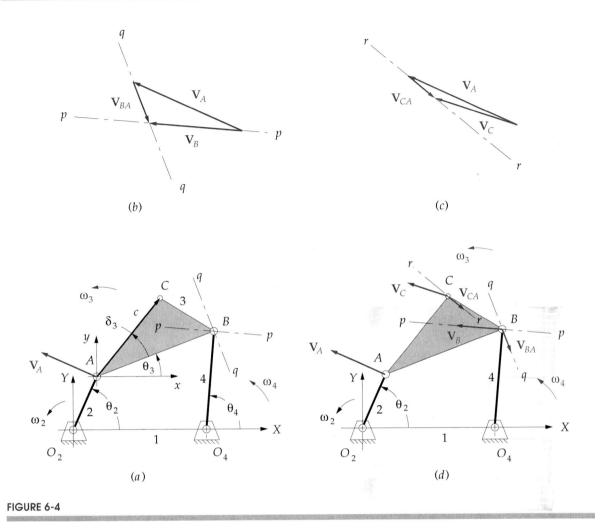

FIGURE 6-4

Graphical solution for velocities in a pin-jointed linkage

Figure 6-4 shows a fourbar linkage in a particular position. We wish to solve for the angular velocities of links 3 and 4 (ω_3, ω_4) and the linear velocities of points A, B and C ($\mathbf{V}_A$, $\mathbf{V}_B$, $\mathbf{V}_C$). Point C represents any general point of interest. Perhaps C is a coupler point. The solution method is valid for any point on any link. To solve this problem we need to know the *lengths of all the links*, the *angular positions of all the links*, and the *instantaneous input velocity of any one driving link or driving point*. Assuming we have designed this linkage, we will know or can measure the link lengths. We must also first do a **complete position analysis** to find the link angles θ_3 and θ_4 given the input link's position θ_2. This can be done by any of the methods in Chapter 4. In general we must solve these problems in stages, first for link positions, then for velocities, and finally for accelerations. For the following example, we will assume that a complete position analysis has been done and that the input is to link 2 with known θ_2 and ω_2 for this one "freeze frame" position of the moving linkage.

✎ EXAMPLE 6-1

Graphical Velocity Analysis for One Position of Linkage.

Problem: Given θ_2, θ_3, θ_4, ω_2, find ω_3, ω_4, $\mathbf{V}_A$, $\mathbf{V}_B$, $\mathbf{V}_C$ by graphical methods.

Solution: (see Figure 6-4)

1   Start at the end of the linkage about which you have the most information. Calculate the magnitude of the velocity of point A using scalar equation 6.7.

$$v_a = (AO_2)\omega_2 \tag{a}$$

2   Draw the velocity vector $\mathbf{V}_A$ with its length equal to its magnitude v_A at some convenient scale with its root at point A and its direction perpendicular to the radius AO_2. Its sense is the same as that of ω_2 as shown in Figure 6-4a.

3   Move next to a point about which you have some information. Note that the direction of the velocity of point B is predictable since it is pivoting in pure rotation about point O_4. Draw the construction line pp through point B perpendicular to BO_4, to represent the direction of $\mathbf{V}_B$ as shown in Figure 6-4a.

4   Write the velocity difference vector equation 6.5 for point B versus point A.

$$\mathbf{V}_B = \mathbf{V}_A + \mathbf{V}_{BA} \tag{b}$$

We will use point A as the reference point to find $\mathbf{V}_B$ because A is in the same link as B and we have already solved for $\mathbf{V}_A$. Any vector equation can be solved for two unknowns. Each term has two parameters, namely magnitude and direction. There are then potentially six unknowns in this equation, two per term. We must know four of them to solve it. We know both magnitude and direction of $\mathbf{V}_A$ and the direction of $\mathbf{V}_B$. We need to know one more parameter.

5   The term $\mathbf{V}_{BA}$ represents the velocity of B with respect to A. If we assume that the link BA is rigid, then there can be no component of $\mathbf{V}_{BA}$ which is directed along the line BA, because point B cannot move toward or away from point A without shrinking or stretching the rigid link! Therefore the direction of $\mathbf{V}_{BA}$ must be perpendicular to the line BA. Draw construction line qq through point B and perpendicular to BA to represent the direction of $\mathbf{V}_{BA}$, as shown in Figure 6-4a.

6   Now the vector equation can be solved graphically by drawing a vector diagram as shown in Figure 6-4b. Either drafting tools or a CAD package is necessary for this step. First draw velocity vector $\mathbf{V}_A$ carefully to some scale, maintaining its direction. (It is drawn twice size in the figure.) The equation in step 4 says to add $\mathbf{V}_{BA}$ to $\mathbf{V}_A$, so draw a line parallel to line qq across the tip of $\mathbf{V}_A$. The resultant, or left side of the equation, must close the vector diagram, from the tail of the first vector drawn ($\mathbf{V}_A$) to the tip of the last, so draw a line parallel to pp across the tail of $\mathbf{V}_A$. The intersection of these lines parallel to pp and qq defines the lengths of $\mathbf{V}_B$ and $\mathbf{V}_{BA}$. The senses of the vectors are determined from reference to the equation. $\mathbf{V}_A$ was added to $\mathbf{V}_{BA}$, so they must be arranged tip to tail. $\mathbf{V}_B$ is the resultant, so it must be from the tail of the first to the tip of the last. The resultant vectors are shown in Figure 6-4b and d.

7   The angular velocities of links 3 and 4 can be calculated from equation 6.7:

$$\omega_4 = \frac{v_B}{BO_4} \qquad \text{and} \qquad \omega_3 = \frac{v_{BA}}{BA} \qquad (c)$$

Note that the velocity difference term $\mathbf{V}_{BA}$ represents the rotational component of velocity of link 3 due to ω_3. This must be true if point B cannot move toward or away from point A. The only velocity difference they can have, one to the other, is due to rotation of the line connecting them. You may think of point B on the line BA rotating about point A as a center, or point A on the line AB rotating about B as a center. The rotational velocity ω of any body is a "free vector" which has no particular point of application to the body. It exists everywhere on the body.

8   Finally we can solve for $\mathbf{V}_C$, again using equation 6.5. We select any point in link 3 for which we know the absolute velocity to use as the reference, such as point A.

$$\mathbf{V}_C = \mathbf{V}_A + \mathbf{V}_{CA} \qquad (d)$$

In this case, we can calculate the magnitude of $\mathbf{V}_{CA}$ from equation 6.7 as we have already found ω_3,

$$v_{ca} = c\omega_3 \qquad (e)$$

Since both $\mathbf{V}_A$ and $\mathbf{V}_{CA}$ are known, the vector diagram can be directly drawn as shown in Figure 6-4c. $\mathbf{V}_C$ is the resultant which closes the vector diagram. Figure 6-4d shows the calculated velocity vectors on the linkage diagram. Note that the velocity difference vector $\mathbf{V}_{CA}$ is perpendicular to line CA (along line rr) for the same reasons as discussed in step 7 above.

The above example contains some interesting and significant principles which deserve further emphasis. Equation 6.5a is repeated here for discussion.

$$\mathbf{V}_P = \mathbf{V}_A + \mathbf{V}_{PA} \qquad (6.5a)$$

This equation represents the *absolute* velocity of some general point P referenced to the origin of the global coordinate system. The right side defines it as the sum of the absolute velocity of some other reference point A in the same system and the velocity difference (or relative velocity) of point P versus point A. This equation could also be written as:

Velocity = Translation component + Rotation component

These are the same two components of motion defined by Chasles' theorem, and introduced for displacement in Section 4.3. Chasles' theorem holds for velocity as well. These two components of motion, translation, and rotation, are independent of one another. If either is zero in a particular example, the complex motion will reduce to one of the special cases of pure translation or pure rotation. When both are present, the total velocity is merely their vector sum.

Let us review what was done in Example 6-1 in order to extract the general strategy for solution of this class of problem. We started at the input side of the mechanism, as that is where the driving angular velocity is defined. We first looked for a point (A) for which the motion was pure rotation so that one of the terms in equation 6.5 would be zero. (We could as easily have looked for a point in pure translation to bootstrap the solution.) We then solved for the absolute velocity of that point ($\mathbf{V}_A$) using equations 6.5 and 6.7. *(Steps 1 and 2)*

We then used the point (A) just solved for as a reference point to define the translation component in equation 6.5 written for a new point (B). Note that we needed to choose a second point (B) which was in the same rigid body as the reference point (A) which we had already solved and about which we could predict some aspect of the new point's (B's) velocity. In this example, we knew the direction of the velocity $\mathbf{V}_B$. In general this condition will be satisfied by any point on a link which is jointed to ground (as is link 4). In this example, we could not have solved for point C until we solved for B, because point C is on a floating link for which point we do not yet know the velocity direction. *(Steps 3 and 4)*

To solve the equation for the second point (B), we also needed to recognize that the rotation component of velocity is directed perpendicular to the line connecting the two points in the link (B and A in the example). You **will always know the direction of the rotation component** in equation 6.5 *if it represents a velocity difference* (CASE 1) **situation**. *If the rotation component relates two points in the **same rigid body**, then that velocity difference component is always perpendicular to the line connecting those two points* (see Figure 6-2). This will be true regardless of the two points selected. But, *this is not true in a* CASE 2 *situation* (see Figure 6-3). *(Steps 5 and 6)*

Once we found the absolute velocity ($\mathbf{V}_B$) of a second point on the same link (CASE 1) we could solve for the angular velocity of that link. (Note that points A and B are both on link 3 and the velocity of point O_4 is zero.) Once the angular velocities of all the links were known, we could solve for the linear velocity of any point (such as C) in any link using Equation 6.5. To do so, we had to understand the concept of angular velocity as a **free vector**, meaning that it exists everywhere on the link at any given instant. It has no particular center. *It has an infinity of potential centers.* The link simply *has an angular velocity,* just as does a frisbee thrown and spun across the lawn.

All points on a *frisbee*, if spinning while flying, obey equation 6.5. Left to its own devices, the frisbee will spin about its center of gravity (*CG*), which is close to the center of its circular shape. But if you are an expert frisbee player, (and have rather pointed fingers) you can imagine catching that flying frisbee between your two index fingers in some off center location (not at the *CG*), such that the frisbee continues to spin about your fingertips. In this somewhat farfetched example of championship frisbee play, you will have taken the translation component of the frisbee's motion to zero, but its independent rotation component will still be present. Moreover, it will now be spinning about a different center (your fingers) than it was in flight (its *CG*). Thus this *free vector* of angular velocity (ω) is happy to attach itself to any point on the body. The body still has the same ω, regardless of the assumed center of rotation. It is this property that allows us to solve equation 6.5 for literally **any point** on a rigid body in complex motion **referenced to any other point** on that body. *(Steps 7 and 8)*.

---

[1] Note that this *graph* is not a plot of points on an *x,y* coordinate system. Rather it is a *linear graph* from the fascinating branch of mathematics called *graph theory,* which is itself a branch of topology. Linear graphs are often used to depict interrelationships between various phenomena. They have many applications in kinematics especially as a way to classify linkages and to find isomers.

6.3   INSTANT CENTERS OF VELOCITY

The definition of an **instant center** of velocity is *a point, common to two bodies in plane motion, which point has the same instantaneous velocity in each body.* Instant centers are sometimes also called *centros* or *poles.* Since it takes two bodies or links to create an instant center (*IC*), we can easily predict the quantity of instant centers to expect from any collection of links. The combination formula for n things taken r at a time is:

$$C = \frac{n(n-1)(n-2)\cdots(n-r+1)}{r!} \qquad (6.8a)$$

For our case $r = 2$ and it reduces to:

$$C = \frac{n(n-1)}{2} \qquad (6.8b)$$

From equation 6.8b we can see that a fourbar linkage has six instant centers, a sixbar has 15 and an eightbar has 28.

Figure 6-5 shows a fourbar linkage in an arbitrary position. It also shows a **linear graph**[1] which is useful for keeping track of which *ICs* have been found. This particular graph can be created by drawing a circle on which we mark off as many points as there are links in our assembly. We will then draw a line between the dots representing the link pairs each time we find an instant center. The resulting linear graph is the set of lines connecting the dots. It does not include the circle which was used only to place the dots. This graph is actually a geometric solution to equation 6.8b, since connecting all the points in pairs gives all the possible combinations of points taken two at a time.

Some *ICs* can be found by inspection, using only the definition of the instant center. Note in Figure 6-5a that the four pin joints each satisfy the definition. They clearly must have the same velocity in both links at all times. These have been labelled $I_{1,2}$, $I_{2,3}$, $I_{3,4}$, and $I_{1,4}$. The order of the subscripts is immaterial. Instant center $I_{2,1}$ is the same as $I_{1,2}$. These pin joint *ICs* are sometimes called "permanent" instant centers as they remain in the same location for all positions of the linkage. In general, instant centers will move to new locations as the linkage changes position, thus the adjective *instant.* In this fourbar example there are two more *ICs* to be found. It will help to use the Aronhold-Kennedy theorem[2], also called *Kennedy's rule,* to locate them.

Kennedy's rule:

*Any three bodies in plane motion will have exactly three instant centers, and **they will lie on the same straight line**.*

The first part of this rule is just a restatement of equation 6.8b for $n = 3$. It is the second clause in this rule that is most useful. Note that this rule does ***not*** require that the three bodies be connected in any way. We can use this rule, in conjunction with the linear graph, to find the remaining *ICs* which are not obvious from inspection. Figure 6.5b shows the construction necessary to find instant center $I_{1,3}$. Figure 6-5c shows the construction necessary to find instant center $I_{2,4}$. The following example describes the procedure in detail.

[2] Discovered independently by Aronhold in Germany, in 1872, and by Kennedy in England, in 1886.

✎⬚EXAMPLE 6-2

Finding all Instant Centers for a Fourbar Linkage.

Problem:　　Given a fourbar linkage in one position, find all *ICs* by graphical methods.

Solution:　　(see Figure 6-5)

1　Draw a circle with all links numbered around the circumference as shown in Figure 6-5a.

2　Locate as many *ICs* as possible by inspection. All pin joints will be permanent *ICs*. Connect the link numbers on the circle to create a linear graph and record those *ICs* found, as shown in Figure 6-5a.

3　Identify a link combination on the linear graph for which the *IC* has not been found and draw a dotted line connecting those two link numbers. Identify two triangles on the graph which each contain the dotted line and whose other two sides are solid lines representing *ICs* already found. On the graph in Figure 6-5b, link numbers 1 and 3 have been connected with a dotted line. This line forms one triangle with sides 13, 34, 14 and another with sides 13, 23, 12. These triangles define trios of *ICs* which obey **Kennedy's rule**. Thus *ICs* 13, 34 and 14 **must lie on the same straight line**. Also *ICs* 13, 23 and 12 will **lie on a different straight line**.

4　On the linkage diagram draw a line through the two known *ICs* which form a trio with the unknown *IC*. Repeat for the other trio. In Figure 6-5b, a line has been drawn through $I_{1,2}$ and $I_{2,3}$ and extended. $I_{1,3}$ must lie on this line. Another line has been drawn through $I_{1,4}$ and $I_{3,4}$ and extended to intersect the first line. By Kennedy's rule, instant center $I_{1,3}$ must also lie on this line, so their intersection is $I_{1,3}$.

5　Connect link numbers 2 and 4 with a dotted line on the linear graph as shown in Figure 6-5c. This line forms one triangle with sides 24, 23, 34 and another with sides 24, 12, 14. These sides represent trios of *ICs* which obey Kennedy's rule. Thus *ICs* 24, 23 and 34 must lie on the same straight line. Also *ICs* 24, 21 and 14 lie on a different straight line.

6　On the linkage diagram draw a line through the two known *ICs* which form a trio with the unknown *IC*. Repeat for the other trio. In Figure 6-5c, a line has been drawn through $I_{1,2}$ and $I_{1,4}$ and extended. $I_{2,4}$ must lie on this line. Another line has been drawn through $I_{2,3}$ and $I_{3,4}$ and extended to intersect the first line. By Kennedy's rule, instant center $I_{2,4}$ must also lie on this line, so their intersection is $I_{2,4}$.

7　If there were more links, this procedure would be repeated until all *ICs* were found.

　　　　The presence of slider joints makes finding the instant centers a little more subtle as is shown in the next example. Figure 6-6a shows a **fourbar slider-crank linkage**. Note that there are only three pin joints in this linkage. All pin joints are *permanent instant centers*. But the joint between links 1 and 4 is a rectilinear, sliding full joint. A sliding joint is kinematically equivalent to an infinitely long link, "pivoted" at infinity. Figure 6-6b shows a nearly equivalent pin-jointed version of the slider-crank in which link 4 is a very long rocker. Point *B* now swings

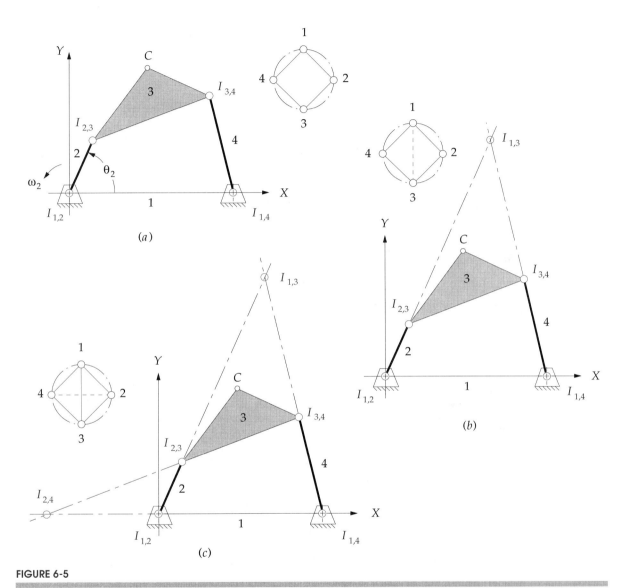

FIGURE 6-5

Locating instant centers in the pin-jointed linkage

through a shallow arc which is nearly a straight line. It is clear in Figure 6-6b that, in this linkage, $I_{1,4}$ is at pivot O_4. Now imagine increasing the length of this long, link 4 rocker even more. In the limit, link 4 approaches infinite length, the pivot O_4 approaches infinity along the line which was originally the long rocker and the arc motion of point B approaches a straight line. Thus, *a slider joint will have its instant center at infinity along a line perpendicular to the direction of sliding* as shown in Figure 6-6a.

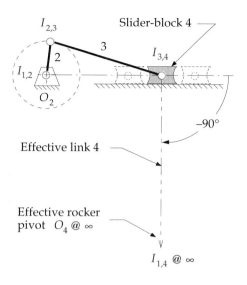

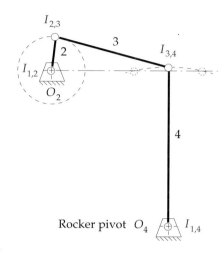

(*a*) Slider-crank linkage   (*b*) Crank-rocker linkage

FIGURE 6-6

Rectilinear slider's instant center is at infinity

EXAMPLE 6-3

Finding all Instant Centers for a Slider-Crank Linkage.

Problem: Given a slider-crank linkage in one position, find all *ICs* by graphical meth-
                 ods.

Solution: (see Figure 6-7)

1   Draw a circle with all links numbered around the circumference as shown in Figure 6-7a.

2   Locate all *ICs* possible by inspection. All pin joints will be permanent *ICs* The slider
    joint's instant center will be at infinity along a line perpendicular to the axis of sliding.
    Connect the link numbers on the circle to create a linear graph and record those *ICs*
    found, as shown in Figure 6-7a.

3   Identify a link combination on the linear graph for which the *IC* has not been found
    and draw a dotted line connecting those two link numbers. Identify two triangles on
    the graph which each contain the dotted line and whose other two sides are solid lines
    representing *ICs* already found. In the graph on Figure 6-7b, link numbers 1 and 3
    have been connected with a dotted line. This line forms one triangle with sides 13,
    34, 14 and another with sides 13, 23, 12. These sides represent trios of *ICs* which obey
    Kennedy's rule. Thus *ICs* 13, 34 and 14 must lie on the same straight line. Also *ICs* 13,
    23, and 12 lie on a different straight line.

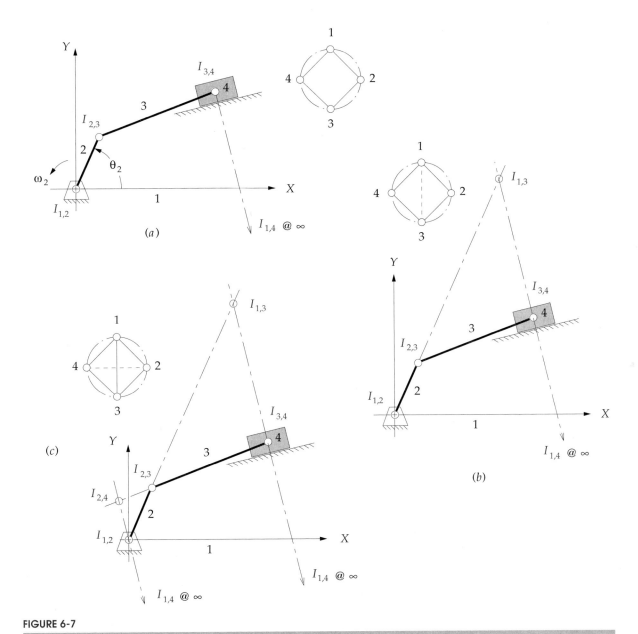

FIGURE 6-7

Locating instant centers in the slider-crank linkage

4 On the linkage diagram draw a line through the two known *IC*s which form a trio
   with the unknown *IC*. Repeat for the other trio. In Figure 6-7b, a line has been drawn
   from $I_{1,2}$ through $I_{2,3}$ and extended. $I_{1,3}$ must lie on this line. Another line has been
   drawn from $I_{1,4}$ (at infinity) through $I_{3,4}$ and extended to intersect the first line. By
   Kennedy's rule, instant center $I_{1,3}$ must also lie on this line, so their intersection is $I_{1,3}$.

5   Connect link numbers 2 and 4 with a dotted line on the graph as shown in Figure 6-7c. This line forms one triangle with sides 24, 23, 34 and another with sides 24, 12, 14. These sides also represent trios of *ICs* which obey Kennedy's rule. Thus *ICs* 24, 23 and 34 must lie on the same straight line. Also *ICs* 24, 12 and 14 lie on a different straight line.

6   On the linkage diagram draw a line through the two known *ICs* which form a trio with the unknown *IC*. Repeat for the other trio. In Figure 6-7c, a line has been drawn from $I_{1,2}$ to intersect $I_{1,4}$, and extended. Note that the only way to "intersect" $I_{1,4}$ at infinity is to draw a line parallel to the line $I_{3,4}\,I_{1,4}$ since all parallel lines intersect at infinity. Instant center $I_{2,4}$ must lie on this parallel line. Another line has been drawn through $I_{2,3}$ and $I_{3,4}$ and extended to intersect the first line. By Kennedy's rule, instant center $I_{2,4}$ must also lie on this line, so their intersection is $I_{2,4}$.

7   If there were more links, this procedure would be repeated until all *ICs* were found.

The procedure in this slider example is identical to that used in the pin-jointed fourbar, except that it is complicated by the presence of instant centers located at infinity.

In Section 2.9 and Figure 2-10c we showed that a cam-follower mechanism is really a fourbar linkage in disguise. As such it will also possess instant centers. The presence of the half joint in this, or any linkage, makes the location of the instant centers a little more complicated. We have to recognize that the instant center between any two links will be along a line that is perpendicular to the *relative velocity* vector between the links at the half joint, as shown in the following example. Figure 6-8 shows the same cam-follower mechanism as in Figure 2-14. The effective links 2, 3, and 4 are also shown.

✏️EXAMPLE 6-4

Finding all Instant Centers for a Cam-Follower Mechanism.

Problem: Given a cam and follower in one position, find all *ICs* by graphical methods.

Solution: (see Figure 6-8)

1   Draw a circle with all links numbered around the circumference as shown in Figure 6-8b. In this case there are only three links and thus only three *ICs* to be found as shown by equation 6.8. Note that the links are numbered 1, 2, and 4. The missing link 3 is the variable length effective coupler.

2   Locate all *ICs* possible by inspection. All pin joints will be permanent *ICs*. The two fixed pivots $I_{1,2}$ and $I_{1,4}$ are the only pin joints here. Connect the link numbers on the circle to create a linear graph and record those *ICs* found, as shown in Figure 6-8b. The only link combination on the linear graph for which the *IC* has not been found is $I_{2,4}$, so draw a dotted line connecting those two link numbers.

3   Kennedy's rule says that all three *ICs* must lie on the same straight line; thus the remaining instant center $I_{2,4}$ must lie on the line $I_{1,2}\,I_{1,4}$ extended. Unfortunately in this example, we have too few links to find a second line on which $I_{2,4}$ must lie.

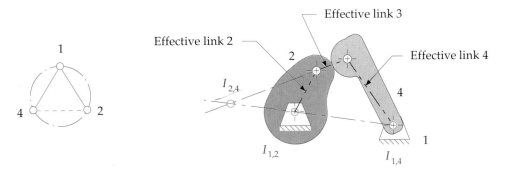

(b) The linkage graph

(c) The effective linkage equivalent

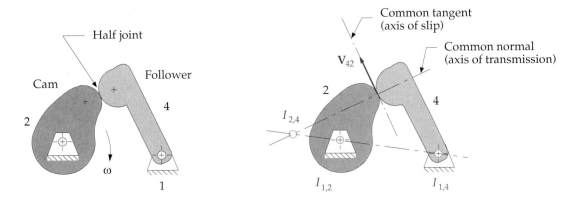

(a) The cam and follower

(d) Finding I_{24} without using the effective linkage

FIGURE 6-8

Locating instant centers in the cam-follower

4   On the linkage diagram draw a line through the two known *ICs* which form a trio with the unknown *IC*. In Figure 6-8c, a line has been drawn from $I_{1,2}$ through $I_{1,4}$ and extended. This is, of course, link 1. By Kennedy's rule, $I_{2,4}$ must lie on this line.

5   Looking at Figure 6-8c which shows the effective links of the equivalent fourbar linkage for this position, we can extend effective link 3 until it intersects link 1 extended. Just as in the "pure" fourbar linkage, instant center 2,4 lies on the intersection of links 1 and 3 extended (see Example 6-2).

6   Figure 6-8d shows that it is not necessary to construct the effective fourbar linkage to find $I_{2,4}$. Note that the **common tangent** to links 2 and 4 at their contact point (the half joint) has been drawn. This line is also called the **axis of slip** because it is the line along which all relative (slip) velocity will occur between the two links. Thus the velocity of link 4 versus 2, $\mathbf{V}_{42}$, is directed along the axis of slip. Instant center $I_{2,4}$ must therefore lie along a line perpendicular to the common tangent, called the **common normal**. Note that this line is the same as the effective link 3 line in Figure 6-8c.

6.4   VELOCITY ANALYSIS WITH INSTANT CENTERS

Once the *ICs* have been found, they can be used to do a very rapid graphical velocity analysis of the linkage. Note that, depending on the particular position of the linkage being analyzed, some of the *ICs* may be very far removed from the links. For example, if links 2 and 4 are nearly parallel, their extended lines will intersect at a point far away, and not be practically available for velocity analysis. Figure 6-9 shows the same linkage as Figure 6-5 with $I_{1,3}$ located and labelled. From the definition of the instant center, both links sharing the instant center will have identical velocity at that point. Instant center $I_{1,3}$ involves the coupler (link 3) which is in complex motion, and the ground link 1, which is stationary. All points on link 1 have zero velocity in the global coordinate system, which is embedded in link 1. Therefore, $I_{1,3}$ must have zero velocity at this instant. If $I_{1,3}$ has zero velocity, then it can be considered to be an instantaneous "fixed pivot" about which link 3 is in pure rotation with respect to link 1. A moment later, $I_{1,3}$ will move to a new location and link 3 will be "pivoting" about a new instant center.

The velocity of point A is shown on Figure 6-9. The magnitude of $\mathbf{V}_A$ can be computed from equation 6.7. Its direction and sense can be determined by inspection as was done in Example 6-1. Note that point A is also instant center $I_{2,3}$. It has the same velocity as part of link 2 and as part of link 3. Since link 3 is effectively pivoting about $I_{1,3}$ at this instant, the angular velocity ω_3 can be found by rearranging equation 6.7:

$$\omega_3 = \frac{v_A}{\left(AI_{1,3}\right)} \tag{6.9a}$$

Once w_3 is known, the magnitude of $\mathbf{V}_B$ can also be found from equation 6.7:

$$v_B = \left(BI_{1,3}\right)\omega_3 \tag{6.9b}$$

Once $\mathbf{V}_B$ is known, w_4 can also be found from equation 6.7:

$$\omega_4 = \frac{v_B}{\left(BO_4\right)} \tag{6.9c}$$

Finally, the magnitude of $\mathbf{V}_C$ (or the velocity of any other point on the coupler) can be found from equation 6.7:

$$v_C = \left(CI_{1,3}\right)\omega_3 \tag{6.9d}$$

Note that equations 6.7 and 6.9 provide only the **scalar magnitude** of these velocity vectors. We have to determine their **direction** from the information in the scale diagram (Figure 6-9). Since we know the location of $I_{1,3}$, which is an instantaneous "fixed" pivot for link 3, all of that link's absolute velocity vectors for this instant will be **perpendicular to their radii from $I_{1,3}$ to the point in question**. $\mathbf{V}_B$ and $\mathbf{V}_C$ can be seen to be perpendicular to their radii from $I_{1,3}$. Note that $\mathbf{V}_B$ is also perpendicular to the radius from O_4 because B is also pivoting about that point as part of link 4.

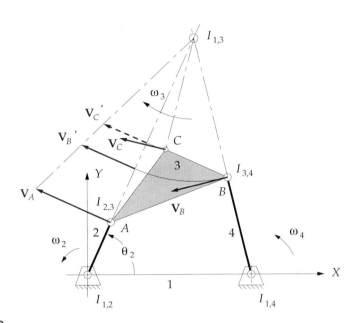

FIGURE 6-9

Velocity analysis using instant centers

A rapid graphical solution to equations 6.9 is shown in the figure. Arcs centered at $I_{1,3}$ are swung from points B and C to intersect line $AI_{1,3}$. The magnitudes of velocities $\mathbf{V}_B'$ and $\mathbf{V}_C'$ are found from the vectors drawn perpendicular to that line at the intersections of the arcs and line $AI_{1,3}$. The lengths of the vectors are defined by the line from the tip of $\mathbf{V}_A$ to the instant center $I_{1,3}$. These vectors can then be slid along their arcs back to points B and C, maintaining their tangency to the arcs.

Thus, we have in only a few steps found all the same velocities that were found in the more tedious method of Example 6-1. The instant center method is a quick graphical method to analyze velocities, but it will only work if the instant centers are in reachable locations for the particular linkage position analyzed. However, the graphical method using the velocity difference equation shown in Example 6-1 will always work, regardless of linkage position.

Angular Velocity Ratio

The **angular velocity ratio** VR is defined as *the output angular velocity divided by the input angular velocity.* For a fourbar mechanism this is expressed as:

$$VR = \frac{\omega_4}{\omega_2} \qquad (6.10)$$

We can derive this ratio for any linkage by constructing a **pair of effective links** as shown in Figure 6-10a. The definition of **effective link pairs** is *two lines, mutually parallel, drawn through the fixed pivots and intersecting the coupler extended.* These are shown as O_2A' and O_4B' in Figure 6-10a. Note that there is an infinity of possible effective link pairs. They must be parallel to one another but may make any angle with link 3. In the figure they are shown perpendicular to link 3 for convenience in the derivation to follow. The angle between links 2 and 3 is shown as ν. The transmission angle between links 3 and 4 is μ. We will now derive an expression for the angular velocity ratio using these effective links, the actual link lengths, and angles ν and μ.

From geometry:

$$O_2A' = (O_2A)\sin \nu \qquad\qquad O_4B' = (O_4B)\sin \mu \qquad\qquad (6.11a)$$

From equation 6.7

$$v_{A'} = (O_2A')\omega_2 \qquad\qquad (6.11b)$$

The component of velocity $v_{A'}$ lies along the link AB. Just as with a two-force member in which a force applied at one end transmits only its component that lies along the link to the other end, this velocity component can be transmitted along the link to point B. This is sometimes called the **principle of transmissibility**. We can then equate these components at either end of the link.

$$v_{A'} = v_{B'} \qquad\qquad (6.11c)$$

Then:

$$O_2A'\omega_2 = O_4B'\omega_4 \qquad\qquad (6.11d)$$

rearranging:

$$\frac{\omega_4}{\omega_2} = \frac{O_2A'}{O_4B'} \qquad\qquad (6.11e)$$

and substituting:

$$\frac{\omega_4}{\omega_2} = \frac{O_2A\sin \nu}{O_4B\sin \mu} = VR \qquad\qquad (6.11f)$$

Note in equation 6.11f that as angle ν goes through zero, the angular velocity ratio will be zero regardless of the values of ω_2 or the link lengths, and thus ω_4 will be zero. When angle ν is zero, links 2 and 3 will be colinear and thus be in their toggle positions. We learned in Section 3.3 that the limiting positions of link 4 are defined by these toggle conditions. We should expect that the velocity of link 4 will be zero when it has come to the end of its travel. An even more interesting situation obtains if we allow angle μ to go to zero. Equation 6.11f shows that ω_4 **will go to infinity** when $\mu = 0$, regardless of the values of ω_2 or the link lengths. We clearly cannot allow μ to reach zero. In fact, we learned in Section 3.3 that we should keep this transmission angle μ above about 35 degrees to maintain good quality of motion and force transmission.

Figure 6-10b shows the same linkage as in Figure 6-10a, but the effective links have now been drawn so that they are not only parallel but are colinear, and thus lie on top of one another. Both intersect the extended coupler at the same point, which is instant center $I_{2,4}$. So, A' and B' of Figure 6-10a are now coincident at $I_{2,4}$.

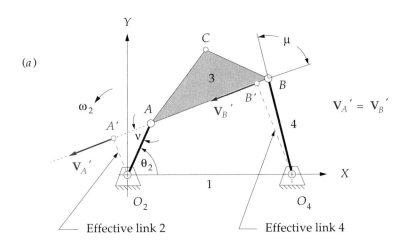

(a)

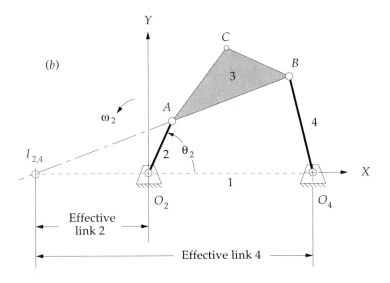

(b)

FIGURE 6-10

Effective links and the angular velocity ratio

This allows us to write an equation for the **angular velocity ratio** in terms of the distances from the fixed pivots to instant center $I_{2,4}$.

$$\frac{\omega_4}{\omega_2} = \frac{O_2 I_{2,4}}{O_4 I_{2,4}} \qquad (6.11g)$$

Thus, the instant center $I_{2,4}$ can be used to determine the **angular velocity ratio**.

Mechanical Advantage

The power in a mechanical system is defined as:

$$Power = force \times velocity$$

For a rotational system:

$$Power = torque \times angular\ velocity$$

The power flows through a passive system and:

$$Power\ in = Power\ out + losses$$

Mechanical efficiency can be defined as:

$$Efficiency = Power\ out\ /\ Power\ in$$

Linkage systems can be very efficient if they are well made with low friction bearings on all pivots. Losses are often less than 10%. For simplicity in the following analysis we will assume that the losses are zero. Then, letting T_2 and ω_2 represent input torque and velocity, and T_4 and ω_4 represent output torque and velocity:

$$Power\ in = T_2\omega_2$$
$$Power\ out = T_4\omega_4$$
$$Power\ out = Power\ in \qquad (6.12)$$
$$T_4\omega_4 = T_2\omega_2$$
$$\frac{\omega_4}{\omega_2} = \frac{T_2}{T_4}$$

Note that the **torque ratio** (T_4/T_2) is the inverse of the angular velocity ratio.

Mechanical advantage (MV) can be defined as:

$$MV = \frac{Output\ Torque}{Input\ Torque}$$

$$MV = \frac{T_4}{T_2} \qquad (6.13a)$$

Substituting equation 6.12:

$$MV = \frac{\omega_2}{\omega_4} \qquad (6.13b)$$

and substituting equation 6.11f:

$$MV = \frac{O_4 B \sin\mu}{O_2 A \sin\nu} \qquad (6.13c)$$

See Figure 6-10 and compare equation 6.13c to equation 6.11f and its discussion under **angular velocity ratio** above. Equation 6.13c shows that the mechanical advantage responds to changes in angles ν and μ in opposite fashion to

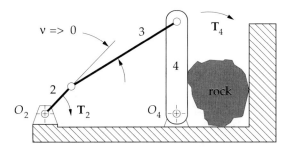

FIGURE 6-11

"Rock-crusher" toggle mechanism

that of the angular velocity ratio. If the transmission angle μ goes to zero (which we don't want it to do), the output torque also goes to zero regardless of the amount of input torque applied (because there will then be no moment arm for the force to act on). But, when angle v goes to zero (which it can and does, twice per cycle in a Grashof linkage), the mechanical advantage becomes infinite! This is the principle of a rock-crusher mechanism as shown in Figure 6-11. A quite moderate torque applied to link 2 can generate a huge torque on link 4 to crush the rock. Of course, we cannot expect to achieve the theoretical output of infinite torque magnitude, as the strengths of the links and joints will limit the maximum forces and torques obtainable. Another common example of a linkage which takes advantage of this infinite ratio of torque at the toggle position is a ViseGrip™ locking pliers.

These two ratios, **angular velocity ratio** and **mechanical advantage**, provide useful, dimensionless **indices of merit** by which we can judge the relative quality of various linkage designs which may be proposed as solutions.

Using Instant Centers in Linkage Design

In addition to providing a quick numerical velocity analysis, instant center analysis more importantly gives the designer a remarkable overview of the linkage's global behavior. It is quite difficult to mentally visualize the complex motion of a "floating" coupler link even in a simple fourbar linkage, unless you build a model or run a computer simulation. Because this complex coupler motion in fact reduces to an instantaneous pure rotation about the instant center $I_{1,3}$, finding that center allows the designer to visualize the motion of the coupler as a pure rotation. One can literally *see* the motion and the directions of velocities of any points of interest by relating them to the instant center. It is only necessary to draw the linkage in a few positions of interest, showing the instant center locations for each position.

Figure 6-12 shows a practical example of how this visual, qualitative analysis technique could be applied to the design of an automobile rear suspension system. Most automobile suspension mechanisms are either fourbar linkages or fourbar slider-cranks, with the wheel assembly carried on the coupler (as was also shown in Figure 3-19). Figure 6-12a shows a rear suspension design from a domestic car of 1970's vintage which was later redesigned because of a disturbing tendency to

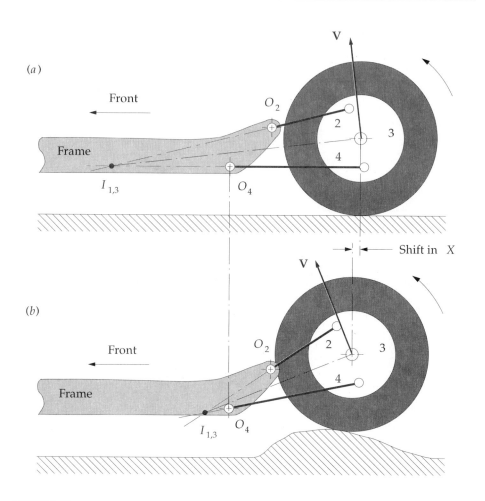

FIGURE 6-12

"Bump steer" due to shift in instant center location

"bump steer," i.e., turn the rear axle when hitting a bump on one side of the car. The figure is a view looking from the center of the car outward, showing the fourbar linkage which controls the up and down motion of one side of the rear axle and one wheel. Links 2 and 4 are pivoted to the frame of the car which is link 1. The wheel and axle assembly is rigidly attached to the coupler, link 3. Thus the wheel assembly has complex motion in the vertical plane. Ideally, one would like the wheel to move up and down in a straight vertical line when hitting a bump. Figure 6-12b shows the motion of the wheel and the new instant center ($I_{1,3}$) location for the situation when one wheel has hit a bump. The velocity vector for the center of the wheel in each position is drawn perpendicular to its radius from $I_{1,3}$. You can see that the wheel center has a significant horizontal component of motion as it moves up over the bump. This horizontal component causes the wheel center on that side of the car to move forward while it moves upward, thus turning the axle (about a vertical axis) and steering the car with the rear wheels in the same way that you steer a toy wagon.

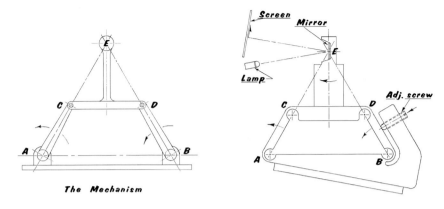

FIGURE 6-13

An optical adjustment linkage (*Reprinted with permission*)

Viewing the path of the instant center over some range of motion gives a clear picture of the behavior of the coupler link. The undesirable behavior of this suspension linkage system could have been predicted from this simple instant center analysis before ever building the mechanism.

Another practical example of the effective use of instant centers in linkage design is shown in Figure 6-13, which is an optical adjusting mechanism used to position a mirror and allow a small amount of rotational adjustment.[1] A more detailed account of this design case study[2] is provided in Chapter 18. The designer, K. Towfigh, recognized that $I_{1,3}$ is an instantaneous "fixed pivot" and will allow very small pure rotations about that point with very small translational error. He then designed a one-piece, plastic fourbar linkage whose "pin-joints" are thin webs of plastic which flex to allow slight rotation. He then placed the mirror on the coupler at $I_{1,3}$. Even the fixed link 1 is the same piece as the "moveable links" and has a small set screw to provide the adjustment. A simple and elegant design.

6.5   CENTRODES

Figure 6-14 illustrates the fact that the successive positions of an instant center (or **Centro**) form a path of their own. *This path, or locus of the instant center is called the* **centrode**. Since there are two links needed to create an instant center, there will be two centrodes associated with any one instant center. These are formed by projecting the path of the instant center first on one link and then on the other. Figure 6-14a shows the locus of instant center $I_{1,3}$ as projected onto link 1. Because link 1 is stationary, or fixed, this is called the **fixed centrode**. By temporarily inverting the mechanism and fixing link 3 as the ground link, as shown in Figure 6-14b, we can move link 1 as the coupler and project the locus of $I_{1,3}$ onto link 3. In the original linkage, link 3 was the moving coupler, so this is called the **moving centrode**. Figure 6-14c shows the original linkage with both fixed and moving centrodes superposed.

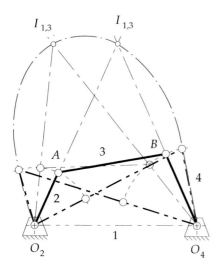

a. The fixed centrode

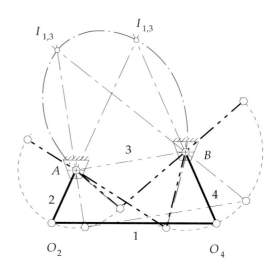

b. The moving centrode

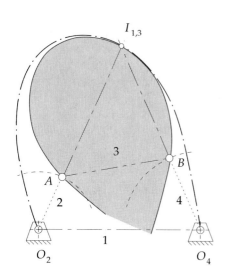

c. The centrodes in contact

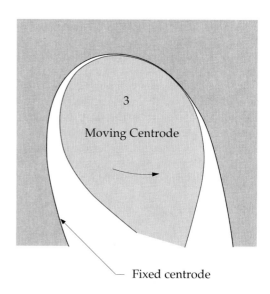

d. Roll the moving centrode against the fixed centrode to produce the same coupler motion as original linkage.

FIGURE 6-14

Open-loop fixed and moving centrodes (or polodes) of a fourbar linkage

The definition of the instant center says that both links have the same velocity at that point, at that instant. Link 1 has zero velocity everywhere, as does the fixed centrode. So, as the linkage moves, the moving centrode must roll against the fixed centrode without slipping. If you cut the fixed and moving centrodes out of metal, as shown in Figure 6-14d, and roll the moving centrode (which is link 3) against the fixed centrode (which is link 1), the complex motion of link 3 will be identical to that of the original linkage. *All of the coupler curves of points on link 3 will have the same path shapes as in the original linkage.* We now have, in effect, a "linkless" fourbar linkage, really one composed of two bodies which have these centrode shapes rolling against one another. Links 2 and 4 have been eliminated. Note that the example shown in Figure 6-14 is a non-Grashof fourbar. The lengths of its centrodes are limited by the double-rocker toggle positions.

All instant centers of a linkage will have centrodes. If the links are directly connected by a joint, such as $I_{2,3}$, $I_{3,4}$, $I_{1,2}$ and $I_{1,4}$, their fixed and moving centrodes will degenerate to a point at that location on each link. The most interesting centrodes are those involving links not directly connected to one another such as $I_{1,3}$ and $I_{2,4}$. If we look at the double-crank linkage in Figure 6-15a in which links 2 and 4 both revolve fully, we see that the centrodes of $I_{1,3}$ form closed curves. The motion of link 3

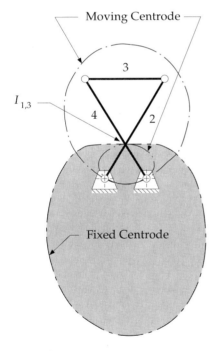

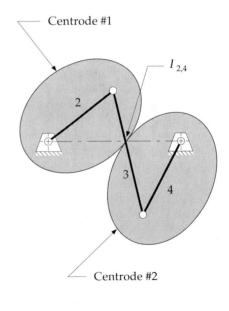

(*a*) Closed loop centrodes of I_{13} for a
Grashof double crank linkage

(*b*) Ellipsoidal centrodes of I_{24}
for a special case Grashof
anti-parallelogram linkage

FIGURE 6-15

Closed-loop fixed and moving centrodes

with respect to link 1 could be duplicated by causing these two centrodes to roll against one another without slipping. Note that there are two loops to the moving centrode. Both must roll on the single-loop fixed centrode to complete the motion of the equivalent double-crank linkage.

We have so far dealt largely with the instant center $I_{1,3}$. Instant center $I_{2,4}$ involves two links which are each in pure rotation and not directly connected to one another. If we use a special case Grashof linkage with the links crossed (sometimes called an **antiparallelogram** linkage), the centrodes of $I_{2,4}$ become ellipses as shown in Figure 6-15b. To guarantee no slip, it will probably be necessary to put meshing teeth on each centrode. We then will have a pair of elliptical, **noncircular gears**, or *gearset*, which gives the *same output motion as the original double-crank linkage* and will have the *same variations in the angular velocity ratio and mechanical advantage as the linkage* had. Thus we can see that *gearsets are also just fourbar linkages in disguise*. Noncircular gears find much use in machinery, such as printing presses, where rollers must be speeded and slowed with some pattern during each cycle or revolution. More complicated shapes of noncircular gears are analogous to cams and followers in that the equivalent fourbar linkage must have variable length links. **Circular gears** are just a special case of noncircular gears which give a **constant angular velocity ratio** and are widely used in all machines. Gears and gearsets will be dealt with in more detail in Chapter 10.

In general, centrodes of crank rockers and double rockers will be open curves with asymptotes. Centrodes of double-crank linkages will be closed curves. Program FOURBAR will calculate and draw the fixed and moving centrodes for any linkage input to it. Input the datafiles FIG6-14 and FIG6-15A and B into program FOURBAR to see the centrodes of these linkage drawn as the linkages rotate.

A "Linkless" Linkage

A common example of a mechanism made of centrodes is shown in Figure 6-16a. You have probably rocked in a *Boston* or *Hitchcock* rocking chair and experienced the soothing motions that it delivers to your body. You may have also rocked in a *platform* rocker as shown in Figure 6-16b and noticed that its motion did not feel as soothing.

There are good kinematic reasons for the difference. The platform rocker has a fixed pin joint between the seat and the base (floor). Thus all parts of your body are in pure rotation along concentric arcs. You are in effect riding on the rocker of a linkage.

The Boston rocker has a shaped (curved) base, or "runners," which rolls against the floor. These runners are usually *not* circular arcs. They have a higher-order curve contour. They are, in fact, **moving centrodes**. The floor is the **fixed centrode**. When one is rolled against the other, the chair and its occupant experience coupler curve motion. Every part of your body travels along a different sixth-order coupler curve which provides smooth accelerations and velocities and feels better than the cruder second-order (circular) motion of the platform rocker. Our ancestors, who carved these rocking chairs, probably had never heard of fourbar linkages and centrodes, but they knew intuitively how to create comfortable motions.

Coupler motion

Moving centrode

Fixed centrode

(a) Boston rocker

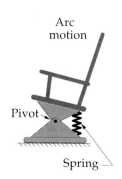

Arc motion

Pivot

Spring

(b) Platform rocker

FIGURE 6-16

Some rocking chairs are centrodes of a fourbar linkage

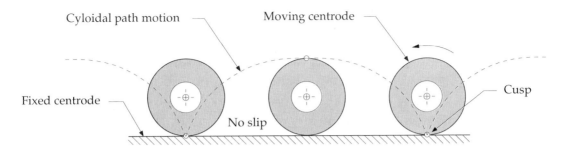

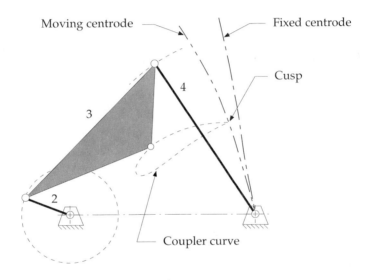

(*a*)   Cycloidal motion of a circular, moving centrode rolling on a straight, fixed centrode

(*b*) Coupler curve cusps exist only on the moving centrode

FIGURE 6-17

Examples of centrodes

Cusps

Another example of a centrode which you probably use frequently is the path of
the tire on your car or bicycle. As your tire rolls against the road without slipping,
the road becomes a fixed centrode and the circumference of the tire is the moving
centrode. The tire is, in effect, the coupler of a linkless fourbar linkage. All points
on the contact surface of the tire move along cycloidal coupler curves and pass
through a cusp of zero velocity when they reach the fixed centrode at the road
surface as shown in Figure 6-17a. All other points on the tire and wheel assembly travel along coupler curves which do not have cusps. This last fact is a clue
to a means to identify coupler points which will have cusps in their coupler curve.
If a coupler point is chosen to be on the moving centrode at one extreme of its path motion

(i.e., at one of the positions of $I_{1,3}$), then it will have a cusp in its coupler curve. Figure 6-17b shows a coupler curve of such a point, drawn with program FOURBAR . The right end of the coupler path touches the moving centrode and as a result has a cusp at that point. So, if you desire a cusp in your coupler motion, many are available. Simply choose a coupler point on the moving centrode of link 3. Read the diskfile FIG6-17B into program Fourbar to animate that linkage with its coupler curve or centrodes. Note in Figure 6-14 that choosing any location of instant center $I_{1,3}$ on the coupler as the coupler point will provide a cusp at that point.

6.6   VELOCITY OF SLIP

When there is a sliding joint between two links and neither one is the ground link, the velocity analysis is more complicated. Figure 6-18 shows an inversion of the four-bar slider-crank mechanism in which the sliding joint is floating, i.e., not grounded. To solve for the velocity at the sliding joint A, we have to recognize that there is more than one point A at that joint. There is a point A as part of link 2 (A_2), a point A as part of link 3 (A_3), and a point A as part of link 4 (A_4). This is a CASE 2 situation in which we have at least two points belonging to different links but occupying the same location at a given instant. Thus, the **relative velocity** equation 6.6 will apply. We can usually solve for the velocity of at least one of these points directly from the known input information using equation 6.7. It and equation 6.6 are all that are needed to solve for everything else. In this example link 2 is the driver, and θ_2 and ω_2 are given for the "freeze frame" position shown. We wish to solve for ω_4, the angular velocity of link 4, and also for the velocity of slip at the joint labelled A.

In Figure 6-18 the **axis of slip** is shown to be tangent to the slider motion and is the line along which all sliding occurs between links 3 and 4. The **axis of transmission** is defined to be perpendicular to the axis of slip and pass through the slider joint at A. This *axis of transmission is the **only line** along which we can transmit motion or force across the slider joint, except for friction.* We will assume friction to be negligible in this example. Any force or velocity vector applied to point A can be resolved into two components along these two axes which provide a *translating and rotating, local coordinate system* for analysis at the joint. The component along the axis of transmission will do useful work at the joint. But, the component along the axis of slip does no work, except *friction work.*

✍️ EXAMPLE 6-5

Graphical Velocity Analysis at a Sliding Joint.

Problem: Given θ_2, θ_3, θ_4, ω_2, find ω_3, ω_4, $\mathbf{V}_A$, by graphical methods.

Solution: (see Figure 6-18)

1   Start at the end of the linkage about which you have the most information. Calculate the magnitude of the velocity of **point A as part of link 2** (A_2) using scalar equation 6.7.

$$v_{A_2} = (AO_2)\omega_2 \qquad\qquad (a)$$

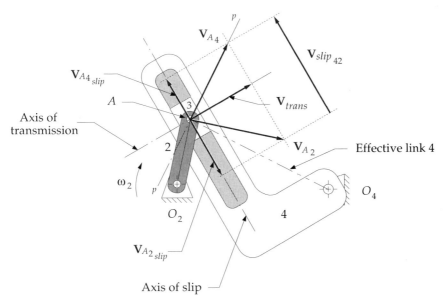

FIGURE 6-18

Velocity of slip and velocity of transmission (note that the applied ω is negative as shown)

2   Draw the velocity vector $\mathbf{V}_{A_2}$ with its length equal to its magnitude v_{A_2} at some convenient scale and with its root at point A and its direction perpendicular to the radius AO_2. Its sense is the same as that of ω_2 as is shown in Figure 6-18.

3   Draw the **axis of slip** and **axis of transmission** through point A.

4   Project $\mathbf{V}_{A_2}$ onto the axis of slip and onto the axis of transmission to create the components $\mathbf{V}_{A_{2slip}}$ and $\mathbf{V}_{trans}$ of $\mathbf{V}_{A_2}$ on the axes of slip and transmission respectively. Note that the **transmission component** is shared by all true velocity vectors at this point, as it is the only component which can transmit across the joint.

5   Note that link 3 is pin-jointed to link 2, so $\mathbf{V}_{A_3} = \mathbf{V}_{A_2}$.

6   Note that the direction of the velocity of point $\mathbf{V}_{A_4}$ is predictable since all points on link 4 are pivoting in pure rotation about point O_4. Draw the line pp through point A and perpendicular to the effective link 4, AO_4. Line pp is the direction of velocity $\mathbf{V}_{A_4}$.

7   Construct the true magnitude of velocity vector $\mathbf{V}_{A_4}$ by extending the projection of the **transmission component** $\mathbf{V}_{trans}$ until it intersects line pp.

8   Project $\mathbf{V}_{A_4}$ onto the axis of slip to create the **slip component** $\mathbf{V}_{A_{4slip}}$.

9   Write the relative velocity vector equation 6.6 for the **slip components** of point A_2 versus point A_4.

$$V_{slip_{42}} = V_{A_{4slip}} - V_{A_{2slip}} \qquad\qquad (b)$$

10   The angular velocities of links 3 and 4 are identical because they share the slider joint and must rotate together. They can be calculated from equation 6.7:

$$\omega_4 = \omega_3 = \frac{V_{A_4}}{AO_4} \qquad\qquad (c)$$

Instant center analysis can also be used to solve sliding joint velocity problems graphically.

✎]EXAMPLE 6-6

Graphical Velocity Analysis in the Fourbar Inverted Slider-Crank Mechanism using Instant Centers.

Problem: Given θ_2, θ_3, θ_4, ω_2, find ω_3, ω_4, V_A, by graphical methods.

Solution: (see Figure 6-19)

1   Start at the end of the linkage about which you have the most information. Calculate the magnitude of the velocity of **point A as part of link 2** (A_2) using scalar equation 6.7.

$$v_{A_2} = \left(AO_2\right)\omega_2 \qquad\qquad (a)$$

2   Draw the velocity vector $\mathbf{V}_{A_2}$ with its length equal to its magnitude v_{A_2} at some convenient scale and with its root at point A and its direction perpendicular to the radius AO_2. Its sense is the same as that of ω_2 as is shown in Figure 6-19. Note that link 3 is pin-jointed to link 2, so $\mathbf{V}_{A_3} = \mathbf{V}_{A_2}$.

3   Find the instant centers of the linkage as shown in Figure 6-19.

4   Define a point (B) on the slider block for analysis. Draw the **axis of slip** and **axis of transmission** through point B. Note that point B is a multiple point, belonging to both link 3 and link 4, and has different linear velocities in each.

5   Project $\mathbf{V}_{A_2}$ onto the axis of slip to create the orthogonal component $\mathbf{V}_{A_{3slip}}$ along link 3. Translate this slip component along link 3 and place it at point B. Rename it $\mathbf{V}_{B_{3slip}}$.

6   The direction of the true velocity of point B as part of link 3 ($\mathbf{V}_{B_3}$) is along a line perpendicular to the radius from $I_{1,3}$ to B. Construct a perpendicular to $\mathbf{V}_{B_{3slip}}$ at its tip and create $\mathbf{V}_{B_3}$.

7   Project $\mathbf{V}_{B_3}$ onto the axis of transmission to create the component $\mathbf{V}_{trans}$. Note that the **transmission component** is shared by all true velocity vectors at this point, as it is the only component which can transmit across the joint.

8   Note that the direction of the velocity of point $\mathbf{V}_{B_4}$ is predictable since all points on link 4 are pivoting in pure rotation about point O_4. Construct a line in the direction of $\mathbf{V}_{B_4}$ perpendicular to the effective link 4. Construct the true magnitude of velocity vector $\mathbf{V}_{B_4}$ by extending the projection of the **transmission component $\mathbf{V}_{trans}$** until it intersects the line of $\mathbf{V}_{B_4}$.

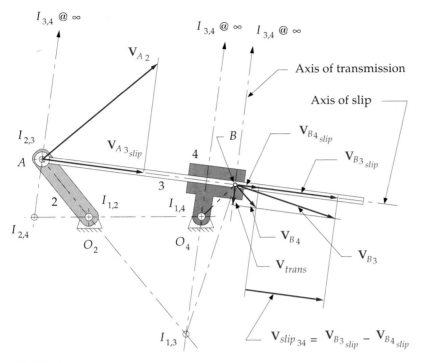

FIGURE 6-19

Graphical velocity analysis of an inverted slider-crank linkage

9 Project $\mathbf{V}_{B_4}$ onto the axis of slip to create the **slip component** $\mathbf{V}_{B4slip}$.

10 The total slip velocity at B is the difference between the two slip components. Write the relative velocity vector equation 6.6 for the **slip components** of point B_3 versus point B_4.

$$V_{slip34} = V_{B3slip} - V_{B4slip} \qquad (b)$$

11 The angular velocities of links 3 and 4 are identical because they share the slider joint and must rotate together. They can be calculated from equation 6.7:

$$\omega_4 = \omega_3 = \frac{V_{B_4}}{AO_4} \qquad (c)$$

The above examples show how a sliding joint linkage can be solved graphically for velocities at one position. In the next section, we will develop the general solution using algebraic equations to solve the same type of problem.

6.7   ANALYTIC SOLUTIONS FOR VELOCITY ANALYSIS

The Fourbar Pin-Jointed Linkage

The position equations for the fourbar pin-jointed linkage were derived in Section 4.5. The linkage was shown in Figure 4-7 and is shown again in Figure 6-20 on which we also show an input angular velocity ω_2 applied to link 2. This ω_2 can be a time-varying input velocity. The vector loop equation is shown in equations 4.5a and 4.5c, repeated here for your convenience.

$$\mathbf{R}_2 + \mathbf{R}_3 - \mathbf{R}_4 - \mathbf{R}_1 = 0 \qquad (4.5a)$$

As before, we substitute the complex number notation for the vectors, denoting their scalar lengths as a, b, c, d as shown in Figure 6-20a.

$$a e^{j\theta_2} + b e^{j\theta_3} - c e^{j\theta_4} - d e^{j\theta_1} = 0 \qquad (4.5c)$$

To get an expression for velocity, differentiate equation 4.5c with respect to time.

$$j a e^{j\theta_2} \frac{d\theta_2}{dt} + j b e^{j\theta_3} \frac{d\theta_3}{dt} - j c e^{j\theta_4} \frac{d\theta_4}{dt} = 0 \qquad (6.14a)$$

But,

$$\frac{d\theta_2}{dt} = \omega_2; \qquad \frac{d\theta_3}{dt} = \omega_3; \qquad \frac{d\theta_4}{dt} = \omega_4 \qquad (6.14b)$$

and:

$$j a \omega_2 e^{j\theta_2} + j b \omega_3 e^{j\theta_3} - j c \omega_4 e^{j\theta_4} = 0 \qquad (6.14c)$$

Note that the θ_1 term has dropped out because that angle is a constant and thus its derivative is zero. Note also that equation 6.14 is, in fact the **relative velocity** or **velocity difference equation**.

$$\mathbf{V}_A + \mathbf{V}_{BA} - \mathbf{V}_B = 0 \qquad (6.15a)$$

where:

$$\mathbf{V}_A = j a \omega_2 e^{j\theta_2}$$
$$\mathbf{V}_{BA} = j b \omega_3 e^{j\theta_3} \qquad (6.15b)$$
$$\mathbf{V}_B = j c \omega_4 e^{j\theta_4}$$

Please compare equations 6.15 to equations 6.3, 6.5, and 6.6. This equation is solved graphically in the vector diagram of Figure 6-20b.

We now need to solve equation 6.14 for ω_3 and ω_4, knowing the input velocity ω_2, the link lengths, and all link angles. Thus the position analysis derived in Section 4.5 must be done first to determine the link angles before this velocity analysis can be completed. We wish to solve equation 6.14 to get expressions in this form:

$$\omega_3 = f(a,b,c,d,\theta_2,\theta_3,\theta_4,\omega_2); \qquad \omega_4 = g(a,b,c,d,\theta_2,\theta_3,\theta_4,\omega_2) \qquad (6.16)$$

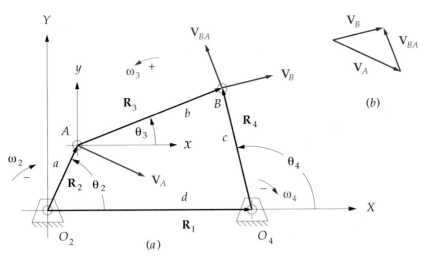

FIGURE 6-20

Position vector loop for a fourbar linkage showing velocity vectors for a negative (cw) ω_2

The strategy of solution will be the same as was done for the position analysis. First, substitute the Euler identity from equation 4.4a in each term of equation 6.14c:

$$ja\,\omega_2\left(\cos\theta_2 + j\sin\theta_2\right) + jb\,\omega_3\left(\cos\theta_3 + j\sin\theta_3\right) - jc\,\omega_4\left(\cos\theta_4 + j\sin\theta_4\right) = 0 \quad (6.17a)$$

Multiply through by the operator j:

$$a\,\omega_2\left(j\cos\theta_2 + j^2\sin\theta_2\right) + b\,\omega_3\left(j\cos\theta_3 + j^2\sin\theta_3\right) - c\,\omega_4\left(j\cos\theta_4 + j^2\sin\theta_4\right) = 0 \quad (6.17b)$$

The cosine terms have become the imaginary, or y-directed terms, and because $j^2 = -1$, the sine terms have become real or x-directed.

$$a\,\omega_2\left(-\sin\theta_2 + j\cos\theta_2\right) + b\,\omega_3\left(-\sin\theta_3 + j\cos\theta_3\right) - c\,\omega_4\left(-\sin\theta_4 + j\cos\theta_4\right) = 0 \quad (6.17c)$$

We can now separate this vector equation into its two components by collecting all real and all imaginary terms separately:

real part (x component):

$$-a\,\omega_2\sin\theta_2 - b\,\omega_3\sin\theta_3 + c\,\omega_4\sin\theta_4 = 0 \qquad\qquad (6.17d)$$

imaginary part (y component):

$$a\,\omega_2\cos\theta_2 + b\,\omega_3\cos\theta_3 - c\,\omega_4\cos\theta_4 = 0 \qquad\qquad (6.17e)$$

Note that the j's have cancelled in equation 6.17e. We can solve these two equations, 6.17d and 6.17e, simultaneously by direct substitution to get:

$$\omega_3 = \frac{a\omega_2}{b}\frac{\sin\left(\theta_4 - \theta_2\right)}{\sin\left(\theta_3 - \theta_4\right)} \qquad\qquad (6.18a)$$

$$\omega_4 = \frac{a\omega_2}{c}\frac{\sin(\theta_2 - \theta_3)}{\sin(\theta_4 - \theta_3)} \tag{6.18b}$$

Once we have solved for ω_3 and ω_4, we can then solve for the linear velocities by substituting the Euler identity into equations 6.15,

$$\mathbf{V}_A = ja\,\omega_2(\cos\theta_2 + j\sin\theta_2) = a\,\omega_2(-\sin\theta_2 + j\cos\theta_2) \tag{6.19a}$$

$$\mathbf{V}_{BA} = jb\,\omega_3(\cos\theta_3 + j\sin\theta_3) = b\,\omega_3(-\sin\theta_3 + j\cos\theta_3) \tag{6.19b}$$

$$\mathbf{V}_B = jc\,\omega_4(\cos\theta_4 + j\sin\theta_4) = c\,\omega_4(-\sin\theta_4 + j\cos\theta_4) \tag{6.19c}$$

where the real and imaginary terms are the x and y components, respectively. Equations 6.18 and 6.19 provide a complete solution for the angular velocities of the links and the linear velocities of the joints in the pin-jointed fourbar linkage. Note that there are also two solutions to this velocity problem, corresponding to the open and crossed branches of the linkage. They are found by the substitution of the open or crossed branch values of θ_3 and θ_4 obtained from equations 4.10 and 4.13 into equations 6.18 and 6.19. Figure 6-20a shows the open branch.

The Fourbar Slider-Crank

The position equations for the fourbar offset slider-crank linkage (inversion #1) were derived in Section 4.6. The linkage was shown in Figure 4-9 and is shown again in Figure 6-21a on which we also show an input angular velocity ω_2 applied to link 2. This ω_2 can be a time-varying input velocity. The vector loop equation 4.14 is repeated here for your convenience.

$$\mathbf{R}_2 - \mathbf{R}_3 - \mathbf{R}_4 - \mathbf{R}_1 = 0 \tag{4.14a}$$

$$ae^{j\theta_2} - be^{j\theta_3} - ce^{j\theta_4} - de^{j\theta_1} = 0 \tag{4.14b}$$

Differentiate equation 4.14b with respect to time noting that $a, b, c, \theta_1,$ and θ_4 are constant but the length of link d varies with time in this inversion.

$$ja\,\omega_2 e^{j\theta_2} - jb\,\omega_3 e^{j\theta_3} - \dot{d} = 0 \tag{6.20a}$$

The term d dot is the linear velocity of the slider block. Equation 6.20a is the velocity difference equation (6.5) and can be written in that form.

$$\mathbf{V}_A - \mathbf{V}_{AB} - \mathbf{V}_B = 0$$

or:

$$\mathbf{V}_A = \mathbf{V}_B + \mathbf{V}_{AB}$$

but: (6.20b)

$$\mathbf{V}_{AB} = -\mathbf{V}_{BA}$$

then:

$$\mathbf{V}_B = \mathbf{V}_A + \mathbf{V}_{BA}$$

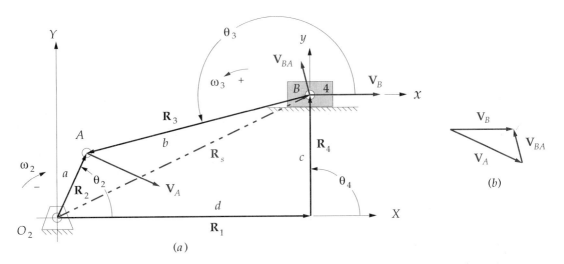

FIGURE 6-21

Position vector loop for a fourbar slider-crank linkage showing velocity vectors for a negative (cw) ω_2

Equation 6.20 is identical in form to equations 6.5 and 6.15a. Note that because we arranged the position vector $\mathbf{R}_3$ in Figures 4-9 and 6-21 with its root at point B, directed from B to A, its derivative represents the velocity difference of point A with respect to point B, the opposite of that in the previous fourbar example. Compare this also to equation 6.15b noting that its vector $\mathbf{R}_3$ is directed from A to B. Figure 6-21b shows the vector diagram of the graphical solution to equation 6.20b.

Substitute the Euler equivalent, equation 4.4a, in equation 6.20a,

$$ja\,\omega_2\left(\cos\theta_2 + j\sin\theta_2\right) - jb\,\omega_3\left(\cos\theta_3 + j\sin\theta_3\right) - \dot{d} = 0 \qquad (6.21a)$$

simplify,

$$a\,\omega_2\left(-\sin\theta_2 + j\cos\theta_2\right) - b\,\omega_3\left(-\sin\theta_3 + j\cos\theta_3\right) - \dot{d} = 0 \qquad (6.21b)$$

and separate into real and imaginary components.

real part (x component):

$$-a\,\omega_2\sin\theta_2 + b\,\omega_3\sin\theta_3 - \dot{d} = 0 \qquad (6.21c)$$

imaginary part (y component):

$$a\,\omega_2\cos\theta_2 - b\,\omega_3\cos\theta_3 = 0 \qquad (6.21d)$$

These are two simultaneous equations in the two unknowns, d dot and ω_3. Equation 6.21d can be solved for ω_3 and substituted into 6.21c to find d dot.

$$\omega_3 = \frac{a\,\cos\theta_2}{b\,\cos\theta_3}\,\omega_2 \qquad (6.22a)$$

$$\dot{d} = -a\,\omega_2\sin\theta_2 + b\,\omega_3\sin\theta_3 \qquad (6.22b)$$

The absolute velocity of point A and the velocity difference of point A versus point B are found from equation 6.20:

$$\mathbf{V}_A = a\,\omega_2\left(-\sin\theta_2 + j\cos\theta_2\right) \qquad (6.23a)$$

$$\mathbf{V}_{AB} = b\,\omega_3\left(-\sin\theta_3 + j\cos\theta_3\right) \qquad (6.23b)$$

$$\mathbf{V}_{BA} = -\mathbf{V}_{AB} \qquad (6.23c)$$

The Fourbar Inverted Slider-Crank

The position equations for the fourbar inverted slider-crank linkage were derived in Section 4.7. The linkage was shown in Figure 4-10 and is shown again in Figure 6-22 on which we also show an input angular velocity ω_2 applied to link 2. This ω_2 can vary with time. The vector loop equations 4.14 shown above are valid for this linkage as well.

All slider linkages will have at least one link whose effective length between joints varies as the linkage moves. In this inversion the length of link 3 between points A and B, designated as b, will change as it passes through the slider block on link 4. To get an expression for velocity, differentiate equation 4.14b with respect to time noting that a, c, d, and θ_1, are constant and b varies with time.

$$ja\,\omega_2 e^{j\theta_2} - jb\,\omega_3 e^{j\theta_3} - \dot{b}e^{j\theta_3} - jc\,\omega_4 e^{j\theta_4} = 0 \qquad (6.24)$$

The value of db/dt will be one of the variables to be solved for in this case and is the b *dot* term in the equation. Another variable will be ω_4, the angular velocity of link 4. Note, however, that we also have an unknown in ω_3, the angular velocity of link 3. This is a total of three unknowns. Equation 6.24 can only be solved for two unknowns. Thus we require another equation to solve the system. There is a fixed relationship between angles θ_3 and θ_4, shown as γ in Figure 6-22 and defined in equation 4.18, repeated here:

$$\theta_3 = \theta_4 + \gamma \qquad (4.18)$$

Differentiate it with respect to time to obtain:

$$\omega_3 = \omega_4 \qquad (6.25)$$

We wish to solve equation 6.24 to get expressions in this form:

$$\omega_3 = \omega_4 = f\left(a,b,c,d,\theta_2,\theta_3,\theta_4,\omega_2\right)$$

$$\qquad (6.26)$$

$$\frac{db}{dt} = \dot{b} = g\left(a,b,c,d,\theta_2,\theta_3,\theta_4,\omega_2\right)$$

Substitution of the Euler identity (Eq 4.4a) into equation 6.24 yields:

$$ja\,\omega_2\left(\cos\theta_2 + j\sin\theta_2\right) - jb\,\omega_3\left(\cos\theta_3 + j\sin\theta_3\right)$$
$$-\dot{b}\left(\cos\theta_3 + j\sin\theta_3\right) - jc\,\omega_4\left(\cos\theta_4 + j\sin\theta_4\right) = 0 \qquad (6.27a)$$

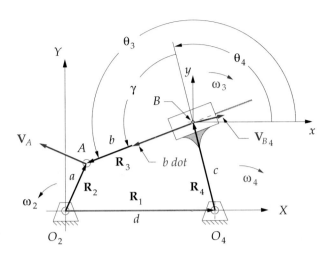

FIGURE 6-22

Velocity analysis of inversion #3 of the slider-crank fourbar linkage

Multiply by the operator j and substitute ω_4 for ω_3 from equation 6.25:

$$a\omega_2\left(-\sin\theta_2 + j\cos\theta_2\right) - b\omega_4\left(-\sin\theta_3 + j\cos\theta_3\right)$$
$$-\dot{b}\left(\cos\theta_3 + j\sin\theta_3\right) - c\omega_4\left(-\sin\theta_4 + j\cos\theta_4\right) = 0 \qquad (6.27b)$$

We can now separate this vector equation into its two components by collecting all real and all imaginary terms separately:

real part (x component):

$$-a\omega_2\sin\theta_2 + b\omega_4\sin\theta_3 - \dot{b}\cos\theta_3 + c\omega_4\sin\theta_4 = 0 \qquad (6.28a)$$

imaginary part (y component):

$$a\omega_2\cos\theta_2 - b\omega_4\cos\theta_3 - \dot{b}\sin\theta_3 - c\omega_4\cos\theta_4 = 0 \qquad (6.28b)$$

We will now collect terms and rearrange equations 6.28 to isolate one unknown on the left side.

$$\dot{b}\cos\theta_3 = -a\omega_2\sin\theta_2 + \omega_4\left(b\sin\theta_3 + c\sin\theta_4\right) \qquad (6.29a)$$
$$\dot{b}\sin\theta_3 = a\omega_2\cos\theta_2 - \omega_4\left(b\cos\theta_3 + c\cos\theta_4\right) \qquad (6.29b)$$

Either equation can be solved for b dot and the result substituted in the other. Solving equation 6.29a:

$$\dot{b} = \frac{-a\omega_2\sin\theta_2 + \omega_4\left(b\sin\theta_3 + c\sin\theta_4\right)}{\cos\theta_3} \qquad (6.30a)$$

Substitute in equation 6.29b and simplify:

$$\omega_4 = \frac{a\omega_2 \cos(\theta_2 - \theta_3)}{b + c\cos(\theta_4 - \theta_3)} \qquad (6.30b)$$

Equation 6.30a provides the **velocity of slip** at point B. Equation 6.30b gives the **angular velocity** of link 4. Note that we can substitute $-\gamma = \theta_4 - \theta_3$ from equation 4.18 into equation 6.30b to further simplify it. Note that $\cos(-\gamma) = \cos(\gamma)$.

$$\omega_4 = \frac{a\omega_2 \cos(\theta_2 - \theta_3)}{b + c\cos\gamma} \qquad (6.30c)$$

The **velocity of slip** from equation 6.30a is always directed along the **axis of slip** as shown in Figures 6-19 and 6-22. There is also a component orthogonal to the axis of slip called the **velocity of transmission**. This lies along the **axis of transmission** which is the only line along which any useful work can be transmitted across the sliding joint. All energy associated with motion along the slip axis is converted to heat and lost.

The absolute linear velocity of point A is found from equation 6.23a. We can find the absolute velocity of point B on link 4 since ω_4 is now known. From equation 6.15b:

$$\mathbf{V}_{B_4} = jc\,\omega_4 e^{j\theta_4} = c\,\omega_4\left(-\sin\theta_4 + j\cos\theta_4\right) \qquad (6.31)$$

6.8   VELOCITY ANALYSIS OF THE GEARED FIVEBAR LINKAGE

The position loop equation for the geared fivebar mechanism was derived in Section 4.8 and is repeated here. See Figure P6-4 for notation.

$$a e^{j\theta_2} + b e^{j\theta_3} - c e^{j\theta_4} - d e^{j\theta_5} - f e^{j\theta_1} = 0 \qquad (4.23b)$$

Differentiate this with respect to time to get an expression for velocity.

$$a\omega_2 j e^{j\theta_2} + b\omega_3 j e^{j\theta_3} - c\omega_4 j e^{j\theta_4} - d\omega_5 j e^{j\theta_5} = 0 \qquad (6.32a)$$

Substitute the Euler equivalents:

$$a\omega_2 j\left(\cos\theta_2 + j\sin\theta_2\right) + b\omega_3 j\left(\cos\theta_3 + j\sin\theta_3\right)$$
$$-c\omega_4 j\left(\cos\theta_4 + j\sin\theta_4\right) - d\omega_5 j\left(\cos\theta_5 + j\sin\theta_5\right) = 0 \qquad (6.32b)$$

Note that the angle θ_5 is defined in terms of θ_2, the gear ratio λ, and the phase angle ϕ.

$$\theta_5 = \lambda\theta_2 + \phi \qquad (4.23c)$$

Differentiate with respect to time:

$$\omega_5 = \lambda\omega_2 \qquad (6.32c)$$

Since a complete position analysis must be done before a velocity analysis, we will assume that the values of θ_5 and ω_5 have been found and will leave these equations in terms of θ_5 and ω_5.

Separating the real and imaginary terms in equation 6.32b:

real:

$$-a\omega_2 \sin\theta_2 - b\omega_3 \sin\theta_3 + c\omega_4 \sin\theta_4 + d\omega_5 \sin\theta_5 = 0 \tag{6.32d}$$

imaginary:

$$a\omega_2 \cos\theta_2 + b\omega_3 \cos\theta_3 - c\omega_4 \cos\theta_4 - d\omega_5 \cos\theta_5 = 0 \tag{6.32e}$$

The only two unknowns are ω_3 and ω_4. Either equation 6.32d or 6.32e can be solved for one unknown and the result substituted in the other. The solution for ω_3 is:

$$\omega_3 = -\frac{2\sin\theta_4\left[a\omega_2 \sin(\theta_2 - \theta_4) + d\omega_5 \sin(\theta_4 - \theta_5)\right]}{b\left[\cos(\theta_3 - 2\theta_4) - \cos\theta_3\right]} \tag{6.33a}$$

The angular velocity ω_4 can be found from equation 6.32d using ω_3.

$$\omega_4 = \frac{a\omega_2 \sin\theta_2 + b\omega_3 \sin\theta_3 - d\omega_5 \sin\theta_5}{c\sin\theta_4} \tag{6.33b}$$

With all link angles and angular velocities known, the linear velocities of the pin joints can be found from:

$$\mathbf{V}_A = a\omega_2\left(-\sin\theta_2 + j\cos\theta_2\right) \tag{6.33c}$$

$$\mathbf{V}_{BA} = b\omega_3\left(-\sin\theta_3 + j\cos\theta_3\right) \tag{6.33d}$$

$$\mathbf{V}_C = d\omega_5\left(-\sin\theta_5 + j\cos\theta_5\right) \tag{6.33e}$$

$$\mathbf{V}_B = \mathbf{V}_A + \mathbf{V}_{BA} \tag{6.33f}$$

6.9 VELOCITY OF ANY POINT ON A LINKAGE

Once the angular velocities of all the links are found it is easy to define and calculate the velocity of *any point on any link* for any input position of the linkage. Figure 6-23 shows the fourbar linkage with its coupler, link 3, enlarged to contain a coupler point P. The crank and rocker have also been enlarged to show points S and U which might represent the centers of gravity of those links. We want to develop algebraic expressions for the velocities of these (or any) points on the links.

To find the velocity of point S, draw the position vector from the fixed pivot O_2 to point S. This vector, $\mathbf{R}_{SO_2}$ makes an angle δ_2 with the vector $\mathbf{R}_{AO_2}$. The angle δ_2 is completely defined by the geometry of link 2 and is constant. The position vector for point S is then:

$$\mathbf{R}_{SO_2} = \mathbf{R}_S = se^{j(\theta_2 + \delta_2)} = s\left[\cos(\theta_2 + \delta_2) + j\sin(\theta_2 + \delta_2)\right] \tag{4.25}$$

Differentiate this position vector to find the velocity of that point.

$$\mathbf{V}_S = jse^{j(\theta_2 + \delta_2)}\omega_2 = s\omega_2\left[-\sin(\theta_2 + \delta_2) + j\cos(\theta_2 + \delta_2)\right] \tag{6.34}$$

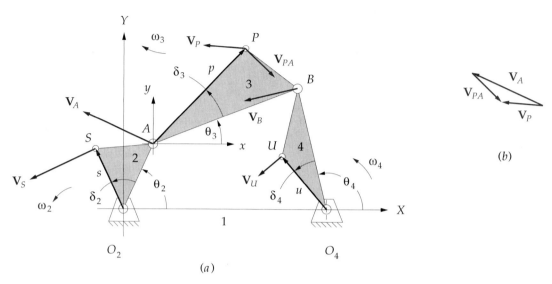

FIGURE 6-23

Finding the velocities of points on the links

The position of point U on link 4 is found in the same way, using the angle δ_4 which is a constant angular offset within the link. The expression is:

$$\mathbf{R}_{UO_4} = ue^{j(\theta_4 + \delta_4)} = u\left[\cos(\theta_4 + \delta_4) + j\sin(\theta_4 + \delta_4)\right] \tag{4.26}$$

Differentiate this position vector to find the velocity of that point.

$$\mathbf{V}_U = jue^{j(\theta_4 + \delta_4)}\omega_4 = u\omega_4\left[-\sin(\theta_4 + \delta_4) + j\cos(\theta_4 + \delta_4)\right] \tag{6.35}$$

The velocity of point P on link 3 can be found from the addition of two velocity vectors, such as $\mathbf{V}_A$ and $\mathbf{V}_{PA}$. $\mathbf{V}_A$ is already defined from our analysis of the link velocities. $\mathbf{V}_{PA}$ is the velocity difference of point P with respect to point A. Point A is chosen as the reference point because angle θ_3 is defined in a local co-ordinate system whose origin is at A. Position vector $\mathbf{R}_{PA}$ is defined in the same way as $\mathbf{R}_S$ or $\mathbf{R}_U$ using the internal link offset angle δ_3 and the angle of link 3, θ_3. This was done in equation 4.27.

$$\mathbf{R}_{PA} = pe^{j(\theta_3 + \delta_3)} = p\left[\cos(\theta_3 + \delta_3) + j\sin(\theta_3 + \delta_3)\right] \tag{4.27a}$$

$$\mathbf{R}_P = \mathbf{R}_A + \mathbf{R}_{PA} \tag{4.27b}$$

Differentiate this position vector to find the velocity of that point.

$$\mathbf{V}_{PA} = jpe^{j(\theta_3 + \delta_3)}\omega_3 = p\omega_3\left[-\sin(\theta_3 + \delta_3) + j\cos(\theta_3 + \delta_3)\right] \tag{6.36a}$$

$$\mathbf{V}_P = \mathbf{V}_A + \mathbf{V}_{PA} \tag{6.36b}$$

Please compare equation 6.36 with equations 6.5 and 6.15. It is, again, the velocity difference equation.

6.10   REFERENCES

1   **Towfigh, K.,** "The Four-Bar Linkage as an Adjustment Mechanism," Proc. of the Applied Mechanisms Conference, Oklahoma State Univ., 1969, pp. 27-1 to 27-4.

2   **Wood, G. A.,** "Educating for Creativity in Engineering," Proc. of the 85th Annual ASEE Conference, Univ. of N. Dakota, June 1977.

6.11   PROBLEMS

6-1   A ship is steaming due north at 20 knots (nautical miles per hour). A submarine is laying in wait 1/2 mile due west of the ship. The sub fires a torpedo on a course of 85 degrees. The torpedo travels at a constant speed of 30 knots. Will it strike the ship? If not, by how many nautical miles will it miss? Hint: Use the relative velocity equation and solve graphically or analytically.

6-2   A point is at a 6.5 inch radius on a body in pure rotation with $\omega = 100$ rad/sec. The rotation center is at the origin of a coordinate system. When the point is at position A, its position vector makes a 45° angle with the X axis. At position B, its position vector makes a 75° angle with the X axis. Draw this system to some convenient scale and:

    a.   Write an expression for the particle's velocity vector in position A using complex number notation, in both polar and cartesian forms.

    b.   Write an expression for the particle's velocity vector in position B using complex number notation, in both polar and cartesian forms.

    c.   Write a vector equation for the velocity difference between point B and A. Substitute the complex number notation for the vectors in this equation and solve for the position difference numerically.

    d.   Check the result of part c with a graphical method.

6-3   Repeat problem 6-2 considering points A and B to be on separate bodies rotating about the origin with ω's of –50 (A) and + 75 rad/sec (B). Find their relative velocity.

*6-4   The link lengths, coupler point location, and the values of θ_2 and ω_2 for the same fourbar linkages as used for position analysis in Chapter 4 are redefined in Table P6-1, which is the same as Table P4-1. The general linkage configuration and terminology are shown in Figure P6-1. *For the row(s) assigned,* draw the linkage to scale and **graphically** find the velocities of the pin joints A and B, and of instant centers $I_{1,3}$ and $I_{2,4}$. Then calculate ω_3 and ω_4 and find the velocity of point P.

*6-5   Repeat problem 6-4 except solve by the analytical vector loop method of Section 6.7.

*6-6   The link lengths and the values of θ_2 and ω_2 for some noninverted, offset fourbar slider-crank linkages are defined in Table P6-2. The general linkage configuration and terminology are shown in Figure P6-2. *For the row(s) assigned,* draw the linkage to scale and **graphically** find the velocities of the pin joints A and B and the velocity of slip at the sliding joint.

---

* Answers in Appendix E

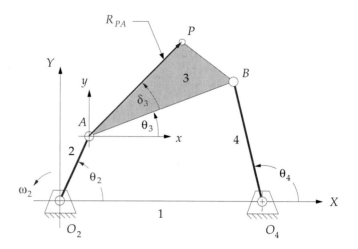

FIGURE P6-1

Configuration and terminology for the pin-jointed fourbar linkage

*6-7   Repeat problem 6-6 using an analytical method.

*6-8   The link lengths and the values of θ_2, ω_2, and γ for some inverted fourbar slider-crank linkages are defined in Table P6-3. The general linkage configuration and terminology are shown in Figure P6-3. *For the row(s) assigned*, draw the linkage to scale and **graphically** find the velocities of the pin joints A and B and velocity of slip at the sliding joint.

TABLE P6-1   Data for Fourbar Analysis Problems

Row	Link 1	Link 2	Link 3	Link 4	θ_2	ω_2	R_{pa}	δ_3
a	6	2	7	9	30	10	6	30
b	7	9	3	8	85	−12	9	25
c	3	10	6	8	45	−15	10	80
d	8	5	7	6	25	24	5	45
e	8	5	8	6	75	−50	9	300
f	5	8	8	9	15	−45	10	120
g	6	8	8	9	25	100	4	300
h	20	10	10	10	50	−65	6	20
i	4	5	2	5	80	25	9	80
j	20	10	5	10	33	25	1	0
k	4	6	10	7	88	−80	10	330
l	9	7	10	7	60	−90	5	180
m	9	7	11	8	50	75	10	90
n	9	7	11	6	120	15	15	60

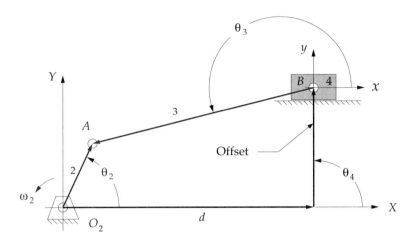

FIGURE P6-2

Configuration and terminology for a fourbar offset slider-crank linkage

*6-9   Repeat problem 6-8 using an analytical method.

*6-10 The link lengths, gear ratio (λ), phase angle (ϕ), and the values of θ_2 and ω_2 for some geared fivebar linkages are defined in Table P6-4. The general linkage configuration and terminology are shown in Figure P6-4. *For the row(s) assigned,* draw the linkage to scale and graphically find ω_3 and ω_4.

*6-11 Repeat problem 6-10 using an analytical method.

6-12   Find all the instant centers of the linkages shown in Figure P6-5.

6-13   Find all the instant centers of the linkages shown in Figure P6-6.

6-14   Trace or measure the link geometry in Figure P6-7 and find ω_2, ω_3, ω_4, V_{A4}, V_{trans}, and V_{slip} for the position shown with $V_{A2} = 20$ units/sec in direction shown. Link 2 is 2.5 units long.

TABLE P6-2   Data for Noninverted Slider-Crank Problems

Row	Link 2	Link 3	Offset	θ_2	ω_2
a	1.4	4	1	45	10
b	2	6	−3	60	−12
c	3	8	2	−30	−15
d	3.5	10	1	120	24
e	5	20	−5	225	−50
f	3	13	0	100	−45
g	7	25	10	330	100

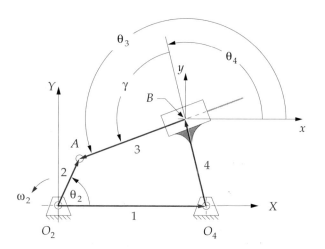

FIGURE P6-3

Configuration and terminology for inversion #3 of the fourbar slider-crank linkage

6-15 A plane is flying due south at 500 mph at 35,000 feet altitude, straight and level. A second plane is initially 40 miles due east of the first plane, also at 35,000 feet altitude, flying straight and level and traveling at 550 mph.

    a.   Determine the compass angle at which the second plane would be on a collision course with the first.

    b.   How long will it take for the second plane to catch the first?

    Hint: Use the relative velocity equation.

6-16 Trace or measure the link geometry in Figure P6-5a and find ω_3, V_A, V_B, and V_C for the position shown for $\omega_2 = 15$ rad/sec in direction shown,

    a.   Using the velocity difference graphical method.

    b.   Using the instant center graphical method.

    c.   Using the analytic method.

6-17 Trace or measure the link geometry in Figure P6-5c and find ω_3, ω_4, V_A, V_B, and V_C for the position shown for $\omega_2 = 15$ rad/sec in direction shown,

    a.   Using the velocity difference graphical method.

TABLE P6-3   **Data for Inverted Slider-Crank Problems**

Row	Link 1	Link 2	Link 4	γ	θ_2	ω_2
a	6	2	4	90	30	10
b	7	9	3	75	85	-15
c	3	10	6	45	45	24
d	8	5	3	60	25	-50
e	8	4	2	30	75	-45
f	5	8	8	90	150	100

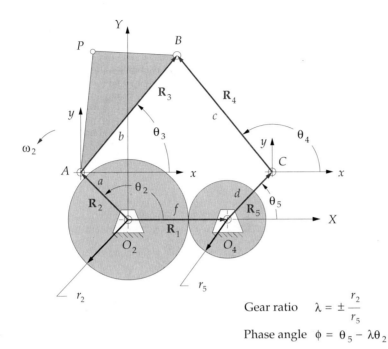

Gear ratio $\quad \lambda = \pm \dfrac{r_2}{r_5}$

Phase angle $\quad \phi = \theta_5 - \lambda\theta_2$

FIGURE P6-4

Configuration and terminology for the geared fivebar linkage

    b.   Using the instant center graphical method.

    c.   Using the analytic method. (Hint: create an effective linkage for the position shown and analyze as a pin-jointed fourbar.)

6-18   Repeat problem 6-16 for the mechanism in Figure P6-5f. There is no ω_2 in this linkage. Let $V_A = 10$ in/sec.

TABLE P6-4   **Data for Geared Fivebar Linkage Problems**

Row	Link 1	Link 2	Link 3	Link 4	Link 5	λ	ϕ	$\omega 2$	θ_2
a	6	1	7	9	4	2	30	10	60
b	6	5	7	8	4	−2.5	60	−12	30
c	3	5	7	8	4	−0.5	0	−15	45
d	4	5	7	8	4	−1	120	24	75
e	5	9	11	8	8	3.2	−50	−50	−39
f	10	2	7	5	3	1.5	30	−45	120
g	15	7	9	11	4	2.5	−90	100	75
h	12	8	7	9	4	−2.5	60	−65	55
i	9	7	8	9	4	−4	120	25	100

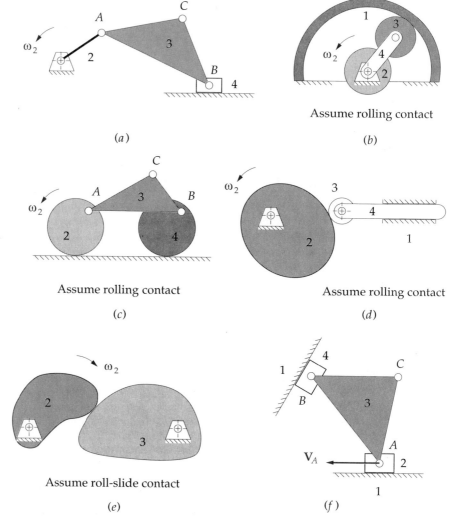

(a)

Assume rolling contact

(b)

Assume rolling contact

(c)

Assume rolling contact

(d)

Assume roll-slide contact

(e)

(f)

FIGURE P6-5

Velocity analysis and instant center problems

6-19 Trace or measure the link geometry in Figure P6-5d and find $\mathbf{V}_4$, $\mathbf{V}_{trans}$, and $\mathbf{V}_{slip}$ for the position shown with $\omega_2 = 20$ rad/sec in the direction shown.

6-20 Trace or measure the link geometry in Figure P6-5e and find ω_3, $\mathbf{V}_{trans}$, and $\mathbf{V}_{slip}$ for the position shown for $\omega_2 = 10$ rad/sec in the direction shown.

6-21 Trace or measure the link geometry in Figure P6-6b and find, for the position shown, the velocity ratio $V_{I5,6}/V_{I2,3}$ and the mechanical advantage from link 2 to link 6.

    a.   Using the velocity difference graphical method.

    b.   Using the instant center graphical method.

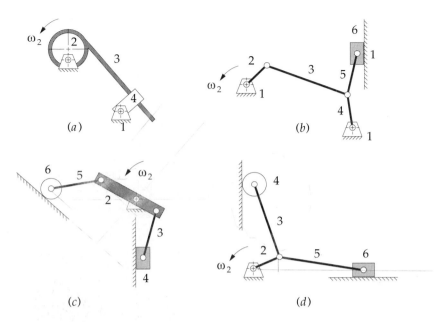

(a)

(b)

(c)

(d)

FIGURE P6-6

Mechanical advantage and instant center problems

6-22   Repeat problem 6-21 for the mechanism in Figure P6-6d.

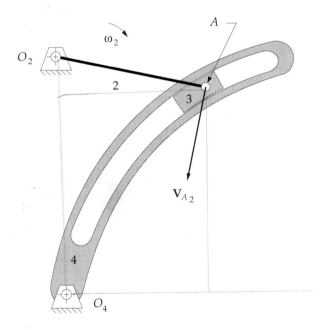

FIGURE P6-7

Velocity analysis at a sliding joint

Chapter **7**

ACCELERATION ANALYSIS

Take it to warp five, Mr. Sulu
CAPTAIN KIRK

7.0 INTRODUCTION

Once a velocity analysis is done, the next step is to determine the accelerations of all links and points of interest in the mechanism or machine. We need to know the accelerations to calculate the dynamic forces from $\mathbf{F} = m\mathbf{a}$. The dynamic forces will contribute to the stresses in the links and other components. Many methods and approaches exist to find accelerations in mechanisms. We will examine only a few of these methods here. We will first develop a manual graphical method, which is often useful as a check on the more complete and accurate analytical solution. Then we will derive the analytical solution for accelerations in the fourbar and inverted slider-crank linkages as examples of the general vector loop equation solution to acceleration analysis problems.

7.1 DEFINITION OF ACCELERATION

Acceleration is defined as *the rate of change of velocity with respect to time*. Velocity ($\mathbf{V}$, ω) is a vector quantity and so is acceleration. Accelerations can be **angular** or **linear**. **Angular acceleration** will be denoted as α and **linear acceleration** as $\mathbf{A}$.

$$\alpha = \frac{d\omega}{dt}; \qquad \mathbf{A} = \frac{d\mathbf{V}}{dt} \qquad (7.1)$$

Figure 7-1 shows a link PA in pure rotation, pivoted at point A in the xy plane. We are interested in the acceleration of point P when the link is subjected to an angular velocity ω and an angular acceleration α, which need not have the same sense. The link's position is defined by the position vector $\mathbf{R}$, and the velocity of point P is $\mathbf{V}_{PA}$. These vectors were defined in equations 6.2 and 6.3 which are repeated here for convenience. (See also Figure 6-1.)

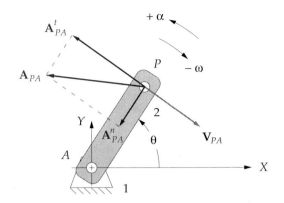

FIGURE 7-1

Acceleration of a link in pure rotation with a positive (ccw) α_2 and a negative (cw) ω_2

$$\mathbf{R}_{PA} = p e^{j\theta} \tag{6.2}$$

$$\mathbf{V}_{PA} = \frac{d\mathbf{R}_{PA}}{dt} = p j e^{j\theta} \frac{d\theta}{dt} = p\omega\, j e^{j\theta} \tag{6.3}$$

where p is the scalar length of the vector $\mathbf{R}_{PA}$. We can easily differentiate equation 6.3 to obtain an expression for the acceleration of point P:

$$\mathbf{A}_{PA} = \frac{d\mathbf{V}_{PA}}{dt} = \frac{d\left(p\omega\, j e^{j\theta}\right)}{dt}$$

$$\mathbf{A}_{PA} = j p \left(e^{j\theta} \frac{d\omega}{dt} + j\omega e^{j\theta} \frac{d\theta}{dt} \right) \tag{7.2}$$

$$\mathbf{A}_{PA} = p\alpha\, j e^{j\theta} - p\omega^2\, e^{j\theta}$$

$$\mathbf{A}_{PA} = \mathbf{A}_{PA}^{t} + \mathbf{A}_{PA}^{n}$$

Note that there are two functions of time in equation 6.3, θ and ω. Thus there are two terms in the expression for acceleration, the tangential component of acceleration $\mathbf{A}_{PA}^{t}$ involving α, and the normal (or centripetal) component $\mathbf{A}_{PA}^{n}$ involving ω^2. As a result of the differentiation, the tangential component is multiplied by the (constant) complex operator j. This causes a rotation of this acceleration vector through 90 degrees with respect to the original position vector. (See also Figure 4-5b.) This 90° rotation is nominally positive, or counterclockwise. However, the tangential component is also multiplied by α, which may be either positive or negative. As a result, the tangential component of acceleration will be **rotated 90 degrees** from the angle θ of the *position vector* **in a direction dictated by the sign of** α. This is just mathematical verification of what you already knew, namely that *tangential acceleration is always in a direction perpendicular to the radius of rotation and is thus tangent to the path of motion* as shown in Figure 7-1. The normal, or centripetal acceleration component is multiplied by j^2, or -1. This directs *the centripetal component at 180 degrees to the angle θ of the original position* *vector*, i.e., toward the center (centripetal means *toward the center*). The total accelera-tion $\mathbf{A}_{PA}$ of point P is the vector sum of the tangential $\mathbf{A}_{PA}^{t}$ and normal $\mathbf{A}_{PA}^{n}$ com-ponents as shown in Figure 7-1 and equation 7.2.

Substituting the Euler identity (Eq. 4.4a) into equation 7.2 gives us the real and imaginary (or x and y) components of the acceleration vector.

$$\mathbf{A}_{PA} = p\alpha\left(-\sin\theta + j\cos\theta\right) - p\omega^2\left(\cos\theta + j\sin\theta\right) \tag{7.3}$$

The acceleration $\mathbf{A}_{PA}$ in Figure 7-1 can be referred to as an **absolute acceler-ation** since it is referenced to A, which is the origin of the global coordinate axes in that system. As such, we could have referred to it as $\mathbf{A}_P$, with the absence of the sec-ond subscript implying reference to the global coordinate system.

Figure 7-2a shows a different and slightly more complicated system in which the pivot A is no longer stationary. It has a known linear acceleration $\mathbf{A}_A$ as part of the translating carriage, link 3. If α is unchanged, the acceleration of point P versus A will be the same as before, but $\mathbf{A}_{PA}$ can no longer be considered an absolute acceleration. It is now an **acceleration difference** and **must** carry the second sub-script as $\mathbf{A}_{PA}$. The absolute acceleration $\mathbf{A}_P$ must now be found from the **accelera-tion difference** equation whose graphical solution is shown in Figure 7-2b:

$$\mathbf{A}_P = \mathbf{A}_A + \mathbf{A}_{PA}$$

$$\tag{7.4}$$

$$\left(\mathbf{A}_P^t + \mathbf{A}_P^n\right) = \left(\mathbf{A}_A^t + \mathbf{A}_A^n\right) + \left(\mathbf{A}_{PA}^t + \mathbf{A}_{PA}^n\right)$$

Note the similarity of equation 7.4 to the **velocity difference equation** (equation 6.5). Note also that the solution for $\mathbf{A}_P$ in equation 7.4 can be found either by adding the resultant vector $\mathbf{A}_{PA}$ or its normal and tangential compo-nents, $\mathbf{A}_{PA}^n$ and $\mathbf{A}_{PA}^t$ to the vector $\mathbf{A}_A$ in Figure 7-2b. The vector $\mathbf{A}_A$ has a zero normal component in this example because link 3 is in pure translation.

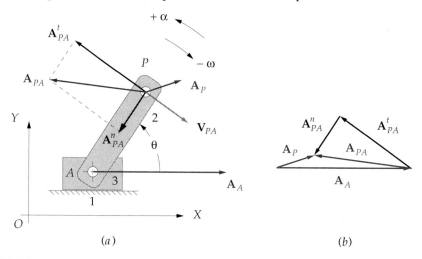

(a)   (b)

FIGURE 7-2

Acceleration difference in a system with a positive (*ccw*) α_2 and a negative (*cw*) ω_2

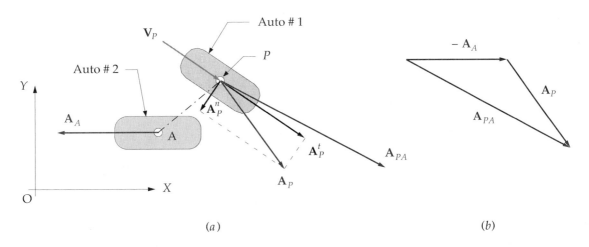

FIGURE 7-3

Relative acceleration

Figure 7-3 shows two independent bodies P and A, which could be two automobiles, moving in the same plane. Auto #1 is turning and accelerating into the path of auto #2, which is decelerating to avoid a crash. If their independent accelerations $\mathbf{A}_P$ and $\mathbf{A}_A$ are known, their **relative acceleration** $\mathbf{A}_{PA}$ can be found from equation 7.4 arranged algebraically as:

$$\mathbf{A}_{PA} = \mathbf{A}_P - \mathbf{A}_A \qquad (7.5)$$

The graphical solution to this equation is shown in Figure 7-3b.

As we did for velocity analysis, we give these two cases different names despite the fact that the same equation applies. Repeating the definition from Section 6.1, modified to refer to acceleration:

CASE 1:   Two points in the same body => *acceleration difference*

CASE 2:   Two points in different bodies => *relative acceleration*

7.2   GRAPHICAL ACCELERATION ANALYSIS

The comments made in regard to graphical velocity analysis in Section 6.2 apply as well to graphical acceleration analysis. Historically, graphical methods were the only practical way to solve these acceleration analysis problems. With some practice, and with proper tools such as a drafting machine or CAD package, one can fairly rapidly solve for the accelerations of particular points in a mechanism for any one input position by drawing vector diagrams. However, if accelerations for many positions of the mechanism are to be found, each new position requires a completely new set of vector diagrams be drawn. Very little of the work done to solve for the accelerations at position 1 carries over to position 2, etc. This is an even more tedious process than that for graphical velocity analysis because there are more components to draw. Nevertheless, this method still has more than historical value as it can provide a quick check on the results from a computer program solution. Such a check needs only be done for a few positions to prove the validity of the program.

To solve any acceleration analysis problem graphically, we need only three equations, equation 7.4 and equations 7.6 (which are merely the scalar magnitudes of the terms in equation 7.2):

$$\left|\mathbf{A}^t\right| = A^t = r\alpha$$

$$\left|\mathbf{A}^n\right| = A^n = r\omega^2$$

(7.6)

Note that the scalar equations 7.6 define only the **magnitudes** (A^t, A^n) of the components of acceleration of any point in rotation. In a CASE 1 graphical analysis, the **directions** of the vectors due to the centripetal and tangential components of the acceleration difference must be understood from equation 7.2 to be perpendicular to and along the radius of rotation, respectively. Thus, if the center of rotation is known or assumed, the directions of the acceleration difference components due to that rotation are known and their senses will be consistent with the angular velocity ω and angular acceleration α of the body.

Figure 7-4 shows a fourbar linkage in one particular position. We wish to solve for the angular accelerations of links 3 and 4 (α_3, α_4), and the linear accelerations of points A, B, and C, ($\mathbf{A}_A$, $\mathbf{A}_B$, $\mathbf{A}_C$). Point C represents any general point of interest such as a coupler point. The solution method is valid for any point on any link. To solve this problem we need to know the *lengths of all the links*, the *angular positions of all the links*, the *angular velocities of all the links*, and the *instantaneous input acceleration of any one driving link or driving point*. Assuming that we have designed this linkage, we will know or can measure the link lengths. We must also first do a **complete position and velocity analysis** to find the link angles θ_3 and θ_4, and angular velocities ω_3 and ω_4 given the input link's position θ_2, input angular velocity ω_2, and input acceleration α_2. This can be done by any of the methods in Chapters 4 and 6. In general we must solve these problems in stages, first for link positions, then for velocities, and finally for accelerations. For the following example, we will assume that a complete position and velocity analysis has been done, and that the input is to link 2 with known θ_2, ω_2, and α_2 for this one "freeze frame" position of the moving linkage.

EXAMPLE 7-1

Graphical Acceleration Analysis for One Position of a Fourbar Linkage.

Problem: Given θ_2, θ_3, θ_4, ω_2, ω_3, ω_4, α_2, find α_3, α_4, $\mathbf{A}_A$, $\mathbf{A}_B$, $\mathbf{A}_P$ by graphical methods.

Solution: (see Figure 7-4)

1 Start at the end of the linkage about which you have the most information. Calculate the magnitudes of the centripetal and tangential components of acceleration of point *A* using scalar equations 7.6.

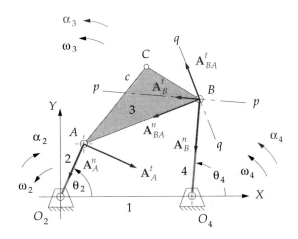

(a) Vector construction

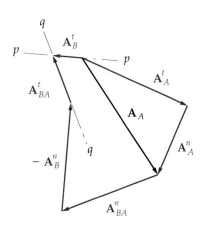

(b) Vector polygon (2x size)

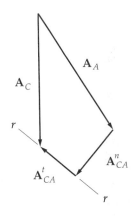

(c) Vector polygon (2x size)

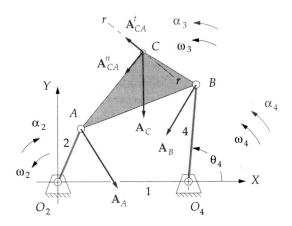

(d) Resultant vectors

FIGURE 7-4

Graphical solution for acceleration in a pin-jointed linkage with a negative (cw) α_2 and a positive (ccw) ω_2

$$A_A^n = (AO_2)\omega_2^2; \qquad\qquad A_A^t = (AO_2)\alpha_2 \qquad\qquad (a)$$

2   On the linkage diagram, Figure 7-4a, draw the acceleration component vectors A_A^n and A_A^t with their lengths equal to their magnitudes at some convenient scale. Place their roots at point A with their directions respectively along and perpendicular to the radius AO_2. The sense of A_A^t is defined by that of α_2 (according to the right-hand rule), and the sense of A_A^n is the opposite of that of the position vector R_A as shown in Figure 7-4a.

3   Move next to a point about which you have some information, such as B on link 4. Note that the directions of the tangential and normal components of acceleration of point B

are predictable since this link is in pure rotation about point O_4. Draw the construction line pp through point B perpendicular to BO_4, to represent the direction of $\mathbf{A}_B^t$ as shown in Figure 7-4a.

4   Write the acceleration difference vector equation 7.4 for point B versus point A.

$$\mathbf{A}_B = \mathbf{A}_A + \mathbf{A}_{BA} \qquad\qquad (b)$$

Substitute the normal and tangential components for each term:

$$\left(\mathbf{A}_B^t + \mathbf{A}_B^n\right) = \left(\mathbf{A}_A^t + \mathbf{A}_A^n\right) + \left(\mathbf{A}_{BA}^t + \mathbf{A}_{BA}^n\right) \qquad\qquad (c)$$

We will use point A as the reference point to find $\mathbf{A}_B$ because A is in the same link as B and we have already solved for $\mathbf{A}_A^t$ and $\mathbf{A}_A^n$. Any vector equation can be solved for two unknowns. Each term has two parameters, namely magnitude and direction. There are then potentially twelve unknowns in this equation, two per term. We must know ten of them to solve it. We know both the magnitudes and directions of $\mathbf{A}_A^t$ and $\mathbf{A}_A^n$ and the directions of $\mathbf{A}_B^t$ and $\mathbf{A}_B^n$ which are along line pp and line BO_4 respectively. We can also calculate the magnitude of $\mathbf{A}_B^n$ from equation 7.6 since we know ω_4. This provides seven known values. We need to know three more parameters to solve the equation.

5   The term $\mathbf{A}_{BA}$ represents the acceleration difference of B with respect to A. This has two components. The normal component $\mathbf{A}_{BA}^n$ is directed along the line BA because we are using point A as the reference center of rotation for the free vector ω_3, and its magnitude can be calculated from equation 7.6. The direction of $\mathbf{A}_{BA}^t$ must then be perpendicular to the line BA. Draw construction line qq through point B and perpendicular to BA to represent the direction of $\mathbf{A}_{BA}^t$ as shown in Figure 7-4a. The calculated magnitude and direction of component $\mathbf{A}_{BA}^n$ and the known direction of $\mathbf{A}_{BA}^t$ provide the needed additional three parameters.

6   Now the vector equation can be solved graphically by drawing a vector diagram as shown in Figure 7-4b. Either drafting tools or a CAD package is necessary for this step. The strategy is to first draw all vectors for which we know both magnitude and direction, being careful to arrange their senses according to equation 7.4.

   First draw acceleration vectors $\mathbf{A}_A^t$ and $\mathbf{A}_A^n$ tip to tail, carefully to some scale, maintaining their directions. (They are drawn twice size in the figure.) Note that the sum of these two components is the vector $\mathbf{A}_A$. The equation in step 4 says to add $\mathbf{A}_{BA}$ to $\mathbf{A}_A$. We know $\mathbf{A}_{BA}^n$, so we can draw that component at the end of $\mathbf{A}_A$. We also know $\mathbf{A}_B^n$, but this component is on the left side of equation 7.4, so we must subtract it. Draw the negative (opposite sense) of $\mathbf{A}_B^n$ at the end of $\mathbf{A}_{BA}^n$.

   This exhausts our supply of components for which we know both magnitude and direction. Our two remaining knowns are the directions of $\mathbf{A}_B^t$ and $\mathbf{A}_{BA}^t$ which lie along the lines pp and qq respectively. Draw a line parallel to line qq across the tip of the vector representing *minus* $\mathbf{A}_B^n$. The resultant, or left side of the equation, must close the vector diagram, from the tail of the first vector drawn ($\mathbf{A}_A$) to the tip of the last, so draw a line parallel to pp across the tail of $\mathbf{A}_A$. The intersection of these lines parallel to pp and qq defines the lengths of $\mathbf{A}_B^t$ and $\mathbf{A}_{BA}^t$. The senses of these vectors are determined from reference to equation 7.6. Vector $\mathbf{A}_A$ was added to $\mathbf{A}_{BA}$, so their components must be arranged tip to tail. Vector $\mathbf{A}_B$ is the resultant, so its component $\mathbf{A}_B^t$ must be from the tail of the first to the tip of the last. The resultant vectors are shown in Figure 7-4b and d.

7   The angular accelerations of links 3 and 4 can be calculated from equation 7.6:

$$\alpha_4 = \frac{A_B^t}{BO_4} \qquad\qquad \alpha_3 = \frac{A_{BA}^t}{BA} \qquad\qquad (d)$$

Note that the acceleration difference term $\mathbf{A}_{BA}^t$ represents the rotational component of acceleration of link 3 due to α_3. The rotational acceleration α of any body is a "**free vector**" which has no particular point of application to the body. It exists everywhere on the body.

8   Finally we can solve for $\mathbf{A}_C$ using equation 7.4 again. We select any point in link 3 for which we know the absolute velocity to use as the reference, such as point A.

$$\mathbf{A}_C = \mathbf{A}_A + \mathbf{A}_{CA} \qquad\qquad (e)$$

In this case, we can calculate the magnitude of $\mathbf{A}_{CA}^t$ from equation 7.6 as we have already found α_3,

$$A_{CA}^t = c\alpha_3 \qquad\qquad (f)$$

The magnitude of the component $\mathbf{A}_{CA}^n$ can be found from equation 7.6 using ω_3.

$$A_{CA}^n = c\omega_3^2 \qquad\qquad (g)$$

Since both $\mathbf{A}_A$ and $\mathbf{A}_{CA}$ are known, the vector diagram can be directly drawn as shown in Figure 7-4c. Vector $\mathbf{A}_C$ is the resultant which closes the vector diagram. Figure 7-4d shows the calculated acceleration vectors on the linkage diagram.

The above example contains some interesting and significant principles which deserve further emphasis. Equation 7.4 is repeated here for discussion.

$$\mathbf{A}_P = \mathbf{A}_A + \mathbf{A}_{PA}$$

$$(7.4)$$

$$\left(\mathbf{A}_P^t + \mathbf{A}_P^n\right) = \left(\mathbf{A}_A^t + \mathbf{A}_A^n\right) + \left(\mathbf{A}_{PA}^t + \mathbf{A}_{PA}^n\right)$$

This equation represents the *absolute* acceleration of some general point P referenced to the origin of the global coordinate system. The right side defines it as the sum of the absolute acceleration of some other reference point A in the same system and the acceleration difference (or relative acceleration) of point P versus point A. These terms are then further broken down into their normal (centripetal) and tangential components which have definitions as shown in equation 7.2.

Let us review what was done in Example 7-1 in order to extract the general strategy for solution of this class of problem. We started at the input side of the mechanism, as that is where the driving angular acceleration α_2 was defined. We first looked for a point (A) for which the motion was pure rotation. We then solved for the absolute acceleration of that point ($\mathbf{A}_A$) using equations 7.4 and 7.6 by breaking $\mathbf{A}_A$ into its normal and tangential components. *(Steps 1 and 2)*

We then used the point (A) just solved for as a reference point to define the translation component in equation 7.4 written for a new point (B). Note that we needed to choose a second point (B) which was in the same rigid body as the reference point (A) which we had already solved, and about which we could predict some aspect of the new point's (B's) acceleration components. In this example, we knew the direction of the component $\mathbf{A}_B^t$, though we did not yet know its magnitude. We could also calculate both magnitude and direction of the centripetal component, $\mathbf{A}_B^n$, since we knew ω_3 and the link length. In general this situation will obtain for any point on a link which is jointed to ground (as is link 4). In this example, we could not have solved for point C until we solved for B, because point C is on a floating link for which we do not yet know the angular acceleration or absolute acceleration direction. *(Steps 3 and 4)*

To solve the equation for the second point (B), we also needed to recognize that the tangential component of the acceleration difference $\mathbf{A}_{BA}^t$ is always directed perpendicular to the line connecting the two related points in the link (B and A in the example). In addition, you will always know the magnitude and direction of the centripetal acceleration components in equation 7.4 ***if it represents an acceleration difference*** (CASE 1) **situation**. *If the two points are in the **same rigid body**, then that acceleration difference centripetal component has a magnitude of $r\omega^2$ and is always directed along the line connecting the two points, pointing toward the reference point as the center* (see Figure 7-2). These observations will be true regardless of the two points selected. But, *note this is not true in a* CASE 2 *situation* as shown in Figure 7-3a where the normal component of acceleration of auto #2 is **not** directed along the line connecting points A and P. *(Steps 5 and 6)*

Once we found the absolute acceleration ($\mathbf{A}_B$) of a second point on the same link (CASE 1) we could solve for the angular acceleration of that link. (Note that points A and B are both on link 3 and the acceleration of point O_4 is zero.) Once the angular accelerations of all the links were known, we could solve for the linear acceleration of any point (such as C) in any link using equation 7.4. To do so, we had to understand the concept of angular acceleration as a **free vector**, which means that it exists everywhere on the link at any given instant. It has no particular center. *It has an infinity of potential centers.* The link simply *has an angular acceleration.* It is this property that allows us to solve equation 7.4 for literally **any point** on a rigid body in complex motion **referenced to any other point** on that body. *(Steps 7 and 8)*

7.3   ANALYTIC SOLUTIONS FOR ACCELERATION ANALYSIS

The Fourbar Pin-Jointed Linkage

The position equations for the fourbar pin-jointed linkage were derived in Section 4.5. The linkage was shown in Figure 4-7 and is shown again in Figure 7-5a on which we also show an input angular acceleration α_2 applied to link 2. This input angular acceleration α_2 may vary with time. The vector loop equation was shown in equations 4.5a and 4.5c, repeated here for your convenience.

$$\mathbf{R}_2 + \mathbf{R}_3 - \mathbf{R}_4 - \mathbf{R}_1 = 0 \qquad (4.5a)$$

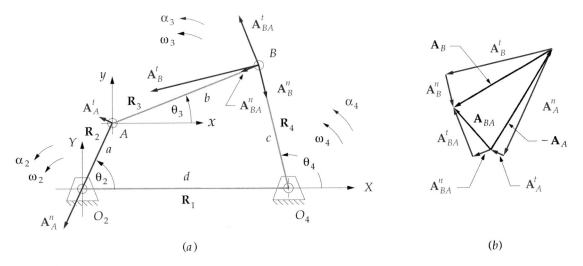

FIGURE 7-5

Position vector loop for a fourbar linkage showing acceleration vectors

As before, we substitute the complex number notation for the vectors, denoting their scalar lengths as a, b, c, d as shown in Figure 7-5.

$$ae^{j\theta_2} + be^{j\theta_3} - ce^{j\theta_4} - de^{j\theta_1} = 0 \qquad (4.5c)$$

In Section 6.7, we differentiated equation 4.5c versus time to get an expression for velocity which is repeated here.

$$ja\,\omega_2 e^{j\theta_2} + jb\,\omega_3 e^{j\theta_3} - jc\,\omega_4 e^{j\theta_4} = 0 \qquad (6.14c)$$

We will now differentiate equation 6.14c versus time to obtain an expression for accelerations in the linkage. Each term in equation 6.14c contains two functions of time, θ and ω. Differentiating with the chain rule in this example will result in two terms in the acceleration expression for each term in the velocity equation.

$$\left(j^2 a\omega_2^2\, e^{j\theta_2} + ja\alpha_2\, e^{j\theta_2} \right) + \left(j^2 b\omega_3^2\, e^{j\theta_3} + jb\alpha_3\, e^{j\theta_3} \right) - \left(j^2 c\,\omega_4^2\, e^{j\theta_4} + jc\,\alpha_4\, e^{j\theta_4} \right) = 0 \quad (7.7a)$$

Simplifying and grouping terms:

$$\left(a\alpha_2\, je^{j\theta_2} - a\omega_2^2\, e^{j\theta_2} \right) + \left(b\alpha_3\, je^{j\theta_3} - b\omega_3^2\, e^{j\theta_3} \right) - \left(c\,\alpha_4\, je^{j\theta_4} - c\,\omega_4^2\, e^{j\theta_4} \right) = 0 \qquad (7.7b)$$

Compare the terms grouped in parentheses with equations 7.2. Equation 7.7 contains the tangential and normal components of the accelerations of points A and B and of the acceleration difference of B to A. Note that these are the same relationships which we used to solve this problem graphically in Section 7.2 above. Equation 7.7 is, in fact, the **acceleration difference equation** 7.4 which, with the labels used here, is:

$$\mathbf{A}_A + \mathbf{A}_{BA} - \mathbf{A}_B = 0 \qquad (7.8a)$$

where:

$$\mathbf{A}_A = \left(\mathbf{A}_A^t + \mathbf{A}_A^n\right) = \left(a\alpha_2\, je^{j\theta_2} - a\omega_2^2\, e^{j\theta_2}\right)$$

$$\mathbf{A}_{BA} = \left(\mathbf{A}_{BA}^t + \mathbf{A}_{BA}^n\right) = \left(b\alpha_3\, je^{j\theta_3} - b\omega_3^2\, e^{j\theta_3}\right) \qquad (7.8b)$$

$$\mathbf{A}_B = \left(\mathbf{A}_B^t + \mathbf{A}_B^n\right) = \left(c\alpha_4\, je^{j\theta_4} - c\omega_4^2\, e^{j\theta_4}\right)$$

The vector diagram in Figure 7-5b shows these components and is a graphical solution to equation 7.8a. The vector components are also shown acting at their respective points on Figure 7-5a.

We now need to solve equation 7.7 for α_3 and α_4, knowing the input angular acceleration α_2, the link lengths, all link angles, and angular velocities. Thus, the position analysis derived in Section 4.5 and the velocity analysis from Section 6.7 must be done first to determine the link angles and angular velocities before this acceleration analysis can be completed. We wish to solve equation 7.8 to get expressions in this form:

$$\alpha_3 = f\left(a,b,c,d,\theta_2,\theta_3,\theta_4,\omega_2,\omega_3,\omega_4,\alpha_2\right) \qquad (7.9a)$$

$$\alpha_4 = g\left(a,b,c,d,\theta_2,\theta_3,\theta_4,\omega_2,\omega_3,\omega_4,\alpha_2\right) \qquad (7.9b)$$

The strategy of solution will be the same as was done for the position and velocity analysis. First, substitute the Euler identity from equation 4.4a in each term of equation 7.7:

$$\left[a\alpha_2\, j\left(\cos\theta_2 + j\sin\theta_2\right) - a\omega_2^2\left(\cos\theta_2 + j\sin\theta_2\right)\right]$$
$$+\left[b\alpha_3\, j\left(\cos\theta_3 + j\sin\theta_3\right) - b\omega_3^2\left(\cos\theta_3 + j\sin\theta_3\right)\right] \qquad (7.10a)$$
$$-\left[c\alpha_4\, j\left(\cos\theta_4 + j\sin\theta_4\right) - c\omega_4^2\left(\cos\theta_4 + j\sin\theta_4\right)\right] = 0$$

Multiply by the operator j and rearrange:

$$\left[a\alpha_2\left(-\sin\theta_2 + j\cos\theta_2\right) - a\omega_2^2\left(\cos\theta_2 + j\sin\theta_2\right)\right]$$
$$+\left[b\alpha_3\left(-\sin\theta_3 + j\cos\theta_3\right) - b\omega_3^2\left(\cos\theta_3 + j\sin\theta_3\right)\right] \qquad (7.10b)$$
$$-\left[c\alpha_4\left(-\sin\theta_4 + j\cos\theta_4\right) - c\omega_4^2\left(\cos\theta_4 + j\sin\theta_4\right)\right] = 0$$

We can now separate this vector equation into its two components by collecting all real and all imaginary terms separately:

real part (x component):

$$-a\alpha_2 \sin\theta_2 - a\omega_2^2 \cos\theta_2 - b\alpha_3 \sin\theta_3 - b\omega_3^2 \cos\theta_3 + c\alpha_4 \sin\theta_4 + c\omega_4^2 \cos\theta_4 = 0 \qquad (7.11a)$$

imaginary part (y component):

$$a\alpha_2 \cos\theta_2 - a\omega_2^2 \sin\theta_2 + b\alpha_3 \cos\theta_3 - b\omega_3^2 \sin\theta_3 - c\alpha_4 \cos\theta_4 + c\omega_4^2 \sin\theta_4 = 0 \qquad (7.11b)$$

Note that the j's have cancelled in equation 7.11b. We can solve equations 7.11a and 7.11b simultaneously to get:

$$\alpha_3 = \frac{CD - AF}{AE - BD} \tag{7.12a}$$

$$\alpha_4 = \frac{CE - BF}{AE - BD} \tag{7.12b}$$

where:

$$
\begin{aligned}
A &= c \sin \theta_4 \\
B &= b \sin \theta_3 \\
C &= a\alpha_2 \sin \theta_2 + a\omega_2^2 \cos \theta_2 + b\omega_3^2 \cos \theta_3 - c\omega_4^2 \cos \theta_4 \\
D &= c \cos \theta_4 \\
E &= b \cos \theta_3 \\
F &= a\alpha_2 \cos \theta_2 - a\omega_2^2 \sin \theta_2 - b\omega_3^2 \sin \theta_3 + c\omega_4^2 \sin \theta_4
\end{aligned}
\tag{7.12c}
$$

Once we have solved for α_3 and α_4, we can then solve for the linear accelerations by substituting the Euler identity into equations 7.8b,

$$\mathbf{A}_A = a\alpha_2 \left(-\sin \theta_2 + j \cos \theta_2\right) - a\omega_2^2 \left(\cos \theta_2 + j \sin \theta_2\right) \tag{7.13a}$$

$$\mathbf{A}_{BA} = b\alpha_3 \left(-\sin \theta_3 + j \cos \theta_3\right) - b\omega_3^2 \left(\cos \theta_3 + j \sin \theta_3\right) \tag{7.13b}$$

$$\mathbf{A}_B = c\alpha_4 \left(-\sin \theta_4 + j \cos \theta_4\right) - c\omega_4^2 \left(\cos \theta_4 + j \sin \theta_4\right) \tag{7.13c}$$

where the real and imaginary terms are the x and y components, respectively. Equations 7.12 and 7.13 provide a complete solution for the angular accelerations of the links and the linear accelerations of the joints in the pin-jointed fourbar linkage.

The Fourbar Slider-Crank

The first inversion of the offset slider-crank has its slider block sliding against the ground plane as shown in Figure 7-6a. Its accelerations can be solved for in similar manner as was done for the pin-jointed fourbar.

The position equations for the fourbar offset slider-crank linkage (inversion #1) were derived in Section 4.6. The linkage was shown in Figures 4-9 and 6-21 and is shown again in Figure 7-6a on which we also show an input angular acceleration α_2 applied to link 2. This α_2 can be a time-varying input acceleration. The vector loop equation 4.14 is repeated here for your convenience.

$$\mathbf{R}_2 - \mathbf{R}_3 - \mathbf{R}_4 - \mathbf{R}_1 = 0 \tag{4.14a}$$

$$ae^{j\theta_2} - be^{j\theta_3} - ce^{j\theta_4} - de^{j\theta_1} = 0 \tag{4.14b}$$

In Section 6.7 we differentiated equation 4.14b with respect to time noting that a, b, c, θ_1 and θ_4 are constant but the length of link d varies with time in this inversion.

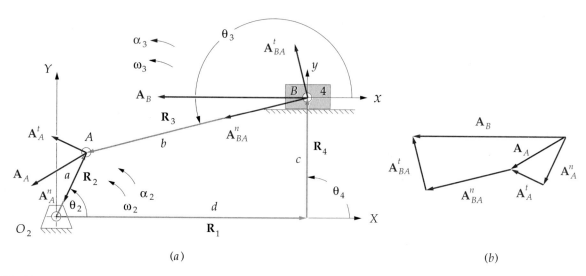

FIGURE 7-6

Position vector loop for a fourbar slider-crank linkage showing acceleration vectors

$$ja\,\omega_2 e^{j\theta_2} - jb\,\omega_3 e^{j\theta_3} - \dot{d} = 0 \qquad (6.20a)$$

The term d dot is the linear velocity of the slider block. Equation 6.20a is the velocity difference equation.

We now will differentiate equation 6.20a with respect to time to get an expression for acceleration in this inversion of the slider-crank mechanism.

$$\left(ja\,\alpha_2 e^{j\theta_2} + j^2 a\,\omega_2^2 e^{j\theta_2}\right) - \left(jb\,\alpha_3 e^{j\theta_3} + j^2 b\,\omega_3^2 e^{j\theta_3}\right) - \ddot{d} = 0 \qquad (7.14a)$$

Simplifying:

$$\left(a\,\alpha_2\, je^{j\theta_2} - a\,\omega_2^2 e^{j\theta_2}\right) - \left(b\,\alpha_3\, je^{j\theta_3} - b\,\omega_3^2 e^{j\theta_3}\right) - \ddot{d} = 0 \qquad (7.14b)$$

Note that equation 7.14 is again the acceleration difference equation:

$$\mathbf{A}_A - \mathbf{A}_{AB} - \mathbf{A}_B = 0$$
$$\mathbf{A}_{BA} = -\mathbf{A}_{AB} \qquad (7.15a)$$
$$\mathbf{A}_B = \mathbf{A}_A + \mathbf{A}_{BA}$$

$$\mathbf{A}_A = \left(\mathbf{A}_A^t + \mathbf{A}_A^n\right) = \left(a\alpha_2\, je^{j\theta_2} - a\omega_2^2\, e^{j\theta_2}\right)$$
$$\mathbf{A}_{BA} = \left(\mathbf{A}_{BA}^t + \mathbf{A}_{BA}^n\right) = \left(b\alpha_3\, je^{j\theta_3} - b\omega_3^2\, e^{j\theta_3}\right) \qquad (7.15b)$$
$$\mathbf{A}_B = \mathbf{A}_B^t = \ddot{d}$$

Note that in this mechanism, link 4 is in pure translation and so has zero ω_4 and zero α_4. The acceleration of link 4 has only a "tangential" component of acceleration along its path.

The two unknowns in the vector equation 7.14 are the angular acceleration of link 3, α_3, and the linear acceleration of link 4, d double dot. To solve for them, substitute the Euler identity,

$$a\alpha_2\left(-\sin\theta_2 + j\cos\theta_2\right) - a\omega_2^2\left(\cos\theta_2 + j\sin\theta_2\right)$$
$$-b\alpha_3\left(-\sin\theta_3 + j\cos\theta_3\right) + b\omega_3^2\left(\cos\theta_3 + j\sin\theta_3\right) - \ddot{d} = 0 \qquad (7.16a)$$

and separate the real (x) and imaginary (y) components:

real part (x component):

$$-a\alpha_2\sin\theta_2 - a\omega_2^2\cos\theta_2 + b\alpha_3\sin\theta_3 + b\omega_3^2\cos\theta_3 - \ddot{d} = 0 \qquad (7.16b)$$

imaginary part (y component):

$$a\alpha_2\cos\theta_2 - a\omega_2^2\sin\theta_2 - b\alpha_3\cos\theta_3 + b\omega_3^2\sin\theta_3 = 0 \qquad (7.16c)$$

Equation 7.16c can be solved directly for α_3 and the result substituted in equation 7.16b to find d double dot.

$$\alpha_3 = \frac{a\alpha_2\cos\theta_2 - a\omega_2^2\sin\theta_2 + b\omega_3^2\sin\theta_3}{b\cos\theta_3} \qquad (7.16d)$$

$$\ddot{d} = -a\alpha_2\sin\theta_2 - a\omega_2^2\cos\theta_2 + b\alpha_3\sin\theta_3 + b\omega_3^2\cos\theta_3 \qquad (7.16e)$$

The other linear accelerations can be found from equation 7.15b and are shown in the vector diagram of Figure 7-6b.

Coriolis Acceleration

The examples used for acceleration analysis above have involved only pin-jointed linkages or the inversion of the slider-crank in which the slider block has no rotation. When a sliding joint is present on a rotating link, an additional component of acceleration will be present, called the **Coriolis component**, after its discoverer. Figure 7-7a shows a simple, two link system consisting of a link with a radial slot, and a slider block free to slip within that slot.

The instantaneous location of the block is defined by a position vector ($\mathbf{R}_P$) referenced to the global origin at the link center. *This vector is both rotating and changing length as the system moves.* As shown this is a two degree of freedom system. The **two inputs to the system** are the angular acceleration (α) of the link and the relative linear slip velocity ($\mathbf{V}_{Pslip}$) of the block versus the disk. The angular velocity ω is a result of the time history of the angular acceleration. The situation shown, with a counterclockwise α and a clockwise ω, implies that earlier in time the link had been accelerated up to a clockwise angular velocity and is now being slowed down. The transmission component of velocity ($\mathbf{V}_{Ptrans}$) is a result of the ω of the link acting at the radius R_P whose magnitude is p.

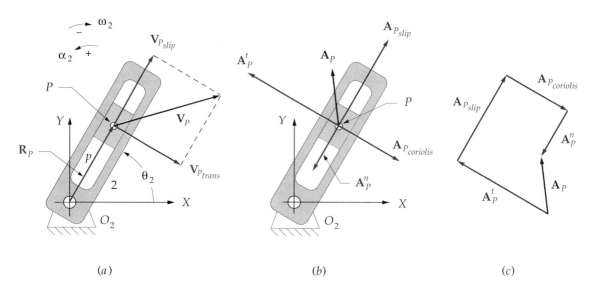

FIGURE 7-7

The Coriolis component of acceleration shown in a system with a positive (ccw) α_2 and a negative (cw) ω_2

We show the situation in Figure 7-7 at one instant of time. However, the equations to be derived will be valid for all time. We want to determine the acceleration at the center of the block (P) under this combined motion of rotation and sliding. To do so we first write the expression for the position vector $\mathbf{R}_P$ which locates point P.

$$\mathbf{R}_P = pe^{j\theta_2} \qquad (7.17)$$

Note that there are two functions of time in equation 7.17, p and θ. When we differentiate versus time we get two terms in the velocity expression:

$$\mathbf{V}_P = p\omega_2 je^{j\theta_2} + \dot{p}e^{j\theta_2} \qquad (7.18a)$$

These are the transmission component and the slip component of velocity.

$$\mathbf{V}_P = \mathbf{V}_{P_{trans}} + \mathbf{V}_{P_{slip}} \qquad (7.18b)$$

The $p\omega$ term is the transmission component and is directed at 90 degrees to the axis of slip which, in this example, is coincident with the position vector $\mathbf{R}_P$. The p dot term is the **slip component** and is directed along the **axis of slip** in the same direction as the position vector in this example. Their vector sum is $\mathbf{V}_P$ as shown in Figure 7-7a.

To get an expression for acceleration, we must differentiate equation 7.18 versus time. Note that the transmission component has **three** functions of time in it, p, ω and θ. The chain rule will yield three terms for this one term. The slip component of velocity contains two functions of time, p and θ, yielding two terms in the derivative for a total of five terms, two of which turn out to be the same.

$$\mathbf{A}_P = \left(p\alpha_2 je^{j\theta_2} + p\omega_2^2 j^2 e^{j\theta_2} + \dot{p}\omega_2 je^{j\theta_2} \right) + \left(\dot{p}\omega_2 je^{j\theta_2} + \ddot{p}e^{j\theta_2} \right) \qquad (7.19a)$$

Simplifying and collecting terms:

$$\mathbf{A}_P = p\alpha_2 je^{j\theta_2} - p\omega_2^2 e^{j\theta_2} + 2\dot{p}\omega_2 je^{j\theta_2} + \ddot{p}e^{j\theta_2} \qquad (7.19b)$$

These terms represent the following components:

$$\mathbf{A}_P = \mathbf{A}_{P_{tangential}} + \mathbf{A}_{P_{normal}} + \mathbf{A}_{P_{coriolis}} + \mathbf{A}_{P_{slip}} \qquad (7.19c)$$

Note that the Coriolis term has appeared in the acceleration expression as a result of the differentiation simply because the length of the vector p is a function of time. The Coriolis component magnitude is twice the product of the velocity of slip (Eq. 7.18) and the angular velocity of the link containing the slider slot. Its direction is rotated 90 degrees from that of the original position vector $\mathbf{R}_P$ either clockwise or counterclockwise depending on the sense of ω. (Note that we chose to align the position vector $\mathbf{R}_P$ with the axis of slip which can always be done regardless of the location of the center of rotation. See Figure 7-6.) All four components from equation 7.19 are shown acting on point P in Figure 7-7b. The total acceleration $\mathbf{A}_P$ is the vector sum of the four terms as shown in Figure 7-7c. Note that the normal acceleration term in equation 7.19b is negative in sign, so it becomes a subtraction when substituted in equation 7.19c.

This Coriolis component of acceleration will always be present when there is a velocity of slip associated with any member which also has an angular velocity. In the absence of either of those two factors the Coriolis component will be zero. You have probably experienced Coriolis acceleration if you have ever ridden on a carousel or merry-go-round. If you attempted to walk radially from the outside to the inside (or vice versa) while the carousel was turning, you were thrown sideways by the inertial force due to the Coriolis acceleration. You were the *slider block* in Figure 7-7, and your *slip velocity* combined with the rotation of the carousel created the Coriolis component. As you walked from a large radius to a smaller one, your tangential velocity had to change to match that of the new location of your foot on the spinning carousel. Any change in velocity requires an acceleration to accomplish. It was the *"ghost of Coriolis"* that pushed you sideways on that carousel.

Another example of the Coriolis component is its effect on weather systems. Large objects which exist in the earth's lower atmosphere, such as hurricanes, span enough area to be subject to significantly different velocities at their northern and southern extremities. The atmosphere turns with the earth. The earth's surface tangential velocity due to its angular velocity varies from zero at the poles to a maximum of about 1000 MPH at the equator. The winds of a storm system are attracted toward the low pressure at its center. These winds have a slip velocity with respect to the surface, which in combination with the earth's ω, creates a Coriolis component of acceleration on the moving air masses. This Coriolis acceleration causes the inrushing air to rotate about the center, or "eye" of the storm system. This rotation will be counterclockwise in the northern hemisphere and clockwise in the southern hemisphere. The movement of the entire storm system from south to north also creates a Coriolis component which will tend to deviate the storm's track eastward, though this effect is often overridden by the forces due to other large air masses such as high-pressure systems which can deflect a storm. These complicated factors make it difficult to predict a large storm's true track.

Note that in the analytic solution presented here, the Coriolis component will be accounted for automatically as long as the differentiations are correctly done. However, when doing a graphical acceleration analysis one must be on the alert to recognize the presence of this component, calculate it, and include it in the vector diagrams when its two constituents $\mathbf{V}_{slip}$ and ω are both nonzero.

The Fourbar Inverted Slider-Crank

The position equations for the fourbar inverted slider-crank linkage were derived in Section 4.7. The linkage was shown in Figures 4-10 and 6-22a and is shown again in Figure 7-8a on which we also show an input angular acceleration α_2 applied to link 2. This α_2 can vary with time. The vector loop equations 4.14 shown above are valid for this linkage as well.

All slider linkages will have at least one link whose effective length between joints varies as the linkage moves. In this inversion the length of link 3 between points A and B, designated as b, will change as it passes through the slider block on link 4. In Section 6.7 we got an expression for velocity, by differentiating equation 4.14b with respect to time noting that $a, c, d,$ and θ_1 are constant and b varies with time.

$$ja\,\omega_2 e^{j\theta_2} - jb\,\omega_3 e^{j\theta_3} - \dot{b}e^{j\theta_3} - jc\,\omega_4 e^{j\theta_4} = 0 \tag{6.24}$$

Differentiating this with respect to time will give an expression for accelerations in this inversion of the slider-crank mechanism.

$$\left(ja\,\alpha_2 e^{j\theta_2} + j^2 a\,\omega_2^2 e^{j\theta_2}\right) - \left(jb\,\alpha_3 e^{j\theta_3} + j^2 b\,\omega_3^2 e^{j\theta_3} + j\dot{b}\,\omega_3 e^{j\theta_3}\right)$$
$$-\left(\ddot{b}e^{j\theta_3} + j\dot{b}\,\omega_3 e^{j\theta_3}\right) - \left(jc\,\alpha_4 e^{j\theta_4} + j^2 c\,\omega_4^2 e^{j\theta_4}\right) = 0 \tag{7.20a}$$

Simplifying and collecting terms:

$$\left(a\,\alpha_2\,je^{j\theta_2} - a\,\omega_2^2 e^{j\theta_2}\right) - \left(b\,\alpha_3\,je^{j\theta_3} - b\,\omega_3^2 e^{j\theta_3} + 2\dot{b}\,\omega_3\,je^{j\theta_3} + \ddot{b}e^{j\theta_3}\right)$$
$$-\left(c\,\alpha_4\,je^{j\theta_4} - c\,\omega_4^2 e^{j\theta_4}\right) = 0 \tag{7.20b}$$

Equation 7.20 is in fact the acceleration difference equation (Eq. 7.4) and can be written in that notation as shown in equation 7.21.

$$\mathbf{A}_A - \mathbf{A}_{AB} - \mathbf{A}_B = 0$$

but: $\mathbf{A}_{BA} = -\mathbf{A}_{AB}$ (7.21a)

and: $\mathbf{A}_B = \mathbf{A}_A + \mathbf{A}_{BA}$

$$\mathbf{A}_A = \mathbf{A}_{A_{tangential}} + \mathbf{A}_{A_{normal}}$$
$$\mathbf{A}_{AB} = \mathbf{A}_{AB_{tangential}} + \mathbf{A}_{AB_{normal}} + \mathbf{A}_{AB_{coriolis}} + \mathbf{A}_{AB_{slip}} \tag{7.21b}$$
$$\mathbf{A}_B = \mathbf{A}_{B_{tangential}} + \mathbf{A}_{B_{normal}}$$

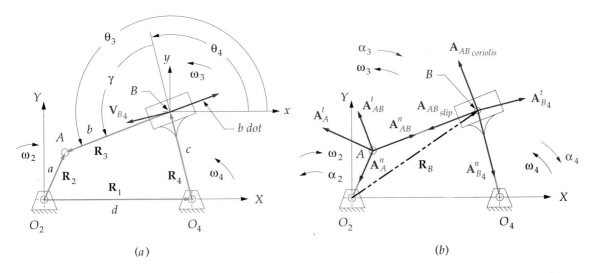

FIGURE 7-8

Acceleration analysis of inversion #3 of the fourbar slider crank driven with a positive (*ccw*) α_2 and a negative (*cw*) ω_2

$$\mathbf{A}_{A_{tangential}} = a\,\alpha_2\,je^{j\theta_2} \qquad \mathbf{A}_{A_{normal}} = -a\,\omega_2^2 e^{j\theta_2}$$

$$\mathbf{A}_{B_{tangential}} = c\,\alpha_4\,je^{j\theta_4} \qquad \mathbf{A}_{B_{normal}} = -c\,\omega_4^2 e^{j\theta_4}$$

$$\mathbf{A}_{AB_{tangential}} = b\,\alpha_3\,je^{j\theta_3} \qquad \mathbf{A}_{AB_{normal}} = -b\,\omega_3^2 e^{j\theta_3} \qquad (7.21c)$$

$$\mathbf{A}_{AB_{coriolis}} = 2\dot{b}\,\omega_3\,je^{j\theta_3} \qquad \mathbf{A}_{AB_{slip}} = \ddot{b}e^{j\theta_3}$$

Because this sliding link also has an angular velocity, there will be a non-zero Coriolis component of acceleration at point *B* which is the 2 *b dot* term in equation 7.20. Since a complete velocity analysis was done before doing this acceleration analysis, the Coriolis component can be readily calculated at this point, knowing both ω and $\mathbf{V}_{slip}$ from the velocity analysis.

The *b double dot* term in equation 7.21a is the *slip component of acceleration*. This is one of the variables to be solved for in this acceleration analysis. Another variable to be solved for is α_4, the angular acceleration of link 4. Note, however, that we also have an unknown in α_3, the angular acceleration of link 3. This is a total of three unknowns. Equation 7.20 can only be solved for two unknowns. Thus we require another equation to solve the system. There is a fixed relationship between angles θ_3 and θ_4, shown as γ in Figure 7-8 and defined in equation 4.18, repeated here:

$$\theta_3 = \theta_4 + \gamma \qquad (4.18)$$

Differentiate it twice with respect to time to obtain:

$$\omega_3 = \omega_4; \qquad \alpha_3 = \alpha_4 \qquad (7.22)$$

We wish to solve equation 7.20 to get expressions in this form:

$$\alpha_3 = \alpha_4 = f\left(a,b,\dot{b},c,d,\theta_2,\theta_3,\theta_4,\omega_2,\omega_3,\omega_4,\alpha_2\right) \qquad (7.23a)$$

$$\frac{d^2 b}{dt^2} = \ddot{b} = g\left(a,b,\dot{b},c,d,\theta_2,\theta_3,\theta_4,\omega_2,\omega_3,\omega_4,\alpha_2\right) \qquad (7.23b)$$

Substitution of the Euler identity (Eq. 4.4a) into equation 7.20 yields:

$$a\,\alpha_2\,j\left(\cos\theta_2 + j\sin\theta_2\right) - a\,\omega_2^2\left(\cos\theta_2 + j\sin\theta_2\right)$$
$$-b\,\alpha_3\,j\left(\cos\theta_3 + j\sin\theta_3\right) + b\,\omega_3^2\left(\cos\theta_3 + j\sin\theta_3\right)$$
$$-2\dot{b}\,\omega_3\,j\left(\cos\theta_3 + j\sin\theta_3\right) - \ddot{b}\left(\cos\theta_3 + j\sin\theta_3\right) \qquad (7.24a)$$
$$-c\,\alpha_4\,j\left(\cos\theta_4 + j\sin\theta_4\right) + c\,\omega_4^2\left(\cos\theta_4 + j\sin\theta_4\right) = 0$$

Multiply by the operator j and substitute α_4 for α_3 from equation 7.22:

$$a\,\alpha_2\left(-\sin\theta_2 + j\cos\theta_2\right) - a\,\omega_2^2\left(\cos\theta_2 + j\sin\theta_2\right)$$
$$-b\,\alpha_4\left(-\sin\theta_3 + j\cos\theta_3\right) + b\,\omega_3^2\left(\cos\theta_3 + j\sin\theta_3\right)$$
$$-2\dot{b}\,\omega_3\left(-\sin\theta_3 + j\cos\theta_3\right) - \ddot{b}\left(\cos\theta_3 + j\sin\theta_3\right) \qquad (7.24b)$$
$$-c\,\alpha_4\left(-\sin\theta_4 + j\cos\theta_4\right) + c\,\omega_4^2\left(\cos\theta_4 + j\sin\theta_4\right) = 0$$

We can now separate this vector equation 7.24b into its two components by collecting all real and all imaginary terms separately:

real part (x component):

$$-a\,\alpha_2\sin\theta_2 - a\,\omega_2^2\cos\theta_2 + b\,\alpha_4\sin\theta_3 + b\,\omega_3^2\cos\theta_3$$
$$+2\dot{b}\,\omega_3\sin\theta_3 - \ddot{b}\cos\theta_3 + c\,\alpha_4\sin\theta_4 + c\,\omega_4^2\cos\theta_4 = 0 \qquad (7.25a)$$

imaginary part (y component):

$$a\,\alpha_2\cos\theta_2 - a\,\omega_2^2\sin\theta_2 - b\,\alpha_4\cos\theta_3 + b\,\omega_3^2\sin\theta_3$$
$$-2\dot{b}\,\omega_3\cos\theta_3 - \ddot{b}\sin\theta_3 - c\,\alpha_4\cos\theta_4 + c\,\omega_4^2\sin\theta_4 = 0 \qquad (7.25b)$$

Note that the j's have cancelled in equation 7.25b. We can solve equations 7.25 simultaneously for the two unknowns, α_4 and b *double dot*. The solution is:

$$P = a\,\alpha_2\cos\left(\theta_3 - \theta_2\right)$$
$$Q = a\,\omega_2^2\sin\left(\theta_3 - \theta_2\right)$$
$$R = c\,\omega_4^2\sin\gamma$$
$$S = 2\dot{b}\,\omega_3$$
$$T = b + c\cos\gamma$$
$$\gamma = \theta_3 - \theta_4$$

$$\alpha_4 = \frac{P + Q - R - S}{T} \qquad (7.26a)$$

$$K = -a\omega_2^2 \left[b\cos(\theta_3 - \theta_2) + c\cos(\theta_4 - \theta_2) \right]$$

$$L = a\alpha_2 \left[b\sin(\theta_3 - \theta_2) + c\sin(\theta_4 - \theta_2) \right]$$

$$M = \omega_4{}^2 \left(b^2 + c^2 + 2bc\cos\gamma \right)$$

$$N = 2\dot{b}c\omega_4 \sin\gamma$$

$$\ddot{b} = \frac{K + L + M + N}{T} \qquad (7.26b)$$

Equation 7.26a provides the **angular acceleration** of link 4. Equation 7.26b provides the **acceleration of slip** at point B. Once these variables are solved for, the linear accelerations at points A and B in the linkage of Figure 7-8 can be found by substituting the Euler identity into equations 7.21.

$$\mathbf{A}_A = a\alpha_2 \left(-\sin\theta_2 + j\cos\theta_2 \right) - a\omega_2^2 \left(\cos\theta_2 + j\sin\theta_2 \right) \qquad (7.27a)$$

$$\mathbf{A}_{BA} = b\alpha_3 \left(\sin\theta_3 - j\cos\theta_3 \right) + b\omega_3^2 \left(\cos\theta_3 + j\sin\theta_3 \right) \qquad (7.27b)$$

$$\mathbf{A}_B = 2\dot{b}\,\omega_3 \left(\sin\theta_3 - j\cos\theta_3 \right) - \ddot{b} \left(\cos\theta_3 + j\sin\theta_3 \right)$$
$$-c\,\alpha_4 \left(\sin\theta_4 - j\cos\theta_4 \right) - c\,\omega_4^2 \left(\cos\theta_4 + j\sin\theta_4 \right) \qquad (7.27c)$$

These components of these vectors are shown in Figure 7-8b.

7.4 ACCELERATION ANALYSIS OF THE GEARED FIVEBAR LINKAGE

The velocity equation for the geared fivebar mechanism was derived in Section 6.8 and is repeated here. See Figure P7-4 for notation.

$$a\omega_2 je^{j\theta_2} + b\omega_3 je^{j\theta_3} - c\omega_4 je^{j\theta_4} - d\omega_5 je^{j\theta_5} = 0 \qquad (6.32a)$$

Differentiate this with respect to time to get an expression for acceleration.

$$\left(a\alpha_2 je^{j\theta_2} - a\omega_2^2 e^{j\theta_2} \right) + \left(b\alpha_3 je^{j\theta_3} - b\omega_3^2 e^{j\theta_3} \right)$$
$$- \left(c\alpha_4 je^{j\theta_4} - c\omega_4^2 e^{j\theta_4} \right) - \left(d\alpha_5 je^{j\theta_5} - d\omega_5^2 e^{j\theta_5} \right) = 0 \qquad (7.28a)$$

Substitute the Euler equivalents:

$$a\alpha_2 \left(-\sin\theta_2 + j\cos\theta_2 \right) - a\omega_2^2 \left(\cos\theta_2 + j\sin\theta_2 \right)$$
$$+ b\alpha_3 \left(-\sin\theta_3 + j\cos\theta_3 \right) - b\omega_3^2 \left(\cos\theta_3 + j\sin\theta_3 \right)$$
$$- c\alpha_4 \left(-\sin\theta_4 + j\cos\theta_4 \right) + c\omega_4^2 \left(\cos\theta_4 + j\sin\theta_4 \right)$$
$$- d\alpha_5 \left(-\sin\theta_5 + j\cos\theta_5 \right) + d\omega_5^2 \left(\cos\theta_5 + j\sin\theta_5 \right) = 0 \qquad (7.28b)$$

Note that the angle θ_5 is defined in terms of θ_2, the gear ratio λ, and the phase angle ϕ. This relationship and its derivatives are:

$$\theta_5 = \lambda\theta_2 + \phi; \qquad \omega_5 = \lambda\omega_2; \qquad \alpha_5 = \lambda\alpha_2 \qquad (7.28c)$$

Since a complete position and velocity analysis must be done before an acceleration analysis, we will assume that the values of θ_5 and ω_5 have been found and will leave these equations in terms of θ_5, ω_5, and α_5.

Separating the real and imaginary terms in equation 7.28b:

real:

$$-a\,\alpha_2\sin\theta_2 - a\,\omega_2^2\cos\theta_2 - b\,\alpha_3\sin\theta_3 - b\,\omega_3^2\cos\theta_3$$
$$+c\,\alpha_4\sin\theta_4 + c\,\omega_4^2\cos\theta_4 + d\,\alpha_5\sin\theta_5 + d\,\omega_5^2\cos\theta_5 = 0 \qquad (7.28d)$$

imaginary:

$$a\,\alpha_2\cos\theta_2 - a\,\omega_2^2\sin\theta_2 + b\,\alpha_3\cos\theta_3 - b\,\omega_3^2\sin\theta_3$$
$$-c\,\alpha_4\cos\theta_4 + c\,\omega_4^2\sin\theta_4 - d\,\alpha_5\cos\theta_5 + d\,\omega_5^2\sin\theta_5 = 0 \qquad (7.28e)$$

The only two unknowns are α_3 and α_4. Either equation 7.28d or 7.28e can be solved for one unknown and the result substituted in the other. The solution for α_3 is:

$$\alpha_3 = \frac{\left[\begin{array}{l} -a\omega_2^2\cos(\theta_2 - \theta_4) - a\alpha_2\sin(\theta_2 - \theta_4) \\ -b\omega_3^2\cos(\theta_3 - \theta_4) + d\omega_5^2\cos(\theta_5 - \theta_4) \\ +d\alpha_5\sin(\theta_5 - \theta_4) + c\omega_4^2 \end{array}\right]}{b\sin(\theta_3 - \theta_4)} \qquad (7.29a)$$

The angle α_4 can be found from equation 7.28e using α_3.

$$\alpha_4 = \frac{\left(\begin{array}{l} a\,\alpha_2\cos\theta_2 - a\,\omega_2^2\sin\theta_2 + b\,\alpha_3\cos\theta_3 \\ -b\,\omega_3^2\sin\theta_3 + c\,\omega_4^2\sin\theta_4 \\ -d\,\alpha_5\cos\theta_5 + d\,\omega_5^2\sin\theta_5 \end{array}\right)}{c\cos\theta_4} \qquad (7.29b)$$

With all link angles, angular velocities, and angular accelerations known, the linear accelerations for the pin joints can be found from:

$$\mathbf{A}_A = a\,\alpha_2\left(-\sin\theta_2 + j\cos\theta_2\right) - a\,\omega_2^2\left(\cos\theta_2 + j\sin\theta_2\right) \qquad (7.29c)$$

$$\mathbf{A}_{BA} = b\,\alpha_3\left(-\sin\theta_3 + j\cos\theta_3\right) - b\,\omega_3^2\left(\cos\theta_3 + j\sin\theta_3\right) \qquad (7.29d)$$

$$\mathbf{A}_C = c\,\alpha_5\left(-\sin\theta_5 + j\cos\theta_5\right) - c\,\omega_5^2\left(\cos\theta_5 + j\sin\theta_5\right) \qquad (7.29e)$$

$$\mathbf{A}_B = \mathbf{A}_A + \mathbf{A}_{BA} \qquad (7.29f)$$

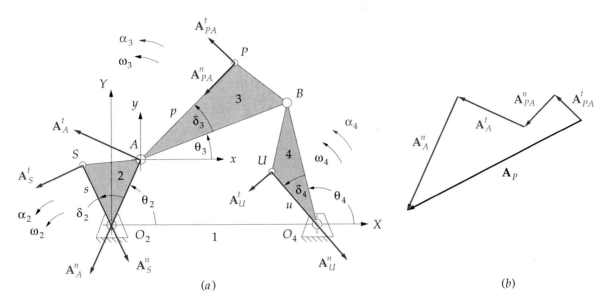

FIGURE 7-9

Finding the accelerations of any points on the links

7.5 ACCELERATION OF ANY POINT ON A LINKAGE

Once the angular accelerations of all the links are found it is easy to define and calculate the acceleration of *any point on any link* for any input position of the linkage. Figure 7-9 shows the fourbar linkage with its coupler, link 3, enlarged to contain a coupler point P. The crank and rocker have also been enlarged to show points S and U which might represent the centers of gravity of those links. We want to develop algebraic expressions for the accelerations of these (or any) points on the links.

To find the acceleration of point S, draw the position vector from the fixed pivot O_2 to point S. This vector $\mathbf{R}_{SO_2}$ makes an angle δ_2 with the vector $\mathbf{R}_{AO_2}$. This angle δ_2 is completely defined by the geometry of link 2 and is constant. The position vector for point S is then:

$$\mathbf{R}_{SO_2} = \mathbf{R}_S = se^{j(\theta_2+\delta_2)} = s\left[\cos(\theta_2+\delta_2)+j\sin(\theta_2+\delta_2)\right] \qquad (4.25)$$

We differentiated this position vector in Section 6.9 to find the velocity of that point. The equation is repeated here for your convenience.

$$\mathbf{V}_S = jse^{j(\theta_2+\delta_2)}\omega_2 = s\omega_2\left[-\sin(\theta_2+\delta_2)+j\cos(\theta_2+\delta_2)\right] \qquad (6.34)$$

We can differentiate again versus time to find the acceleration of point S.

$$\mathbf{A}_S = s\alpha_2\, je^{j(\theta_2+\delta_2)} - s\omega_2^2\, e^{j(\theta_2+\delta_2)}$$
$$= s\alpha_2\left[-\sin(\theta_2+\delta_2) + j\cos(\theta_2+\delta_2)\right] \qquad (7.30)$$
$$-s\omega_2^2\left[\cos(\theta_2+\delta_2) + j\sin(\theta_2+\delta_2)\right]$$

The position of point U on link 4 is found in the same way, using the angle δ_4 which is a constant angular offset within the link. The expression is:

$$\mathbf{R}_{UO_4} = ue^{j(\theta_4+\delta_4)} = u\left[\cos(\theta_4+\delta_4) + j\sin(\theta_4+\delta_4)\right] \qquad (4.26)$$

We differentiated this position vector in Section 6.9 to find the velocity of that point. The equation is repeated here for your convenience.

$$\mathbf{V}_U = jue^{j(\theta_4+\delta_4)}\omega_4 = u\omega_4\left[-\sin(\theta_4+\delta_4) + j\cos(\theta_4+\delta_4)\right] \qquad (6.35)$$

We can differentiate again versus time to find the acceleration of point U.

$$\mathbf{A}_U = u\alpha_4\, je^{j(\theta_4+\delta_4)} - u\omega_4^2\, e^{j(\theta_4+\delta_4)}$$
$$= u\alpha_4\left[-\sin(\theta_4+\delta_4) + j\cos(\theta_4+\delta_4)\right] \qquad (7.31)$$
$$-u\omega_4^2\left[\cos(\theta_4+\delta_4) + j\sin(\theta_4+\delta_4)\right]$$

The acceleration of point P on link 3 can be found from the addition of two acceleration vectors, such as $\mathbf{A}_A$ and $\mathbf{A}_{PA}$. Vector $\mathbf{A}_A$ is already defined from our analysis of the link accelerations. $\mathbf{A}_{PA}$ is the acceleration difference of point P with respect to point A. Point A is chosen as the reference point because angle θ_3 is defined at a local coordinate system whose origin is at A. Position vector $\mathbf{R}_{PA}$ is defined in the same way as $\mathbf{R}_U$ or $\mathbf{R}_S$, using the internal link offset angle δ_3 and the angle of link 3, θ_3. We previously analyzed this position vector and differentiated it in Section 6.9 to find the velocity difference of that point with respect to point A. Those equations are repeated here for your convenience.

$$\mathbf{R}_{PA} = pe^{j(\theta_3+\delta_3)} = p\left[\cos(\theta_3+\delta_3) + j\sin(\theta_3+\delta_3)\right] \qquad (4.27a)$$

$$\mathbf{R}_P = \mathbf{R}_A + \mathbf{R}_{PA} \qquad (4.27b)$$

$$\mathbf{V}_{PA} = jpe^{j(\theta_3+\delta_3)}\omega_3 = p\omega_3\left[-\sin(\theta_3+\delta_3) + j\cos(\theta_3+\delta_3)\right] \qquad (6.36a)$$

$$\mathbf{V}_P = \mathbf{V}_A + \mathbf{V}_{PA} \qquad (6.36b)$$

We can differentiate equation 6.36 again versus time to find $\mathbf{A}_{PA}$, the acceleration of point P versus A. This vector can then be added to the vector $\mathbf{A}_A$ already found to define the absolute acceleration $\mathbf{A}_P$ of point P.

$$\mathbf{A}_P = \mathbf{A}_A + \mathbf{A}_{PA} \qquad (7.32a)$$

where:

$$\mathbf{A}_{PA} = p\alpha_3\, je^{j(\theta_3+\delta_3)} - p\omega_3^2\, e^{j(\theta_3+\delta_3)}$$

$$= p\alpha_3\left[-\sin(\theta_3+\delta_3)+j\cos(\theta_3+\delta_3)\right] \qquad (7.32b)$$

$$-p\omega_3^2\left[\cos(\theta_3+\delta_3)+j\sin(\theta_3+\delta_3)\right]$$

Please compare equation 7.32 with equation 7.4. It is again the acceleration difference equation. Note that this equation applies to **any point** on **any link** at any position for which the positions and velocities are defined. It is a general solution for any rigid body.

7.6   HUMAN TOLERANCE OF ACCELERATION

It is interesting to note that the human body does not sense velocity, except with the eyes, but is very sensitive to acceleration. Riding in an automobile, in the daylight, one can see the scenery passing by and have a sense of motion. But, traveling at night in a commercial airliner at a 500 MPH constant velocity, we have no sensation of motion as long as the flight is smooth. What we will sense in this situation is any change in velocity due to atmospheric turbulence, takeoffs, or landings. The semi-circular canals in the inner ear are sensitive accelerometers which report to us on any accelerations which we experience. You have no doubt also experienced the sensation of acceleration when riding in an elevator, and starting, stopping, or turning in an automobile. Accelerations produce dynamic forces on physical systems, as expressed in Newton's second law, **F**=m**a**. Force is proportional to acceleration, for a constant mass. The dynamic forces produced within the human body in response to acceleration can be harmful if excessive. The human body is, after all, not rigid. It is a loosely coupled bag of water and tissue, most of which is quite internally mobile. Accelerations in the headward or footward directions will tend to either starve or flood the brain with blood as this liquid responds to Newton's law and effectively moves within the body in a direction opposite to the imposed acceleration as it lags the motion of the skeleton. Lack of blood supply to the brain causes blackout; excess blood supply causes redout. Either results in death if sustained for a long enough period.

A great deal of research has been done, largely by the military and NASA, to determine the limits of human tolerance to sustained accelerations in various directions. Figure 7-10 shows data developed from such tests.[1] The units of linear acceleration were defined in Table 1-4 as in/sec², ft/sec², or m/sec². Another common unit for acceleration is the g, defined as the acceleration due to gravity, which on Earth at sea level is approximately 386.4 in/sec², 32.2 ft/sec², or 9.81 m/sec². The g is a very convenient unit to use for accelerations involving the human as we live in a 1 g environment. Our weight, felt on our feet or buttocks, is defined by our mass times the acceleration due to gravity or mg. Thus an imposed acceleration of 1 g above the baseline of our gravity, or 2 g's, will be felt as a doubling of our weight. At 6 g's we would feel six times as heavy as normal and would have great difficulty even moving our arms against that acceleration. Figure 7-10 shows that the body's tolerance of acceleration is a function of its direction versus the body, its magnitude, and its duration. Note also that the data used for this chart were developed from tests on young, healthy military personnel in prime physical condition. The general

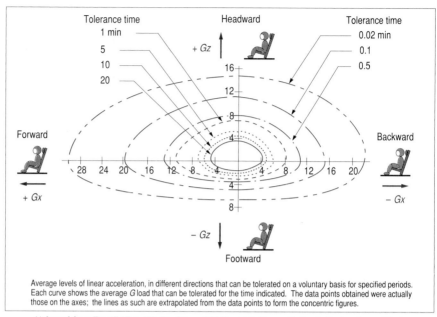

Average levels of linear acceleration, in different directions that can be tolerated on a voluntary basis for specified periods. Each curve shows the average G load that can be tolerated for the time indicated. The data points obtained were actually those on the axes; the lines as such are extrapolated from the data points to form the concentric figures.

(Adapted from Fig. 17-17, p. 505, Sanders and McCormick, *Human Factors in Engineering and Design*, 6th ed., McGraw-Hill Inc, New York, reprinted with permission.)

FIGURE 7-10

Human tolerance of acceleration

population, children and elderly in particular, should not be expected to be able to withstand such high levels of acceleration. Since much machinery is designed for human use, these acceleration tolerance data should be of great interest and value to the machine designer. Several references dealing with these human factors data are provided in the bibliography to Chapter 1.

Another useful benchmark when designing machinery for human occupation is to attempt to relate the magnitudes of accelerations which you commonly experience to the calculated values for your potential design. Table 7-1 lists some approximate levels of acceleration, in g's, which humans can experience in everyday life. Your own experience of these will help you develop a "feel" for the values of acceleration which you encounter in designing machinery intended for human occupation.

Note that machinery which does not carry humans is limited in its acceleration levels only by considerations of the stresses in its parts. These stresses are often generated in large part by the dynamic forces due to accelerations. The range of acceleration values in such machinery is so wide that it is not possible to comprehensively define any guidelines for the designer as to acceptable or unacceptable levels of acceleration. If the moving mass is small, then very large numerical values of acceleration are reasonable. If the mass is large, the dynamic stresses which the materials can sustain may limit the allowable accelerations to low values. Unfortunately, the designer does not usually know how much acceleration is too much

TABLE 7-1   Common Values of Acceleration in Human Activities

Gentle acceleration in an automobile	0.1 g's
Jet aircraft on takeoff	0.3 g's
Hard acceleration in an automobile	0.5 g's
Panic stop in an automobile	0.7 g's
Fast cornering in an automobile	0.8 g's
Roller coaster	3.5 g's
F-16 Air Force jet	9.0 g's

in his design until he has completed it to the point of calculating stresses in the parts. This usually requires a fairly complete and detailed design. If the stresses turn out to be too high and are due to dynamic forces, then the only recourse is to iterate back through the design process and reduce the accelerations and or masses in the design. This is one reason that the design process is a circular and not a linear one.

As one point of reference, the acceleration of the piston in a small, four cylinder economy car engine (about 1.5 liters displacement) at idle speed is about 40 g's. At highway speeds the piston acceleration may be as high as 700 g's. At the engine's top speed of 6000 RPM the peak piston acceleration is 2000 g's! As long as you're not riding on the piston, this is acceptable. These engines last a long time in spite of the high accelerations they experience. One key factor is the choice of low-mass, high-strength materials for the moving parts to both keep the dynamic forces down at these high accelerations and to enable them to tolerate high stresses.

7.7   JERK

No, not you! The **time derivative of acceleration** is called *jerk*, *pulse*, or *shock*. The name is apt, as it conjures the proper image of this phenomenon. **Jerk** is *the time rate of change of acceleration.* Force is proportional to acceleration. Rapidly changing acceleration means a rapidly changing force. Rapidly changing forces tend to "jerk" the object about! You have probably experienced this phenomenon when riding in an automobile. If the driver is inclined to "jack-rabbit" starts and accelerates violently away from the traffic light, you will suffer from large jerk because your acceleration will go from zero to a large value quite suddenly. But, when Jeeves, the chauffeur, is driving the *Rolls*, he always attempts to minimize jerk by accelerating gently and smoothly, so that *Madame* is entirely unaware of the change.

Controlling and minimizing jerk in machine design is often of interest, especially if low vibration is desired. Large magnitudes of jerk will tend to excite the natural frequencies of vibration of the machine or structure to which it is attached and cause increased vibration and noise levels. Jerk control is of greater interest in the design of cams than of linkages, and we will investigate it in more detail in Chapter 9 on cam design.

The procedure for calculating the jerk in a linkage is a straightforward extension of the methods shown for acceleration analysis. Let angular jerk be represented by:

$$\varphi = \frac{d\alpha}{dt} \tag{7.33a}$$

and linear jerk by:

$$\mathbf{J} = \frac{d\mathbf{A}}{dt} \tag{7.33b}$$

To solve for jerk in a fourbar linkage, for example, the vector loop equation for acceleration (Eq. 7.7) is differentiated versus time. Refer to Figure 7-5 for notation.

$$
\begin{aligned}
-a\omega_2^3 je^{j\theta_2} - 2a\omega_2\alpha_2 e^{j\theta_2} + a\alpha_2\omega_2 j^2 e^{j\theta_2} + a\varphi_2 je^{j\theta_2} \\
-b\omega_3^3 je^{j\theta_3} - 2b\omega_3\alpha_3 e^{j\theta_3} + b\alpha_3\omega_3 j^2 e^{j\theta_3} + b\varphi_3 je^{j\theta_3} \\
+c\omega_4^3 je^{j\theta_4} + 2c\omega_4\alpha_4 e^{j\theta_4} - c\alpha_4\omega_4 j^2 e^{j\theta_4} - c\varphi_4 je^{j\theta_4} = 0
\end{aligned}
\tag{7.34a}
$$

Collect terms and simplify:

$$
\begin{aligned}
-a\omega_2^3 je^{j\theta_2} - 3a\omega_2\alpha_2 e^{j\theta_2} + a\varphi_2 je^{j\theta_2} \\
-b\omega_3^3 je^{j\theta_3} - 3b\omega_3\alpha_3 e^{j\theta_3} + b\varphi_3 je^{j\theta_3} \\
+c\omega_4^3 je^{j\theta_4} + 3c\omega_4\alpha_4 e^{j\theta_4} - c\varphi_4 je^{j\theta_4} = 0
\end{aligned}
\tag{7.34b}
$$

Substitute the Euler identity and separate into x and y components:

real part (x component):

$$
\begin{aligned}
a\omega_2^3 \sin\theta_2 - 3a\omega_2\alpha_2 \cos\theta_2 - a\varphi_2 \sin\theta_2 \\
+b\omega_3^3 \sin\theta_3 - 3b\omega_3\alpha_3 \cos\theta_3 - b\varphi_3 \sin\theta_3 \\
-c\omega_4^3 \sin\theta_4 + 3c\omega_4\alpha_4 \cos\theta_4 + c\varphi_4 \sin\theta_4 = 0
\end{aligned}
\tag{7.35a}
$$

imaginary part (y component):

$$
\begin{aligned}
-a\omega_2^3 \cos\theta_2 - 3a\omega_2\alpha_2 \sin\theta_2 + a\varphi_2 \cos\theta_2 \\
-b\omega_3^3 \cos\theta_3 - 3b\omega_3\alpha_3 \sin\theta_3 + b\varphi_3 \cos\theta_3 \\
+c\omega_4^3 \cos\theta_4 + 3c\omega_4\alpha_4 \sin\theta_4 - c\varphi_4 \cos\theta_4 = 0
\end{aligned}
\tag{7.35b}
$$

These can be solved simultaneously for φ_3 and φ_4 which are the only unknowns. The driving angular jerk, φ_2, if nonzero, must be known in order to solve the system. All the other factors in equations 7.35 are defined or have been calculated from the position, velocity and acceleration analyses. To simplify these expressions we will set the known terms to temporary constants.

In equation 7.35a, let:

$A = a\omega_2^3 \sin\theta_2$	$D = b\omega_3^3 \sin\theta_3$	$G = 3c\omega_4\alpha_4 \cos\theta_4$
$B = 3a\omega_2\alpha_2 \cos\theta_2$	$E = 3b\omega_3\alpha_3 \cos\theta_3$	$H = c\sin\theta_4$
$C = a\varphi_2 \sin\theta_2$	$F = c\omega_4^3 \sin\theta_4$	$K = b\sin\theta_3$

$$\tag{7.36a}$$

Equation 7.35a then reduces to:

$$\varphi_3 = \frac{A - B - C + D - E - F + G + H}{K}\varphi_4 \qquad (7.36b)$$

Note that equation 7.36b defines angle φ_3 in terms of angle φ_4. We will now simplify equation 7.35b and substitute equation 7.36b into it.

In equation 7.35b, let:

$$
\begin{array}{lll}
L = a\omega_2^3 \cos\theta_2 & P = b\omega_3^3 \cos\theta_3 & S = c\omega_4^3 \cos\theta_4 \\
M = 3a\omega_2\alpha_2 \sin\theta_2 & Q = 3b\omega_3\alpha_3 \sin\theta_3 & T = 3c\omega_4\alpha_4 \sin\theta_4 \quad (7.37a) \\
N = a\varphi_2 \cos\theta_2 & R = b\cos\theta_3 & U = c\cos\theta_4
\end{array}
$$

Equation 7.35b then reduces to:

$$R\varphi_3 - U\varphi_4 - L - M + N - P - Q + S + T = 0 \qquad (7.37b)$$

Substituting equation 7.36b in equation 7.35b:

$$R\varphi_4\left(\frac{A - B - C + D - E - F + G + H}{K}\right) - U\varphi_4 - L - M + N - P - Q + S + T = 0 \quad (7.38)$$

The solution is:

$$\varphi_4 = \frac{KN - KL - KM - KP - KQ + AR - BR - CR + DR - ER - FR + GR + KS + KT}{KU - HR} \qquad (7.39)$$

The result from equation 7.39 can be substituted into equation 7.36b to find φ_3. Once the angular jerk values are found, the linear jerk at the pin joints can be found from:

$$
\begin{array}{l}
J_A = -a\omega_2^3 je^{j\theta_2} - 3a\omega_2\alpha_2 e^{j\theta_2} + a\varphi_2 je^{j\theta_2} \\
J_{BA} = -b\omega_3^3 je^{j\theta_3} - 3b\omega_3\alpha_3 e^{j\theta_3} + b\varphi_3 je^{j\theta_3} \qquad (7.40) \\
J_B = -c\omega_4^3 je^{j\theta_4} - 3c\omega_4\alpha_4 e^{j\theta_4} + c\varphi_4 je^{j\theta_4} = 0
\end{array}
$$

The same approach as used in Section 7.4 to find the acceleration of any point on any link can be used to find the linear jerk at any point.

$$J_P = J_A + J_{PA} \qquad (7.41)$$

The jerk difference equation 7.41 can be applied to any point on any link if we let P represent any arbitrary point on any link and A represent any reference point on the same link for which we know the value of the jerk vector. Note that if you substitute equations 7.40 into 7.41, you will get equation 7.34.

7.8   LINKAGES OF N BARS

The same analysis techniques presented here for position, velocity, acceleration, and jerk, using the fourbar and fivebar linkage as the examples, can be extended to more complex assemblies of links. Multiple vector loop equations can be written around a linkage of arbitrary complexity. The resulting vector equations can be differentiated and solved simultaneously for the variables of interest. In some cases, the solution will require simultaneous solution of a set of nonlinear equations. A root-finding algorithm such as the Newton-Rapheson method will be needed to solve these more complicated cases. A computer is necessary. An equation solver software package such as *TK Solver* which will do an iterative root-finding solution will be a useful aid to the solution of any of these analysis problems, including the examples shown here.

7.9   REFERENCES

1   **Sanders**, M.S., and McCormick, E.J., *Human Factors in Engineering and Design*, 6th ed., McGraw-Hill Co., New York, 1987, p. 505.

7.10   PROBLEMS

7-1   A point at a 6.5 inch radius is on a body which is in pure rotation with $\omega = 100$ rad/sec and a constant $\alpha = -500$ rad/sec^2 at point A. The rotation center is at the origin of a coordinate system. When the point is at position A, its position vector makes a 45° angle with the X axis. It takes 0.01 sec to reach point B. Draw this system to some convenient scale, calculate the θ and ω of position B and:

    a.   Write an expression for the particle's acceleration vector in position A using complex number notation, in both polar and cartesian forms.

    b.   Write an expression for the particle's acceleration vector in position B using complex number notation, in both polar and cartesian forms.

    c.   Write a vector equation for the acceleration difference between points B and A. Substitute the complex number notation for the vectors in this equation and solve for the acceleration difference numerically.

    d.   Check the result of part c with a graphical method.

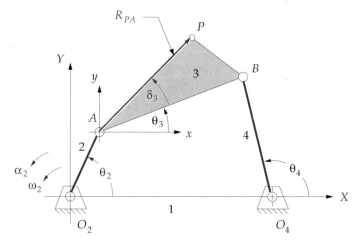

FIGURE P7-1

Configuration and terminology for the pin-jointed fourbar linkage

TABLE P7-1 · Data for Fourbar Linkage Analysis Problems

Row	Link 1	Link 2	Link 3	Link 4	θ_2	ω_2	α_2	R_{pa}	δ_3
a	6	2	7	9	30	10	0	6	30
b	7	9	3	8	85	– 12	5	9	25
c	3	10	6	8	45	– 15	– 10	10	80
d	8	5	7	6	25	24	– 4	5	45
e	8	5	8	6	75	– 50	10	9	300
f	5	8	8	9	15	– 45	50	10	120
g	6	8	8	9	25	100	18	4	300
h	20	10	10	10	50	– 65	25	6	20
i	4	5	2	5	80	25	– 25	9	80
j	20	10	5	10	33	25	– 40	1	0
k	4	6	10	7	88	– 80	30	10	330
l	9	7	10	7	60	– 90	20	5	180
m	9	7	11	8	50	75	– 5	10	90
n	9	7	11	6	120	15	– 65	15	60

7-2 In problem 7-1 let A and B represent points on separate, rotating bodies both having the given ω and α at $t = 0$, $\theta_A = 45°$, and $\theta_B = 120°$. Find their relative acceleration.

*7-3 The link lengths, coupler point location, and the values of θ_2, ω_2, and α_2 for the same fourbar linkages as used for position and velocity analysis in Chapters 4 and 6 are redefined in Table P7-1, which is the same as Table P6-1. The general

---

* Answers in Appendix E

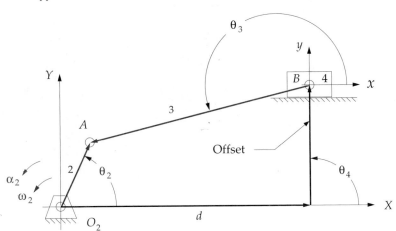

FIGURE P7-2

Configuration and terminology for a fourbar offset slider-crank linkage

TABLE P7-2   Data for Noninverted Slider-Crank Problems

Row	Link 2	Link 3	Offset	θ_2	ω_2	α_2
a	1.4	4	1	45	10	0
b	2	6	– 3	60	– 12	5
c	3	8	2	– 30	– 15	– 10
d	3.5	10	1	120	24	– 4
e	5	20	– 5	225	– 50	10
f	3	13	0	100	– 45	50
g	7	25	10	330	100	18

linkage configuration and terminology are shown in Figure P7-1. *For the row(s) assigned,* draw the linkage to scale and **graphically** find the accelerations of points A and B. Then calculate α_3 and α_4 and the acceleration of point P.

*7-4   Repeat problem 7-3 except solve by the analytical vector loop method of Section 7.3.

*7-5   The link lengths and offset and the values of θ_2, ω_2, and α_2 for some noninverted, offset fourbar slider-crank linkages are defined in Table P7-2. The general linkage configuration and terminology are shown in Figure P7-2. *For the row(s) assigned,* draw the linkage to scale and **graphically** find the accelerations of the pin joints A and B and the acceleration of slip at the sliding joint.

*7-6   Repeat problem 7-5 using an analytical method.

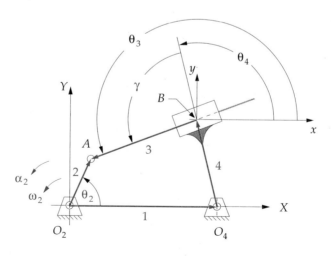

FIGURE P7-3

Configuration and terminology for inversion #3 of the fourbar slider-crank linkage

TABLE P7-3   Data for Inverted Slider-Crank Problems

Row	Link 1	Link 2	Link 4	γ	θ_2	ω_2	α_2
a	6	2	4	90	30	10	– 25
b	7	9	3	75	85	–15	– 40
c	3	10	6	45	45	24	30
d	8	5	3	60	25	– 50	20
e	8	4	2	30	75	– 45	– 5
f	5	8	8	90	150	100	– 65

*7-7   The link lengths and the values of θ_2, ω_2, and γ for some inverted fourbar slider-crank linkages are defined in Table P7-3. The general linkage configuration and terminology are shown in Figure P7-3. *For the row(s) assigned,* find the accelerations of the pin joints A and B and the acceleration of slip at the sliding joint. Solve by the analytical vector loop method of Section 7.3 for the open configuration of the linkage.

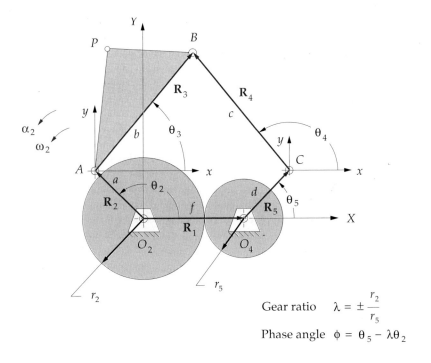

$$\text{Gear ratio} \quad \lambda = \pm \frac{r_2}{r_5}$$

$$\text{Phase angle} \quad \phi = \theta_5 - \lambda\theta_2$$

FIGURE P7-4

Configuration and terminology for the geared fivebar linkage

TABLE P7-4   Data for Geared Fivebar Linkage Problems

Row	Link 1	Link 2	Link 3	Link 4	Link 5	λ	ϕ	θ_2	ω_2	α_2
a	6	1	7	9	4	2	30	60	10	0
b	6	5	7	8	4	-2.5	60	30	-12	5
c	3	5	7	8	4	-0.5	0	45	-15	-10
d	4	5	7	8	4	-1	120	75	24	-4
e	5	9	11	8	8	3.2	-50	-39	-50	10
f	10	2	7	5	3	1.5	30	120	-45	50
g	15	7	9	11	4	2.5	-90	75	100	18
h	12	8	7	9	4	-2.5	60	55	-65	25
i	9	7	8	9	4	-4	120	100	25	-25

*7-8   Repeat problem 7-7 for the crossed configuration of the linkage.

*7-9   The link lengths, gear ratio (λ), phase angle (ϕ), and the values of θ_2, ω_2, and α_2 for some geared fivebar linkages are defined in Table P7-4. The general linkage configuration and terminology are shown in Figure P7-4. *For the row(s) assigned,* find α_3 and α_4.

7-10   An automobile driver took a curve too fast. The car spun out of control about its *CG* and slid off the road in a northeasterly direction. The friction of the skidding tires provided a 0.25 *g* linear deceleration. The car rotated at 100 rpm. When the car hit the tree head-on at 30 mph, it took 0.1 second to come to rest.

   a.   What was the acceleration experienced by the child seated on the middle of the rear seat, 2 feet behind the car's *CG*, just prior to impact?

   b.   What force did the 100-lb. child exert on her seatbelt harness as a result of the acceleration, just prior to impact?

   c.   Assuming a constant deceleration during the 0.1 seconds of impact, what was the magnitude of the average deceleration felt by the passengers in that interval?

7-11   For the row(s) assigned in Table P7-1, find the angular jerk of links 3 and 4 and the linear jerk of the pin joint between link 3 and 4 (point *B*). Assume an angular jerk of zero on link 2. The linkage configuration and terminology are shown in Figure P7-1.

7-12   You are riding on a carousel which is rotating at a constant 15 RPM. It has an inside radius of 3 feet and an outside radius of 10 feet. You begin to run from the inside to the outside along a radius. Your peak velocity with respect to the carousel is 5 mph and occurs at a radius of 7 feet. What is your maximum Coriolis acceleration magnitude and its direction with respect to the carousel?

Chapter 8

COMPUTER PROGRAMS

I really hate this damned machine;
I wish that they would sell it.
It never does quite what I want
But only what I tell it.
FROM THE FORTUNE DATABASE, BERKELY UNIX

8.0 INTRODUCTION

There are several computer programs provided, on disk, with this text. Programs FOURBAR, FIVEBAR, and SIXBAR will be discussed in detail in this chapter. Program MATRIX was introduced in Chapter 5. Programs DYNACAM, DYNAFOUR, and ENGINE will be discussed in later chapters after the mathematical theory on which they are based has been derived and described.

Programs FOURBAR, FIVEBAR, and SIXBAR are based on the mathematics derived in Chapters 4 through 7 and use the equations presented therein to solve for position, velocity, and acceleration in pin-jointed linkages of the variety described in the particular program's name. Much of the description of general program operation and function in this chapter will also be applicable to the programs DYNACAM, MATRIX, DYNAFOUR, and ENGINE as all are based on a common format of input menus, and have similar choices for display of output data in the form of tables and plots. All the programs are designed to be user friendly and reasonably "crashproof." The author encourages reports of any "bugs" or problems encountered in their use.

LEARNING TOOLS All the programs provided with this text are designed to be learning tools to aid in the understanding of the relevant subject matter and *are specifically not intended to be used for commercial purposes in the design of hardware* **and should not be so used**. It is quite possible to obtain inappropriate (but mathematically correct) results to any problem solved with these programs, due to incorrect or inappropriate input of data. In other words, the user is expected to understand the kinematic and dynamic theory underlying the program's structure and to also understand the mathematics on which the program's algorithms are based. This information on the underlying theory and mathematics is derived and described in the noted chapters of this text. Most equations used in the programs are presented or derived in this textbook.

DISCLAIMER Commercial software for use in design or analysis needs to have built-in safeguards against the possibility of the user providing incorrect, inappropriate, or ridiculous values for input variables, in order to guard against erroneous results due to user ignorance or inexperience. **The programs with this text are** *not commercial software* **and deliberately do not contain such safeguards against improper input data**, on the premise that to do so would "short circuit" the student's learning process. We learn most from our failures. These programs provide a consequence-free environment to explore failure of your designs "on paper" and in the process come to a more thorough and complete understanding of the subject matter. **The author and publisher are not responsible for any damages which may result from the use or misuse of these programs**.

BRUTE FORCE AND IGNORANCE The very rapid computation speed of these programs allows the student to explore a much larger number and variety of potential solutions to more realistic and comprehensive problems than could be accomplished using only hand calculator solutions of these complicated systems of equations. This is both an advantage and a danger. The advantage is that the student can use the programs like a "flight simulator" to "fly" his potential design solutions through their paces with no consequences from a "crash" of the design not yet built. If the student diligently attempts to interpret the program's results and relate them to the relevant theory, a more thorough understanding of the kinematics and dynamics can result. On the other hand, there is a great temptation to use these programs with "brute force and ignorance" (BFI) to somewhat randomly try solutions without regard to what the theory and equations are telling you, and hope that somehow a useable solution will "pop out." The student who succumbs to this temptation will not obtain much benefit from the exercise and will probably have a poor design result. This situation is probably best summed up in the following comment from a student who had suffered through the course in kinematics using these programs to design solutions to three project problems like those listed at the end of Chapter 3.

. . . The computer, with its immense benefits for the engineer, can also be a hindrance if one does not first develop a thorough understanding of the theory upon which a particular program is based. An over-reliance on the computer can leave one "computer smart" and "engineering stupid." The BFI approach becomes increasingly tempting when it can be employed with such ease. . . . BRIAN KIMBALL

Smart student!

Use these computer programs wisely. Avoid *brute force and ignorance*! **Engineer** your solutions and understand the theory behind them.

8.1   HARDWARE REQUIREMENTS

These programs all run on any MS DOS[1] computer that is fully IBM™ compatible.[2] Some will run with only 256K bytes of random access memory (RAM).[3] Others require at least 512K bytes of RAM.[4] One floppy disk drive is needed. No hard disk drive is necessary but can be used and will speed operation.[5] The graphics output requires at least an IBM CGA[6] graphics emulation capacity but does not **require** a color monitor. All **CGA** graphics output will be monochrome even on a color monitor, as the color capability is being used instead to obtain higher resolution on the plots. The higher resolution **EGA** and **VGA** graphics screens are supported by these programs and will display in color if you have a color monitor. Text output will print on any printer. A dot matrix printer with graphics screen dump[7] ability is needed to get hard copy of the plots. The MS DOS *Graphics.Com* utility program[8] supplied with your DOS must be loaded into memory for the graphics dump ability to be enabled. If the SHIFT PRINTSCREEN command does not work from these programs, exit the program and type GRAPHICS from the DOS prompt with your MS DOS system disk in the drive. Then rerun the program. You may want to put the GRAPHICS command in your AUTOEXEC.BAT file which runs when you start the computer.

The disk supplied with this text is a 3.5-inch high density 1.4 megabyte floppy. If your computer disk drive does not read this size or format, your school's computer center can probably convert it to a useable format for you. There are several batch files on the disk which will copy the files to another medium. The programs can be copied to a hard drive or to other floppies by using these batch files which will create the proper subdirectories and transfer the files. You will need eight 5.25-inch 360K byte floppies to hold the entire contents of this disk. See the README files on the disk for instructions once you have it in the proper type of disk drive. If all attempts to copy and run the programs fail, contact the author.

8.2   RUNNING THE PROGRAMS

In the following descriptions of user interaction with the programs, prompts which are printed on the screen by the computer will be placed in **BOLD CAPS**. Responses which you must type will be in *italics*. Note that the responses are case insensitive, meaning any mix of lower and upper case in your response should be understandable to the computer. References to keys on the keyboard will be in SMALL CAPS.

Each disk contains a file called README which can be read by typing *Type Readme* at the DOS prompt for the drive in which the disk is placed. For example, if the disk is in drive **A:**, the prompt and your response will look like: **A:>**type readme. (Note the case insensitivity.) The contents of the text file README will then be typed on the screen. It contains the information contained in the next paragraph about how to start the programs running. If you want a hard copy of the contents of this file (or any other text file on disk), type **A:>***print readme*. The command **Print** sends output to the attached printer provided that it is turned on!

The programs are run by typing the name of the program, FOURBAR, FIVEBAR, or SIXBAR (or other) at the DOS prompt for the drive in which the disk contain-

This program deals only in "pure" numbers without regard for units.

It is your responsibility to ensure that all input data is consistent with you choice of units. Your choice only affects labels on plots and tables

Time is in seconds and angles in degrees for all systems used here.

System	Length	Force	Mass	
1 –	inches	pounds	pound-sec^2/inch	(blobs)
2 –	feet	pounds	pound-sec^2/foot	(slugs)
3 –	meters	newtons	kilograms	(MKS or SI system)
4 –	cm	dynes	grams	(CGS system)

<Return> Accepts Value of   1

Choice ?

FIGURE 8-1

Units selection menu common to all programs

ing the programs is placed. For example, if the disk is in drive **A:**, the prompt and your response will look like: **A:>**_fourbar_. Note that the character following the **A:** may be different on different machines, and may be more than one character. If the programs are on the hard drive **C:**, the prompt and your response will look like: **C:>** _fourbar_.

In response to your command, the program will be loaded from disk into RAM and run. A copyright notice and educational use disclaimer will first be displayed. The first time you run the program on a given computer it will prompt you for the type of graphics display you have in that computer, (CGA, EGA, or VGA) and whether you have a color or monochrome monitor. If you choose monochrome, you will also be asked to select black-on-white or white-on-black display. The latter is needed for graphics screen dumps to the printer, while the former may be more pleasant to view on the screen. It will also ask you to choose a mode of angle display, one of modulo 360°, modulo ±180° or absolute. The next screen asks for a choice of the system of units to be used, as shown in Figure 8-1. The choices should be self-explanatory. See Section 1.8 and Table 1-4 for a review of unit systems if confused. Note that these input screens, like most in these programs, provide _default options_ which are selected by hitting the RETURN or ENTER key. These choices can later be changed from the help menu within the program. Your choices will be written to a **configuration file** on the default disk drive called _Programname_.CON. These settings will become the defaults. On subsequent startups of the program this configuration file will be automatically read and the defaults set. When you change any settings from the **Help** menu, it will rewrite the _Programname_.CON file to disk with the latest settings. If you are un-

```
  Fourbar Linkage Analyzer   – Version X.x – mm/dd/yy

        1 –  INPUT        the linkage data

        2 –  DISPLAY      the linkage data

        3 –  CALCULATE    the results

        4 –  PRINT        the results

        5 –  PLOT         the results

        6 –  ANIMATE      the linkage

        7 –  HELP         screens

      <RET> –  QUIT PROGRAM

                            CHOICE ?
```

FIGURE 8-2

The main menu for program FOURBAR is similar to the other programs' main menus

sure what settings to make, just accept the defaults. You can later experiment with new settings from the **Help** menu. The worst problem that an inappropriate choice can cause is to require the reentry of your data.

The program will display a version number and ask the user to type his or her *Name*. This information will be used to annotate all tables and plots printed from the program. It then requests a *Design Number*. Any number may be input. A null response (the RETURN key) will default the design number to 1. This number will also be used to annotate tables and plots and will be automatically updated each time you change your input data within any program session. This feature provides a convenient means to identify output for different designs after many have been done. When restarting a new session after having stopped the program, simply type a number that is one greater than the highest design number reached in the prior session. This will allow you to return to an earlier design in the iteration process at any time and identify all the plots and tables associated with it.

Finally, the main menu, Figure 8-2, is reached. The programs are completely menu-driven. A choice from one menu frequently brings up a submenu which requires additional choices until the task is accomplished. Most menus provide a default response chosen with the RETURN or ENTER key. Improper responses are trapped and re-requested. The menus are very similar among all the programs. The three linkage kinematics programs discussed in this chapter (FOUR-BAR, FIVEBAR, SIXBAR) have nearly identical menus. We will use program FOURBAR as the example in the figures except where there are significant differences. These differences will be discussed in separate sections devoted to the other programs.

SELECT INPUT METHOD

1 – TYPE DATA   from KEYBOARD

2 – READ DATA   from DISKFILE

3 – Create GRASHOF   test linkage

4 – Create HOEKENS   straight-line linkage

5 – Create ROBERTS   straight-line linkage

6 – Create WATTS   straight-line linkage

7 – Create CHEBYCHEV   straight-line linkage

8 – SYNTHESIZE   a linkage for multiple positions

9 – Choose a COGNATE   of current linkage

<Ret> – EXIT

Choice ?

FIGURE 8-3

Input data menu for program FOURBAR

8.3   PROGRAM OPERATION

Figure 8-2 shows the main menu. It is, in general, used in numerical order, with some exceptions. Note that to quit the program hit RETURN from this main menu. You will be asked to confirm this decision. Only the letter *Y* (either upper or lower case) in response to the confirmation question will stop the program. Any other response will return you to the main menu. This is to avoid inadvertent loss of data.

1 - Input Data

Input Data will bring up the submenu shown in Figure 8-3. This provides several choices, the first of which is **Type data from keyboard**. This is the choice you will make to put in the necessary numerical data needed to do the calculations. Once you are used to the program this will be your most common choice from this menu. Many of the other choices are built-in examples which allow you to experiment with the program using linkages which will work and illustrate program function. You should choose one or more of these and experiment until you are comfortable with the program. Note that in the FOURBAR program, the examples include some of the straight line linkages discussed in Section 3.8.

TYPE DATA FROM KEYBOARD starts a sequence of input screens which request the link lengths in the order, input link 2, coupler link 3, output link 4, and ground link 1. Note they are not input in numerical order. The link numbering and the axis system used are shown in Figure 8-4a. Link lengths are defined as

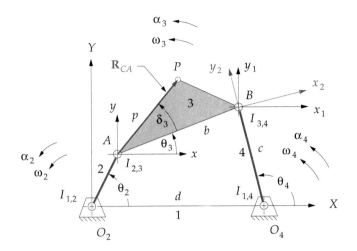

(*a*) Linkage data for program FOURBAR

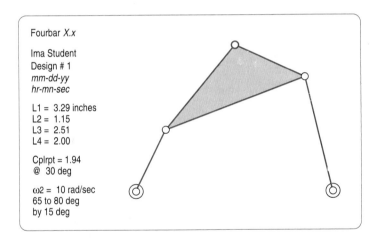

(*b*) Linkage animated in program FOURBAR

FIGURE 8-4

Linkage data for program FOURBAR and linkage animated in program

the pin-to-pin distances and are labelled as a, b, c, d in the figure. The X or real axis is constrained to lie along link 1, defined by the instant centers $I_{1,2}$ and $I_{1,4}$ which are also labelled as points O_2 and O_4 in the figure. Instant center $I_{1,2}$, the driver crank pivot, is the origin of the global coordinate system.

It might at first seem an overly restrictive constraint to force the X axis to lie on link 1. Many linkages will have their ground link at some angle other than zero. However, reorienting the linkage after designing and analyzing it merely involves rotating the piece of paper on which it is drawn to the desired final angle of the ground link. More formally, it means a rotation of the coordinate system. In effect,

SELECT TYPE OF LINKAGE SYNTHESIS DESIRED

1 – Two-Position Motion Synthesis

2 – Three-Position Motion Synthesis

3 – Three-Position Motion Synthesis - Specified Pivots

4 – Circle-Point and Center-Point Circles

<Ret> – EXIT

Choice ?

FIGURE 8-5

Submenu for multiposition linkage synthesis

the actual final angle of the ground link must be added or subtracted from all angles of links and vectors calculated in our aligned axis system.

In addition to the link lengths, you must supply the location of one coupler point on link 3 to find that point's coupler curve positions, velocities, and accelerations. This point is located by a position vector rooted at $I_{2,3}$ (point A) and directed to the coupler point P of interest which can be anywhere on link 3. The program requires that you input the polar coordinates of this vector which are labelled p and δ_3 in Figure 8-4a. The program asks for the **distance from $I_{2,3}$ to coupler point**, which is p, and the **angle coupler point makes with link 3** which is δ_3. Note that angle δ_3 is not referenced to either the global coordinate system XY or to the local coordinate system xy at point A. Rather, it is referenced to the line AB which is the pin-to-pin edge of link 3. Angle δ_3 is a property of link 3 and is embedded in it. The angle which locates vector $\mathbf{R}_{CA}$ in the x,y coordinate system is the sum of angle δ_3 and angle θ_3. This addition is done in the program, after θ_3 is calculated for each position of the input crank. Also see Section 4.8.

Note that the coordinate system, dimensions, angles, and nomenclature in Figure 8-4a are completely consistent with that of Figures 4-6, 4-7, 4-8, and 4-13 which were used in the derivation of the equations used in this FOURBAR program. Figure 8-4b shows the FOURBAR program's depiction of the linkage in Figure 8-4a. Read the diskfile FIG8-4B into program Fourbar to animate this linkage.

READ DATA FROM DISKFILE Item 2 on the input menu allows you to read data from a disk. Files which have been previously stored on disk (using item 2 on the main menu) can be recalled into the program for further analysis and display.

Ima Student   Design # 1   *mm/dd/yy*   at *hr:mn*

Link #	Length in Inches
1	9.38
2	2
3	8.63
4	7

Coupler Pt.   = 10 at 50 Degrees

Open/Crossed   = OPEN

Start Alpha2   = 0 Radians / Sec ^ 2

Start Omega2   = 1 Radians / Sec

Start Theta2   = 0 Degrees

Final Theta2   = 360 Degrees

Delta Theta2   = 10 Degrees

FIGURE 8-6

The "Display Data" screen

This selection will bring up a series of prompts, the first of which asks for the disk drive letter (a, b, c, or d) on which you have placed the file to be read. The program will add the mandatory colon to the chosen drive letter. If you choose a hard drive (**C:** or **D:**), it will ask for the subdirectory pathname which can be null (blank). (To cancel a selected subdirectory, type *none*.) You may also see the directory of the selected drive. It will show only those files which have the proper suffix for the program. Finally it will ask for the filename to be read and then input the data into the program. You will then have to **Calculate** (see below) to generate the output data.

EXAMPLE LINKAGES Items 3 through 7 on the input menu provide built-in example linkages which you can use to familiarize yourself with the program's operation. Many diskfiles are also provided which reproduce various examples and figures from this text. These are noted where the topics are discussed and can be input as described above to gain familiarity with the program.

SYNTHESIZE LINKAGES Item 8 on the input menu allows you to synthesize a fourbar linkage for two or three positions by the methods derived in Chapter 5. Figure 8-5 shows the submenu of four choices. Item 1 does either of the two-position synthesis methods described in Section 5.3. Item 2 does the general three-position synthesis method described in Section 5.7. Item 3 does three-position synthesis with specified fixed pivots as described in Section 5.9. Item 4 creates the circle-point and center-point circles, as defined in Section 5.10, for later plotting. These synthesis routines solve the equations derived in the noted sections when you input the required data and then place the synthesized link lengths and relevant data into the same variables as would have been defined by direct input of link lengths. Thus all other program features operate on the synthesized linkage just as they would for one whose data were typed in.

```
                    SELECT OUTPUT DEVICE
                    ────────────────────────

                    0  –  Print to Screen Only

                    1  –  Print to Printer Only

                    2  –  Print to a Disk File
                          (For later import to this program,
                          or to a word processor, to Lotus
                          or other spreadsheet program)


                    <Return> Accepts Value of  0

                                        Choice ?
```

FIGURE 8-7

Display/Print submenu

COGNATE LINKAGES Item 9 on the input menu allows you to **Choose a Cognate** of the current linkage as described in Section 3.7. Whenever a linkage is input to the program by any method (i.e., using items 1 to 8 in Figure 8-3), the program automatically calculates the dimensions of its two cognates. These can be switched to, calculated, and investigated at any time. The original linkage is always cognate #1. However, if you switch to another cognate and subsequently step through **Type data from Keyboard** with that data, you will redefine the current cognate as number 1 and recompute the dimensions of **its** two cognates.

2 - Display Data

After the input data is supplied, the program calculates and displays the Grashof condition for the linkage, and then displays the data which were input for this linkage as shown in Figure 8-6. The second item on the main menu **Display Data**, will bring up the same display screen at any time to allow checking of the current input data. A hard copy of this (or any) screen can be made at any time by using SHIFT PRINTSCRN. Alternatively the data in Figure 8-6 can be sent to the printer by choice from the submenu shown in Figure 8-7 which also allows the writing of the data to a diskfile.

WRITE TO DISK When option 2, **Print to a Disk File** is chosen, the programs will typically write two files to the disk drive. The first one is the same as Figure 8-6. The second is a file containing the data needed for reinput to the program from the **Readfile** option on the input menu (see Figure 8-3). Program FOURBAR

CALCULATE FOR :

    1 – One-Position

    2 – Time-Steps   (constant Alpha2)

    3 – Crank-Angles   (constant Omega2)

    0 – Abort this calculation

        <Return> Accepts Value of 3

             Choice ?

FIGURE 8-8

The calculation submenu

also writes a third file which is an input file for program FIVEBAR containing data for the equivalent geared fivebar linkage to the current fourbar as defined in Section 3.7. These are ASCII text files and so can be imported into most word processors for inclusion in reports and into many spreadsheet programs for further calculations if desired.

3 - Calculate

Main menu item 3, **Calculate**, will ask whether the linkage is assembled in the open or crossed configuration as defined in Section 4.5 and Figure 4-8. It then brings up a submenu which gives three choices as shown in Figure 8-8.

ONE POSITION will calculate position, velocity, and acceleration for any one specified input position θ_2, input angular velocity ω_2, and angular acceleration α_2. These parameters are supplied in response to prompts on input screens.

TIME STEPS requires input of a start time, finish time, and a time step, all in seconds. The value for α_2 (which must be either a constant or zero) and the initial position θ_2 and initial velocity ω_2 of link 2 at time zero must also be supplied. The program will then calculate all linkage parameters for each time step by applying the specified acceleration, which of course will change the angular velocity of the driver link with time. This is a transient analysis. The linkage will make as many revolutions of the driver crank as is necessary to run for the specified time. This choice is more appropriate for Grashof linkages, unless very short time durations are specified, as a non-Grashof linkage will quickly reach its toggle positions.

PRINT SELECT
_____

1 – Link Angles

2 – Link Omegas

3 – Link Alphas

4 – Point Velocities

5 – Point Accelerations

6 – Coupler-Point Coordinates

7 – Velocity Ratio

8 – Mechanical Advantage

<Ret> – EXIT

Choice ?

FIGURE 8-9

The print selection menu

CRANK ANGLES assumes that the angular acceleration α_2 of input link 2 is zero, thus ω_2 is constant. The values of initial and final crank angle θ_2, angle step $\Delta\theta_2$, and the constant input crank velocity ω_2 are requested with prompts. The program will then compute all linkage parameters for each angle step. This is a steady-state analysis and is suitable for either Grashof or non-Grashof linkages provided that the total linkage excursion is limited in the latter case.

If a position is encountered (in either the **Time Step** or **Crank Angle** calculations) which cannot be reached by the links, the mathematical result will be an attempt to take the square root of a negative number (See Section 4.5 and equation 4.10b). The program will skip the calculation for these positions and report to the screen that **Links do not connect for Theta2=xx**. The calculated parameters will all be arbitrarily forced to zero at those positions and the computation will continue at the next step. Note that a combination of successive **Time Step**, **Crank Angle,** and **Time Step** analyses can be used to simulate the startup, steady-state and the deceleration phase of a system, for a complete analysis.

4 - Print Results

Once a linkage has been calculated, the resulting data can be printed in tabular form. Figure 8-9 shows the **Print Select** submenu which comes up when **Print** is chosen from the main menu. The available choices allow printing of link angles (including transmission angles), angular velocities, angular accelerations, linear velocities and accelerations of points A, B, and C in Figure 8-4a and the XY coordinates of the coupler curve generated by point P. The angular velocity ratio and

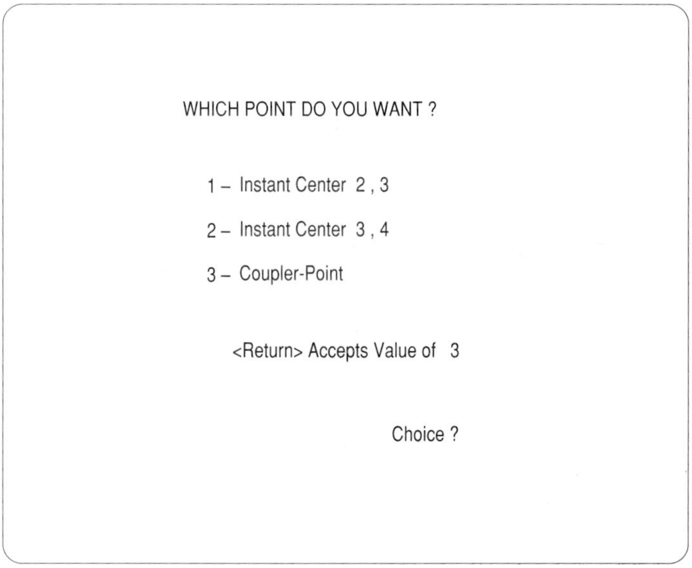

WHICH POINT DO YOU WANT ?

1 – Instant Center 2 , 3

2 – Instant Center 3 , 4

3 – Coupler-Point

<Return> Accepts Value of 3

Choice ?

FIGURE 8-10

Selection of the linkage point for output

mechanical advantage of the linkage for each calculated position as defined in equations 6.11 and 6.13 can also be printed. Any choice from the print select menu will request selection of either the CRT screen, the printer, or a diskfile as the output device with the same screen as shown in Figure 8-7.

The maximum number of positions which can be calculated is limited by array dimensions and varies with each program. This number is noted on the program's startup screen. For program FOURBAR, the maximum number is 180. Depending on your choice of start, finish, and step values of either time or angle, you will have some maximum number of calculated values available, necessarily less than 180 in this example. Your choice of **Print Step Size** determines how many of the calculated values are printed. A choice of one will print every calculated value, a choice of ten will print every tenth calculated value, etc. Note that you can examine linkage output with very high precision by applying the maximum number of calculated positions over a small, local range of input motion.

Choosing item 4 or 5 from the **Print Select** menu (Figure 8-9) also requires the selection of the point on the linkage for which you want to print the data, as shown in Figure 8-10. The points are identified in the menu as instant center $I_{2,3}$, $I_{3,4}$ and the coupler point. These labels correspond, in the order listed, to points A, B, and P in Figure 8-4a.

Figure 8-11 shows a table of link angles printed from FOURBAR. The first column is the transmission angle from equation 4.28. Column 2 is angle θ_2 which is the input value, as this was a **Crank Angle** calculation. Columns 3 and 4 are angles θ_3 and θ_4 which have been calculated from equations 4.10 and 4.13. The table is

FOURBAR *X.x*   Ima Student   Design # 2   *mm/dd/yy*   at   *hr:mn*

Angles in Degrees

Trans. Angle	Link 2	Link 3	Link 4
23.074	0.000	78.463	101.537
28.702	30.000	51.855	80.557
40.536	60.000	39.732	80.268
53.130	90.000	36.870	90.000
63.896	120.000	38.945	102.841
71.163	150.000	44.522	115.686
73.740	180.000	53.130	126.870
71.163	210.000	64.314	135.478
63.896	240.000	77.159	141.055
53.130	270.000	90.000	143.130
40.536	300.000	99.732	140.268
28.702	330.000	99.443	128.145
23.074	360.000	78.463	101.537

Table #____ Title_____

FIGURE 8-11

Table of link angles as printed by the program

annotated at the top with the user name, the design number, and the date and time of the printing. The bottom of the table has a line for table number and title

FOURBAR *X.x*   Ima Student   Design # 2   *mm/dd/yy*   at   *hr:mn*

Acceleration of Coupler Pt. - Inches/Sec^2
Referenced to Non-Rotating Global Coordinate System

Crank Angle	Magnitude	Angle in Degrees
0.0	13,268.072	270.000
30.0	16,310.163	169.199
60.0	6,797.073	155.112
90.0	2,304.887	167.471
120.0	658.601	195.580
150.0	42.708	217.654
180.0	138.889	90.000
210.0	42.708	322.354
240.0	658.601	344.420
270.0	2,304.887	12.529
300.0	6,797.073	24.888
330.0	16,310.163	10.801
360.0	13,268.072	270.000

Table #____ Title_____

FIGURE 8-12

Table of accelerations of a coupler point on a linkage as printed by the programs

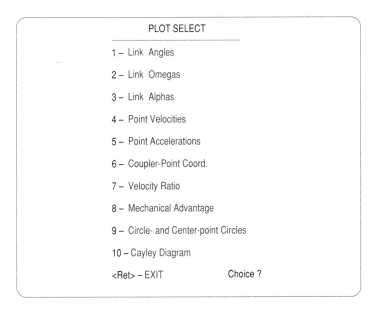

PLOT SELECT

1 – Link Angles

2 – Link Omegas

3 – Link Alphas

4 – Point Velocities

5 – Point Accelerations

6 – Coupler-Point Coord.

7 – Velocity Ratio

8 – Mechanical Advantage

9 – Circle- and Center-point Circles

10 – Cayley Diagram

<Ret> – EXIT   Choice ?

FIGURE 8-13

The plot selection menu

so that it can be used in an engineering report. The units used are defined in the table. The angular velocity and angular acceleration tables are similarly arranged.

Figure 8-12 shows a table of linear accelerations of the coupler point as printed by the program. The acceleration vectors are presented in polar form, i.e., magnitude and direction. The units are defined in the table. The header and footer are the same as in the other tables (Figure 8-11). The printed table of linear velocities has the same format.

5 - Plot Results

The **Plot Select** menu is shown in Figure 8-13. The same data that can be printed can also be plotted. However, only one link's data can be plotted at a time. The submenu in Figure 8-14 allows choice of the link for which angles, angular velocities, or angular accelerations are to be plotted. Figure 8-15 shows an example plot of the angular velocity of link 3. A sidebar contains the user name, design number, date, time, link lengths, coupler point location, and input position and velocity information. These are commonly called cartesian plots as they plot y against x on cartesian axes. The independent variable on the x axis in these cases is either time or angle depending on the choice made under **Calculate** above. The variable on the y axis is the one selected from the plot menu. Though these angular velocities are true vectors, they are all directed along the Z axis in this two-dimensional system. Thus their magnitudes alone can be plotted on cartesian axes and compared as their directions are constant, known, and the same.

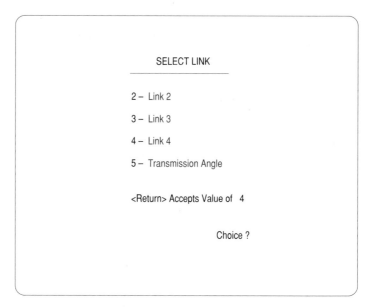

SELECT LINK

2 – Link 2

3 – Link 3

4 – Link 4

5 – Transmission Angle

<Return> Accepts Value of 4

Choice ?

FIGURE 8-14

The link selection menu

Plots of the linear velocities and linear accelerations require a different treatment than the cartesian plots used for the angular vector parameters. Their directions are not the same but vary with time or linkage position. One way to represent these linear vectors would be to make two cartesian plots, one for the magnitude and one for the angle of the vector at each time or angle step. Alternatively, the x and y cartesian components of the vector at each time or angle step could be presented as a pair of cartesian plots. This again requires two plots per

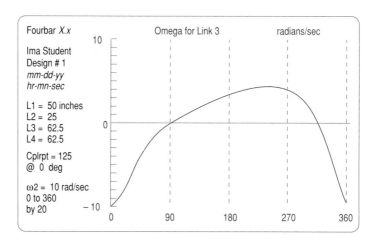

FIGURE 8-15

Cartesian plot of angular velocity for the coupler, link 3, of a fourbar linkage vector. Either of these methods has the disadvantage of being difficult to interpret. A better method is to present the vectors on a **polar plot**.

Figure 8-16 shows a **polar plot** of the velocity of point A in Figure 8-4a. The polar plot is made with respect to the **local axis system** x,y at point A as shown in Figure 8-4a. This local x,y axis system translates with point A as it moves but remains always parallel to the global axis system X,Y. By plotting the linear vectors on this moving axis system we can see both their magnitude and direction at each time or angle step, since we are attaching the roots of all the vectors to the moving point (A) at which they are acting.

In our example, ω_2 is constant, so the magnitude of $\mathbf{V}_A$ is also constant from equation 6.7. Only the direction of the velocity vector at point A changes with θ_2. This can be seen in Figure 8-16. The lines plotted radially outward from the origin represent individual velocity vectors for each time or angle step plotted. The vector root is at the origin. The large solid circle at the tip of one vector marks it as the velocity vector that goes with the first value of θ_2 calculated. In this case the initial θ_2 is $0°$ and ω_2 is counterclockwise. Thus, the velocity of point A at $\theta_2 = 0$ is directed $90°$ upward, as it should be from equation 6.3. The smaller circles at the ends of some of the other vectors represent their arrowheads and are placed on every fifth vector, corresponding to every fifth time or angle step. The vector corresponding to the second value of the independent variable calculated has a small filled circle at its tip. In this example, the angle step is $20°$, so the circled vectors are 100 input crank degrees (θ_2) apart (at 90 and 190°).

These plots can be done either with or without a pause between the plotting of each vector. Without the pause invoked, the plot will occur too quickly for the eye to detect the order in which they are drawn. Pause allows one to see their order because a keypress is required between the drawing of each vector. The current value of the independent variable (driver crank angle θ_2) is also printed to the screen with each pause.

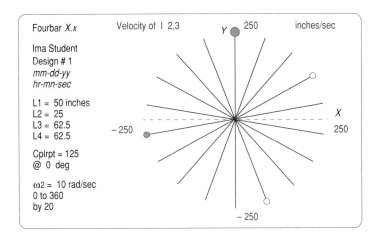

FIGURE 8-16

Polar plot of velocity at instant center I 2,3

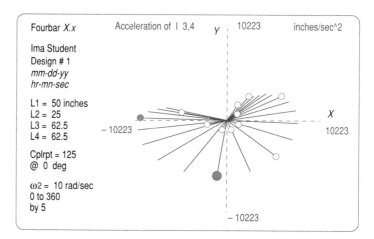

FIGURE 8-17

Type 1 polar plot of the acceleration of instant center I 3,4

Figure 8-17 shows a polar plot of the accelerations at point B in Figure 8-4a, for 5° increments of crank angle. Here the vectors are varying in both magnitude and direction and their polar plot creates a distinct pattern. We can see that the largest acceleration magnitudes are approximately in the negative x direction and relatively small magnitudes exist in the positive y direction. In fact the lines which represent the vectors in this figure add little information to the plot. We are most concerned with the locations of the tips of these vectors with

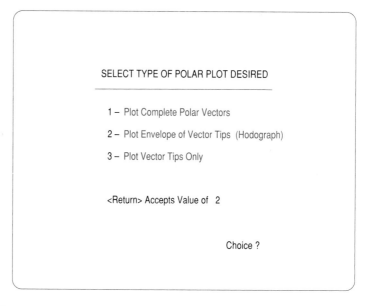

FIGURE 8-18

Selection of vector display mode

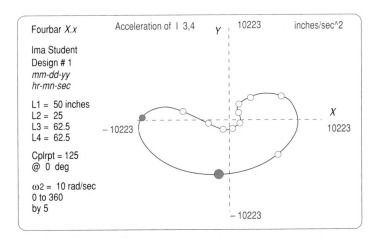

FIGURE 8-19

Type 2 polar plot (hodograph) of the acceleration at instant center I 3,4

respect to the origin. The programs allow alternate presentations of these polar plots as shown in Figure 8-18, which is a submenu of choices for all polar plots. Figures 8-16 and 8-17 were drawn using option 1 from this menu, **Plot Complete Polar Vectors**. Figure 8-19 shows the same data as Figure 8-17 but uses selection 2 from Figure 8-18, which plots the **Envelope of Vector Tips**. This plot connects the tips of the vectors with a line to give a more complete picture of the envelope of acceleration at this point. This type of plot is sometimes called a **hodograph**.

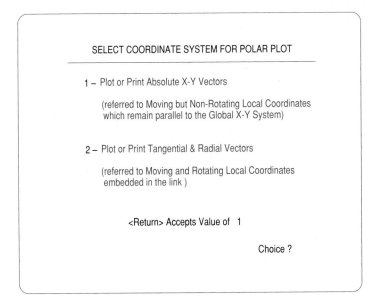

FIGURE 8-20

Selection of coordinate system for vector display

The third choice in Figure 8-18 plots the vector tips without any connecting lines. Any of these three alternate presentations of the polar plot is equally valid. The data are the same in all cases. The choice of presentation is largely a matter of individual preference.

When point B or point P (see Figure 8-4) is chosen for an acceleration polar plot, an additional option for data display is presented. This option does not apply to velocity plots. Figure 8-20 shows the submenu for this option. The choice is between two different sets of local coordinate systems as shown at point B in Figure 8-4a. Coordinate system x_1, y_1 at point B behaves in the same way as the local coordinate system x, y at point A; that is, it travels with point B but remains parallel to the world coordinate system X, Y at all times. This is choice 1 on the menu.

Coordinate system x_2, y_2 also travels with point B as its origin, but it is embedded in link 4 and rotates with that link, continuously changing its orientation with respect to the world coordinate system X, Y. This is choice 2 on the menu, and its purpose is to allow the plotting and printing of tangential and radial components of acceleration on the link. This is of value if a bending stress analysis is wanted. The dynamic force component due to tangential acceleration perpendicular to the link will create a bending moment. The radial component will create tension or compression. Actually, we will want to do the analysis at the center of gravity of the link for this problem rather than at its pin joint. Program DYNAFOUR, to be discussed in a later chapter, allows this more complete analysis to be done. Choice 1 in Figure 8-20 will be the most useful in our present kinematic analysis and is the default.

```
         <Return> Accepts Value of  N

                  Pause Between Plots ? <Y/N>

         <Return> Accepts Value of  N

                  Trace on (Show all positions) ? <Y/N>

         <Return> Accepts Value of  N

                  Suppress Links to see Coupler Curve ? <Y/N>
    _____

      Fixed centrode is brown, w/filled dots; moving centrode is blue, w/open circles

         <Return> Accepts Value of  N

                  Show Centrodes instead of Coupler Curve on Link Plot ? <Y/N>

         <Return> Accepts Value of  Y

                  Do you want to Autoscale for animation Y/N ?
```

FIGURE 8-21

Selection of animation parameters

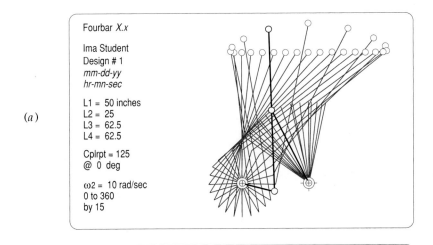

(a)

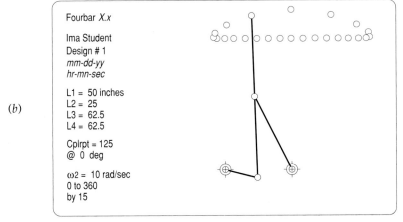

(b)

FIGURE 8-22

Animation with trace on (a) and with trace off (b)

6 - Animate

Item 6 on the main menu allows a graphical animation of the linkage to be done. Three parameters must be set from prompted input screens as shown in Figure 8-21. The first parameter toggles the PAUSE BETWEEN PLOTS option on/off. When on, a keystroke is required to draw each new position. The second parameter turns the TRACE feature on or off. TRACE ON keeps all positions of the linkage on the screen so that the pattern of motions can be seen as shown in Figure 8-22a. TRACE OFF erases all prior positions, showing only the current position as it cycles the linkage through all positions. This gives a dynamic view of the linkage behavior as shown in Figure 8-22b. SUPPRESS LINKS TO SEE CURVE, when on, will draw only the coupler curve (or centrodes) and not the links. This option is of value when the coupler curve lies within the links' excursions and they obscure its shape.

CENTRODES   Program Fourbar also calculates and draws the fixed and moving centrodes as defined in Section 6.5 and shown in Figure 6-15a. These loci of the instant centers can be substituted for the coupler curve with the SHOW CENTRODES instead of coupler curve option. Different colors are used to distinguish the fixed from the moving centrode. For monochrome plots, one is drawn with open, one with filled circles. The centrodes are drawn with their point of common tangency located at the first position calculated. Thus you can orient them anywhere by choosing the start angle for the calculation appropriately.

The linkage animation plot is normally autoscaled to fit the screen based on the size of the linkage and its coupler curves (but not of the centrodes as they can go to infinity). In some cases you may want to turn off autoscaling as when you wish to print two plots of slightly different linkages at the same scale for later manual superposition. The animation control menu in Figure 8-21 allows this. Turning off autoscaling will retain the most recent scale factor used. When on, it will rescale each plot to fit the screen. The same sidebar of linkage and user information is provided on the animation plots as on the others. A hard copy of any plot can be obtained by using the SHIFT PRINTSCRN keystroke combination. All plots are provided with a footer labelled for figure number and title so that they can be used as report illustrations.

7 - Help

Limited help in program operation is available from this selection. The submenu is shown in Figure 8-23. It should be self-explanatory. The last three items allow the program configuration parameters to be reset while the program is running. For example, when running on a VGA-equipped computer with color monitor, you will want to see the plots in high-resolution color on the screen. But, unless you have DOS 4.0 or higher you cannot dump these screens to a printer. So, change the screen resolution to CGA and replot the data. Then do a SHIFT PRINTSCRN to get the hard copy from the printer and switch back to VGA from the help menu to again have color screen plots. Even with DOS 4.0, which can dump EGA and VGA plots to the printer, you will have to replot in monochrome with white lines on a black background to get a noninverted screen dump. Whenever you change these parameters from the help menu, the configuration file will be rewritten to disk and the next time you start the program it will "wake up" in the last mode it was in when shut down.

8.4   FIVEBAR DIFFERENCES

Program FIVEBAR is very similar to FOURBAR. Its main menu is identical to FOURBAR's as shown in Figure 8-2.

FIVEBAR Input Data

The input submenu (Figure 8-24) has fewer built-in examples than FOURBAR, but the first two selections, **Type data from keyboard** and **Read data from DiskFile,** are the same. Because this program deals with the geared fivebar mechanism (GFBM), it

HELP IS AVAILABLE ON THESE TOPICS

1 – General Program Use

2 – Getting Hardcopy Graphics Output

3 – Program Ordering Information

4 – Copyright Notice

5 – Changing Screen Resolution & Clr/Mono

6 – Changing Angle Display Range

7 – Changing Units System

<Ret> – EXIT

Choice ?

FIGURE 8-23

Program help menu

requires more input data than the fourbar. Five link lengths must be supplied, in the order: driver crank (link 2), first coupler (link 3), second coupler (link 4), driven crank (link 5), and ground (link 1). These are as shown in Figure 8-25a which also shows

SELECT INPUT METHOD

1 – TYPE DATA from KEYBOARD

2 – READ DATA from DISKFILE

3 – Create Test Linkage with Gear Ratio = – 1

4 – Create Test Linkage with Gear Ratio = + 3

<Ret> – EXIT

Choice ?

FIGURE 8-24

Input menu for program FIVEBAR

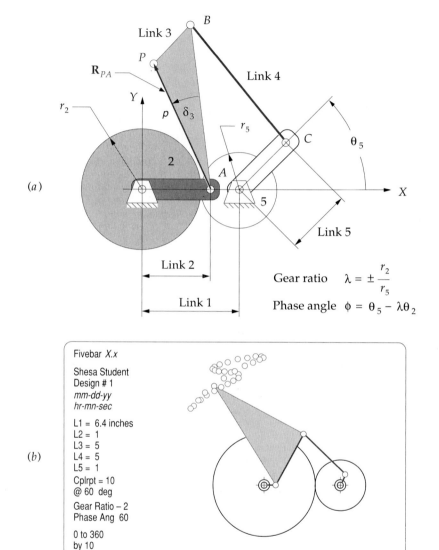

(a)

Gear ratio $\lambda = \pm \dfrac{r_2}{r_5}$

Phase angle $\phi = \theta_5 - \lambda\theta_2$

(b)

Fivebar X.x

Shesa Student
Design # 1
mm-dd-yy
hr-mn-sec

L1 = 6.4 inches
L2 = 1
L3 = 5
L4 = 5
L5 = 1
Cplrpt = 10
@ 60 deg

Gear Ratio – 2
Phase Ang 60

0 to 360
by 10

FIGURE 8-25

Input data for program FIVEBAR (a) and animation from the program (b)

the other two linkage parameters that must be defined as input, namely, the gear ratio (λ) and the phase angle (ϕ). These last two parameters are also defined in equation 4.23.

The phase angle is defined as the angle of link 5 when link 2 is at 0° as shown in Figure 8-25a. Note that the gear ratio as shown in the figure is a *negative ratio* because the external gears turn in opposite directions. The addition of an idler gear as shown in Figure 3-23b will create a positive gear ratio. It is also worth noting that the gear ratios defined in the *ZNH Atlas of Geared Fivebar Linkages* (see Figure 3-20 and ref. 8,

Chap. 3) are the reciprocal of the gear ratio in program FIVEBAR. So, when transferring data from this atlas to the program, the gear ratio must be inverted in order to get the same linkage as shown in the atlas. Otherwise the coupler curve will be a mirror image of the one in the atlas.

The coupler point P is defined in the same way as in the fourbar linkage. See Section 8.3 for a detailed explanation. A coupler point can only be placed on link 3 in this program. If you want a coupler point on link 4, mirror your linkage and renumber the links to put the coupler point on link 3.

Other FIVEBAR Main Menu Items

All of the other main menu items operate identically to FOURBAR except for the choice of an additional link and additional point for output data. Figure 8-25b shows the animation of a fivebar linkage from program FIVEBAR. Read the diskfile FIG8-25A into program Fivebar to animate this linkage.

8.5   SIXBAR DIFFERENCES

The SIXBAR program is generally similar to the FOURBAR program. It will analyze the **Watt's II** and the **Stephenson's III** linkage isomers as defined in Figure 2-14. These are two of the five distinct sixbar isomers. The **Watt's II** mechanism is shown in Figure 8-26a with the program's input parameters defined. The **Stephenson's III** mechanism is shown in Figure 8-27a with its input parameters defined. Note that the program divides the sixbar into two stages of fourbar linkages, asking first for stage 1 information, then for stage 2. Stage 1 is the left half of the mechanism as shown in Figures 8-26a and 8-27a. Stage 2 is the right half. The X axis of the global coordinate system is defined by instant centers $I_{1,2}$ and $I_{1,4}$ with its origin at $I_{1,2}$. The third fixed pivot $I_{1,6}$ can be anywhere in the plane. Its coordinates must be supplied as input. The main menu for SIXBAR is the same as in FOURBAR.

SIXBAR Input Data

Figure 8-28 shows the input submenu for SIXBAR. The built-in examples include some linkages discussed in previous chapters. Selection 3 is similar to the linkage designed in Example 3-11 and shown in Figure 3-25. Selection 5 shows an alternate approach to the problem presented in Example 3-12.

Selection 1 requires that more data be input than for the fourbar because of the more complicated nature of the sixbar linkage. The first input screen asks for a linkage name to annotate the plots. The next asks for a choice of Watt's or Stephenson's linkage. Then follows a sequence of input screens which ask for the the link lengths of the first stage of the linkage, followed by a request for the second stage data. These differ for the Watt's and Stephenson's linkages.

WATT'S II LINKAGE For the Watt's linkage (Figure 8-26a) the stage 1 data is requested in the order: crank, first coupler, first rocker, and ground link segment from

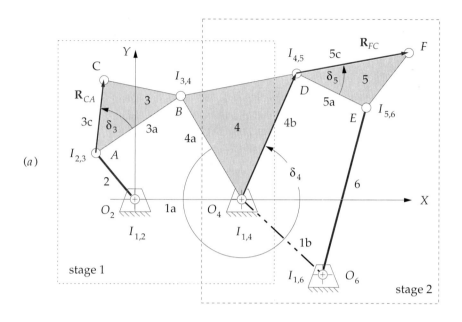

(a)

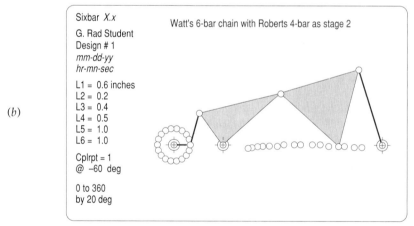

(b)

FIGURE 8-26

Input data for program SIXBAR - Watt's linkage (a) and animation from the program (b)

instant centers $I_{1,2}$ to $I_{1,4}$. These correspond to links 2, 3a, 4a, and 1a, respectively, as labelled in the figure. The program then asks for the link lengths in stage 2 of the linkage, in the order: second crank, second coupler, second rocker, corresponding respectively to links 4b, 5a, and 6 in Figure 8-26a. The angle δ_4 that the second crank (4b) makes with the first rocker (4a) is also requested. Note that this angle obeys the right-hand rule as do all angles in these programs.

Two coupler points are allowed to be defined in this linkage, one on link 3 and one on link 5. The method of location is by polar coordinates of a position vector embedded in the link as was done for the fourbar and fivebar linkages.

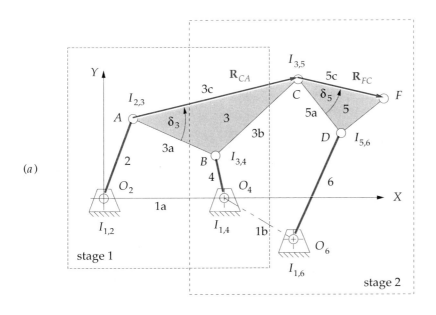

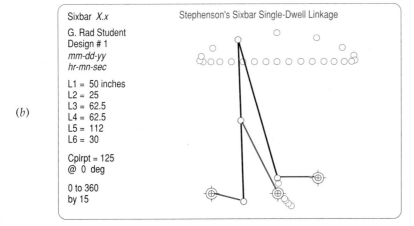

FIGURE 8-27

Input data for program SIXBAR - Stephenson's linkage (*a*) and program animation (*b*)

The first coupler point *C* is on link 3 and is defined in the same way as in the fourbar. The program requests the length (3c) of its position vector $\mathbf{R}_{CA}$ and the angle δ_3 which that vector makes with line 3a in Figure 8-26a. The second coupler point *F* is on link 5 and is defined with a position vector $\mathbf{R}_{FD}$ rooted at instant center $I_{4,5}$. The program requests the length (5c) of this position vector $\mathbf{R}_{FD}$ and the angle δ_5 which that vector makes with line 5a.

The *X* and *Y* components of the location of the third fixed pivot $I_{1,6}$ are then requested. These are with respect to the global *X,Y* axis system whose origin is at $I_{1,2}$. Figure 8-26b shows a Watt's linkage from program SIXBAR.

Table 8-1   Equations Used in Program Fourbar

Variable	Equation
θ_3	4.13
θ_4	4.10
ω_3	6.18a
ω_4	6.18b
α_3	7.12a
α_4	7.12b
V_A	6.19a
V_B	6.19c
V_P	6.36
A_A	7.13a
A_B	7.13c
A_P	7.32
Transmission Angle	4.28
Velocity Ratio	6.10, 6.11
Mechanical Advantage	6.13
Coupler Point Coord.	4.27
3-Pos Linkage Synthesis	5.25, to 5.31, 5.2
3-Pos Synthesis w/Pivots	5.34 plus 5.25, to 5.31, 5.2
Circle, Center Point Circles	5.25, to 5.31

STEPHENSON'S III LINKAGE The stage 1 data for the Stephenson's linkage is input in a similar way to the Watt's version. The first stage's crank, first coupler, first rocker, and ground link segment from instant center $I_{1,2}$ to $I_{1,4}$ correspond to links 2, 3a, 4, and 1a, respectively, as shown in Figure 8-27a. The program then asks for the link lengths in stage 2 of the linkage, in the order: second coupler, second rocker, corresponding respectively to links 5a and 6 in Figure 8-27a.

Note that, unlike the Watt's example above, there is no "second crank" in the Stephenson's linkage because the second stage is driven by the coupler (link 3) of the first stage. In this program link 5 is constrained to be connected to link 3 at link 3's defined coupler point C which then becomes instant center $I_{3,5}$. The data for this is requested in similar format to the Watt's linkage, namely the length of the position vector $\mathbf{R}_{CA}$ (line 3c) and its angle δ_3. The second coupler point location on link 5 is defined, as before, by position vector $\mathbf{R}_{FC}$ with length 5c and angle δ_5.

The X and Y components of the location of the third fixed pivot $I_{1,6}$ are then requested. These are with respect to the global X,Y axis system. Figure 8-27b shows a Stephenson's linkage from program SIXBAR.

Other SIXBAR Main Menu Items

All of the other menu items operate identically to FOURBAR except for the choice
of two additional links and three more points for output data. These differences
should be self-explanatory from the menu prompts.

8.6   EQUATIONS USED IN THE PROGRAMS

Tables 8-1 and 8-2 are lists of all the equations used in programs FOURBAR and
FIVEBAR identified by their equation numbers in this text. Program SIXBAR uses
similar equations derived by the same methods as explained in previous chapters,
but which are not specifically derived or presented herein. The student should
study these equations in concert with the use of programs FOURBAR and FIVEBAR in
order to understand the kinematic theory and mathematics on which the programs
are based. The understanding gained can be extrapolated to the other program.
See Figure 8-4a for FOURBAR notation and Figure 8-25a for FIVEBAR notation. The
notation for program SIXBAR will be found in Figures 8-26a and 8-27a.

Table 8-2   Equations Used in Program Fivebar

Variable	Equation
θ_3	4.24i
θ_4	4.24h
θ_5	4.23c
ω_3	6.33a
ω_4	6.33b
α_3	7.14a
α_4	7.14b
V_A	6.33c
V_B	6.33f
V_P	6.36
A_A	7.29c
A_B	7.29f
A_P	7.32
Transmission Angle	4.28
Velocity Ratio	6.10, 6.11
Mechanical Advantage	6.13
Coupler Point Coord.	4.27

SELECT INPUT METHOD

1 – TYPE DATA from KEYBOARD

2 – READ DATA from DISKFILE

3 – Create SIXBAR Test Linkage

4 – Create SINGLE-DWELL linkage

5 – Create DOUBLE-DWELL linkage

6 – Create ROBERTS straight line linkage

7 – Create WATT'S   straight line linkage

<Ret> – EXIT

Choice ?

FIGURE 8-28

Input menu for program SIXBAR

8.7   ENDNOTES

[1] Micro Soft Disk Operating System. This is the "supervisory program" which is automatically loaded into memory from disk when you turn on an IBM-compatible computer. It controls all operations involving the disk drives, printers, etc.

[2] The degree of compatibility with the IBM standard varies widely among brands of computers. Not all will run these programs without problems. Only an actual test will determine this. Graphics incompatibilities are the most common problems.

[3] Random access memory (RAM) is the main memory of the computer and is measured in bytes. It takes 2 bytes to express a single-integer number in these machines, and more to express a real number.

[4] DYNAFOUR, DYNACAM, and ENGINE.

[5] When copying the programs to a hard drive, be sure to copy all the programs on the disk, as some are needed at startup of the programs.

[6] The Color Graphics Adapter (CGA) is the hardware driver in the computer which "paints" the picture on the screen. The resolution needed for plots requires the color driver even though output may be monochrome.

[7] A graphics screen dump is obtained by holding the SHIFT key and pressing the PRINTSCREEN key simultaneously. Whatever is on the CRT screen will be duplicated on the printer, whether text or graphics. Not all printers can perform this function.

[8] This program is included on the MS DOS system disks which are provided with the computer. You can make it load automatically at startup by including the one word command GRAPHICS in your AUTOEXEC.BAT file on your startup disk. See your MS DOS manual for instruction in creating or modifying the AUTOEXEC.BAT file.

Chapter 9

CAM DESIGN

It is much easier to design than to perform
SAMUEL JOHNSON

9.0   INTRODUCTION

Cam-follower systems are frequently used in all kinds of machines. The valves in your automobile engine are opened by cams. Machines used in the manufacture of many consumer goods are full of cams. Compared to linkages, cams are easier to design to give a specific output function, but they are much more difficult and expensive to make than a linkage. Cams are a form of degenerate fourbar linkage in which the coupler link has been replaced by a half joint as shown in Figure 9-1. This topic was discussed in Section 2.9 on linkage transformation (see also Figure 2-10). For any one instantaneous position of cam and follower, we can substitute an effective linkage which will, for that instantaneous position, have the same motion as the original. In effect, the cam-follower is a fourbar linkage with variable length (effective) links. It is this conceptual difference that makes the cam-follower such a flexible and useful **function generator**. We can specify virtually any output function we desire and quite likely create a curved surface on the cam to generate that function in the motion of the follower. We are not limited to fixed length links as we were in linkage synthesis. The cam-follower is an extremely useful mechanical device, without which the machine designer's tasks would be more difficult to accomplish. But, as with everything else in engineering, there are tradeoffs. These will be discussed in later sections.

This chapter will present the proper approach to designing a cam-follower system, and in the process also present some less than proper designs as examples of the problems which inexperienced cam designers often get into. Theoretical considerations of the mathematical functions commonly used for cam curves will be discussed. Methods for the derivation of custom polynomial functions, to suit any set of boundary conditions, will be presented. The task of sizing the cam with considerations of pressure angle and radius of curvature will be addressed, and

TABLE 9-1   Notation Used in This Chapter

t = time, seconds

θ = camshaft angle, degrees or radians

ω = camshaft angular velocity, radians/sec

β = total angle of any segment, rise, fall or dwell, degrees or radians

h = total lift (rise or fall) of any one segment, length units

s or S = follower displacement, length units

$v = ds/d\theta$ = follower velocity, length/radian

$V = dS/dt$ = follower velocity, length/sec

$a = dv/d\theta$ = follower acceleration, length/radian^2

$A = dV/dt$ = follower acceleration, length/sec^2

$j = da/d\theta$ = follower jerk, length/radian^3

$J = dA/dt$ = follower jerk, length/sec^3

$s\ v\ a\ j$ refer to the group of diagrams, length units versus radians

$S\ V\ A\ J$ refer to the group of diagrams, length units versus time

R_b = base circle radius, length units

R_p = prime circle radius, length units

R_f = roller follower radius, length units

ε = eccentricity of cam follower, length units

ϕ = pressure angle, degrees or radians

ρ = radius of curvature of cam surface, length units

ρ_{pitch} = radius of curvature of pitch curve, length units

ρ_{min} = min imum radius of curvature of pitch curve or cam surface, length units

manufacturing processes and their limitations discussed. The computer program DYNACAM will be used throughout the chapter as a tool to present and illustrate design concepts and solutions. A user manual for this program is included as the last section in the chapter. The reader can skip ahead to that section at any time without loss of continuity in order to become familiar with the program's operation.

9.1   CAM TERMINOLOGY

Cam-follower systems can be classified in several ways: by *type of follower motion*, either **translating** or **rotating** (oscillating); by type of cam, radial, cylindrical, three-dimensional; by *type of joint closure*, either **force-** or **form**-closed; by *type of follower*, **curved** or **flat**, **rolling** or **sliding**; by *type of motion constraints*, **critical extreme position** (CEP), **critical path motion** (CPM); by *type of motion program*, **rise-fall** (RF), **rise-fall-dwell** (RFD), **rise-dwell-fall-dwell** (RDFD). We will now discuss each of these classification schemes in more detail.

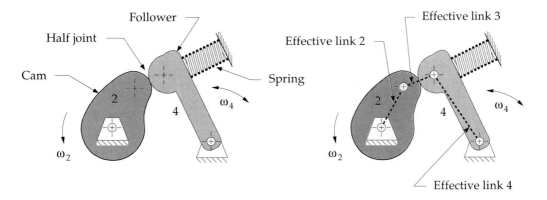

(*a*) Oscillating cam-follower has an effective pin-jointed fourbar equivalent

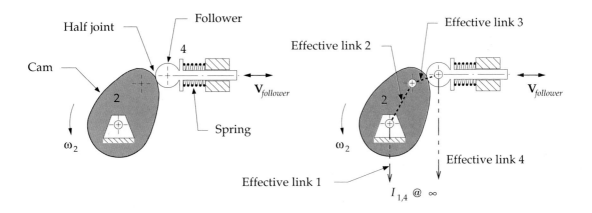

(*b*) Translating cam-follower has an effective fourbar slider-crank equivalent

FIGURE 9-1

Effective linkages in the cam-follower

Type of Follower Motion

Figure 9-1a shows a system with an oscillating, or **rotating, follower**. Figure 9-1b shows a **translating follower**. These are analogous to the crank-rocker fourbar and the slider-crank fourbar linkages, respectively. The choice between these two forms of the cam-follower is usually dictated by the type of output motion desired. If true rectilinear translation is required, then the translating follower is dictated. If pure rotation output is needed, then the oscillator is the obvious choice. There are advantages to each of these approaches, separate from their motion characteristics, depending on the type of follower chosen. These will be discussed in a later section.

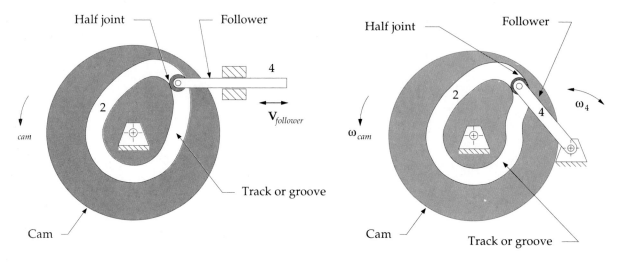

(*a*) Cam with translating follower

(*b*) Cam with oscillating follower

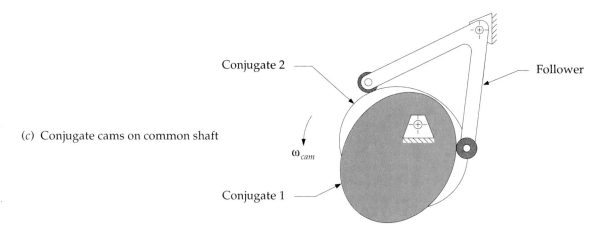

(*c*) Conjugate cams on common shaft

FIGURE 9-2

Form-closed cam-follower systems

Type of Joint Closure

Force and form closure were discussed in Section 2.3 on the subject of joints and have the same meaning here. **Force closure**, as shown in Figure 9-1, *requires an external force be applied to the joint* in order to keep the two links, cam and follower, physically in contact. This force is usually provided by a spring. This force, defined as positive in a direction which closes the joint, cannot be allowed to become negative. If it does, the links have lost contact because a *force-closed joint can only push, not pull*. **Form closure**, as shown in Figure 9-2, *closes the joint by geometry*. No external force is required.

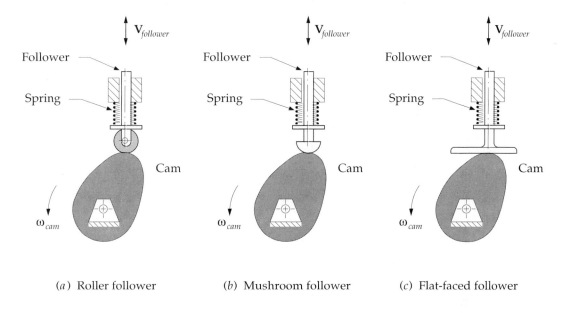

(*a*) Roller follower   (*b*) Mushroom follower   (*c*) Flat-faced follower

FIGURE 9-3

Three common types of cam followers

There are really two cam surfaces in this arrangement, one surface on each side of the follower. Each surface pushes, in its turn, to drive the follower in both directions.

Figure 9-2a and b shows track or groove cams which capture a single follower in the groove and both push and pull on the follower. Figure 9-2c shows another variety of form-closed cam-follower arrangement, called **conjugate cams**. There are two cams fixed on a common shaft which are mathematical conjugates of one another. Two roller followers, attached to a common arm, are each pushed in opposite directions by the conjugate cams. When form-closed cams are used in automobile or motorcycle engine valve trains, they are called **desmodromic** cams. There are advantages and disadvantages to both force- and form-closed arrangements which will be discussed in a later section.

Type of Follower

Follower, in this context, refers only to that part of the follower link which contacts the cam. Figure 9-3 shows three common arrangements, **flat-faced, mushroom** (curved), and **roller**. The roller follower has the advantage of lower (rolling) friction than the sliding contact of the other two but can be more expensive. **Flat-faced followers** can package smaller than roller followers for some cam designs and are often favored for that reason as well as cost for automotive valve trains. **Roller followers** are more frequently used in production machinery where their ease of replacement and availability from bearing manufacturers' stock in any quantities are advantages. Grooved or track cams require roller followers. Roller followers are essentially ball or roller bearings with customized mounting details. Figure 9-5a shows two common types of commercial roller followers. Flat-faced or **mushroom followers** are

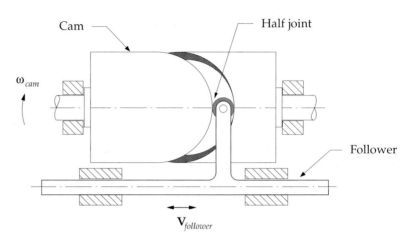

FIGURE 9-4

Axial, cylindrical or barrel cam with form-closed, translating follower

usually custom designed and manufactured for each application. For high-volume applications such as automobile engines, the quantities are high enough to warrant a custom-designed follower.

Type of Cam

The direction of the follower's motion relative to the axis of rotation of the cam determines whether it is a **radial** or **axial** cam. All cams shown in Figures 9-1 to 9-3 are radial cams because the follower motion is generally in a radial direction. Open **radial cams** are also called **plate cams**.

Figure 9-4 shows an **axial cam** whose follower moves parallel to the axis of cam rotation. This arrangement is also called a **face** cam if open (force-closed) and a **cylindrical** or **barrel** cam if grooved or ribbed (form-closed).

Figure 9-5b shows a selection of cams of various types. Clockwise from the lower left, they are: an open (force-closed) axial or face cam; an axial grooved (track) cam (form-closed) with external gear; an open radial, or plate cam (force-closed); a ribbed axial cam (form-closed); an axial grooved (barrel) cam.

A **three-dimensional cam** or **camoid** (not shown) is a combination of radial and axial cams. It is a two degree of freedom system. The two inputs are rotation of the cam about its axis and translation of the cam along its axis. The follower motion is a function of both inputs. The follower tracks along a different portion of the cam depending on the axial input.

Type of Motion Constraints

There are two general categories of motion constraint, **critical extreme position** (CEP - also called endpoint specification) and **critical path motion** (CPM). **Critical extreme position** refers to the case in which the design specifications define the start

(*a*) Commercial roller followers

Courtesy of McGill Manufacturing Co.
South Bend, Indiana

(*b*) Commercial cams of various types

Courtesy of The Ferguson Co.
St. Louis Missouri

FIGURE 9-5

Cams and roller followers

and finish positions of the follower (i.e., extreme positions) but do not specify any constraints on the path motion between the extreme positions. This case is discussed in sections 9.3 and 9.4 and is the easier of the two to design as the designer has great freedom to choose the cam functions which control the motion between extremes. **Critical path motion** is a more constrained problem than CEP because the path motion, and/or one or more of its derivatives are defined over all or part of the interval of motion. This is analogous to **function generation** in the linkage design case except that with a cam we can achieve a continuous output function for the follower. Section 9.6 discusses this CPM case. It may only be possible to create an approximation of the specified function and still maintain suitable dynamic behavior.

Type of Motion Program

The motion programs **rise-fall** (RF), **rise-fall-dwell** (RFD), and **rise-dwell-fall-dwell** (RDFD) all refer mainly to the CEP case of motion constraint and in effect define how many dwells are present in the full cycle of motion, either none (RF), one (RFD) or more than one (RDFD). **Dwells,** defined as *no output motion for a specified period of input motion*, are an important feature of cam-follower systems because it is very easy to create exact dwells in these mechanisms. The cam-follower is the design type of choice whenever a dwell is required. We saw in Section 3.9 how to design dwell linkages and found that at best we could obtain only an approximate

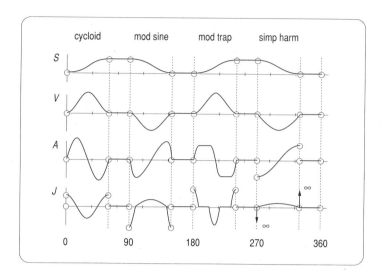

FIGURE 9-6

Cycloidal, modified sine, modified trapezoid and simple harmonic motion functions

dwell. The resulting single- or double-dwell linkages tend to be quite large for their output motion and are somewhat difficult to design. (See program SIXBAR for some built-in examples of these dwell linkages.) Cam-follower systems tend to be more compact than linkages for the same output motion.

If your need is for a **rise-fall** (RF) CEP motion, with no dwell, then you should really be considering a crank-rocker linkage rather than a cam-follower to obtain all the linkage's advantages over cams of reliability, ease of construction, and lower cost which were discussed in Section 2.15. If your needs for compactness outweigh those considerations, then the choice of a cam-follower in the RF case may be justified. Also, if you have a CPM design specification, and the motion or its derivatives are defined over the interval, then a cam-follower system is the logical choice in the RF case.

The **rise-fall-dwell** (RFD) and **rise-dwell-fall-dwell** (RDFD) cases are obvious choices for cam-followers for the reasons discussed above. However, each of these two cases has its own set of constraints on the behavior of the cam functions at the interfaces between the segments which control the rise, the fall, and the dwells. In general, we must match the **boundary conditions** (BC's) of the functions and their derivatives at all interfaces between the segments of the cam. This topic will be thoroughly discussed in the following sections.

9.2   *S V A J* DIAGRAMS

The first task faced by the cam designer is to select the mathematical functions to be used to define the motion of the follower. The easiest approach to this process is to "linearize" the cam, i.e., "unwrap it" from its circular shape and consider it

Camshaft Constant Omega = 38.0 Radians/Sec
Calculated Maximum Lift = 0.4 Inches
Number of Cam Segments = 8

Segment Number	Function used	Start Angle	End Angle	Delta Angle
1	Cycloid rise	0	60	60
2	Dwell	60	90	30
3	ModSine fall	90	150	60
4	Dwell	150	180	30
5	ModTrap rise	180	240	60
6	Dwell	240	270	30
7	SimpHarm fall	270	330	60
8	Poly5	330	360	30

FIGURE 9-7

Cam programs defined in DYNACAM for the cam functions shown in Figure 9-6

as a function plotted on cartesian axes. We plot the displacement function s, its first derivative velocity v, its second derivative acceleration a, and its third derivative jerk j, all on aligned axes as a function of camshaft angle θ as shown in Figure 9-6. Note that we can consider the independent variable in these plots to be either time t or shaft angle θ, as we know the constant angular velocity ω of the camshaft and can easily convert from angle to time and vice versa.

$$\theta = \omega t \qquad (9.1)$$

Figure 9-6 shows the $s\,v\,a\,j$ curves for example cam #5 in program DYNACAM (also see Figure 9-69). Figure 9-6 shows the whole cam, i.e., 360 degrees of camshaft rotation. This particular cam has eight segments, RDFDRDFD as shown in Figure 9-7. We will begin any cam design by sketching the $s\,v\,a\,j$ diagrams to define the general shapes of the functions. Particular functions will then be chosen for each segment and their actual shapes, magnitudes and interactions at the interfaces between segments investigated with program DYNACAM.

9.3 DOUBLE-DWELL CAM DESIGN - CHOOSING S V A J FUNCTIONS

Many cam design applications require multiple dwells. The double-dwell case is quite common. Perhaps a **double-dwell** cam is driving a part feeding station on a production machine that makes toothpaste. This hypothetical cam's follower is fed an empty toothpaste tube (during the low dwell), then moves the empty

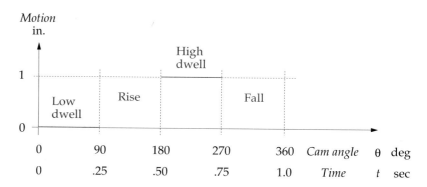

FIGURE 9-8

Cam timing diagram

tube into a loading station (during the rise), holds the tube absolutely still in a **critical extreme position** (CEP) while toothpaste is squirted into the open bottom of the tube (during the high dwell), and then retracts the filled tube back to the starting (zero) position and holds it in this other critical extreme position where another mechanism (during the low dwell) picks the tube up and carries it to the next operation, which might be to seal the bottom of the tube. A similar cam could be used to feed, align, and retract the tube at the bottom-sealing station as well.

Cam specifications such as this are often depicted on a timing diagram as shown in Figure 9-8 which is a graphical representation of the specified events in the machine cycle. A **machine's cycle** is defined as *one revolution of its master driveshaft*. In a complicated machine, such as our toothpaste maker, there will be a **timing diagram** for each subassembly in the machine. The time relationships among all subassemblies are defined by their timing diagrams which are all drawn on a common time axis. Obviously all these operations must be kept in precise synchrony and time phase for the machine to work.

This simple example in Figure 9-8 is a critical extreme position (CEP) case, because nothing is specified about the functions to be used to get from the low dwell position (one extreme) to the high dwell position (other extreme). The designer is free to choose any function that will do the job. Note that these specifications contain only information about the displacement function. The higher derivatives are not specifically constrained in this example. We will now use this problem to investigate several different ways to meet the specifications.

EXAMPLE 9-1

Naive Cam Design - A Bad Cam

Problem: Consider the following cam design CEP specification:

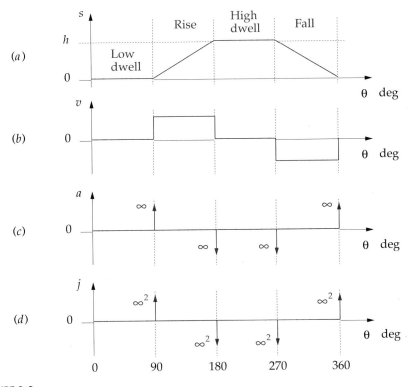

FIGURE 9-9

The *s v a j* diagrams of a "bad" cam design

dwell	at zero displacement for 90 degrees (low dwell)
rise	1 inch in 90 degrees
dwell	at 1 inch for 90 degrees (high dwell)
fall	1 inch in 90 degrees.
cam ω	2π radians/sec = 1 rev/sec

Solution:

1 The naive or inexperienced cam designer might proceed with a design as shown in Figure 9-9a. Taking the given specifications literally, it is tempting to merely "connect the dots" on the timing diagram to create the displacement (*s*) diagram. (After all, when we wrap this *s* diagram around a circle to create the actual cam, it will look quite smooth despite the sharp corners on the *s* diagram.) The mistake our beginning designer is making here is to ignore the effect on the higher derivatives of the displacement function which results from this simplistic approach.

2 Figure 9-9 b, c, and d shows the problem. Note that we have to treat each segment of the cam (rise, fall, dwell) as a separate entity in developing mathematical functions for the cam. Taking the rise segment (#2) first, the displacement function in Figure 9-9a during this portion is a straight line, or first-degree polynomial. The general equation for a straight line is:

$$y = mx + b \qquad (9.2)$$

where m is the slope of the line and b is the y intercept. Substituting variables appropriate to this example in equation 9.2, angle θ replaces the independent variable x, and the displacement s replaces the dependent variable y. By definition, the constant slope m of the displacement is the velocity constant K_v.

3   For the rise segment, the y intercept b is zero because the **low dwell position is taken as zero displacement by convention**. Equation 9.2 then becomes:

$$s = K_v \theta \qquad (9.3)$$

4   Differentiating with respect to θ gives a function for velocity during the rise.

$$v = K_v = \text{constant} \qquad (9.4)$$

5   Differentiating again with respect to θ gives a function for acceleration during the rise.

$$a = 0 \qquad (9.5)$$

This seems too good to be true (and it is). Zero acceleration means zero dynamic force. This cam appears to have no dynamic forces or stresses in it!

Figure 9-9 shows what is really happening here. If we return to the displacement function and graphically differentiate it twice, we will observe that, from the definition of the derivative as the instantaneous slope of the function, the acceleration is in fact zero **during the interval**. *But, at the boundaries of the interval*, where rise meets low dwell on one side and high dwell on the other, note that *the velocity function is multivalued. There are discontinuities at these boundaries*. The effect of these discontinuities is to create a portion of the velocity curve which has **infinite slope** and zero duration. This results in the *infinite spikes of acceleration* shown at those points.

These spikes are more properly called **Dirac delta functions**. Infinite acceleration cannot really be obtained, as it requires infinite force. Clearly the dynamic forces will be very large at these boundaries and will create high stresses and rapid wear. In fact, if this cam were built and run at any significant speeds, the sharp corners on the displacement diagram which are creating these theoretical infinite accelerations would be quickly worn to a smoother contour by the unsustainable stresses generated in the materials. **This is an unacceptable design**.

The unacceptability of this design is reinforced by the **jerk** diagram which shows theoretical values of **infinity squared** at the discontinuities. The problem has been engendered by an inappropriate choice of displacement function. In fact, the cam designer should not be as concerned with the displacement function as with its higher derivatives.

The Fundamental Law of Cam Design

Any cam designed for operation at other than very low speeds must be designed with the following constraints:

The cam function must be continuous through the first and second derivatives of displacement across the entire interval (360 degrees).

corollary:

The jerk function must be finite across the entire interval (360 degrees).

In any but the simplest of cams, the cam motion program cannot be defined by a single mathematical expression, but rather must be defined by several separate functions, each of which defines the follower behavior over one segment, or piece, of the cam. These are sometimes called *piecewise functions*. These functions must have **third-order continuity** (the function plus two derivatives) at all boundaries. **The displacement, velocity and acceleration functions must have no discontinuities in them**.

If any discontinuities exist in the acceleration function, then there will be infinite spikes, or Dirac delta functions, appearing in the derivative of acceleration, jerk. Thus the corollary merely restates the fundamental law of cam design. Our naive designer failed to recognize that by starting with a low-degree (linear) polynomial as the displacement function, discontinuities would appear in the upper derivatives.

Polynomial functions are one of the best choices for cams as we shall shortly see, but they do have one fault that can lead to trouble in this application. Each time they are differentiated, they reduce by one degree. Eventually, after enough differentiations, polynomials degenerate to zero degree (a constant value) as the velocity function in Figure 9-9b demonstrates. Thus, by starting with a first-degree polynomial as a displacement function, it was inevitable that discontinuities would soon appear in its derivatives.

In order to obey the fundamental law of cam design, one must start with at least a third-degree polynomial (cubic) as the displacement function. This will degenerate to a first-degree function in the acceleration. The jerk function will have discontinuities, and the (unnamed) derivative of jerk will have infinite spikes in it. This is acceptable, as the jerk is still finite.

Simple Harmonic Motion (SHM)

Our naive cam designer recognized his mistake in choosing a straight-line function for the displacement. He also remembered a family of functions he had met in a calculus course which have the property of remaining continuous throughout any number of differentiations. These are the harmonic functions. On repeated differentiation, sine becomes cosine, which becomes negative sine, which becomes negative cosine, etc., ad infinitum. One never runs out of derivatives with the harmonic family of curves. In fact, differentiation of a harmonic function really only amounts to a 90° phase shift of the function. It is though, as you differentiated it, you cut out, with a scissors, a different portion of the same continuous sine wave function, which is defined from minus infinity to plus infinity. The equations of simple harmonic motion (SHM) for a rise motion are:

$$s = \frac{h}{2}\left[1 - \cos\left(\pi\frac{\theta}{\beta}\right)\right] \qquad (9.6a)$$

$$v = \frac{\pi}{\beta}\frac{h}{2}\sin\left(\pi\frac{\theta}{\beta}\right) \qquad (9.6b)$$

$$a = \frac{\pi^2}{\beta^2}\frac{h}{2}\cos\left(\pi\frac{\theta}{\beta}\right) \qquad (9.6c)$$

$$j = -\frac{\pi^3}{\beta^3}\frac{h}{2}\sin\left(\pi\frac{\theta}{\beta}\right) \qquad (9.6d)$$

where h is the total rise, or lift, θ is the camshaft angle, and β is the total angle of the rise interval.

We have here introduced a notation to simplify the expressions. The independent variable in our cam functions is θ, the camshaft angle. The period of any one segment is defined as the angle β. Its value can, of course, be different for each segment. We normalize the independent variable θ by dividing it by the period of the segment β. Both θ and β are measured in radians (or both in degrees). The value of θ/β will then vary from 0 to 1 over any segment. It is a dimensionless ratio. Equations 9.6 define simple harmonic motion and its derivatives for this rise segment in terms of θ/β.

This family of harmonic functions appears, at first glance, to be well suited to our cam design problem above. If we define the displacement function to be one of the harmonic functions, we should not "run out of derivatives" before reaching the acceleration.

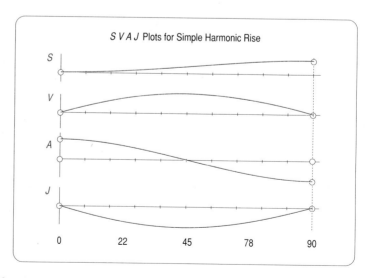

FIGURE 9-10

Simple harmonic motion with dwells has discontinuous acceleration

✍️ EXAMPLE 9-2

Sophomoric[1] Cam Design - Simple Harmonic Motion - Still a Bad Cam

Problem: Consider the same cam design CEP specification as in Example 9-1:

dwell	at zero displacement for 90 degrees (low dwell)
rise	1 inch in 90 degrees
dwell	at 1 inch for 90 degrees (high dwell)
fall	1 inch in 90 degrees.
cam ω	2π radians/sec = 1 rev/sec

Solution:

1   Figure 9-10 shows a full-rise **simple harmonic** function[2] applied to the rise segment of our cam design problem.

2   Note that the velocity function is continuous, as it matches the zero velocity of the dwells at each end.

3   The acceleration function, however, is **not** continuous. It is a half-period cosine curve and has nonzero values at start and finish.

4   Unfortunately, the dwell functions, which adjoin this rise on each side, have zero acceleration as can be seen in Figure 9-6. Thus there are **discontinuities in the acceleration at each end of the interval** which uses this simple harmonic displacement function.

5   This violates the fundamental law of cam design and creates **infinite spikes of jerk** at the ends of this fall interval. **This is also an unacceptable design**.

What went wrong? While it is true that harmonic functions are differentiable ad infinitum, we are not dealing here with single harmonic functions. Our cam function over the entire interval is a **piecewise function**, (Figure 9-6) made up of several segments, some of which may be dwell portions or other functions. A dwell will always have zero velocity and zero acceleration. Thus we must match the dwells' zero values at the ends of those derivatives of any nondwell segments that adjoin them. The simple harmonic displacement function, when used with dwells, does **not** satisfy the fundamental law of cam design. Its second derivative, acceleration, is nonzero at its ends and thus does not match the dwells required in this example.

The only case in which the simple harmonic displacement function will satisfy the fundamental law is the non quick-return RF case, i.e., rise in 180° and fall in 180° with no dwells. Then the cam becomes an eccentric as shown in Figure 9-11. As a single continuous (not piecewise) function, its derivatives are continuous also. Figure 9-12 shows the displacement (in inches) and acceleration functions (in *g*'s) of the eccentric cam in Figure 9-11 as actually measured on the follower. The

---

[1] **Sophomoric**, from sophomore, *def. wise fool*, from the Greek, *sophos* = *wisdom*, *moros* = *fool*.
[2] Though this is actually a half-period cosine wave, we will call it a *full-rise* (or *full-fall*) simple harmonic function to differentiate it from the *half-rise* (and *half-fall*) simple harmonic function which is actually a quarter-period cosine (see Section 9.6).

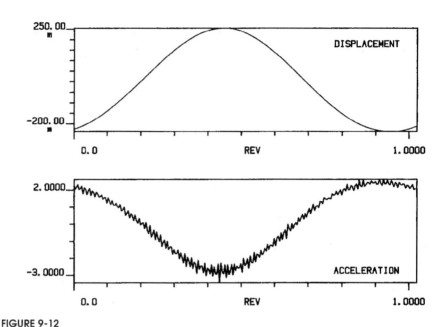

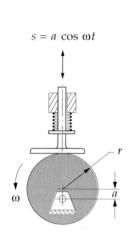

FIGURE 9-11

An eccentric cam
has simple harmonic
motion

FIGURE 9-12

Displacement and acceleration as measured on the follower of an eccentric cam

noise, or "ripple", on the acceleration curve is due to small, unavoidable, manu-facturing errors. Manufacturing limitations will be discussed in a later section.

Cycloidal Displacement

The two bad examples of cam design described above should lead the cam de-signer to the conclusion that consideration only of the displacement function when designing a cam is erroneous. The better approach is to start with consid-eration of the higher derivatives, especially acceleration. The acceleration func-tion, and to a lesser extent the jerk function, should be the principal concern of the designer. In some cases, especially when the mass of the follower train is large, or when there is a specification on velocity, that function must be carefully designed as well.

With this in mind, we will redesign the cam for the same example speci-fications as above. This time we will start with the acceleration function. The harmonic family of functions still have advantages which make them attractive for these applications. Figure 9-13 shows a full-period sinusoid applied as the acceleration function. It meets the constraint of zero magnitude at each end to match the dwell segments which adjoin it. The equation for a sine wave is:

$$a = C \sin\left(2\pi \frac{\theta}{\beta} \right) \tag{9.7}$$

We have again normalized the independent variable θ by dividing it by the period of the segment β with both θ and β measured in radians. The value of

θ/β ranges from 0 to 1 over any segment and is a dimensionless ratio. Since we want a full-cycle sine wave, we must multiply the argument by 2π. The argument of the sine function will then vary between 0 and 2π regardless of the value of β. The constant C defines the amplitude of the sine wave.

Integrate to obtain velocity,

$$a = \frac{dv}{d\theta} = C\sin\left(2\pi\frac{\theta}{\beta}\right)$$

$$\int dv = \int C\sin\left(2\pi\frac{\theta}{\beta}\right)d\theta \qquad (9.8)$$

$$v = -C\frac{\beta}{2\pi}\cos\left(2\pi\frac{\theta}{\beta}\right) + k_1$$

where k_1 is the constant of integration. To evaluate k_1, substitute the boundary condition $v = 0$ at $\theta = 0$, since we must match the zero velocity of the dwell at that point. The constant of integration is then:

$$k_1 = C\frac{\beta}{2\pi}$$

and: $\qquad\qquad\qquad\qquad\qquad\qquad\qquad\qquad\qquad\qquad\qquad\qquad$ (9.9)

$$v = C\frac{\beta}{2\pi}\left[1 - \cos\left(2\pi\frac{\theta}{\beta}\right)\right]$$

Note that substituting the boundary values at the other end of the interval, $v = 0$, $\theta = \beta$, will give the same result for k_1. Integrate again to obtain displacement:

$$v = \frac{ds}{d\theta} = C\frac{\beta}{2\pi}\left[1 - \cos\left(2\pi\frac{\theta}{\beta}\right)\right]$$

$$\int ds = \int\left\{C\frac{\beta}{2\pi}\left[1 - \cos\left(2\pi\frac{\theta}{\beta}\right)\right]\right\}d\theta \qquad (9.10)$$

$$s = C\frac{\beta}{2\pi}\theta - C\frac{\beta^2}{4\pi^2}\sin\left(2\pi\frac{\theta}{\beta}\right) + k_2$$

To evaluate k_2, substitute the boundary condition $s = 0$ at $\theta = 0$, since we must match the zero displacement of the dwell at that point. To evaluate the amplitude constant C, substitute the boundary condition $s = h$ at $\theta = \beta$, where h is the maximum follower rise (or lift) required over the interval and is a constant for any one cam specification.

$$k_2 = 0$$

$$C = 2\pi\frac{h}{\beta^2} \qquad (9.11)$$

Substituting the value of the constant C in equation 9.7 for acceleration gives:

$$a = 2\pi \frac{h}{\beta^2} \sin\left(2\pi \frac{\theta}{\beta}\right)$$ (9.12a)

Differentiating with respect to θ gives the expression for jerk.

$$j = 4\pi^2 \frac{h}{\beta^3} \cos\left(2\pi \frac{\theta}{\beta}\right)$$ (9.12b)

Substituting the values of the constants C and k_1, in equation 9.9 for velocity gives:

$$v = \frac{h}{\beta}\left[1 - \cos\left(2\pi \frac{\theta}{\beta}\right)\right]$$ (9.12c)

This velocity function is the sum of a negative cosine term and a constant term. The coefficient of the cosine term is equal to the constant term. This results in a velocity curve which starts and ends at zero and reaches a maximum magnitude at $\beta/2$ as can be seen in Figure 9-13. Substituting the values of the constants C, k_1 and k_2 in equation 9.10 for displacement gives:

$$s = h\left[\frac{\theta}{\beta} - \frac{1}{2\pi} \sin\left(2\pi \frac{\theta}{\beta}\right)\right]$$ (9.12d)

Note that this displacement expression is the sum of a straight line of slope h and a negative sine wave. The sine wave is, in effect, "wrapped around" the straight line as can be seen in Figure 9-13. Equation 9.12d is the expression for a cycloid. This cam function is referred to either as **cycloidal displacement** or **sinusoidal acceleration**.

In the form presented, with θ (in radians) as the independent variable, the units of equation 9.12d are length, of equation 9.12c length/radian, of equation 9.12a length/radian^2, and of equation 9.12b length/radian^3. To convert these equations to a time base, multiply velocity v by the camshaft angular velocity ω (in radians/sec), multiply acceleration a by ω^2, and jerk j by ω^3.

✐ EXAMPLE 9-3

Junior Cam Design - Cycloidal Displacement - An Acceptable Cam

Problem: Consider the same cam design CEP specification as in Examples 9-1 and 9-2:

dwell	at zero displacement for 90 degrees (low dwell)
rise	1 inch in 90 degrees
dwell	at 1 inch for 90 degrees (high dwell)
fall	1 inch in 90 degrees.
cam ω	2π radians/sec = 1 rev/sec

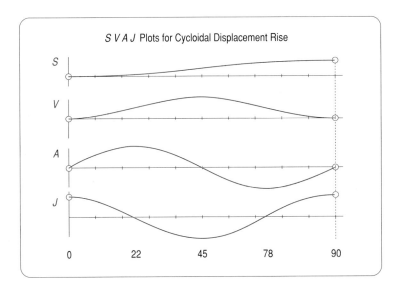

FIGURE 9-13

Sinusoidal acceleration gives cycloidal displacement

Solution:

1   The cycloidal displacement function is an acceptable one for this double-dwell cam specification. Its derivatives are continuous through the acceleration function as shown in Figure 9-13.

2   The jerk curve in Figure 9-13 is discontinuous at its boundaries but is of finite magnitude, and this is acceptable.

3   The velocity is smooth and matches the zeros of the dwell at each end.

4   The only drawback to this function is that it has relatively large magnitudes of peak acceleration and peak velocity compared to some other possible functions for the double-dwell case.

The reader is encouraged to input the diskfile EX9-3 to program DYNACAM to investigate this example in more detail.

Combined Functions

Dynamic force is proportional to acceleration. We generally would like to minimize dynamic forces, and thus should be looking to minimize the magnitude of the acceleration function as well as to keep it continuous. Kinetic energy is proportional to velocity squared. We also would like to minimize stored kinetic energy, especially with large mass follower trains, and so are concerned with the magnitude of the velocity function as well.

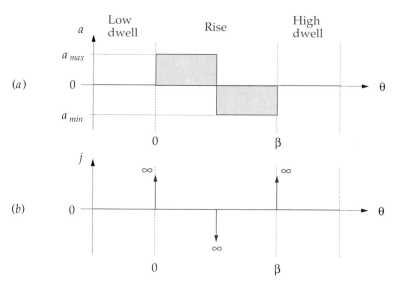

FIGURE 9-14

Constant acceleration gives infinite jerk

CONSTANT ACCELERATION If we wish to minimize the peak value of the magnitude of the acceleration function for a given problem, the function that would best satisfy this constraint is the square wave as shown in Figure 9-14. This function is also called **constant acceleration**. The square wave has the property of minimum peak value for a given area in a given interval. However, this function is not continuous. It has discontinuities at the beginning, middle, and end of the interval, so, by itself, **this is *unacceptable* as a cam acceleration function**.

TRAPEZOIDAL ACCELERATION The square wave's discontinuities can be removed by simply "knocking the corners off" the square wave function and creating the **trapezoidal acceleration** function shown in Figure 9-15a. The area lost from the "knocked off corners" must be replaced by increasing the peak magnitude above that of the original square wave in order to maintain the required specifications on lift and duration. But, this increase in peak magnitude is small, and the theoretical maximum acceleration can be significantly less than the theoretical peak value of the sinusoidal acceleration (cycloidal displacement) function. One disadvantage of this trapezoidal function is its very discontinuous jerk function, as shown in Figure 9-15b. Ragged jerk functions such as this tend to excite vibratory behavior in the follower train. The cycloidal's sinusoidal acceleration has a relatively smoother cosine jerk function with only two discontinuities in the interval and is preferable to the trapezoid's square waves of jerk. But the cycloidal's theoretical peak acceleration will be larger, which is not desirable. So, some tradeoffs must be made in selecting the cam functions.

MODIFIED TRAPEZOIDAL ACCELERATION An improvement can be made to the trapezoidal acceleration function by substituting pieces of sine waves for the

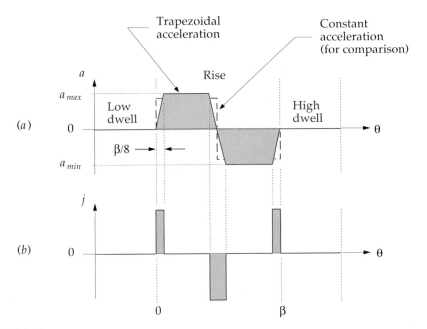

FIGURE 9-15

Trapezoidal acceleration gives finite jerk

sloped sides of the trapezoids as shown in Figure 9-16. This function is called the **modified trapezoidal acceleration** curve. This function is a marriage of the sine acceleration and constant acceleration curves. Conceptually, a full period sine wave is cut into fourths and "pasted into" the square wave to provide a smooth transition from the zeros at the endpoints, to the maximum and minimum peak values, and to make the transition from maximum to minimum in the center of the interval. The portions of the total segment period (β) used for the sinusoidal parts of the function can be varied. The most common arrangement is to cut the square wave at $\beta/8$, $3\beta/8$, $5\beta/8$, and $7\beta/8$ to insert the pieces of sine wave as shown in Figure 9-16. The $s\ v\ a\ j$ formulae for that arrangement of a modified trapezoidal rise are:

for $\quad 0 \le \theta < \dfrac{1}{8}\beta$

$$s = h\left[0.38898448\frac{\theta}{\beta} - 0.0309544\sin\left(4\pi\frac{\theta}{\beta}\right)\right]$$

$$v = 0.38898448\frac{h}{\beta}\left[1 - \cos\left(4\pi\frac{\theta}{\beta}\right)\right]$$

$$a = 4.888124\frac{h}{\beta^2}\sin\left(4\pi\frac{\theta}{\beta}\right)$$

$$j = 61.425769\frac{h}{\beta^3}\cos\left(4\pi\frac{\theta}{\beta}\right)$$

(9.13a)

(a) Take a
    sine wave

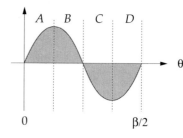

(b) Split the
    sine wave
    apart

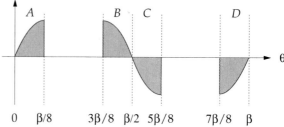

(c) Take a
    constant
    acceleration
    square wave

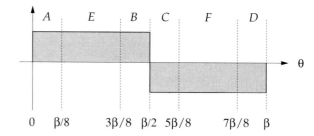

(d) Combine
    the two

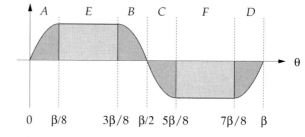

(e) Modified
    trapezoidal
    acceleration

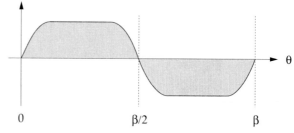

FIGURE 9-16

Creating the modified trapezoidal acceleration function

for $\quad \dfrac{1}{8}\beta \le \theta < \dfrac{3}{8}\beta$

$$s = h\left[2.44406184\left(\dfrac{\theta}{\beta}\right)^2 - 0.22203097\left(\dfrac{\theta}{\beta}\right) + 0.00723407 \right]$$

$$v = \dfrac{h}{\beta}\left[4.888124\left(\dfrac{\theta}{\beta}\right) - 0.22203097 \right]$$

$$a = 4.888124\dfrac{h}{\beta^2}$$

$$j = 0$$

(9.13b)

for $\quad \dfrac{3}{8}\beta \le \theta < \dfrac{5}{8}\beta$

$$s = h\left[1.6110154\dfrac{\theta}{\beta} - 0.0309544\sin\left(4\pi\dfrac{\theta}{\beta} - \pi \right) - 0.3055077 \right]$$

$$v = \dfrac{h}{\beta}\left[1.6110154 - 0.38898448\cos\left(4\pi\dfrac{\theta}{\beta} - \pi \right) \right]$$

$$a = 4.888124\dfrac{h}{\beta^2}\sin\left(4\pi\dfrac{\theta}{\beta} - \pi \right)$$

$$j = 61.425769\dfrac{h}{\beta^3}\cos\left(4\pi\dfrac{\theta}{\beta} - \pi \right)$$

(9.13c)

for $\quad \dfrac{5}{8}\beta \le \theta < \dfrac{7}{8}\beta$

$$s = h\left[-2.44406184\left(\dfrac{\theta}{\beta}\right)^2 + 4.6660917\left(\dfrac{\theta}{\beta}\right) - 1.2292648 \right]$$

$$v = \dfrac{h}{\beta}\left[-4.888124\left(\dfrac{\theta}{\beta}\right) + 4.6660917 \right]$$

$$a = -4.888124\dfrac{h}{\beta^2}$$

$$j = 0$$

(9.13d)

for $\dfrac{7}{8}\beta \le \theta \le \beta$

$$s = h\left[0.6110154 + 0.38898448\frac{\theta}{\beta} + 0.0309544\sin\left(4\pi\frac{\theta}{\beta} - 3\pi\right)\right]$$

$$v = 0.38898448\frac{h}{\beta}\left[1 + \cos\left(4\pi\frac{\theta}{\beta} - 3\pi\right)\right] \qquad (9.13e)$$

$$a = -4.888124\frac{h}{\beta^2}\sin\left(4\pi\frac{\theta}{\beta} - 3\pi\right)$$

$$j = -61.425769\frac{h}{\beta^3}\cos\left(4\pi\frac{\theta}{\beta} - 3\pi\right)$$

The modified trapezoidal function defined above is one of many combined functions created for cams by piecing together various functions, while being careful to match the values of the s, v, and a curves at all the interfaces between the joined functions. It has the advantage of relatively low theoretical peak acceleration, and reasonably rapid, smooth transitions at beginning and end of the interval. Note that in the form presented, with θ (in radians) as the independent variable, the units of the expressions in equations 9.13 are length, length/radian, length/radian^2, and length/radian^3 for s, v, a, j respectively. To convert equations 9.13 to a time base, multiply velocity v by the camshaft angular velocity ω (in radians/sec), multiply acceleration a by ω^2, and jerk j by ω^3.

The modified trapezoidal cam function is a popular and often used program for double-dwell cams. Its $S\ V\ A\ J$ curves are shown in Figure 9-17. The

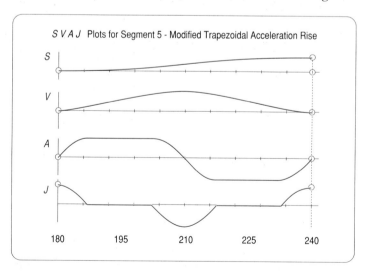

$S\ V\ A\ J$ Plots for Segment 5 - Modified Trapezoidal Acceleration Rise

FIGURE 9-17

Modified trapezoidal acceleration

relative theoretical peak magnitudes of acceleration for the cycloidal (sine acceleration) and modified trapezoidal cases calculated for a common total rise and duration (β) can be seen in Figure 9-18, which is the example from selection 6 on the input menu in program DYNACAM. The cycloidal curve (segment 1) has a theoretical peak acceleration which is approximately 1.3 times that of the modified trapezoid's peak value (segment 5) for the same cam specification. However, their peak velocity values are the same, so each will store the same peak kinetic energy in the follower train. The other two nondwell segments in Figure 9-18 are a simple harmonic fall (segment 7) and a modified sinusoidal acceleration fall (segment 3).

MODIFIED SINUSOIDAL ACCELERATION The sine acceleration curve (cycloidal displacement) has the advantage of smoothness (less ragged jerk curve) compared to the modified trapezoid but has higher theoretical peak acceleration. By combining two harmonic (sinusoid) curves of different frequencies, we can retain some of the smoothness characteristics of the cycloid and also reduce the peak acceleration. As an added bonus we will find that the peak velocity is also lower than in either the cycloidal or modified trapezoid. Figure 9-19 shows how the modified sine acceleration curve is made up of pieces of two sinusoid functions, one of higher frequency than the other. The first and last quarter of the high-frequency (short period, $\beta/2$) sine curve is used for the first and last eighths of the combined function. The center half of the low-frequency (long period, $3\beta/2$) sine wave is used to fill in the center three-fourths of the combined curve. Obviously, the magnitudes of the two curves and their derivatives must be matched at their interfaces in order to avoid discontinuities.

The equations for the **modified sine** curve for a rise of height h over a period β, with the functions joined at the $\beta/8$ and $7\beta/8$ points are as follows:

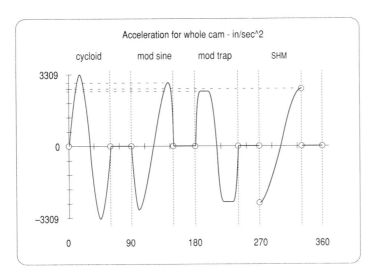

FIGURE 9-18

Relative peak accelerations for cycloidal, mod sine, mod trap, and simple harmonic motion

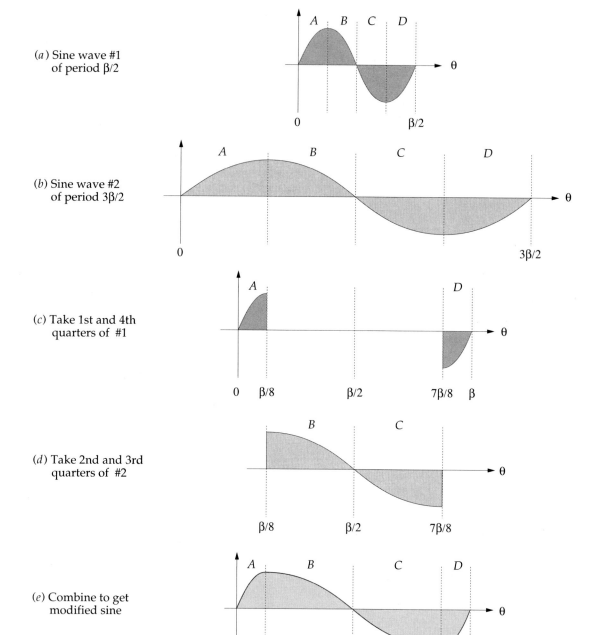

(*a*) Sine wave #1
of period β/2

(*b*) Sine wave #2
of period 3β/2

(*c*) Take 1st and 4th
quarters of #1

(*d*) Take 2nd and 3rd
quarters of #2

(*e*) Combine to get
modified sine

FIGURE 9-19

Creating the modified sine acceleration function

for $\quad 0 \le \theta < \frac{1}{8}\beta$

$$s = h\left[0.43990085\frac{\theta}{\beta} - 0.0350062\sin\left(4\pi\frac{\theta}{\beta}\right)\right]$$

$$v = 0.43990085\frac{h}{\beta}\left[1 - \cos\left(4\pi\frac{\theta}{\beta}\right)\right]$$

$$a = 5.5279571\frac{h}{\beta^2}\sin\left(4\pi\frac{\theta}{\beta}\right)$$

$$j = 69.4663577\frac{h}{\beta^3}\cos\left(4\pi\frac{\theta}{\beta}\right)$$

(9.14a)

for $\quad \frac{1}{8}\beta \le \theta < \frac{7}{8}\beta$

$$s = h\left[0.28004957 + 0.43990085\frac{\theta}{\beta} - 0.31505577\cos\left(\frac{4\pi}{3}\frac{\theta}{\beta} - \frac{\pi}{6}\right)\right]$$

$$v = 0.43990085\frac{h}{\beta}\left[1 + 3\sin\left(\frac{4\pi}{3}\frac{\theta}{\beta} - \frac{\pi}{6}\right)\right]$$

$$a = 5.5279571\frac{h}{\beta^2}\cos\left(\frac{4\pi}{3}\frac{\theta}{\beta} - \frac{\pi}{6}\right)$$

$$j = -23.1553\frac{h}{\beta^3}\sin\left(\frac{4\pi}{3}\frac{\theta}{\beta} - \frac{\pi}{6}\right)$$

(9.14b)

for $\quad \frac{7}{8}\beta \le \theta \le \beta$

$$s = h\left\{0.56009915 + 0.43990085\frac{\theta}{\beta} + 0.0350062\sin\left[2\pi\left(2\frac{\theta}{\beta} - 1\right)\right]\right\}$$

$$v = 0.43990085\frac{h}{\beta}\left\{1 - \cos\left[2\pi\left(2\frac{\theta}{\beta} - 1\right)\right]\right\}$$

$$a = 5.5279571\frac{h}{\beta^2}\sin\left[2\pi\left(2\frac{\theta}{\beta} - 1\right)\right]$$

$$j = 69.4663577\frac{h}{\beta^3}\cos\left[2\pi\left(2\frac{\theta}{\beta} - 1\right)\right]$$

(9.14c)

Figure 9-20 shows a comparison of the shapes and relative magnitudes of five cam acceleration programs including the cycloidal, modified trapezoid,

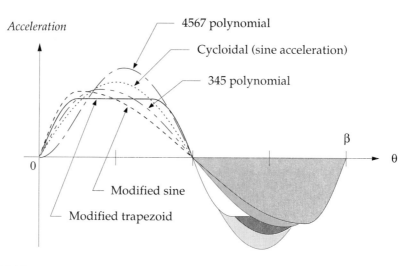

FIGURE 9-20

Comparison of five double-dwell cam acceleration functions

and modified sine acceleration curves. The peak value of acceleration for the modified sine is between the cycloidal and modified trapezoid. Table 9-2 lists the peak values of acceleration, velocity, and jerk for these functions in terms of the total rise h and period β. Figure 9-21 compares the jerk curves for the same functions. The modified sine jerk is somewhat less ragged than the modified trapezoid jerk but not as smooth as that of the cycloid, which is a full-period cosine.

TABLE 9-2   Maximum Velocity Acceleration and Jerk Factors

Program	Max. Veloc.	Max. Accel.	Max. Jerk	Comments
Constant accel.	$2.000h/\beta$	$4.000h/\beta^2$	infinite	∞ jerk - not acceptable
Harmonic disp.	$1.571h/\beta$	$4.945h/\beta^2$	infinite	∞ jerk - not acceptable
Trapezoid accel.	$2.000h/\beta$	$5.300h/\beta^2$	$44h/\beta^3$	Not as good as mod. trap.
Mod. trap. accel.	$2.000h/\beta$	$4.888h/\beta^2$	$61h/\beta^3$	Low accel but rough jerk
Mod. sine. accel.	$1.760h/\beta$	$5.528h/\beta^2$	$69h/\beta^3$	Low veloc - good accel.
3-4-5 Poly. disp.	$1.875h/\beta$	$5.777h/\beta^2$	$60h/\beta^3$	Good compromise.
Cycloidal disp.	$2.000h/\beta$	$6.283h/\beta^2$	$40h/\beta^3$	Smooth accel. & jerk
4-5-6-7 Poly. disp.	$2.188h/\beta$	$7.526h/\beta^2$	$52h/\beta^3$	Smooth jerk - high accel

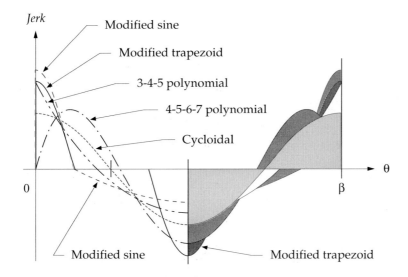

FIGURE 9-21

Comparison of five double-dwell cam jerk functions

Figure 9-22 compares their velocity curves. Note that the peak velocity of the modified sine is the lowest of the five functions shown. This is the principal advantage of the modified sine acceleration curve and the reason it is usually chosen for applications in which the follower mass is very large.

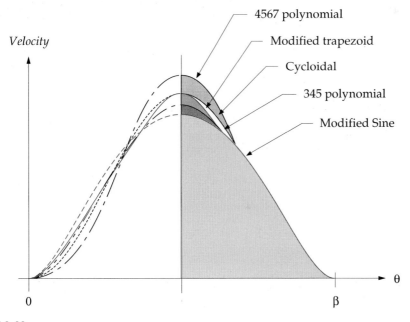

FIGURE 9-22

Comparison of five double-dwell cam velocity functions

An example of such an application is shown in Figure 9-23 which is an indexing table drive used for automated assembly lines. The round indexing table (not shown) is mounted on the tapered vertical spindle and driven as part of the follower train by a form-closed barrel cam which moves it through some angular displacement, and then holds the table still in a dwell (called a "stop") while an assembly operation is performed on the workpiece carried on the table. These indexers may have from three to many stops, each corresponding to an index position. The table (not shown) is solid steel and may be several feet in diameter, thus its mass is large. To minimize the stored kinetic energy, which must be dissipated each time the table is brought to a stop, the manufacturers often use the modified sine program on these multidwell cams, because of its lower peak velocity.

Let us again try to improve the double-dwell cam example with these combined functions of modified trapezoid and modified sine acceleration.

EXAMPLE 9-4

Senior Cam Design - Combined Functions - Better Cams

Problem: Consider the same cam design CEP specification as in Examples 9-1 to 9-3:

dwell	at zero displacement for 90 degrees (low dwell)
rise	1 inch in 90 degrees
dwell	at 1 inch for 90 degrees (high dwell)
fall	1 inch in 90 degrees.
cam ω	2π radians/sec = 1 rev/sec

Solution:

1   The modified trapezoidal function is an acceptable one for this double-dwell cam specification. Its derivatives are continuous through the acceleration function as shown in Figures 9-17, 9-20, and 9-22.

2   The modified trapezoidal jerk curve in Figures 9-17 and 9-21 is discontinuous at its boundaries but is of finite magnitude, and this is acceptable.

3   The modified trapezoidal velocity in Figures 9-17 and 9-22 is smooth and matches the zeros of the dwell at each end.

4   The advantage of this modified trapezoidal function is that it has relatively lower magnitude of theoretical peak acceleration compared to the cycloidal but its peak velocity is identical to the cycloidal.

5   The modified sinusoid function is also an acceptable one for this double-dwell cam specification. Its derivatives are also continuous through the acceleration function as shown in Figures 9-20 and 9-22.

6   The modified sine jerk curve in Figure 9-21 is discontinuous at its boundaries but is of finite magnitude and is smoother than that of the modified trapezoid.

7   The modified sine velocity (Figure 9-22) is smooth, matches the zeros of the dwell at each end, and is lower in peak magnitude than either the cycloid of modified trapezoidal.

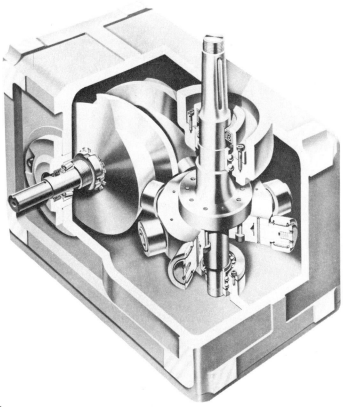

FIGURE 9-23

Multi-stop rotary indexer (table removed) - *Courtesy of The Ferguson Co., St. Louis Mo.*

This is an advantage for high-mass follower systems as it reduces the kinetic energy. This, coupled with a peak acceleration lower than the cycloidal (but higher than the modified trapezoidal), is its chief advantage.

The reader is encouraged to run Example 4 from the input menu of program DYNACAM to investigate this example in more detail.

Figure 9-24 shows the displacement curves for these three cam programs. Note how little difference there is between the displacement curves despite the large differences in their acceleration waveforms in Figure 9-20. This is evidence of the smoothing effect of the integration process. Differentiating any two functions will exaggerate their differences. Integration tends to mask their differences. It is nearly impossible to recognize these very differently behaving cam functions by looking only at their displacement curves. This is further evidence of the folly of our earlier naive approach to cam design which dealt exclusively with the displacement function. The cam designer must be concerned with the higher derivatives of displacement. The displacement function is primarily of value to the manufacturer of the cam who needs its coordinate information in order to cut the cam.

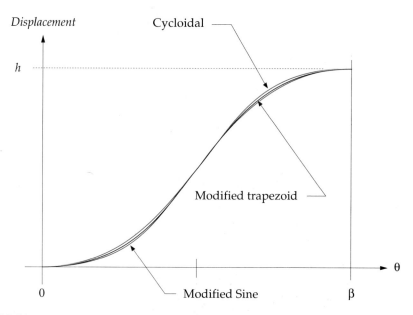

FIGURE 9-24

Comparison of three double-dwell cam displacement functions

FALL FUNCTIONS We have used only the rise portion of the cam for these examples. The fall is handled similarly. The rise functions presented here are applicable to the fall with slight modification. To convert rise equations to fall equations, it is only necessary to subtract the rise displacement function s from the maximum lift h, and to negate the higher derivatives, $v, a,$ and j.

SUMMARY This section has attempted to present an approach to the selection of appropriate double-dwell cam functions, using the common rise-dwell-fall-dwell cam as the example, and to point out some of the pitfalls awaiting the cam designer. The particular functions described are only a few of the ones that have been developed for this double-dwell case over many years, by many designers, but they are probably the most used and most popular among cam designers. Most of them are also included in program DYNACAM. There are many tradeoffs to be considered in selecting a cam program for any application, some of which have already been mentioned, such as function continuity, peak values of velocity and acceleration, and smoothness of jerk. There are many other tradeoffs still to be discussed in later sections of this chapter, involving the sizing and the manufacturability of the cam.

9.4   SINGLE-DWELL CAM DESIGN - CHOOSING $S V A J$ FUNCTIONS

Many applications in machinery require a **single-dwell** cam program, **rise-fall-dwell** (RFD). Perhaps a single-dwell cam is needed to lift and lower a roller which carries a moving paper web on a production machine that makes envelopes. This cam's follower lifts the paper up to one critical extreme position at the right time to contact a roller which applies a layer of glue to the envelope flap. Without dwelling in the up position, it immediately retracts the web back to the starting (zero) position and holds it in this other critical extreme position (low dwell) while the rest of the envelope passes by. It repeats the cycle for the next envelope as it comes by. Another common example of a single-dwell application is the cam which opens the valves in your automobile engine. This lifts the valve open on the rise, immediately closes it on the fall, and then keeps the valve closed in a dwell while the combustion takes place.

If we attempt to use the same type of cam programs as were defined for the double-dwell case for a single-dwell application, we will achieve a solution which may work but is not optimal. We will nevertheless do so here as an example in order to point out the problems that result. Then we will redesign the cam to eliminate those problems.

EXAMPLE 9-5

Using Cycloidal Motion for Single-Dwell

Problem: Consider the following single-dwell cam specification:

rise	1 inch in 90 degrees
fall	1 inch in 90 degrees
dwell	at zero displacement for 180 degrees (low dwell)
cam ω	15 radians/sec

Solution:

1   Figure 9-25 shows a cycloidal displacement rise and separate cycloidal displacement fall applied to this single-dwell example. Note that the displacement (s) diagram looks acceptable in that it moves the follower from the low to the high position and back in the required intervals.

2   The velocity (v) also looks acceptable in shape in that it takes the follower from zero velocity at the low dwell to zero again at the maximum displacement, where the glue is applied.

3   Figure 9-26 shows the acceleration function for this solution. Its maximum value is about 573 in/sec^2.

4   The problem is that this acceleration curve has an **unnecessary return to zero** at the end of the rise. It is unnecessary because the acceleration during the first part of the fall is also negative. It would be better to keep it in the negative region at the end of the rise.

5   This unnecessary oscillation to zero in the acceleration causes the jerk to have more abrupt changes and discontinuities. The only real justification for taking the acceleration to zero is the need to change its sign (as is the case halfway through the rise or fall) or to match an adjacent segment which has zero acceleration.

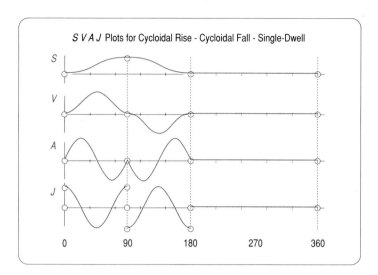

FIGURE 9-25

Cycloidal motion (or any double-dwell program) is a poor choice for the single-dwell case

The reader is encouraged to input the diskfile EX9-5 to program DYNACAM to investigate this example in more detail.

For the single-dwell case we would like a function for the rise which does not return its acceleration to zero at the end of the interval. The function for the

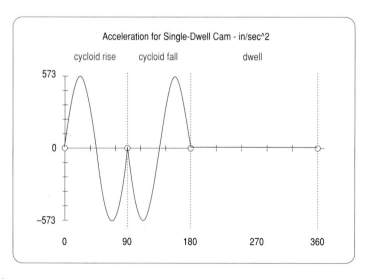

FIGURE 9-26

Sinusoidal acceleration of cycloid used with single-dwell has undesirable return to zero

fall should begin with the same nonzero acceleration value as ended the rise and then be zero at its terminus to match the dwell. One function which meets those criteria is the **double harmonic** which gets its name from its two cosine terms, one of which is a half period harmonic and the other a full period wave. The equations for the double harmonic functions are:

for the rise:

$$s = \frac{h}{2}\left\{\left[1 - \cos\left(\pi\frac{\theta}{\beta}\right)\right] - \frac{1}{4}\left[1 - \cos\left(2\pi\frac{\theta}{\beta}\right)\right]\right\}$$

$$v = \frac{\pi}{\beta}\frac{h}{2}\left[\sin\left(\pi\frac{\theta}{\beta}\right) - \frac{1}{2}\sin\left(2\pi\frac{\theta}{\beta}\right)\right]$$

$$a = \frac{\pi^2}{\beta^2}\frac{h}{2}\left[\cos\left(\pi\frac{\theta}{\beta}\right) - \cos\left(2\pi\frac{\theta}{\beta}\right)\right] \qquad (9.15a)$$

$$j = -\frac{\pi^3}{\beta^3}\frac{h}{2}\left[\sin\left(\pi\frac{\theta}{\beta}\right) - 2\sin\left(2\pi\frac{\theta}{\beta}\right)\right]$$

for the fall:

$$s = \frac{h}{2}\left\{\left[1 + \cos\left(\pi\frac{\theta}{\beta}\right)\right] - \frac{1}{4}\left[1 - \cos\left(2\pi\frac{\theta}{\beta}\right)\right]\right\}$$

$$v = -\frac{\pi}{\beta}\frac{h}{2}\left[\sin\left(\pi\frac{\theta}{\beta}\right) + \frac{1}{2}\sin\left(2\pi\frac{\theta}{\beta}\right)\right]$$

$$a = -\frac{\pi^2}{\beta^2}\frac{h}{2}\left[\cos\left(\pi\frac{\theta}{\beta}\right) + \cos\left(2\pi\frac{\theta}{\beta}\right)\right] \qquad (9.15b)$$

$$j = \frac{\pi^3}{\beta^3}\frac{h}{2}\left[\sin\left(\pi\frac{\theta}{\beta}\right) + 2\sin\left(2\pi\frac{\theta}{\beta}\right)\right]$$

Note that these double harmonic functions should **never** be used for the double-dwell case because their acceleration is nonzero at one end of the interval.

EXAMPLE 9-6

Double Harmonic Motion for Single-Dwell

Problem: Consider the same single-dwell cam specification as in example 9-5:

rise	1 inch in 90 degrees
fall	1 inch in 90 degrees
dwell	at zero displacement for 180 degrees (low dwell)
cam ω	15 radians/sec

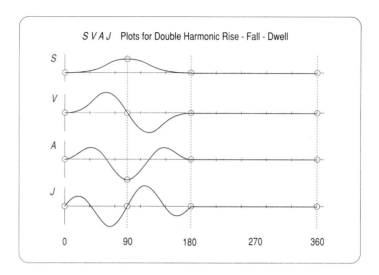

FIGURE 9-27

Double harmonic motion can be used for the single-dwell case

Solution:

1.   Figure 9-27 shows a double harmonic rise and a double harmonic fall. This is example 3 in program DYNACAM.

2.   Note that the acceleration of this double harmonic function does not return to zero at the end of the rise as also shown in Figure 9-28. This makes it more suitable for a single-dwell case in that respect.

3.   The jerk function (Figure 9-27) is quite smooth compared to the cycloidal solution.

4.   Unfortunately, the peak negative acceleration is 900 in/sec^2, nearly twice that of the cycloidal solution. This is a smoother function but will develop higher dynamic forces.

      The reader is encouraged to run example cam #3 from the input menu of program DYNACAM to investigate this example in more detail. Neither of the solutions in Examples 9-5 and 9-6 is optimal. We will revisit this example problem after introducing polynomial functions and redesign it yet again to improve both its smoothness and its acceleration magnitude.

      SUMMARY This section has attempted to present an approach to the selection of appropriate single-dwell cam functions, and to point out some of their limitations. The particular functions described (cycloidal and double harmonic) are only a few of the ones that have been developed for this single-dwell case. We will visit this single-dwell problem again in the next section and develop a superior solution using other techniques.

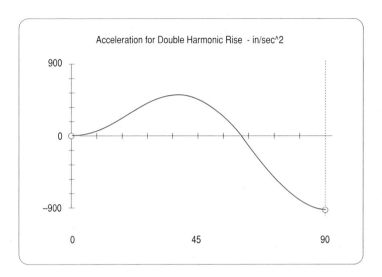

Acceleration for Double Harmonic Rise - in/sec^2

FIGURE 9-28

The acceleration of the double harmonic function is asymmetric

9.5 POLYNOMIAL FUNCTIONS

The class of polynomial functions is perhaps the most versatile that can be used for cam design. They can be tailored to most any design specification. They are not limited to single- or double-dwell applications. They are adaptable to any situation. The general form of a polynomial function is:

$$s = C_0 + C_1 x + C_2 x^2 + C_3 x^3 + C_4 x^4 + C_5 x^5 + C_6 x^6 \cdots + C_n x^n \qquad (9.16)$$

where s is follower displacement, x is the independent variable, and in our case will be replaced by either θ/β or time t. The constant coefficients C_n are the unknowns to be solved for in our development of the particular polynomial equation to suit our design specification. The degree of a polynomial is defined as the highest power present in any term. Note that a polynomial of degree n will have $n + 1$ terms because there is an x^0 or constant term with coefficient C_0, as well as coefficients through and including C_n.

We structure a polynomial cam design problem by deciding how many boundary conditions (BC's) we want to specify on the $s\ v\ a\ j$ diagrams. The number of BC's then determines the degree of the resulting polynomial. We can write an independent equation for each BC by substituting it into equation 9.16 or one of its derivatives. We will then have a system of linear equations which can be solved for the unknown coefficients $C_0 \ldots C_n$. If k represents the number of chosen boundary conditions, there will be k equations in k unknowns $C_0 \ldots C_n$ and the **degree** of the polynomial will be $n = k - 1$. The **order** of the n-degree polynomial is equal to the number of terms, k.

Double-Dwell Applications of Polynomials

THE 3-4-5 POLYNOMIAL Let us return to the double-dwell problem of Section 9.3 and solve it with polynomial functions. Many different polynomial solutions are possible. We will start with the simplest one possible for the double-dwell case.

✐ EXAMPLE 9-7

The 3-4-5 Polynomial for the Double-Dwell Case

Problem: Consider the same cam design CEP specification as in Examples 9-1 to 9-4:

dwell	at zero displacement for 90 degrees (low dwell)
rise	1 inch in 90 degrees
dwell	at 1 inch for 90 degrees (high dwell)
fall	1 inch in 90 degrees.
cam ω	2π radians/sec = 1 rev/sec

Solution:

1   To satisfy the fundamental law of cam design the values of the rise (and fall) functions at their boundaries with the dwells must match with no discontinuities in, at a minimum, s, v, and a.

2   Figure 9-29 shows the axes for the $s\,v\,a\,j$ diagrams on which the known data have been drawn. The dwells are the only fully defined segments at this stage. The requirement for continuity through the acceleration defines a minimum of **six boundary conditions** for the rise segment and six more for the fall in this problem. They are shown as filled circles on the plots. For generality, we will let the specified total rise be represented by the variable h. The minimum set of required BC's for this example is then:

for the rise:

$$\text{when} \quad \theta = 0; \qquad \text{then} \quad s = 0, \qquad v = 0, \qquad a = 0$$
$$\text{(9.17a)}$$
$$\text{when} \quad \theta = \beta_1; \qquad \text{then} \quad s = h, \qquad v = 0, \qquad a = 0$$

for the fall:

$$\text{when} \quad \theta = 0; \qquad \text{then} \quad s = h, \qquad v = 0, \qquad a = 0$$
$$\text{(9.17b)}$$
$$\text{when} \quad \theta = \beta_2; \qquad \text{then} \quad s = 0, \qquad v = 0, \qquad a = 0$$

3   We will use the rise for an example solution. (The fall is a similar derivation.) We have six BC's on the rise. This requires six terms in the equation. The highest term will be fifth degree. We will use the normalized angle θ/β as our independent variable, as before. Because our boundary conditions involve velocity and acceleration as well as displacement, we need to differentiate equation 9.16 versus θ to obtain expressions into which we can substitute those BC's. Rewriting equation 9.16 to fit these constraints, and differentiating twice, we get:

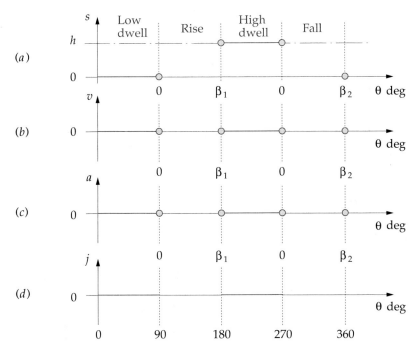

FIGURE 9-29

Minimum boundary conditions for the double dwell case

$$s = C_0 + C_1\left(\frac{\theta}{\beta}\right) + C_2\left(\frac{\theta}{\beta}\right)^2 + C_3\left(\frac{\theta}{\beta}\right)^3 + C_4\left(\frac{\theta}{\beta}\right)^4 + C_5\left(\frac{\theta}{\beta}\right)^5 \qquad (9.18a)$$

$$v = \frac{1}{\beta}\left[C_1 + 2C_2\left(\frac{\theta}{\beta}\right) + 3C_3\left(\frac{\theta}{\beta}\right)^2 + 4C_4\left(\frac{\theta}{\beta}\right)^3 + 5C_5\left(\frac{\theta}{\beta}\right)^4\right] \qquad (9.18b)$$

$$a = \frac{1}{\beta^2}\left[2C_2 + 6C_3\left(\frac{\theta}{\beta}\right) + 12C_4\left(\frac{\theta}{\beta}\right)^2 + 20C_5\left(\frac{\theta}{\beta}\right)^3\right] \qquad (9.18c)$$

4   Substitute the boundary conditions $\theta = 0$, $s = 0$ into equation 9.18a:

$$0 = C_0 + 0 + 0 + \cdots$$
$$C_0 = 0 \qquad (9.19a)$$

5   Substitute $\theta = 0$, $v = 0$ into equation 9.18b:

$$0 = \frac{1}{\beta}[C_1 + 0 + 0 + \cdots]$$
$$C_1 = 0 \qquad (9.19b)$$

6 Substitute $\theta = 0, a = 0$ into equation 9.18c:

$$0 = \frac{1}{\beta^2}\left[C_2 + 0 + 0 + \cdots\right]$$
$$C_2 = 0 \qquad\qquad (9.19\text{c})$$

7 Substitute $\theta = \beta, s = h$ into equation 9.18a:

$$h = C_3 + C_4 + C_5 \qquad\qquad (9.19\text{d})$$

8 Substitute $\theta = \beta, v = 0$ into equation 9.18b:

$$0 = \frac{1}{\beta}\left[3C_3 + 4C_4 + 5C_5\right] \qquad\qquad (9.19\text{e})$$

9 Substitute $\theta = \beta, a = 0$ into equation 9.18c:

$$0 = \frac{1}{\beta^2}\left[6C_3 + 12C_4 + 20C_5\right] \qquad\qquad (9.19\text{f})$$

10 Three of our unknowns are found to be zero, leaving three unknowns to be solved for, C_3, C_4, C_5. Equations 9.19d, 9.19e, and 9.19f can be solved simultaneously to get:

$$C_3 = 10h; \qquad\qquad C_4 = -15h; \qquad\qquad C_5 = 6h \qquad\qquad (9.19\text{g})$$

11 The equation for this cam design's displacement is then:

$$s = h\left[10\left(\frac{\theta}{\beta}\right)^3 - 15\left(\frac{\theta}{\beta}\right)^4 + 6\left(\frac{\theta}{\beta}\right)^5\right] \qquad\qquad (9.20)$$

12 The expressions for velocity and acceleration can be obtained by substituting the values of C_3, C_4, and C_5 into equations 9.18b and 9.18c. This function is referred to as the **3-4-5 polynomial**, after its exponents.

Figure 9-30 shows the resulting $s\ v\ a\ j$ diagrams for a **3-4-5 polynomial rise** function from example cam 5 in program DYNACAM. Note that the acceleration is continuous but the jerk is not, because we did not place any constraints on the boundary values of the jerk function. It is also interesting to note that the acceleration waveform looks very similar to the sinusoidal acceleration of the cycloidal function in Figure 9-13. Segment 5 in Figure 9-31 shows the relative peak accelerations of this 3-4-5 polynomial compared to three other functions with the same h and β. The cam in Figure 9-31 also has a modified sine rise (segment 1), cycloidal fall (segment 3), and 4-5-6-7 polynomial fall (segment 7) on it which allows direct comparisons of the relative peak values of the accelerations of these functions. Table 9-2 also provides a numerical comparison, in general terms, of the maximum velocity, acceleration and jerk for these functions. The reader is encouraged to input the diskfile EX9-7 to program DYNACAM to investigate this example in more detail.

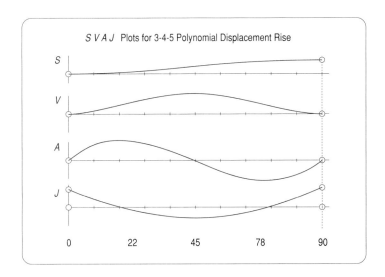

FIGURE 9-30

3-4-5 polynomial rise - Its acceleration is very similar to the sinusoid of cycloidal motion

THE 4-5-6-7 POLYNOMIAL We left the jerk unconstrained in the previous example. We will now redesign the cam for the same specifications but will also constrain the jerk function to be zero at both ends of the rise. It will then match the dwells in the jerk function with no discontinuities. This gives eight boundary conditions and yields a seventh-degree polynomial. The solution procedure to find the eight unknown coefficients is identical to that used in the previous example. Write the polynomial with the appropriate number of terms. Differentiate it to get expressions for all orders of boundary conditions. Substitute the boundary conditions and solve the resulting set of simultaneous equations. This problem reduces to four equations in four unknowns, as the coefficients C_0, C_1, C_2, and C_3 turn out to be zero. It is not necessary to solve these sets of equations by hand, as any matrix solving calculator, program MATRIX, or program DYNACAM will do the solution for you. The details of program DYNACAM's use are covered in a subsequent section, but it is only necessary to supply the desired boundary conditions to the program and the coefficients will be computed. The student is encouraged to do so and examine all the example problems presented here with the DYNACAM program. For this example the resulting displacement equation for the rise is:

$$s = h\left[35\left(\frac{\theta}{\beta}\right)^4 - 84\left(\frac{\theta}{\beta}\right)^5 + 70\left(\frac{\theta}{\beta}\right)^6 - 204\left(\frac{\theta}{\beta}\right)^7\right] \tag{9.21}$$

This is known as the **4-5-6-7 polynomial**, after its exponents. Figure 9-32 shows the $s\ v\ a\ j$ diagrams for this function. Compare them to the 3-4-5 polynomial in Figures 9-30 and 9-31. Note that the acceleration of the 4-5-6-7 starts off slowly, with zero slope (as we demanded with our zero jerk BC), and as a result goes to a larger peak value of acceleration in order to replace the missing area in the leading edge.

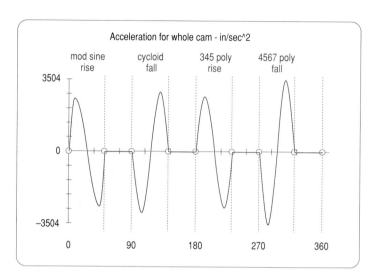

FIGURE 9-31

Modified sine, cycloidal, 345 polynomial and 4567 polynomial acceleration functions

This **4-5-6-7 polynomial** function has the advantage of smoother jerk for better vibration control, compared to the **3-4-5 polynomial**, the **cycloidal** and all other functions so far discussed (except the double harmonic), but pays a stiff price in the form of significantly higher acceleration than all those functions. Figure 9-31 and Table 9-2 show the relative values of peak acceleration for five of the double-dwell functions discussed.

Single-Dwell Applications of Polynomials

Let us return to the single-dwell example used in the previous section and attempt to solve it with a polynomial cam function. Restating the original problem for convenience:

rise	1 inch in 90 degrees
fall	1 inch in 90 degrees
dwell	at zero displacement for 180 degrees (low dwell)
cam ω	15 radians/sec

To solve this with a polynomial we must decide on a suitable set of boundary conditions. We must also first decide how many segments to divide the cam cycle into. The problem statement seems to imply three segments, a rise, a fall, and a dwell. We could use those three segments to create the functions, but a better approach is to use only **two segments**, one for the rise-fall combined, and one for the dwell. *As a general rule we would like to minimize the number of segments in our polynomial cam functions.* Any dwell requires its own segment. So, the minimum number possible in this case is two segments.

Another rule of thumb is that *we would like to minimize the number of boundary conditions specified,* because the degree of the polynomial is tied to the number of BC's. As the degree of the function increases, so will the number of its

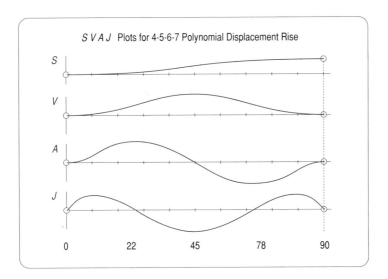

FIGURE 9-32

4-5-6-7 polynomial rise - Its jerk is piecewise continuous with the dwells

inflection points and its number of **minima and maxima**. The polynomial derivation process will guarantee that the function will pass through all specified BC's but says nothing about the function's behavior between the BC's. **A high-degree function may have undesirable oscillations between BC's.**

With these assumptions we can select a set of boundary conditions for a trial solution. First we will restate the problem to reflect our two-segment configuration:

EXAMPLE 9-8

Designing a Polynomial for the Single-Dwell Case

Problem: Redefine the CEP specification from Examples 9-5 and 9-6

rise	1 inch in 90° and **fall** 1 inch in 90° for a total of 180°
dwell	at zero displacement for 180° (low dwell)
cam ω	15 radians/sec

Solution:

1   Figure 9-33 shows the minimum set of seven BC's for this problem. The dwell on either side of the combined rise-fall segment has zero values of $s, v, a,$ and j. The fundamental law of cam design requires that we match these zero values, through the acceleration function, at each end of the rise-fall segment.

2   These then account for six BC's; $s, v, a = 0$ at each end of the rise-fall segment.

3   We also must specify a value of displacement at the 1 inch peak of the rise which occurs at $\theta = 90$ degrees. This is the seventh BC.

Segment number	Function used	Start angle	End angle	Delta angle
1	Poly 6	0	180	180

Boundary Conditions Imposed

Function	Theta	% Beta	Boundary Cond.
Displ	0	0	0
Veloc	0	0	0
Accel	0	0	0
Displ	180	1	0
Veloc	180	1	0
Accel	180	1	0
Displ	90	.5	1

FIGURE 9-33

Boundary conditions for a single-dwell polynomial application

4   Figure 9-33 shows the BC's as displayed in program DYNACAM. This set of seven BC's will lead to a sixth-degree polynomial.

5   Figure 9-34 shows the coefficients of the displacement polynomial which result from the simultaneous solution of the polynomial equations for the BC's in Figure 9-33 as calculated by the program. For generality we have substituted the variable h for the specified 1 inch rise. The function turns out to be a 3-4-5-6 polynomial whose equation is:

$$s = h\left[64\left(\frac{\theta}{\beta}\right)^3 - 192\left(\frac{\theta}{\beta}\right)^4 + 192\left(\frac{\theta}{\beta}\right)^5 - 64\left(\frac{\theta}{\beta}\right)^6 \right]$$

(9.22)

Figure 9-35 shows the $s\,v\,a\,j$ diagrams for this solution, and Figure 9-36 shows the acceleration curve of the rise-fall segment with its maximum values noted. Compare these acceleration and $s\,v\,a\,j$ curves to the double harmonic and cycloidal solutions to the same problem in Section 9.4 (Figures 9-25 to 9-28). Note that this sixth-degree polynomial function is as smooth as the double harmonic functions (Example 9-6) and does not unnecessarily return the acceleration to zero at the top of the rise as does the cycloidal (Example 9-5). The polynomial has a peak acceleration of 547 in/sec², which is less than that of either the cycloidal or double harmonic solution. This 3-4-5-6 polynomial is a superior solution to either of those presented for the same problem in Section 9.4 and is an example of how polynomial functions can be easily tailored to particular design specifications. The reader is encouraged to input the diskfile EX9-8 to program DYNACAM to investigate this example in more detail.

Segment number	Function used	Start angle	End angle	Delta angle
1	Poly 6	0	180	180

Equation Resulting

Exponent	Coefficient
0	0
1	0
2	0
3	64
4	-192
5	192
6	-64

FIGURE 9-34

Coefficients of the sixth-degree polynomial for a single-dwell application

SUMMARY This section has presented polynomial functions as the most versatile approach of those shown to virtually any cam design problems. It is only since the development and general availability of computers that these functions have become practical to use, as the computation to solve the simultaneous equations is often beyond hand calculation abilities. With the availability of a design aid to solve the equations such as program DYNACAM, polynomials have become a practical and preferable way to solve many cam design problems.

9.6   CRITICAL PATH MOTION (CPM)

Probably the most common application of **critical path motion** (CPM) specifications in production machinery design is the need for **constant velocity motion**. There are two general types of automated production machinery in common use, **intermittent motion** assembly machines and **continuous motion** assembly machines.

Intermittent motion assembly machines carry the manufactured goods from work station to work station, stopping the workpiece or subassembly at each station while another operation is performed upon it. The throughput speed of this type of automated production machine is typically limited by the dynamic forces which are due to accelerations and decelerations of the mass of the moving parts of the machine and its workpieces.

Continuous motion assembly machines never allow the workpiece to stop and thus are capable of higher throughput speeds. All operations are performed on a moving target. Any tools which operate on the product have to "chase" the moving assembly line to do their job. Since the assembly line (often

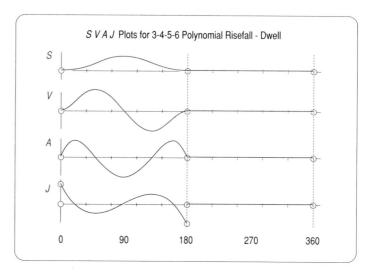

FIGURE 9-35

Sixth-degree 3-4-5-6 polynomial function for two-segment risefall - single-dwell cam

a conveyor belt or chain) is moving at some constant velocity, there is a need for mechanisms to provide constant velocity motion, matched exactly to the conveyor, in order to carry the tools alongside for a long enough time to do their job. These cam driven "chaser" mechanisms must then return the tool quickly to its start position in time to meet the next part or subassembly on the conveyor (quick-return). There is a motivation in manufacturing to convert from intermittent motion machines to continuous motion in order to increase production rates.

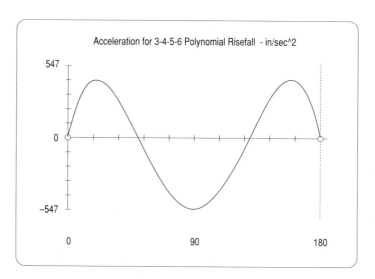

FIGURE 9-36

Single-dwell 3-4-5-6 polynomial acceleration

Thus there is considerable demand for this type of constant velocity mechanism. The cam-follower system is well suited to this problem and the polynomial cam function is particularly adaptable to the task.

Polynomials Used for Critical Path Motion

⬛EXAMPLE 9-9

Designing a Polynomial for Constant Velocity Critical Path Motion

Problem:　Consider the following statement of a Critical Path Motion (CPM) problem:

Accelerate　the follower from zero to 10 in/sec
Maintain　a constant velocity of 10 in/sec for 0.5 sec
Decelerate　the follower to zero velocity
Return　the follower to start position
Cycle time　exactly 1 second

Solution:

1　This unstructured problem statement is typical of real design problems as was discussed in Chapter 1. No information is given as to the means to be used to accelerate or decelerate the follower or even as to the portions of the available time to be used for those tasks. A little reflection will cause the engineer to recognize that the specification on total cycle time in effect defines the camshaft velocity to be its reciprocal or **one revolution per second**. Converted to appropriate units, this is an angular velocity of 2π radians/sec.

2　The constant velocity portion uses half of the total period of one second in this example. The designer must next decide how much of the remaining 0.5 second to devote to each other phase of the required motion.

3　The problem statement seems to imply that four segments are needed. Note that the designer has to somewhat arbitrarily select the lengths of the individual segments

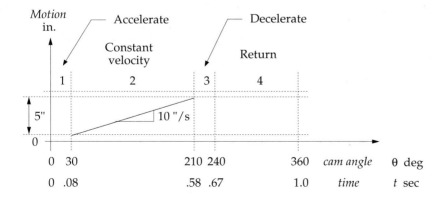

FIGURE 9-37

Constant velocity cam timing diagram

```
Camshaft Constant Omega    =    6.28   Radians/Sec
Calculated Maximum Lift     =    0      Inches
Number of Cam Segments      =    4


Segment    Function    Start    End      Delta
number     used        angle    angle    angle

   1       Poly3          0       30       30
   2       Poly1         30      210      180
   3       Poly3        210      240       30
   4       Poly5        240      360      120
```

FIGURE 9-38

Assumed segment divisions for constant velocity problem

(except the constant velocity one). Some iteration may be required to optimize the result. Program DYNACAM makes the iteration process quick and easy, however.

4   Assuming four segments, the timing diagram in Figure 9-37 shows an acceleration phase, a constant velocity phase, a deceleration phase, and a return phase, labelled as segments 1 through 4.

5   The segment angles (β's) are assumed, for a first approximation, to be 30° for segment 1, 180° for segment 2, 30° for segment 3 and 120° for segment 4 as shown in Figure 9-38. These angles may need to be adjusted in later iterations, except for segment 2 which is rigidly constrained in the specifications.

6   Figure 9-39 shows a tentative $s\,v\,a\,j$ diagram. The solid circles indicate a set of boundary conditions which will constrain the continuous function to these specifications. These are, for segment 1:

$$\text{when}\quad \theta = 0°; \qquad s = 0, \qquad v = 0, \qquad none$$

$$(a)$$

$$\text{when}\quad \theta = 30°; \qquad none, \qquad v = 10, \qquad a = 0$$

7   Note that the displacement at $\theta = 30°$ is left unspecified. The resulting polynomial function will provide us with the values of displacement at that point, which can then be used as a boundary condition for the next segment, in order to make the overall functions continuous as required. The acceleration at $\theta = 30°$ must be zero in order to match that of the constant velocity segment 2. The acceleration at $\theta = 0$ is left unspecified. The resulting value will be used later to match the end of the last segment's acceleration.

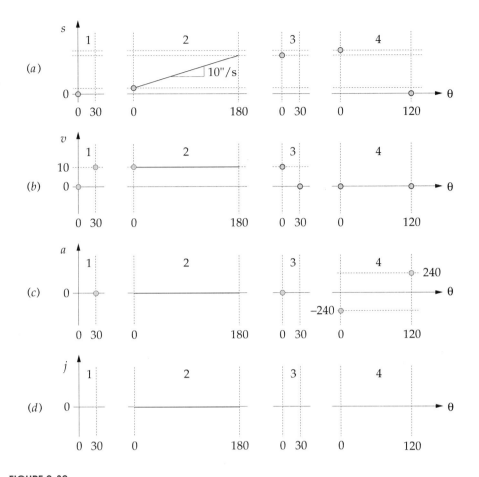

FIGURE 9-39

A possible set of boundary conditions for the four-segment constant velocity solution

8 Putting these four BC's for segment 1 into program DYNACAM yields a cubic function whose *s v a j* plots are shown in Figure 9-40. Its equation is:

$$s = 0.83376\left(\frac{\theta}{\beta}\right)^2 - 0.27792\left(\frac{\theta}{\beta}\right)^3 \tag{9.23a}$$

Figure 9-41 shows the maximum displacement which occurs at $\theta = 30°$. This will be used as one BC for segment 2.   The entire set for segment 2 is:

$$\text{when } \theta = 30°; \qquad s = 0.556, \qquad v = 10 \tag{b}$$

$$\text{when } \theta = 210°; \qquad none, \qquad none$$

9 Note that, in the derivations and in the DYNACAM program each segment's local angles run from zero to the β for that segment. Thus, segment 2's local angles are 0° to 180°,

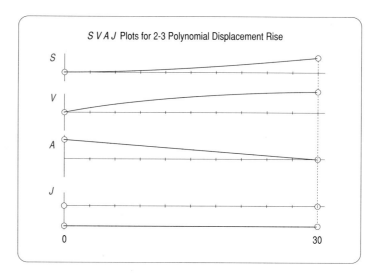

FIGURE 9-40

Segment one for the four-segment solution to the constant velocity problem

which correspond to 30° to 210° globally in this example. We have left the displacement, velocity and acceleration at the end of segment 2 unspecified. They will be determined by the computation.

Segment number	Function used	Start angle	End angle	Delta angle
1	Poly 3	0	30	30

Shaft omega = 6.28 rad/sec   Total lift = 0.55584

Angle degrees	Displacement inches
0	0.000
10	0.082
20	0.288
30	0.556

Maximum value =   0.556

FIGURE 9-41

Displacement at end of segment one

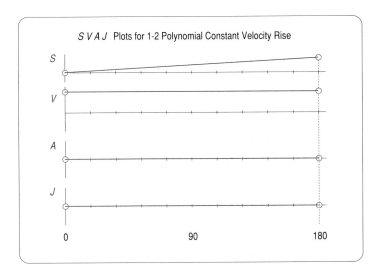

S V A J Plots for 1-2 Polynomial Constant Velocity Rise

FIGURE 9-42

Segment two for the four-segment solution to the constant velocity problem

10 Since this is a constant velocity segment, its integral, the displacement function, must be a polynomial of degree one, i.e., a straight line. If we specify more than two BC's we will get a function of higher degree than one which **will** pass through the specified endpoints but may also oscillate between them and deviate from the desired constant velocity. Thus we can *only* provide two BC's, a slope and an intercept, as defined in equation 9.2. We must also provide at least one displacement boundary condition in order to compute the coefficient C_0 from equation 9.16. Specifying the two BC's at only one end of the interval is perfectly acceptable. The equation for segment 2 is:

$$s = 5\left(\frac{\theta}{\beta}\right) + 0.556 \tag{9.23b}$$

11 Figure 9-42 shows the displacement and velocity plots of segment 2. The acceleration and jerk are both zero. The resulting displacement at $\theta = 210°$ is 5.556.

12 The displacement at the end of segment 2 is now known from its equation. The four boundary conditions for segment 3 are then:

$$\text{when} \quad \theta = 210°; \quad s = 5.556, \quad v = 10, \quad a = 0$$

$$\tag{c}$$

$$\text{when} \quad \theta = 240°; \quad \textit{none}, \quad v = 0, \quad \textit{none}$$

13 This generates a cubic displacement function which can be seen in Figure 9-43. Its equation is:

$$s = -0.27792\left(\frac{\theta}{\beta}\right)^3 + 0.83376\left(\frac{\theta}{\beta}\right) + 5.556 \tag{9.23c}$$

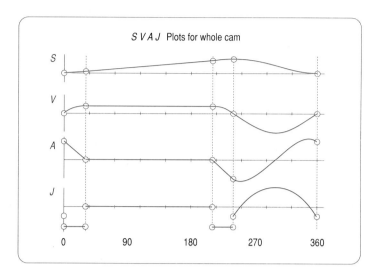

FIGURE 9-43

Four-segment solution to the constant velocity problem

14   The boundary conditions for the last segment 4, are now defined, as they must match those of the end of segment 3 and the beginning of segment 1. The displacement at the end of segment 3 is found from the computation in DYNACAM to be $s = 6.112$ at $\theta = 240°$ and the acceleration at that point is –239.9. We left the acceleration at the beginning of segment 1 unspecified. From the second derivative of the equation for displacement in that segment we find that the acceleration is 239.9 at $\theta = 0°$. The BC's for segment 4 are then:

$$\text{when}\quad \theta = 240°; \qquad s = 6.112, \qquad v = 0, \qquad a = -239.9$$

$$(d)$$

$$\text{when}\quad \theta = 360°; \qquad s = 0, \qquad v = 0, \qquad a = 239.9$$

15   The equation for segment 4 is then:

$$s = -9.9894\left(\frac{\theta}{\beta}\right)^5 + 24.9735\left(\frac{\theta}{\beta}\right)^4 - 7.7548\left(\frac{\theta}{\beta}\right)^3 - 13.3413\left(\frac{\theta}{\beta}\right)^2 + 6.112 \qquad (9.23d)$$

16   Figure 9-43 shows the $s\,v\,a\,j$ plots for the complete cam. It obeys the fundamental law of cam design because the piecewise functions are continuous through the acceleration. The maximum value of acceleration is 257 in/sec². The maximum negative velocity is –29.4 in/sec. We now have four piecewise and continuous functions, equations 9.23, which will meet the performance specifications for this problem.

     The reader is encouraged to read the diskfile EX9-9 into program DYNACAM to investigate this example in more detail.

While this design is acceptable, it can be improved. One useful strategy in designing polynomial cams is to minimize the number of segments, provided that this does not result in functions of such high degree that they misbehave between boundary conditions. Another strategy is to always start with the segment for which you have the most information. In this example, the constant velocity portion is the most constrained, and must be a separate segment, just as a dwell must be a separate segment. The rest of the cam motion exists only to return the follower to the constant velocity segment for the next cycle. If we start by designing the constant velocity segment, it may be possible to complete the cam with only one additional segment. We will now redesign this cam, to the same specifications but with only two segments as shown in Figure 9-44.

✐ EXAMPLE 9-10

Designing an Optimum Polynomial for Constant Velocity Critical Path Motion

Problem: Redefine the problem statement of Example 9-9 to have only two segments.

Maintain   a constant velocity of 10 in/sec for 0.5 sec
Decelerate and **accelerate** follower to constant velocity
Cycle time exactly 1 second

Solution:

1   The BC's for the first, constant velocity, segment will be similar to our previous solution except for the global values of its angles and the fact that we will start at zero displacement. They are:

$$\text{when} \quad \theta = 0°; \qquad s = 0, \qquad v = 10$$

$$(a)$$

$$\text{when} \quad \theta = 180°; \qquad none, \qquad none$$

2   The displacement and velocity plots for this segment are identical to those in Figure 9-42 except that the displacement starts at zero. The equation for this segment 1 is:

$$s = 5\left(\frac{\theta}{\beta}\right) \qquad (9.24a)$$

3   The program calculates the displacement at the end of segment 1 to be 5.00 inches. This defines that BC for segment 2. The set of BC's for segment 2 is then:

$$\text{when} \quad \theta = 180°; \qquad s = 5.00, \qquad v = 10, \qquad a = 0$$

$$(b)$$

$$\text{when} \quad \theta = 360°; \qquad s = 0, \qquad v = 10, \qquad a = 0$$

The equation for segment 2 is:

$$s = -60\left(\frac{\theta}{\beta}\right)^5 + 150\left(\frac{\theta}{\beta}\right)^4 - 100\left(\frac{\theta}{\beta}\right)^3 + 5\left(\frac{\theta}{\beta}\right)^1 + 5 \qquad (9.24b)$$

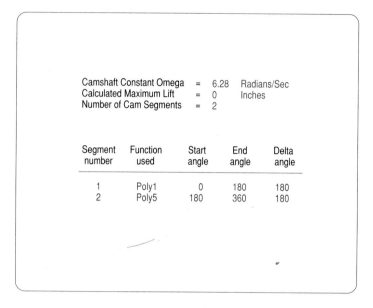

Camshaft Constant Omega	=	6.28	Radians/Sec
Calculated Maximum Lift	=	0	Inches
Number of Cam Segments	=	2	

Segment number	Function used	Start angle	End angle	Delta angle
1	Poly1	0	180	180
2	Poly5	180	360	180

FIGURE 9-44

Two-segment solution for constant velocity example problem

4   The *s v a j* diagrams for this design are shown in Figure 9-45. Note that they are much smoother than the four-segment design. The maximum acceleration in this example is now 230 in/sec², and the maximum negative velocity is −27.5 in/sec. These are both less than in the previous design.

5   The fact that our displacement in this design contains negative values as shown in the Figure 9-45 *s* diagram is of no concern. This is due to our starting with the beginning of the constant velocity portion as zero displacement. The follower has to go to a negative position in order to have distance to accelerate up to speed again. We will simply shift the displacement coordinates by that negative amount to make the cam. To do this simply calculate the displacement coordinates for the cam. Note the value of the largest negative displacement. Add this value to the displacement boundary conditions for all segments and recalculate the cam functions with DYNACAM. (Do not change the BC's for the higher derivatives.) The finished cam's displacement profile will be shifted up such that its minimum value will now be zero.

So, not only do we now have a smoother cam but the dynamic forces and stored kinetic energy are both lower. Note that we did not have to make any assumptions about the portions of the available nonconstant velocity time to be devoted to speeding up or slowing down. This all happened automatically from our choice of only two segments and the specification of the minimum set of necessary boundary conditions. This is clearly a superior design to the previous attempt and is in fact an optimal polynomial solution to the given specifications. The reader is encouraged to read the diskfile EX9-10 into program DYNACAM to investigate this example in more detail.

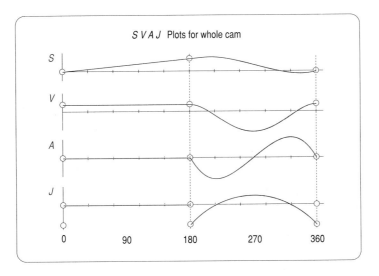

FIGURE 9-45

Two-segment solution to the constant velocity problem

Half-Period Harmonic Family Functions

The full-rise simple harmonic, the full-rise cycloidal, and the full-rise modified sine functions are generally suited only to the double-dwell cases as they have zero acceleration at each end. However, pieces of these functions can be used to match other functions such as constant velocity segments in a similar fashion to that used to piece together the modified sine from two harmonics of different frequency. The full-rise functions mentioned above (except simple harmonic) contain one complete period in their velocity and acceleration. The simple harmonic (Figure 9-10) does not have zero acceleration at its ends, but the modified sine's and cycloid's accelerations (Figure 9-13) do start and end at zero. In order to match a nonzero velocity as in the example above, we could use half of any of these harmonic family functions and design them to match the desired constant velocity of the adjacent segment.

As an example of this approach, Figure 9-46a shows the $s\ v\ a\ j$ functions for a **half-cycloid** rise function #1 which has zero velocity at the beginning of the interval and nonzero velocity at the end. Note that its displacement starts at zero and ends at some positive value, but its acceleration is zero at both extremes. This makes it possible to mate this function to a constant velocity segment and match both the desired velocity and its zero acceleration at the boundary. The total displacement required of the half-cycloid will "come out in the wash" when the required boundary conditions of velocity and duration are applied to the particular case. The equations for this half-cycloid #1 are:

$$s = L\left[\frac{\theta}{\beta} - \frac{1}{\pi}\sin\left(\pi\frac{\theta}{\beta}\right)\right]$$

(9.25a)

$$v = \frac{L}{\beta}\left[1 - \cos\left(\pi\frac{\theta}{\beta}\right)\right] \qquad (9.25b)$$

$$a = \frac{\pi L}{\beta^2}\sin\left(\pi\frac{\theta}{\beta}\right) \qquad (9.25c)$$

$$j = \frac{\pi^2 L}{\beta^3}\cos\left(\pi\frac{\theta}{\beta}\right) \qquad (9.25d)$$

Figure 9-46b shows the *s v a j* functions for the **half-cycloid** rise function #2 with nonzero velocity at the beginning of the interval and zero velocity at the end. The equations for this half-cycloid #2 are:

$$s = L\left[\frac{\theta}{\beta} + \frac{1}{\pi}\sin\left(\pi\frac{\theta}{\beta}\right)\right] \qquad (9.26a)$$

$$v = \frac{L}{\beta}\left[1 + \cos\left(\pi\frac{\theta}{\beta}\right)\right] \qquad (9.26b)$$

$$a = -\frac{\pi L}{\beta^2}\sin\left(\pi\frac{\theta}{\beta}\right) \qquad (9.26c)$$

$$j = -\frac{\pi^2 L}{\beta^3}\cos\left(\pi\frac{\theta}{\beta}\right) \qquad (9.26d)$$

For a fall instead of a rise, subtract the rise displacement expressions from the total rise *L* and negate all the higher derivatives.

To fit these functions to a particular constant velocity situation, solve either equation 9.25b or 9.26b (depending on which function is desired) for the value of *L* which results from the specification of the known constant velocity *v* to be matched at $\theta = \beta$ or $\theta = 0$. You will have to choose a value of β for the interval of this half-cycloid which is appropriate to the problem. In our example above, the value of $\beta = 30°$ used for the first segment of the four-piece polynomial could be tried as a first iteration. Once *L* and β are known, all the functions are defined.

The same approach can be taken with the **modified sine** and the **simple harmonic** functions. Either half of their full-rise functions can be sized to match with a constant velocity segment. The half-modified sine function mated with a constant velocity segment has the advantage of low peak velocity, useful with large inertia loads. When matched to a constant velocity, the half-simple harmonic has the same disadvantage of infinite jerk as its full-rise counterpart does when matched to a dwell, so it is not recommended.

We will now solve the previous constant velocity example problem using half-cycloid, constant velocity and full-fall modified sine functions.

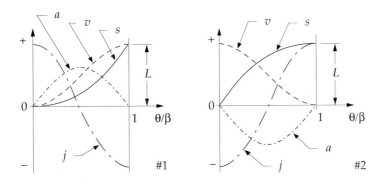

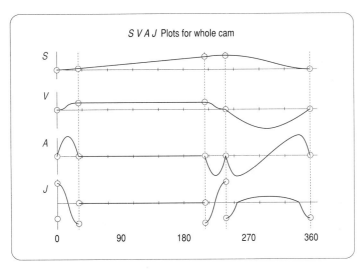

(*a*) Half-cycloidal rise motions

(*b*) Constant velocity with half-cycloidal rise transitions

FIGURE 9-46

Half-cycloidal functions

EXAMPLE 9-11

Using Half-Cycloids to Match Constant Velocity Critical Path Motion.

Problem: Consider the same problem statement as Example 9-9.

Accelerate   the follower from zero to 10 in/sec
Maintain   a constant velocity of 10 in/sec for 0.5 sec
Decelerate   the follower to zero velocity
Return   the follower to start position
Cycle time   exactly 1 second

Solution:

1   We must express the specified constant velocity in units of length/radian. The angular velocity is 2π radians/sec.

$$v = 10\frac{in}{sec}\left(\frac{1}{2\pi}\frac{sec}{rad}\right) = \frac{5}{\pi}\frac{in}{rad} \qquad\qquad (a)$$

2   The constant velocity portion uses half of the total period of one second, or π radians, in this example. The designer must decide how much of the remaining 0.5 second to devote to each other phase of the required motion. The segment angles (β's) are assumed, for a first approximation, to be 25° for segment 1, 180° for segment 2, 25° for segment 3, and 130° for segment 4. These angles may need to be adjusted in later iterations to balance and minimize the accelerations (except for segment 2 which is rigidly constrained in the specifications).

3   The segments will consist of:

Segment	β, radians	Function	Motion
1	0.43633	Half-cycloid #1	Rise
2	3.14159	1° polynomial	Rise
3	0.43633	Half-cycloid #2	Rise
4	2.26893	Modified sine	Fall

4   To determine the total rise L of the half-cycloid needed to match the specified constant velocity, solve equation 9.25b for L at $\theta = \beta$ where it must match the specified constant velocity v.

$$L = \frac{\beta v}{1 - \cos\left(\pi\dfrac{\theta}{\beta}\right)} = \frac{0.43633\left(\dfrac{5}{\pi}\right)}{1 - \cos\left(\pi\dfrac{\beta}{\beta}\right)} = 0.34722 \qquad\qquad (b)$$

5   Substitute this value of L in equation 9.25a to get the displacement for the first segment:

$$s = 0.3472\left[\frac{\theta}{\beta} - \frac{1}{\pi}\sin\left(\pi\frac{\theta}{\beta}\right)\right] \qquad\qquad (c)$$

6   The constant velocity segment is found in the same way as was done in Example 9-9. The initial displacement for segment 2 in this case is the value of L and the equation for segment 2 is:

$$s = 5\left(\frac{\theta}{\beta}\right) + 0.3472 \qquad\qquad (d)$$

The total lift within this segment is 5 inches as before.

7   Segment 3 is a half-cycloid #2. Its coefficient L is identical to that of segment 1 because
    we used the same β for both. But, we must offset it by the sum of the displacement of
    segments 2 and 3 or $L + 5$. Program DYNACAM provides for the specification of this
    offset. The lift for this segment is 0.3472 and the offset is 5.3472. The equation for
    segment 3 (from Eq. 9.26a) is then:

$$s = 0.3472\left[\frac{\theta}{\beta} + \frac{1}{\pi}\sin\left(\pi\frac{\theta}{\beta}\right)\right] + 5.3472 \qquad (e)$$

8   Segment 4 is a full-period modified sine to return the follower from its maximum dis-
    placement of $h = 0.3472 + 5.0 + 0.3472 = 5.6944$. See equation 9.14.

9   The complete set of data needed to compute these functions in (or out of) DYNACAM is:

Segment	β, degrees	Lift	Offset
1	25	0.3472	0
2	180	5.0[*]	0.3472[*]
3	25	0.3472	5.3472
4	130	5.6944	0

[*] Note that these are boundary conditions on the polynomial. Lift and offset are not
     defined for polynomials in DYNACAM as the boundary conditions determine them.

10 The resulting $s\,v\,a\,j$ diagrams are shown in Figure 9-46b. The peak acceleration is 241 in/sec^2
    and peak velocity is –28 in/sec.

These results are nearly as low as the values from the two-segment polynomial solution in Example 9-10. The factor that makes this an inferior cam design to
Example 9-10 is the unnecessary returns to zero in the acceleration waveform.
This creates a more "ragged" jerk function which will increase vibration problems. The polynomial solution is superior in this case as it often is in cam design.
The reader is encouraged to read the diskfile EX9-11 into program DYNACAM to
investigate this example in more detail.

9.7   SIZING THE CAM - PRESSURE ANGLE AND RADIUS OF CURVATURE

Once the $s\,v\,a\,j$ functions have been defined, the next step is to size the cam. There
are two major factors which affect cam size, the **pressure angle** and the **radius of
curvature**. Both of these involve either the **base circle radius** on the cam (R_b)
when using flat-faced followers, or the **prime circle radius** on the cam (R_p) when
using roller or curved followers.

The base circle's and prime circle's centers are at the center of rotation of
the cam. The base circle is defined as *the smallest circle which can be drawn tangent*

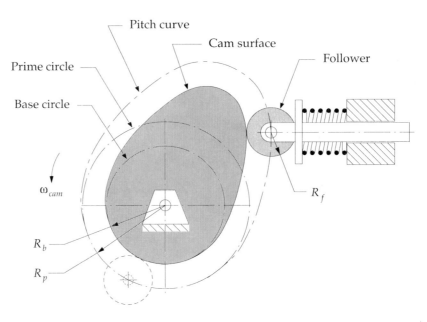

FIGURE 9-47

Base circle, prime circle and pitch curve of a radial cam with roller follower

to the physical cam surface as shown in Figure 9-47. All radial cams will have a base circle, regardless of the follower type used.

The prime circle is only applicable to cams with roller followers or radiused (mushroom) followers and is measured to the center of the follower. The **prime circle** is defined as *the smallest circle which can be drawn tangent to the locus of the centerline of the follower* as shown in Figure 9-47. *The locus of the centerline of the follower* is called the **pitch curve**. Cams with roller followers are in fact defined for manufacture with respect to the pitch curve rather than with respect to the cam's physical surface. Cams with flat-faced followers must be defined for manufacture with respect to their physical surface, as there is no pitch curve.

The process of creating the physical cam from the s diagram can be visualized conceptually by imagining the s diagram to be cut out of a flexible material such as rubber. The x axis of the s diagram represents the circumference of a circle, which could be either the **base circle**, or the **prime circle**, around which we will "wrap" our "rubber" s diagram. We are free to choose the initial length of our s diagram's x axis, though the height of the displacement curve is fixed by the cam displacement function we have chosen. In effect we will choose the base or prime circle radius as a design parameter and stretch the length of the s diagram's axis to fit the circumference of the chosen circle.

Pressure Angle - Roller Followers

The **pressure angle** is defined as shown in Figure 9-48. It is the complement of the transmission angle which was defined for linkages in previous chapters and has a

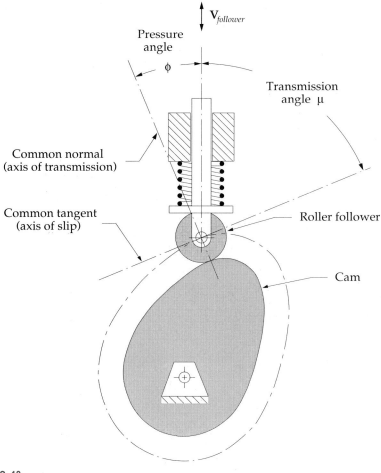

FIGURE 9-48

Cam pressure angle

similar meaning with respect to cam-follower operation. By convention, the pressure angle is used for cams, rather than the transmission angle. Force can only be transmitted from cam to follower or vice versa along the **axis of transmission** which is perpendicular to the **axis of slip**, or common tangent.

PRESSURE ANGLE The **pressure angle** ϕ is *the angle between the direction of motion (velocity) of the follower and the direction of the axis of transmission*. When $\phi = 0$, all the transmitted force goes into motion of the follower and none into slip velocity. When ϕ becomes 90° there will be no motion of the follower. As a rule of thumb, we would like the pressure angle to be between zero and about 30 degrees for translating followers to avoid excessive side load on the sliding follower. If the follower is oscillating on a pivoted arm, a pressure angle up to about 35° is acceptable. Values of ϕ greater than this will increase the follower sliding or pivot friction to undesirable levels and may tend to jam a translating follower in its guides.

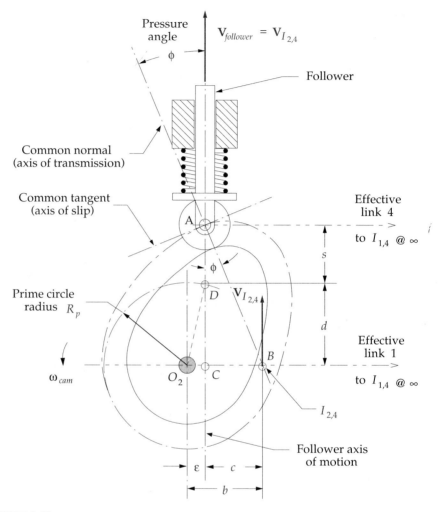

FIGURE 9-49

Geometry for the derivation of the equation for pressure angle

ECCENTRICITY Figure 9-49 shows the geometry of a cam and translating roller follower in an arbitrary position. This shows the general case in that the axis of motion of the follower does not intersect the center of the cam. There is an **eccentricity** ε defined as *the perpendicular distance between the follower's axis of motion and the center of the cam*. Often this eccentricity ε will be zero, making it an **aligned follower**, which is the special case.

In the figure, the axis of transmission is extended to intersect effective link 1, which is the ground link. (See Section 9.0 and Figure 9-1 for a discussion of effective links in cam systems.) This intersection is instant center $I_{2,4}$ (labelled B) which, by definition, has the same velocity in link 2 (the cam) and in link 4 (the follower). Because link 4 is in pure translation, all points on it have identical velocities $V_{follower}$, which are equal to the velocity of $I_{2,4}$ in link 2. We can write an expression for the velocity of $I_{2,4}$ in terms of cam angular velocity and the radius b from cam center to $I_{2,4}$,

$$V_{I_{2,4}} = b\omega = \dot{s} \qquad (9.27)$$

where s is the instantaneous displacement of the follower from the s diagram and s dot is its time derivative in units of length/sec. (Note that capital S V A J denote time-based variables rather than functions of cam angle.)

But:
$$\dot{s} = \frac{ds}{dt}$$

and:
$$\frac{ds}{dt}\frac{d\theta}{d\theta} = \frac{ds}{d\theta}\frac{d\theta}{dt} = \frac{ds}{d\theta}\omega = v\omega$$

so:
$$b\omega = v\omega$$

then:
$$b = v \qquad (9.28)$$

This is an interesting relationship which says that the **distance b to the instant center $I_{2,4}$ is numerically equal to the velocity of the follower** v in units of length/radian as derived in previous sections. We have reduced this expression to pure geometry, independent of the angular velocity ω of the cam.

Note that we can express the distance b in terms of the prime circle radius R_p and the eccentricity ε, by the construction shown in the figure. Swing the arc of radius R_p until it intersects the axis of motion of the follower at point D. This defines the length of line d from effective link 1 to this intersection. This is constant for any chosen prime circle radius R_p. Points A, C and $I_{2,4}$ form a right triangle whose upper angle is the pressure angle ϕ and whose vertical leg is $(s + d)$, where s is the instantaneous displacement of the follower. From this triangle:

$$c = (s+d)\tan\phi \qquad (9.29a)$$

and

$$b = (s+d)\tan\phi + \varepsilon$$

Then from equation 9.28,
$$v = (s+d)\tan\phi + \varepsilon \qquad (9.29b)$$

and from triangle CDO_2,
$$d = \sqrt{R_P^2 - \varepsilon^2} \qquad (9.29c)$$

Substituting equation 9.29c into equation 9.29b and solving for ϕ gives an expression for pressure angle in terms of displacement s, velocity v, eccentricity ε, and the prime circle radius R_p.

$$\phi = \arctan \frac{v - \varepsilon}{s + \sqrt{R_P^2 - \varepsilon^2}} \qquad (9.29d)$$

The velocity v in this expression is in units of length/radian and all other quantities are in compatible length units. We have typically defined s and v by this stage of the cam design process and wish to manipulate R_p and ε to get an acceptable maximum pressure angle ϕ. As R_p is increased ϕ will be reduced. The only constraints against large values of R_p are the practical ones of package size and cost. Often there will be some upper limit on the size of the cam-follower package dictated by its surroundings. There will always be a cost constraint and bigger = heavier = more expensive.

Choosing a Prime Circle Radius

Both R_p and ε are within a transcendental expression in equation 9.29d, so they cannot be conveniently solved for directly. The simplest approach is to assume a trial value for R_p and an initial eccentricity of zero, and use program DYNACAM, your own program, or an equation solver such as *TKSolver*™ to quickly calculate the values of ϕ for the entire cam, and then adjust R_p and repeat the calculation until an acceptable arrangement is found. Figure 9-50 shows the calculated pressure angles for example #5 in DYNACAM. Note the similarity in shape to the velocity functions for the same cam in Figure 9-6, as that term is dominant in equation 9.29d.

USING ECCENTRICITY If a suitably small cam cannot be obtained with acceptable pressure angle, then eccentricity can be introduced to change the pressure angle. Using eccentricity to control pressure angle has its limitations. A **positive value of eccentricity** will *decrease the pressure angle on the rise* but will *increase it on the fall*. **Negative eccentricity does the reverse.** This is of little value with a form-

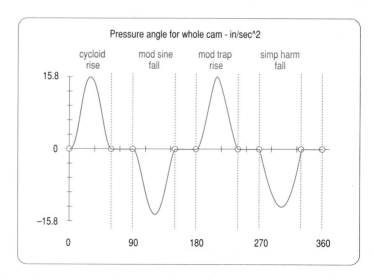

FIGURE 9-50

Pressure angle functions are similar in shape to the velocity functions

closed (groove or track) cam, as it is driving the follower in both directions. For a force-closed cam with spring return, you can sometimes afford to have a larger pressure angle on the fall than on the rise because the stored energy in the spring is attempting to speed up the camshaft on the fall, whereas the cam is storing that energy in the spring on the rise. The limit of this technique can be the degree of overspeed attained with a larger pressure angle on the fall. The variations in cam angular velocity that result may be unacceptable.

The most value gained from adding eccentricity to a follower comes in situations where the cam program is asymmetrical and significant differences exist (with no eccentricity) between maximum pressure angles on rise and fall. Introducing eccentricity can balance the pressure angles in this situation and create a smoother running cam.

If adjustments to R_p or ε do not yield acceptable pressure angles, the only recourse is to return to an earlier stage in the design process and redefine the problem. Less lift or more time to rise or fall will reduce the causes of the large pressure angle. Design is, after all, an iterative process.

Overturning Moment - Flat-Faced Follower

Figure 9-51 shows a translating, flat-faced follower running against a radial cam. The pressure angle can be seen to be zero for all positions of cam and follower. This seems to be giving us something for nothing, which can't be true. As the contact point moves left and right, the point of application of the force between cam and follower moves with it. There is an overturning moment on the follower associated with this off-center force which tends to jam the follower in its guides, just as did too large a pressure angle in the roller follower case. In this case, we would like to keep the cam as small as possible in order to minimize the moment arm of the force. Eccentricity will affect the average value of the moment, but the peak-to-peak variation of the moment about that average is unaffected by eccentricity. Considerations of too-large pressure angle do not limit the size of this cam, but other factors do. The minimum radius of curvature (see below) of the cam surface must be kept large enough to avoid undercutting. This is true regardless of the type of follower used.

Radius of Curvature - Roller Follower

The **radius of curvature** is a *mathematical property of a function*. Its value and use is not limited to cams but has great significance in their design. The concept is simple. No matter how complicated a curve's shape may be, nor how high the degree of the function which describes it, it will have an instantaneous radius of curvature at every point on the curve. These radii of curvature will have instantaneous centers (which may be at infinity), and the radius of curvature of any function is itself a function which can be computed and plotted. For example, the radius of curvature of a straight line is infinity everywhere; that of a circle is a constant value. A parabola has a constantly changing radius of curvature which approaches infinity along the parabola's asymptotes. A cubic curve will have radii of curvature that are sometimes positive (convex) and sometimes negative (concave). The higher the degree of a function, in general, the more potential variety in its radius of curvature.

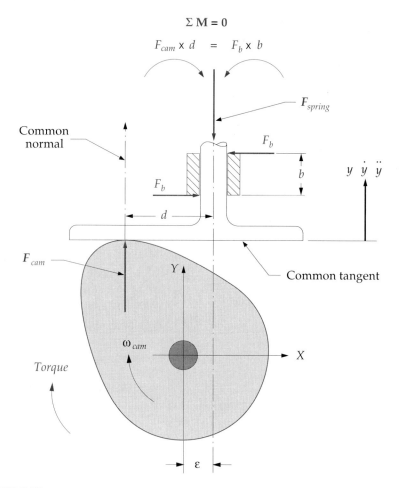

$$\Sigma \, M = 0$$

$$F_{cam} \times d \;=\; F_b \times b$$

FIGURE 9-51

Overturning moment on a flat-faced follower

Cam contours are usually functions of high degree. When they are wrapped around their base or prime circles, they may have portions which are concave, convex, or flat. Infinitesimally short flats of infinite radius will occur at all inflection points on the cam surface where it changes from concave to convex or vice versa.

The radius of curvature of the finished cam is of concern regardless of the follower type, but the concerns are different for different followers. Figure 9-52 shows an obvious problem with a roller follower whose own (constant) radius of curvature R_f is too large to follow the locally smaller concave (negative) radius $-\rho$ on the cam.

A more subtle problem occurs when the roller follower radius R_f is larger than the smallest positive (convex) local radius $+\rho$ on the cam. This problem is called **undercutting** and is depicted in Figure 9-53. Recall that for a roller follower cam, the cam contour is actually defined as the locus of the center of the roller follower, or

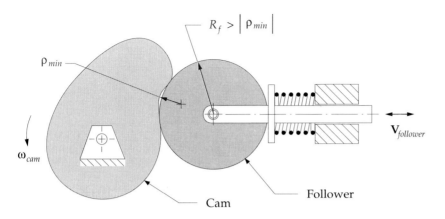

FIGURE 9-52

Mismatch of radii of curvature between cam and roller follower

the **pitch curve**. The machinist is given these x,y coordinate data (on computer tape or disk) and also told the radius of the follower R_f. The machinist will then cut the cam with a cutter of the same effective radius as the follower, following the pitch curve coordinates with the center of the cutter.

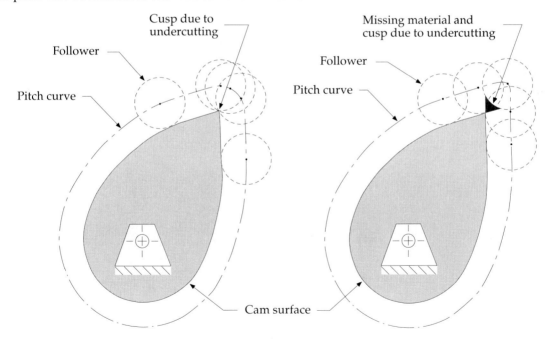

(a) Radius of curvature of pitch curve equals the radius of the roller follower

(b) Radius of curvature of pitch curve is less than the radius of the roller follower

FIGURE 9-53

Small positive radius of curvature can cause undercutting

Figure 9-53a shows the situation in which the follower (cutter) radius R_f is at one point exactly equal to the minimum convex radius of curvature of the cam ($+\rho_{min}$). The cutter creates a perfect sharp point, or **cusp**, on the cam surface. This cam will not run very well at speed! Figure 9-53b shows the situation in which the follower (cutter) radius is greater than the minimum convex radius of curvature of the cam. The cutter now **undercuts** or removes material needed for cam contours in different locations and also creates a sharp point or cusp on the cam surface. This cam no longer has the same displacement function you so carefully designed.

The rule of thumb is to keep the absolute value of the minimum radius of curvature ρ_{min} of the cam pitch curve preferably at least 2 to 3 times as large as the radius of the roller follower R_f.

$$|\rho_{min}| >> R_f \tag{9.30}$$

A derivation for radius of curvature can be found in any calculus text. For our case of a roller follower, we can write the equation for the radius of curvature of the pitch curve of the cam as:

$$\rho_{pitch} = \frac{\left[(R_P + s)^2 + v^2\right]^{\frac{3}{2}}}{(R_P + s)^2 + 2v^2 - a(R_P + s)} \tag{9.31}$$

In this expression, s, v, and a are the displacement, velocity, and acceleration of the cam program as defined in a previous section. Their units are length, length/radian, and length/radian², respectively. R_p is the prime circle radius. **Do not confuse** this *prime circle radius R_p* with the *radius of curvature, ρ_{pitch}*. R_p is a **constant value** which you choose as a design parameter and ρ_{pitch} is the constantly changing radius of curvature which results from your design choices.

Also do not confuse R_p, the *prime circle radius* with R_f, the *radius of the roller follower*. See Figures 9-47 and 9-55 for definitions. You can choose the value of R_f to suit the problem, so you might think that it is simple to satisfy equation 9.30 by just selecting a roller follower with a small value of R_f. Unfortunately it is more complicated than that, as a small roller follower may not be strong enough to withstand the dynamic forces from the cam. The radius of the pin on which the roller follower pivots is substantially smaller than R_f because of the space needed for roller or ball bearings within the follower. The topic of dynamic forces will be addressed in later chapters at which point we will revisit this problem.

We can solve equation 9.31 for ρ_{pitch} since we know s, v, and a for all values of θ and can choose a trial R_p. If the pressure angle has already been calculated, the R_p found for its acceptable values should be used to calculate ρ_{pitch} as well. If a suitable follower radius cannot be found which satisfies equation 9.30 for the minimum values of ρ_{pitch} calculated from equation 9.31, then further iteration will be needed, possibly including a redefinition of the cam specifications.

Program DYNACAM calculates ρ_{pitch} for all values of θ for a user supplied prime circle radius R_p. Figure 9-54 shows ρ_{pitch} for the program's example cam #5. Note that this cam has both positive and negative radii of curvature. The large values of radius of curvature are truncated at arbitrary levels on the plot as

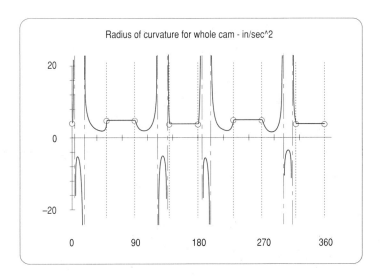

FIGURE 9-54

Radius of curvature of a four-dwell cam

they are heading to infinity at the inflection points between convex and concave portions. Note that the radii of curvature go out to positive infinity and return from negative infinity or vice versa at these inflection points (perhaps after a round trip through the universe?).

Once an acceptable prime circle radius and roller follower radius are determined based on pressure angle and radius of curvature considerations, the cam can be drawn in finished form and subsequently manufactured. Figure 9-55 shows the profile of the cam from example # 4 on DYNACAM's input menu. The acceleration plots for this eight-segment, four-dwell RDFDRDFD cam are shown in Figure 9-31 on p. 336. Its radii of curvature are shown in Figure 9-54.

The cam surface contour is swept out by the envelope of follower positions just as the cutter will create the cam in metal. The sidebar shows the parameters for the design, which is an acceptable one. The ρ_{min} is 1.7 times R_f and the pressure angles are less than 30°. The contours on the cam surface appear smooth, with no sharp corners. Figure 9-56 shows the same cam with only one change. The radius of follower R_f has been made the same as the minimum radius of curvature, ρ_{min}. The sharp corners or cusps in several places indicate that undercutting has occurred. This has now become an **unacceptable cam**, *simply because of a roller follower that is too large.*

The coordinates for the cam contour, measured to the locus of the center of the roller follower, or the **pitch curve** as shown in Figure 9-55, are defined by the following expressions, referenced to the center of rotation of the cam. See Figure 9-49 for nomenclature. The subtraction of the cam input angle θ from 2π is necessary because the relative motion of the follower versus the cam is opposite to that of the cam versus the follower. In other words, to define the contour of the centerline of the follower's path around a stationary cam, we must move the follower (and also the cutter to make the cam) in the opposite direction of cam rotation.

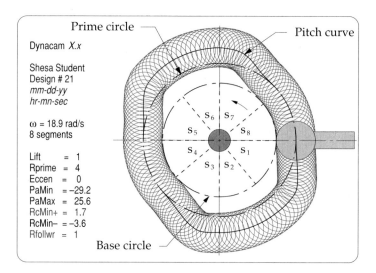

FIGURE 9-55

Radial plate cam profile is generated by the locus of the roller follower (or cutter)

$$x = \cos\lambda\sqrt{(d+s)^2 + \varepsilon^2}$$
$$y = \sin\lambda\sqrt{(d+s)^2 + \varepsilon^2}$$

(9.32)

where:

$$\lambda = (2\pi - \theta) - \arctan\left(\frac{\varepsilon}{d+s}\right)$$

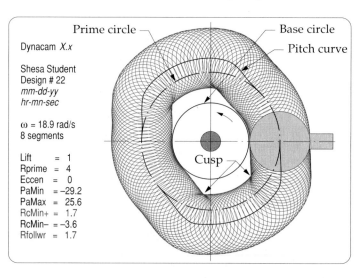

FIGURE 9-56

Cusps formed by undercutting due to radius of follower $R_f \geq$ cam radius of curvature ρ

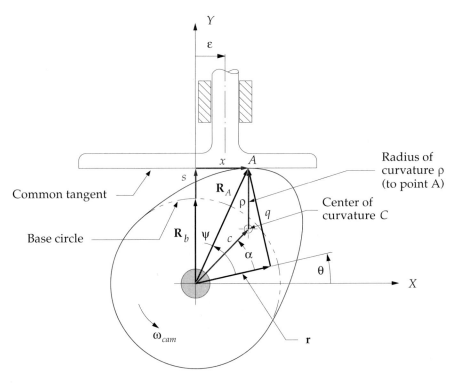

FIGURE 9-57

Geometry for derivation of radius of curvature and cam contour with flat-faced follower

Radius of Curvature - Flat-Faced Follower

The situation with a flat-faced follower is different to that of a roller follower. A negative radius of curvature on the cam cannot be accommodated with a flat-faced follower. The flat follower obviously cannot follow a concave cam. Undercutting will occur when the radius of curvature becomes negative if a cam with that condition is made.

Figure 9-57 shows a cam and flat-faced follower in an arbitrary position. The origin of the global XY coordinate system is placed at the cam's center of rotation and the X axis is defined parallel to the common tangent, which is the surface of the flat follower. The vector $\mathbf{r}$ is attached to the cam, rotates with it, and serves as the reference line to which the cam angle θ is measured from the X axis. The point of contact A is defined by the position vector $\mathbf{R}_A$. The instantaneous center of curvature is at C and the radius of curvature is ρ. R_b is the radius of the base circle and s is the displacement of the follower for angle θ. The eccentricity is ε.

We can define the location of contact point A from two vector loops (in complex notation).

$$\mathbf{R}_A = x + j(R_b + s)$$

and

$$\mathbf{R}_A = ce^{j(\theta+\alpha)} + j\rho$$

so:

$$ce^{j(\theta+\alpha)} + j\rho = x + j(R_b + s) \qquad (9.33a)$$

Substitute the Euler equivalent (Eq. 4.4a) in equation 9.33a and separate the real and imaginary parts.

real:

$$c\cos(\theta+\alpha) = x \qquad (9.33b)$$

imaginary:

$$c\sin(\theta+\alpha) + \rho = R_b + s \qquad (9.33c)$$

The center of curvature C is **stationary** on the cam meaning that the magnitudes of c and ρ, and angle α do not change for small changes in cam angle θ. (These values are not constant but are at stationary values. Their first derivatives with respect to θ are zero, but their higher derivatives are not zero.)

Differentiating equation 9.33a with respect to θ then gives:

$$jce^{j(\theta+\alpha)} = \frac{dx}{d\theta} + j\frac{ds}{d\theta} \qquad (9.34)$$

Substitute the Euler equivalent (Eq. 4.4a) in equation 9.34 and separate the real and imaginary parts.

real:

$$-c\sin(\theta+\alpha) = \frac{dx}{d\theta} \qquad (9.35)$$

imaginary:

$$c\cos(\theta+\alpha) = \frac{ds}{d\theta} = v \qquad (9.36)$$

Inspection of equations 9.33b and 9.36 shows that:

$$x = v \qquad (9.37)$$

This is an interesting relationship that says the x position of the contact point between cam and follower is numerically equal to the velocity of the follower in length/radian. This means that the v diagram gives a direct measure of the necessary minimum facewidth of the flat follower.

$$facewidth > v_{max} - v_{min} \qquad (9.38)$$

If the velocity function is asymmetric then the follower will have to be also, in order for it not to fall off the cam.

Differentiating equation 9.37 with respect to θ gives:

$$\frac{dx}{d\theta} = \frac{dv}{d\theta} = a \tag{9.39}$$

Equations 9.33c and 9.35 can be solved simultaneously and equation 9.39 substituted in the result to yield:

$$\rho = R_b + s + a \tag{9.40}$$

BASE CIRCLE Note that equation 9.40 defines the radius of curvature in terms of the base circle radius and both the displacement and acceleration functions from the $s\,v\,a\,j$ diagrams only. Because ρ cannot be allowed to become negative with a flat-faced follower, we can formulate a relationship from this equation which will predict the minimum base circle radius R_b needed to avoid undercutting. The only factor on the right side of equation 9.40 which can be negative is the acceleration, a. We have defined s to be always positive, as is R_b. Therefore the worst case for undercutting will occur when a is at its **largest negative value**, a_{min}, whose value we know from the a diagram. The minimum base circle radius can then be defined as:

$$R_{b_{min}} > \rho_{min} - s_{@a_{min}} - a_{min} \tag{9.41}$$

Note that the value of s in this equation is taken at the cam angle θ corresponding to that of a_{min}. Because the value of a_{min} is negative and it is also negated in equation 9.41, it dominates the expression. To use this relationship, we must choose some minimum radius of curvature ρ_{min} for the cam surface as a design parameter. Since the Hertzian contact stresses at the contact point are a function of local radius of curvature, that criterion can be used to select ρ_{min}. That topic is beyond the scope of this text and will not be further explored here. See reference 1 for further information on contact stresses.

CAM CONTOUR For a flat-faced follower cam, the coordinates of the physical cam surface must be provided to the machinist as there is no pitch curve to work to. Figure 9-57 shows two orthogonal vectors, **r** and **q**, which define the cartesian coordinates of contact point A between cam and follower with respect to a rotating axis coordinate system embedded in the cam. Vector **r** is the rotating "x" axis of this embedded coordinate system. Angle ψ defines the position of vector $\mathbf{R}_A$ in this system. Two vector loop equations can be written and equated to define the coordinates of all points on the cam surface as a function of cam angle θ.

$$\mathbf{R}_A = x + j(R_b + s)$$

and

$$\mathbf{R}_A = re^{j\theta} + qe^{j\left(\theta + \frac{\pi}{2}\right)}$$

so:

$$re^{j\theta} + qe^{j\left(\theta + \frac{\pi}{2}\right)} = x + j(R_b + s) \tag{9.42}$$

Divide both sides by $e^{j\theta}$:

$$r + jq = xe^{-j\theta} + j(R_b + s)e^{-j\theta} \qquad (9.43)$$

Separate into real and imaginary components and substitute v for x from equation 9.37:

real (x component):

$$r = (R_b + s)\sin\theta + v\cos\theta \qquad (9.44a)$$

imaginary (y component):

$$q = (R_b + s)\cos\theta - v\sin\theta \qquad (9.44b)$$

Equations 9.44 can be used to machine the cam for a flat-faced follower. These x,y components are in the rotating coordinate system that is embedded in the cam.

Note that none of the equations developed above for this case involve the **eccentricity**, ε. It is only a factor in cam size when a roller follower is used. It does not affect the geometry of a flat follower cam.

Figure 9-58 shows the result of trying to use a flat-faced follower on a cam with negative radius of curvature. If the follower could contact the cam at all the points necessary to control the follower to the function in the s diagram, the cam surface would be as developed by the envelope of straight lines. However, these loci of the follower face are cutting into cam contours which are needed for other cam angles. The line running through the forest of follower loci is the theoretical cam contour needed for this design. The undercutting can be clearly seen as the crescent-shaped missing pieces at four places between the cam contour and the follower loci.

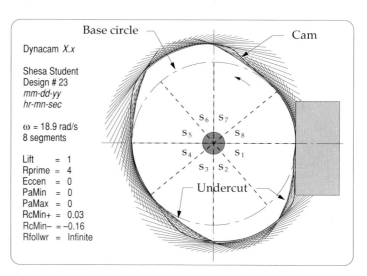

FIGURE 9-58

Undercutting due to negative radius of curvature used with flat-faced follower

SUMMARY The task of sizing a cam is an excellent example of the need for and value of iteration in design. Rapid recalculation of the relevant equations with a tool such as program DYNACAM makes it possible to quickly and painlessly arrive at an acceptable solution while balancing the often conflicting requirements of pressure angle and radius of curvature constraints. In any cam, either the pressure angle or radius of curvature considerations will dictate the minimum size of the cam. Both factors must be checked. The choice of follower type, either roller or flat-faced, makes a big difference in the cam geometry. Cam programs which generate negative radii of curvature are unsuited to the flat-faced type of follower unless very large base circles are used to force ρ to be positive everywhere.

9.8 CAM MANUFACTURING CONSIDERATIONS

The preceding sections illustrate that there are a number of factors to consider when designing a cam. A great deal of care in design is necessary to obtain a good compromise of all factors, some of which conflict. Once the cam design is complete a whole new set of considerations must be dealt with that involve manufacturing the cam. After all, if your design cannot be successfully machined in metal in a way that truly represents the theoretical functions chosen, their benefits will not be realized. Unlike linkages, which are very easy to make, cams are a challenge to manufacture properly.

Cams are usually made from strong, hard materials such as **medium to high carbon steels** (case- or through-hardened) or **cast ductile iron** or **grey cast iron** (case-hardened). Cams for low loads and speeds or marine applications are sometimes made of brass or bronze. Even plastic cams are used in such applications as washing machine timers where the cam is merely tripping a switch at the right time. We will concentrate on the higher load-speed situations here for which steel or cast/ductile iron are the only practical choices. These materials range from fairly difficult to very difficult to machine depending on the alloy. At a minimum, a reasonably accurate milling machine is needed to make a cam. A computer controlled machining center is far preferable and is most often the choice for serious cam production.

Cams are typically milled with rotating cutters that in effect "tear" the metal away leaving a less than perfectly smooth surface at a microscopic level. For a better finish and better geometric accuracy, the cam can be ground after milling away most of the unneeded material. Heat treatment is usually necessary to get sufficient hardness to prevent rapid wear. Steel cams are typically hardened to about Rockwell Rc 50-55. Heat treatment introduces some geometric distortion. The grinding is done after heat treatment to correct the contour as well as to improve the finish. The grinding step nearly doubles the cost of an already expensive part, so it is often skipped in order to save money. A hardened but unground cam will have some heat distortion error despite accurate milling before hardening. There are several methods of cam manufacture in common use as shown in Table 9-3.

Geometric Generation

Geometric generation refers to the continuous "sweeping out" of a surface as in turning a cylinder on a lathe. This is perhaps the ideal way to make a cam because it

TABLE 9-3   Common Cam Manufacturing Methods

1   Geometric Generation

2   Manual or **N**umerical **C**ontrol (**NC**) machining to cam coordinates (plunge-cutting)

3   **C**ontinuous **N**umerical **C**ontrol (**CNC**) with Linear Interpolation (**LI**)

4   **C**ontinuous **N**umerical **C**ontrol (**CNC**) with Circular Interpolation (**CI**)

5   **A**nalog **D**uplication (**AD**) of a hand-dressed master cam

creates a truly continuous surface with an accuracy limited only by the quality of the machine and tools used. Unfortunately there are very few types of cams that can be made by this method. The most obvious one is the eccentric cam (Figure 9-11 on p. 310) which can be turned and ground on a lathe. A cycloid can also be geometrically generated. Very few other curves can. The presence of dwells makes it extremely difficult to apply this method. Thus, it is seldom used for cams. However, when it can be, as in the case of the eccentric cam of Figure 9-11, the resulting acceleration, though not perfect, is very close to the theoretical cosine wave as seen in Figure 9-12 (p. 310). This eccentric cam was made by turning and grinding on a high-quality lathe. This is the best that can be obtained in cam manufacture. Note that the displacement function is virtually perfect. The errors are only visible in the more sensitive acceleration function measurement.

Manual or NC Machining to Cam Coordinates (Plunge-Cutting)

Computer-aided manufacturing (CAM) has become the virtual standard for high accuracy machining in the U.S. Numerical Control (NC) machinery comes in many types. Lathes, milling machines, grinders, etc., are all available with on-board computers which control either the position of the workpiece, the tool or both. The simplest type of NC machine moves the tool (or workpiece) to a specified x,y location and then drives the tool (say a drill) down through the workpiece to make a hole. This process is repeated as much as necessary to create the part. This simple process is referred to as NC to distinguish it from CNC or Continuous Numerical Control.

This NC process is sometimes used for cam manufacture, and even for master cams as described below. It is, in fact, merely a computerized version of the old manual method of cam milling, which is often called **plunge-cutting** to refer to plunging the spinning milling cutter down through the workpiece. This is not the best way to machine a cam because it leaves "scallops" on the surface as shown in Figure 9-59, due to the fact that the machinist can only *plunge* at a discrete number of positions around the cam. In effect, the displacement function which we developed has to be "discretized" or sampled at some finite number of places around the cam. Practicality limits this digitizing process to increments of about 1/2 to 1 degree. With an NC process the increment might be reduced to 1/4 or 1/10 degree. At some point "diminishing returns" will set in as the machine's ability to resolve positions spaced too closely will limit the accuracy. Standard milling machines can be expected to give accuracies in the 0.001 inch tolerance range. Tooling quality machining centers, jig borers and grinders can be as much as two to ten times that accurate (0.0005 inch down to 0.0001 inch tolerance).

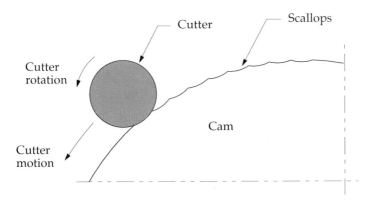

FIGURE 9-59

Plunge-cutting a cam leaves scallops on its surface

The scallops that are left on the cam after plunge-cutting have to be removed by hand-dressing with files and grindstones. This obviously introduces more error. Even if the bottoms of the scallops at the sample increments were exactly correct, all points between are subject to the vagaries of hand work. The chance of exactly achieving the designed *s v a j* functions, especially the higher derivatives, with this manufacturing method is slight.

Continuous Numerical Control with Linear Interpolation

In a CNC machine, the tool is in constant contact with the workpiece, always cutting, while the computer controls the movement of the workpiece from position to position as stored in its memory. This is a **continuous cutting process** as opposed to the discrete one of NC. However, the cam displacement function must still be discretized or sampled at some angular increment. The common increments are $1/4, 1/2$ and $1°$. Since the machine only has information about the x,y locations of these 360, 720, or 1440 points around the cam, it has to figure out how to get from one point to another while cutting. The most common method used to "fill in" the missing data is **linear interpolation** (LI). The machine's computer calculates the straight line between each pair of data points and then drives the cutter (or workpiece) so as to stay as close to that line as it can. If it could do this perfectly (which it can't), we would get a continuous first-degree approximation to the displacement function. Figure 9-60 shows what would happen to the higher derivatives if this happened. We would be back to our "bad cam by naive designer" of Section 9.3 with infinite pulses of acceleration regardless of the actual function selected!

Fortunately, the dynamics of the cutting process which are a function of speeds, feeds, tool sharpness, tool chatter, deflection of the spindle, etc., all conspire to prevent the formation of a series of distinct "flats" which would give the derivatives shown in Figure 9-60. Rather, the kind of acceleration curve that actually results from a cam which was milled (but not ground) on a very high-quality CNC machining center, using $1°$ linear interpolation, is as shown in Figure 9-61. The program

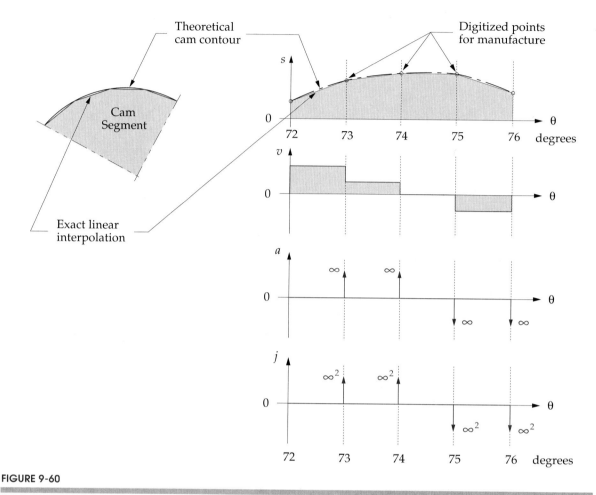

FIGURE 9-60

Theoretical result of linear interpolation of cam surface cut with Continuous Numerical Control (CNC)

is a simple harmonic eccentric with no dwells. The dynamic curves were measured with instrumentation on the roller follower while the cam was running at 600 RPM on a custom designed cam dynamic test fixture (CDTF).[3] The displacement is quite accurate, but the acceleration has a significant amount of vibratory noise present which distorts the function from its theoretical cosine waveshape. The acceleration is in g's. Compare the peak-to-peak values in Figure 9-61 (8 g's) with the same cam design made with geometric generation in Figure 9-12, (5 g's). The error is of the order of 3 g's on a base of 5 g's. These errors in the acceleration are due to a combination of manufacturing factors as described above. It was found that the use of 1/4, 1/2 or 1° digitization increments made no statistically significant difference in the fidelity of the actual acceleration function to its theoretical waveform.[2]

The physical fidelity of the cam surface of the same 1° linear interpolated eccentric cam to a geometrically generated (turned and ground) reference cam can

[3] This research was supported by the National Science Foundation under grants MEA-82-10865 and MSM-85-12913, and by AMP Incorporated, Harrisburg, Pa.

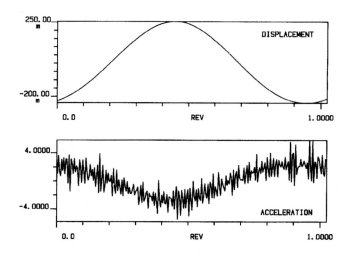

FIGURE 9-61

Displacement and acceleration of eccentric cam made with 1° linear interpolation CNC

be seen in Figure 9-62 which is an enlarged section of a portion of the cams as measured to 0.0005 inch accuracy. The axes are calibrated in inches. These figures show that linear interpolated CNC is a reasonably accurate method of cam manufacture.

Continuous Numerical Control with Circular Interpolation

This process is similar to CNC with linear interpolation except that a **circular interpolation** (CI) algorithm is used between data points. This theoretically allows a sparser (thus smaller) data base in the machine which can be an advantage. The

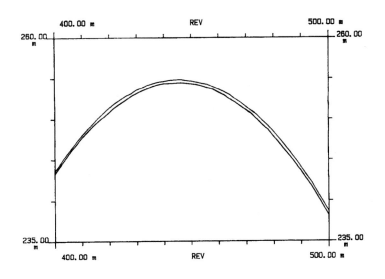

FIGURE 9-62

Contours of 2 cams - one turned and ground - one milled with 1° linear interpolation CNC

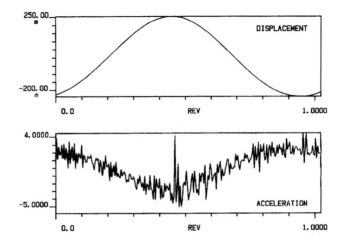

FIGURE 9-63

Displacement and acceleration of eccentric cam made with circular interpolation CNC

computer tries to fit a circle arc to as many adjacent data points as possible without exceeding a user-selected error band around the actual displacement function. This reduces the (variable) number of fitted data points to three, a radius and its center coordinates. Most CNC machines have a built-in algorithm to generate (cut) circle arcs quickly and efficiently, as this is a common requirement in normal machining.

Figure 9-63 shows the same cam design as in Figures 9-12 and 9-61, made on the same machining center from the same bar of steel, but with circular interpolation (CI). The error in acceleration is more than the turned-ground (TG) cam but less than the linear interpolated (LI) one. Figure 9-64 shows the cam contour of the CI cam compared to the TG cam. Based on dynamic performance the circular interpolated cam has lower acceleration error than the linear interpolated cam and the dif-

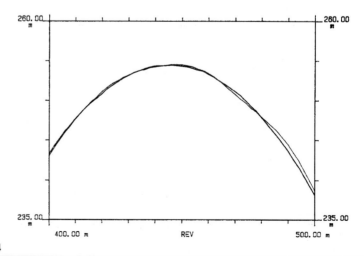

FIGURE 9-64

Contours of 2 cams - one turned and ground - one milled with circular interpolation CNC

ference is statistically significant. The "blip" in the middle of the period is due to the slight ridge formed at the point where the cutter starts and stops its continuous sweep around the cam contour. This will be present regardless of the CNC algorithm used.

Analog Duplication

The last method listed, analog duplication, involves the creation of a master cam, sometimes at larger than full scale, which is subsequently used in a cam duplicating machine to turn out large quantities of the finished cams. Most automotive cam-shafts are made by this method. It is the most economical where production quantities are high. A cam duplicating machine will have two spindles. The master cam will be mounted on one. The workpiece will be placed on the other. A dummy cutter mounted on one crank of a parallelogram linkage (see Figure 2-17a on p. 47) is used as a follower against the master. The actual cutter is mounted on the other crank of the parallelogram linkage. A multiplication ratio can be introduced between the dummy and the cutter to size the finished cam versus the master. As the master and slave slowly turn together, synchronous and in phase, the dummy cutter follows the master's contour and the workpiece is cut to match. This process can be done with either milling (cutting) or grinding of the cam surface. Typically the cam is rough cut first, and then heat treated and reground to finished size. Some cams are left as-milled with no post heat-treat grinding.

This analog duplication method can obviously create a cam that is only as good as the master cam at best. Some errors will be introduced in the duplicating process due to deflections of the tool or machine parts, but the quality of the master cam ultimately limits the quality of the finished cams. The master cam is typically made by one of the other methods listed in Table 9-3, each of which has its limitations. The master cam usually requires some hand-dressing with files or hand grindstones to smooth its surface. A plunge cut cam requires a lot of hand dressing, the CNC cams less so. If hand dressing is done, it will result in a very smooth surface but the chances that the resulting contour is an accurate representation of the designed $s\ v\ a\ j$ functions, especially the higher derivatives, is slim. Thus the finished cam may not be an accurate representation of the design.

Figure 9-65 shows the same cam design as in Figure 9-63, made on the same machining center from the same bar of steel, but analog duplicated from a hand-dressed, plunge-cut master. This represents the worst case in terms of manufacturing error. The error in acceleration is more than any of the other cams. Figure 9-66 shows the cam contour of the analog milled cam compared to the reference turned-ground cam. It is much less accurate than either of the CNC versions. Based on dynamic performance, the analog milled cam from a hand-dressed, plunge-cut master has higher acceleration error than any other cam tested and the difference is statistically significant. If the master cam were made by a more accurate method (as is the case with automotive cams), the accuracy of the production cam would be much greater.

Actual Cam Performance Compared to Theoretical Performance

The relative peak accelerations of several common double-dwell cam functions were discussed in Section 9.3 and summarized in Table 9-2. That discussion also empha-

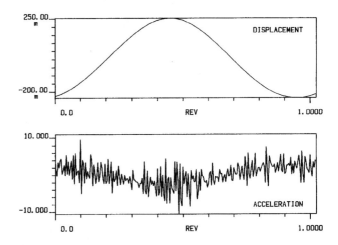

FIGURE 9-65

Displacement and acceleration of eccentric cam made with analog duplication of a
hand-dressed master cam

sized the importance of a smooth jerk function for minimizing vibrations. The
theoretical differences between peak accelerations of different cam functions will be
altered by the presence of vibratory noise in the actual acceleration waveforms.
This noise will be due in part to errors introduced in the manufacturing process,
as discussed above, but there will also be inherent differences due to the degree to
which the jerk function excites vibrations in the cam-follower train. These vibrations
will be heavily influenced by the structural dynamic characteristics of the follower
train itself. In general, a very stiff and massive follower train will vibrate less than a
light and flexible one, but the presence of frequencies in the cam function that are near
the natural frequencies of the follower train will exacerbate the problem.

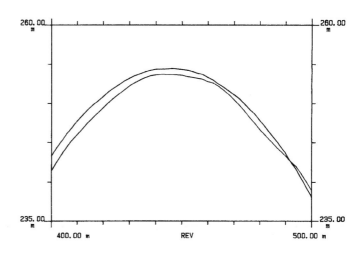

FIGURE 9-66

Contours of 2 cams - one turned and ground - one milled with analog duplication of a
hand-dressed master cam

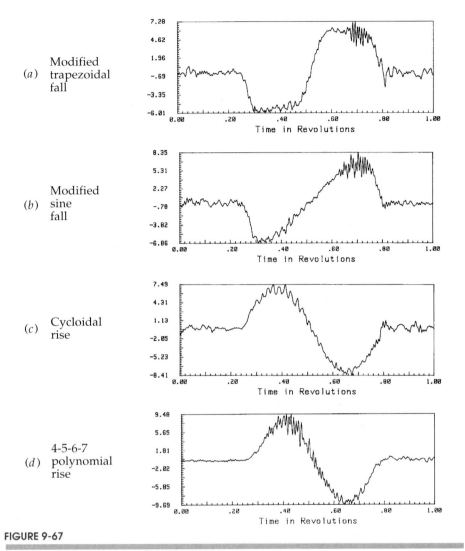

(a) Modified trapezoidal fall

(b) Modified sine fall

(c) Cycloidal rise

(d) 4-5-6-7 polynomial rise

FIGURE 9-67

Actual measured follower acceleration curves of four, double-dwell programs

Figure 9-67 shows the actual acceleration waveforms of four common double-dwell cam programs, **modified trapezoid**, **modified sine**, **cycloidal**, and **4-5-6-7 polynomial**, which were ground on the same four-dwell cam as can be seen as example #5 in program DYNACAM and in Figure 9-55. These waveforms were measured with an accelerometer mounted on the stiff follower train of a cam dynamic test fixture (CDTF) specially designed for low vibration and noise. These test cams were made by a specialty cam manufacturer using linear interpolation CNC with a 1/4° digitizing increment on very high quality machining and grinding equipment. After hardening, the cam contour was ground to a surface finish with an average roughness (R_a) of 0.125 microns. Comparing these functions on the same cam, made on the same production machinery, and run against the same follower system gives a means to measure their actual differences in relative accelerations.

TABLE 9-4   Actual Peak Acceleration Compared to the Theoretical
for Various Double-Dwell Cam Programs (in g's)

Cam Program	Theor. A_{pp} g's	Actual A_{pp} g's	Ratio A_{mf} Eq. 9.45	Theor.* Accel. Factor	Actual Accel. Factor
Mod. trap. Accel.	11.78	13.3	1.13	$4.888h/\beta^2$	$5.5h/\beta^2$
Mod. sine. Accel.	13.32	15.2	1.14	$5.528h/\beta^2$	$6.3h/\beta^2$
Cycloidal disp.	15.14	15.9	1.05	$6.283h/\beta^2$	$6.6h/\beta^2$
4-5-6-7 poly. disp.	18.14	19.2	1.06	$7.526h/\beta^2$	$8.0h/\beta^2$

* From Table 9-2.

All four acceleration waveforms have some amount of vibratory noise present which is seen as oscillations or ripples on the curves. Compare these curves to their exact theoretical equivalents in Figure 9-20 on p. 322 and in program DYNACAM. Table 9-4 compares the expected theoretical peak-to-peak acceleration $A_{pp_{theoretical}}$ for each cam (column 2) with its actual, measured peak-to-peak acceleration $A_{pp_{actual}}$ (column 3).   The error is expressed in column 4 as an acceleration multiplier factor A_{mf} defined as:

$$A_{mf} = \frac{A_{pp_{actual}}}{A_{pp_{theoreticall}}}$$

(9.45)

Column five in Table 9-4 shows the maximum acceleration factors for these four functions taken from Table 9-2. The last (sixth) column shows the actual maximum acceleration values based on these test data and is the product of columns four and five.

These values of actual maximum acceleration show smaller differences between the functions than are predicted by their theoretical waveforms. The modified trapezoid, which has the lowest theoretical acceleration, has a 13% noise penalty due to vibration. The modified sine has a 14% noise penalty, while the cycloidal and 4-5-6-7 polynomial functions have only 5 to 6% noise. This is due to the fact that the last two functions have smoother jerk waveforms than the first two. The cycloidal waveform, with its cosine jerk function, is a good choice for high-speed operations. Its acceleration is actually only about 19% greater than the modified trapezoid's and 5% more than the modified sine, rather than the 28% and 14% differentials predicted by the theoretical peak values.

9.9   PRACTICAL DESIGN CONSIDERATIONS

The cam designer is often faced with many confusing decisions, especially at an early stage of the design process. Many early decisions, often made somewhat arbitrarily and without much thought, can have significant and costly consequences later in the design. The following is a discussion of some of the tradeoffs involved with such decisions in the hope that it will provide the cam designer with some guidance in making these decisions.

Translating or Oscillating Follower?

There are many cases, especially early in a design, when either translating or rotating motion could be accommodated as output from the cam. Approximate straight-line motion is often adequate and can be obtained from a large-radius rocker follower. The rocker or oscillating follower has one significant advantage over the translating follower when a roller follower is used. A translating follower is free to rotate about its axis of translation and may need to have some antirotation guiding (such as a keyway) provided to prevent misalignment of a roller follower with the cam.

Conversely, the oscillating follower will keep the roller follower aligned in the same plane as the cam with no guiding except its own pivot. Also, the pivot friction in an oscillating follower typically has a small moment arm compared to the moment of the force from the cam on the follower arm. The friction force on a translating follower has a one-to-one geometric relationship with the cam force. This can have a larger parasitic effect on the system.

On the other hand, translating flat-faced followers are often deliberately arranged with their axis slightly out of the plane of the cam in order to create a rotation about their own axis due to the frictional moment resulting from the offset. The flat follower will then precess around its own axis and distribute the wear over its entire face surface. This is common practice in automotive valve cams that use flat-faced followers or "tappets".

Force or Form-Closed?

A form-closed (track or groove) cam is more expensive to make than a force-closed (open) cam simply because there are two surfaces to machine and grind. Also, heat treating will often distort the track of a form-closed cam, narrowing or widening it such that the roller follower will not fit properly. This virtually requires post heat-treat grinding for track cams in order to resize the slot. An open (force-closed) cam will also distort on heat-treating but may still be useable without grinding.

FOLLOWER JUMP The principal advantage of a form-closed (track) cam is that it does not need a return spring, and thus can be run at higher speeds than a force-closed cam whose spring and follower mass will go into resonance at some speed, causing potentially destructive follower jump. This phenomenon will be investigated in Chapter 17 on cam dynamics. High-speed automobile and motorcycle racing engines often use form-closed (desmodromic) valve cam trains to allow higher engine rpm without incurring valve "float," or **follower jump**.

CROSSOVER SHOCK Though the lack of a return spring can be an advantage, it comes, as usual, with a tradeoff. In a form-closed (track) cam there will be **crossover shock** each time the acceleration changes sign. Crossover shock describes the impact force that occurs when the follower suddenly jumps from one side of the track to the other as the dynamic force (ma) reverses sign. There is no flexible spring in this system to absorb the force reversal as in the force-closed case. The high impact forces at crossover cause noise, high stresses, and local wear. Also, the roller follower has to reverse direction at each crossover which causes sliding and accelerates follower wear. Studies have shown that roller followers running against a well-lubricated open radial cam have slip rates of less than 1%.[3]

Radial or Axial Cam?

This choice is largely dictated by the overall geometry of the machine for which the cam is being designed. If the follower must move parallel to the camshaft axis, then an axial cam is dictated. If there is no such constraint, a radial cam is probably a better choice simply because it is a less complicated, thus cheaper, cam to manufacture.

Roller or Flat-Faced Follower?

The roller follower is an easier choice from a cam design standpoint simply because it allows negative radius of curvature on the cam. This allows more variety in the cam program. Also, for any production quantities, the roller follower has the advantage of being available from several manufacturers in any quantity from one to a million. For low quantities it is not usually economical to design and build your own custom follower. In addition, replacement roller followers can be obtained from suppliers on short notice when repairs are needed. They are also not particularly expensive even in small quantities.

Perhaps the largest users of flat-faced followers are automobile engine makers. Their quantities are high enough to allow any custom design they desire. It can be made or purchased economically in large quantity and can be less expensive than a roller follower in that case. Also with engine valve cams, a flat follower can save space over a roller. However some manufacturers are switching to roller followers in auto engines to reduce friction and improve fuel mileage. Diesel engines have long used roller followers (tappets) as have racers who "hop-up" engines for high performance.

Cams used in automated production line machinery use stock roller followers almost exclusively. The ability to quickly change a worn follower for a new one taken from the stockroom without losing much production time on the "line" is a strong argument in this environment. Roller followers come in several varieties (see Figure 9-5a, p. 301). They are based on roller or ball bearings. Plain bearing versions are also available for low-noise requirements. The outer surface, which rolls against the cam can be either cylindrical or spherical in shape. The "crown" on the spherical follower is slight, but it guarantees that the follower will ride near the center of a flat cam regardless of the accuracy of alignment of the axes of rotation of cam and follower. If a cylindrical follower is chosen and care is not taken to align the axes of cam and roller follower, the follower will ride on one edge and rapid wear of the follower will result.

Commercial roller followers are typically made of high carbon alloy steel such as 52100 and hardened to Rockwell Rc 60-62. The 52100 alloy is well suited to thin sections that must be heat-treated to a uniform hardness. Because the roller makes many revolutions for each cam rotation, its wear rate may be higher than the cam. Chrome plating the follower can markedly improve its life. Chrome is harder than steel at about Rc 70. Steel cams are typically hardened to a range of Rc 50 - 55.

To Dwell or Not to Dwell?

The need for a dwell is usually clear from the problem specifications. If the follower must be held stationary for any time, then a dwell is required. Some cam designers tend to insert dwells in situations where they are not specifically needed for follower stasis, in a mistaken belief that this is preferable to providing a rise-return motion when that is what is really needed. If the designer is attempting to use a double-dwell program in a single-dwell case, then perhaps his motivation to "let the vibrations settle out" by providing a "short dwell" at the end of the motion is justified. However, he probably should be using another cam program, perhaps a polynomial tailored to the specifications. Taking the acceleration to zero, whether for an instant or for a "short dwell", is generally unnecessary and undesirable. (See Examples 9-5, 9-6, and 9-8.) A dwell should be used only when the follower is required to be stationary for some measurable time. Moreover, if you do not need any dwell at all, consider using a linkage instead. They are a lot easier and cheaper to manufacture.

To Grind or Not to Grind?

Most production machinery cams are used as-milled, not ground. Most, if not all, automotive valve cams are ground. The reasons are largely due to cost and quantity considerations. There is no question that a ground cam is superior to a milled cam. The question in each case is whether the advantage gained is worth the cost. In small quantities, as are typical of production machinery, grinding about doubles the cost of a cam. The advantages in terms of smoothness and quietness of operation, and of wear, are not in the same ratio as the cost difference. A well-machined cam can perform nearly as well as a well-ground cam and better than a poorly ground cam.[3]

Automotive cams are made in large quantity, run at very high speed, and are expected to last for a very long time with minimal maintenance. This is a very challenging specification. It is a great credit to the engineering of these cams that they very seldom fail in 100,000 miles or more of operation. These cams are made on specialized equipment which keeps the cost of their grinding to a minimum.

To Lubricate or Not to Lubricate?

Cams need lots of lubrication. Automotive cams are literally drowned in a flow of engine oil. Many production machine cams run immersed in an oil bath. These are reasonably happy cams. Others are not so fortunate. Cams which operate in close proximity to the product on an assembly machine in which oil would cause contamination of the product (food products, personal products) often are run dry. Camera mechanisms, which are full of linkages and cams, are often run dry. Lubricant would eventually find its way to the film.

Unless there is some good reason to eschew lubrication, a cam-follower should be provided with a generous supply of clean lubricant, preferably a hypoid-type oil containing additives for boundary lubrication conditions. The geometry of a cam-follower joint (half-joint) is among the worst possible from a lubrication standpoint. Unlike a journal bearing, which tends to trap a film of lubricant within the joint, the half-joint is continually trying to squeeze the lubricant out of itself. This results in a boundary lubrication state in which some metal-to-metal contact will occur. Lubricant must be continually resupplied to the joint. Another purpose of the liquid lubricant is to remove the heat of friction from the joint. If run dry, significantly higher material temperatures will result, with accelerated wear and possible early failure.

9.10   PROGRAM DYNACAM

This program operates in a similar way to programs FOURBAR, FIVEBAR and SIXBAR which were described in Chapter 8. Please refer to that chapter for information on hardware requirements and on loading and running the program. The main menu for DYNACAM is shown in Figure 9-68. The first five items operate similarly to FOURBAR.

1 - Input Data

The **Input Data** menu (Figure 9-69) has several built-in examples, some of which have already been used in this chapter (#'s 3, 4 and 5). The examples on the input menu include a single-dwell cam using double harmonic functions, a double-dwell cam using modified trapezoid and modified sine functions and two four-dwell cams with various combinations of functions including polynomials. The student should exercise these examples to become familiar with the program.

KEYBOARD The first choice on the input menu, **Type Data from Keyboard** allows input of data specific to your problem. It will first request the camshaft angular velocity with a warning that it must be positive in sign (*ccw*) for this program.

```
Main Menu - Program DYNACAM  Version X.x   mm/dd/yy

        1  –  INPUT         problem setup data

        2  –  DISPLAY       problem setup data

        3  –  COMPUTE       s-v-a-j for cam program

        4  –  PRINT         any of the results

        5  –  PLOT          any of the results

        6  –  COMPUTE       pressure angles

        7  –  COMPUTE       radius of curvature

        8  –  COMPUTE       dynamic force and torque

        9  –  DRAW          the cam profile

       10  –  HELP

    <Ret>  –  QUIT          the program        CHOICE?
```

FIGURE 9-68

The main menu for program DYNACAM

(A clockwise angular velocity is tantamount to viewing the cam from the back side. Its shape will be the same.) You must then supply the number of segments in the cam. Segments are typically dwells, rises, falls, or combinations of the last two (for polynomials). The angle (β), in degrees, of each segment must then be supplied. If these β's do not sum to 360°, you will be asked to reenter the data. A code letter is then requested for each segment, one of: R = rise, F = fall, D = dwell or P = polynomial. These are used in the program to properly compute the data but can be overridden in some cases when the cam computation is later done. A polynomial can always be substituted for any other choice at the time of computation. To change a rise to a fall or vice versa, it is necessary to return to this menu and redo that choice.

LIFT A value will be requested for the total lift, h, required for any non-dwell segment. This must always be entered as a **positive value**, even if it is for a fall segment. The program will automatically take care of its direction based on the supplied codes for the segments.

OFFSET A value for offset is requested for nondwell and nonpolynomial segments. **Offset** *should not be confused with eccentricity.* Offset is used to start a rise or fall at other than the base or prime circle for a "staircase effect".

For example, if you wish to rise 2 inches in segment 1, dwell in segment 2, then rise again for another inch in segment 3, the offset for segment 1 is zero, and for segment 3 is 2 inches. The offset is always measured as a positive value, up from the zero displacement baseline. So, to offset a fall segment 0.5 inch below the top of a 3-inch rise, the offset would be 2.5 inches.

A common use of a nonzero offset is to properly position a half-rise or half-fall function within the total rise (see Example 9-11). Normally the offset should be

```
                      SELECT INPUT METHOD
          ────────────────────────────────────────────
          (Note: any selection will overwrite existing data)


          1 −  TYPE DATA from KEYBOARD

          2 −  READ DATA from DISKFILE

          3 −  Example Cam  -  Single-Dwell

          4 −  Example Cam  -  Double-Dwell

          5 −  Example Cam  -  Four-Dwell

          6 −  Example Cam  -  AMP Four-Dwell

                 <Ret> − EXIT

                          Choice ?
```

FIGURE 9-69

Input menu for program DYNACAM

left at zero unless one of the special cases noted is desired. An **offset is never used with polynomial functions because the boundary conditions take care of positioning the function.**

DISK FILES The second choice on the input menu, **Read Data from Disk-File** allows input of previously stored data files from disk. Its use is the same as in the FOURBAR program (see Chapter 8).

2 - Display Data

The current cam configuration can be seen at any time with this choice. All the data input in step 1 above will be displayed as shown in Figure 9-38 on p. 342. The segment angles can be visually checked for correctness. As calculations are later done, the information in this display screen will be updated as shown in Figure 9-7 on p. 303. The one-letter codes under "Function used" will change to show the type of function used, such as "*poly5*" or "*cycloid fall*", etc. Also, the calculated maximum lift will have a nonzero value in it after calculations are done. A hard copy of this or any display screen can be obtained at any time with Shift PrtScrn or by selecting printer from the menu.

3 - Compute S V A J

The computations must be done one segment at a time. After choosing the segment, the **Select Program Menu** in Figure 9-70 comes up. There are five standard cam programs (items 1 to 5) all of which are defined elsewhere in this chapter, a polynomial choice (item 6), and a dwell choice (item 7). The five standard programs

Segment number	Function used	Start angle	End angle	Delta angle
1	R	0	50	50

Select program for this segment

1 – Simple Harmonic

2 – Cycloidal

3 – Modified Sine

4 – Modified Trapezoid

5 – Double Harmonic

6 – Any Polynomial

7 – Dwell

Choice?

FIGURE 9-70

The calculate menu

use the r/f flag set in step 1 to calculate the function as a rise or fall, negating the specified lift for a fall segment. A polynomial can be either a rise or a fall, or **both**, so does not pay attention to the r/f flag. Thus a polynomial selection will override any r/f choice made in step 1.

DWELLS The dwell is easiest to calculate. You supply the displacement during the dwell. It does the rest.

STANDARD FUNCTIONS The standard cam programs (items 1 to 5) calculate automatically based on the previously specified total lift and the r/f flag for that segment. Items 1 and 2, the simple harmonic and the cycloidal, allow a choice between full-wave and half-wave versions. The full waves will give a complete rise or fall. The half waves are for matching to other functions, such as constant velocity, within a segment. (See Example 9-11.)

POLYNOMIALS The polynomial choice requires more information from the user. The total number of boundary conditions for the segment (interval) must be supplied. A set of input screens will then appear, in sequence. One is shown in Figure 9-71. These will ask for boundary conditions at each end of the segment, first at $\theta = 0$, then at $\theta = \beta$. For each end, four requests will be made, for s, v, a, j conditions in that order. You must supply either a number corresponding to the desired boundary constraint, hit the **enter** or **return** key if you wish to accept the default value or type n to have **no constraint** on the function at that point. There will be no default responses for these data until after your first set of BC's have been input.

If we use the boundary conditions in equation 9.17a from Example 9-7 for discussion, the responses would be: 0, 0, 0, n, 1, 0, 0, n, where n means no constraint. If there are still any unfilled boundary conditions after the ends of the segment have

Segment number	Function used	Start angle	End angle	Delta angle
5	poly5	180	230	50

at Theta = 0 - Local to Segment # 5

Enter Boundary Condition - Use the letter 'n' (for none) if want NO constraint

<Return> accepts value of 2

Enter DISPLACEMENT condition ?

FIGURE 9-71

Boundary condition input for polynomial functions

You have 3 of 9 conditions still unfilled

Doing boundary condition number 7

  0 – Displacement

  1 – Velocity

  2 – Acceleration

  3 – Jerk

Enter DEGREE of function you want to constrain ?

Enter value of THETA at constraint ?

Enter VALUE of Boundary Constraint ?

FIGURE 9-72

Defining boundary conditions within the interval

been taken care of, the input screen shown in Figure 9-72 appears. You must specify the function you wish to place the boundary constraint on, the value of local segment angle θ (in degrees) at which you want to constrain it, and the value of the constraint. **Note that the angles are local to each segment, running always from zero to β.** If, for example, you want to put a constraint at the middle of the segment, the value of θ will be $\beta/2$, regardless of where in the 360° circumference of the whole cam it might actually be. **These computations are always done within one segment.** *When calculating, the program is temporarily ignorant of the existence of anything outside of the current segment.*

When all conditions have been supplied, it calculates the coefficients of the polynomial by a Gauss-Jordan reduction method with partial pivoting. All computations are done in double-precision for accuracy. If an inconsistent set of conditions is sent to the solver, an error message will appear. If the solution succeeds, it calculates *s v a j* for the segment. You must repeat the calculation process for each segment, including dwells. Results for any calculated segment can be printed and plotted at any time, even prior to any other segments being calculated.

4 - Print Results

The **Print Select** menu is shown in Figure 9-73. At this stage, items 1 to 4, 9 and 10 are available for printing. We have yet to calculate the others. Any selection from 1 to 9 brings up a sequence of input screens which allow selection of any one segment or the entire cam for output. Figure 9-41 on p. 344 shows a table of displacements printed from this menu.

Item 10 will print three screens in succession which are of interest primarily for polynomial segments (though they print more limited data for other segments).

```
╭─────────────────────────────────────────────────────────╮
│                    PRINT SELECT MENU                      │
│                                                           │
│              1 –  Displacement                            │
│                                                           │
│              2 –  Velocity                                │
│                                                           │
│              3 –  Acceleration                            │
│                                                           │
│              4 –  Jerk                                    │
│                                                           │
│              5 –  Pressure Angle                          │
│                                                           │
│              6 –  Radius of Curvature                     │
│                                                           │
│              7 –  Dynamic Forces                          │
│                                                           │
│            _ 8 –  Camshaft Torque                         │
│                                                           │
│              9 –  S - V - A - J  Table                    │
│                                                           │
│             10 –  Boundary Conditions                     │
│                                                           │
│           <Ret> –  EXIT                                   │
│                                       Choice ?            │
╰─────────────────────────────────────────────────────────╯
```

FIGURE 9-73

DYNACAM print select menu (similar to the plot select menu)

The first screen (Figure 9-33 on p. 338) shows the boundary conditions which were applied to that segment. The second screen (Figure 9-34 on p. 339) shows the coefficients of the polynomial which results from those boundary conditions. The third screen (Figure 9-74) is a table of the actual calculated values at all segment boundaries (as opposed to what you asked for). All function values are shown whether or not they were constrained. They are grouped so that adjacent segment function values are together. This table allows a rapid check for discontinuities at segment boundaries that will violate the fundamental law of cam design. Note that the values shown for the second end of one segment and the first end of the next segment are all identical pairs of values, through the acceleration function, indicating no discontinuities.

5 - Plot Results

The **Plot Select** menu (not shown) is identical to the **Print Select** menu except for item 10 which gives a combined plot of pressure angle and radius of curvature on one screen. Many plots from this menu have been presented in this chapter. Table 9-5 lists some figures which were made with various selections from the plot menu.

6 - Compute Pressure Angle

This choice requires you to specify the type of follower used, roller or flat-faced. The prime or base circle desired for the cam and any eccentricity of the follower are requested. The computation can be done for any one segment or for the entire cam at once. The maximum and minimum values are reported to the screen, and all values are stored for later plotting or printing. See Section 9.7 for the equations and discussion.

Segment number	Function used	Start angle	End angle	Delta angle
1	Poly6	0	120	120
2	Dwell	120	360	240

Boundary Values Resulting

Segment	End	S	V	A	J
1	1	0	10	0	0
1	2	5	10	0	0
2	1	5	10	0	− 4793.9
2	2	0	10	0	4793.9
1	1	0	10	0	0

FIGURE 9-74

Boundary value table from DYNACAM

7 - Compute Radius of Curvature

These input screens are identical to those for the pressure angle calculation. If the choice is for a flat-faced follower, a minimum desired radius of curvature is requested as defined in section 9.7. From this value a minimum base circle radius needed to keep the radius of curvature positive is reported to the screen. You may enter any value of base circle radius, above or below this minimum. See Section 9.7 for a discussion of the consequences.

8 - Compute Dynamic Force and Torque

Discussion of this topic will be taken up in a later chapter after the principles of dynamic force analysis have been presented.

TABLE 9-5   Figures from the DYNACAM Plot Menu

Menu Item	Function	Figure	Page
1	Displacement	9-42	345
3	Acceleration	9-36	340
5	Pressure angle	9-50	358
6	Radius of curvature	9-54	363
9	s v a j plot	9-45	349

9 - Draw Cam Profile

This selection will draw the finished cam as shown in Figures 9-55, 9-56, and 9-58 on pp. 364-368. All calculations must have been done, including pressure angle and radius of curvature, before this option can be used. The minimum radius of curvature is displayed and the roller follower diameter is requested if that choice had been made in steps 6 and 7 above. Otherwise the minimum width of the flat follower is reported to the screen, and the cam is drawn.

Note in Figure 9-55 that the curved arrow indicates the direction of cam rotation (fixed as *ccw* in the program). The initial position of a roller follower at cam angle $\theta = 0$ is shown as a filled circle with rectangular stem, and a flat-faced follower as a filled rectangle. Any eccentricity would show as a shift up or down of the filled follower with respect to the X axis through the cam center. The smallest filled circle on the cam centerline represents the camshaft. The smallest unfilled circle is the base circle. The prime circle is somewhat obscured by the repeated drawings of the follower diameter, which sweep out the cam surface. The pitch curve is drawn along the locus of the roller follower centers. The radial lines that form pieces of pie within the base circle, represent the segments of the cam. They are numbered **clockwise** around the circumference! If the cam turns counterclockwise, the first segment (here a rise) is in the **fourth** quadrant. The sidebar contains the user name, date, time, design number, and pertinent data on the cam design, similar to the sidebar in the linkage programs.

10 - Help

Limited help in program operation is available from this selection. The help submenu is shown as Figure 8-23 on p. 287. It should be self-explanatory. The last three items allow the program configuration parameters to be reset while the program is running. For example, when running on a VGA equipped computer with color monitor, you will want to see the plots in high resolution color on the screen. But, unless you have DOS 4.0 or higher you cannot dump these screens to a printer. So, change the screen resolution to CGA and replot the data. Then do a SHIFT PRINTSCRN to get the hard copy from the printer and switch back to VGA from the help menu to have color screen plots.

Even with DOS 4.0 which can dump EGA and VGA plots to the printer you will have to replot in monochrome with white lines on a black background to get a noninverted screen dump. Whenever you change these parameters from the help menu, the configuration file will be rewritten to disk, and the next time you start the program it will "wake up" in the last mode it was in when shut down.

If you have trouble with your printer executing a graphics dump of a plot, make sure that the GRAPHICS.COM program has been loaded (run) at startup of the computer. Also, try plotting in the lower resolution CGA mode to make screen dumps as some printers may have trouble handling the higher resolution screens.

<ret> Quit

Hitting the return key will bring up a request for confirmation of your decision to exit the program. Only a Y or y response will cause a termination of execution. All data in memory will be lost on exit unless it was previously saved to a diskfile.

9.11 REFERENCES

1   **Roark, R.**, and **Young, W.**, *Formulas for Stress and Strain*, 5th ed. McGraw-Hill Inc., New York, 1975, pp. 513-530.

2   **Norton, R. L.**, "Effect of Cam Manufacturing Method on Dynamic Performance," *Mechanism and Machine Theory*, Pergamon Press, London, vol. 23, no. 3, 1988, pp. 191-208.

3   **Norton, R. L.**, "Analysis of the Effects of Manufacturing Methods and Heat Treatment on the Performance of Double-Dwell Cams," *Mechanism and Machine Theory*, Pergamon Press, London, vol. 23, no. 6, 1988, pp. 461-474.

9.12 PROBLEMS

Programs DYNACAM *and* MATRIX *may be used to solve these problems where appropriate.*

9-1   Design a double-dwell cam to move a follower from 0 to 2.5" in 60°, dwell for 120°, fall 2.5" in 30° and dwell for the remainder. The total cycle must take 4 seconds. Choose suitable programs for rise and fall to minimize accelerations. Plot the *s v a j* diagrams.

9-2   Design a double-dwell cam to move a follower from 0 to 1.5" in 45°, dwell for 150°, fall 1.5" in 90° and dwell for the remainder. The total cycle must take 6 seconds. Choose suitable programs for rise and fall to minimize velocities. Plot the *s v a j* diagrams.

9-3   Design a single-dwell cam to move a follower from 0 to 2" in 60°, fall 2" in 90° and dwell for the remainder. The total cycle must take 2 seconds. Choose suitable programs for rise and fall to minimize accelerations. Plot the *s v a j* diagrams.

9-4   Design a three dwell cam to move a follower from 0 to 2.5" in 40°, dwell for 100°, fall 1.5" in 90°, dwell for 20°, fall 1" in 30° and dwell for the remainder. The total cycle must take 10 seconds. Choose suitable programs for rise and fall to minimize velocities. Plot the *s v a j* diagrams.

9-5   Design a four dwell cam to move a follower from 0 to 2.5" in 40°, dwell for 100°, fall 1.5" in 90°, dwell for 20°, fall 0.5" in 30°, dwell for 40°, fall 0.5" in 30° and dwell for the remainder. The total cycle must take 15 seconds. Choose suitable programs for rise and fall to minimize accelerations. Plot the *s v a j* diagrams.

9-6   Size the cam from Problem 9-1 for a 1" radius roller follower considering pressure angle and radius of curvature. Use eccentricity only if necessary to balance those functions. Plot both those functions. Draw the cam profile. Repeat for a flat-faced follower. Which would you use?

9-7   Size the cam from Problem 9-2 for a 1.5" radius roller follower considering pressure angle and radius of curvature. Use eccentricity only if necessary to balance those functions. Plot both those functions. Draw the cam profile. Repeat for a flat-faced follower. Which would you use?

9-8   Size the cam from Problem 9-3 for a 0.5" radius roller follower considering pressure angle and radius of curvature. Use eccentricity only if necessary to balance those functions. Plot both those functions. Draw the cam profile. Repeat for a flat-faced follower. Which would you use?

9-9   Size the cam from Problem 9-4 for a 2" radius roller follower considering pres-
        sure angle and radius of curvature. Use eccentricity only if necessary to balance
        those functions. Plot both those functions. Draw the cam profile. Repeat for a
        flat-faced follower. Which would you use?

9-10   Size the cam from Problem 9-5 for a 0.5" radius roller follower considering pres-
        sure angle and radius of curvature. Use eccentricity only if necessary to balance
        those functions. Plot both those functions. Draw the cam profile. Repeat for a
        flat-faced follower. Which would you use?

9-11   A high friction, high inertia load is to be driven. We wish to keep peak velocity
        low. Combine segments of half period cycloidal displacements with a constant
        velocity segment on both rise and fall to reduce the maximum velocity below that
        obtainable with a full period modified sine acceleration alone (i.e., with no con-
        stant velocity portion). Rise is 1" in 90°, dwell for 60°, fall in 50°, dwell for remain-
        der. Compare the two designs and comment. Use an ω of one for comparison.

9-12   A constant velocity of 0.4 in/sec must be matched for 1.5 seconds. Then the follow-
        er must return to your choice of start point and dwell for 2 seconds. The total cycle
        time is 6 seconds. Design a cam for a follower radius of 0.75" and a maximum
        pressure angle of 30° absolute value.

9-13   A constant velocity of 0.25 in/sec must be matched for 3 seconds. Then the follow-
        er must return to your choice of start point and dwell for 3 seconds. The total
        cycle time is 12 seconds. Design a cam for a follower radius of 1.25" and a max-
        imum pressure angle of 35° absolute value.

9-14   A constant velocity of 2 in/sec must be matched for 1 second. Then the follower
        must return to your choice of start point. The total cycle time is 2.75 seconds. De-
        sign a cam for a follower radius of 0.5" and a maximum pressure angle of 25° ab-
        solute value.

9.13   PROJECTS

*These larger-scale project statements deliberately lack detail and structure and are loosely de-
fined. Thus, they are similar to the kind of "identification of need" or problem statement commonly encountered in engineering practice. It is left to the student to structure the problem
through background research and to create a clear goal statement and set of task specifications
before attempting to design a solution. This design process is spelled out in Chapter 1 and
should be followed in all of these examples. All results should be documented in a professional engineering report. (See the bibliography of Chapter 1 for information on report writing.)*

P9-1   A timing diagram for a halogen headlight filament insertion device is shown in
        the Figure P9-1. Four points are specified. Point A is the start of rise. At B the
        grippers close to grab the filament from its holder. The filament enters its socket
        at C and is fully inserted at D. The high dwell from D to E holds the filament sta-
        tionary while it is soldered in place. The follower returns to its start position
        from E to F. From F to A the follower is stationary while the next bulb is indexed
        into position. It is desirable to have low to zero velocity at point B where the grip-
        pers close on the fragile filament. The velocity at C should not be so high as to
        "bend the filament in the breeze." Design and size a complete cam-follower sys-
        tem to do this job.

Timing Diagram

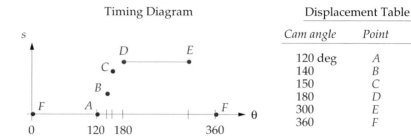

Displacement Table		
Cam angle	Point	s
120 deg	A	0
140	B	2
150	C	3
180	D	3.5
300	E	3.5
360	F	0

FIGURE P9-1

Data for cam design project

P9-2 A cam-driven pump to simulate human aortic pressure is needed to serve as a consistent, repeatable pseudo-human input to the operating room computer monitoring equipment, in order to test it daily. Figure P9-2 shows a typical aortic pressure curve and a pump pressure-volume characteristic. Design a cam to drive the piston and give as close an approximation to the aortic pressure curve shown as can be obtained without violating the fundamental law of cam design. Simulate the dichrotic notch as best you can.

P9-3 An athletic footwear manufacturer wants a device to test rubber heels for their ability to withstand millions of cycles of force similar to that which a walking human's foot applies to the ground. Figure P9-3 shows a typical walker's force-time function and a pressure-volume curve for a piston-accumulator. Design a cam-follow-

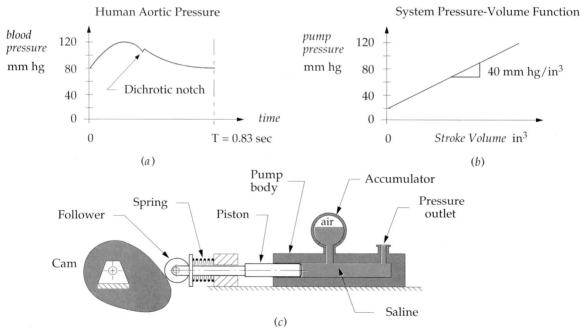

FIGURE P9-2

Data for cam design project

er system to drive the piston in a way that will create a force-time function on the heel similar to the one shown. Choose suitable piston diameters at each end.

P9-4 A fluorescent light bulb production machine moves 5500 lamps per hour through a 550°C oven on a chain conveyor which is in constant motion. The lamps are on 2 inch centerlines. The bulbs must be sprayed internally with a tin oxide coating as they leave the oven, still hot. This requires a cam driven device to track the bulbs at constant velocity for the 0.5 second required to spray them. The spray guns will fit on a 6 x 10 inch table. The spray creates hydrochloric acid, so all exposed parts must be resistant to that environment. The spray head transport device will be driven from the conveyor chain by a shaft having a 28-tooth sprocket in mesh with the chain. Design a complete spray gun transport assembly to these specifications.

P9-5 A 30-foot-tall drop tower is being used to study the shape of water droplets as they fall through air. A camera is to be carried by a cam-operated linkage which will track the droplet's motion from the 9-foot to the 10-foot point in its fall (measured from release point at the top of the tower). The drops are released every 1/2 second. Every drop is to be filmed. Design a cam and linkage which will track these droplets, matching their velocities and accelerations in the 1-foot filming window.

P9-6 A device is needed to accelerate a 3000 lb. vehicle into a barrier with constant velocity, to test its 5 MPH bumpers. The vehicle will start at rest, move forward, and have constant velocity for the last part of its motion before striking the barrier with the specified velocity. Design a cam-follower system to do this. The vehicle will leave contact with your follower just prior to the crash.

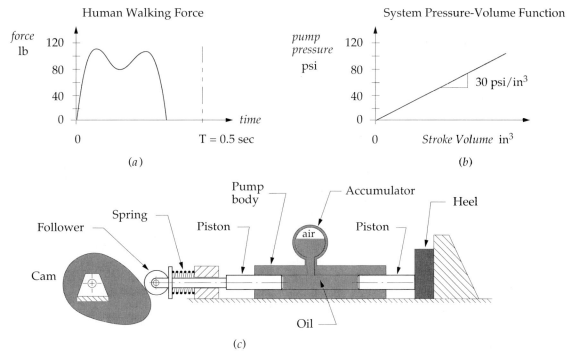

FIGURE P9-3

Data for cam design project

Chapter 10

GEAR TRAINS

Cycle and epicycle,
orb in orb
JOHN MILTON, PARADISE LOST

10.0 INTRODUCTION

Gear trains are widely used in all kinds of mechanisms and machines, from can openers to aircraft carriers. Whenever a change in the speed or torque of a rotating device is needed, a gear train or one of its cousins, the belt or chain drive mechanism, will usually be used. This chapter will explore the theory of gear tooth action and the design of these ubiquitous devices for motion control. The calculations involved are trivial compared to those for cams or linkages. The shape of gear teeth has become quite standardized for good kinematic reasons which we will explore.

Gears of various sizes and styles are readily available from many manufacturers. Assembled gearboxes for particular ratios are also stock items. The kinematic design of gear trains is principally involved with the selection of appropriate ratios and gear diameters. A complete gear train design will necessarily involve considerations of strength of materials and the complicated stress states to which gear teeth are subjected. This text will not deal with the stress analysis aspects of gear design. There are many texts which do. Some are listed in the bibliography at the end of this chapter. This chapter will discuss the kinematics of gear tooth theory, gear types, and the kinematic design of gearsets and gear trains of simple, compound, reverted, and epicyclic types. Chain and belt drives will also be discussed. Examples of the use of these devices will be presented as well.

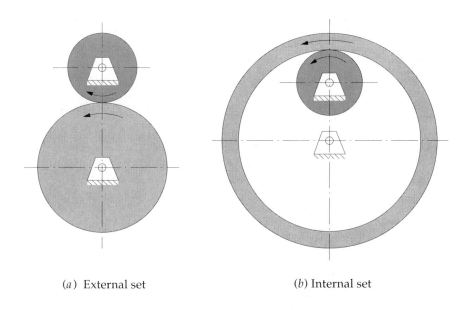

(*a*) External set   (*b*) Internal set

FIGURE 10-1

Rolling cylinders

10.1 ROLLING CYLINDERS

The simplest means of transferring rotary motion from one shaft to another is a pair of rolling cylinders. They may be an external set of rolling cylinders as shown in Figure 10-1a or an internal set as in Figure 10-1b. Provided that sufficient friction is available at the rolling interface, this mechanism will work quite well. There will be no slip between the cylinders until the maximum available frictional force at the joint is exceeded by the demands of torque transfer.

A variation on this mechanism is what causes your car or bicycle to move along the road. Your tire is one rolling cylinder and the road the other (very large radius) one. Friction is all that prevents slip between the two, and it works well unless the friction coefficient is reduced by the presence of ice or other slippery substances. In fact, some early automobiles had rolling cylinder drives inside the transmission, as do some present-day snowblowers and garden tractors which use a rubber-coated wheel rolling against a steel disk to transmit power from the engine to the wheels.

A variant on the rolling cylinder drive is the flat or vee belt as shown in Figure 10-2. This mechanism also transfers power through friction and is capable of quite large power levels, provided enough belt cross section is provided. Friction belts are used in a wide variety of applications from small sewing machines to the alternator drive on your car, to multihorsepower generators and pumps. Whenever absolute phasing is not required and power levels are moderate, a friction belt drive may be the best choice. They are relatively quiet running, require no lubrication, and are inexpensive compared to gears and chain drives.

FIGURE 10-2

A two-groove vee belt drive - *Courtesy of T. B. Wood's Sons Co., Chambersburg Pa.*

Both rolling cylinders and belt (or chain) drives have effective linkage equivalents as shown in Figure 10-3. These effective linkages are valid only for one instantaneous position but nevertheless show that these devices are just another variation of the fourbar linkage in disguise.

The principal drawbacks to the rolling cylinder drive (or smooth belt) mechanism are its relatively low torque capability and the possibility of slip. Some drives require absolute phasing of the input and output shafts for timing purposes. A common example is the valve train drive in an automobile engine. The valve cams must be kept in phase with the piston motion or the engine will not run properly. A smooth belt or rolling cylinder drive from crankshaft to camshaft would not guarantee correct phasing. In this case some means of preventing slip is needed.

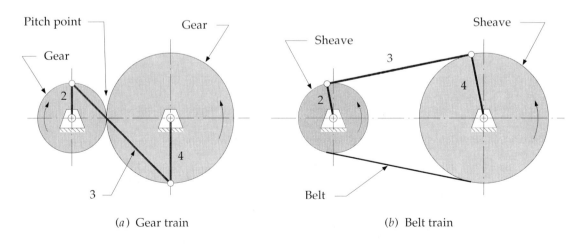

(*a*) Gear train   (*b*) Belt train

FIGURE 10-3

Gear and belt trains each have an equivalent fourbar linkage for any instantaneous position

This usually means adding some meshing teeth to the rolling cylinders. They then become gears as shown in Figure 10-4 and are together called a *gearset*. When two gears are placed in mesh to form a gearset such as this one, it is conventional to refer to the smaller of the two gears as the *pinion* and to the other as the *gear*.

10.2   THE FUNDAMENTAL LAW OF GEARING

Conceptually, teeth of any shape will do to prevent gross slip. Old water-powered mills and windmills used wooden gears whose teeth were merely round wooden pegs stuck into the rims of the cylinders. Even ignoring the crudity of construction of these early examples of gearsets, there was no possibility of smooth velocity transmission because the geometry of the tooth "pegs" violated the **fundamental law of gearing** which states that *the angular velocity ratio between the gears of a gearset must remain constant throughout the mesh*. The angular velocity ratio (*VR*) referred to in this law is the same one that we derived for the fourbar linkage in Section 6.4 and equation 6.10. It is equal to the ratio of the radius of the input gear to that of the output gear.

$$VR = \frac{\omega_{out}}{\omega_{in}} = \pm \frac{r_{in}}{r_{out}} \qquad (10.1a)$$

$$MV = \frac{\omega_{in}}{\omega_{out}} = \pm \frac{r_{out}}{r_{in}} \qquad (10.1b)$$

The **torque ratio** or **mechanical advantage** (*MV*) was shown in equation 6.13 to be the reciprocal of the velocity ratio (*VR*); thus a gearset is essentially a device to exchange torque for velocity or vice versa. The most common application is to reduce velocity and increase torque to drive heavy loads as in your automobile transmission. Other applications require an increase in velocity, for which a reduction in torque must be accepted. In either case, it is usually desirable to maintain a constant ratio between the gears as they rotate. Any variation in ratio will show up as oscillation in the output velocity and torque even if the input is constant with time.

The radii in equation 10.1 are those of the rolling cylinders to which we are adding the teeth. The positive or negative sign accounts for internal or external cylinder sets as defined in Figure 10-1. An external set reverses the direction of rotation between the cylinders and requires the negative sign. An internal gearset or a belt or chain drive will have the same direction of rotation on input and output shafts and require the positive sign in equation 10.1. The surfaces of the rolling cylinders will become the **pitch circles,** and their diameters the **pitch diameters** of the gears. The contact point between the cylinders lies on the line of centers as shown in Figure 10-3, and this point is called the **pitch point**.

In order for the fundamental law of gearing to be true, the gear tooth contours on mating teeth must be conjugates of one another. There is an infinite number of possible conjugate pairs that could be used, but only a few curves have seen practical application as gear teeth. The **cycloid** still is used as a tooth form in watches and clocks, but most other gears use the **involute** curve for their shape.

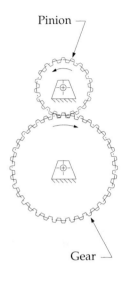

Pinion

Gear

FIGURE 10-4

An external gearset

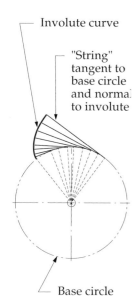

Involute curve

"String" tangent to base circle and normal to involute

Base circle

The Involute Tooth Form

The involute is a curve which can be generated by unwrapping a taut string from a cylinder as shown in Figure 10-5. Note the following about this involute curve:

The string is always tangent to the cylinder.

The center of curvature of the involute is always at the point of tangency of the string with the cylinder.

A tangent to the involute is then always normal to the string, which is the instantaneous radius of curvature of the involute curve.

Figure 10-6 shows two involutes on separate cylinders in contact or "in mesh." These represent gear teeth. The cylinders from which the strings are unwrapped are called the **base circles** of the respective gears. Note that the base circle is necessarily smaller than the pitch circle, as the gear tooth must project both below and above the surface (pitch circle) of the original rolling cylinder and the *involute only exists outside of the base circle.*

The geometry at this tooth-tooth interface is similar to that of a cam-follower joint as was defined in Figure 9-48 on p. 355. There is a common tangent to both curves at the contact point, and a common normal, perpendicular to the common tangent. Note that the common normal is, in fact, the "strings" of both involutes, which are colinear. Thus the **common normal**, which is also the **axis of transmission**, always passes through the pitch point regardless of where in the mesh the two teeth are contacting. Figure 10-7 shows the same two involute tooth

FIGURE 10-5

Development of the involute curve

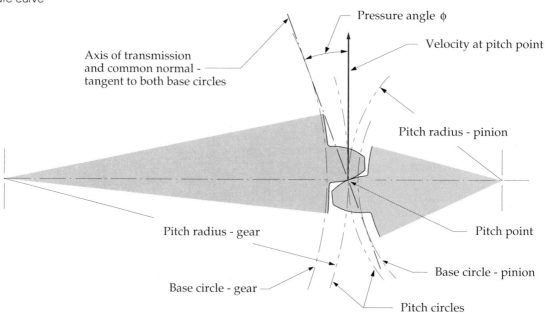

Pressure angle ϕ

Velocity at pitch point

Axis of transmission and common normal - tangent to both base circles

Pitch radius - pinion

Pitch radius - gear

Pitch point

Base circle - gear

Base circle - pinion

Pitch circles

Contact geometry of involute teeth and pressure angle

forms in two other positions, just beginning contact and about to leave contact. The common normals of both these contact points still pass through the same pitch point. It is this property of the involute that causes it to obey the fundamental law of gearing. The ratio of the driving gear radius to the driven gear radius remains constant as the teeth move into and out of mesh.

From this observation of the behavior of the involute we can restate the **fundamental law of gearing** in a more kinematically formal way as: *the common normal of the tooth profiles, at all contact points within the mesh, must always pass through a fixed point on the line of centers, called the pitch point.* The gearset's velocity ratio will then be a constant defined by the ratio of the respective radii of the gears to the pitch point.

Changing Center Distance

When involute teeth (or any teeth) have been cut into a cylinder, with respect to a particular base circle, to create a single gear, we do not yet have a pitch circle. The pitch circle only comes into being when we mate this gear with another to create a pair of gears, or gearset. There will be some range of center-to-center distances over which we can achieve a mesh between the gears. There will also be an ideal center distance that will give us the nominal pitch diameters for which the gears were designed. However, the limitations of the manufacturing process give a low probability that we will be able to exactly achieve this ideal center distance in every case. More likely, there will be some error in the center distance, even if small.

What will happen to the adherence to the fundamental law of gearing if there is error in the location of the gear centers? If the gear tooth form is **not** an involute, then an error in center distance will violate the fundamental law, and there will be variation, or "ripple," in the output velocity. The output angular velocity will not be constant for a constant input velocity. However, **with an involute tooth form,**

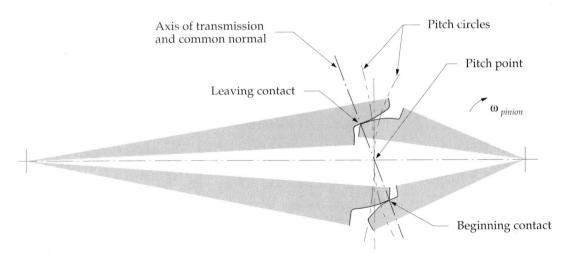

Axis of transmission
and common normal

Pitch circles

Pitch point

Leaving contact

ω_{pinion}

Beginning contact

FIGURE 10-7

All contact points in the mesh lie on the common normal which passes through the pitch point with involute teeth *center distance errors do not affect the velocity ratio.* This is the principal advantage of the involute over all other possible tooth forms and the reason why it is nearly universally used for gear teeth. Figure 10-8 shows what happens when the center distance is varied on an involute gearset. Note that the common normal still goes through a pitch point, common to all contact points within the mesh. Only the pressure angle is affected by the change in center distance.

Pressure Angle

The **pressure angle** in a gearset is defined exactly as it was for the cam and follower. It is the angle between the axis of transmission (common normal) and the direction of velocity at the point of contact (the pitch point) and is shown in Figure 10-6.

Figure 10-8 also shows the pressure angles for two different center distances. As the center distance increases, so will the pressure angle and vice versa. This is one result of a change, or error, in center distance when using involute teeth.

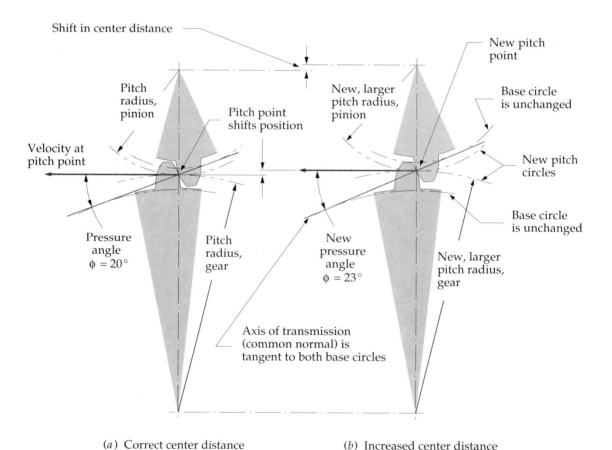

(*a*) Correct center distance   (*b*) Increased center distance

FIGURE 10-8

Changing center distance of involute gears changes only the pressure angle and pitch diameters

Note that the fundamental law of gearing still holds in the modified center distance case. The common normal is still tangent to the two base circles and still goes through the pitch point. The pitch point has moved, but in proportion to the move of the center distance and the gear radii. The velocity ratio is unchanged despite the shift in center distance. In fact, the velocity ratio of involute gears is fixed by the ratio of the base circle diameters, which are unchanging once the gear is cut.

The pressure angles of gearsets are standardized at a few values by the gear manufacturers. These are defined at the nominal center distance for the gearset as cut. The standard values are 14.5, 20 and 25° with 20° being the most commonly used. Any custom pressure angle can be made, but its expense over the available stock gears with standard pressure angles would be hard to justify. Special cutters would have to be made. Gears to be run together must be cut to the same nominal pressure angle.

Backlash

Another factor affected by changing center distance (*CD*) is backlash. Increasing the *CD* will increase the backlash and vice versa. **Backlash** is defined as *the clearance between mating teeth measured along the circumference of the pitch circle.* Manufacturing tolerances preclude a zero clearance, as all teeth cannot be exactly the same dimensions, and all must mesh. So, there must be some small difference between the tooth thickness and the space width, which are shown in Figure 10-9. As long as the gearset is run with a nonreversing torque, the backlash should not be a problem. But, whenever the torque changes sign, the teeth will move from contact on one side to the other. The backlash gap will be traversed and the teeth will impact with noticeable noise. This is the same phenomenon as crossover shock in the form-closed cam. As well as increasing stresses and wear, backlash can cause undesirable positional error in some applications.

In servomechanisms, where motors are driving, for example, the control surfaces on an aircraft, backlash can cause potentially destructive "hunting" in which the control system tries in vain to correct the positional errors due to the backlash "slop" in the mechanical drive system. Such applications need **anti-backlash gears** which are really two gears back-to-back on the same shaft which can be rotated slightly at assembly with respect to one another, and then fixed so as to take up the backlash. In less critical applications, such as the propellor drive on a boat, backlash on reversal will not even be noticed.

The *American Gear Manufacturers Association* (**AGMA**), defines standards for gear design and manufacture. They define a spectrum of quality and tolerances ranging from the lowest (3) to the highest precision (16). Obviously the cost of the gear will be a function of this quality index.

10.3   GEAR TOOTH NOMENCLATURE

Figure 10-9 shows two teeth of a gear with the standard nomenclature defined. **Pitch circle** and **base circle** have been defined above. The tooth height is defined by the **addendum** (*added on*) and the **dedendum** (*subtracted from*) which are referenced to the

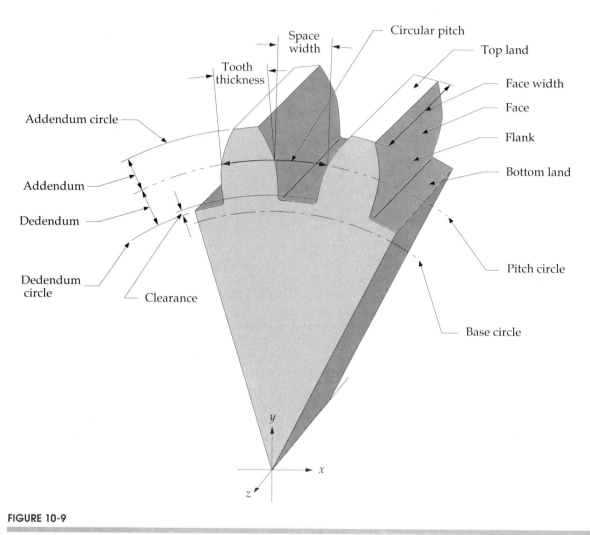

FIGURE 10-9

Gear tooth nomenclature

nominal pitch circle. The dedendum is slightly larger than the addendum to provide a small amount of **clearance** between the tip of one mating tooth (**addendum
circle**) and the bottom of the tooth space of the other (**dedendum circle**). The
tooth thickness is measured at the pitch circle, and the tooth **space width** is
slightly larger than the tooth thickness. The difference between these two dimensions is the **backlash**. The **face width** of the tooth is measured along the axis of
the gear. The **circular pitch** is the arc length along the pitch circle circumference
measured from a point on one tooth to the same point on the next. The circular
pitch defines the tooth size. The other tooth dimensions are standardized based
on that dimension as shown in Table 10-1. The definition of **circular pitch** p_c is:

$$p_c = \frac{\pi d}{N}$$

where: (10.2)

$$d = \text{pitch diameter}$$
$$N = \text{number of teeth}$$

The units of p_c are inches or millimeters. A more convenient way to define tooth size is to relate it to the diameter of the pitch circle rather than its circumference. The **diametral pitch** p_d is:

$$p_d = \frac{N}{d}$$ (10.3)

The units of p_d are reciprocal inches, or number of teeth per inch. This measure is only used in U.S. specification gears. The SI system, used for metric gears, defines a parameter called the **module** which is *the reciprocal of diametral pitch* with pitch diameter measured in millimeters.

$$m = \frac{d}{N}$$ (10.4)

The units of the module are millimeters. Unfortunately, metric gears are not interchangeable with U.S. gears, despite both being involute tooth forms, as their standards for tooth sizes are different. In the United States, gear tooth sizes are specified by diametral pitch. The velocity ratio of the gearset can be put into a more convenient form by substituting equation 10.3 into equation 10.1, noting that the diametral pitch of meshing gears must be the same.

$$VR = \pm \frac{r_{in}}{r_{out}} = \pm \frac{d_{in}}{d_{out}} = \pm \frac{N_{in}}{N_{out}}$$ (10.5)

TABLE 10-1   AGMA Full-Depth Gear Tooth Specifications

Quantity	Coarse Pitch ($p_d < 20$)	Fine Pitch ($p_d \geq 20$)
Pressure angle ϕ	20° or 25°	20°
Addendum a	$1.000/p_d$	$1.000/p_d$
Dedendum d	$1.250/p_d$	$1.250/p_d$
Working depth h_k	$2.000/p_d$	$2.000/p_d$
Min. whole depth h_t	$2.250/p_d$	$(2.200/ p_d) + 0.002$ in
Circular tooth thickness t	$1.571/p_d$	$1.571/p_d$
Fillet radius of basic rack r_f	$0.300/p_d$	Not standardized
Min. basic clearance c	$0.250/p_d$	$(0.200/p_d) + 0.002$ in
Clearance (shaved or ground teeth) c	$0.350/p_d$	$(0.350/p_d) + 0.002$ in
Min. width of top land t_o	$0.250/p_d$	Not standardized

Thus the **velocity ratio** can be computed from the number of teeth on the meshing gears, which are integers. Note that a minus sign implies an external gearset and a positive sign an internal gearset as shown in Figure 10-1on p. 395. Table 10-1 shows the definitions of dimensions of standard, full-depth gear teeth as defined by the AGMA.

10.4   INTERFERENCE AND UNDERCUTTING

The involute tooth form is only defined outside of the base circle. In some cases, the dedendum will be large enough to extend below the base circle. If so, then the portion of tooth below the base circle will not be an involute and will interfere with the tip of the tooth on the mating gear, which is an involute. If the gear is cut with a standard gear shaper or a "hob," the cutting tool will also interfere with the portion of tooth below the base circle and will cut away the interfering material. This results in an undercut tooth as shown in Figure 10-10. This undercutting weakens the tooth by removing material at its root. The maximum moment and maximum shear from the tooth loaded as a cantilever beam both occur in this region. Severe undercutting will promote early tooth failure.

This undercutting problem can be prevented simply by avoiding gears with too few teeth. If a gear has a large number of teeth, they will be small compared to its diameter. As the number of teeth is reduced for a fixed diameter gear, the teeth must become larger. At some point, the dedendum will exceed the radial distance between the base circle and the pitch circle, and interference will occur. Table 10-2 shows the minimum number of teeth required for no undercutting as a function of pressure angle and also the minimum number usually recommended for an acceptable level of undercutting.

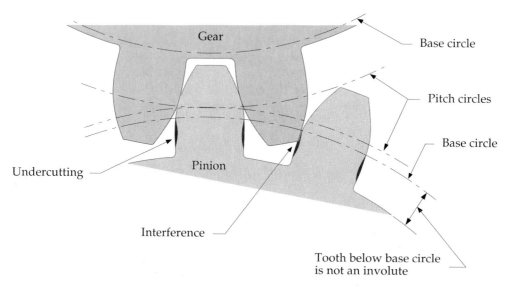

FIGURE 10-10

Interference and undercutting

TABLE 10-2   Minimum Number of Teeth to Avoid Undercutting

Pressure Angle	Min. for No Undercut	Min. Recommended
14.5	32	20
20	18	14
25	12	10

Spur, Helical, and Herringbone Gears

SPUR GEARS are ones in which the *teeth are parallel to the axis of the gear.* This is the simplest and least expensive form of gear to make. Spur gears can only be meshed if their axes are parallel. Figure 10-11 shows a spur gear.

HELICAL GEARS are ones in which the teeth are at a helix angle ψ with respect to the axis of the gear as shown in Figure 10-12. Figure 10-13 shows a pair of opposite-hand **helical gears** in mesh. Their axes are parallel. Two **crossed helical gears** of the same hand can be meshed with their axes at an angle as shown in Figure 10-14. The helix angles can be designed to accommodate any skew angle between the nonintersecting shafts.

Helical gears are more expensive than spur gears but offer some advantages. They run quieter than spur gears because of the smoother and more gradual contact between their angled surfaces as the teeth come into mesh. Spur gear teeth mesh along their entire face width at once. The sudden impact of tooth on tooth causes vibrations which are heard as a "whine" which is characteristic of spur gears but is absent with helicals. Also, for the same gear diameter and diametral pitch, a helical gear is stronger due to the slightly thicker tooth form in a plane perpendicular to the axis of rotation.

HERRINGBONE GEARS are formed by joining two helical gears of identical pitch and diameter but of opposite hand on the same shaft. These two sets of teeth are often cut on the same gear blank. The advantage compared to a helical gear is the internal cancellation of its axial thrust loads since each "hand" half of the herring-

FIGURE 10-11

A spur gear

Courtesy of Martin Sprocket and Gear Co. Arlington, Texas

FIGURE 10-13

Parallel helical gears

Courtesy of Boston Gear, Division of IMO Industries, Quincy, Mass.

(*a*) Herringbone gear

Helix angle
ψ

(*b*) Helical gear

FIGURE 10-12

A helical gear and a herringbone gear

FIGURE 10-14

Crossed helical gears

*Courtesy of the Boston
Gear Division of IMO
Industries, Quincy, Mass.*

FIGURE 10-15

A worm and worm gear
(or worm wheel)

*Courtesy of Martin
Sprocket and Gear Co.
Arlington, Texas*

bone gear has an oppositely directed thrust load. Thus no thrust bearings are needed other than to locate the shaft axially. This type of gear is much more expensive than a helical gear and tends to be used in large, high-power applications such as ship drives, where the frictional losses from axial loads would be prohibitive. A herringbone gear is shown in Figure 10-12. Its face view is the same as the helical gear's.

EFFICIENCY   A spur gearset can be 98 to 99% efficient. The helical gearset is less efficient than the spur gearset due to sliding friction along the helix angle. They also present a reaction force along the axis of the gear, which the spur gear does not. Thus helical gearsets must have thrust bearings as well as radial bearings on their shafts to prevent them from pulling apart along the axis. Some friction losses occur in the thrust bearings as well. A parallel helical gearset will be about 96 to 98% efficient and a crossed helical set only 50 to 90% efficient. The parallel helical set (opposite hand but same helix angle) has line contact between the teeth and can handle high loads at high speeds. The crossed helical set has point contact and a large sliding component which limit its application to light load situations.

If the gearsets have to be shifted in and out of mesh while in motion, spur gears are a better choice than helical, as the helix angle interferes with the axial shifting motion. (Herringbone gears of course cannot be axially disengaged.) Truck transmissions often use spur gears for this reason, whereas automobile (standard) transmissions use helical, constant mesh gears for quiet running, and have a synchromesh mechanism to allow shifting. These transmission applications will be described in a later section.

Worms and Worm Gears

If the helix angle is increased sufficiently, the result will be a **worm**, which has only one tooth wrapped continuously around its circumference a number of times, analogous to a screw thread. This worm can be meshed with a special **worm gear** (or **worm wheel**), whose axis is perpendicular to that of the worm as shown in Figure 10-15. Because the driving worm typically has only one tooth, the ratio of the gearset is equal to one over the number of teeth on the worm gear (see equation 10.5). These teeth are not involutes over their entire face which means that the center distance must be maintained accurately to guarantee conjugate action.

Worms and wheels are made and replaced as matched sets. These worm gearsets have the advantage of very high gear ratios in a small package and can carry very high loads especially in their single or double enveloping forms. **Single enveloping** means that the worm gear teeth are wrapped around the worm. **Double enveloping** sets also wrap the worm around the gear, resulting in an hourglass-shaped worm. Both of these techniques increase the surface area of contact between worm and wheel, increasing load carrying capacity and also cost. One tradeoff in any wormset is very high sliding and thrust loads which make the wormset rather inefficient at 40 to 85% efficiency.

Perhaps the major advantage of the wormset is that it can be designed to be impossible to **backdrive**. A spur or helical gearset can be driven from either shaft, as a velocity step-up or step-down device. While this may be desirable in many cases, if the load being driven must be held in place after the power is shut off, the spur or helical gearset will not do. They will "backdrive." This makes them unsuitable for such applications as a jack to raise a car unless a brake is added to the design to hold the load. The wormset, on the other hand, can only be driven from the worm. The friction can be large enough to prevent it being backdriven from the worm wheel. Thus it can be used without a brake in load-holding applications such as jacks and hoists.

Rack and Pinion

If the diameter of the base circle of a gear is increased without limit, the base circle will become a straight line. If the "string" wrapped around this base circle to generate the involute were still in place after the base circle's enlargement to an infinite radius, the string would be pivoted at infinity and would generate an involute that is a straight line. This linear gear is called a **rack**. Its teeth are trapezoids, yet are true involutes. This fact makes it easy to create a cutting tool to generate involute teeth on circular gears, by accurately machining a rack and hardening it to cut teeth in other gears. Rotating the gear blank with respect to the rack cutter while moving the cutter axially back and forth across the gear blank will shape, or develop, a true involute tooth on the circular gear.

Figure 10-16 shows a **rack and pinion**. The most common application of this device is in rotary to linear motion conversion or vice versa. It can be back-driven, so it requires a brake if used to hold a load. An example of its use is in **rack-and-pinion steering** in automobiles. The pinion is attached to the bottom end of the steering column and turns with the steering wheel. The rack meshes with the pinion and is free to move left and right in response to your angular input at the steering wheel. The rack is also one link in a multibar linkage which converts the linear translation of the rack to the proper amount of angular motion of a rocker link attached to the front wheel assembly to steer the car.

Bevel and Hypoid Gears

BEVEL GEARS For right-angle drives, crossed helical gears or a wormset can be used. For any angle between the shafts, including 90°, bevel gears may be the solution. Just as spur gears are based on rolling cylinders, **bevel gears** are based on rolling cones as shown in Figure 10-17. The angle between the axes of the cones and the included

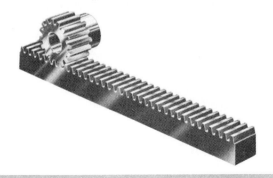

FIGURE 10-16

A rack and pinion - *Courtesy of Martin Sprocket and Gear Co., Arlington, Texas*

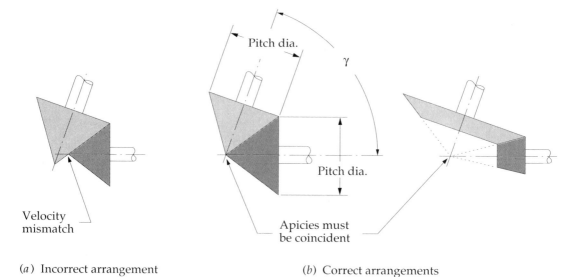

(a) Incorrect arrangement   (b) Correct arrangements

FIGURE 10-17

Bevel gears are based on rolling cones

angles of the cones can be any compatible values as long as the apices of the cones intersect. If they did not intersect, there would be a mismatch of velocity at the interface. The apex of each cone has zero radius, thus zero velocity. All other points on the cone surface will have nonzero velocity. The velocity ratio of the bevel gears is defined by equation 10.1 using the pitch diameters at any common point of intersection of cone diameters.

SPIRAL BEVEL GEARS If the teeth are parallel to the axis of the gear, it will be a straight bevel gear as shown in Figure 10-18. If the teeth are angled with respect to the axis, it will be a **spiral bevel gear** (see Figure 10-19), analogous to a helical gear. The cone axes and apices must intersect in both cases. The advantages and disadvantages of straight bevel and spiral bevel gears are similar to those of the spur gear and helical gear, respectively, regarding strength, quietness, and cost. Bevel gear teeth are not involutes but are based on an "octoid" tooth curve. They must be purchased or replaced in pairs (gearsets) as they are not universally interchangeable, and their center distances must be accurately maintained.

HYPOID GEARS If the axes between the gears are nonparallel and also nonintersecting, bevel gears cannot be used. **Hypoid gears** will accommodate this geometry. Hypoid gears are based on rolling hyperboloids of revolution as shown in Figure 10-20. (The term hypoid is a contraction of hyperboloid.) The tooth form is not an involute. These hypoid gears are used in the final drive of front-engined, rear wheel drive automobiles, in order to lower the axis of the driveshaft below the center of the rear axle to reduce the "driveshaft hump" in the back seat.

FIGURE 10-18

Straight bevel gears

Courtesy of Martin Sprocket and Gear Co., Arlington, Texas

Noncircular Gears

Noncircular gears are based on the rolling centrodes of a Grashof double-crank four-bar linkage. Centrodes are the loci of the instant centers of the linkage and were described in Section 6.5. Figure 6-15b on p. 209 shows a pair of centrodes that could be used for noncircular gears. Teeth would be added to their circumferences in the same way that we add teeth to rolling cylinders for circular gears. The teeth then act to guarantee no slip. Of course, the velocity ratio of noncircular gears is not constant. That is their purpose, to provide a time-varying output function in response to a constant velocity input. Their instantaneous velocity ratio is defined by equation 6.11b. These devices are used in a variety of rotating machinery such as printing presses where variation in the angular velocity of rollers is required on a cyclical basis.

FIGURE 10-19

Spiral bevel gears

Courtesy of the Boston Gear Division of IMO Industries, Quincy, Mass.

Belt and Chain Drives

VEE BELTS A **vee belt** drive is shown in Figure 10-2 on p. 396. Vee belts are made of elastomers (synthetic rubber) reinforced with synthetic or metallic cords for strength. The pulleys, or *sheaves*, have a matching vee-groove which helps to grip the belt as belt tension jams the belt into the vee. Flat belts running on flat and crowned pulleys are still used in some applications as well. As discussed above, slip is possible with this device and phasing cannot be guaranteed.

SYNCHRONOUS (TIMING) BELTS The **synchronous belt** solves the phasing problem while retaining the advantages of quiet running of vee belts and can cost less than gears or chains. Figure 10-21a shows a synchronous (or toothed) belt and its special gearlike pulleys or sheaves. These belts are made of a rubberlike material but

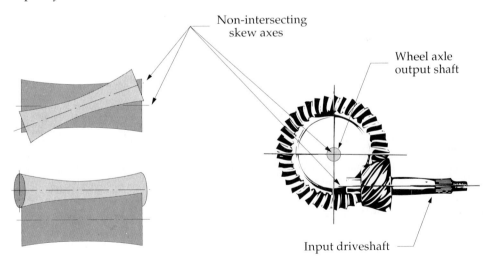

(*a*) Rolling hyperboloids of revolution

(*b*) Automotive hypoid final drive gears
Courtesy General Motors Co., Detroit MI

FIGURE 10-20

Hypoid gears are based on hyperboloids of revolution

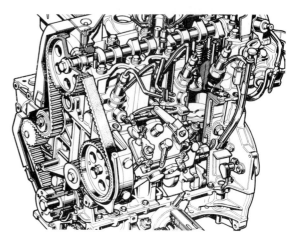

(a) Standard synchronous belt

Courtesy of T. B. Wood's Sons Co.,
Chambersburg PA

(b) Engine valve cam drive

Courtesy of Chevrolet Division,
General Motors Co., Detroit MI

FIGURE 10-21

Toothed synchronous belts and sprockets

are reinforced with steel or synthetic cords for higher strength and have molded-in teeth which fit in the grooves of the pulleys for positive drive. They are capable of fairly high torque and power transmission levels and are frequently used to drive automotive engine camshafts as shown in Figure 10-21b. They are more expensive than conventional vee belts. Manufacturers' catalogs provide detailed information on sizing both vee and synchronous belts for various applications.

CHAIN DRIVES are often used for applications where positive drive (phasing) is needed and large torque requirements or high temperature levels preclude the use of timing belts. When the input and output shafts are far apart, chain drive may be the most economical choice. Conveyor systems often use chain drives to carry the work along the assembly line. Steel chain can be run through many (but not all) hostile chemical or temperature environments. Many types and styles of chain have been designed for various applications ranging from the common roller chain as used on your bicycle or motorcycle, to more expensive "silent chain" designs used for camshaft drives in expensive automobile engines. Figure 10-22 shows a two-row roller chain used for a slightly different task, as a coupling for two in-line shafts. Figure 10-23 shows a typical sprocket for roller chain. Note that the sprocket teeth are not the same shape as gear teeth and are not involutes. The sprocket tooth shape is dictated by the need to match the contour of the portion of chain which nestles in the grooves. In this case the roller chain has cylindrical pins which engage the sprocket.

One unique limitation of chain drive is something called "**chordal action.**" The links of the chain form a set of chords when wrapped around the circumference of the sprocket. As these links enter and leave the sprocket, they impart a "jerky" motion to the driven shaft which causes some variation, or ripple, on the output

FIGURE 10-22

Roller chain used as a
shaft coupling

Courtesy of Martin
Sprocket and Gear Co.,
Arlington, Texas

velocity. Roller chain does not exactly obey the fundamental law of gearing. If very accurate, constant output velocity is required, a chain drive may not be the best choice.

10.6 SIMPLE GEAR TRAINS

A gear train is any collection of two or more meshing gears. A simple gear train is one in which each shaft carries only one gear, the most basic, two-gear example of which is shown in Figure 10-4. The *velocity ratio* (sometimes called *train ratio*) of this gearset is given by equation 10.1a. Figure 10-24 shows a simple gear train with five gears in series. Equation 10.6 derives the expression for the train's velocity ratio:

$$VR = \left(-\frac{N_2}{N_3}\right)\left(-\frac{N_3}{N_4}\right)\left(-\frac{N_4}{N_5}\right)\left(-\frac{N_5}{N_6}\right) = +\frac{N_2}{N_6} \tag{10.6}$$

Each gearset potentially contributes to the overall train ratio, but in any case of a simple (series) train, the numerical effects of all gears except the first and last cancel out. The train ratio of a simple train is always just the ratio of the first gear over the last. Only the sign of the overall ratio is affected by the intermediate gears which are called *idlers* because no power is typically taken from their shafts. If all gears in the train are external and there is an even number of gears in the train, the output direction will be opposite that of the input. If there is an odd number of external gears in the train, the output will be in the same direction as the input. Thus a single, external, idler gear *of any diameter* can be used to change the direction of the output gear without affecting its velocity.

A **single gearset** of spur, helical, or bevel gears *is usually limited to a ratio of about* **10:1** simply because the gearset will become very large, expensive, and hard to package above that ratio if the pinion is kept above the minimum numbers of teeth shown in Table 10-2. If the need is to get a larger train ratio than can be obtained with a single gearset, it is clear from equation 10.6 that the simple train will be of no help.

It is common practice to insert a single idler gear to change direction, but more than one idler is superfluous. There is little justification for designing a gear train as is shown in Figure 10-24. If the need is to connect two shafts which are far apart, a simple train of many gears could be used but will be much more expensive than a chain or belt drive for the same application. Gears are not cheap.

FIGURE 10-23

A roller chain sprocket

Courtesy of Martin Sprocket and Gear Co., Arlington, Texas

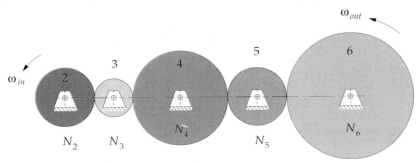

FIGURE 10-24

A simple gear train

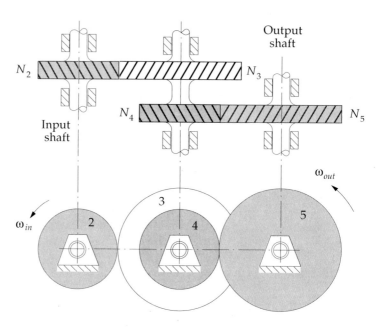

FIGURE 10-25

A compound gear train

10.7   COMPOUND GEAR TRAINS

To get a train ratio of greater than about 10:1 with spur, helical, or bevel gears (or any combination thereof) it is necessary to **compound the train** (unless an epicyclic train is used - see next section). A **compound train** *is one in which at least one shaft carries more than one gear.* This will be a parallel or series-parallel arrangement, rather than the pure series connections of the simple gear train. Figure 10-25 shows a compound train of four gears, two of which, gears 3 and 4, are fixed on the same shaft and thus have the same angular velocity.

The train ratio is now:

$$VR = \left(-\frac{N_2}{N_3} \right)\left(-\frac{N_4}{N_5} \right)$$ (10.7a)

This can be generalized for any number of gears in the train as:

$$VR = \pm \frac{\text{product of number of teeth on driver gears}}{\text{product of number of teeth on driven gears}}$$ (10.7b)

Note that these intermediate ratios do not cancel and the overall train ratio is the product of the ratios of parallel gearsets. Thus a larger ratio can be obtained in a compound gear train despite the approximately 10:1 limitation on individual gearset ratios. The plus or minus sign in equation 10.7b depends on the number and type of meshes in the train, whether external or internal. Writing the expression in the form of equation 10.7a and carefully noting the sign of each mesh ratio in the expression will result in the correct algebraic sign for the overall train ratio.

Design of Compound Trains

If presented with a completed design of a compound gear train such as that in Figure 10-25, it is a trivial task to apply equation 10.7 and determine the train ratio. It is not so simple to do the inverse, namely, design a compound train for a specified train ratio.

✎ EXAMPLE 10-1

Compound Gear Train Design

Problem: Design a compound train for an exact train ratio of 180:1. Find a combination of gears which will give that ratio.

Solution:

1   The first step is to determine how many stages, or gearsets, are necessary. Simplicity is the mark of good design, so try the smallest possibility first. Take the square root of 180, which is 13.42. So, two stages each of that ratio will give approximately 180:1. However, this is larger than our design limit of 10:1 for each stage, so try three stages. The cube root of 180 is 5.646, well within 10, so three stages will do.

2   If we can find some integer ratio of gear teeth which will yield 5.646:1, we can simply use three of them to design our gearbox. Using a lower limit of 12 teeth for the pinion and trying several possibilities we get the gearsets shown in Table 10-3 as possibilities.

3   The number of gear teeth must obviously be an integer. The closest to an integer in Table 10-3 is the 79.05 result. Thus a 79:14 gearset comes closest to the desired ratio. Applying this ratio to all three stages will yield a train ratio of $(79/14)^3 = 179.68{:}1$, which is close to 180:1. This may be an acceptable solution provided that the gearbox is not being used in a timing application. If the purpose of this gearbox is to step down the motor speed for a crane hoist, for example, an approximate ratio will be adequate.

4   Many gearboxes are used in production machinery to drive camshafts or linkages from a master driveshaft and must have the exact ratio needed or else the driven device will eventually get out of phase with the rest of the machine. If that were the case in this example, then the solution found above would not be good enough. We will need to

TABLE 10-3   Possible Gearsets for 180:1 Three-Stage Compound Train

Gearset Ratio	x	Pinion Teeth	=	Gear Teeth
5.646	x	12	=	67.75
5.646	x	13	=	73.40
5.646	x	14	=	79.05
5.646	x	15	=	84.69

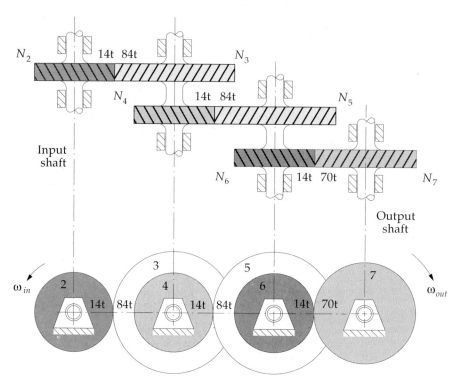

FIGURE 10-26

Three stage compound gear train for 1 to 180 train ratio

redesign it for exactly 180:1. Since our overall train ratio is an integer, it will be simplest to look for integer gearset ratios. Thus we need three integer factors of 180. The first solution above gives us a reasonable starting point in the cube root of 180, which is 5.65. If we round this up (or down) to an integer, we may be able to find a suitable combination.

5   Two compounded stages of 6:1 together give 36:1. Dividing 180 by 36 gives 5. Thus the stages shown in Table 10-4 are a possible exact solution.

This solution, shown in Figure 10-26, meets our design criteria. It has the correct, exact ratio, the stages are all less than 10:1 and no pinion has less than 14 teeth, which controls undercutting to an acceptable level.

TABLE 10-4   **Exact Solution for 180:1 Three-Stage Compound Train**

Gearset Ratio	x	Pinion Teeth	=	Gear Teeth
6	x	14	=	84
6	x	14	=	84
5	x	14	=	70

Design of Reverted Compound Trains

In the preceding example the input and output shaft locations are in different locations. This may well be acceptable or even desirable in some cases, depending on other packaging constraints in the overall machine design. Such a gearbox, whose *input and output shafts are not coincident*, is called a **nonreverted compound train**. In some cases, such as automobile transmissions, it is desirable or even necessary to have the *output shaft concentric with the input shaft*. This is referred to as "reverting the train" or "bringing it back onto itself." The design of a **reverted compound train** is more complicated because of the additional constraint that the center distances of the stages must be equal.

✒️ EXAMPLE 10-2

Reverted Gear Train Design

Problem: Design a reverted compound train for an exact train ratio of 18:1.

Solution:

1   Though it is not at all necessary to have integer gearset ratios in a compound train (only integer tooth numbers), if the train ratio is an integer, it is easier to design with integer ratio gearsets.

2   The square root of 18 is 4.24, well within our 10:1 limitation. So two stages will suffice in this gearbox.

3   If we could form two identical stages, each with a ratio equal to the square root of the overall train ratio, the train would be reverted by default. Table 10-5 shows that there are no reasonable combinations of tooth ratios which will give the exact square root needed. Moreover, this square root is not a rational number, so we cannot get an exact solution by this approach.

TABLE 10-5   Possible Gearsets for an 18:1 Two-Stage Reverted Compound Train

Gearset Ratio	x	Pinion Teeth	=	Gear Teeth
4.24264	x	12	=	50.91
4.24264	x	13	=	55.15
4.24264	x	14	=	59.40
4.24264	x	15	=	63.64
4.24264	x	16	=	67.88
4.24264	x	17	=	72.12
4.24264	x	18	=	76.37
4.24264	x	19	=	80.61
4.24264	x	20	=	84.85

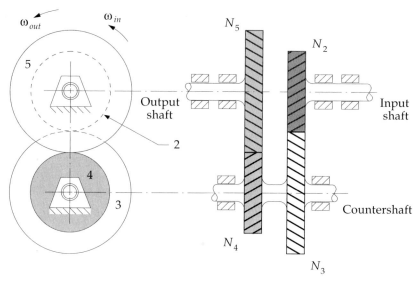

FIGURE 10-27

A reverted compound gear train

4   Instead, let's factor the train ratio. All numbers in the factors 9 x 2 and 6 x 3 are < 10, so they are acceptable on that basis. It is probably better to have the ratios of the two stages closer in value to one another for packaging reasons, so the 6 x 3 choice will be tried.

5   Figure 10-27 shows a two-stage reverted train. Note that, unlike the nonreverted train in Figure 10-25, the input and output shafts are now in-line and cantilevered; thus each must have double bearings on one end for moment support and a good bearing ratio as was defined in Section 2.15.

6   Equation 10.7 states the relationship for its train ratio. In addition, we have the constraint that the center distances of the stages must be equal. We can state this in terms of gear tooth numbers provided that the diametral pitch of both stages is the same.

Summing pitch radii:

$$r_2 + r_3 = r_4 + r_5 \tag{10.8a}$$

and expressing it in terms of pitch diameters:

$$d_2 + d_3 = d_4 + d_5 \tag{10.8b}$$

Substituting equation 10.3 and cancelling the diametral pitch p_d:

$$N_2 + N_3 = N_4 + N_5 = K \tag{10.8c}$$

where K is a constant to be determined.

7   We wish to solve equations 10.7 and 10.8 simultaneously. We can separate the terms in equation 10.7 and set them each equal to one of the stage ratios chosen for this design.

$$\frac{N_2}{N_3} = \frac{1}{6}$$

$$N_3 = 6N_2 \tag{10.9a}$$

$$\frac{N_4}{N_5} = \frac{1}{3}$$
$$N_5 = 3N_4 \qquad\qquad (10.9b)$$

8 Separating the terms in equation 10.8c:

$$N_2 + N_3 = K \qquad\qquad (10.10a)$$
$$N_4 + N_5 = K \qquad\qquad (10.10b)$$

9 Substituting equation 10.9a in equation 10.10a and equation 10.9b in equation 10.10b yields:

$$N_2 + 6N_2 = K = 7N_2 \qquad\qquad (10.11a)$$
$$N_4 + 3N_4 = K = 4N_4 \qquad\qquad (10.11b)$$

10 To make equations 10.11 compatible, K must be set to at least the lowest common multiple of 7 and 4, which is 28. This yields values of $N_2 = 4$ teeth and $N_4 = 7$ teeth.

11 Since a four-tooth gear will have unacceptable undercutting, we need to increase our value of K sufficiently to make the smallest pinion large enough.

12 A value of $K = 28 \times 3 = 84$ will increase the four-tooth gear to a twelve-tooth gear which is acceptable for a $25°$ pressure angle. With this assumption of $K = 84$, equations 10.11 and 10.9 can be solved to give:

$$N_2 = 12 \qquad\qquad N_3 = 72$$
$$\qquad\qquad\qquad\qquad\qquad (10.11c)$$
$$N_4 = 21 \qquad\qquad N_5 = 63$$

which is a viable solution for this reverted train.

The same procedure outlined here can be applied to the design of reverted trains involving several stages such as the helical gearbox in Figure 10-28.

10.8 EPICYCLIC OR PLANETARY GEAR TRAINS

The conventional gear trains described in the previous sections are all one degree of freedom (*DOF*) devices. Another class of gear train has wide application, the **epicyclic or planetary train**. This is a two degree of freedom device. Two inputs are needed to obtain a predictable output. In some cases, such as the automotive differential, one input is provided (the driveshaft) and two frictionally coupled outputs are obtained (the two driving wheels).

Figure 10-29a shows a conventional, one *DOF* gearset in which link 1 is immobilized as the ground link. Figure 10-29b shows the same gearset with link 1 now free to rotate as an **arm** which connects the two gears. Now only the joint O_2 is grounded and the system $DOF = 2$. This has become an **epicyclic** train with a **sun**

FIGURE 10-28

A three stage reverted compound commercial gearbox
Courtesy of Boston Gear Div. of IMO Industries, Quincy, Mass.

gear and a **planet gear** orbiting around the sun, held in orbit by the **arm**. Two inputs are required. Typically, the arm and the sun gear will each be driven in some direction at some velocity. In many cases, one of these inputs will be zero velocity, i.e., a brake applied to either the arm or the sun gear. Note that a zero velocity input to the arm merely makes a conventional train out of the epicyclic train as shown in Figure 10-29a. Thus the conventional gear train is simply a special case of the more complex epicyclic train, in which the arm is held stationary.

In this simple example of an epicyclic train, the only gear left to take an output from, after putting inputs to sun and arm, is the planet. It is a bit difficult to get a useable output from this orbiting gear as its pivot is moving. A more useful configuration is shown in Figure 10-30 to which a ring gear has been added. This

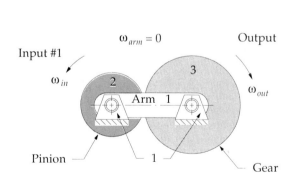

(*a*) Conventional gearset

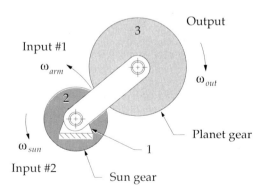

(*b*) Planetary or epicyclic gearset

FIGURE 10-29

Conventional gearsets are special cases of planetary or epicyclic gearsets

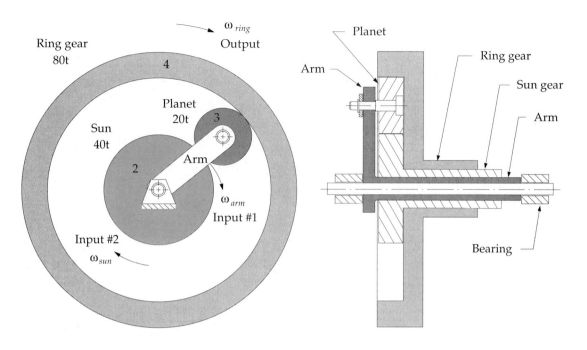

FIGURE 10-30

Planetary gearset with ring gear used as output

ring gear meshes with the planet and pivots at O_2, so it can be easily tapped as the output member. Most planetary trains will be arranged with ring gears to bring the planetary motion back to a grounded pivot. Note how the sun gear, ring gear, and arm are all brought out as concentric hollow shafts so that each can be accessed to tap its angular velocity and torque either as an input or an output.

While it is relatively easy to visualize the power flow through a conventional gear train and observe the directions of motion for its member gears, it is very difficult to determine the behavior of a planetary train by observation. We must do the necessary calculations to determine its behavior and may be surprised at the often counterintuitive results. Since the gears are rotating with respect to the arm and the arm itself has motion, we have a velocity difference problem here which requires equation 6.5 be applied to this problem. Rewriting the velocity difference equation in terms of angular velocities specific to this system, we get:

$$\omega_{gear} = \omega_{arm} + \omega_{gear/arm} \qquad (10.12)$$

Equations 10.12 and 10.5 are all that are needed to solve for the velocities in an epicyclic train, provided that the tooth numbers and two input conditions are known.

The Tabular Method

One approach to the analysis of velocities in an epicyclic train is to create a table which represents equation 10.12 for each gear in the train.

✐ EXAMPLE 10-3

Epicyclic Gear Train Design by the Tabular Method

Problem: Consider the train in Figure 10-30 which has the following tooth numbers and initial conditions:

Sun gear	N_2 = 40-tooth external gear
Planet gear	N_3 = 20-tooth external gear
Ring gear	N_4 = 80-tooth internal gear

Input to arm	200 RPM clockwise
Input to sun	100 RPM clockwise

We wish to find the absolute output angular velocity of the ring gear.

Solution:

1   The solution table is set up with a column for each term in equation 10.12 and a row for each gear in the train. It will be most convenient if we can arrange the table so that meshing gears occupy adjacent rows. The table for this method, prior to data entry, is shown in Figure 10-31.

2   Note that the gear ratios are shown straddling the rows of gears to which they apply. The gear ratio column is placed next to the column containing the velocity differences $\omega_{gear/arm}$ because the gear ratios only apply to the velocity difference. The gear ratios **cannot be directly applied to the absolute velocities** in the ω_{gear} column.

3   The solution strategy is simple but is fraught with opportunities for careless errors. Note that we are solving a vector equation with scalar algebra and the signs of the terms denote the sense of the ω vectors which are all directed along the Z axis. Great care must be taken to get the signs of the input velocities and of the gear ratios correct in the table, or the answer will be wrong. Some gear ratios may be negative if they involve external gearsets, and some will be positive if they involve an internal gear. We have both types in this example.

4   The first step is to enter the known data as shown in Figure 10-32 which in this case are the arm velocity (in all rows) and the absolute velocity of gear 2 in column 1. The gear ratios can also be calculated and placed in their respective locations. Note that these ratios should be calculated for each gearset in a consistent manner, following the power flow through the train. That is, starting at gear 2 as the driver, it drives gear 3

	1	2	3	
Gear #	ω_{gear} =	ω_{arm} +	$\omega_{gear/arm}$	*Gear ratio*

FIGURE 10-31

Table for the solution of planetary gear trains

	1	2	3	
Gear #	ω_{gear} =	ω_{arm} +	$\omega_{gear/arm}$	Gear ratio
2	−100	−200		
				−40/20
3		−200		
				+20/80
4		−200		

FIGURE 10-32

Given data for planetary gear train from example 10-3 placed in solution table

directly. This makes its ratio $-N_2/N_3$, or input over output, not the reciprocal. *This ratio is negative because the gearset is external.* Gear 3 in turn drives gear 4 so its ratio is $+N_3/N_4$. *This is a positive ratio because of the internal gear.*

5 Once any one row has two entries, the value for its remaining column can be calculated from equation 10.12. Once any one value in the velocity difference column (column 3) is found, the gear ratios can be applied to calculate all other values in that column. Finally, the remaining rows can be calculated from equation 10.12 to yield the absolute velocities of all gears in column 1. These computations are shown in Figure 10-33 which completes the solution.

6 The overall train value for this example can be calculated from the table and is, from arm to ring gear +1.25:1 and from sun gear to ring gear +2.5:1.

In this example, the arm velocity was given. If it is to be found as the output, then it must be entered in the table as an unknown, x, and the equations solved for that unknown.

FERGUSON'S PARADOX Epicyclic trains have several advantages over conventional trains among which are higher train ratios in smaller packages, reversion by default, and simultaneous, concentric, bidirectional outputs available from a single unidirectional input. These features make planetary trains popular as automatic transmissions in automobiles and trucks, etc.

	1	2	3	
Gear #	ω_{gear} =	ω_{arm} +	$\omega_{gear/arm}$	Gear ratio
2	−100	−200	+100	
				−40/20
3	−400	−200	−200	
				+20/80
4	−250	−200	−50	

FIGURE 10-33

Solution for planetary gear train from example 10-3

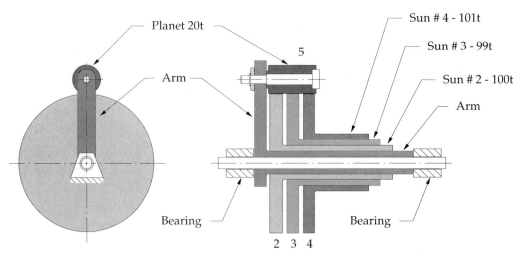

FIGURE 10-34

Ferguson's paradox compound planetary gear train

The so-called **Ferguson's paradox** of Figure 10-34 illustrates all these features of the planetary train. It is a **compound epicyclic train** with one 20-tooth planet gear (gear 5) carried on the arm and meshing simultaneously with three sun gears. These sun gears have 100 teeth (gear 2), 99 teeth (gear 3), and 101 teeth (gear 4), respectively. The center distances between all sun gears and the planet are the same despite the slightly different pitch diameters of each sun gear. This is possible because of the properties of the involute tooth form as described in Section 10.2. Each sun gear will run smoothly with the planet gear. Each gearset will merely have a slightly different pressure angle.

✐ EXAMPLE 10-4

Analyzing Ferguson's Paradox by the Tabular Method

Problem: Consider the Ferguson's paradox train in Figure 10-34 which has the following tooth numbers and initial conditions:

Sun gear # 2	N_2= 100-tooth external gear
Sun gear # 3	N_3= 99-tooth external gear
Sun gear # 4	N_4= 101-tooth external gear
Planet gear	N_5 = 20-tooth external gear
Input to sun # 2	0 RPM
Input to arm	100 RPM counterclockwise

Sun gear 2 is fixed to the frame, thus providing one input (zero velocity) to the system. The arm is driven at 100 RPM counterclockwise as the second input. Find the angular velocities of the two outputs which are available from this compound train, one from gear 3 and one from gear 4, both of which are free to rotate on the main shaft.

Gear #	1 $\omega_{gear} =$	2 $\omega_{arm} +$	3 $\omega_{gear/arm}$	Gear ratio
2	0	+100		
				−100/20
5		+100		
				−20/99
3		+100		
5		+100		
				−20/101
4		+100		

FIGURE 10-35

Given data for Ferguson's paradox planetary gear train from example 10-4

Solution:

1 The tabular solution for this train is set up in Figure 10-35 which shows the given data. Note that the row for gear 5 is repeated for clarity in applying the gear ratio between gears 5 and 4.

2 The known input values of velocity are the arm angular velocity and the zero absolute velocity of gear 2.

3 The gear ratios in this case are all negative because of the external gear sets and their values reflect the direction of power flow from gear 2 to 5, then 5 to 3 and 5 to 4 in the second branch.

4 Figure 10-36 shows the calculated values added to the table. Note that for a **counterclockwise** 100 RPM input to the arm, we get a *counterclockwise* 1 RPM output from gear 4 and a *clockwise* 1 RPM output from gear 3, simultaneously.

Gear #	1 $\omega_{gear} =$	2 $\omega_{arm} +$	3 $\omega_{gear/arm}$	Gear ratio
2	0	+100	−100	
				−100/20
5	+600	+100	+500	
				−20/99
3	−1.01	+100	−101.01	
5	+600	+100	+500	
				−20/101
4	+0.99	+100	−99.01	

FIGURE 10-36

Solution to Ferguson's paradox planetary gear train from example 10-4

This result accounts for the use of the word **paradox** to describe this train. Not only do we get a much larger ratio (100:1) than we could from a conventional train with gears of 100 and 20 teeth, but we have our choice of output directions!

Automotive automatic transmissions use compound planetary trains, which are always in mesh, and which give different ratio forward speeds, plus reverse, by simply engaging and disengaging brakes on different members of the train. The brake provides zero velocity input to one train member. The other input is from the engine. The output is thus modified by the application of these internal brakes in the transmission according to the selection of the operator (*Drive, Reverse, Neutral* etc.).

The Formula Method

It is not necessary to tabulate the solution to an epicyclic train. The velocity difference formula can be solved directly for the train ratio. We can rearrange equation 10.12 to solve for the velocity difference term. Then, let ω_F represent the angular velocity of the first gear in the train (chosen at either end), and ω_L represent the angular velocity of the last gear in the train (at the other end).

For the first gear in the system:

$$\omega_{F/arm} = \omega_F - \omega_{arm} \tag{10.13a}$$

For the last gear in the system:

$$\omega_{L/arm} = \omega_L - \omega_{arm} \tag{10.13b}$$

Dividing the last by the first:

$$\frac{\omega_{L/arm}}{\omega_{F/arm}} = \frac{\omega_L - \omega_{arm}}{\omega_F - \omega_{arm}} = VR \tag{10.13c}$$

This gives an expression for the overall train value VR. The left-most side of equation 10.13c involves only the velocity difference terms which are relative to the arm. This fraction is equal to the ratio of the products of tooth numbers of the gears from first to last in the train as defined in equation 10.7b which can be substituted for the left-most side of equation 10.13c.

$$\pm \frac{\text{product of number of teeth on driver gears}}{\text{product of number of teeth on driven gears}} = \frac{\omega_L - \omega_{arm}}{\omega_F - \omega_{arm}} \tag{10.14}$$

This equation can be solved for any one of the variables on the right side provided that the other two are defined as the two inputs to this two *DOF* train. Either the velocities of the arm plus one gear must be known or the velocities of two gears, the first and last, as so designated, must be known. Another limitation of this method is that both the first and last gears chosen must be pivoted to ground (not orbiting), and there must be a path of meshes connecting them, which may include orbiting planet gears. Let us use this method to again solve the Ferguson's paradox of the previous example.

✍️ EXAMPLE 10-5

Analyzing Ferguson's Paradox by the Formula Method

Problem: Consider the same Ferguson's paradox train as in Example 10-4 which has
the following tooth numbers and initial conditions (See Figure 10-34):

Sun gear #2	N_2 = 100-tooth external gear
Sun gear #3	N_3 = 99-tooth external gear
Sun gear #4	N_4 = 101-tooth external gear
Planet gear	N_5 = 20-tooth external gear
Input to sun #2	0 RPM
Input to arm	100 RPM counterclockwise

Sun gear 2 is fixed to the frame, thus providing one input (zero velocity) to
the system. The arm is driven at 100 RPM counterclockwise as the second
input. Find the angular velocities of the two outputs which are available
from this compound train, one from gear 3 and one from gear 4, both of
which are free to rotate on the main shaft.

Solution:

1   We will have to apply equation 10.14 twice, once for each output gear. Taking gear 3
as the last gear in the train with gear 2 as the first, we have:

$$N_2 = 100 \qquad N_3 = 99 \qquad N_5 = 20$$

(10.15a)

$$\omega_{arm} = +100 \qquad \omega_F = 0 \qquad \omega_L = ?$$

2   Substituting in equation 10.14 we get:

$$\left(-\frac{N_2}{N_5}\right)\left(-\frac{N_5}{N_3}\right) = \frac{\omega_L - \omega_{arm}}{\omega_F - \omega_{arm}}$$

$$\left(-\frac{100}{20}\right)\left(-\frac{20}{99}\right) = \frac{\omega_3 - 100}{0 - 100}$$

(10.15b)

$$\omega_3 = -1.01$$

3   Now taking gear 4 as the last gear in the train with gear 2 as the first, we have:

$$N_2 = 100 \qquad N_4 = 101 \qquad N_5 = 20$$

(10.15c)

$$\omega_{arm} = +100 \qquad \omega_F = 0 \qquad \omega_L = ?$$

4   Substituting in equation 10.14 we get:

$$\left(-\frac{N_2}{N_5}\right)\left(-\frac{N_5}{N_4}\right) = \frac{\omega_L - \omega_{arm}}{\omega_F - \omega_{arm}}$$

$$\left(-\frac{100}{20}\right)\left(-\frac{20}{101}\right) = \frac{\omega_4 - 100}{0 - 100}$$

(10.15d)

$$\omega_4 = +0.99$$

These are the same results as were obtained with the tabular method.

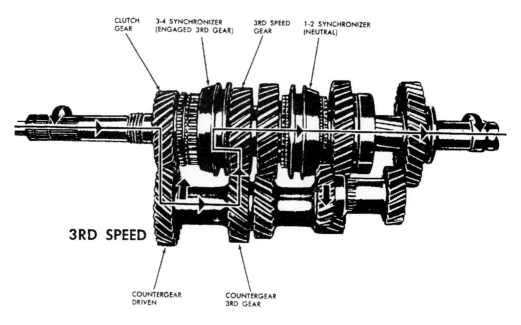

CLUTCH   3-4 SYNCHRONIZER   3RD SPEED   1-2 SYNCHRONIZER
GEAR   (ENGAGED 3RD GEAR)   GEAR   (NEUTRAL)

3RD SPEED

COUNTERGEAR   COUNTERGEAR
DRIVEN   3RD GEAR

FIGURE 10-37

Four speed manual synchromesh automobile transmission
from Crouse, Automotive Mechanics, 8th ed., p. 480, McGraw-Hill, 1980, New York, NY. Reprinted with permission

10.9   TRANSMISSIONS

COMPOUND REVERTED GEAR TRAINS are commonly used in manual (nonautomatic) automotive transmissions to provide user-selectable ratios between the engine and the drive wheels for torque multiplication (mechanical advantage). These gearboxes usually have from three to six forward speeds and one reverse. Most modern transmissions of this type use helical gears for quiet operation. These gears are **not** moved into and out of engagement when shifting from one speed to another except for reverse. Rather, the desired ratio gears are selectively locked to the output shaft by synchromesh mechanisms as in Figure 10-37 which shows a four-speed, manually-shifted, synchromesh automotive transmission.

The input shaft is at top left. The input gear is always in mesh with the leftmost gear on the countershaft at the bottom. This countershaft has several gears integral with it, each of which meshes with a different output gear that is freewheeling on the output shaft. The output shaft is concentric with the input shaft , making this a reverted train, but the input and output shafts only connect through the gears on the countershaft except in "top gear" (fourth speed), for which the input and output shafts are directly coupled together with a synchromesh clutch for a 1:1 ratio.

The synchromesh clutches are beside each gear on the output shaft and are partially hidden by the shifting collars which move them left and right in response to the driver's hand on the shift lever. These clutches act to lock one gear

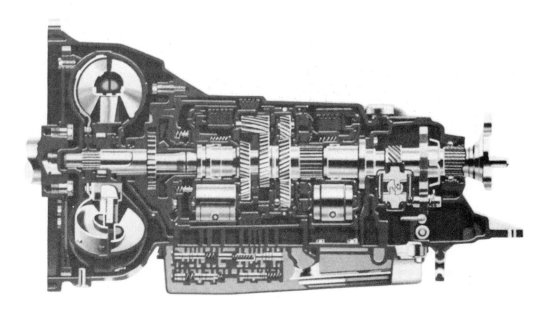

FIGURE 10-38

Four speed automatic automobile transmission
Courtesy of Mercedes Benz of North America Inc.

to the output shaft at a time to provide a power path from input to output of a particular ratio. The arrows on the figure show the power path for third-speed forward, which is engaged. Reverse gear, on the lower right, engages an idler gear which is physically shifted into and out of mesh at standstill.

PLANETARY OR EPICYCLIC TRAINS are commonly used in automatic shifting automotive transmissions as shown in Figure 10-38. At the left is a turbine-like fluid coupling between engine and transmission, called a **torque converter**. This device allows sufficient slip in the coupling fluid to let the engine idle with the transmission engaged and the wheels stopped. As the vehicle is allowed to move, the engine-driven turbine blades, running in oil, transmit torque through the oil to the turbine blades attached to the transmission input shaft. This is one input to the multi-*DOF* transmission. Different ratios are chosen within the transmission based on a combination of driver selection, road speed, engine load, and other factors which are automatically monitored.

The epicyclic gearsets can be seen near the center of the transmission. They are controlled by hydraulically operated clutches and brakes within the transmission that impart zero velocity (second) inputs to various elements of the train to create one of four forward velocity ratios plus reverse in this particular example. Automatic transmissions can have any number of ratios. Automotive examples typically have from two to four speeds forward. Truck and bus automatic transmissions may have more.

10.10 BIBLIOGRAPHY

1   **Roark**, **R.**, **and Young**, W., *Formulas for Stress and Strain*, 5th ed., McGraw-Hill Inc, New York, 1965.

2   **Shigley**, **J.**, **and Mischke**, **C.**, *Mechanical Engineering Design*, 5th ed., McGraw-Hill Inc, New York, 1989.

10.11 PROBLEMS

*10-1   A 22-tooth gear has AGMA standard full-depth involute teeth with diametral pitch of 4. Calculate the pitch diameter, circular pitch, addendum, dedendum, tooth thickness, and clearance.

10-2   A 40-tooth gear has AGMA standard full-depth involute teeth with diametral pitch of 10. Calculate the pitch diameter, circular pitch, addendum, dedendum, tooth thickness, and clearance.

10-3   A 30-tooth gear has AGMA standard full-depth involute teeth with diametral pitch of 12. Calculate the pitch diameter, circular pitch, addendum, dedendum, tooth thickness, and clearance.

10-4   Using any available string, some tape, a pencil, and a drinking glass or tin can, generate and draw an involute curve on a piece of paper. With your protractor, show that all normals to the curve are tangent to the base circle.

*10-5   A spur gearset has pitch diameters of 4.5 and 12 inches. What is the largest tooth size, in terms of diametral pitch, that can be used without having any interference and undercutting?
   a.   For a 20° pressure angle.
   b.   For a 25° pressure angle. (Note that diametral pitch need not be an integer.)

*10-6   Design a *simple*, spur gear train for a ratio of –9:1 and diametral pitch of 7.6. Specify pitch diameters and numbers of teeth. Draw a scale layout.

*10-7   Design a *simple*, spur gear train for a ratio of +8:1 and diametral pitch of 6. Specify pitch diameters and numbers of teeth. Draw a scale layout.

10-8   Design a *simple*, spur gear train for a ratio of –7:1 and diametral pitch of 8. Specify pitch diameters and numbers of teeth. Draw a scale layout.

10-9   Design a *simple*, spur gear train for a ratio of +6.5:1 and diametral pitch of 5. Specify pitch diameters and numbers of teeth. Draw a scale layout.

*10-10   Design a *compound*, spur gear train for a ratio of –70:1 and diametral pitch of 10. Specify pitch diameters and numbers of teeth. Draw a scale layout.

10-11   Design a *compound*, spur gear train for a ratio of 50:1 and diametral pitch of 8. Specify pitch diameters and numbers of teeth. Draw a scale layout.

*10-12   Design a *compound*, spur gear train for a ratio of 150:1 and diametral pitch of 6. Specify pitch diameters and numbers of teeth. Draw a scale layout.

10-13   Design a *compound*, spur gear train for a ratio of –250:1 and diametral pitch of 9. Specify pitch diameters and numbers of teeth. Draw a scale layout.

* Answers in Appendix E

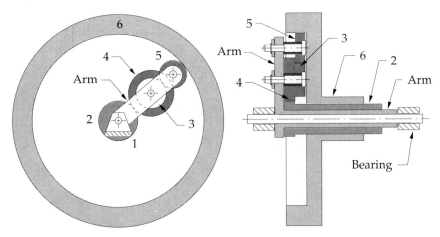

FIGURE P10-1

Planetary gearset for problem 10-25

*10-14 Design a compound, *reverted*, spur gear train for a ratio of 30:1 and diametral pitch of 10. Specify pitch diameters and numbers of teeth. Draw a scale layout.

10-15 Design a compound, *reverted*, spur gear train for a ratio of 40:1 and diametral pitch of 8. Specify pitch diameters and numbers of teeth. Draw a scale layout.

*10-16 Design a compound, *reverted*, spur gear train for a ratio of 75:1 and diametral pitch of 15. Specify pitch diameters and numbers of teeth. Draw a scale layout.

10-17 Design a compound, *reverted*, spur gear train for a ratio of 7:1 and diametral pitch of 4. Specify pitch diameters and numbers of teeth. Draw a scale layout.

10-18 Design a compound, *reverted*, spur gear train for a ratio of 12:1 and diametral pitch of 6. Specify pitch diameters and numbers of teeth. Draw a scale layout.

*10-19 Design a compound, reverted, spur gear *transmission* which will give two shiftable ratios of +3:1 forward and –4.5:1 reverse with diametral pitch of 6. Specify pitch diameters and numbers of teeth. Draw a scale layout. See Figure 10-37

TABLE P10-1   Data for Figure P10-1

row	N_2	N_3	N_4	N_5	N_6	ω_2	ω_6	ω_{arm}
a	30	25	45	50	200	?	20	– 50
b	30	25	45	50	200	30	?	– 90
c	30	25	45	50	200	50	0	?
d	30	25	45	30	160	?	40	– 50
e	30	25	45	30	160	50	?	– 75
f	30	25	45	30	160	50	0	?

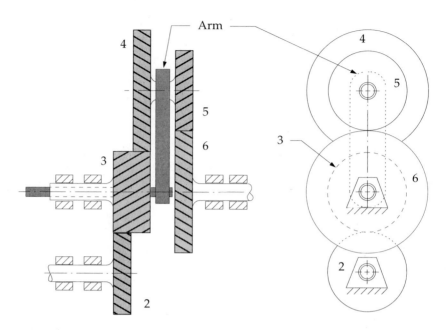

FIGURE P10-2

Compound planetary gear train for problem 10-26

10-20 Design a compound, reverted, spur gear *transmission* which will give two shift-able ratios of +5:1 forward and –3.5:1 reverse with diametral pitch of 6. Specify pitch diameters and numbers of teeth. Draw a scale layout. See Figure 10-37

*10-21 Design a compound, reverted, spur gear *transmission* which will give three shiftable ratios of +6:1, +3.5:1 forward and –4:1 reverse with diametral pitch of 8. Specify pitch diameters and numbers of teeth. Draw a scale layout. See Figure 10-37

10-22 Design a compound, reverted, spur gear *transmission* which will give three shiftable ratios of +4.5:1, +2.5:1 forward and –3.5:1 reverse with diametral pitch of 5. Specify pitch diameters and numbers of teeth. Draw a scale layout. See Figure 10-37

TABLE P10-2   Data for Figure P10-2

row	N_2	N_3	N_4	N_5	N_6	ω_2	ω_6	ω_{arm}
a	50	25	45	30	40	?	20	– 50
b	30	35	55	40	50	30	?	– 90
c	40	20	45	30	35	50	0	?
d	25	45	35	30	50	?	40	– 50
e	35	25	55	35	45	30	?	– 75
f	30	30	45	40	35	40	0	?

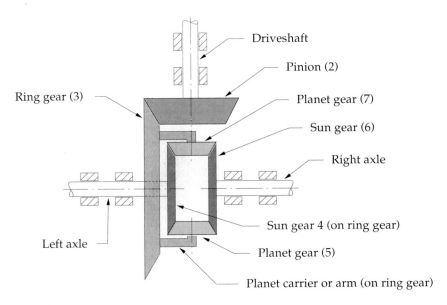

Driveshaft

Pinion (2)

Ring gear (3)

Planet gear (7)

Sun gear (6)

Right axle

Left axle

Sun gear 4 (on ring gear)

Planet gear (5)

Planet carrier or arm (on ring gear)

FIGURE P10-3

Automotive differential planetary gear train for problem 10-27

10-23 Design the rolling cones for a –3:1 ratio and a 60° included angle between the shafts. Draw a scale layout.

10-24 Design the rolling cones for a –4.5:1 ratio and a 40° included angle between the shafts. Draw a scale layout.

*10-25 Figure P10-1 shows a compound planetary gear train (not to scale). Table P10-1 gives data for gear numbers of teeth and input velocities. For the row(s) assigned, find the variable represented by a question mark.

*10-26 Figure P10-2 shows a compound planetary gear train (not to scale). Table P10-2 gives data for gear numbers of teeth and input velocities. For the row(s) assigned, find the variable represented by a question mark.

*10-27 Figure P10-3 shows a planetary gear train used in an automotive rear-end differential (not to scale). The car has wheels with a 15-inch rolling radius and is moving forward in a straight line at 50 mph. The engine is turning 2000 rpm. The transmission is in direct drive (1:1) with the driveshaft.

    a.   What is the rear wheels' rpm and the gear ratio between ring and pinion?

    b.   As the car hits a patch of ice, the right wheel speeds up to 800 RPM. What is the speed of the left wheel? *Hint: The average of both wheels' rpm is a constant.*

10-28 Design a speed reducing planetary gearbox to be used to lift a 5 ton load 50 feet with a motor that develops 20 lb-ft of torque at its operating speed of 1750 RPM. The available winch drum has no more than a 16-inch diameter when full of its steel cable. The speed reducer should be no larger in diameter than the winch drum. Gears of no more than about 75 teeth are desired, and diametral pitch needs to be no smaller than 6 to stand the stresses. Make multiview sketches of your design and show all calculations. How long will it take to raise the load with your design?

End
of
Part I

PART II

The entire world of machinery ...
is inspired by the play of organs of
reproduction. The designer animates
artificial objects by simulating the
movements of animals engaged in
propagating the species. Our
machines are Romeos of steel
and Juliets of cast iron.

J. COHEN, *Human Robots in Myth*
and Science, Allen & Unwin,
London, 1966, p. 67.

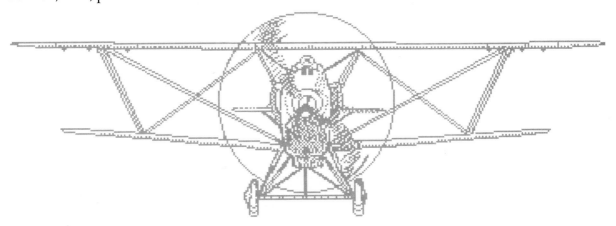

DYNAMICS OF
MACHINERY

Chapter 11

DYNAMICS FUNDAMENTALS

He has half the deed done
who has made a beginning
HORACE, 65-8 BC

11.0 INTRODUCTION

Part I of this text has dealt with the **kinematics** of mechanisms while temporarily ignoring the forces present in those mechanisms. This second part will address the problem of determining the forces present in moving mechanisms and machinery. This topic is called **kinetics** or **dynamic force analysis**. We will start with a brief review of some fundamentals needed for dynamic analysis. It is assumed that the reader has had an introductory course in dynamics. If that topic is rusty, one can review it by referring to reference 1 or to any other text on the subject.

11.1 NEWTON'S LAWS OF MOTION

Dynamic force analysis involves the application of **Newton's** three **laws of motion** which are:

1 *A body at rest tends to remain at rest and a body in motion at constant velocity will tend to maintain that velocity unless acted upon by an external force.*

2 *The time rate of change of momentum of a body is equal to the magnitude of the applied force and acts in the direction of the force.*

3 *For every action force there is an equal and opposite reaction force.*

The second law is expressed in terms of rate of change of *momentum*, $\mathbf{M} = m\mathbf{v}$, where m is mass and $\mathbf{v}$ is velocity. Mass is assumed to be constant in this analysis.

The time rate of change of $m\mathbf{v}$ is, of course, $m\mathbf{a}$, so this second law is expressed in the familiar equation:

$$\mathbf{F} = m\mathbf{a} \tag{11.1}$$

We can differentiate between two subclasses of dynamics problems depending upon which quantities are known and which are to be found. The **"forward dynamics problem"** is the one in which we know everything about the forces and/or torques being exerted on the system, and we wish to determine the accelerations, velocities, and displacements which result from the application of those forces and torques. This subclass is typical of the problems you probably encountered in an introductory dynamics course, such as determining the acceleration of a block sliding down a plane, acted upon by gravity. Given $\mathbf{F}$ and m, solve for $\mathbf{a}$.

The second subclass of dynamics problem, called the **"inverse dynamics problem"** is one in which we know the (desired) accelerations, velocities, and displacements to be imposed upon our system and wish to solve for the magnitudes and directions of the forces and torques which are necessary to provide the desired motions and which result from them. This inverse dynamics case is sometimes also called **kinetostatics**. Given $\mathbf{a}$ and m, solve for $\mathbf{F}$.

Whichever subclass of problem is addressed, it is important to realize that they are both dynamics problems. Each merely solves $\mathbf{F} = m\mathbf{a}$ for a different variable. To do so we must first review some fundamental geometric and mass properties which are needed for the calculations.

11.2   DYNAMIC MODELS

It is often convenient in dynamic analysis to create a simplified model of a complicated part. These models are sometimes considered to be a collection of point masses connected by massless rods. For a model of a rigid body to be dynamically equivalent to the original body, three things must be true:

1   *The mass of the model must equal that of the original body.*

2   *The center of gravity must be in the same location as that of the original body.*

3   *The mass moment of inertia must equal that of the original body.*

11.3   MASS

Mass is not weight! Mass is an invariant property of a rigid body. The weight of the same body varies depending on the gravitational system in which it sits. See Section 1.8 for a discussion of the use of proper mass units in various measuring systems. We will assume the mass of our parts to be constant in our calculations. For most earth-bound machinery, this is a reasonable assumption. The rate at which an automobile or bulldozer loses mass due to fuel consumption, for example, is slow enough to be ignored when calculating dynamic forces over short time spans. However, this would not be a safe assumption for a vehicle such as the space shuttle, whose mass changes rapidly and drastically during liftoff.

When designing machinery, we must first do a complete kinematic analysis of our design, as described in Part I of this text, in order to obtain information about the accelerations of the moving parts. We next want to use Newton's second law to calculate the dynamic forces. But to do so we need to know the masses of all the moving parts which have these known accelerations. These parts do not exist yet! As with any design problem, we lack sufficient information at this stage of the design to accurately determine the best sizes and shapes of the parts. We must estimate the masses of the links and other parts of the design in order to make a first pass at the calculation. We will then have to iterate to better and better solutions as we generate more information. See Section 1.4 on the design process to review the use of iteration in design.

A first estimate of your parts' masses can be obtained by assuming some reasonable shapes and sizes for all the parts and choosing appropriate materials. Then calculate the volume of each part and multiply its volume by the material's **mass density** (not weight density) to obtain a first approximation of its mass. These mass values can then be used in Newton's equation. The densities of some common engineering materials can be found in Appendix A.

How will we know whether our chosen sizes and shapes of links are even acceptable, let alone optimal? Unfortunately, we will not know until we have carried the computations all the way through a complete stress and deflection analysis of the parts. It is often the case, especially with long, thin elements such as shafts or slender links, that the deflections of the parts under their dynamic loads will limit the design even at low stress levels. In some cases the stresses will be excessive.

We will probably discover that the parts fail under the dynamic forces. Then we will have to go back to our original assumptions about the shapes, sizes, and materials of these parts, redesign them, and repeat the force, stress, and deflection analyses. Design is, unavoidably, an **iterative process**.

The topic of stress and deflection analysis is beyond the scope of this text and will not be further discussed here. It is mentioned only to put our discussion of dynamic force analysis into context. We are analyzing these dynamic forces primarily to provide the information needed to do the stress and deflection analyses on our parts! It is also worth noting that, unlike a static force situation in which a failed design might be fixed by adding more mass to the part to strengthen it, to do so in a dynamic force situation can have a deleterious effect. More mass with the same acceleration will generate even higher forces and thus higher stresses! The machine designer often needs to remove mass (in the right places) from parts in order to reduce the stresses and deflections due to $\mathbf{F} = m\mathbf{a}$. Thus the designer needs to have a good understanding of both material properties and stress and deflection analysis to properly shape and size parts for minimum mass while maximizing the strength and stiffness needed to withstand the dynamic forces.

11.4 MASS MOMENT AND CENTER OF GRAVITY

When the mass of an object is distributed over some dimensions, it will possess a moment with respect to any axis of choice. Figure 11-1 shows a mass of general shape in an xyz axis system. A differential element of mass is also shown. The

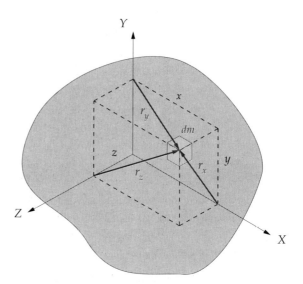

FIGURE 11-1

A generalized mass in a 3-D coordinate system

mass moment (first moment of mass) of the differential element is equal to the **product of its mass and its distance** from the axis of interest. With respect to the x, y, and z axes these are:

$$dM_x = r_x^1 dm = \sqrt{(y^2 + z^2)} dm \qquad (11.2a)$$

$$dM_y = r_y^1 dm = \sqrt{(x^2 + z^2)} dm \qquad (11.2b)$$

$$dM_z = r_z^1 dm = \sqrt{(x^2 + y^2)} dm \qquad (11.2c)$$

The radius from the axis of interest to the differential element is shown with an exponent of 1 to emphasize the reason for this property being called the first moment of mass. To obtain the mass moment of the entire body we integrate each of these expressions.

$$M_x = \int \sqrt{(y^2 + z^2)} dm \qquad (11.3a)$$

$$M_y = \int \sqrt{(x^2 + z^2)} dm \qquad (11.3b)$$

$$M_z = \int \sqrt{(x^2 + y^2)} dm \qquad (11.3c)$$

If the mass moment with respect to a particular axis is numerically zero, then that axis passes through the **center of gravity (CG)** of the object. By definition the summation of first moments about all axes through the center of gravity is zero. We will need to locate the CG of all moving bodies in our designs because the linear acceleration component of each body is calculated acting at that point.

We can determine the *CG* location for any body with respect to any selected axis system by setting equations 11.3 each equal to zero and solving them simultaneously for the values of x, y, and z.

It is often convenient to model a complicated shape as several interconnected simple shapes whose individual geometries allow easy computation of their masses and the locations of their local *CGs*. The global *CG* can then be found from the summation of the first moments of these simple shapes set equal to zero. Appendix B contains formulas for the volumes and locations of centers of gravity of some common shapes.

Figure 11-2 shows a simple model of a mallet broken into two cylindrical parts, the handle and the head, which have masses m_h and m_d, respectively. The individual centers of gravity of the two parts are at l_d and $l_h/2$, respectively, with respect to the axis ZZ. We want to find the location of the composite center of gravity of the mallet with respect to ZZ. Summing first moments of the individual components about ZZ and setting them equal to the moment of the entire mass about ZZ.

$$\sum M_{ZZ} = m_h \frac{l_h}{2} + m_d l_d = (m_h + m_d)d \tag{11.3d}$$

This equation can be solved for the distance d along the X axis, which, in this symmetrical example, is the only dimension of the composite *CG* not discernable by inspection. The y and z components of the composite *CG* are both zero.

$$d = \frac{m_h \dfrac{l_h}{2} + m_d l_d}{(m_h + m_d)} \tag{11.3e}$$

11.5   MASS MOMENT OF INERTIA (SECOND MOMENT OF MASS)

Newton's law applies to systems in rotation as well as to those in translation. The rotational form of Newton's second law is:

$$\mathbf{T} = I\alpha \tag{11.4}$$

where $\mathbf{T}$ is torque, α is angular acceleration, and I is mass moment of inertia.

Mass moment of inertia is referred to some axis of rotation, quite often one through the *CG*. Refer again to Figure 11-1 which shows a mass of general shape and an *xyz* axis system. A differential element of mass is also shown. The **mass moment of inertia** of the differential element is equal to the **product of its mass and the square of its distance** from the axis of interest. With respect to the x, y, and z axes they are:

$$dI_x = r_x^2 dm = (y^2 + z^2)dm \tag{11.5a}$$
$$dI_y = r_y^2 dm = (x^2 + z^2)dm \tag{11.5b}$$
$$dI_z = r_z^2 dm = (x^2 + y^2)dm \tag{11.5c}$$

The exponent of two on the radius term gives this property its other name of **second moment of mass**. To obtain the mass moments of inertia of the entire body we integrate each of these expressions.

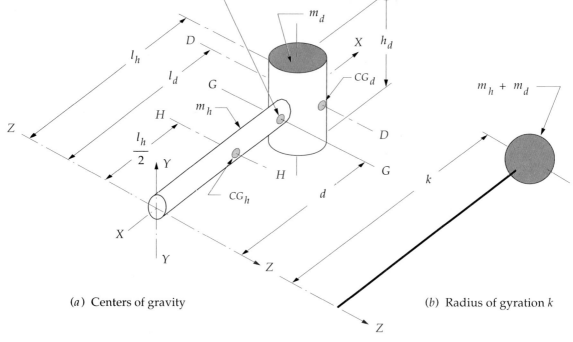

(a) Centers of gravity   (b) Radius of gyration k

FIGURE 11-2

Dynamic models, composite center of gravity and radius of gyration of a mallet

$$I_x = \int (y^2 + z^2)\,dm \qquad\qquad (11.6a)$$

$$I_y = \int (x^2 + z^2)\,dm \qquad\qquad (11.6b)$$

$$I_z = \int (x^2 + y^2)\,dm \qquad\qquad (11.6c)$$

While it is fairly intuitive to appreciate the physical significance of the first moment of mass, it is more difficult to do the same for the second moment, or moment of inertia.

Consider equation 11.4. It says that torque is proportional to angular acceleration, and the constant of proportionality is this moment of inertia, I. Picture a common hammer or mallet as depicted in Figure 11-2. The head, made of steel, has large mass compared to the light wooden handle. When gripped properly, at the end of the handle, the radius to the mass of the head is large. Its contribution to the total I of the mallet is proportional to the square of the radius from the axis of rotation (your wrist at axis ZZ) to the head. Thus it takes considerably more torque to swing (and thus angularly accelerate) the mallet when it is held properly than if held near the head. As a child you probably chose to hold a hammer close to its head because you lacked the strength to provide the larger torque it needed when held properly. You also found it ineffective in driving nails when held close to the head because you were unable to store very much **kinetic energy** in it. In a translating system kinetic energy is:

$$KE = \frac{1}{2}mv^2 \qquad (11.7a)$$

and in a rotating system kinetic energy is:

$$KE = \frac{1}{2}I\omega^2 \qquad (11.7b)$$

Thus the kinetic energy stored in the mallet is also proportional to its moment of inertia I and to ω^2. So, you can see that holding the mallet close to its head reduces the I and lowers the energy available for driving the nail.

Moment of inertia then is one indicator of the ability of the body to store rotational kinetic energy and is also an indicator of the amount of torque that will be needed to rotationally accelerate the body. Unless you are designing a device intended for the storage and transfer of large amounts of energy (punch press, drop hammer, rock crusher etc.) you will probably be trying to minimize the moments of inertia of your rotating parts. Just as mass is a measure of resistance to linear acceleration, moment of inertia is a measure of resistance to angular acceleration. A large I will require a large driving torque, and thus a larger and more powerful motor to obtain the same acceleration. Later we will see how to make moment of inertia work for us in rotating machinery by using flywheels with large I. The units of moment of inertia can be determined by doing a unit balance on either equation 11.4 or equation 11.7 and are shown in Table 1-4. In the **ips** system they are lb-in-sec^2.

11.6 PARALLEL AXIS THEOREM (TRANSFER THEOREM)

The moment of inertia of a body with respect to any axis *(ZZ)* can be expressed as the sum of its moment of inertia about an axis *(GG)* parallel to *ZZ* through its *CG*, and the square of the perpendicular distance between those parallel axes.

$$I_{ZZ} = I_{GG} + md^2 \qquad (11.8)$$

where *ZZ* and *GG* are parallel axes, *GG* goes through the *CG* of the body or assembly, m is the mass of the body or assembly, and d is the perpendicular distance between the parallel axes. This property is most useful when computing the moment of inertia of a complex shape which has been broken into a collection of simple shapes as shown in Figure 11-2a which represents a simplistic model of a mallet. The mallet is broken into two cylindrical parts, the handle and the head, which have masses m_h and m_d, and radii r_h and r_d, respectively. The expressions for the mass moments of inertia of a cylinder with respect to axes through its *CG* can be found in Appendix B and are for the handle about its *CG* axis *HH*:

$$I_{HH} = \frac{m_h\left(3r_h^2 + l_h^2\right)}{12} \qquad (11.9a)$$

and for the head about its *CG* axis *DD*:

$$I_{DD} = \frac{m_d\left(3r_d^2 + h_d^2\right)}{12} \qquad\qquad (11.9b)$$

Using the parallel axis theorem to transfer the moment of inertia to the axis ZZ at the end of the handle:

$$I_{ZZ} = \left[I_{HH} + m_h\left(\frac{l_h}{2}\right)^2\right] + \left[I_{DD} + m_d l_d^2\right] \qquad\qquad (11.9c)$$

11.7   RADIUS OF GYRATION

The **radius of gyration** of a body is defined as the radius at which the entire mass of the body could be concentrated such that the resulting model will have the same moment of inertia as the original body. The mass of this model must be the same as that of the original body. Let I_{ZZ} represent the mass moment of inertia about ZZ from equation 11.9 and m the mass of the original body. From the parallel axis theorem, a concentrated mass m at a radius k will have a moment of inertia:

$$I_{ZZ} = mk^2 \qquad\qquad (11.10a)$$

Since we want I_{ZZ} to be equal to the original moment of inertia, the required **radius of gyration** at which we will concentrate the mass m is then:

$$k = \sqrt{\frac{I_{ZZ}}{m}} \qquad\qquad (11.10b)$$

Note that this property of radius of gyration allows the construction of an even simpler dynamic model of the system in which all the system mass is concentrated in a "point mass" at the end of a massless rod of length k. Figure 11-2b shows such a model of the mallet in Figure 11-2a.

By comparing equation 11.10a with equation 11.8, it can be seen that the radius of gyration k will always be larger than the radius to the composite CG of the original body.

$$I_{CG} + md^2 = I_{ZZ} = mk^2 \qquad\qquad \therefore k > d \qquad\qquad (11.10c)$$

Appendix B contains formulas for the moments of inertia and radii of gyration of some common shapes.

11.8   CENTER OF PERCUSSION

The **center of percussion** is a point on a body which, when struck with a force, will have associated with it another point called the **center of rotation** at which there will be a zero reaction force. You have probably experienced the result of "missing the center of percussion" when you hit a baseball or softball with the wrong spot on the bat. The "right place on the bat" to hit the ball is the center of percussion associated with the point that your hands grip the bat (the center of rotation). Hitting the ball at other than the center of percussion results in a stinging force

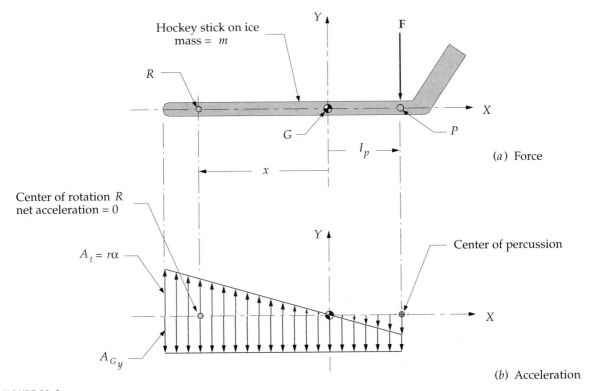

(a) Force

(b) Acceleration

FIGURE 11-3

Center of percussion and center of rotation

being delivered to your hands. Hit the right spot and you feel no force (nor pain). The center of percussion is sometimes called the "sweet spot" on a bat, tennis racquet, or golf club. In the case of our mallet example, a center of percussion at the head corresponds to a center of rotation near the end of the handle, and the handle is usually contoured to encourage gripping it there.

The explanation of this phenomenon is quite simple. To make the example two dimensional and eliminate the effects of friction, consider a hockey stick of mass m laying on the ice as shown in Figure 11-3a. Strike it a sharp blow at point P with a force $\mathbf{F}$ perpendicular to the stick axis. The stick will begin to travel across the ice in complex planar motion, both rotating and translating. Its complex motion at any instant can be considered as the superposition of two components: pure translation of its center of gravity G in the direction of $\mathbf{F}$ and pure rotation about that point G. Set up an embedded coordinate system centered at G with the X axis along the stick in its initial position as shown. The translating component of acceleration of the CG resulting from the force $\mathbf{F}$ is (from Newton's law)

$$A_{G_y} = \frac{F}{m} \qquad (11.11a)$$

and the angular acceleration is:

$$\alpha = \frac{T}{I_G} \tag{11.11b}$$

where I_G is its mass moment of inertia about the Z axis (out of the page) through the CG. But torque is also:

$$T = F l_p \tag{11.11c}$$

where l_p is the distance along the X axis from point G to point P so:

$$\alpha = \frac{F l_p}{I_G} \tag{11.11d}$$

The total linear acceleration at any point along the stick will be the sum of the linear acceleration A_{G_y} of the CG and the tangential component ($r\alpha$) of the angular acceleration as shown in Figure 11-3b.

$$A_{y_{total}} = A_{G_y} + r\alpha$$

$$= \frac{F}{m} + x\left(\frac{F l_p}{I_G}\right) \tag{11.12}$$

where x is the distance to any point along the stick. Equation 11.12 can be set equal to zero and solved for the value of x for which the $r\alpha$ component exactly cancels the A_{G_y} component. This will be the **center of rotation** at which there is no translating acceleration, and thus no linear dynamic force. The solution for x when $A_{y_{total}} = 0$ is:

$$x = -\frac{I_G}{m l_p} \tag{11.13a}$$

and substituting equation 11.10b:

$$x = -\frac{k^2}{l_p} \tag{11.13b}$$

where the radius of gyration k is calculated with respect to the axis ZZ through the CG.

Note that this relationship between the center of percussion and the center of rotation involves only geometry and mass properties. The magnitude of the applied force is irrelevant, but its location l_p completely determines x. Thus there is **not** just one center of percussion on a body. Rather there will be pairs of points. For every point (center of percussion) at which a force is applied there will be a corresponding center of rotation at which the reaction force felt will be zero. This center of rotation need not fall within the physical length of the body however. Consider the value of x predicted by equation 11.13b if you strike the body at its CG!

11.9 SOLUTION METHODS

Dynamic force analysis can be done by any of several methods. Two will be discussed here, **superposition** and **linear simultaneous equation solution**. Both methods require that the system be linear.

These dynamic force problems typically have a large number of unknowns and thus have multiple equations to solve. The method of superposition attacks the problem by solving for parts of the solution and then adding (superposing) the partial results together to get the complete result. For example, if there are two loads applied to the system, we solve independently for the effects of each load, and then add the results. In effect we solve an N variable system by doing sequential calculations on parts of the problem. It can be thought of as a "serial processing" approach.

Another method writes all the relevant equations for the entire system as a set of linear simultaneous equations. These equations can then be solved simultaneously to obtain the results. This can be thought of as a "parallel processing" approach. A convenient approach to the solution of sets of simultaneous equations is to put them in a standard matrix form and use a numerical matrix solver to obtain the answers. Matrix solvers are built into most engineering and scientific pocket calculators. Some spreadsheet packages and equation solvers will also do a matrix solution. A brief introduction to matrix solution of simultaneous equations was presented in Section 5.5. Section 5.6 describes the use of the computer program MATRIX, included on disk with this text. This program allows the rapid calculation of the solution to systems of up to 40 simultaneous equations. Please refer to the sections in Chapter 5 to review these calculation procedures and the use of program MATRIX. Reference 2 also provides an introduction to matrix algebra.

We will use both superposition and simultaneous equation solution to various dynamic force analysis problems in the remaining chapters. Both have their place, and one can serve as a check on the results from the other. So it is useful to be familiar with more than one approach. Historically, superposition was the only practical method for systems involving large numbers of equations until computers became available to solve large sets of simultaneous equations. Now the simultaneous equation solution method is more popular.

11.10 THE PRINCIPLE OF D'ALEMBERT

Newton's second law (equations 11.1 and 11.4) are all that are needed to solve any dynamic force system by the newtonian method. Jean le Rond d'Alembert (1717-1783), a French mathematician, rearranged Newton's equations to create a "quasi-static" situation from a dynamic one. D'Alembert's versions of equations 11.1 and 11.4 are:

$$\sum \mathbf{F} - ma = 0$$

$$\sum \mathbf{T} - I\alpha = 0$$

(11.14)

All d'Alembert did was to move the terms from the right side to the left, changing their algebraic signs in the process as required. These are obviously still the same equations as 11.1 and 11.4, algebraically rearranged. The motivation for this algebraic manipulation was to make the system look like a statics problem in which, for equilibrium, all forces and torques must sum to zero. Thus, this is sometimes called a quasi-static problem when expressed in this form. The premise is that by placing an "inertia force" equal to $-ma$ and an "inertia torque" equal to $-I\alpha$ on our free-body diagrams, the system will then be in a state of "dynamic equilibrium" and can be solved by the familiar methods of statics. These inertia forces and torques are equal in magnitude, opposite in sense, and along the same line of action as ma and $I\alpha$. This was a useful and popular approach which made the solution of dynamic force analysis problems somewhat easier when graphical vector solutions were the methods of choice.

With the availability, literally in your pocket or on your desk, of calculators and computers which can solve the simultaneous equations for these problems, there is now little motivation to labor through the complicated tedium of a graphical force analysis. It is for this reason that graphical force analysis methods are not presented in this text. However, d'Alembert's concept of "inertia forces and torques" still has, at a minimum, historical value and, in many instances, can prove useful in understanding what is going on in a dynamic system. Moreover, the concept of inertia force has entered the popular lexicon and is often used in a lay context when discussing motion. Thus we present a simple example of its use here and will use it again in our discussion of dynamic force analysis later this text where it helps us to understand some topics such as balancing and superposition.

The popular term **centrifugal force**, used by laypersons everywhere to explain why a mass on a rope keeps the rope taut when swung in a circle, is in fact a d'Alembert inertial force. Figure 11-4a shows such a mass, being rotated at the end of a flexible but inextensible cord at a constant angular velocity ω and constant radius r. Figure 11-4b shows "pure" free-body diagrams of both members in this system, the ground link (1) and the rotating link (2). The only real force acting on link 2 is the force of link 1 on 2, $\mathbf{F}_{12}$. Since angular acceleration is zero in this example, the acceleration acting on the link is only the $r\omega^2$ component, which is a **centripetal acceleration**, i.e., directed *toward the center*. The force at the pin from Newton's equation 11.1 is then:

$$\mathbf{F}_{12} = mr\omega^2 \qquad\qquad (11.15a)$$

Note that this force is directed toward the center, so it is a *centripetal* not a *centrifugal* (away from center) force. The force $\mathbf{F}_{21}$ which link 2 exerts on link 1 can be found from Newton's third law and is obviously equal and opposite to $\mathbf{F}_{12}$.

$$\mathbf{F}_{21} = -\mathbf{F}_{12} \qquad\qquad (11.15b)$$

Thus it is the reaction force on link 1 which is centrifugal, not the force on link 2. Of course it is this reaction force that your hand (link 1) feels, and this gives rise to the popular conception of something pulling centrifugally on the rotating weight. Now let us look at this through d'Alembert's eyes. Figure 11-4c shows another set of free-body diagrams done according to the principle of d'Alembert. Here we show a negative $m\mathbf{a}$ inertia force applied to the mass on link 2. The force at the pin from d'Alembert's equation is:

$$\mathbf{F}_{12} - mr\omega^2 = 0$$

$$\mathbf{F}_{12} = mr\omega^2 \qquad\qquad (11.15c)$$

Not surprisingly, the result is the same as equation 11.15a, as it must be. The only difference is that the free-body diagram shows an "inertia force" applied

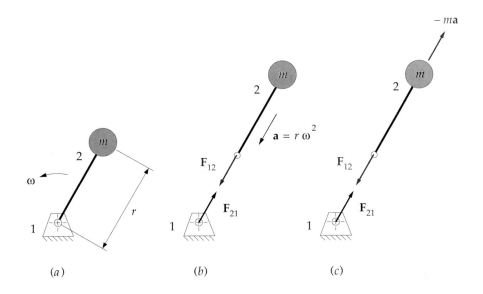

$\mathbf{a} = r\,\omega^2$

(a)　　　　(b)　　　　(c)

FIGURE 11-4

Centripetal and centrifugal forces

to the rotating mass on link 2. This is the centrifugal force of popular repute which takes the blame for keeping the cord taut.

Clearly, any problem can be solved for the right answer no matter how we may algebraically rearrange the correct equations. So, if it helps our understanding to think in terms of these inertia forces being applied to a dynamic system, we will do so. When dealing with the topic of balancing, this approach does, in fact, help to visualize the effects of the balance masses on the system.

11.11 ENERGY METHODS - VIRTUAL WORK

The newtonian methods of dynamic force analysis described above have the advantage of providing complete information about all interior forces at pin joints as well as about the external forces and torques on the system. One consequence of this fact is the relative complexity of their application which requires the simultaneous solution of large systems of equations. Other methods are available for the solution of these problems which are easier to implement but give less information. Energy methods of solution are of this type. Only the external, work producing, forces and torques are found by these methods. The internal joint forces are not computed. One chief value of the energy approach is its use as a quick check on the correctness of the newtonian solution for input torque. Usually we are forced to use the more complete newtonian solution in order to obtain force information at pin joints so that pins and links can be analyzed for failure due to stress.

The **Law of Conservation of Energy** states that energy can neither be created nor destroyed, only converted from one form to another. Most machines are designed specifically to convert energy from one form to another in some controlled fashion. Depending on the efficiency of the machine, some portion of the input energy will be converted to heat which cannot be completely recaptured. But large quantities of energy will typically be stored temporarily within the machine in both potential and kinetic form. It is not uncommon for the magnitude of this internally stored energy, on an instantaneous basis, to far exceed the magnitude of any useful external work being done by the machine.

Work is defined as *the dot product of force and displacement.* It can be positive, negative, or zero and is a scalar quantity.

$$W = \mathbf{F} \cdot \mathbf{R} \tag{11.16a}$$

Since the forces at the pin joints between the links have no relative displacement associated with them, they do no work on the system, and thus will not appear in the work equation. The work done by the system plus losses is equal to the energy delivered to the system.

$$E = W + Losses \tag{11.16b}$$

Pin-jointed linkages with low friction bearings at the pivots can have very high efficiencies, above 90%. Thus it is not unreasonable, for a first approximation in designing such a mechanism, to assume the losses to be zero. **Power** is the time rate of change of energy:

$$P = \frac{dE}{dt} \tag{11.16c}$$

Since we are assuming the machine member bodies to be rigid, only a change of position of the CG's of the members will alter the stored potential energy in the system. The gravitational forces of the members in moderate to high speed machinery often tend to be dwarfed by the dynamic forces from the accelerating masses. For these reasons we will ignore the weights and the gravitational potential energy and consider only the kinetic energy in the system for this analysis. The time rate of change of the kinetic energy stored within the system for linear and angular motion, respectively is then:

$$\frac{d\left(\frac{1}{2}m\mathbf{v}^2\right)}{dt} = m\mathbf{a} \cdot \mathbf{v} \tag{11.17a}$$

and:

$$\frac{d\left(\frac{1}{2}I\omega^2\right)}{dt} = I\alpha \cdot \omega \tag{11.17b}$$

These are, of course, expressions for power in the system, equivalent to:

$$P = \mathbf{F} \cdot \mathbf{v} \tag{11.17c}$$

and:

$$P = \mathbf{T} \cdot \omega \qquad (11.17d)$$

The rate of change of energy in the system at any instant must balance between that which is externally supplied and that which is stored within the system (neglecting losses). Equations 11.17a and 11.17b represent change in the energy stored in the system, and equations 11.17c and 11.17d represent change in energy passing into or out of the system. In the absence of losses, these two must be equal in order to conserve energy. We can express this relationship as a summation of all the delta energies (or power) due to each moving element (or link) in the system.

$$\sum_{k=2}^{n} \mathbf{F}_k \cdot \mathbf{v}_k + \sum_{k=2}^{n} \mathbf{T}_k \cdot \omega_k = \sum_{k=2}^{n} m_k \mathbf{a}_k \cdot \mathbf{v}_k + \sum_{k=2}^{n} I_k \alpha_k \cdot \omega_k \qquad (11.18a)$$

The subscript k represents each of the n links or moving elements in the system, starting with link 2 because link 1 is the stationary ground link. Note that all the angular and linear velocities and accelerations in this equation must have been calculated, for all positions of the mechanism of interest, from a prior kinematic analysis. Likewise, the masses and mass moments of inertia of all moving links must be known.

If we use the principle of d'Alembert to rearrange this equation, we can more easily "name" the terms for discussion purposes.

$$\sum_{k=2}^{n} \mathbf{F}_k \cdot \mathbf{v}_k + \sum_{k=2}^{n} \mathbf{T}_k \cdot \omega_k - \sum_{k=2}^{n} m_k \mathbf{a}_k \cdot \mathbf{v}_k - \sum_{k=2}^{n} I_k \alpha_k \cdot \omega_k = 0 \qquad (11.18b)$$

The first two terms in equation 11.18b represent, respectively, the change in energy due to all **external forces** and all **external torques** applied to the system. These would include any forces or torques from other mechanisms which impinge upon any of these links and also includes the driving torque. The second two terms represent, respectively, the change in energy due to all **inertial forces** and all **inertial torques** present in the system. These last two terms define the change in stored kinetic energy in the system at each time step. The only unknown in this equation when properly set up is the **driving torque** (or driving force) to be supplied by the mechanism's motor or actuator. This driving torque (or force) is then the only variable which can be solved for with this approach. The internal joint forces are not present in the equation as they do no net work on the system.

Equation 11.18b is sometimes called the **virtual work equation**, which is something of a misnomer, as it is in fact a **power equation**. When this analysis approach is applied to a statics problem, there is no motion. The term **virtual work** comes from the concept of each force causing an infinitesimal, or virtual, displacement of the static system element to which it is applied over an infinitesimal delta time. The dot product of the force and the virtual displacement is the virtual work. In the limit, this becomes the instantaneous power in the system. We will present an example of the use of this method of virtual work in the next chapter along with examples of the newtonian solution applied to linkages in motion.

11.12 REFERENCES

1 **Beer, F. P.**, and **Johnson, E. R.**, *Vector Mechanics for Engineers, Statics and Dynamics*, McGraw-Hill Inc., New York, 1984.

2 **Jennings**, A., *Matrix Computation for Engineers and Scientists*, John Wiley and Sons, New York, 1977.

11.13 PROBLEMS

*11-1 The mallet shown in Figure 11-2 has the following specifications: The steel head is 1" dia by 3" tall; the wood handle is 1.25" dia. and 10" long. Find the location of its composite CG, and its moment of inertia and radius of gyration about axis ZZ.

*11-2 Repeat problem 11.1 using a wooden mallet head of 2" dia.

11-3 Calculate the location of the composite center of gravity, the mass moment of inertia and the radius of gyration with respect to the specified axis, for whichever of the following commonly available items that are assigned. (Note these are not short problems.)

a. A good-quality writing pen, about the pivot point at which you grip it to write. (How does placing the cap on the upper end of the pen affect these parameters when you write?)

b. Two table knives, one metal and one plastic, about the pivot axis when held for cutting. Compare the calculated results and comment on what they tell you about the dynamic usability of the two knives (ignore sharpness considerations).

c. A ball-peen hammer (available for inspection in any university machine shop), about the center of rotation (after you calculate its location for the proper center of percussion).

d. A baseball bat (see the coach) about the center of rotation (after you calculate its location for the proper center of percussion).

e. A cylindrical coffee mug, about the handle hole.

*11-4 Set up these equations in matrix form. Use program MATRIX or a calculator which has matrix math ability to solve them.

a.
$$
\begin{aligned}
-5x & -2y & +12z & -w & = -9 \\
x & +3y & -2z & +4w & = 10 \\
-x & -y & +z & & = -7 \\
3x & -3y & +7z & +9w & = -6
\end{aligned}
$$

b.
$$
\begin{aligned}
3x & -5y & +17z & -5w & = -5 \\
-2x & +9y & -14z & +6w & = 22 \\
-x & -y & & -2w & = 13 \\
4x & -7y & +8z & +4w & = -9
\end{aligned}
$$

---

* Answers in Appendix E

Chapter 12

DYNAMIC FORCE ANALYSIS

Don't force it!
Use a bigger hammer
ANONYMOUS

12.0 INTRODUCTION

When kinematic synthesis and analysis have been used to define a geometry and set of motions for a particular design task, it is logical and convenient to then use a **kinetostatic**, or **inverse dynamics**, solution to determine the forces and torques in the system. We will take that approach here and concentrate on solving for the forces and torques that result from, and are required to drive, our kinematic system in such a way as to provide the designed accelerations. Numerical examples are presented throughout this chapter. These examples are also provided as diskfiles for input to either program MATRIX or DYNAFOUR. MATRIX is described in Chapter 5. DYNAFOUR is described in a later section of this chapter which can be read out of order without any loss of continuity. The student is encouraged to read the noted diskfiles into the programs and investigate the examples in more detail. The diskfile names are noted in the discussion of each example.

12.1 NEWTONIAN SOLUTION METHOD

Dynamic force analysis can be done by any of several methods. The one which gives the most information about forces internal to the mechanism requires only the use of Newton's law as defined in equations 11.1 and 11.4. These can be written as a summation of all forces and torques in the system.

$$\sum \mathbf{F} = m\mathbf{a} \qquad\qquad \sum \mathbf{T} = I_G \alpha \qquad\qquad (12.1a)$$

It is also convenient to separately sum force components in X and Y directions, with the coordinate system chosen for convenience. The torques in our two dimensional system are all in the Z direction. This lets us break the two vector equations into three scalar equations:

$$\sum F_x = ma_x \qquad\qquad \sum F_y = ma_y \qquad\qquad \sum T = I_G\alpha \qquad\qquad (12.1b)$$

These three equations must be written for each moving body in the system which will lead to a set of linear simultaneous equations for any system. The set of simultaneous equations can most conveniently be solved by a matrix method as was shown in Chapter 5. These equations do not account for the gravitational force (weight) on a link unless the constant gravitational acceleration is added (vectorially) to the kinematic acceleration for each position. If the kinematic accelerations are large compared to gravity, which is often the case, then the weight forces can be ignored in the dynamic analysis. If the machine members are very massive or moving slowly with small kinematic accelerations, or both, the weight of the members may need to be included in the analysis. As we will see, the weight can conveniently be treated as an external force acting on the CG of the member at a constant angle.

12.2   SINGLE LINK IN PURE ROTATION

As a simple example of this solution procedure, consider the single link in pure rotation shown in Figure 12-1a. In any of these kinetostatic dynamic force analysis problems, the kinematics of the problem must first be fully defined. That is, the angular accelerations of all rotating members and the linear accelerations of the CGs of all moving members must be found for all positions of interest. The mass of each member and the mass moment of inertia I_G with respect to each member's CG must also be known. In addition there may be external forces or torques applied to any member of the system. These are all shown in the figure.

While this analysis can be approached in many ways, it is useful for the sake of consistency to adopt a particular arrangement of coordinate systems and stick with it. We present such an approach here which, if carefully followed, will tend to minimize the chances of error. The reader may wish to develop his or her own approach once the principles are understood. The underlying mathematics is invariant, and one can choose coordinate systems for convenience. The vectors which are acting on the dynamic system in any loading situation are the same at a particular time regardless of how we may decide to resolve them into components for the sake of computation. The solution result will be the same.

We will first set up a nonrotating, local coordinate system on each moving member, located at its CG. (In this simple example we have only one moving member.) All externally applied forces, whether due to other connected members or to other systems must then have their points of application located in this local coordinate system. Figure 12-1b shows a free-body diagram of the moving link 2. The pin joint at O_2 has a force $\mathbf{F}_{12}$ due to the mating link 1, the x and y components of which are F_{12x} and F_{12y}. These subscripts are read "force of link 1 on 2" in the x or y direction. This subscript notation scheme will be used consistently to indicate which of the "action-reaction" pair of forces at each joint is being solved for.

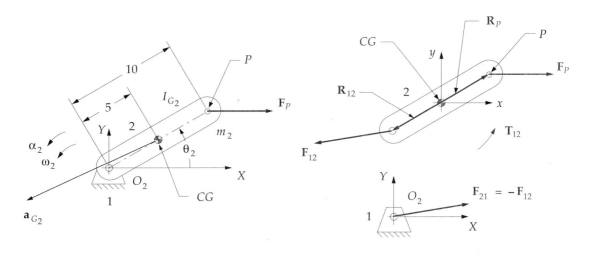

(a) Kinematic diagram   (b) Force (free-body) diagrams

FIGURE 12-1

Dynamic force analysis of a single link in pure rotation

There is also an externally applied force $\mathbf{F}_P$ shown at point P, with components F_{Px} and F_{Py}. The points of application of these forces are defined by position vectors $\mathbf{R}_{12}$ and $\mathbf{R}_P$, respectively. These position vectors are defined with respect to the local coordinate system at the CG of the member. We will need to resolve them into x and y components. There will have to be a source torque available on the link to drive it at the kinematically defined accelerations. This is one of the unknowns to be solved for. The source torque is the torque delivered *from the ground to the driver link 2* and so is labelled $\mathbf{T}_{12}$. The other two unknowns in this example are the force components at the pin joint F_{12x} and F_{12y}.

We have three unknowns and three equations, so the system can be solved. Equations 12.1 can now be written for the moving link 2. Any applied forces or torques whose directions are known must retain the proper signs on their components. We will assume all unknown forces and torques to be positive. Their true signs will "come out in the wash."

$$\sum \mathbf{F} = \mathbf{F}_P + \mathbf{F}_{12} = m_2 \mathbf{a}_G$$
$$\sum \mathbf{T} = \mathbf{T}_{12} + (\mathbf{R}_{12} \times \mathbf{F}_{12}) + (\mathbf{R}_P \times \mathbf{F}_P) = I_G \alpha \qquad (12.2)$$

The force equation can be broken into its two components. The torque equation contains two cross product terms which represent torques due to the forces applied at a distance from the CG. When these cross products are expanded, the system of equations becomes:

$$F_{P_x} + F_{12_x} = m_2 a_{G_x}$$
$$F_{P_y} + F_{12_y} = m_2 a_{G_y} \tag{12.3}$$
$$T_{12} + \left(R_{12_x} F_{12_y} - R_{12_y} F_{12_x} \right) + \left(R_{P_x} F_{P_y} - R_{P_y} F_{P_x} \right) = I_G \alpha$$

This can be put in matrix form with the coefficients of the unknown variables forming the **A** matrix, the unknown variables the **B** vector, and the constant terms the **C** vector and then solved for **B**.

$$
[\mathbf{A}] \qquad \times \quad [\mathbf{B}] \quad = \qquad [\mathbf{C}]
$$

$$
\begin{bmatrix} 1 & 0 & 0 \\ 0 & 1 & 0 \\ -R_{12_y} & R_{12_x} & 1 \end{bmatrix} \times \begin{bmatrix} F_{12_x} \\ F_{12_y} \\ T_{12} \end{bmatrix} = \begin{bmatrix} m_2 a_{G_x} - F_{P_x} \\ m_2 a_{G_y} - F_{P_y} \\ I_G \alpha - \left(R_{P_x} F_{P_y} - R_{P_y} F_{P_x} \right) \end{bmatrix} \tag{12.4}
$$

Note that the **A** matrix contains all the geometrical information and the **C** matrix contains all the dynamical information about the system. The **B** matrix contains all the unknown forces and torques. We will now present a numerical example to reinforce your understanding of this method.

EXAMPLE 12-1

Dynamic Force Analysis of a Single Link in Pure Rotation. (See Figure 12-1)

Given: The 10-inch-long link shown weighs 4 lb. Its *CG* is on the line of centers at the 5 inch point. Its mass moment of inertia about its *CG* is 0.08 lb-in-sec². Its kinematic data are:

θ_2 deg	ω_2 rad / sec	α_2 rad / sec²	a_{G_2} in / sec²
30	20	15	2001 @ 208°

An external force of 40 lb at 0° is applied at point *P*.

Find: The **force F**$_{12}$ at pin joint O_2 and the driving **torque T**$_{12}$ needed to maintain motion with the given acceleration for this instantaneous position of the link.

Solution:

1   Convert the given weight to proper mass units, in this case blobs:

$$ mass = \frac{weight}{g} = \frac{4 \text{ lb}}{386 \text{ in / sec}^2} = 0.0104 \text{ blobs} \tag{a} $$

2   Set up a local coordinate system at the *CG* of the link and draw all applicable vectors acting on the system as shown in the figure. Draw a free-body diagram as shown.

3   Calculate the *x* and *y* components of the position vectors **R**$_{12}$ and **R**$_P$ in this coordinate system:

$$R_{12} = 5''@\angle 210°; \qquad R_{12_x} = -4.33, \qquad R_{12_y} = -2.50$$

$$\mathbf{R}_P = 5''@\angle 30°; \qquad R_{P_x} = +4.33, \qquad R_{P_y} = +2.50 \qquad (b)$$

4   Calculate the x and y components of the acceleration of the CG in this coordinate system:

$$\mathbf{a}_G = 2001@\angle 208°; \qquad a_{G_x} = -1766.78, \qquad a_{G_y} = -939.41 \qquad (c)$$

5   Calculate the x and y components of the external force at P in this coordinate system:

$$\mathbf{F}_P = 40@\angle 0°; \qquad F_{P_x} = 40, \qquad F_{P_y} = 0 \qquad (d)$$

6   Substitute these given and calculated values into the matrix equation 12.4

$$\begin{bmatrix} 1 & 0 & 0 \\ 0 & 1 & 0 \\ 2.50 & -4.33 & 1 \end{bmatrix} \times \begin{bmatrix} F_{12_x} \\ F_{12_y} \\ T_{12} \end{bmatrix} = \begin{bmatrix} (.01)(-1766.78) - 40 \\ (.01)(-939.41) - 0 \\ (.08)(15) - \{(4.33)(0) - (2.5)(40)\} \end{bmatrix}$$

$$(e)$$

$$\begin{bmatrix} 1 & 0 & 0 \\ 0 & 1 & 0 \\ 2.50 & -4.33 & 1 \end{bmatrix} \times \begin{bmatrix} F_{12_x} \\ F_{12_y} \\ T_{12} \end{bmatrix} = \begin{bmatrix} -57.67 \\ -9.39 \\ 101.2 \end{bmatrix}$$

7   Solve this system either by inverting matrix $\mathbf{A}$ and premultiplying that inverse times matrix $\mathbf{C}$ using a pocket calculator such as the HP-15c or by inputting the values for matrices $\mathbf{A}$ and $\mathbf{C}$ to program MATRIX provided with this text.

Program MATRIX gives the following solution:

$$F_{12_x} = -57.67 \text{ lb}, \qquad F_{21_y} = -9.39 \text{ lb}, \qquad T_{12} = 204.72 \text{ lb} - \text{in} \qquad (f)$$

Converting the force to polar coordinates:

$$\mathbf{F}_{12} = 58.43@\angle 189.25° \qquad (g)$$

Read the diskfile EX12-1 into program MATRIX to exercise this example.

12.3   FORCE ANALYSIS OF A THREEBAR CRANK-SLIDE LINKAGE

When there are more links than one in the assembly, the solution simply requires that the three equations 12.1 be written for each link and then solved simultaneously. Figure 12-2a shows a threebar crank-slide linkage. This linkage has been simplified from the fourbar slider-crank (see Figure 12-4) by replacing the kinematically redundant slider block (link 4) with a half joint as shown. This linkage transformation reduces the number of links to three with no change in degree of freedom (see Section 2.9). Only links 2 and 3 are moving. Link 1 is ground. Thus we should expect to have six equations in six unknowns (three per moving link).

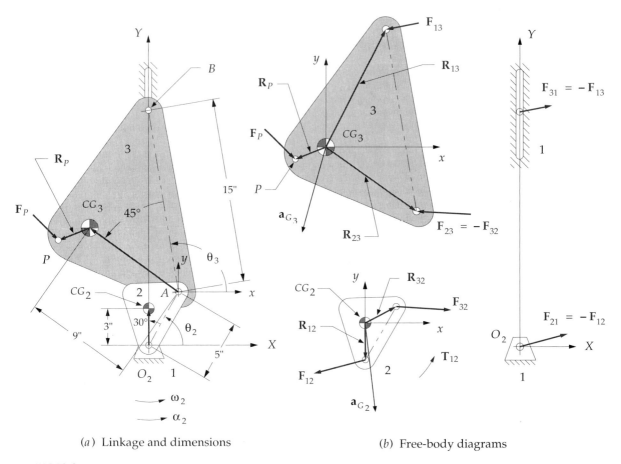

(*a*) Linkage and dimensions   (*b*) Free-body diagrams

FIGURE 12-2

Dynamic force analysis of a slider-crank linkage

Figure 12-2b shows the linkage "exploded" into its three separate links, drawn as free bodies. A kinematic analysis must have been done in advance of this dynamic force analysis in order to determine, for each moving link, its angular acceleration and the linear acceleration of its *CG*. For the kinematic analysis, only the link lengths from pin to pin were required. For a dynamic analysis the mass (*m*) of each link, the location of its *CG*, and its mass moment of inertia (I_G) about that *CG* are also needed.

The *CG* of each link is initially defined by a position vector rooted at one pin joint whose angle is measured with respect to the line of centers of the link. This is the most convenient way to establish the *CG* location since the link line of centers is the kinematic definition of the link. However, we will need to define the link's dynamic parameters and force locations with respect to an axis system located at its *CG*. The position vector locations of all attachment points of other links and points of application of external forces must be defined with respect to this **translating but nonrotating** local *xy* axis system. Note that these kinematic and applied force data must be available for all positions of the linkage for which a force analysis is desired.

In the following discussion and examples, only one linkage position will be addressed. However, the process is identical for each succeeding position and only the calculations must be repeated. Obviously, a computer will be a valuable aid in accomplishing the task.

Link 2 in Figure 12-2b shows forces acting on it at each pin joint, designated $\mathbf{F}_{12}$ and $\mathbf{F}_{32}$. By convention their subscripts denote the force that the adjoining link is exerting *on* the link being analyzed; that is, $\mathbf{F}_{12}$ is the force of 1 *on* 2 and $\mathbf{F}_{32}$ is the force of 3 *on* 2. Obviously there is also an equal and opposite force at each of these pins which would be designated as $\mathbf{F}_{21}$ and $\mathbf{F}_{23}$, respectively. The choice of which of the members of these pairs of forces to be solved for is arbitrary. As long as proper bookkeeping is done, their identities will be maintained.

When we move to link 3, we maintain the same convention of showing forces acting *on* the link in its free-body diagram. Thus at instant center I_{23} we show $\mathbf{F}_{23}$ acting on link 3. However, because we showed force $\mathbf{F}_{32}$ acting at the same point on link 2, this introduces an additional unknown to the problem for which we need an additional equation. The equation is available from Newton's third law:

$$\mathbf{F}_{23} = -\mathbf{F}_{32} \qquad (12.5)$$

Thus we are free to substitute the negative reaction force for any action force at any joint. This has been done on link 3 in the figure in order to reduce the unknown forces at that joint to one, namely $\mathbf{F}_{32}$. The same procedure is followed at each joint with one of the action-reaction forces arbitrarily chosen to be solved for and its negative reaction applied to the mating link.

The naming convention used for the position vectors ($\mathbf{R}_{ap}$) which locate the pin joints with respect to the *CG* in the link's nonrotating local coordinate system is as follows. The first subscript (a) denotes the adjoining link to which the position vector points. The second subscript (p) denotes the parent link to which the position vector belongs. Thus in the case of link 2 in Figure 12-2b, vector $\mathbf{R}_{12}$ locates the attachment point of link 1 to link 2, and $\mathbf{R}_{32}$ the attachment point of link 3 to link 2. Note that in some cases these subscripts will match those of the pin forces shown acting at those points, but where the negative reaction force has been substituted as described above, the subscript order of the force and its position vector will not agree. This can lead to confusion and must be carefully watched for typographical errors when setting up the problem.

Any external forces acting on the links are located in similar fashion with a position vector to a point on the line of application of the force, which point is given the same letter subscript as that of the external force. Link 3 in the figure shows such an external force $\mathbf{F}_P$ acting on it at point P. The position vector $\mathbf{R}_P$ locates that point with respect to the *CG*. It is important to note that the *CG* of each link is consistently taken as the point of reference for all forces acting on that link. Left to its own devices, an unconstrained body in complex motion will spin about its own *CG*; thus we analyze its linear acceleration at that point and apply the angular acceleration about the *CG* as a center.

Equations 12.1 are now written for each moving link. For link 2, with the cross products expanded:

$$F_{12_x} + F_{32_x} = m_2 a_{G_{2x}}$$
$$F_{12_y} + F_{32_y} = m_2 a_{G_{2y}}$$

$$T_{12} + \left(R_{12_x} F_{12_y} - R_{12_y} F_{12_x} \right) + \left(R_{32_x} F_{32_y} - R_{32_y} F_{32_x} \right) = I_{G_2} \alpha_2$$

(12.6a)

For link 3, with the cross products expanded, note the substitution of the reaction force $-\mathbf{F}_{32}$ for $\mathbf{F}_{23}$:

$$F_{13_x} - F_{32_x} + F_{P_x} = m_3 a_{G_{3x}}$$
$$F_{13_y} - F_{32_y} + F_{P_y} = m_3 a_{G_{3y}}$$

$$\left(R_{13_x} F_{13_y} - R_{13_y} F_{13_x} \right) - \left(R_{23_x} F_{32_y} - R_{23_y} F_{32_x} \right) + \left(R_{P_x} F_{P_y} - R_{P_y} F_{P_x} \right) = I_{G_3} \alpha_3$$

(12.6b)

Note also that $\mathbf{T}_{12}$, the source torque, only appears in the equation for link 2 as that is the driver crank to which the motor is attached. Link 3 has no externally applied torque but does have an external force $\mathbf{F}_P$ which might be due to whatever link 3 is pushing on to do its external work.

There are seven unknowns present in these six equations, F_{12x}, F_{12y}, F_{32x}, F_{32y}, F_{13x}, F_{13y}, and T_{12}. But, F_{13y} is due only to friction at the joint between link 3 and link 1. We can write a relation for the friction force f at that interface such as $f = \pm\mu N$, where $\pm\mu$ is a known coefficient of coulomb friction. The friction force always opposes motion. The kinematic analysis will provide the velocity of the link at the sliding joint. The direction of f will always be the opposite of this velocity. Note that μ is a nonlinear function which has a discontinuity at zero velocity, thus at the linkage positions where velocity is zero, the inclusion of μ in these linear equations is not valid. (See Figure 17-2a on p. 626.) In this example, the normal force N is equal to F_{13x} and the friction force f is equal to F_{13y}. For linkage positions with nonzero velocity, we can eliminate F_{13y} by substituting into equation 12.6b,

$$F_{13_y} = \pm\mu \, F_{13_x}$$

(12.6c)

where the sign of μ is taken as the opposite of the sign of the velocity at that point. We are then left with six unknowns in equations 12.6 and can solve them simultaneously. We also rearrange equations 12.6a and 12.6b to put all known terms on the right side.

$$F_{12_x} + F_{32_x} = m_2 a_{G_{2x}}$$
$$F_{12_y} + F_{32_y} = m_2 a_{G_{2y}}$$
$$T_{12} + R_{12_x} F_{12_y} - R_{12_y} F_{12_x} + R_{32_x} F_{32_y} - R_{32_y} F_{32_x} = I_{G_2} \alpha_2$$

(12.6d)

$$F_{13_x} - F_{32_x} = m_3 a_{G_{3x}} - F_{P_x}$$
$$\pm\mu \, F_{13_x} - F_{32_y} = m_3 a_{G_{3y}} - F_{P_y}$$

$$\left(\pm\mu \, R_{13_x} - R_{13_y} \right) F_{13_x} - R_{23_x} F_{32_y} + R_{23_y} F_{32_x} = I_{G_3} \alpha_3 - R_{P_x} F_{P_y} + R_{P_y} F_{P_x}$$

Putting these six equations in matrix form we get:

$$
\begin{bmatrix}
1 & 0 & 1 & 0 & 0 & 0 \\
0 & 1 & 0 & 1 & 0 & 0 \\
-R_{12_y} & R_{12_x} & -R_{32_y} & R_{32_x} & 0 & 1 \\
0 & 0 & -1 & 0 & 1 & 0 \\
0 & 0 & 0 & -1 & \pm\mu & 0 \\
0 & 0 & R_{23_y} & -R_{23_x} & \left(\pm\mu R_{13_x} - R_{13_y}\right) & 0
\end{bmatrix}
\times
\begin{bmatrix}
F_{12_x} \\
F_{12_y} \\
F_{32_x} \\
F_{32_y} \\
F_{13_x} \\
T_{12}
\end{bmatrix}
=
$$

(12.7)

$$
\begin{bmatrix}
m_2 a_{G_{2x}} \\
m_2 a_{G_{2y}} \\
I_{G_2}\alpha_2 \\
m_3 a_{G_{3x}} - F_{P_x} \\
m_3 a_{G_{3y}} - F_{P_y} \\
I_{G_3}\alpha_3 - R_{P_x}F_{P_y} + R_{P_y}F_{P_x}
\end{bmatrix}
$$

This system can be solved by using program MATRIX or any other matrix solving calculator.

As an example of this solution consider the following linkage data.

✎ EXAMPLE 12-2

Dynamic Force Analysis of a Threebar Crank Slide Linkage with Half Joint. (See Figure 12-2)

Given: The 5 inch long crank (link 2) shown weighs 2 lb. Its CG is at 3 inches and 30° from the line of centers. Its mass moment of inertia about its CG is 0.05 lb-in-sec². Its kinematic data are:

θ_2 deg	ω_2 rad / sec	α_2 rad / sec²	a_{G_2} in / sec²
60	30	−10	2700.17 @ −89.4°

The coupler (link 3) is 15 inches long and weighs 4 lb. Its CG is at 9 inches and 45° from the line of centers. Its mass moment of inertia about its CG is 0.10 lb-in-sec². Its kinematic data are:

θ_3 deg	ω_3 rad / sec	α_3 rad / sec²	a_{G_3} in / sec²
99.52	−8.79	−135.65	3447.46 @ 253.6°

The sliding joint on link 3 has a velocity of +96.7 in/sec in the y direction.

There is an external force of 50 lb at − 45°, applied at point P which is located at 2.7 inches and 101° from the CG of link 3, measured in the link's embedded, rotating coordinate system (origin at I_{23} and x axis from I_{23} to I_{34}). The coefficient of friction μ is -0.2, the negative sign being used because of that point's positive velocity.

Find: The **forces** F_{12}, F_{32}, F_{13} at the joints and the driving **torque** T_{12} needed to maintain motion with the given acceleration for this instantaneous position of the link.

Solution:

1   Convert the given weights to proper mass units, in this case blobs:

$$mass_{link2} = \frac{weight}{g} = \frac{2 \text{ lb}}{386 \text{ in} / \sec^2} = 0.0052 \text{ blobs} \qquad (a)$$

$$mass_{link3} = \frac{weight}{g} = \frac{4 \text{ lb}}{386 \text{ in} / \sec^2} = 0.0104 \text{ blobs} \qquad (b)$$

2   Set up a local xy coordinate system at the *CG* of each link and draw all applicable vectors acting on that system as shown in the figure. Draw a free-body diagram of each moving link as shown.

3   Calculate the x and y components of the position vectors R_{12}, R_{32}, R_{23}, R_{13}, and R_P in this coordinate system:

$$
\begin{array}{lll}
R_{12} = 3.00 \ @ \angle \ 270.0°; & R_{12_x} = 0, & R_{12_y} = -3.0 \\
R_{32} = 2.83 \ @ \angle \ 28.0°; & R_{32_x} = 2.500, & R_{32_y} = 1.333 \\
R_{23} = 9.00 \ @ \angle \ 324.5°; & R_{23_x} = 7.329, & R_{23_y} = -5.224 \\
R_{13} = 10.72 \ @ \angle \ 63.14°; & R_{13_x} = 4.843, & R_{13_y} = 9.563 \\
R_P = 2.70 \ @ \angle \ 201.0°; & R_{P_x} = -2.521, & R_{P_y} = -0.968 \\
\end{array}
\qquad (c)
$$

Note that these position vector angles are all measured with respect to the global coordinate system.

4   Calculate the x and y components of the acceleration of the *CGs* of all moving links in this coordinate system:

$$
\begin{array}{lll}
a_{G_2} = 2700.17 \ @ \angle \ -89.4°; & a_{G_{2x}} = 28.28, & a_{G_{2y}} = -2700
\end{array}
$$

$$
\begin{array}{lll}
a_{G_3} = 3447.46 \ @ \angle \ 253.6°; & a_{G_{3x}} = -973.36, & a_{G_{3y}} = -3307
\end{array}
\qquad (d)
$$

5   Calculate the x and y components of the external force at P in this coordinate system:

$$F_P = 50 @ \angle -45°; \qquad F_{P_x} = 35.36, \qquad F_{P_y} = -35.36 \qquad (e)$$

6   Substitute these given and calculated values into the matrix equation 12.7.

$$
\begin{bmatrix}
1 & 0 & 1 & 0 & 0 & 0 \\
0 & 1 & 0 & 1 & 0 & 0 \\
3 & 0 & -1.333 & 2.5 & 0 & 1 \\
0 & 0 & -1 & 0 & 1 & 0 \\
0 & 0 & 0 & -1 & -0.2 & 0 \\
0 & 0 & -5.224 & -7.329 & [(-0.2)4.843-(9.563)] & 0
\end{bmatrix}
\times
\begin{bmatrix}
F_{12_x} \\
F_{12_y} \\
F_{32_x} \\
F_{32_y} \\
F_{13_x} \\
T_{12}
\end{bmatrix}
=
$$

$$(f)$$

$$
\begin{bmatrix}
(.005)(28.28) \\
(.005)(-2700) \\
(.05)(-10) \\
(.01)(-973.36)-35.36 \\
(.01)(-3307)-(-35.36) \\
(.1)(-135.65)-(-2.521)(-35.36)+(-.968)(35.36)
\end{bmatrix}
=
\begin{bmatrix}
0.141 \\
-13.500 \\
-.500 \\
-45.094 \\
2.288 \\
-136.936
\end{bmatrix}
$$

7 Solve this system either by inverting matrix **A** and premultiplying that inverse times matrix **C** using a pocket calculator such as the HP-15c, or by inputting the values for matrices **A** and **C** to program MATRIX provided with this text. Program MATRIX gives the following solution:

$$
\begin{bmatrix}
F_{12_x} \\
F_{12_y} \\
F_{32_x} \\
F_{32_y} \\
F_{13_x} \\
T_{12}
\end{bmatrix}
=
\begin{bmatrix}
-39.225 \\
-12.356 \\
38.366 \\
-1.144 \\
-5.728 \\
172.520
\end{bmatrix}
$$

$$(g)$$

Converting the forces to polar coordinates:

$$
\begin{aligned}
\mathbf{F}_{12} &= 40.3 \quad @ \angle \ 198.2° \\
\mathbf{F}_{32} &= 38.4 \quad @ \angle \ -1.4° \\
\mathbf{F}_{13} &= 6.8 \quad @ \angle \ 191.3°
\end{aligned}
$$

$$(h)$$

Read the diskfile EX12-2 into program MATRIX to exercise this example.

12.4 FORCE ANALYSIS OF A FOURBAR LINKAGE

Figure 12-3a shows a fourbar linkage. All dimensions of link lengths, link positions, locations of the links' CGs, linear accelerations of those CGs, and link angular accelerations and velocities have been previously determined from a kinematic analysis. We now wish to find the forces acting at all the pin joints of the linkage for one or more positions. The procedure is exactly the same as that used in the above two examples. This linkage has three moving links. Equation 12.1 provides three equations for any link or rigid body in motion. Thus we should expect to have nine equations in nine unknowns for this problem.

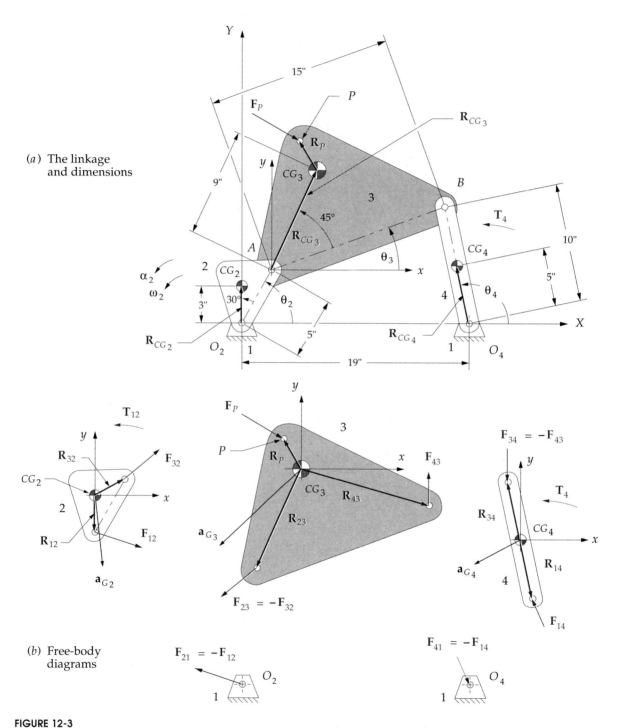

(a) The linkage and dimensions

(b) Free-body diagrams

FIGURE 12-3

Dynamic force analysis of a fourbar linkage

Figure 12-3b shows the free-body diagrams for all links, with all forces shown. Note that an external force $\mathbf{F}_P$ is shown acting on link 3 at point P. Also an external torque $\mathbf{T}_4$ is shown acting on link 4. These external loads are due to some other mechanism (device, person, thing, etc.) pushing or twisting against the motion of the linkage. Any link can have any number of external loads and torques acting on it. Only one external torque and one external force are shown here to serve as examples of how they are handled in the computation. (Note that a more complicated force system, if present, could also be reduced to the combination of a single force and torque on each link.)

To solve for the pin forces it is necessary that these applied external forces and torques be defined for all positions of interest. We will solve for one member of the pair of action-reaction forces at each joint, and also for the driving torque $\mathbf{T}_{12}$ needed to be supplied at link 2 in order to maintain the kinematic state as defined. The force subscript convention is the same as that defined in the previous example. For example, $\mathbf{F}_{12}$ is the force of 1 *on* 2 and $\mathbf{F}_{32}$ is the force of 3 *on* 2. The equal and opposite forces at each of these pins are designated $\mathbf{F}_{21}$ and $\mathbf{F}_{23}$, respectively. All the unknown forces in the figure are shown at arbitrary angles and lengths as their true values are still to be determined.

The linkage kinematic parameters are defined with respect to a global XY system whose origin is at the driver pivot O_2 and whose X axis goes through link 4's fixed pivot O_4. The mass (m) of each link, the location of its CG, and its mass moment of inertia (I_G) about that CG are also needed. The CG of each link is initially defined within each link with respect to a moving and rotating axis system embedded in the link. The origin of this axis system is at one pin joint and its x axis is the line of centers of the link. The CG position within the link is defined by a position vector in this moving and rotating coordinate system. The instantaneous location of the CG can easily be determined for each dynamic link position by adding the angle of the internal CG position vector to the current angle of the link.

We need to define each link's dynamic parameters and force locations with respect to a local, moving, but nonrotating axis system located at its CG as shown on each free-body diagram in Figure 12-3b. The position vector locations of all attachment points of other links and points of application of external forces must be defined with respect to this local xy axis system. These kinematic and applied force data differ for each position of the linkage. In the following discussion and examples, only one linkage position will be addressed. The process is identical for each succeeding position.

Equations 12.1 are now written for each moving link. For link 2, the result is identical to that done for the slider-crank example in equation 12.6a.

$$F_{12_x} + F_{32_x} = m_2 a_{G2_x}$$
$$F_{12_y} + F_{32_y} = m_2 a_{G2_y} \qquad (12.8a)$$
$$T_{12} + \left(R_{12_x} F_{12_y} - R_{12_y} F_{12_x} \right) + \left(R_{32_x} F_{32_y} - R_{32_y} F_{32_x} \right) = I_{G2} \alpha_2$$

For link 3, with substitution of the reaction force $-\mathbf{F}_{32}$ for $\mathbf{F}_{23}$, the result is similar to equation 12.6b with some subscript changes to reflect the presence of link 4.

$$F_{43_x} - F_{32_x} + F_{P_x} = m_3 a_{G3_x}$$
$$F_{43_y} - F_{32_y} + F_{P_y} = m_3 a_{G3_y} \tag{12.8b}$$

$$\left(R_{43_x} F_{43_y} - R_{43_y} F_{43_x}\right) - \left(R_{23_x} F_{32_y} - R_{23_y} F_{32_x}\right) + \left(R_{P_x} F_{P_y} - R_{P_y} F_{P_x}\right) = I_{G3} \alpha_3$$

For link 4, substituting the reaction force $-\mathbf{F}_{43}$ for $\mathbf{F}_{34}$, a similar set of equations 12.1 can be written:

$$F_{14_x} - F_{43_x} = m_4 a_{G4_x}$$
$$F_{14_y} - F_{43_y} = m_4 a_{G4_y} \tag{12.8c}$$

$$\left(R_{14_x} F_{14_y} - R_{14_y} F_{14_x}\right) - \left(R_{34_x} F_{43_y} - R_{34_y} F_{43_x}\right) + T_4 = I_{G4} \alpha_4$$

Note again that $\mathbf{T}_{12}$, the source torque, only appears in the equation for link 2 as that is the driver crank to which the motor is attached. Link 3, in this example, has no externally applied torque (though it could have) but does have an external force $\mathbf{F}_P$. Link 4, in this example, has no external force acting on it (though it could have) but does have an external torque $\mathbf{T}_4$. (The driving link 2 could also have an externally applied force on it though it usually does not.) There are nine unknowns present in these nine equations, $F_{12x}, F_{12y}, F_{32x}, F_{32y}, F_{43x}, F_{43y}, F_{14x}, F_{14y}$ and T_{12}, so we can solve them simultaneously. We rearrange terms in equations 12.8 to put all known constant terms on the right side and then put them in matrix form.

$$
\begin{bmatrix}
1 & 0 & 1 & 0 & 0 & 0 & 0 & 0 & 0 \\
0 & 1 & 0 & 1 & 0 & 0 & 0 & 0 & 0 \\
-R_{12_y} & R_{12_x} & -R_{32_y} & R_{32_x} & 0 & 0 & 0 & 0 & 1 \\
0 & 0 & -1 & 0 & 1 & 0 & 0 & 0 & 0 \\
0 & 0 & 0 & -1 & 0 & 1 & 0 & 0 & 0 \\
0 & 0 & R_{23_y} & -R_{23_x} & -R_{43_y} & R_{43_x} & 0 & 0 & 0 \\
0 & 0 & 0 & 0 & -1 & 0 & 1 & 0 & 0 \\
0 & 0 & 0 & 0 & 0 & -1 & 0 & 1 & 0 \\
0 & 0 & 0 & 0 & R_{34_y} & -R_{34_x} & -R_{14_y} & R_{14_x} & 0
\end{bmatrix}
\times
\begin{bmatrix}
F_{12_x} \\
F_{12_y} \\
F_{32_x} \\
F_{32_y} \\
F_{43_x} \\
F_{43_y} \\
F_{14_x} \\
F_{14_y} \\
T_{12}
\end{bmatrix}
=
\tag{12.9}
$$

$$
\begin{bmatrix}
m_2 a_{G2_x} \\
m_2 a_{G2_y} \\
I_{G2} \alpha_2 \\
m_3 a_{G3_x} - F_{P_x} \\
m_3 a_{G3_y} - F_{P_y} \\
I_{G3} \alpha_3 - R_{P_x} F_{P_y} + R_{P_y} F_{P_x} \\
m_4 a_{G4_x} \\
m_4 a_{G4_y} \\
I_{G4} \alpha_4 - T_4
\end{bmatrix}
$$

This system can be solved by using program MATRIX or any matrix solving calculator. As an example of this solution consider the following linkage data.

✐ EXAMPLE 12-3

Dynamic Force Analysis of a Fourbar Linkage. (See Figure 12-3)

Given: The 5-inch-long crank (link 2) shown weighs 1.5 lb. Its *CG* is at 3 inches at +30° from the line of centers. Its mass moment of inertia about its *CG* is 0.4 lb-in-sec². Its kinematic data are:

θ_2 deg	ω_2 rad / sec	α_2 rad / sec^2	a_{G_2} in / sec^2
60	25	−40	1878.84 @ 273.66°

The coupler (link 3) is 15 inches long and weighs 7.7 lb. Its *CG* is at 9 inches at 45° off the line of centers. Its mass moment of inertia about its *CG* is 1.5 lb-in-sec². Its kinematic data are:

θ_3 deg	ω_3 rad / sec	α_3 rad / sec^2	a_{G_3} in / sec^2
20.92	−5.87	120.9	3697.86 @ 223.5°

There is an external force of 80 lb at 330° on link 3, applied at point *P* which is located 3 inches at 100° from the *CG* of link 3. There is an external torque on link 4 of 120 lb-in. The ground link is 19 inches long. The rocker (link 4) is 10 inches long and weighs 5.8 lb. Its *CG* is at 5 inches at 0° off the line of centers. Its mass moment of inertia about its *CG* is 0.8 lb-in-sec². Its kinematic data are:

θ_4 deg	ω_4 rad / sec	α_4 rad / sec^2	a_{G_4} in / sec^2
104.41	7.93	276.29	1416.8 @ 207.2°

Find: The **forces** F_{12}, F_{32}, F_{43}, F_{14}, at the joints and the driving **torque** T_{12} needed to maintain motion with the given acceleration for this instantaneous position of the link.

Solution:

1   Convert the given weight to proper mass units, in this case blobs:

$$mass_{link2} = \frac{weight}{g} = \frac{1.5 \text{ lb}}{386 \text{ in / sec}^2} = 0.0039 \text{ blobs} \qquad (a)$$

$$mass_{link3} = \frac{weight}{g} = \frac{7.7 \text{ lb}}{386 \text{ in / sec}^2} = 0.0199 \text{ blobs} \qquad (b)$$

$$mass_{link4} = \frac{weight}{g} = \frac{5.8 \text{ lb}}{386 \text{ in / sec}^2} = 0.0150 \text{ blobs} \qquad (c)$$

2   Set up a local *xy* coordinate system at the *CG* of each link and draw all applicable vectors acting on that system as shown in the figure. Draw a free-body diagram of each moving link as shown.

3   Calculate the *x* and *y* components of the position vectors R_{12}, R_{32}, R_{23}, R_{43}, R_{34}, R_{14}, and R_P in this coordinate system. Note that the current value of link 3's position angle (θ_3) must be added to the angles of all position vectors before creating their components, because their angles were originally measured with respect to the link's embedded, rotating coordinate system.

$$R_{12} = 3.00 \ @ \ \angle \ 270.00°; \qquad R_{12_x} = 0.000, \qquad R_{12_y} = -3$$

$$R_{32} = 2.83 \ @ \ \angle \ 28.00°; \qquad R_{32_x} = 2.500, \qquad R_{32_y} = 1.333$$

$$R_{23} = 9.00 \ @ \ \angle \ 245.92°; \qquad R_{23_x} = -3.672, \qquad R_{23_y} = -8.217$$

$$R_{43} = 10.72 \ @ \ \angle \ -15.46°; \qquad R_{43_x} = 10.332, \qquad R_{43_y} = -2.858 \qquad (d)$$

$$R_{34} = 5.00 \ @ \ \angle \ 104.41°; \qquad R_{34_x} = -1.244, \qquad R_{34_y} = 4.843$$

$$R_{14} = 5.00 \ @ \ \angle \ 284.41°; \qquad R_{14_x} = 1.244, \qquad R_{14_y} = -4.843$$

$$R_P = 3.00 \ @ \ \angle \ 120.92°; \qquad R_{P_x} = -1.542, \qquad R_{P_y} = 2.574$$

4 Calculate the x and y components of the acceleration of the CGs of all moving links in this coordinate system:

$$\mathbf{a}_{G_2} = 1878.84 \ @ \ \angle 273.66°; \qquad a_{G_{2x}} = 119.94, \qquad a_{G_{2y}} = -1875.01$$

$$\mathbf{a}_{G_3} = 3697.86 \ @ \ \angle 223.50°; \qquad a_{G_{3x}} = -2682.33, \qquad a_{G_{3y}} = -2545.44 \qquad (e)$$

$$\mathbf{a}_{G_4} = 1416.80 \ @ \ \angle 207.24°; \qquad a_{G_{4x}} = -1259.67, \qquad a_{G_{4y}} = -648.50$$

5 Calculate the x and y components of the external force at P in this coordinate system:

$$\mathbf{F}_{P3} = 80 \ @ \ \angle \ 330°; \qquad F_{P3_x} = 69.28, \qquad F_{P3_y} = -40.00 \qquad (f)$$

6 Substitute these given and calculated values into the matrix equation 12.9

$$
\begin{bmatrix}
1 & 0 & 1 & 0 & 0 & 0 & 0 & 0 & 0 \\
0 & 1 & 0 & 1 & 0 & 0 & 0 & 0 & 0 \\
3 & 0 & -1.333 & 2.5 & 0 & 0 & 0 & 0 & 1 \\
0 & 0 & -1 & 0 & 1 & 0 & 0 & 0 & 0 \\
0 & 0 & 0 & -1 & 0 & 1 & 0 & 0 & 0 \\
0 & 0 & -8.217 & 3.672 & 2.858 & 10.332 & 0 & 0 & 0 \\
0 & 0 & 0 & 0 & -1 & 0 & 1 & 0 & 0 \\
0 & 0 & 0 & 0 & 0 & -1 & 0 & 1 & 0 \\
0 & 0 & 0 & 0 & 4.843 & 1.244 & 4.843 & 1.244 & 0
\end{bmatrix}
\times
\begin{bmatrix}
F_{12_x} \\
F_{12_y} \\
F_{32_x} \\
F_{32_y} \\
F_{43_x} \\
F_{43_y} \\
F_{14_x} \\
F_{14_y} \\
T_{12}
\end{bmatrix}
=
$$

$$(g)$$

$$
\begin{bmatrix}
(0.004)(119.94) \\
(0.004)(-1875.01) \\
(0.4)(-40) \\
(0.02)(-2682.33)-(69.28) \\
(0.02)(-2545.44)-(-40) \\
(1.5)(120.9)-\left[(-1.542)(-40)-(2.574)(69.28)\right] \\
(0.015)(-1259.67) \\
(0.015)(-648.50) \\
(0.8)(276.29)-(120)
\end{bmatrix}
=
\begin{bmatrix}
0.480 \\
-7.500 \\
-16.000 \\
-122.927 \\
-10.909 \\
298.003 \\
-18.895 \\
-9.728 \\
101.032
\end{bmatrix}
$$

7 Solve this system either by inverting matrix **A** and premultiplying that inverse times matrix **C** using a pocket calculator such as the HP-28, or by inputting the values for matrices **A** and **C** to program MATRIX provided with this text. Program MATRIX gives the following solution:

$$
\begin{bmatrix}
F_{12_x} \\
F_{12_y} \\
F_{32_x} \\
F_{32_y} \\
F_{43_x} \\
F_{43_y} \\
F_{14_x} \\
F_{14_y} \\
T_{12}
\end{bmatrix}
=
\begin{bmatrix}
-120.51 \\
-108.21 \\
120.99 \\
100.71 \\
-1.94 \\
89.81 \\
-20.84 \\
80.08 \\
255.04
\end{bmatrix}
\tag{h}
$$

Converting the forces to polar coordinates:

$$
\begin{aligned}
\mathbf{F}_{12} &= 161.96 \ @ \ \angle \ 221.73° \\
\mathbf{F}_{32} &= 157.42 \ @ \ \angle \ 39.77° \\
\mathbf{F}_{43} &= 89.83 \ @ \ \angle \ 91.24° \\
\mathbf{F}_{14} &= 82.75 \ @ \ \angle \ 104.59°
\end{aligned}
\tag{k}
$$

This solves the linkage for one position. A new set of values can be put into the **A** and **C** matrices for each position of interest at which a force analysis is needed. Read the diskfile EX12-3 into program MATRIX to exercise this example. A diskfile of the same name can also be read into program DYNAFOUR which will run the linkage through a series of positions starting with the stated parameters as initial conditions. The linkage will slow to a stop and then run in reverse due to the negative acceleration.

It is worth noting some general observations about this method at this point. The solution is done using cartesian coordinates of all forces and position vectors. Before being placed in the matrices, these vector components must be defined in the global coordinate system or in nonrotating, local coordinate systems, parallel to the global coordinate system, with their origins at the links' CGs. Some of the linkage parameters are normally expressed in such coordinate systems, but some are not, and so must be converted. The kinematic data should all be computed in the global system or in parallel, **nonrotating**, local systems placed at the CGs of individual links. Any external forces on the links must also be defined in the global system.

However, the position vectors that define intralink locations, such as the pin joints versus the CG, or which locate points of application of external forces versus the CG are defined in local, **rotating** coordinate systems embedded in the links. Thus these position vectors must be redefined in a **nonrotating**, parallel system before being used in the matrix. An example of this is vector $\mathbf{R}_p$, which was initially defined as 3 inches at $100°$ in link 3's embedded, **rotating** coordinate system. Note in the example above that its cartesian coordinates for use in the equations were calculated after adding the current value of θ_3 to its angle. This redefined $\mathbf{R}_p$ as 3 inches at $120.92°$ in the **nonrotating** local system. The same was done for position vectors $\mathbf{R}_{12}$, $\mathbf{R}_{32}$, $\mathbf{R}_{23}$, $\mathbf{R}_{43}$, $\mathbf{R}_{34}$, and $\mathbf{R}_{14}$. In each case the **intralink angle** of these vectors (which is independent of linkage position) was added to the current link angle to obtain its position in the xy system at the link's CG. The proper definition of these position vector components is critical to the solution, and it is very easy to make errors in defining them.

To further confuse things, even though the position vector $\mathbf{R}_p$ is initially measured in the link's embedded, rotating coordinate system, the force $\mathbf{F}_p$, which it locates, is not. The force $\mathbf{F}_p$ is not part of the link, as is $\mathbf{R}_p$, but rather is part of the external world, so it is defined in the global system.

12.5   FORCE ANALYSIS OF A FOURBAR SLIDER-CRANK LINKAGE

The approach taken for the pin-jointed fourbar is equally valid for a fourbar slider-crank linkage. The principal difference will be that the slider block will have no angular acceleration. Figure 12-4 shows a fourbar slider-crank with an external force on the slider block, link 4. This is representative of the mechanism used extensively in piston pumps and internal combustion engines. We wish to determine the forces at the joints and the driving torque needed on the crank to provide the specified accelerations. A kinematic analysis must have previously been done in order to determine all position, velocity, and acceleration information for the positions being analyzed. Equations 12.1 are written for each link. For link 2:

$$F_{12_x} + F_{32_x} = m_2 a_{G2_x}$$
$$F_{12_y} + F_{32_y} = m_2 a_{G2_y} \quad\quad (12.10a)$$
$$T_{12} + \left(R_{12_x} F_{12_y} - R_{12_y} F_{12_x} \right) + \left(R_{32_x} F_{32_y} - R_{32_y} F_{32_x} \right) = I_{G2} \alpha_2$$

This is identical to equation 12.8a for the "pure" fourbar linkage. For link 3:

$$F_{43_x} - F_{32_x} = m_3 a_{G3_x}$$
$$F_{43_y} - F_{32_y} = m_3 a_{G3_y} \quad\quad (12.10b)$$
$$\left(R_{43_x} F_{43_y} - R_{43_y} F_{43_x} \right) - \left(R_{23_x} F_{32_y} - R_{23_y} F_{32_x} \right) = I_{G3} \alpha_3$$

This is similar to equation 12.8b, lacking only the terms involving $\mathbf{F}_p$ since there is no external force shown acting on link 3 of our example slider-crank. For link 4:

$$F_{14_x} - F_{43_x} + F_{P_x} = m_4 a_{G4_x}$$
$$F_{14_y} - F_{43_y} + F_{P_y} = m_4 a_{G4_y} \quad\quad (12.10c)$$
$$\left(R_{14_x} F_{14_y} - R_{14_y} F_{14_x} \right) - \left(R_{34_x} F_{43_y} - R_{34_y} F_{43_x} \right) + \left(R_{P_x} F_{P_y} - R_{P_y} F_{P_x} \right) = I_{G4} \alpha_4$$

These contain the external force $\mathbf{F}_p$ shown acting on link 4.

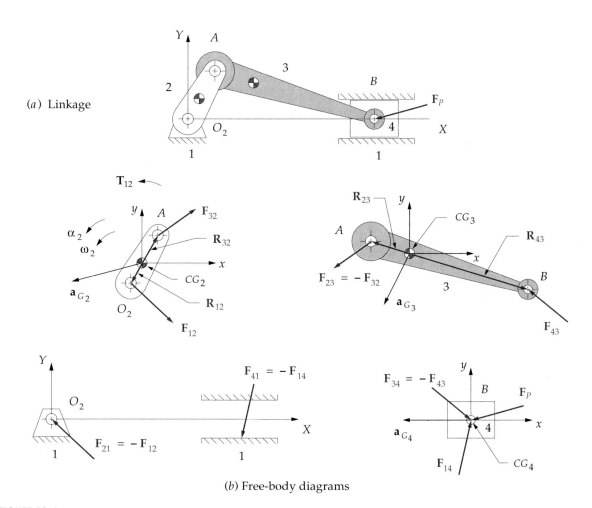

(a) Linkage

(b) Free-body diagrams

FIGURE 12-4

Dynamic force analysis of the fourbar slider-crank linkage

For the inversion of the slider-crank shown, the slider block, or piston, is in pure translation against the stationary ground plane; thus it can have no angular acceleration or angular velocity. Also, the position vectors in the torque equation (Eq. 12.10c) are all zero as the force $\mathbf{F}_p$ acts at the CG. Thus the torque equation for link 4 (third expression in Eq 12.10c) is zero for this inversion of the slider-crank linkage. Its linear acceleration also has no y component.

$$\alpha_4 = 0, \qquad\qquad a_{G_{4y}} = 0 \qquad\qquad (12.10d)$$

The only x directed force that can exist at the interface between links 4 and 1 is friction. Assuming coulomb friction, the x component can be expressed in terms of the y component of force at this interface. We can write a relation for the friction force f at that interface such as $f = \pm\mu N$, where $\pm\mu$ is a known coefficient of friction.

The plus and minus signs on the coefficient of friction are to recognize the fact that the friction force always opposes motion. The kinematic analysis will provide the velocity of the link at the sliding joint. The sign of μ will always be the opposite of the sign of this velocity.

$$F_{14_x} = \pm\mu F_{14_y} \tag{12.10e}$$

Substituting equations 12.10d and 12.10e into the reduced equation 12.10c yields:

$$\pm\mu F_{14_y} - F_{43_x} + F_{P_x} = m_4 a_{G4_x}$$
$$F_{14_y} - F_{43_y} + F_{P_y} = 0 \tag{12.10f}$$

This last substitution has reduced the unknowns to eight, F_{12x}, F_{12y}, F_{32x}, F_{32y}, F_{43x}, F_{43y}, F_{14y}, and T_{12}, thus we need only eight equations. We can now use the eight equations in 12.10a, b, and f to assemble the matrices for solution.

$$
\begin{bmatrix}
1 & 0 & 1 & 0 & 0 & 0 & 0 & 0 \\
0 & 1 & 0 & 1 & 0 & 0 & 0 & 0 \\
-R_{12_y} & R_{12_x} & -R_{32_y} & R_{32_x} & 0 & 0 & 0 & 1 \\
0 & 0 & -1 & 0 & 1 & 0 & 0 & 0 \\
0 & 0 & 0 & -1 & 0 & 1 & 0 & 0 \\
0 & 0 & R_{23_y} & -R_{23_x} & -R_{43_y} & R_{43_x} & 0 & 0 \\
0 & 0 & 0 & 0 & -1 & 0 & \pm\mu & 0 \\
0 & 0 & 0 & 0 & 0 & -1 & 1 & 0
\end{bmatrix}
\times
\begin{bmatrix}
F_{12_x} \\
F_{12_y} \\
F_{32_x} \\
F_{32_y} \\
F_{43_x} \\
F_{43_y} \\
F_{14_y} \\
T_{12}
\end{bmatrix}
=
$$

$$\tag{12.10g}$$

$$
\begin{bmatrix}
m_2 a_{G2_x} \\
m_2 a_{G2_y} \\
I_{G_2}\alpha_2 \\
m_3 a_{G3_x} \\
m_3 a_{G3_y} \\
I_{G_3}\alpha_3 \\
m_4 a_{G4_x} - F_{P_x} \\
-F_{P_y}
\end{bmatrix}
$$

Solution of this matrix equation (Eq. 12.10g) plus equation 12.10e will yield complete dynamic force information for the fourbar slider-crank linkage.

12.6   FORCE ANALYSIS OF THE INVERTED SLIDER-CRANK

Another inversion of the fourbar slider-crank was also analyzed kinematically in Part I. It is shown in Figure 12-5. Link 4 does have an angular acceleration in this inversion. In fact, it must have the same angle, angular velocity, and angular acceleration as link 3 because they are rotationally coupled by the sliding joint. We wish to determine the forces at all pin joints and at the sliding joint as well as the driving

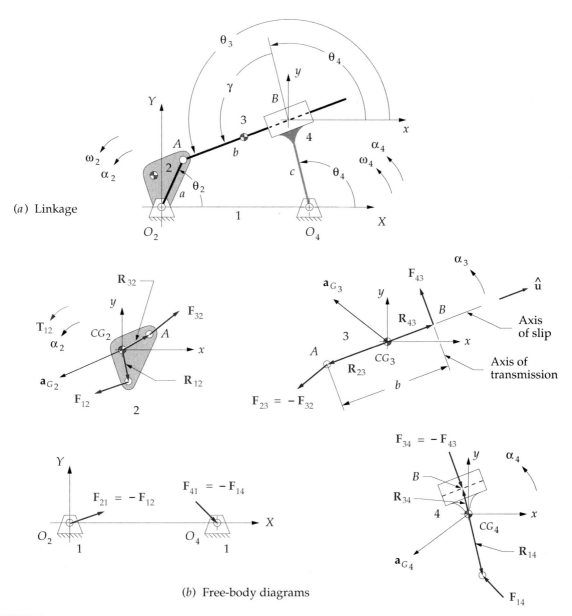

FIGURE 12-5

Dynamic forces in the inverted slider-crank fourbar linkage

torque needed to create the desired accelerations. Each link's joints are located by position vectors referenced to nonrotating local xy coordinate systems at each link's CG as before. The sliding joint is located by the position vector $\mathbf{R}_{43}$ to the center of the slider, point B. The instantaneous position of point B was determined from the kinematic analysis as length b referenced to instant center I_{23} (point A). See Sections 4.7, 6.7, and 7.3 to review the position, velocity, and acceleration analysis of this mechanism. Recall that this mechanism has a nonzero Coriolis component of acceleration. The force between link 3 and link 4 within the sliding joint is distributed along the unspecified length of the slider block. For this analysis the distributed force can be modelled as a force concentrated at point B within the sliding joint. We will neglect friction in this example.

The equations for links 2 and 3 are identical to those for the noninverted slider-crank (Eqs. 12.10a and b). The equations for link 4 are the same as equations 12.10c except for the absence of the terms involving $\mathbf{F}_p$ since no external force is shown acting on link 4 in this example. The slider joint can only transmit force from link 3 to link 4 or vice versa along a line perpendicular to the axis of slip. This line is called the axis of transmission. In order to guarantee that the force $\mathbf{F}_{34}$ or $\mathbf{F}_{43}$ is always perpendicular to the axis of slip, we can write the following relation:

$$\hat{\mathbf{u}} \cdot \mathbf{F}_{43} = 0 \qquad (12.11a)$$

which expands to:

$$u_x F_{43_x} + u_y F_{43_y} = 0 \qquad (12.11b)$$

The dot product of two vectors will be zero when the vectors are mutually perpendicular. The unit vector u *hat* is in the direction of link 3 which is defined from the kinematic analysis as θ_3.

$$u_x = \cos\theta_3, \qquad u_y = \sin\theta_3 \qquad (12.11c)$$

Equation 12.11 provides a tenth equation, but we have only nine unknowns, $F_{12x}, F_{12y}, F_{32x}, F_{32y}, F_{43x}, F_{43y}, F_{14x}, F_{14y}$, and T_{12}, so one of our equations is redundant. Since we must include equation 12.11, we will combine the torque equations for links 3 and 4 rewritten here in vector form and without the external force $\mathbf{F}_p$.

$$\left(\mathbf{R}_{43} \times \mathbf{F}_{43}\right) - \left(\mathbf{R}_{23} \times \mathbf{F}_{32}\right) = I_{G_3}\alpha_3 = I_{G_3}\alpha_4$$

$$(12.12a)$$

$$\left(\mathbf{R}_{14} \times \mathbf{F}_{14}\right) - \left(\mathbf{R}_{34} \times \mathbf{F}_{43}\right) = I_{G_4}\alpha_4$$

Note that the angular acceleration of link 3 is the same as that of link 4 in this linkage. Adding these equations gives:

$$\left(\mathbf{R}_{43} \times \mathbf{F}_{43}\right) - \left(\mathbf{R}_{23} \times \mathbf{F}_{32}\right) + \left(\mathbf{R}_{14} \times \mathbf{F}_{14}\right) - \left(\mathbf{R}_{34} \times \mathbf{F}_{43}\right) = \left(I_{G_3} + I_{G_4}\right)\alpha_4 \qquad (12.12b)$$

Expanding and collecting terms:

$$\left(R_{43_x} - R_{34_x}\right)F_{43_y} + \left(R_{34_y} - R_{43_y}\right)F_{43_x} - R_{23_x}F_{32_y}$$

$$+ R_{23_y}F_{32_x} + R_{14_x}F_{14_y} - R_{14_y}F_{14_x} = \left(I_{G_3} + I_{G_4}\right)\alpha_4 \qquad (12.12c)$$

Equations 12.10a, 12.11b, 12.12c, and the four force equations from equations 12.10b and 12.10c (excluding the external force $\mathbf{F}_P$) give us nine equations in the nine unknowns which we can put in matrix form for solution.

$$
\begin{bmatrix}
1 & 0 & 1 & 0 & 0 & 0 & 0 & 0 & 0 \\
0 & 1 & 0 & 1 & 0 & 0 & 0 & 0 & 0 \\
-R_{12_y} & R_{12_x} & -R_{32_y} & R_{32_x} & 0 & 0 & 0 & 0 & 1 \\
0 & 0 & -1 & 0 & 1 & 0 & 0 & 0 & 0 \\
0 & 0 & 0 & -1 & 0 & 1 & 0 & 0 & 0 \\
0 & 0 & R_{23_y} & -R_{23_x} & \left(R_{34_y}-R_{43_y}\right) & \left(R_{43_x}-R_{34_x}\right) & -R_{14_y} & R_{14_x} & 0 \\
0 & 0 & 0 & 0 & -1 & 0 & 1 & 0 & 0 \\
0 & 0 & 0 & 0 & 0 & -1 & 0 & 1 & 0 \\
0 & 0 & 0 & 0 & u_x & u_y & 0 & 0 & 0
\end{bmatrix}
\times
$$

$$
\begin{bmatrix}
F_{12_x} \\
F_{12_y} \\
F_{32_x} \\
F_{32_y} \\
F_{43_x} \\
F_{43_y} \\
F_{14_x} \\
F_{14_y} \\
T_{12}
\end{bmatrix}
=
\begin{bmatrix}
m_2 a_{G_{2x}} \\
m_2 a_{G_{2y}} \\
I_{G_2}\alpha_2 \\
m_3 a_{G_{3x}} \\
m_3 a_{G_{3y}} \\
\left(I_{G_3}+I_{G_4}\right)\alpha_4 \\
m_4 a_{G_{4x}} \\
m_4 a_{G_{4y}} \\
0
\end{bmatrix}
\qquad (12.13)
$$

12.7 FORCE ANALYSIS OF LINKAGES WITH MORE THAN FOUR BARS

This matrix method of force analysis can easily be extended to more complex assemblages of links. The equations for each link are of the same form. We can create a more general notation for equations 12.1 to apply them to any assembly of n pin-connected links. Let j represent any link in the assembly. Let $i = j - 1$ be the previous link in the chain and $k = j + 1$ be the next link in the chain, then, using the vector form of equations 12.1:

$$
\mathbf{F}_{ij} + \mathbf{F}_{jk} + \sum \mathbf{F}_{ext_j} = m_j\,\mathbf{a}_{G_j} \qquad (12.14a)
$$

$$
\left(\mathbf{R}_{ij}\times\mathbf{F}_{ij}\right)+\left(\mathbf{R}_{jk}\times\mathbf{F}_{jk}\right)+\sum \mathbf{T}_j +\left(\mathbf{R}_{ext_j}\times\sum \mathbf{F}_{ext_j}\right)= I_{G_j}\alpha_j \qquad (12.14b)
$$

where:

$$
j = 2,3,\dots n; \quad i = j-1; \quad k = j+1,\ j \neq n; \qquad \text{if } j = n,\ k = 1
$$

and $\qquad (12.14c)$

$$
\mathbf{F}_{ji} = -\mathbf{F}_{ij}; \qquad \mathbf{F}_{kj} = -\mathbf{F}_{jk};
$$

The sum of forces vector equation 12.14a can be broken into its two xy component equations and then applied, along with the sum of torques equation 12.14b, to each of the links in the chain to create the set of simultaneous equations for solution. Any link may have external forces and/or external torques applied to it. All will have pin forces. Since the nth link in a closed chain connects to the first link, the value of k for the nth link is set to 1. In order to reduce the number of variables to a tractable quantity, substitute the negative reaction forces from equation 12.14c where necessary as was done in the examples above.

When sliding joints are present, it will be necessary to add constraints on the allowable directions of forces at those joints as was done in the inverted slider-crank example above.

12.8   SHAKING FORCES AND SHAKING TORQUE

It is usually of interest to know the net effect of the dynamic forces as felt on the ground plane as this can set up vibrations in the structure which supports the machine. For our simple examples of three- and fourbar linkages, there are only two points at which the dynamic forces can be delivered to link 1, the ground plane. More complicated mechanisms will have more joints with the ground plane. The forces delivered by the moving links to the ground at the fixed pivots O_2 and O_4 are designated $\mathbf{F}_{21}$ and $\mathbf{F}_{41}$ by our subscript convention as defined in Section 12.1. Since we chose to solve for $\mathbf{F}_{12}$ and $\mathbf{F}_{14}$ in our solutions, we simply negate those forces to obtain their equal and opposite counterparts (see also equation 12.5).

$$\mathbf{F}_{21} = -\mathbf{F}_{12} \qquad\qquad \mathbf{F}_{41} = -\mathbf{F}_{14} \qquad\qquad (12.15a)$$

The *sum of all the forces acting on the ground plane* is called the **shaking force ($\mathbf{F}_s$)**. In these simple examples it is equal to:

$$\mathbf{F}_s = \mathbf{F}_{21} + \mathbf{F}_{41} \qquad\qquad (12.15b)$$

The *reaction torque felt by the ground plane* is also called the **shaking torque ($\mathbf{T}_s$)**. This is simply the negative of the source torque $\mathbf{T}_{12}$ which is delivered to the driving link from the ground.

$$\mathbf{T}_s = \mathbf{T}_{21} = -\mathbf{T}_{12} \qquad\qquad (12.15c)$$

The shaking force will tend to move the ground plane back and forth, and the shaking torque will tend to rock the ground plane about the driveline axis. Both will cause vibrations. We are usually looking to minimize the effects of the shaking forces and shaking torques on the frame. This can sometimes be done by balancing, sometimes by the addition of a flywheel to the system, and sometimes by shock mounting the frame to isolate the vibrations from the rest of the assembly. Most often we will use a combination of all three approaches. We will investigate some of these techniques shortly.

12.9   PROGRAM DYNAFOUR

The matrix methods introduced in the preceding sections all provide force and torque information for one position of the linkage assembly as defined by its kinematic and geometric parameters. To do a complete force analysis for multiple positions of a machine merely requires that these computations be repeated with new input data for each position. A computer program is the obvious way to accomplish this.

Program FOURBAR, described in Chapter 8, computes the kinematic parameters for any fourbar linkage over changes in time or driver (crank) angle. Program DY-NAFOUR, also supplied with this text, adds to program FOURBAR the ability to compute the forces and torques concomitant with the linkage kinematics and link geometry. Please refer to Chapter 8 for information on computer hardware requirements and on the use of program FOURBAR. DYNAFOUR operates in a similar manner to that program and their menus are nearly identical. This section will describe those aspects of DYNAFOUR which are different than FOURBAR's.

The main menu is shown in Figure 12-6. It is the same as FOURBAR's except that two additional choices appear, one to balance the linkage and one to calculate for a flywheel. These two operations will be discussed in later sections after their mathematical approach has been presented. The input data menu shown in Figure 12-7 is also the same as the menu in FOURBAR (shown in Figure 8-3 on p. 270), except for two of that program's choices which are not included here. Items 3 through 7 provide the same built-in examples. Item 2 allows reading diskfiles previously stored by DYNAFOUR and also allows FOURBAR diskfiles to be read into this program. Thus the kinematic data for a linkage designed in FOURBAR can be easily transferred into DYNAFOUR for dynamic force analysis. (You will still have to input the mass and mass moment of inertia information from the keyboard as that is not present in FOURBAR.)

Item 1 on the input menu allows the linkage data to be entered from the keyboard. The first six items requested are the same as were required for FOURBAR, namely, the four link lengths and the polar coordinates of a position vector which locates the coupler point of interest with respect to link 3's line of centers. See Figure 8-4 on p. 271 for a definition of this position vector. The program then asks

DYNAFOUR Linkage Analyzer - Version X.x

1 – INPUT   the linkage data

2 – DISPLAY   the linkage data

3 – CALCULATE   the results

4 – PRINT   the results

5 – PLOT   the results

6 – ANIMATE   the linkage

7 – BALANCE   the linkage

8 – FLYWHEEL   calculation

9 – HELP   screens

<RET> – QUIT PROGRAM   CHOICE ?

FIGURE 12-6

Main menu for program DYNAFOUR

SELECT INPUT METHOD

1 – TYPE linkage data from KEYBOARD

2 – READ linkage data from DISKFILE

3 – Create GRASHOF   test linkage

4 – Create HOEKENS   straight line linkage

5 – Create ROBERTS   straight line linkage

6 – Create WATTS   straight line linkage

7 – Create CHEBYCHEV   straight line linkage

<Ret> - EXIT

Choice ?

FIGURE 12-7

Input menu for program DYNAFOUR

for the location of the *CG* of each moving link. These are defined in the same way as the coupler point, by a position vector whose root is at the low-numbered instant center of each link. That is, for link 2 it is $I_{1,2}$; for link 3, $I_{2,3}$; and for link 4, $I_{1,4}$. These vectors are shown in Figure 12-3a, labelled R_{CGi} where i is the link number. Note that the angle of each *CG* vector is measured with respect to a *rotating, local coordinate system* whose origin is at the aforementioned instant center and whose x axis lies along the line of centers of the link. For example, in Figure 12-3a, link 2's *CG* vector is 3 inches at 30°, link 3's is 9 inches at 45° and link 4's is 5 inches at 0°. The program will automatically create the necessary position vectors R_{12}, R_{32}, R_{23}, R_{43}, R_{34}, and R_{14} needed for the dynamic force equations as shown in the free-body diagrams in Figure 12-3b. These position vectors are, of course, recalculated in the *nonrotating local coordinate systems* at the links' *CGs* for each new position of the linkage as the link angles change.

DYNAFOUR next asks for the masses and mass moments of inertia with respect to their CGs of the moving links and then for any external forces or torques which may be applied to links 3 or 4. **If you answer** *no* to the question about these external forces and torques, **the program will set them all to zero** for that link. If you answer *yes*, the screen shown in Figure 12-8 will appear to remind you of the coordinate systems to be used for definition of the **external force** and its location vector. You will then be prompted for the magnitude and direction of the force shown as F_P in Figure 12-3. The direction angle *of the force* must be measured with respect to the *global coordinate system*. The program **will assume that this angle remains constant** for all positions of the linkage analyzed.

You must also supply the magnitude and direction of the **position vector R_P** which locates any point on the force vector F_P. This R_P vector is measured in the

Do you have any external Forces or Torques on Link 3? (y/n) ? y

You must resolve the link's EXTERNAL FORCE system to a
single force acting at some point with respect to EACH LINK's CG.

That force will be considered to be constant and non-rotating
at a FIXED ANGLE with respect to the GLOBAL COORDINATE SYSTEM.

You must also define the POSITION VECTOR (Mag and Angle) from
the link CG to the point of application of the force on that link.

Note that this angle is referenced to the link's embedded, rotating
LOCAL coordinate system whose X axis lies along its line of centers.

There may also be an EXTERNAL TORQUE on the link, independent of
the external forces.

Press Any Key to Continue

FIGURE 12-8

External force coordinate system information

rotating, local coordinate system embedded in the link, as were the CGs of the links. $\mathbf{R}_P$ is *not* measured in the global system. The program takes care of the resolution of these $\mathbf{R}_P$ vectors, for each position of the linkage, into coordinates in the nonrotating local coordinate system at the CG. Note that if you wish to account for the gravitational force on a heavy link, you may do so by applying that weight as an external force acting through the link's CG at 270° in the global system ($\mathbf{R}_P = 0$).[1]

The input data for the linkage can be checked by using the **Display Data** option from the main menu which brings up the screen shown in Figure 12-9. **Calculate** and **Animate** work the same as in FOURBAR except that the larger number of computations take longer and the CG locations are drawn on the links in addition to the coupler point. The program uses equations 12.8 to calculate the forces and torques but provides for both external forces and external torques on links 3 and 4 as shown in equation 12.14. The **Help** screens are the same as in program FOURBAR.

The **Print** and **Plot** menus are identical as shown in Figure 12-10. The first 8 items are the same as those on the corresponding FOURBAR menus. Item 9 allows printing or plotting of the torques on links 1, 2, 3, or 4. Link 2 torque is the source torque $\mathbf{T}_{12}$ found from equation 12.8a. Its plot for an entire revolution of the crank is shown in Figure 12-11. The arithmetic average (mean) and root mean square (RMS) average values of the link torques are also printed and plotted. The "**nonbalanced**" label on the torque and force plots indicates that the linkage has not yet been balanced by the methods provided in main menu item 7. This linkage balancing method will be derived and discussed in the next chapter. The torques displayed for links 3 and 4 are the sum of their applied torque, the couple due to the applied force and their inertial (d'Alembert) torques as defined in Section 11.11. The torque on the ground plane can also be plotted. This torque $\mathbf{T}_{21}$ is the negative of the source torque $\mathbf{T}_{12}$. It is

[1] The latest version of DYNAFOUR also allows the input of a constant-magnitude friction force which acts in the plane of the linkage at the coupler point.

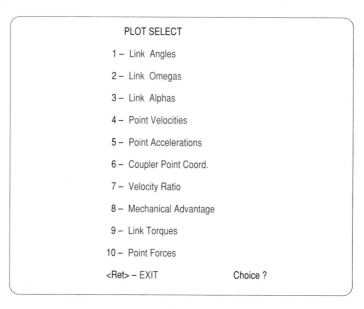

DYNAFOUR   Ima Student Design # 5   *mm-dd-yy at hh:mn*

Link No.	Length Inches	Mass Units	Inertia Units	CG Posit	at Deg	Ext. Force Pounds	at Deg
1	5.5						
2	2.0	.002	.004	1.0	0		
3	6.0	.030	.060	2.5	30	12	270
4	3.0	.010	.020	1.5	0	60	−45

Ext. Force 3 acts at 5 inches and 30 degrees versus CG of Link 3
Ext. Force 4 acts at 5 inches and 90 degrees versus CG of Link 4

Open/Crossed = OPEN   Coupler Pt.   =   3 Inches at 45 degrees
Ext. Torque 3 = − 20 Pound-inches Ext. Torque 4 = 25 Pound-inches

Start Alpha2   =   0 Radians/Sec^2
Start Omega2 =   50 Radians/Sec

Start Theta2   =   0 Degrees
Final Theta2   =   360 Degrees
Delta Theta2   =   2 Degrees

FIGURE 12-9

Display data screen from program DYNAFOUR

the reaction torque felt by link 1, to which the drive motor is attached, and is the shaking torque as defined in Section 12.8.

PLOT SELECT

1 – Link Angles

2 – Link Omegas

3 – Link Alphas

4 – Point Velocities

5 – Point Accelerations

6 – Coupler Point Coord.

7 – Velocity Ratio

8 – Mechanical Advantage

9 – Link Torques

10 – Point Forces

<Ret> – EXIT   Choice ?

FIGURE 12-10

Print and plot menu from DYNAFOUR

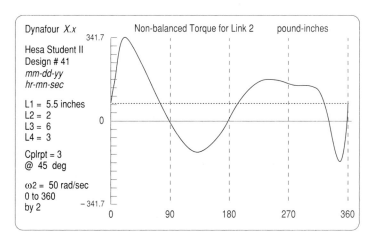

FIGURE 12-11

Input torque curve for a Grashof crank-rocker fourbar linkage from program DYNAFOUR

Item 10 on the **Print** or **Plot** menus brings up the submenu shown in Figure 12-12 which allows choice of the particular force to be displayed. Items 1, 2, 6, and 7 display forces $\mathbf{F}_{32}$, $\mathbf{F}_{43}$, $\mathbf{F}_{12}$, $\mathbf{F}_{14}$, from equations 12.8, in that order. Items 3, 4 and 5 display the d'Alembert inertial forces at the CGs of links 3, 2, and 4, respectively. Item 8 shows the shaking force as defined in Section 12.8. The same style choices are available for plotting the polar diagram of forces as were shown in Figure 8-18 for program FOURBAR. See Section 8.3 for an explanation. Figure 12-13

FIGURE 12-12

Point selection menu from DYNAFOUR

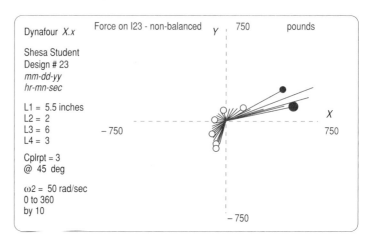

FIGURE 12-13

Polar plot of pin forces on joint I23 of a fourbar linkage from program DYNAFOUR

shows a polar plot of the force $\mathbf{F}_{32}$ at instant center 2,3. Every fifth position has a circle drawn at the tip of the vector. The first and second of these are filled solid.

The **Balance** and **Flywheel** choices will be discussed in later sections. Diskfiles for input to DYNAFOUR containing some of the examples of linkage force analysis presented in this chapter and problem solutions are also on the disks supplied. The student is encouraged to read these into the program in order to view the plots and explore the force analysis in more detail. Small differences in the program's numerical results versus those shown in the text are due to round-off errors in the computations.

12.10 LINKAGE FORCE ANALYSIS BY ENERGY METHODS

In Section 11.12 the method of virtual work was presented. We will now use that approach to solve the linkage from Example 12-3 as a check on its solution by the newtonian method used above. The kinematic data given in Example 12-3 did not include information on the angular velocities of all the links, the linear velocities of the centers of gravities of the links, and the linear velocity of the point P of application of the external force on link 3. These velocity data were not needed for the newtonian solution but are for the virtual work approach and are detailed below. Equation 11.24 is repeated here for convenience.

$$\sum_{k=2}^{n} \mathbf{F}_k \cdot \mathbf{v}_k + \sum_{k=2}^{n} \mathbf{T}_k \cdot \omega_k = \sum_{k=2}^{n} m_k \mathbf{a}_k \cdot \mathbf{v}_k + \sum_{k=2}^{n} I_k \alpha_k \cdot \omega_k \qquad (12.16a)$$

Expanding the summations, still in vector form:

$$\left(\mathbf{F}_{P_3} \cdot \mathbf{v}_{P_3} + \mathbf{F}_{P_4} \cdot \mathbf{v}_{P_4} \right) + \left(\mathbf{T}_{12} \cdot \omega_2 + \mathbf{T}_3 \cdot \omega_3 + \mathbf{T}_4 \cdot \omega_4 \right) =$$

$$\left(m_2 \mathbf{a}_{G_2} \cdot \mathbf{v}_{G_2} + m_3 \mathbf{a}_{G_3} \cdot \mathbf{v}_{G_3} + m_4 \mathbf{a}_{G_4} \cdot \mathbf{v}_{G_4} \right) \qquad (12.16b)$$

$$+ \left(I_{G_2} \alpha_2 \cdot \omega_2 + I_{G_3} \alpha_3 \cdot \omega_3 + I_{G_4} \alpha_4 \cdot \omega_4 \right)$$

Expanding the dot products to create a scalar equation:

$$\left(F_{P_{3x}} v_{P_{3x}} + F_{P_{3y}} v_{P_{3y}} \right) + \left(F_{P_{4x}} v_{P_{4x}} + F_{P_{4y}} v_{P_{4y}} \right) + \left(T_{12} \omega_2 + T_3 \omega_3 + T_4 \omega_4 \right) =$$

$$m_2 \left(a_{G_{2x}} v_{G_{2x}} + a_{G_{2y}} v_{G_{2y}} \right) + m_3 \left(a_{G_{3x}} v_{G_{3x}} + a_{G_{3y}} v_{G_{3y}} \right) \qquad (12.16c)$$

$$+ m_4 \left(a_{G_{4x}} v_{G_{4x}} + a_{G_{4y}} v_{G_{4y}} \right) + \left(I_{G_2} \alpha_2 \omega_2 + I_{G_3} \alpha_3 \omega_3 + I_{G_4} \alpha_4 \omega_4 \right)$$

✎ EXAMPLE 12-4

Analysis of a Fourbar Linkage by the Method of Virtual Work. (See Figure 12-3)

Given: The 5-inch-long crank (link 2) shown weighs 1.5 lb. Its CG is at 3 inches at $+30°$ from the line of centers. Its mass moment of inertia about its CG is 0.4 lb-in-sec^2. Its kinematic data are:

θ_2 deg	ω_2 rad / sec	α_2 rad / sec^2	V_{G_2} in / sec
60	25	−40	193.62 @ 161.17°

The coupler (link 3) is 15 inches long and weighs 7.7 lb. Its CG is at 9 inches at 45° off the line of centers. Its mass moment of inertia about its CG is 1.5 lb-in-sec^2. Its kinematic data are:

θ_3 deg	ω_3 rad / sec	α_3 rad / sec^2	V_{G_3} in / sec
20.92	−5.87	120.9	72.66 @ 145.7°

There is an external force on link 3 of 80 lb at 330°, applied at point P which is located 3 inches @ 100° from the CG of link 3. The linear velocity of that point is 54.03 in/sec at 135.79°.

The rocker (link 4) is 10 inches long and weighs 5.8 lb. Its CG is at 5 inches at 0° off the line of centers. Its mass moment of inertia about its CG is 0.8 lb-in-sec^2. Its kinematic data are:

θ_4 deg	ω_4 rad / sec	α_4 rad / sec^2	V_{G_4} in / sec
104.41	7.93	276.29	155.82 @ 160.26°

There is an external torque on link 4 of 120 lb-in. The ground link is 19 inches long.

Find: The driving **torque** T_{12} needed to maintain motion with the given acceleration for this instantaneous position of the link.

Solution:

1 The torque, angular velocity, and angular acceleration vectors in this two-dimensional problem are all directed along the Z axis, so their dot products each have only one term. Note that in this particular example there is no force F_{P_4} and no torque T_3.

2 The cartesian coordinates of the acceleration data were calculated in Example 12-3 above.

$$\mathbf{a}_{G_2} = 1878.84 \quad @\angle \ 273.66°; \qquad a_{G_{2x}} = 119.94, \qquad a_{G_{2y}} = -1875.01$$

$$\mathbf{a}_{G_3} = 3697.86 \quad @\angle \ 223.50°; \qquad a_{G_{3x}} = -2682.33, \qquad a_{G_{3y}} = -2545.44 \qquad (a)$$

$$\mathbf{a}_{G_4} = 1416.80 \quad @\angle \ 207.24°; \qquad a_{G_{4x}} = -1259.67, \qquad a_{G_{4y}} = -648.50$$

3   The x and y components of the external force at P in this coordinate system were also calculated in Example 12-3:

$$\mathbf{F}_{P_3} = 80 \quad @ \ \angle 330°; \qquad F_{P_{3x}} = 69.28, \qquad F_{P_{3y}} = -40.00 \qquad (b)$$

4   Converting the velocity data for this example to cartesian coordinates:

$$\mathbf{V}_{G_2} = \quad 193.62 \quad @\angle \ 161.17°; \qquad V_{G_{2x}} = -183.26, \qquad V_{G_{2y}} = \ 62.49$$

$$\mathbf{V}_{G_3} - \quad\ \ 72.66 \quad @\angle \ 145.70°; \qquad V_{G_{3x}} = -60.02, \qquad V_{G_{3y}} = \ 40.95$$

$$\mathbf{V}_{G_4} = \quad 155.82 \quad @\angle \ 160.26°; \qquad V_{G_{4x}} = -146.66, \qquad V_{G_{4y}} = \ 52.63 \qquad (c)$$

$$\mathbf{V}_{P_4} = \quad\ \ 54.03 \quad @\angle \ 135.79°; \qquad V_{P_{4x}} = -38.73, \qquad V_{P_{4y}} = \ 37.67$$

5   Substituting the example data into equation 12.16c:

$$\left[(69.28)(-38.73)+(-40)(37.67)\right]+[0]+\left[25T_{12}+(0)+(120)(7.93)\right]=$$

$$\frac{1.5}{386.4}\left[(119.94)(-183.26)+(-1875.01)(62.49)\right]$$

$$+\frac{7.7}{386.4}\left[(-2682.33)(-60.02)+(-2545.44)(40.95)\right] \qquad (d)$$

$$+\frac{5.8}{386.4}\left[(-1259.67)(-146.66)+(-648.50)(52.63)\right]$$

$$+\left[(.4)(-40)(25)+(1.5)(120.9)(-5.87)+(.8)(276.29)(7.93)\right]$$

6   The only unknown in this equation is the input torque $\mathbf{T}_{12}$ which calculates to:

$$\mathbf{T}_{12} = 254.9 \ \hat{\mathbf{k}} \qquad (e)$$

which is the same (to 3 sig. figs.) as the answer of 255 obtained with the other method.

This method of virtual work is useful if a quick answer is needed for the input torque, but it does not give any information about the joint forces.

12.11 CONTROLLING INPUT TORQUE - FLYWHEELS

The typically large variation in accelerations within a mechanism can cause large oscillations in the torque required to drive it at a constant or near constant speed. The peak torques needed may be so high as to require an overly large motor to deliver them. However, the average torque over the cycle, due mainly to losses and external work done, may often be much smaller than the peak torque. We would like to provide some means to smooth out these oscillations in torque during the cycle. This will allow us to size the motor to deliver the average torque rather than the peak torque. One convenient and relatively inexpensive means to this end is the addition of a **flywheel** to the system.

TORQUE VARIATION Figure 12-11 shows the variation in the input torque for a crank-rocker fourbar linkage over one full revolution of the drive crank. It is running at a constant angular velocity of 50 rad/sec. The torque varies a great deal within one cycle of the mechanism, going from a positive peak of 341.7 lb-in to a negative peak of –166.4 lb-in. The **average value of this torque** over the cycle is only 70.2 lb-in, being due to the *external work done plus losses*. This linkage has only a 12-lb external weight force applied to link 3 at the *CG* and a 25 lb-in external torque applied to link 4. These small external loads cannot account for the large variation in input torque required to maintain constant crank speed. What then is the explanation? The large variations in torque are evidence of the kinetic energy that is stored in the links as they move. We can think of the positive pulses of torque as representing energy delivered by the driver (motor) and stored temporarily in the moving links, and the negative pulses of torque as energy attempting to return from the links to the driver. Unfortunately most motors are designed to deliver energy but not to take it back. Thus the "returned energy" has no place to go.

Figure 2-22, reproduced here as Figure 12-14, shows the speed torque characteristic of a permanent magnet (PM) DC electric motor. Other types of motors will have differently shaped functions that relate motor speed to torque, but all drivers (sources) will have some such characteristic curve. As the torque demands on the motor change, its speed must also change according to its inherent characteristic. This means that the torque curve being demanded in Figure 12-11 will be very difficult for a standard motor to deliver without drastic changes in its speed.

The computation of the torque curve in Figure 12-11 was made on the assumption that the crank (thus the motor) speed was a constant value. All the

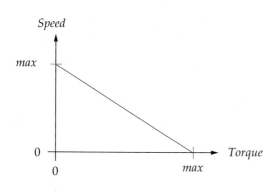

(*a*) *Speed-Torque* characteristic of a PM electric motor

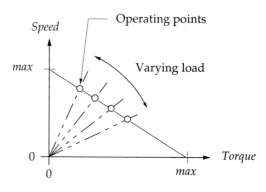

(*b*) Load lines superposed on *Speed-Torque* curve

FIGURE 12-14

DC permanent magnet (PM) motor *Speed-Torque* characteristic

kinematic data used in the force and torque calculation was generated on that basis. With the torque variation shown we would have to use a large-horsepower motor to provide the power required to reach that peak torque at the design speed:

$$Power \;=\; torque \times angular\; velocity$$

$$Peak\; power \;=\; 341.7\; lb\text{-}in \;\times 50\,\frac{rad}{sec} = 17{,}085\,\frac{in\text{-}lb}{sec} = 2.59\;HP$$

The power needed to supply the average torque is much smaller.

$$Average\; power \;=\; 70.2\; lb\text{-}in \times 50\,\frac{rad}{sec} = 3{,}510\,\frac{in\text{-}lb}{sec} = 0.53\;HP$$

It would be extremely inefficient to specify a motor based on the peak demand of the system, as most of the time it will be underutilized. We need something in the system which is capable of storing kinetic energy. One such kinetic energy storage device is called a **flywheel**.

FLYWHEEL ENERGY Figure 12-15 shows a **flywheel**, designed as a flat circular disk, attached to a motor shaft which might also be the driveshaft for the crank of our linkage. The motor supplies a torque magnitude T_M which we would like to be as constant as possible, i.e., to be equal to the average torque T_{avg}. The load (our linkage), on the other side of the flywheel, demands a torque T_L which is time varying as shown in Figure 12-11. The kinetic energy in a rotating system is:

$$E = \frac{1}{2}I\omega^2 \tag{12.17}$$

where I is the moment of inertia of all rotating mass on the shaft. This includes the I of the motor rotor and of the linkage crank plus that of the flywheel. We want to determine how much I we need to add in the form of a flywheel to reduce the speed variation of the shaft to an acceptable level. We begin by writing Newton's law for the free-body diagram in Figure 12-15.

$$\sum T = I\alpha$$

$$T_L - T_M = I\alpha$$

but we want:

$$T_M = T_{avg} \tag{12.18a}$$

so:

$$T_L - T_{avg} = I\alpha$$

substituting:

$$\alpha = \frac{d\omega}{dt} = \frac{d\omega}{dt}\left(\frac{d\theta}{d\theta}\right) = \omega\frac{d\omega}{d\theta}$$

gives:

$$T_L - T_{avg} = I\,\omega\frac{d\omega}{d\theta} \tag{12.18b}$$

$$\left(T_L - T_{avg}\right)d\theta = I\,\omega\,d\omega$$

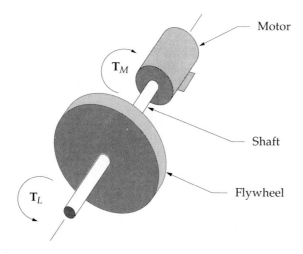

Motor

$\mathbf{T}_M$

Shaft

$\mathbf{T}_L$

Flywheel

FIGURE 12-15

Flywheel on a driveshaft

and integrating:

$$\int_{\theta@\omega_{min}}^{\theta@\omega_{max}} \left(T_L - T_{avg}\right)d\theta = \int_{\omega_{min}}^{\omega_{max}} I\,\omega\,d\omega$$

(12.18c)

$$\int_{\theta@\omega_{min}}^{\theta@\omega_{max}} \left(T_L - T_{avg}\right)d\theta = \frac{1}{2}I\left(\omega_{max}^2 - \omega_{min}^2\right)$$

The left side of this expression represents the change in energy E between the maximum and minimum shaft ω's and is equal to the *area under the torque-time diagram*[1] (Figures 12-11 and 12-16) between those extreme values of ω. The right side of equation 12.18c is the change in energy stored in the flywheel. The only way we can extract energy from the flywheel is to slow it down as shown in equation 12.17. Adding energy will speed it up. Thus it is impossible to obtain exactly constant shaft velocity in the face of changing energy demands by the load. The best we can do is to minimize the speed variation ($\omega_{max} - \omega_{min}$) by providing a flywheel with sufficiently large I.

---

[1] There is often confusion between torque and energy because they appear to have the same units of *lb-in (in-lb)* or *N-m (m-N)*. This leads some students to think that they are the same quantity, but they are not. Torque ≠ energy. The **integral** of torque with respect to angle, measured in radians, **is** equal to energy. This integral has the units of *in-lb-radians*. The radian term is usually omitted since it is in fact unity. Power in a rotating system is equal to torque x angular velocity (measured in *rad/sec*), and the power units are then *(in-lb-rad)/sec*. When power is integrated versus time to get energy, the resulting units are *in-lb-radians*, the same as the integral of torque versus angle. The radians are again usually dropped, contributing to the confusion.

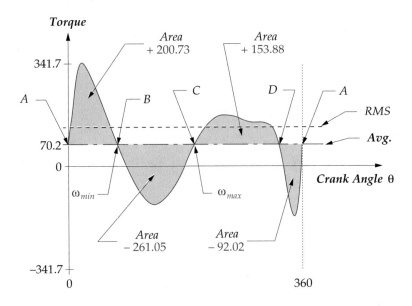

FIGURE 12-16

Integrating the pulses above and below the average value in the input torque function

✎ EXAMPLE 12-5

Determining the Energy Variation in a Torque-Time Function

Given: An input torque-time function which varies over its cycle. Figure 12-16 shows
              the input torque curve from the linkage in Figure 12-11. The torque is vary-
              ing during the 360° cycle about its average value.

Find: The total energy variation over one cycle.

Solution:

1   Calculate the average value of the torque-time function over one cycle. which in this
    case is 70.2 lb-in. (Note that in some cases the average value may be zero.)

2   Note that the **integration** on the left side of equation 12.18c is done **with respect to the
    average line of the torque function**, *not with respect to the θ axis*. (From the definition of
    the average, the sum of positive area above an average line is equal to the sum of nega-
    tive area below that line.) The integration limits in equation 12.18 are from the shaft an-
    gle θ at which the shaft ω is a minimum to the shaft angle θ at which ω is a maximum.

3   The minimum ω will occur after the maximum positive energy has been delivered
    from the motor to the load, i.e., at a point (θ) where the summation of positive ener-
    gy (area) in the torque pulses is at its largest positive value.

4   The maximum ω will occur after the maximum negative energy has been returned
    to the load, i.e., at a point (θ) where the summation of energy (area) in the torque
    pulses is at its largest negative value.

Areas of link 2 torque pulses in order during one cycle

Energy units are Pound - Inches - Radians

Order	Neg Area	Pos Area
1	− 261.05	200.73
2	− 92.02	153.88

FIGURE 12-17

Areas under the torque-time pulses in Figure 12-16

5 To find these locations in θ corresponding to the maximum and minimum ω's and thus find the amount of energy needed to be stored in the flywheel, we need to numerically integrate each pulse of this function from crossover to crossover with the average line. The crossover points in Figure 12-16 have been labelled A, B, C, and D. (Program DYNAFOUR does this integration for you numerically, using a trapezoidal rule. Choose item 8, **Flywheel Calculation**, on the main menu.)

6 The DYNAFOUR program prints the table of areas shown in Figure 12-17. The positive and negative pulses are separately integrated as described above. Reference to the plot of the torque function will indicate whether a positive or negative pulse is the first encountered in a particular case. The first pulse in this example is a positive one.

7 The remaining task is to accumulate these pulse areas beginning at an arbitrary crossover (in this case point A) and proceeding pulse by pulse across the cycle. Table 12-1 shows this process and the result.

8 Note in Table 12-1 that the minimum shaft speed occurs after the largest accumulated positive energy pulse (+200.73 in-lb) has been delivered from the driveshaft to the system. This delivery of energy slows the motor down. The maximum shaft speed occurs after the largest accumulated negative energy pulse (–60.32 in-lb) has been received back from the system by the driveshaft. This return of stored energy will speed up the motor. The total energy variation is the algebraic difference between these two extreme values, which in this example is –261.05 in-lb. This negative energy coming out of the system needs to be absorbed by the flywheel and then returned to the system *during each cycle* to smooth the variations in shaft speed.

TABLE 12-1   Integrating the Torque Function

From	Δ Area = ΔE	Accum. Sum = E	
A to B	+200.73	+200.73	ω_{min} @ B
B to C	−261.05	−60.32	ω_{max} @ C
C to D	+153.88	+93.56	
D to A	−92.02	+1.54	

$$\text{Total } \Delta \text{ Energy} \quad = E \text{ @ } \omega_{max} - E \text{@} \omega_{min}$$
$$= (-60.32) - (+200.73) = -261.05 \text{ in} - \text{lb}$$

SIZING THE FLYWHEEL   We now must determine how large a flywheel is needed to absorb this energy with an acceptable change in speed. The change in shaft speed during a cycle is called its *fluctuation (Fl)* and is equal to:

$$Fl = \omega_{max} - \omega_{min} \tag{12.19a}$$

We can normalize this to a dimensionless ratio by dividing it by the average shaft speed. This ratio is called the *coefficient of fluctuation (k)*.

$$k = \frac{(\omega_{max} - \omega_{min})}{\omega_{avg}} \tag{12.19b}$$

This *coefficient of fluctuation* is a design parameter to be chosen by the designer. It typically is set to a value between 0.01 and 0.05, which correspond to a 1 to 5% fluctuation in shaft speed. The smaller this chosen value, the larger the flywheel will have to be. This presents a design tradeoff. A larger flywheel will add more cost and weight to the system, which factors have to be weighed against the smoothness of operation desired.

We found the required change in energy E by integrating the torque curve

$$\int_{\theta @ \omega_{min}}^{\theta @ \omega_{max}} \left(T_L - T_{avg}\right) d\theta = E \tag{12.20a}$$

and can now set it equal to the right side of equation 12.18c:

$$E = \frac{1}{2} I \left(\omega_{max}^2 - \omega_{min}^2\right) \tag{12.20b}$$

Factoring this expression:

$$E = \frac{1}{2} I \left(\omega_{max} + \omega_{min}\right)\left(\omega_{max} - \omega_{min}\right) \tag{12.20c}$$

If the torque-time function were a pure harmonic then its average value could be expressed exactly as:

$$\omega_{avg} = \frac{\left(\omega_{\max} + \omega_{\min}\right)}{2} \qquad (12.21)$$

Our torque functions will seldom be pure harmonics, but the error introduced by using this expression as an approximation of the average is acceptably small. We can now substitute equations 12.19b and 12.21 into equation 12.20c to get an expression for the mass moment of inertia I_s of the system flywheel needed.

$$E = \frac{1}{2}I\left(2\omega_{avg}\right)\left(k\,\omega_{avg}\right)$$

$$I_s = \frac{E}{k\,\omega_{avg}^2} \qquad (12.22)$$

Equation 12.22 can be used to design the physical flywheel by choosing a desired coefficient of fluctuation k, and using the value of E from the numerical integration of the torque curve (see Table 12-1) and the average shaft ω to compute the needed system I_s. The physical flywheel's mass moment of inertia I_f is then set equal to the required system I_s. But if the moments of inertia of the other rotating elements on the same driveshaft (such as the motor) are known, the physical flywheel's required I_f can be reduced by those amounts.

The most efficient flywheel design in terms of maximizing I_f for minimum material used is one in which the mass is concentrated in its rim and its hub is supported on spokes, like a carriage wheel. This puts the majority of the mass at the largest radius possible and minimizes the weight for a given I_f. Even if a flat, solid circular disk flywheel design is chosen, either for simplicity of manufacture or to obtain a flat surface for other functions (such as an automobile clutch), the design should be done with an eye to reducing weight and thus cost. Since in general, $I = mr^2$, a thin disk of large diameter will need fewer pounds of material to obtain a given I than will a thicker disk of smaller diameter. Dense materials such as cast iron and steel are the obvious choices for a flywheel. Aluminum is seldom used. Though many metals (lead, gold, silver, platinum) are more dense than iron and steel, one can seldom get the accounting department's approval to use them in a flywheel.

Figure 12-18 shows the change in the input torque $\mathbf{T}_{12}$ for the linkage in Figure 12-11 after the addition of a flywheel sized to provide a coefficient of fluctuation of 0.05. The oscillation in torque about the unchanged average value is now 5%, much less than what it was without the flywheel. A much smaller horsepower motor can now be used because the flywheel is available to absorb the energy returned from the linkage during its cycle.

12.12 PRACTICAL CONSIDERATIONS

This chapter has presented some approaches to the computation of dynamic forces in moving machinery. The newtonian approach gives the most information and is necessary in order to obtain the forces at all pin joints so that stress analyses of the members can be done. Its application is really quite straightforward, requiring only the creation of correct free-body diagrams for each member and the application of

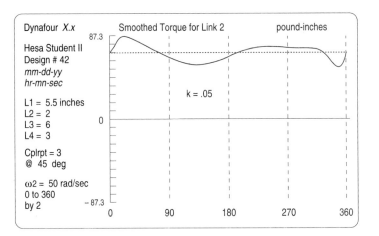

FIGURE 12-18

Input torque curve for the fourbar linkage in Figure 12-11 after smoothing with a flywheel [1]

the two, simple vector equations which express Newton's second law to each free-body. Once these equations are expanded for each member in the system and placed in standard matrix form, their solution (with a computer) is a trivial task.

The real work in designing these mechanisms comes in the determination of the shapes and sizes of the members. In addition to the kinematic data, the force computation requires only the masses, *CG* locations, and mass moments of inertia versus those *CGs* for its completion. These three geometric parameters completely characterize the member for dynamic modelling purposes. Even if the link shapes and materials are completely defined at the outset of the force analysis process (as with the redesign of an existing system), it is a tedious exercise to calculate the dynamic properties of complicated shapes. Current solids modelling *CAD* systems make this step easy by computing these parameters automatically for any part designed within them.

If, however, you are starting from scratch with your design, the *blank-paper syndrome* will inevitably rear its ugly head. A first approximation of link shapes and selection of materials must be made in order to create the dynamic parameters needed for a "first pass" force analysis. A stress analysis of those parts, based on the calculated dynamic forces, will invariably find problems that require changes to the part shapes, thus requiring recalculation of the dynamic properties and recomputation of the dynamic forces and stresses. This process will have to be repeated in circular fashion (*iteration* - see Chapter 1) until an acceptable design is reached. The advantages of using a computer to do these repetitive calculations is obvious and cannot be overstressed. An equation solver program such as *TK Solver* will be a useful aid in this process by reducing the amount of computer programming necessary.

Students with no design experience are often not sure how to approach this process of designing parts for dynamic applications. The following suggestions are offered to get you started. As you gain experience, you will develop your own approach.

[1] The programs calculate the flywheel-smoothed torque by multiplying the raw torque by the chosen coefficient of fluctuation.

It is often useful to create complex shapes from a combination of simple shapes, at least for first approximation dynamic models. For example, a link could be considered to be made up of a hollow cylinder at each pivot end, connected by a rectangular prism along the line of centers. It is easy to calculate the dynamic parameters for each of these simple shapes and then combine them. The steps would be as follows (repeated for each link):

1 Calculate the volume, mass, CG location, and mass moments of inertia with respect to the local CG of each separate part of your built-up link. In our example link these parts would be the two hollow cylinders and the rectangular prism.

2 Find the location of the composite CG of the assembly of the parts into the link by the method shown in Section 11.4 and equation 11.3. See also Figure 11-2.

3 Use the *parallel axis theorem* (Eq. 11.8) to transfer the mass moments of inertia of each part to the common, composite CG for the link. Then add the individual, transferred I's of the parts to get the total I of the link about its composite CG. See Section 11.6.

Steps 1 to 3 will create the link geometry data for each link needed for the dynamic force analysis as derived in this chapter.

4 Do the dynamic force analysis.

5 Do a dynamic stress and deflection analysis of all parts.

6 Redesign the parts and repeat steps 1 to 5 until a satisfactory result is achieved.

Remember that lighter (lower mass) links will have smaller inertial forces on them and thus could have lower stresses despite their smaller cross sections. Also, smaller mass moments of inertia of the links can reduce the driving torque requirements, especially at higher speeds. But be cautious about the dynamic deflections of thin, light links becoming too large. We are assuming rigid bodies in these analyses. That assumption will not be valid if the links are too flexible. Always check the deflections as well as the stresses in your designs.

12.13 PROBLEMS

12-1 Draw free-body diagrams of the links in the geared fivebar linkage shown in Figure 4-11 on p. 127 and write the dynamic equations to solve for all forces plus the driving torque. Assemble the symbolic equations in matrix form for solution.

12-2 Draw free-body diagrams of the links in the sixbar linkage shown in Figure 4-12 on p. 130 and write the dynamic equations to solve for all forces plus the driving torque. Assemble the symbolic equations in matrix form for solution.

*12-3 Table P12-1 shows kinematic and geometric data for several slider-crank linkages of the type and orientation shown in Figure P12-1. The point locations are defined as described in the text. For the row(s) in the table assigned, use the matrix method of Section 12.5 and program MATRIX or a matrix solving calculator to solve for forces and torques at the position shown. Also compute the shaking force and shaking torque. Work in any units system you prefer. Consider μ to be zero. You may check your solution (for $\mu = 0$) by inputting the solution files named P12-3x

---

* Answers in Appendix E

TABLE P12-1   Data for Problem 12-3

Part 1 of 3 Parts to Table P12-1

Row	link 2	link 3	θ_2	θ_3	α_2	α_3	m_2	m_3
a.	4	12	45	103.58	20	– 27.65	.002	.020
b.	3	10	30	105.20	– 5	– 25.47	.050	.100
c.	5	15	260	86.66	15	10.08	.010	.020
d.	6	20	– 75	94.42	– 10	– 10.07	.006	.150
e.	2	8	135	79.79	25	12.91	.001	.004
f.	10	35	120	81.63	– 20	18.63	.150	.300
g.	7	25	– 45	101.37	– 15	– 22.45	.080	.200

Part 2 of 3 Parts to Table P12-1

Row	I_2	I_3	R_{g2} mag	δ_2 ang	R_{g3} mag	δ_3 ang	ag_2 mag	ag_2 ang
a.	.10	.2	2	0	5	0	203.96	213.69
b.	.20	.4	1	20	4	– 30	100.12	232.86
c.	.05	.1	3	– 40	9	50	303.36	31.47
d.	.12	.3	3	120	12	60	301.50	230.71
e.	.30	.8	0.5	30	3	75	51.54	330.96
f.	.24	.6	6	45	15	135	611.88	356.31
g.	.45	.9	4	– 45	10	225	404.47	98.53

Part 3 of 3 Parts to Table P12-1

Row	ω_2	T_3	ag_3 mag	ag_3 ang	F_{P3} mag	F_{P3} ang	R_{P3} mag	δ_{P3} ang
a.	10	20	312.88	243.11	0	0	0	0
b.	10	– 35	247.68	228.57	10	45	4	30
c.	10	– 65	372.07	75.15	32	270	0	0
d.	10	– 12	644.94	91.04	15	180	2	60
e.	10	40	224.66	306.51	6	– 60	2	75
f.	10	– 75	1272.90	305.63	25	270	0	0
g.	10	– 90	732.03	153.15	9	120	5	45

(where x is the row letter) into program DYNAFOUR. These files "fool" that fourbar linkage program by specifying very long links 1 and 4 (10,000 units) to approximate the slider-crank with an equivalent, effective pin-jointed fourbar linkage. See Figure 2-10a on p. 38.

*12-4   Repeat problem 12-3 using the method of virtual work to solve for the input torque on link 2. Additional data for corresponding rows are given in Table P12-2.

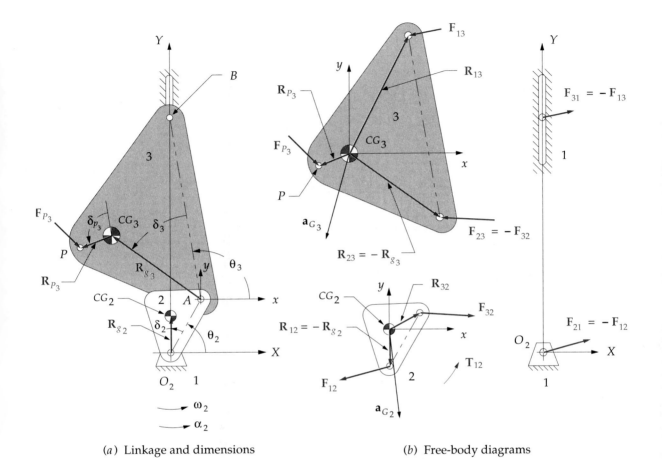

(a) Linkage and dimensions (b) Free-body diagrams

FIGURE P12-1

Linkage for problem 12-3

TABLE P12-2 Data for Problem 12-4

(See Also Table P12-1)

Row	V_{g_2} mag	V_{g_2} ang	V_{g_3} mag	V_{g_3} ang	V_{P_3} mag	V_{P_3} ang	ω_3
a.	20	135	35.23	117.89	35.23	117.89	− 2.43
b.	10	140	25.98	110.24	29.18	99.01	− 1.56
c.	30	310	41.83	313.76	41.83	− 46.26	3.29
d.	30	135	45.76	339.74	45.60	− 27.73	2.91
e.	5	255	15.05	218.04	11.97	− 149.67	− 1.79
f.	60	255	110.74	190.40	110.71	− 169.58	− 2.49
g.	40	0	89.86	47.54	79.92	46.43	2.02

(a) The linkage
and dimensions

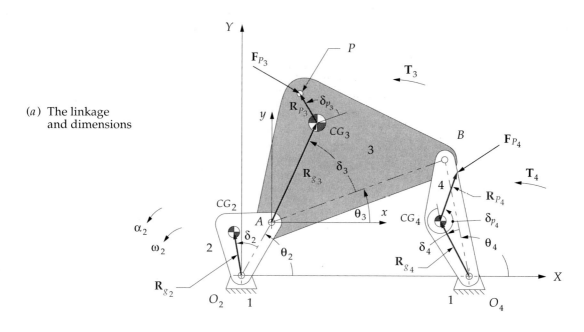

(b) Free-body
diagrams

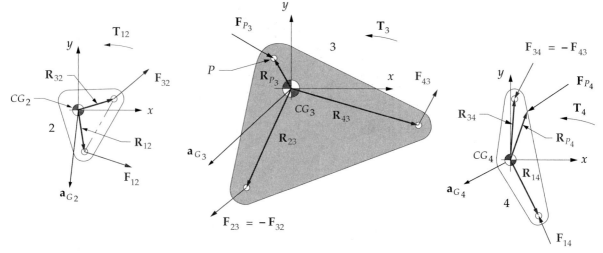

FIGURE P12-2

Linkage for problem 12-5

TABLE P12-3   Data for Problem 12-5

Part 1 of 5 Parts to Table P12-3

Row	link 2	link 3	link 4	link 1	θ_2	θ_3	θ_4	α_2
a.	4	12	8	15	45	24.97	99.30	20
b.	3	10	12	6	30	90.15	106.60	− 5
c.	5	15	14	2	260	128.70	151.03	15
d.	6	19	16	10	− 75	91.82	124.44	− 10
e.	2	8	7	9	135	34.02	122.71	25
f.	17	35	23	4	120	348.08	19.01	− 20
g.	7	25	10	19	65	355.93	27.19	− 15

Part 2 of 5 Parts to Table P12-3

Row	α_3	α_4	m_2	m_3	m_4	I_2	I_3	I_4
a.	75.29	244.43	0.002	0.02	0.10	0.10	0.20	0.50
b.	140.96	161.75	0.050	0.10	0.20	0.20	0.40	0.40
c.	78.78	53.37	0.010	0.02	0.05	0.05	0.10	0.13
d.	− 214.84	− 251.82	0.006	0.15	0.07	0.12	0.30	0.15
e.	71.54	− 14.19	0.001	0.04	0.09	0.30	0.80	0.30
f.	− 101.63	− 150.86	0.150	0.30	0.25	0.24	0.60	0.92
g.	− 251.86	− 661.41	0.080	0.20	0.12	0.45	0.90	0.54

Part 3 of 5 Parts to Table P12-3

Row	T_3	T_4	R_{g2} mag	δ_2 ang	R_{g3} mag	δ_3 ang	R_{g4} mag	δ_4 ang
a.	− 15	25	2	0	5	0	4	30
b.	12	0	1	20	4	− 30	6	40
c.	− 10	20	3	− 40	9	50	7	0
d.	0	30	3	120	12	60	6	− 30
e.	25	40	0.5	30	3	75	2	− 40
f.	0	− 25	6	45	15	135	10	25
g.	0	0	4	− 45	10	225	4	45

*12-5   Table P12-3 shows kinematic and geometric data for several pin-jointed fourbar linkages of the type and orientation shown in Figure P12-2. All have $\theta_1 = 0$. The point locations are defined as described in the text. For the row(s) in the table assigned, use the matrix method of Section 12-4 and program MATRIX or a matrix solving calculator to solve for forces and torques at the position shown. Also com-

Part 4 of 5 Parts to Table P12-3

Row	F_{P3} mag	F_{P3} ang	R_{P3} mag	δ_{P3} ang	R_{P4} mag	δ_{P4} ang	F_{P4} mag	F_{P4} ang
a.	0	0	0	0	8	0	40	– 30
b.	4	30	10	45	12	0	15	– 55
c.	0	0	0	0	14	0	75	45
d.	2	45	15	180	16	0	20	270
e.	9	0	6	– 60	7	0	16	60
f.	0	0	0	0	23	0	23	0
g.	12	– 60	9	120	10	0	32	20

Part 5 of 5 Parts to Table P12-3

Row	ω_2	ag_2 mag	ag_2 ang	ag_3 mag	ag_3 ang	ag_4 mag	ag_4 ang
a.	20	801.00	222.14	1670.18	210.87	979.02	222.27
b.	10	100.12	232.86	989.32	193.91	1032.32	256.52
c.	20	1200.84	37.85	3057.69	23.20	1446.58	316.06
d.	20	1200.87	226.43	4567.65	80.46	1510.34	2.15
e.	20	200.39	341.42	735.20	299.67	69.07	286.97
f.	20	2403.00	347.86	12008.57	308.62	4820.72	242.25
g.	20	1601.12	202.15	2526.60	187.92	3667.70	298.35

pute the shaking force and shaking torque. Work in any units system you prefer. You may check your solution by inputting the solution files named P12-5x (where x is the row letter) into program DYNAFOUR.

*12-6 Repeat problem 12-5 using the method of virtual work to solve for the input torque on link 2. Additional data for corresponding rows are given in Table P12-4.

*12-7 For the row(s) assigned in Table 12-3 (a-f), input the associated diskfile to program DYNAFOUR, calculate the linkage parameters for crank angles from zero to 360° by 5° increments, and design a steel disk flywheel to smooth the input torque using a coefficient of fluctuation of 0.05. Minimize the flywheel weight.

12.14 PROJECTS

The following problem statement applies to all the projects listed.

*These larger-scale project statements deliberately lack detail and structure and are loosely defined. Thus, they are similar to the kind of 'identification of need' or problem statement commonly encountered in engineering practice. It is left to the student to structure the problem through **background research** and to create a **clear goal statement** and set of **task specifications** before attempting to design a solution. This design process is spelled*

TABLE P12-4   Data for Problem 12-6
Part 1 of 2 Parts to Table P12-4

Row	V_{g_2} mag	V_{g_2} ang	V_{g_3} mag	V_{g_3} ang	V_{g_4} mag	V_{g_4} ang
a.	40	135	54.44	145.19	14.23	219.3
b.	10	140	21.46	14.74	45.94	56.6
c.	60	310	191.94	299.70	98.1	241.03
d.	60	135	94.36	353.80	19.03	4.44
e.	10	255	42.89	223.13	11.22	172.71
f.	120	255	618.05	211.39	213.98	134
g.	80	110	84.04	173.73	100.8	162.19

Part 2 of 2 Parts to Table P12-4

Row	V_{P_3} mag	V_{P_3} ang	V_{P_4} mag	V_{P_4} ang	ω_3	ω_4
a.	54.46	145.18	41.42	−160.80	− 5.62	3.56
b.	122.10	40.04	130.52	29.68	− 10.31	− 7.66
c.	191.93	−60.30	296.73	−118.97	16.60	14.13
d.	152.50	−3.13	67.86	26.38	3.90	− 3.17
e.	37.01	−140.37	48.40	−155.86	1.06	5.61
f.	618.03	−148.61	692.09	116.52	18.55	21.40
g.	138.13	−172.97	331.04	129.62	6.62	25.20

out in Chapter 1 and should be followed in all of these examples. All results should be documented in a professional engineering report. See the Bibliography in Chapter 1 for references on report writing.

Some of these project problems are based on the kinematic design projects in Chapter 3. Those kinematic devices can now be designed more realistically with consideration of the dynamic forces that they generate. The strategy in most of the following project problems is to keep the dynamic pin forces and thus the shaking forces to a minimum and also keep the input torque-time curve as smooth as possible to minimize power requirements. **All these problems can be solved with a pin-jointed fourbar linkage**. This fact will allow you to use program DYNAFOUR to do the kinematic and dynamic computations on a large number and variety of designs in a short time. There are infinities of viable solutions to these problems. **Iterate to find the best one!** All links must be designed in detail as to their geometry (mass, moment of inertia, etc.) Determine all pin forces, shaking force, shaking torque, and input horsepower required for your designs.

P12-1   Project 3-1 stated that:

The tennis coach needs a better tennis ball server for practice. This device must fire a sequence of standard tennis balls from one side of a standard tennis court over the net such that they land and bounce within each of the three court areas defined by the court's white lines. *The order and frequency of a ball's landing in any one of the three court areas must be random. The device should operate automatically and unattended except for the refill of balls. It should be capable of firing 50 balls between reloads. The timing of ball releases should vary. For simplicity, a motor-driven pin-jointed linkage design is preferred.*

This project asks you to design such a device to be mounted upon a tripod stand of 5-foot height. Design it, and the stand, for stability against tipover due to the shaking forces and shaking torques which should also be minimized in the design of your linkage. Minimize the input torque required.

P12-2   Project 3-9 stated that:

The "Save the Skeet" foundation has requested a more humane skeet launcher be designed. While they have not yet succeeded in passing legislation to prevent the wholesale slaughter of these little devils, they are concerned about the inhumane aspects of the large accelerations imparted to the skeet as it is launched into the sky for the sportsperson to shoot it down. The need is for a skeet launcher that will smoothly accelerate the clay pigeon onto its desired trajectory.

Design a skeet launcher to be mounted upon a child's "little red wagon." Control your design parameters so as to minimize the shaking forces and torques so that the wagon will remain as nearly stationary as possible during the launch of the clay pigeon.

P12-3   Project 3-10 stated that:

The coin operated "kid bouncer" machines found outside supermarkets typically provide a very unimaginative rocking motion to the occupant. There is a need for a superior "bouncer" which will give more interesting motions while remaining safe for small children.

Design this device for mounting in the bed of a pickup truck. Keep the shaking forces to a minimum and the input torque-time curve as smooth as possible.

P12-4   Project 3-15 stated that:

NASA wants a zero-G machine for astronaut training. It must carry one person and provide a negative 1 G acceleration for as long as possible.

Design this device and its mounting hardware to the ground plane minimizing the dynamic forces and driving torque.

P12-5   Project 3-16 stated that:

The Amusement Machine Co. Inc. wants a portable "WHIP" ride which will give two or four passengers a thrilling but safe ride, and which can be trailed behind a pickup truck from one location to another.

Design this device and its mounting hardware to the truck bed minimizing the dynamic forces and driving torque.

P12-6   Project 3-17 stated that:

The Air Force has requested a pilot training simulator which will give potential pilots exposure to G forces similar to those they will experience in dogfight maneuvers.

Design this device and its mounting hardware to the ground plane minimizing the dynamic forces and driving torque.

P12-7   Project 3-18 stated that:

Cheers needs a better "mechanical bull" simulator for their "yuppie" bar in Boston. It must give a thrilling "bucking bronco" ride but be safe.

Design this device and its mounting hardware to the ground plane minimizing the dynamic forces and driving torque.

P12-8   Gargantuan Motors Inc. is designing a new light military transport vehicle. Their current windshield wiper linkage mechanism develops such high shaking forces when run at its highest speed that the engines are falling out! Design a superior windshield wiper mechanism to sweep the 20 lb armored wiper blade through a 90° arc while minimizing both input torque and shaking forces. The wind load on the blade, perpendicular to the windshield, is 50 lb. The coefficient of friction of the wiper blade on glass is 0.9.

P12-9   The Army's latest helicopter gunship is to be fitted with the Gatling gun, which fires 50-mm-dia by 2-cm-long spent uranium slugs at a rate of 10 rounds per second. The reaction (recoil) force may upset the chopper's stability. A mechanism is needed which can be mounted to the frame of the helicopter and which will provide a synchronous shaking force, 180° out of phase with the recoil force pulses, to counteract the recoil of the gun. Design such a linkage and minimize its torque and power drawn from the aircraft's engine. Total weight of your device should also be minimized.

P12-10   Steel pilings are universally used as foundations for large buildings. These are often driven into the ground by hammer blows from a "pile driver." In certain soils (sandy, muddy) the piles can be "shaken" into the ground by attaching a "vibratory driver" which imparts a vertical, dynamic shaking force at or near the natural frequency of the pile-earth system. The pile can literally be made to "fall into the ground" under optimal conditions. Design a fourbar linkage-based pile shaker mechanism which, when its ground link is firmly attached to the top of a piling (supported from a crane hook), will impart a dynamic shaking force that is predominantly directed along the piling's long, vertical axis. Operating speed should be in the vicinity of the natural frequency of the pile-earth system.

P12-11   Paint can shaker mechanisms are common in paint and hardware stores. While they do a good job of mixing the paint, they are also noisy and transmit their vibrations to the shelves and counters. A better design of the paint can shaker is possible using a balanced fourbar linkage. Design such a portable device to sit on the floor (not bolted down) and minimize the shaking forces and vibrations while still effectively mixing the paint.

Chapter 13

BALANCING

*Moderation is best,
and to avoid all extremes*
PLUTARCH

13.0 INTRODUCTION

Any link or member that is in pure rotation can, theoretically, be perfectly balanced to eliminate all shaking forces and shaking moments. It is accepted design practice to balance all rotating members in a machine unless shaking forces are desired (as in a vibrating shaker mechanism, for example). A rotating member can be balanced either statically or dynamically. Static balance is a subset of dynamic balance. To achieve complete balance requires that dynamic balancing be done. In some cases, static balancing can be an acceptable substitute for dynamic balancing and is generally easier to do.

Rotating parts can, and generally should, be designed to be inherently balanced by their geometry. However, the vagaries of production tolerances guarantee that there will still be some small unbalance in each part. Thus a balancing procedure will have to be applied to each part after manufacture. The amount and location of any imbalance can be measured quite accurately and compensated for by adding or removing material in the correct locations.

In this chapter we will investigate the mathematics of determining and designing a state of static and dynamic balance in rotating elements and also in mechanisms having complex motion, such as the fourbar linkage. The methods and equipment used to measure and correct imbalance in manufactured assemblies will also be discussed. It is quite convenient to use the method of d'Alembert (see Section 11.10) when discussing rotating imbalance, applying inertia forces to the rotating elements, so we will do that.

13.1   STATIC BALANCE

Despite its name, **static balance** *does* apply to things in motion. The unbalanced forces of concern are due to the accelerations of masses in the system. The requirement for **static balance** is simply that *the sum of all forces on the moving system (including d'Alembert inertial forces) must be zero.*

$$\sum \mathbf{F} - m\mathbf{a} = 0 \qquad (13.1)$$

This, of course, is simply a restatement of Newton's law as was discussed in Section 11.11.

Another name for static balance is **single-plane balance**, which means that *the masses which are generating the inertia forces are in, or nearly in, the same plane.* It is essentially a two-dimensional problem. Some examples of common devices which meet this criterion, and thus can successfully be statically balanced, are: a single gear or pulley on a shaft, a bicycle or motorcycle tire and wheel, a thin flywheel, an airplane propellor, an individual turbine blade-wheel (but not the entire turbine). The common denominator among these devices is that they are all short in the axial direction compared to the radial direction, and thus can be considered to exist in a single plane. An automobile tire and wheel is only marginally suited to static balancing as it is reasonably thick in the axial direction compared to its diameter. Despite this fact, auto tires are sometimes statically balanced. More often they are dynamically balanced and will be discussed under that topic.

Figure 13-1a shows a link in the shape of a vee which is part of a linkage. We want to statically balance it. We can model this link dynamically as two point masses m_1 and m_2 concentrated at the local CGs of each "leg" of the link as shown in Figure 13-1b. These point masses each have a mass equal to that of the "leg" they replace and are supported on massless rods at the position ($\mathbf{R}_1$ or $\mathbf{R}_2$) of that leg's CG. We can solve for the required amount and location of a third "balance mass" m_b to be added to the system at some location $\mathbf{R}_b$ in order to satisfy equation 13.1.

Assume that the system is rotating at some constant angular velocity ω. The accelerations of the masses will then be strictly centripetal, and the inertia forces will be centrifugal as shown in the figure. Since the system is rotating, the figure shows a "freeze-frame" image of it. The position at which we "stop the action" for the purpose of drawing the picture and doing the calculations is both arbitrary and irrelevant to the computation. We will set up a coordinate system with its origin at the center of rotation and resolve the inertial forces into components in that system. Writing vector equation 13.1 for this system we get:

$$-m_1\mathbf{R}_1\omega^2 - m_2\mathbf{R}_2\omega^2 - m_b\mathbf{R}_b\omega^2 = 0 \qquad (13.2a)$$

Note that the only forces acting on this system are the inertia forces. For balancing, it does not matter what external forces may be acting on the system. External forces cannot be balanced by making any changes to the system's internal geometry. Note that the ω^2 terms cancel. For balancing, it also does not matter how fast the system is rotating, only that it *is* rotating. (*The ω will determine the magnitudes of these forces, but we are going to force their sum to be zero anyway.*)

Dividing out the ω^2 and rearranging we get:

$$m_b\mathbf{R_b} = -m_1\mathbf{R_1} - m_2\mathbf{R_2} \tag{13.2b}$$

Breaking into x and y components:

$$m_b R_{b_x} = -\left(m_1 R_{1_x} + m_2 R_{2_x}\right)$$

$$\tag{13.2c}$$

$$m_b R_{b_y} = -\left(m_1 R_{1_y} + m_2 R_{2_y}\right)$$

The terms on the right sides are known. We can readily solve for the mR_x and mR_y products needed to balance the system. It will be convenient to convert the results to polar coordinates.

$$\theta_b = \arctan \frac{m_b R_{b_y}}{m_b R_{b_x}}$$

$$\tag{13.2d}$$

$$\theta_b = \arctan \frac{-\left(m_1 R_{1_y} + m_2 R_{2_y}\right)}{-\left(m_1 R_{1_x} + m_2 R_{2_x}\right)}$$

$$R_b = \sqrt{\left(R_{b_x}^2 + R_{b_y}^2\right)}$$

$$m_b R_b = m_b \sqrt{\left(R_{b_x}^2 + R_{b_y}^2\right)}$$

$$m_b R_b = \sqrt{m_b^2\left(R_{b_x}^2 + R_{b_y}^2\right)} \tag{13.2e}$$

$$m_b R_b = \sqrt{m_b^2 R_{b_x}^2 + m_b^2 R_{b_y}^2}$$

$$m_b R_b = \sqrt{\left(m_b R_{b_x}\right)^2 + \left(m_b R_{b_y}\right)^2}$$

The angle at which the balance mass must be placed (with respect to our arbitrarily oriented freeze-frame coordinate system) is θ_b, found from equation 13.2d. Note that the signs of the numerator and denominator of equation 13.2d must be individually maintained and a two-argument arctangent computed in order to obtain θ_b in the correct quadrant. Most calculators and computers will give an arctangent result only between ±90°.

The $m_b R_b$ product is found from equation 13.2e. There is now an infinity of solutions available. We can either select a value for m_b and solve for the necessary radius R_b at which it should be placed, or choose a desired radius and solve for the mass that must be placed there. Packaging constraints may dictate the maximum radius possible in some cases. The balance mass is, of course, confined to the "single plane" of the unbalanced masses.

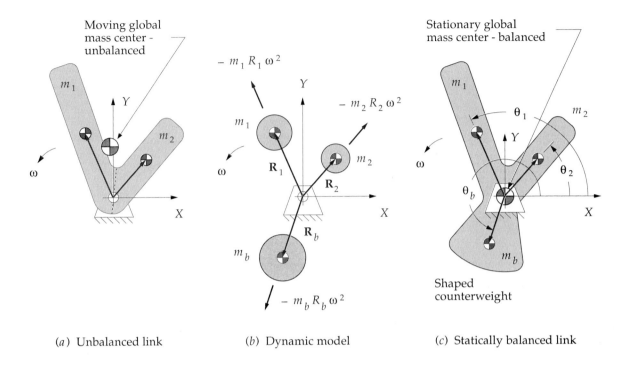

(a) Unbalanced link   (b) Dynamic model   (c) Statically balanced link

FIGURE 13-1

Static balancing a link in pure rotation

Once a combination of m_b and R_b is chosen, it remains to design the physical counterweight. The chosen radius R_b is the distance from the pivot to the CG of whatever shape we create for the counterweight mass. Our simple dynamic model, used to calculate the mR product, assumed a point mass and a massless rod. These ideal devices do not exist. A possible shape for this counterweight is shown in Figure 13-1c. Its mass must be m_b, distributed so as to place its CG at radius R_b at angle θ_b.

**EXAMPLE 13-1**

Static Balancing

Given: The system shown in Figure 13-1 has the following data:

$$m_1 = 1.2 \text{ kg} \qquad\qquad R_1 = 1.135 \text{ m } @\angle 113.4°$$
$$m_2 = 1.8 \text{ kg} \qquad\qquad R_2 = 0.822 \text{ m } @\angle 48.8°$$
$$\omega = 40 \text{ rad / sec}$$

Find: The mass-radius product and its angular location needed to statically balance the system.

Solution:

1   Resolve the position vectors into xy components in the arbitrary coordinate system associated with the freeze-frame position of the linkage chosen for analysis.

$$R_1 = 1.135@\angle 113.4°; \qquad R_{1_x} = -0.451, \qquad R_{1_y} = 1.042$$

$$(a)$$

$$R_2 = 0.822@\angle 48.8°; \qquad R_{2_x} = +0.541, \qquad R_{2_y} = 0.618$$

2   Solve equations 13.2c.

$$m_b R_{b_x} = -m_1 R_{1_x} - m_2 R_{2_x} = -(1.2)(-0.451) - (1.8)(0.541) = -0.433$$

$$(b)$$

$$m_b R_{b_y} = -m_1 R_{1_y} - m_2 R_{2_y} = -(1.2)(1.042) - (1.8)(0.618) = -2.363$$

3   Solve equations 13.2d and 13.2e

$$\theta_b = \arctan \frac{-2.363}{-0.433} = 259.6°$$

$$(c)$$

$$m_b R_b = \sqrt{(-0.433)^2 + (-2.363)^2} = 2.402 \text{ kg-m}$$

4   This mass-radius product of 2.402 kg-m can be obtained with a variety of shapes appended to the assembly. Figure 13-1c shows a particular shape whose CG is at a radius of $R_b = 0.806$ m at the required angle of 259.6°. The mass required for this counterweight design is then:

$$m_b = \frac{2.402 \text{ kg-m}}{0.806 \text{ m}} = 2.980 \text{ kg}$$

$$(d)$$

at a chosen CG radius of:

$$R_b = 0.806 \text{ m}$$

$$(e)$$

Many other shapes are possible. As long as they provide the required mass-radius product at the required angle, the system will be statically balanced. Note that the value of ω was not needed in the calculation.

13.2   DYNAMIC BALANCE

Dynamic balance is sometimes called **two-plane balance**. It requires that two criteria be met. The sum of the forces must be zero (static balance) plus the sum of the moments[1] must also be zero.

[1] We will use the term *Moment* in this text to refer to "turning forces" whose vectors are perpendicular to an axis of rotation or "long axis" of an assembly, and the term *Torque* to refer to "turning forces" whose vectors are parallel to an axis of rotation.

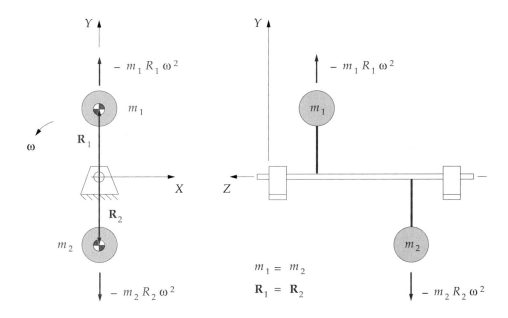

FIGURE 13-2

Balanced forces - unbalanced moment

$$\sum \mathbf{F} = 0$$

$$\sum \mathbf{M} = 0$$

(13.3)

These moments act in planes that include the axis of rotation of the assembly such as planes XZ and YZ in Figure 13-2. The moment's vector direction, or axis, is perpendicular to the assembly's axis of rotation.

Any rotating object or assembly which is relatively long in the axial direction compared to the radial direction requires dynamic balancing for complete balance. It is possible for an object to be statically balanced but not be dynamically balanced. Consider the assembly in Figure 13-2. Two equal masses are at identical radii, 180° apart rotationally, but separated along the shaft length. A summation of $-m\mathbf{a}$ forces due to their rotation will be always zero. However, in the side view, their inertia forces form a couple which rotates with the masses about the shaft. This rocking couple causes a moment on the ground plane, alternately lifting and dropping the left and right ends of the shaft.

Some examples of devices which require dynamic balancing are: rollers, crankshafts, camshafts, axles, clusters of multiple gears, motor rotors, turbines, propellor shafts. The common denominator among these devices is that their mass may be unevenly distributed both rotationally around their axis and also longitudinally along their axis.

To correct dynamic imbalance requires either adding or removing the right amount of mass at the proper angular locations *in two correction planes* separated by some distance along the shaft. This will create the necessary counter forces to statically balance the system and also provide a counter couple to cancel the unbalanced moment. When an automobile tire and wheel is dynamically balanced, the two correction planes are the inner and outer edges of the wheel rim. Correction weights are added at the proper locations in each of these correction planes based on a measurement of the dynamic forces generated by the unbalanced, spinning wheel.

It is always good practice to first statically balance all individual components that go into an assembly, if possible. This will reduce the amount of dynamic imbalance that must be corrected in the final assembly and also reduce the bending moment on the shaft. A common example of this situation is the aircraft turbine which consists of a number of circular turbine wheels arranged along a shaft. Since these spin at high speed, the inertia forces due to any imbalance can be very large. The individual wheels are statically balanced before being assembled to the shaft. The final assembly is then dynamically balanced.

Some devices do not lend themselves to this approach. An electric motor rotor is essentially a spool of copper wire wrapped in a complex pattern around the shaft. The mass of the wire is not uniformly distributed either rotationally or longitudinally, so it will not be balanced. It is not possible to modify the windings' local mass distribution after the fact without compromising electrical integrity. Thus the entire rotor imbalance must be countered in the two correction planes after assembly.

Consider the system of three lumped masses arranged around and along the shaft in Figure 13-3. Assume that, for some reason, they cannot be individually statically balanced within their own planes. We then create two correction planes labelled A and B. In this design example, the unbalanced masses m_1, m_2, m_3 and their radii R_1, R_2, R_3 are known along with their angular locations θ_1, θ_2 and θ_3. We want to dynamically balance the system. A three-dimensional coordinate system is applied with the axis of rotation in the Z direction. Note that the system has again been stopped in an arbitrary freeze-frame position. Angular acceleration is assumed to be zero. The summation of forces is:

$$-m_1\mathbf{R}_1\omega^2 - m_2\mathbf{R}_2\omega^2 - m_3\mathbf{R}_3\omega^2 - m_A\mathbf{R}_A\omega^2 - m_B\mathbf{R}_B\omega^2 = 0 \qquad (13.4a)$$

Dividing out the w^2 and rearranging we get:

$$m_A\mathbf{R}_A + m_B\mathbf{R}_B = -m_1\mathbf{R_1} - m_2\mathbf{R_2} - m_3\mathbf{R_3} \qquad (13.4b)$$

Breaking into x and y components:

$$m_A R_{A_x} + m_B R_{B_x} = -m_1 R_{1_x} - m_2 R_{2_x} - m_3 R_{3_x}$$

$$\qquad (13.4c)$$

$$m_A R_{A_y} + m_B R_{B_y} = -m_1 R_{1_y} - m_2 R_{2_y} - m_3 R_{3_y}$$

This has four unknowns in the form of the $m\mathbf{R}$ products at plane A and the $m\mathbf{R}$ products at plane B. To solve we need the sum of the moments equation which we can take about a point in one of the correction planes such as point O. The moment arm z distances of each force versus plane A are labelled l_1, l_2, l_3, l_B in the figure.

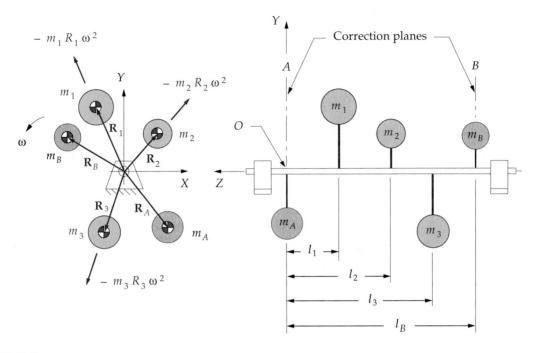

FIGURE 13-3

Two-plane dynamic balancing

$$\left(m_B\mathbf{R}_B\omega^2\right)l_B = -\left(m_1\mathbf{R}_1\omega^2\right)l_1 - \left(m_2\mathbf{R}_2\omega^2\right)l_2 - \left(m_3\mathbf{R}_3\omega^2\right)l_3 \qquad (13.4d)$$

Dividing out the w^2, breaking into x and y components, and rearranging:

moment in the XZ plane (i.e., about the Y axis)

$$m_B R_{B_x} = \frac{-\left(m_1 R_{1_x}\right)l_1 - \left(m_2 R_{2_x}\right)l_2 - \left(m_3 R_{3_x}\right)l_3}{l_B} \qquad (13.4e)$$

moment in the YZ plane (i.e., about the X axis)

$$m_B R_{B_y} = \frac{-\left(m_1 R_{1_y}\right)l_1 - \left(m_2 R_{2_y}\right)l_2 - \left(m_3 R_{3_y}\right)l_3}{l_B} \qquad (13.4f)$$

These can be solved for the $m\mathbf{R}$ products in x and y directions for correction plane B which can then be substituted into equation 13.4c to find the values needed in plane A. Equations 13.2d and 13.2e can then be applied to each correction plane to find the angles at which the balance masses must be placed and the mR products needed in each plane. The physical counterweights can then be designed consistent with the constraints outlined in the section on static balance. Note that the radii R_A and R_B do not have to be the same value.

✏️ EXAMPLE 13-2

Dynamic Balancing

Given: The system shown in Figure 13-3 has the following data:

$$m_1 = 1.2 \text{ kg} \qquad\qquad\qquad R_1 = 1.135 \text{ m @} \angle 113.4°$$
$$m_2 = 1.8 \text{ kg} \qquad\qquad\qquad R_2 = 0.822 \text{ m @} \angle 48.8°$$
$$m_3 = 2.4 \text{ kg} \qquad\qquad\qquad R_3 = 1.04 \text{ m @} \angle 251.4°$$

The z distances in meters from plane A are:

$$l_1 = 0.854, \qquad l_2 = 1.701, \qquad l_3 = 2.396, \qquad l_B = 3.097$$

Find: The mass-radius products and their angular locations needed to dynamically balance the system using the correction planes A and B.

Solution:

1   Resolve the position vectors into xy components in the arbitrary coordinate system associated with the freeze-frame position of the linkage chosen for analysis.

$$R_1 = 1.135 \text{ @ } \angle 113.4°; \qquad R_{1_x} = -0.451, \qquad R_{1_y} = +1.042$$
$$R_2 = 0.822 \text{ @ } \angle 48.8°; \qquad R_{2_x} = +0.541, \qquad R_{2_y} = +0.618 \qquad (a)$$
$$R_3 = 1.040 \text{ @ } \angle 251.4°; \qquad R_{3_x} = -0.332, \qquad R_{3_y} = -0.986$$

2   Solve equations 13.4e for summation of moments about point O.

$$m_B R_{B_x} = \frac{-\left(m_1 R_{1_x}\right)l_1 - \left(m_2 R_{2_x}\right)l_2 - \left(m_3 R_{3_x}\right)l_3}{l_B}$$

$$\hspace{11cm} (b)$$

$$m_B R_{B_x} = \frac{-(1.2)(-0.451)(0.854) - (1.8)(0.541)(1.701) - (2.4)(-0.332)(2.396)}{3.097} = 0.230$$

$$m_B R_{B_y} = \frac{-\left(m_1 R_{1_y}\right)l_1 - \left(m_2 R_{2_y}\right)l_2 - \left(m_3 R_{3_y}\right)l_3}{l_B}$$

$$\hspace{11cm} (c)$$

$$m_B R_{B_y} = \frac{-(1.2)(1.042)(0.854) - (1.8)(0.618)(1.701) - (2.4)(-0.986)(2.396)}{3.097} = 0.874$$

3   Solve equations 13.2d and 13.2e for the mass radius product in plane B.

$$\theta_B = \arctan\frac{0.874}{0.230} = 75.27°$$

$$(d)$$

$$m_B R_B = \sqrt{(0.230)^2 + (0.874)^2} = 0.904 \text{ kg-m}$$

4 Solve equations 13.4c for forces in x and y directions.

$$m_A R_{A_x} = -m_1 R_{1_x} - m_2 R_{2_x} - m_3 R_{3_x} - m_B R_{B_x}$$
$$m_A R_{A_y} = -m_1 R_{1_y} - m_2 R_{2_y} - m_3 R_{3_y} - m_B R_{B_y}$$

$$(e)$$

$$m_A R_{A_x} = -(1.2)(-0.451) - (1.8)(0.541) - (2.4)(-0.332) - 0.230 = 0.133$$
$$m_A R_{A_y} = -(1.2)(1.042) - (1.8)(0.618) - (2.4)(-0.986) - 0.874 = -0.872$$

5 Solve equations 13.2d and 13.2e for the mass radius product in plane A.

$$\theta_A = \arctan\frac{-0.872}{0.133} = -81.35°$$

$$(f)$$

$$m_A R_A = \sqrt{(0.133)^2 + (-0.872)^2} = 0.882 \text{ kg-m}$$

6 These mass-radius products can be obtained with a variety of shapes appended to the assembly in planes A and B. Many shapes are possible. As long as they provide the required mass-radius products at the required angles in each correction plane, the system will be dynamically balanced.

So, when the design is still on the drawing board, these simple analysis techniques can be used to determine the necessary sizes and locations of balance masses for any assembly in pure rotation for which the mass distribution is defined. This two-plane balance method can be used to dynamically balance any system in pure rotation, and all such systems should be balanced unless the purpose of the device is to create shaking forces or moments.

13.3 STATIC BALANCING THE FOURBAR LINKAGE

The rotating links (cranks, rockers) of a linkage can be individually balanced by the methods described above. The effects of the couplers, which are in complex motion, are more difficult to compensate for. Note that the process of statically balancing a rotating link, in effect, forces its mass center (*CG*) to be at its fixed pivot, thus stationary. In other words the condition of **static balance** can also be **defined as** one of *making the mass center stationary.* A coupler has no fixed pivot and thus its mass center is, in general, always in motion.

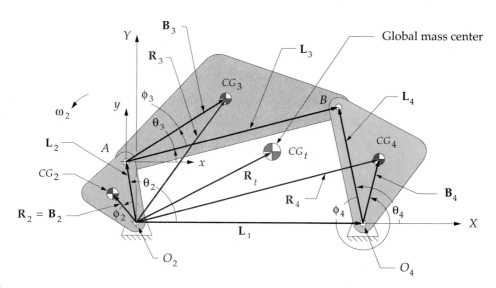

FIGURE 13-4

Static (force) balancing a fourbar linkage

Any mechanism, no matter how complex, will have, for every instantaneous position, a single, overall, *global mass center* located at some particular point. We can calculate its location knowing only the link masses and the locations of the CGs of the individual links at that instant. The global mass center normally changes position as the linkage moves. If we can somehow force this global mass center to be stationary, we will have a state of static balance for the overall linkage.

The Berkof-Lowen method of linearly independent vectors[1] provides us with a means to calculate the magnitude and location of counterweights to be placed on the rotating links which will make the global mass center stationary for all positions. Placement of the proper balance masses on the links will cause the dynamic forces on the fixed pivots to always be equal and opposite, i.e., a couple, thus creating static balance ($\Sigma F = 0$ but $\Sigma M \neq 0$) in the moving linkage.

Figure 13-4 shows a fourbar linkage with its overall global mass center located by the position vector $\mathbf{R}_t$. The individual CGs of the links are located *in the global system* by position vectors $\mathbf{R}_2$, $\mathbf{R}_3$, and $\mathbf{R}_4$ (magnitudes R_2, R_3, R_4), rooted at its origin, the crank pivot O_2. The link lengths are defined by position vectors labelled $\mathbf{L}_1, \mathbf{L}_2, \mathbf{L}_3, \mathbf{L}_4$ (magnitudes l_1, l_2, l_3, l_4), and the local position vectors which locate the CGs *within each link* are $\mathbf{B}_2, \mathbf{B}_3, \mathbf{B}_4$ (magnitudes b_2, b_3, b_4). The angles of the vectors $\mathbf{B}_2, \mathbf{B}_3, \mathbf{B}_4$ are ϕ_2, ϕ_3, ϕ_4 measured internal to the links with respect to the links' lines of centers $\mathbf{L}_2, \mathbf{L}_3, \mathbf{L}_4$. The instantaneous link angles which locate $\mathbf{L}_2, \mathbf{L}_3, \mathbf{L}_4$ in the global system are $\theta_2, \theta_3, \theta_4$. The total mass of the system is simply the sum of the individual link masses:

$$m_t = m_2 + m_3 + m_4 \qquad (13.5a)$$

The total mass moment about the origin must be equal to the sum of the mass moments due to the individual links:

$$\sum M_{O_2} = m_t \mathbf{R}_t = m_2 \mathbf{R}_2 + m_3 \mathbf{R}_3 + m_4 \mathbf{R}_4 \qquad (13.5b)$$

The position of the global mass center is then:

$$\mathbf{R}_t = \frac{m_2 \mathbf{R}_2 + m_3 \mathbf{R}_3 + m_4 \mathbf{R}_4}{m_t} \qquad (13.5c)$$

and from the linkage geometry:

$$\mathbf{R}_2 = b_2 \, e^{j(\theta_2 + \phi_2)} = b_2 \, e^{j\theta_2} e^{j\phi_2}$$
$$\mathbf{R}_3 = l_2 \, e^{j\theta_2} + b_3 \, e^{j(\theta_3 + \phi_3)} = l_2 \, e^{j\theta_2} + b_3 \, e^{j\theta_3} e^{j\phi_3} \qquad (13.5d)$$
$$\mathbf{R}_4 = l_1 \, e^{j\theta_1} + b_4 \, e^{j(\theta_4 + \phi_4)} = l_1 \, e^{j\theta_1} + b_4 \, e^{j\theta_4} e^{j\phi_4}$$

We can solve for the location of the global mass center for any link position for which we know the link angles θ_2, θ_3, θ_4. We want to make this position vector $\mathbf{R}_t$ be a constant. The first step is to substitute equations 13.5d into 13.5b,

$$m_t \mathbf{R}_t = m_2 \left(b_2 \, e^{j\theta_2} e^{j\phi_2} \right) + m_3 \left(l_2 \, e^{j\theta_2} + b_3 \, e^{j\theta_3} e^{j\phi_3} \right) + m_4 \left(l_1 \, e^{j\theta_1} + b_4 \, e^{j\theta_4} e^{j\phi_4} \right) \qquad (13.5e)$$

and rearrange to group the constant terms as coefficients of the time-dependent terms:

$$m_t \mathbf{R}_t = \left(m_4 l_1 \, e^{j\theta_1} \right) + \left(m_2 b_2 e^{j\phi_2} + m_3 l_2 \right) e^{j\theta_2} + \left(m_3 b_3 \, e^{j\phi_3} \right) e^{j\theta_3} + \left(m_4 b_4 \, e^{j\phi_4} \right) e^{j\theta_4} \qquad (13.5f)$$

Note that the terms in parentheses are all constant with time. The only time-dependent terms are the ones containing θ_2, θ_3, and θ_4.

We can also write the vector loop equation for the linkage,

$$l_2 \, e^{j\theta_2} + l_3 \, e^{j\theta_3} - l_4 \, e^{j\theta_4} - l_1 \, e^{j\theta_1} = 0 \qquad (13.6a)$$

and solve it for one of the unit vectors that define a link direction, say link 3:

$$e^{j\theta_3} = \frac{\left(l_1 \, e^{j\theta_1} - l_2 \, e^{j\theta_2} + l_4 \, e^{j\theta_4} \right)}{l_3} \qquad (13.6b)$$

Substitute this into equation 13.5f to eliminate the θ_3 term:

$$m_t \mathbf{R}_t = \left(m_2 b_2 e^{j\phi_2} + m_3 l_2 \right) e^{j\theta_2} + \frac{1}{l_3} \left(m_3 b_3 \, e^{j\phi_3} \right) \left(l_1 \, e^{j\theta_1} - l_2 \, e^{j\theta_2} + l_4 \, e^{j\theta_4} \right)$$
$$+ \left(m_4 b_4 \, e^{j\phi_4} \right) e^{j\theta_4} + \left(m_4 l_1 \, e^{j\theta_1} \right) \qquad (13.7a)$$

and collect terms:

$$m_t \mathbf{R}_t = \left(m_2 b_2 e^{j\phi_2} + m_3 l_2 - m_3 b_3 \frac{l_2}{l_3} e^{j\phi_3} \right) e^{j\theta_2} + \left(m_4 b_4 \, e^{j\phi_4} + m_3 b_3 \frac{l_4}{l_3} e^{j\phi_3} \right) e^{j\theta_4}$$
$$+ m_4 l_1 \, e^{j\theta_1} + m_3 b_3 \frac{l_1}{l_3} e^{j\phi_3} e^{j\theta_1} \qquad (13.7b)$$

This expression gives us the tool to force $\mathbf{R}_t$ to be a constant and make the linkage mass center stationary. For that to be so, the terms in parentheses which multiply the only two time-dependent variables, θ_2 and θ_4, must be forced to be zero. (The fixed link angle θ_1 is a constant.) Thus the requirement for linkage force balance is:

$$\left(m_2 b_2 e^{j\phi_2} + m_3 l_2 - m_3 b_3 \frac{l_2}{l_3} e^{j\phi_3} \right) = 0$$

(13.8a)

$$\left(m_4 b_4 e^{j\phi_4} + m_3 b_3 \frac{l_4}{l_3} e^{j\phi_3} \right) = 0$$

Rearranging to isolate one link's parameters (say link 3) on one side of each of these equations:

$$m_2 b_2 e^{j\phi_2} = m_3 \left(b_3 \frac{l_2}{l_3} e^{j\phi_3} - l_2 \right)$$

(13.8b)

$$m_4 b_4 e^{j\phi_4} = -m_3 b_3 \frac{l_4}{l_3} e^{j\phi_3}$$

We have two equations involving three links. Any one link's parameters can be assumed and the other two solved for. A linkage is typically first designed to satisfy the required motion and packaging constraints before this force balancing procedure is attempted. In that event, the link geometry and masses are already defined, at least in a preliminary way. A useful strategy is to leave the link 3 mass and CG location as originally designed and calculate the necessary masses and CG locations of links 2 and 4 to satisfy these conditions for balanced forces. Links 2 and 4 are in pure rotation, so it is straightforward to add counterweights to them in order to move their CGs to the necessary locations. With this approach, the right sides of equations 13.8b are reducible to numbers for a designed linkage. We want to solve for the mass radius products $m_2 b_2$ and $m_4 b_4$ and also the angular locations of the CGs within the links. Note that the angles ϕ_2 and ϕ_4 in equation 13.8 are measured with respect to the lines of centers of their respective links.

$$\left(m_2 b_2 \right)_x = m_3 \left(b_3 \frac{l_2}{l_3} \cos \phi_3 - l_2 \right)$$

(13.8c)

$$\left(m_2 b_2 \right)_y = m_3 \left(b_3 \frac{l_2}{l_3} \sin \phi_3 \right)$$

$$\left(m_4 b_4 \right)_x = -m_3 b_3 \frac{l_4}{l_3} \cos \phi_3$$

(13.8d)

$$\left(m_4 b_4 \right)_y = -m_3 b_3 \frac{l_4}{l_3} \sin \phi_3$$

Note that these components of the mR product needed to force balance the linkage represent the entire amount needed. If links 2 and 4 are already designed with some individual unbalance (CG not at pivot), then the existing mR product must be subtracted from that found in equations 13.8c and 13.8d in order to determine the size and location of additional counterweights to be added to those links. As we did with the balance of rotating links, any combination of mass and radius that gives the desired product is acceptable. Use equations 13.2d and 13.2e to convert the cartesian mR products in equations 13.8 to polar coordinates. Note that the angle of the mR vector for each link will be referenced to that individual link's line of centers. Design the physical counterweights on the links as discussed in Section 13.1.

13.4 EFFECT OF BALANCING ON INPUT TORQUE

Individually balancing a link which is in pure rotation by the addition of a counterweight will have the side effect of increasing its mass moment of inertia. The "flywheel effect" of the link is increased by this increase in its moment of inertia. Thus the torque needed to accelerate that link will be greater. The input torque will be unaffected by any change in the I of the input crank when it is run at constant angular velocity. But, any rockers in the mechanism will have angular accelerations even when the crank does not. Thus, individually balancing the rockers will tend to increase the required input torque even at constant input crank velocity.

Adding the counterweights to the rotating links, which are necessary to force balance the entire linkage, both increases their mass moment of inertia and also (individually) *unbalances* those rotating links in order to gain the global balance. Thus the CGs of the rotating links will not be at their fixed pivots. Their angular accelerations will then contribute to the torque loading on the linkage. Balancing an entire linkage by this method then can have the side effect of increasing the variation in the required input torque, sometimes quite drastically. A larger flywheel may be needed on a balanced linkage in order to achieve the same coefficient of fluctuation as the unbalanced version of the linkage.

13.5 BALANCING WITH PROGRAM DYNAFOUR

Choosing the **Balance** selection from the main menu in program DYNAFOUR computes equations 13.7 and 13.8 and converts the result to polar coordinates referenced to the link's line of centers. The program has the information on the existing masses and CG locations for the links and uses it to compute the amount of additional counterweight mR product and its angle needed for force balance. The user is asked to select a desired radius at which to place the counterweight on each link (2 and 4). The needed mass for that radius is then computed and placed on the link at the specified radius. Figure 13-5 shows the linkage from diskfile FIG13-5 as animated by the program after balancing. Note the counterweights placed on links 2 and 4 at the calculated locations for complete force balance. Read the diskfile FIG13-5 into program DYNAFOUR, calculate and balance the linkage to see this example.

Once a linkage has been balanced within the program, any request for plots or tables of force or torque information will display the input screen in Figure 13-6 which asks for a choice among the original **unbalanced linkage data**, the **Berkof**

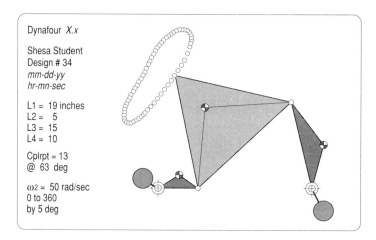

Dynafour *X.x*

Shesa Student
Design # 34
mm-dd-yy
hr-mn-sec

L1 = 19 inches
L2 =   5
L3 = 15
L4 = 10

Cplrpt = 13
@ 63 deg

ω2 = 50 rad/sec
0 to 360
by 5 deg

FIGURE 13-5

Animation of a balanced linkage in program DYNAFOUR

data, or the **net balanced data**. The meanings of the first and last choices should
be obvious. The Berkof choice will display the forces or torques that are being
generated by the added counterweights alone. This gives some indication of the
dynamic effect of adding these balance weights. The net balanced choice is the
vector sum of the first two choices.

Figure 13-7 shows the shaking force generated by this linkage in its unbalanced state. The maximum is 462 lb at 15°. Figure 13-8 shows the shaking

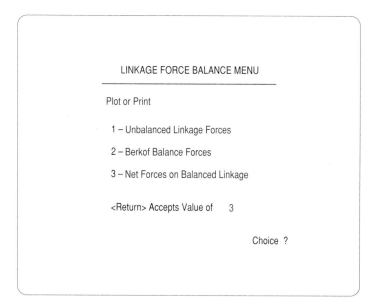

LINKAGE FORCE BALANCE MENU

Plot or Print

1 – Unbalanced Linkage Forces

2 – Berkof Balance Forces

3 – Net Forces on Balanced Linkage

<Return> Accepts Value of   3

Choice ?

FIGURE 13-6

Force balance menu from program DYNAFOUR

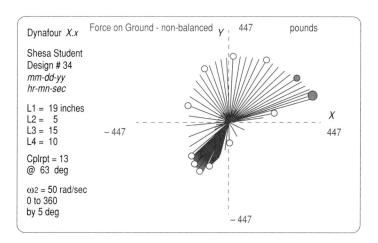

FIGURE 13-7

Polar plot of unbalanced shaking forces on ground plane of a fourbar linkage

force after balancing. It is essentially zero. The small magnitude vectors shown are the result of round-off errors in the computations.

If we look separately at the forces on joints O_2 and O_4 in Figures 13-9 and 13-10 we can see that these forces are not zero. They are, in fact, equal and opposite. Note that after balancing, the pattern of forces at joint 1,2 is the mirror image of the pattern at joint 1,4. The **net shaking force** is the vector sum of these two sets of forces for each time step (see Section 12.8). The equal and opposite pairs of forces acting at the ground pivots at each time step create a time-varying couple on the ground plane which is the **shaking torque**. The fact that these forces can be larger due to the balance weights may increase the shaking torque compared

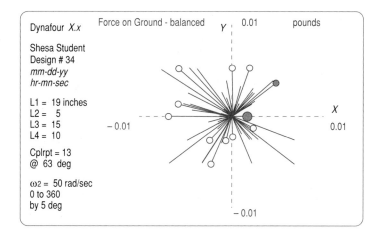

FIGURE 13-8

Polar plot of balanced shaking forces on ground plane of a fourbar linkage

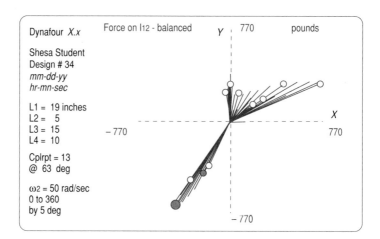

FIGURE 13-9

Polar plot of balanced forces F21 on the ground plane of a fourbar linkage

to its former value in the unbalanced linkage. Thus the tradeoff for reducing the shaking forces to zero may be a larger input torque variation.

Figure 13-11 shows the input torque curve for the unbalanced linkage and Figure 13-12 shows the input torque for the same linkage after complete force-balancing has been done. The peak value of the required drive torque has increased as a result of the force-balancing. The presence of the additional counterweight mass has also created additional forces at all the pin joints (not shown). Thus the stresses in the links and pins may also increase as a result of force-balancing.

Note, however, that the degree of increase in the shaking torque due to force-balancing is dependent upon the choice of radii at which the balance masses

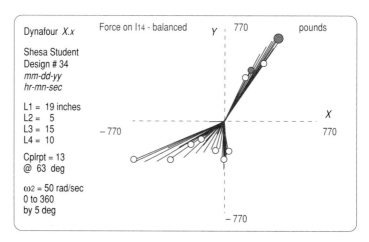

FIGURE 13-10

Polar plot of balanced forces F41 on the ground plane of a fourbar linkage

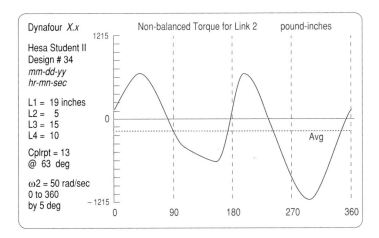

FIGURE 13-11

Input torque curve for an unbalanced fourbar linkage

are placed. The additional mass moment of inertia of the link due to the balance mass is proportional to the square of the radius at which you place the mass. The force balance algorithm only dictates a particular mass-radius product be added. Placing the balance mass at as small a radius as possible will minimize the increase in shaking torque due to force-balancing.

To reduce the torque penalty one may also wish to do less than a complete force balance and accept some shaking force in trade. DYNAFOUR allows this. A choice of complete or partial balance is provided in the balance menus. The latter choice requires the input of the desired mass for the counterweight as well as the radius.

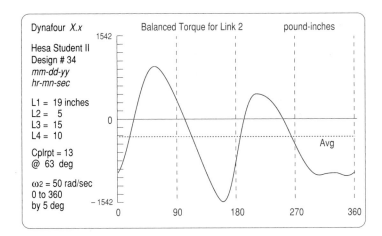

FIGURE 13-12

Input torque curve for a force-balanced fourbar linkage

The calculated mass radius product for complete balance can be printed out any time after performing the balance calculation. Select **Forces** from the print menu and then select **Shaking Forces**. The mR products for links 2 and 4 will be printed at the head of the shaking force table.

13.6   MEASURING AND CORRECTING IMBALANCE

While we can do a great deal to ensure balance when designing a machine, variations and tolerances in manufacturing will preclude even a well-balanced design from being in perfect balance when built. Thus there is need for a means to measure and correct the imbalance in rotating systems. Perhaps the best example assembly to discuss is that of the automobile tire and wheel, with which most readers will be familiar. Certainly the design of this device promotes balance, as it is essentially cylindrical and symmetrical. If manufactured to be perfectly uniform in geometry and homogeneous in material, it should be in perfect balance as is. But typically it is not. The wheel (or rim) is more likely to be close to balanced, as manufactured, than is the tire. The wheel is made of a homogeneous metal and has fairly uniform geometry and cross section. The tire, however, is a composite of synthetic rubber elastomer and fabric cord or metal wire. The whole is compressed in a mold and steam-cured at high temperature. The resulting material varies in density and distribution and its geometry is often distorted in the process of removal from the mold and cooling.

STATIC BALANCING After the tire is assembled to the wheel, the assembly must be balanced to reduce vibration at high speeds. The simplest approach is to statically balance it, though it is not really an ideal candidate for this approach as it is thick axially compared to its diameter. To do so it is typically suspended in a horizontal plane on a cone through its center hole. A bubble level is attached to the wheel and weights are placed at positions around the rim of the wheel until it sits level. These weights are then attached to the rim at those points. This is a single-plane balance and thus can only cancel the unbalanced forces. It has no effect on any unbalanced moments due to uneven distribution of mass along the axis of rotation. It also is not very accurate.

DYNAMIC BALANCING The better approach is to dynamically balance it. This requires a dynamic balancing machine be used. Figure 13-13 shows a schematic of such a device, used for balancing wheels and tires or any other rotating assembly. The assembly to be balanced is mounted temporarily on an axle, called a mandrel, which is supported in bearings within the balancer. These two bearings are each mounted on a suspension which contains a transducer that measures dynamic force. A common type of force transducer contains a piezoelectric crystal which delivers a voltage proportional to the force applied. This voltage is amplified electronically and delivered to circuitry or software which can compute its peak magnitude and the phase angle of that peak with respect to some time reference signal. The reference signal is supplied by a shaft encoder on the mandrel which provides a short duration electrical pulse once per revolution in exactly the same angular location. This encoder pulse triggers the computer to begin processing the force signal. The encoder may also provide some large number of additional pulses equispaced around the shaft circumference (often 1024). These are used to trigger the recording of each data sample from the transducers in exactly the same location around the shaft and to provide a measure of shaft velocity via an electronic counter.

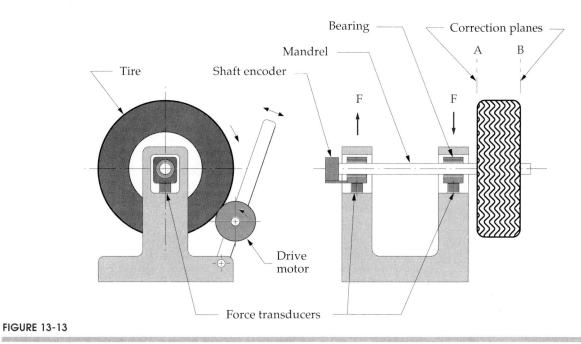

FIGURE 13-13

A dynamic wheel balancer

The assembly to be balanced is then "spun up" to some angular velocity, usually with a friction drive contacting its circumference. The drive torque is then removed and the drive motor stopped, allowing the assembly to "freewheel." (This is to avoid measuring any forces due to imbalances in the drive system.) The measuring sequence is begun and the dynamic forces at each bearing are measured simultaneously and their waveforms stored. Many cycles can be measured and averaged to improve the quality of the measurement. Because forces are being measured at two locations displaced along the axis, both summation of moment and summation of force data are computed.

The force signals are sent to a built-in computer for processing and computation of the needed balance masses and locations. The data needed from the measurements are the magnitudes of the peak forces and the angular locations of those peaks with respect to the shaft encoder's reference angle (which corresponds to a known point on the wheel). The axial locations of the wheel rim's inside and outside edges (the correction planes) with respect to the balance machine's transducer locations are provided to the machine's computer by operator measurement. From these data the net unbalanced force and net unbalanced moment can be calculated since the distance between the measured bearing forces is known. The mass-radius products needed in the correction planes on each side of the wheel can then be calculated from equations 13.3 in terms of the mR product of the balance weights. The correction radius is that of the wheel rim. The balance masses and angular locations are calculated for each correction plane to put the system in dynamic balance. Weights having the needed mass are clipped onto the inside and outside wheel rims (which are the correction planes in this case), at the proper angular locations. The result is a fairly accurately dynamically balanced tire and wheel.

13.7   REFERENCES

1   Berkof, R. S. and Lowen, G. G., "A New Method for Completely Force Balancing Simple Linkages," *J. Eng. Ind. Trans. ASME*, ser. B, vol. 91, no. 1, February 1969, pp. 21-26.

13.8   PROBLEMS

*13-1   A system of two coplanar arms on a common shaft, as shown in Figure 13-1, is to be designed. For the row(s) assigned in Table P13-1, find the shaking force of the linkage when run unbalanced and design a counterweight to statically balance the system. Work in any consistent units system you prefer.

13-2   The minute hand on Big Ben weighs 40 lb and is 10 feet long. Its *CG* is 4 feet from the pivot. Calculate the *mR* product and angular location needed to statically balance this link and design a physical counterweight, positioned close to the center. Select material and design the detailed shape of the counterweight which is of 2 inch uniform thickness in the *Z* direction.

13-3   A "**V** for victory" advertising sign is being designed to be oscillated about the apex of the V, on a billboard, as the rocker of a fourbar linkage. The angle between the legs of the **V** is 20°. Each leg is 8 feet long and 1.5 feet wide. Material is 0.25 inch-thick aluminum. Design the V link for static balance.

13-4   A three-bladed ceiling fan has 1.5 feet by 0.25 feet equispaced rectangular blades that nominally weigh 2 lb each. Manufacturing tolerances will cause the blade weight to vary up to plus or minus 5%. The mounting accuracy of the blades will vary the location of the *CG* versus the spin axis by plus or minus 10% of the blades' diameters. Calculate the weight of the largest steel counterweight needed at a 2 inch radius to statically balance the worst case blade assembly.

*13-5   A system of three noncoplanar weights is arranged on a shaft generally as shown in Figure 13-3. For the dimensions from the row(s) assigned in Table P13-2, find the shaking forces and shaking moment when run unbalanced at 100 rpm and specify the *mR* product and angle of the counterweights in correction planes *A* and *B* needed to dynamically balance the system. The correction planes are 20 units apart. Work in any consistent units system you prefer.

TABLE P13-1   Data for Problem 13-1

row	m_1	m_2	R_1	R_2
a.	0.20	0.40	1.25 @ 30°	2.25 @ 120°
b.	2.00	4.36	3.00 @ 45°	9.00 @ 320°
c.	3.50	2.64	2.65 @ 100°	5.20 @ −60°
d.	5.20	8.60	7.25 @ 150°	6.25 @ 220°
e.	0.96	3.25	5.50 @ −30°	3.55 @ 120°

TABLE P13-2   Data for Problem 13-5

row	m_1	m_2	m_3	l_1	l_2	l_3	R_1	R_2	R_3
a.	0.20	0.40	1.24	2	8	17	1.25 @ 30°	2.25 @ 120°	5.50 @ −30°
b.	2.00	4.36	3.56	5	7	16	3.00 @ 45°	9.00 @ 320°	6.25 @ 220°
c.	3.50	2.64	8.75	4	9	11	2.65 @ 100°	5.20 @ −60°	1.25 @ 30°
d.	5.20	8.60	4.77	7	12	16	7.25 @ 150°	6.25 @ 220°	9.00 @ 320°
e.	0.96	3.25	0.92	1	3	18	5.50 @ −30°	3.55 @ 120°	2.65 @ 100°

*13-6   A wheel and tire assembly has been run at 100 RPM on a dynamic balancing machine as shown in Figure 13-13. The force measured at the left bearing had a peak of 2.2 N at a phase angle of 45° with respect to the zero reference angle on the tire. The force measured at the right bearing had a peak of 1.2 N at a phase angle of –120° with respect to the reference zero on the tire. The center distance between the two bearings on the machine is 50 cm. The left edge of the wheel rim is 10 cm from the centerline of the closest bearing. The wheel is 18 cm wide at the rim. Calculate the size and location with respect to the tire's zero reference angle, of balance weights needed on each side of the rim to dynamically balance the tire assembly. The wheel rim is 15 inches in diameter.

*13-7   Repeat problem 13-6 for measured forces of 3.1 N at a phase angle of –60° with respect to the reference zero on the tire, measured at the left bearing and 4.2 N at a phase angle of 150 degrees with respect to the reference zero on the tire measured at the right bearing. The wheel is 16 inches in diameter.

*13-8   Table P12-3 in Chapter 12 shows geometry and kinematic data for several four-bar linkages.

    a.   For the row(s) from Table P12-3 assigned in this problem, calculate the size and angular locations of the counterbalance mass-radius products needed on links 2 and 4 to completely force balance the linkage by the method of Berkof and Lowen. Check your manual calculation with program DYNAFOUR.

    b.   Calculate the input torque for the linkage both with and without the added balance weights and compare the results. Use DYNAFOUR.

* Answers in Appendix E

Chapter 14

ENGINE DYNAMICS

*I have always thought that the substitution of the
internal combustion machine for the horse marked
a very gloomy milestone in the progress of mankind.*
WINSTON S. CHURCHILL

14.0 INTRODUCTION

The previous chapters have introduced analysis techniques for the determination of dynamic forces, moments, and torques in machinery. Shaking forces, moments, and their balancing have also been discussed. We will now attempt to integrate all these dynamic considerations into the design of a common device, the slider-crank linkage as used in the internal combustion engine. This deceptively simple mechanism will be found to be actually quite complex in terms of the dynamic considerations necessary to its design for high-speed operation. Thus it will serve as an excellent example of the application of the dynamics concepts just presented. We will not address the thermodynamic aspects of the internal combustion engine beyond defining the combustion forces which are necessary to drive the device. Many other texts, such as those listed in the bibliography at the end of this chapter, deal with the very complex thermodynamic and fluid dynamic aspects of this ubiquitous device. We will concentrate only on its kinematics and mechanical dynamics aspects. It is not our intention to make an "engine designer" of the student so much as to apply dynamic principles to a realistic design problem of general interest and also to convey the complexity and fascination involved in the design of an apparently simple dynamic device.

Some students may have had the opportunity to disassemble and service an internal combustion engine, but many will have never done so. Thus we will begin with very fundamental descriptions of engine design and operation. The program ENGINE, supplied with this text, is designed to reinforce and amplify the concepts presented. It will perform all the tedious computations necessary to

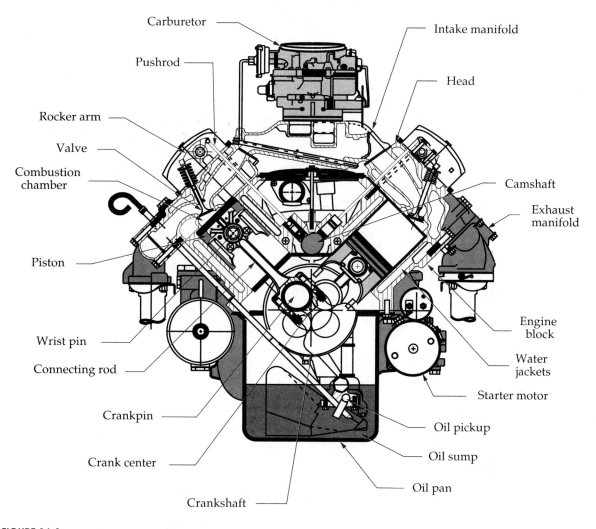

FIGURE 14-1

Cutaway cross section of a vee-eight engine
Adapted from a drawing by Lane Thomas, Western Carolina University, Dept. of Industrial Education - with permission

provide the student with dynamic information for design choices and tradeoffs. The student is encouraged to use this program concurrent with a reading of the text. Many examples and illustrations within the text will be generated with this program and reference will frequently be made to it. A user manual for program ENGINE is provided in Chapter 16. That chapter can be read or referred to out of sequence, with no loss in continuity, in order to gain familiarity with the program's operation. Examples used in Chapters 14 and 15 which deal with engine dynamics are provided as diskfiles which can be read into program ENGINE for student observation and exercise. These are noted in the text. Most of the figures made from program ENGINE plots in this chapter are of the built-in example engine which is #3 on the input menu of the program.

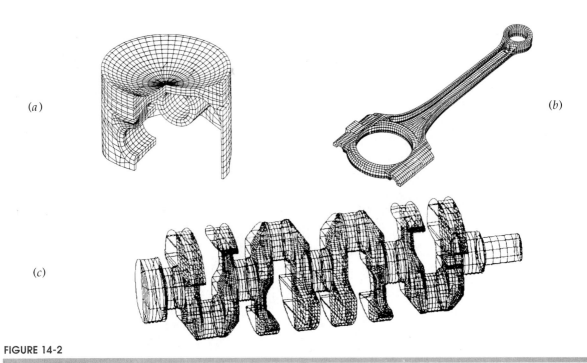

FIGURE 14-2

Finite element models of an engine piston (a), connecting rod (b) and crankshaft (c). *Courtesy of General Motors Co.*

14.1   ENGINE DESIGN

Figure 14-1 shows a detailed cross section of an internal combustion (IC) engine. The basic mechanism consists of a crank, a connecting rod (coupler), and piston (slider). Since this figure depicts a **multicylinder vee-eight** engine configuration, there are four cranks arranged on a crankshaft, and eight sets of connecting rods and pistons, four in the left bank of cylinders and four in the right bank. Only two piston-connecting rod assemblies are visible in this view. The others are behind those shown. Figure 14-2 shows finite-element models of a piston, connecting rod and crankshaft for a four-cylinder inline engine. The most usual arrangement is an inline engine with cylinders all in a common plane. Three-, four-, five- and six-cylinder **inline engines** are in production the world over. **Vee engines** in four-, six-, eight-, ten- and twelve-cylinder versions are also in production, with vee six and vee eight being the most popular configurations. The geometric arrangements of the crankshaft and cylinders have a significant effect on the dynamic condition of the engine. We will explore these effects of multicylinder arrangements in the next chapter. At this stage we wish to deal only with the design of a **single-cylinder** engine. After optimizing the geometry and dynamic condition of one cylinder, we will be ready to assemble combinations of cylinders into multicylinder configurations.

A schematic of the basic **one-cylinder slider-crank** mechanism and the terminology for its principal parts is shown in Figure 14-3. Note that it is "back-driven" compared to the linkages we have been analyzing in previous chapters. That is, the explosion of the combustible mixture in the cylinder drives the piston down, turning

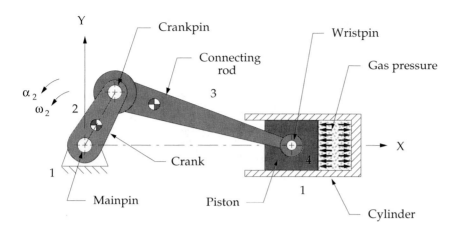

FIGURE 14-3

Fourbar slider-crank mechanism for single-cylinder internal combustion (IC) engine

the crank. The crank torque that results is ultimately delivered to the drive wheels of the vehicle through a transmission (see Section 10.9) to propel the car, motorcycle, or other device. The same slider-crank mechanism is also used "forward-driven", by motor-driving the crank and taking the output energy from the piston end. It is then called a **piston pump** and is used to compress air, pump gasoline and well water, etc.

In the IC engine of Figure 14-3, it should be fairly obvious that at most we can only expect energy to be delivered from the exploding gases to the crank during the down-stroke half of the cycle. The piston must return from bottom dead center (BDC) to top dead center (TDC) on its own momentum before it can receive another push from the next explosion. In fact, some rotational kinetic energy must be stored in the crankshaft merely to carry it through the TDC and BDC points as the moment arm for the gas force at those points is zero. This is why an IC engine must be "spun-up" with a hand crank, pull rope, or starter motor to get it running.

There are two common combustion cycles in use in internal combustion engines, the **Clerk two-stroke cycle** and the **Otto four-stroke cycle**, named after their nineteenth century inventors. The four-stroke cycle is most common in automobile, truck, and stationary gasoline engines. The two-stroke cycle is used in motorcycles, outboard motors, chain saws, and other applications where its better power-to-weight ratio outweighs its drawbacks of higher pollution levels and poor fuel economy compared to the four-stroke.

FOUR-STROKE CYCLE The **Otto four-stroke cycle** is shown in Figure 14-4. It takes four full strokes of the piston to complete one Otto cycle. A piston stroke is defined as its travel from TDC to BDC or the reverse. Thus there are two-strokes per 360° crank revolution and it takes 720° of crankshaft rotation to complete one four-stroke cycle. This engine requires at least two valves per cylinder, one for intake and one for exhaust. For discussion, we can start the cycle at any point as it repeats every two crank revolutions. Figure 14-4a shows the **intake stroke** which starts with the piston at TDC. A mixture of fuel and air is drawn into the cylinder from the induc-

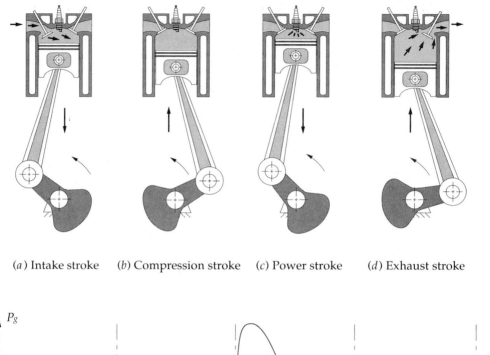

(a) Intake stroke   (b) Compression stroke   (c) Power stroke   (d) Exhaust stroke

(e) The gas pressure curve

FIGURE 14-4

The Otto four-stroke combustion cycle

tion system (the carburetor and intake manifold in Figure 14-1) as the piston descends to BDC, increasing the volume of the cylinder and creating a slight negative pressure.

During the **compression stroke** in Figure 14-4b, all valves are closed and the gas is compressed as the piston travels from BDC to TDC. Slightly before TDC, a spark is ignited to explode the compressed gas. The pressure from this explosion builds very quickly and pushes the piston down from TDC to BDC during the **power stroke** shown in Figure 14-4c. The exhaust valve is opened and the piston's **ex-**

haust stroke from BDC to TDC (Figure 14-4d) pushes the spent gases out of the cylinder into the exhaust manifold (see also Figure 14-1) and thence to the catalytic converter for cleaning before being dumped out the tailpipe. The cycle is then ready to repeat with another intake stroke. The valves are opened and closed at the right times in the cycle by a camshaft which is driven in synchrony with the crankshaft by gears, chain, or toothed belt drive. (See Figure 10-21.) Figure 14-4e shows the gas pressure curve for one cycle. With a one-cylinder Otto cycle engine, power is delivered to the crankshaft, at most, 25% of the time as there is only 1 power stroke per 2 revolutions.

TWO-STROKE CYCLE The **Clerk two-stroke cycle** is shown in Figure 14-5. This engine does not need any valves, though to increase its efficiency it is sometimes provided with a passive (pressure differential operated) one at the intake port. It does not have a camshaft or valve train or cam drive gears to add weight and bulk to the engine. As its name implies, it requires only two-strokes, or 360°, to complete its cycle. There is a passageway, called a transfer port, between the combustion chamber above the piston and the crankcase below. There is also an exhaust port in the side of the cylinder. The piston acts to sequentially block or expose these ports as it moves up and down. The crankcase is sealed and mounts the carburetor on it, serving also as the intake manifold.

Starting at TDC, the two-stroke cycle proceeds as follows: The spark plug ignites the fuel-air charge, compressed on the previous revolution. The expansion of the burning gases drives the piston down, delivering torque to the crankshaft. Partway down, the piston uncovers the exhaust port, allowing the burnt (and also some unburnt) gases to begin to escape to the exhaust system.

As the piston descends, it compresses the charge of fuel-air mixture in the sealed crankcase. The piston blocks the intake port preventing blowback through the carburetor. As the piston clears the transfer port in the cylinder wall, its downward motion pushes the new fuel-air charge up through the transfer port to the combustion chamber. The momentum of the exhaust gases leaving the chamber on the other side helps pull in the new charge as well.

The piston passes BDC and starts up, pushing out the remaining exhaust gases, starting to compress the new charge, and sucking in a new charge of air and fuel to the crankcase from the carburetor. All ports are closed by the piston as it ascends, allowing full compression of the new charge. Slightly before TDC, the spark is ignited and the cycle repeats as the piston passes TDC.

Clearly, this Clerk cycle is not as efficient as the Otto cycle in which each event is more cleanly separated from the others. Here there is much mixing of the various phases of the cycle. Unburned hydrocarbons are exhausted in larger quantities. This accounts for the poor fuel economy and dirty emissions of the Clerk engine.[1] It is nevertheless popular in applications where low weight is paramount.

Lubrication is also more difficult in the two-stroke engine than in the four-stroke as the crankcase is not available as an oil sump. Thus the lubricating oil must be mixed with the fuel. This further increases the emissions problem compared to the Otto cycle engine which burns raw gasoline and pumps its lubricating oil separately throughout the engine.

---

[1] Current research and development is under way to clean up the emissions of the two-stroke engine by using fuel injection and compressed air scavenging of the cylinders. These efforts may yet bring this potentially more powerful engine design into compliance with air quality specifications.

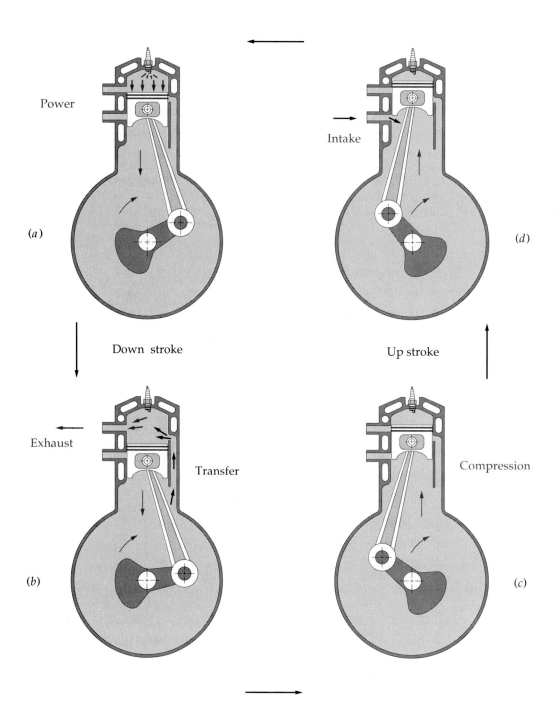

FIGURE 14-5

The Clerk two-stroke combustion cycle

DIESEL CYCLE The **diesel cycle** can be either two-stroke or four-stroke. It is a **compression-ignition** cycle. No spark is needed to ignite the mixture. The air is compressed in the cylinder by a factor of about 14 to 15 (versus 8 to 10 in the spark engine), and a low volatility fuel is injected into the cylinder just before TDC. The heat of compression causes the explosion. Diesel engines are larger and heavier than spark ignition engines for the same power output because the higher pressures and forces at which they operate require stronger, heavier parts. Two-stroke cycle Diesel engines are quite common. Diesel fuel is a better lubricant than gasoline.

GAS FORCE In all the engines discussed here, the useable **output torque** is created from the explosive gas pressure generated within the cylinder either once or twice per two revolutions of the crank, depending on the cycle used. The magnitude and shape of this explosion pressure curve will vary with the engine design, stroke cycle, fuel used, speed of operation, and other factors related to the thermodynamics of the system. For our purpose of analyzing the mechanical dynamics of the system, we need to keep the gas pressure function consistent while we vary other design parameters in order to compare the results of our mechanical design changes. For this purpose, program ENGINE has been provided with a built-in **gas pressure curve** whose peak value is about 600 psi and whose shape is similar to the curve from a real engine. Figure 14-6 shows the **gas force curve** that results from the built-in gas pressure function in program ENGINE applied to a piston of particular area, for both two- and four-stroke engines. Changes in piston area will obviously affect the gas force magnitude for this consistent pressure function, but *no changes in engine design parameters input to this program will change its built-in gas pressure curve.* To see this gas force curve, run program ENGINE and select any one of the example engines from the input menu. **Calculate**, and then plot **Gas Force**.

14.2 SLIDER-CRANK KINEMATICS

In Chapters 4, 6, 7, and 12 we developed general equations for the exact solution of the positions, velocities, accelerations, and forces in the pin-jointed fourbar linkage, and also for two inversions of the **slider-crank linkage**, using vector equations. We could again apply that method to the analysis of the "standard" slider-crank linkage, used in the majority of internal combustion engines, as shown in Figure 14-7. Note that its slider motion has been aligned with the X axis. This is a "nonoffset" slider-crank, because the slider axis extended passes through the crank pivot. It also has its slider block translating against the stationary ground plane; thus there will be no Coriolis component of acceleration (see Section 7-3).

The simple geometry of this particular inversion of the slider-crank mechanism allows a very straightforward approach to the exact analysis of its slider's position, velocity, and acceleration, using only plane trigonometry and scalar equations. Because of this method's simplicity and to present an alternative solution approach we will analyze this device again.

Let the crank radius be l. The conrod length be l. The angle of the crank is θ and the angle that the conrod makes with the X axis is ϕ. For any constant crank angular velocity ω, the crank angle $\theta = \omega t$. The instantaneous piston position is x. Two right triangles rqs and lqu are constructed. Then from geometry:

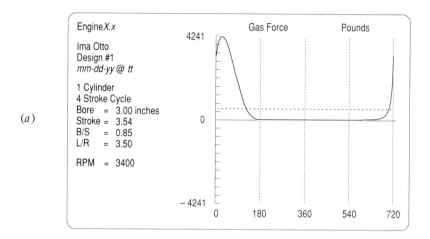

(a)

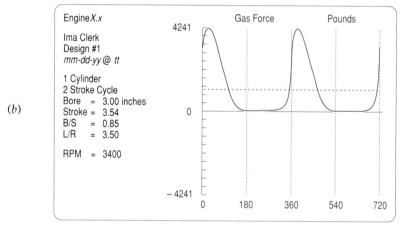

(b)

FIGURE 14-6

Gas force functions in the two-stroke and four-stroke cycle engines

$$q = r \sin \theta = l \sin \phi$$
$$\theta = \omega t \tag{14.1a}$$

$$\sin \phi = \frac{r}{l} \sin \omega t$$

$$s = r \cos \omega t$$
$$u = l \cos \phi \tag{14.1b}$$
$$x = s + u = r \cos \omega t + l \cos \phi$$

$$\cos \phi = \sqrt{1 - \sin^2 \phi} = \sqrt{1 - \left(\frac{r}{l} \sin \omega t\right)^2} \tag{14.1c}$$

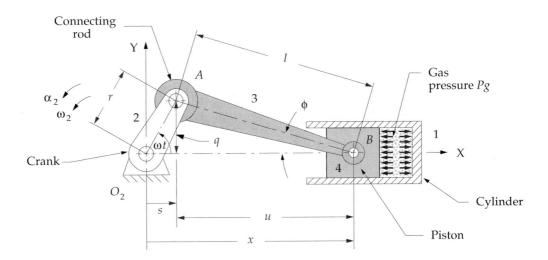

(a) The linkage geometry

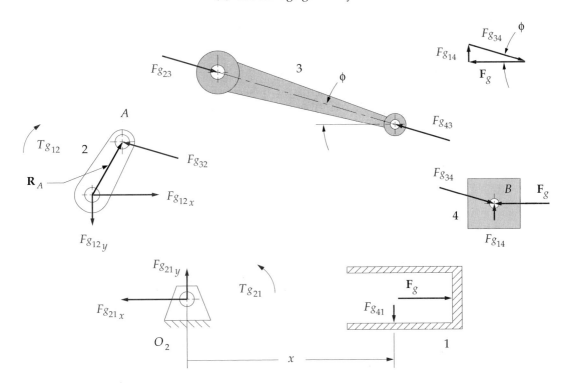

(b) Free-body diagrams

FIGURE 14-7

Fourbar slider-crank linkage position and gas force analysis

$$x = r\cos\omega t + l\sqrt{1 - \left(\frac{r}{l}\sin\omega t\right)^2} \tag{14.1d}$$

Equation 14.1d is an exact expression for the piston position x as a function of r, l, and ωt. This can be differentiated versus time to obtain exact expressions for the velocity and acceleration of the piston. For a steady-state analysis we will assume ω to be constant.

$$\dot{x} = -r\omega\left[\sin\omega t + \frac{r}{2l}\frac{\sin 2\omega t}{\sqrt{1 - \left(\frac{r}{l}\sin\omega t\right)^2}}\right] \tag{14.1e}$$

$$\ddot{x} = -r\omega^2\left\{\cos\omega t - \frac{r\left[l^2\left(1 - 2\cos^2\omega t\right) - r^2\sin^4\omega t\right]}{\left[l^2 - (r\sin\omega t)^2\right]^{\frac{3}{2}}}\right\} \tag{14.1f}$$

Equations 14.1 can easily be solved with a computer for all values of ωt needed. But, it is rather difficult to look at equation 14.1f and visualize the effects of changes in the design parameters r and l on the acceleration. It would be useful if we could derive a simpler expression, even if approximate, that would allow us to more easily predict the results of design decisions involving these variables. To do so, we will use the binomial theorem to expand the radical in equation 14.1d for piston position to put the equations for position, velocity and acceleration in simpler, approximate forms which will shed some light on the dynamic behavior of the mechanism.

The general form of the binomial theorem is:

$$(a+b)^n = a^n + na^{n-1}b + \frac{n(n-1)}{2!}a^{n-2}b^2 + \frac{n(n-1)(n-2)}{3!}a^{n-3}b^3 + \cdots \tag{14.2a}$$

The radical in equation 14.1d is:

$$\sqrt{1 - \left(\frac{r}{l}\sin\omega t\right)^2} = \left[1 - \left(\frac{r}{l}\sin\omega t\right)^2\right]^{\frac{1}{2}} \tag{14.2b}$$

where, for the binomial expansion:

$$a = 1 \qquad b = -\left(\frac{r}{l}\sin\omega t\right)^2 \qquad n = \frac{1}{2} \tag{14.2c}$$

It expands to:

$$1 - \frac{1}{2}\left(\frac{r}{l}\sin\omega t\right)^2 + \frac{1}{8}\left(\frac{r}{l}\sin\omega t\right)^4 - \frac{1}{16}\left(\frac{r}{l}\sin\omega t\right)^6 + \cdots \qquad (14.2d)$$

or:

$$1 - \left(\frac{r^2}{2l^2}\right)\sin^2\omega t + \left(\frac{r^4}{8l^4}\right)\sin^4\omega t - \left(\frac{r^6}{16l^6}\right)\sin^6\omega t + \cdots \qquad (14.2e)$$

Each nonconstant term contains the **crank-conrod ratio** r/l to some power. Applying some engineering common sense to the depiction of the slider-crank in Figure 14-7a, we can see that if r/l were greater than 1 the crank could not make a complete revolution. In fact if r/l even gets close to 1, the piston will hit the fixed pivot O_2 before the crank completes its revolution. If r/l is as large as $1/2$, the transmission angle $(\pi/2 - \phi)$ will be too small (see Sections 3.3 and 4.10) and the linkage will not run well. A practical upper limit on the value of r/l is about $1/3$. Most slider-crank linkages will have this **crank-conrod ratio** somewhere between $1/3$ and $1/5$ for smooth operation. If we substitute this practical upper limit of $r/l = 1/3$ into equation 14.2e we get:

$$1 - \left(\frac{1}{18}\right)\sin^2\omega t + \left(\frac{1}{648}\right)\sin^4\omega t - \left(\frac{1}{11,664}\right)\sin^6\omega t + \cdots$$
$$(14.2f) \qquad\qquad\qquad (14.2f)$$
$$1 - 0.05556\sin^2\omega t + 0.00154\sin^4\omega t - 0.00009\sin^6\omega t + \cdots$$

Clearly we can drop all terms after the second with very small error. Substituting this approximate expression for the radical in equation 14.1d gives an approximate expression for piston displacement with only a fraction of one percent error.

$$x \cong r\cos\omega t + l\left[1 - \left(\frac{r^2}{2l^2}\right)\sin^2\omega t\right] \qquad (14.3a)$$

Substitute the trigonometric identity:

$$\sin^2\omega t = \frac{1 - \cos 2\omega t}{2} \qquad (14.3b)$$

and simplify:

$$x \cong l - \frac{r^2}{4l} + r\left(\cos\omega t + \frac{r}{4l}\cos 2\omega t\right) \qquad (14.3c)$$

Differentiate for velocity of the piston (with constant ω):

$$\dot{x} \cong -r\omega\left(\sin\omega t + \frac{r}{2l}\sin 2\omega t\right) \qquad (14.3d)$$

Differentiate again for acceleration (with constant ω):

$$\ddot{x} \cong -r\omega^2\left(\cos\omega t + \frac{r}{l}\cos 2\omega t\right) \qquad (14.3e)$$

The process of binomial expansion has, in this particular case, led us to Fourier series approximations of the exact expressions for the piston displacement, velocity, and acceleration. Fourier[2] showed that any periodic function can be approximated by a series of sine and cosine terms of integer multiples of the independent variable. Recall that we dropped the fourth, sixth and subsequent power terms from the binomial expansion, which would have provided $\cos 4\omega t$, $\cos 6\omega t$ etc. terms in this expression. These multiple angle functions are referred to as the **harmonics** of the fundamental $\cos \omega t$ term. The $\cos \omega t$ term repeats once per crank revolution and is called the fundamental frequency or the **primary component**. The second harmonic, $\cos 2\omega t$, repeats twice per crank revolution and is called the **secondary component**. The higher harmonics were dropped when we truncated the series. The constant term in the displacement function is the **DC component** or **average value**. The complete function is the sum of its harmonics. The Fourier series form of the expressions for displacement and its derivatives lets us see the relative contributions of the various harmonic components of the functions. This approach will prove to be quite valuable when we attempt to dynamically balance an engine design.

Program ENGINE calculates the position, velocity, and acceleration of the piston according to equations 14.3c, d, and e. Figure 14-8a, b, and c shows these functions for example engine #3 in the program as plotted for constant crank ω over two full revolutions. The acceleration curve shows the effects of the second harmonic term most clearly because that term's coefficient is larger than its correspondent in either of the other two functions. The fundamental ($-\cos \omega t$) term gives a pure harmonic function with a period of 360°. This fundamental term dominates the function as it has the largest coefficient in equation 14.3e. The flat top and slight dip in the positive peak acceleration of Figure 14-8c is caused by the $\cos 2\omega t$ second harmonic adding or subtracting from the fundamental. Note the very high value of peak acceleration of the piston even at the midrange engine speed of 3400 RPM. It is 747 g's! At 6000 RPM this increases to nearly 1300 g's. This is a moderately sized engine, of 3 inch bore and 3.5 inch stroke, with 25 cubic inch (0.4 liter) displacement per cylinder.

SUPERPOSITION We will now analyze the dynamic behavior of the single-cylinder engine based on the approximate kinematic model developed above. Since we have several sources of dynamic excitation to deal with, we will use the method of superposition to separately analyze them and then combine their effects. We will first consider the **forces and torques** which are due to the presence of the **explosive gas forces** in the cylinder, which drive the engine. Then we will analyze the **inertia forces and torques** that result from the high-speed motion of the elements. The total force and torque state of the machine at any instant will be the sum of these components. Finally we will look at the **shaking forces and torques** on the ground plane and the **pin forces** within the linkage that result from the combination of applied and dynamic forces on the system.

---

[2] Baron Jean Baptiste Joseph Fourier (1768-1830) published the description of the mathematical series which bears his name in *The Analytic Theory of Heat* in 1822. The Fourier series is widely used in harmonic analysis of all types of physical systems. Its general form is:

$$y = \frac{a_0}{2} + (a_1 \cos x + b_1 \sin x) + (a_2 \cos 2x + b_2 \sin 2x) + \cdots + (a_n \cos nx + b_n \sin nx)$$

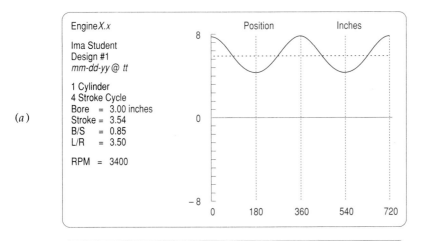

(a)

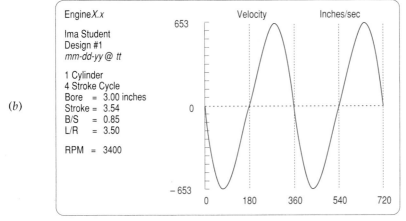

(b)

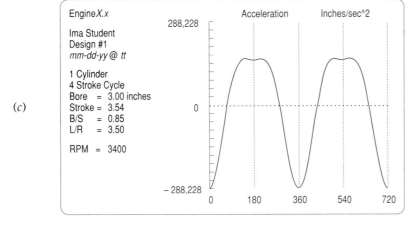

(c)

FIGURE 14-8

Position, velocity, and acceleration functions for a single-cylinder engine

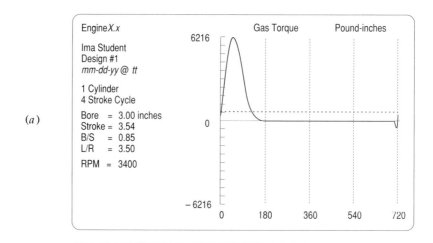

(*a*)

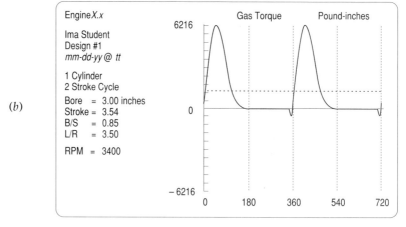

(*b*)

FIGURE 14-9

Gas torque for (*a*) four-stroke and (*b*) two-stroke single-cylinder engines

14.3   GAS FORCE AND GAS TORQUE

The **gas force** is due to the gas pressure from the exploding fuel-air mixture impinging on the top of the piston surface as shown in Figure 14-3. Let F_g = gas force, P_g = gas pressure, A_p = area of piston, and B = bore of cylinder, which is also equal to the piston diameter. Then:

$$\mathbf{F}_g = -P_g A_p \,\hat{\mathbf{i}}; \qquad\qquad A_p = \frac{\pi}{4}B^2$$

$$(14.4)$$

$$\mathbf{F}_g = -\frac{\pi}{4}P_g B^2 \,\hat{\mathbf{i}}$$

The negative sign is due to the choice of engine orientation in the coordinate system of Figure 14-3. The **gas pressure** P_g in this expression is a function of crank angle ωt and is defined by the thermodynamics of the engine. A typical **gas pressure curve** for a four-stroke engine is shown in Figure 14-4. The **gas force curve** shape is identical to that of the gas pressure curve as they differ only by a constant multiplier, the piston area A_p. Figure 14-6 shows the approximation of the gas force curve used in program ENGINE for both four- and two-stroke engines.

The **gas torque** in Figure 14-9 is due to the gas force acting at a moment arm about the crank center O_2 in Figure 14-7. This moment arm varies from zero to a maximum as the crank rotates. The distributed gas force over the piston surface has been resolved to a single force acting through the mass center of link 4 in the free-body diagrams of Figure 14-7b. The concurrent force system at point B is resolved in the vector diagram showing that:

$$\mathbf{F}_{g14} = F_g \tan \phi \, \hat{\mathbf{j}} \tag{14.5a}$$

$$\mathbf{F}_{g34} = F_g \, \hat{\mathbf{i}} - F_g \tan \phi \, \hat{\mathbf{j}} \tag{14.5b}$$

From the free-body diagrams in Figure 14-7 we can see that:

$$\mathbf{F}_{g41} = -\mathbf{F}_{g14}$$
$$\mathbf{F}_{g43} = -\mathbf{F}_{g34}$$
$$\mathbf{F}_{g23} = -\mathbf{F}_{g43}$$
$$\mathbf{F}_{g32} = -\mathbf{F}_{g23}$$

so:

$$\mathbf{F}_{g32} = -\mathbf{F}_{g34} = -F_g \, \hat{\mathbf{i}} + F_g \tan \phi \, \hat{\mathbf{j}} \tag{14.5c}$$

The **driving torque** $\mathbf{T}_{g21}$ at link 2 due to the gas force can be found from the cross product of the position vector to point A and the force at point A.

$$\mathbf{T}_{g21} = \mathbf{R}_A \times \mathbf{F}_{g32} \tag{14.6a}$$

This expression can be expanded and will involve the crank length r and the angles θ and ϕ as well as the gas force $\mathbf{F}_g$. Note from the free-body diagram for link 1 that we can also express the torque in terms of the forces $\mathbf{F}_{g14}$ or $\mathbf{F}_{g41}$ which act always perpendicular to the motion of the slider (neglecting friction), and the distance x, which is their instantaneous moment arm about O_2. The reaction torque $\mathbf{T}_{g12}$ due to the gas force trying to rock the ground plane is:

$$\mathbf{T}_{g12} = F_{g41} \cdot x \, \hat{\mathbf{k}} \tag{14.6b}$$

If you have ever abruptly opened the throttle of a running automobile engine while working on it, you probably noticed the engine move to the side as it rocked in its mounts from the reaction torque. The driving torque $\mathbf{T}_{g21}$ is the negative of this reaction torque.

$$\mathbf{T}_{g21} = -\mathbf{T}_{g12}$$

$$\mathbf{T}_{g21} = -F_{g41} \cdot x \, \hat{\mathbf{k}} \tag{14.6c}$$

and:
$$F_{g14} = -F_{g41}$$

(14.6d)

so:
$$\mathbf{T}_{g21} = F_{g14} \cdot x \ \hat{\mathbf{k}}$$

Equation 14.6d gives us an expression for **gas torque** which involves the displacement of the piston x for which we have already derived equation 14.3c. Substituting equation 14.3c for x and the magnitude of equation 14.5a for F_{g14} we get:

$$\mathbf{T}_{g21} = \left(F_g \tan \phi\right)\left[1 - \frac{r^2}{4l} + r\left(\cos \omega t + \frac{r}{4l}\cos 2\omega t\right)\right]\hat{\mathbf{k}}$$

(14.6e)

Equation 14.6e contains the conrod angle ϕ as well as the independent variable, crank angle ωt. We would like to have an expression which involves only ωt. We can substitute an expression for $\tan \phi$ generated from the geometry of Figure 14-7a.

$$\tan \phi = \frac{q}{u} = \frac{r \sin \omega t}{l \cos \phi}$$

(14.7a)

Substitute equation 14.1c for $\cos \phi$:

$$\tan \phi = \frac{r \sin \omega t}{l\sqrt{1 - \left(\dfrac{r}{l}\sin \omega t\right)^2}}$$

(14.7b)

The radical in the denominator can be expanded using the binomial theorem as was done in equations 14.2, and the first two terms retained for a good approximation to the exact expression,

$$\frac{1}{\sqrt{1 - \left(\dfrac{r}{l}\sin \omega t\right)^2}} \cong 1 + \frac{r^2}{2l^2}\sin^2 \omega t$$

(14.7c)

giving:

$$\tan \phi \cong \frac{r}{l}\sin \omega t \left(1 + \frac{r^2}{2l^2}\sin^2 \omega t\right)$$

(14.7d)

Substitute this into equation 14.6e for the gas torque:

$$\mathbf{T}_{g21} \cong F_g \left[\frac{r}{l}\sin \omega t\left(1 + \frac{r^2}{2l^2}\sin^2 \omega t\right)\right]\left[1 - \frac{r^2}{4l} + r\left(\cos \omega t + \frac{r}{4l}\cos 2\omega t\right)\right]\hat{\mathbf{k}}$$

(14.8a)

Expand this expression and neglect any terms containing the conrod crank ratio r/l raised to any power greater than one since these will have very small coefficients as was seen in equation 14.2. This results in a simpler, but even more approximate expression for the gas torque:

$$\mathbf{T}_{g21} \cong F_g \, r \sin \omega t \left(1 + \frac{r}{l}\cos \omega t\right)$$

(14.8b)

Note that the **exact value** of this **gas torque** can always be calculated from equations 14.1d, 14.5a and 14.6d in combination, or from the expansion of equation 14.6a, if you require a more accurate answer. For design purposes the approximate equation 14.8b will usually be adequate. Program ENGINE calculates the gas torque using equation 14.8b and its built-in gas pressure curve to generate the gas force function. Plots of the gas torque for two- and four-stroke cycles are shown in Figure 14-9. Note the similarity in shape to that of the gas force curve in Figure 14-6. Note also that the two-stroke has theoretically twice the power available as the four-stroke, all other factors equal, because there are twice as many torque pulses per unit time. The poorer efficiency of the two-stroke reduces this theoretical advantage, however.

14.4 EQUIVALENT MASSES

To do a complete dynamic force analysis on any mechanism we need to know the geometric properties (mass, center of gravity, mass moment of inertia) of the moving links as was discussed in previous chapters (see Sections 11.3 to 11.8 and Chapter 12). This is easy to do if the link already is designed in detail and its dimensions are known. When designing the mechanism from scratch, we typically do not yet know that level of detail about the links' geometries. But we must nevertheless make some estimate of their geometric parameters in order to begin the iteration process which will eventually converge on a detailed design.

In the case of this slider-crank mechanism, the **crank** is in **pure rotation** and the **piston** is in **pure translation**. By assuming some reasonable geometries and materials we can make approximations of their dynamic parameters. Their kinematic motions are easily determined. We have already derived expressions for the piston motion in equations 14.3. Further, if we balance the rotating crank, as described and recommended in the previous chapter, then the CG of the crank will be motionless at its center O_2 and will not contribute to the dynamic forces. We will do this in a later section.

The conrod is in complex motion. To do an exact dynamic analysis as was derived in Section 12.5, we need to determine the linear acceleration of its CG for all positions. At the outset of the design, the conrod's CG location is not accurately defined. To "bootstrap" the design, we need a simplified model of this connecting rod which we can later refine, as more dynamic information is generated about our engine design. The requirements for a dynamically equivalent model were stated in Section 11.2 and are repeated here as Table 14-1 for your convenience.

If we could model our still-to-be-designed conrod as two, lumped, point masses, concentrated one at the crankpin (point A in Figure 14-7), and one at the

TABLE 14-1   Requirements for Dynamic Equivalence

1   The mass of the model must equal that of the original body.

2   The center of gravity must be in the same location as that of the original body.

3   The mass moment of inertia must equal that of the original body.

wristpin (point B in Figure 14-7), we would at least know what the motions of these lumps are. The lump at A would be in pure rotation as part of the crank, and the lump at point B would be in pure translation as part of the piston. These lumped, point masses have no dimension and are assumed to be connected with a magical, massless but rigid rod.[3]

DYNAMICALLY EQUIVALENT MODEL Figure 14-10a shows a typical conrod. Figure 14-10b shows a generic two-mass model of the conrod. One mass m_t is located at distance l_t from the CG of the original rod and the second mass m_p at distance l_p from the CG. The mass of the original part is m_3, and its moment of inertia about its CG is I_{G3}. Expressing the three requirements for dynamic equivalence from Table 14-1 mathematically in terms of these variables, we get:

$$m_p + m_t = m_3 \tag{14.9a}$$

$$m_p l_p = m_t l_t \tag{14.9b}$$

$$m_p l_p^2 + m_t l_t^2 = I_{G3} \tag{14.9c}$$

There are four unknowns in these three equations, m_p, l_p, m_t, l_t, which means we must choose a value for any one variable to solve the system. Let us choose the distance l_t equal to the distance to the wrist pin, l_b as shown in Figure 14-10c. This will put one mass at a desired location. Solving equations 14.9a and 14.9b simultaneously with that substitution gives expressions for the two lumped masses:

$$m_p = m_3 \frac{l_b}{l_p + l_b}$$

$$\tag{14.9d}$$

$$m_b = m_3 \frac{l_p}{l_p + l_b}$$

Substituting equation 14.9d into 14.9c gives a relation between l_p and l_b:

$$m_3 \frac{l_b}{l_p + l_b} l_p^2 + m_3 \frac{l_p}{l_p + l_b} l_b^2 = I_{G3} = m_3 l_p l_b \tag{14.9e}$$

$$l_p = \frac{I_{G3}}{m_3 l_b}$$

Please refer to Section 11.8 and equation 11.13 which define the *center of percussion* and its geometric relationship to a corresponding *center of rotation*. Equation 14.9e is the same as equation 11.13 (except for sign which is due to an arbitrary choice of the link's orientation in the coordinate system). The distance l_p is the location of the center of percussion corresponding to a center of rotation at l_b. Thus our second mass m_p must be placed at the link's **center of percussion** P versus point B for exact dynamic equivalence. The masses must be as defined in equation 14.9d.

[3] These lumped mass models have to be made with *very* special materials. *Unobtainium 206* has the property of infinite mass density, thus occupies no space and can be used for "point masses." *Unobtainium 208* has infinite stiffness and zero mass, and thus can be used for rigid but "massless rods".

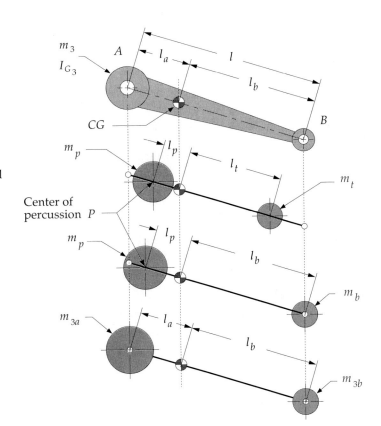

a. Original connecting rod

b. Generic two mass model

Center of
percussion P

c. Exact dynamic model

d. Approximate model

FIGURE 14-10

Lumped mass dynamic models of a connecting rod

The geometry of the typical conrod, as shown in Figures 14-2 and 14-10a, is large at the crankpin end (A) and small at the wristpin end (B). This puts the CG close to the "big end." The **center of percussion** P will be even closer to the big end than is the CG. For this reason, we can place the second lumped mass, that belongs at P, at point A with relatively small error in our dynamic model's accuracy. This approximate model is adequate for our initial design calculations. Once a viable design geometry is established, we will have to do a complete and exact force analysis with the methods of Chapter 12 before considering the design complete.

Making this substitution of distance l_a for l_p and renaming the lumped masses at those distances m_{3a} and m_{3b}, to reflect both their identity with link 3 and with points A and B, we rewrite equations 14.9d.

$$\text{let} \quad l_p = l_a$$

then:

$$m_{3a} = m_3 \frac{l_b}{l_a + l_b} \qquad\qquad (14.10a)$$

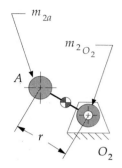

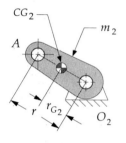

FIGURE 14-11

Lumped mass model
of the crank

and:

$$m_{3b} = m_3 \frac{l_a}{l_a + l_b} \qquad (14.10b)$$

These define the amounts of the total conrod mass to be placed at each end, to approximately dynamically model that link. Figure 14-10d shows this dynamic model. In the absence of any data on the shape of the conrod at the outset of a design, preliminary dynamic force information can be obtained by using the rule of thumb of placing two-thirds of the conrod's mass at the crankpin and one-third at the wristpin.

STATICALLY EQUIVALENT MODEL We can create a similar lumped mass model of the crank. Even though we intend to balance the crank before we are done, for generality we will initially model it *unbalanced* as shown in Figure 14-11. Its *CG* is located at some distance r_{G_2} from the pivot, O_2, on the line to the crankpin, *A*. We would like to model it as a lumped mass at *A* on a massless rod pivoted at O_2. If our principal concern is with a steady-state analysis, then the crank velocity ω will be held constant. An absence of angular acceleration on the crank allows a statically equivalent model to be used because the equation $\mathbf{T} = I\alpha$ will be zero regardless of the value of I. A **statically equivalent model** needs only to have equivalent mass and equivalent first moments as shown in Table 14-2. The moments of inertia need not match. We model it as two lumped masses, one at point *A* and one at the fixed pivot O_2. Writing the two requirements for static equivalence from Table 14-2:

$$m_2 = m_{2a} + m_{2O_2}$$
$$m_{2a}r = m_2 r_{G_2} \qquad (14.11)$$
$$m_{2a} = m_2 \frac{r_{G_2}}{r}$$

The lumped mass m_{2a} can be placed at point *A* to represent the unbalanced crank. The second lumped mass, at the fixed pivot O_2, is not necessary to any calculations as that point is stationary.

These simplifications lead to the lumped parameter model of the slider-crank linkage shown in Figure 14-12. The crankpin, point *A*, has two masses concentrated at it, the equivalent mass of the crank m_{2a} and the portion of conrod m_{3a}. Their sum is m_A. At the wristpin, point *B*, two masses are also concentrated, the piston mass m_4 and the remaining portion of the conrod mass m_{3b}. Their sum is m_B. This model has masses which are either in pure rotation (m_A) or in pure translation (m_B), so it is very easy to dynamically analyze.

TABLE 14-2   Requirements for Static Equivalence

1   The mass of the model must equal that of the original body.

2   The center of gravity must be in the same location as that of the original body.

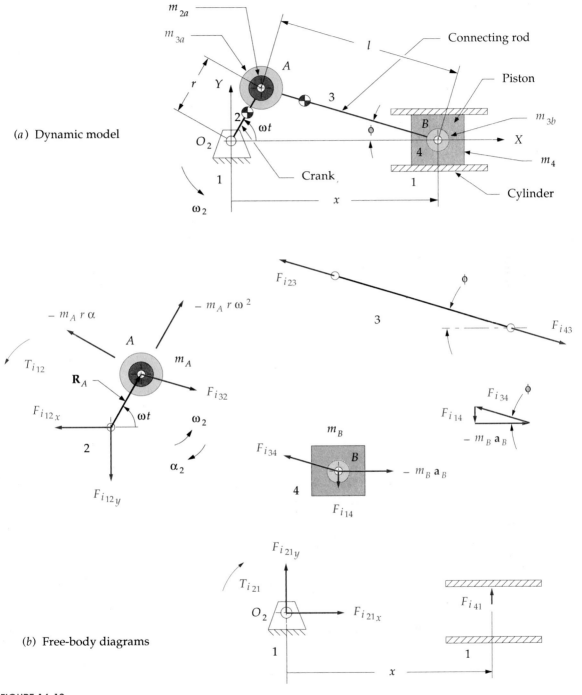

(a) Dynamic model

(b) Free-body diagrams

FIGURE 14-12

Lumped mass dynamic model of the slider-crank

$$m_A = m_{2a} + m_{3a}$$

(14.12)

$$m_B = m_{3b} + m_4$$

VALUE OF MODELS *The value of constructing simple, lumped mass models of complex systems increases with the complexity of the system being designed.* It makes little sense to spend large amounts of time doing sophisticated, detailed analyses of designs which are so ill-defined at the outset that their conceptual viability is as yet unproven. Better to get a reasonably approximate and rapid answer that tells you the concept needs to be rethought, than to spend a greater amount of time reaching the same conclusion to more decimal places.

14.5 INERTIA AND SHAKING FORCES

The simplified, lumped mass model of Figure 14-12 can be used to develop expressions for the forces and torques due to the accelerations of the masses in the system. The method of d'Alembert is of value in visualizing the effects of these moving masses on the system and on the ground plane. Accordingly, the free-body diagrams of Figure 14-12b show the d'Alembert inertia forces acting on the masses at points A and B. Friction is again ignored. The acceleration for point B is given in equation 14.3e. The acceleration of point A in pure rotation is obtained by differentiating the position vector $\mathbf{R}_A$ twice, assuming a constant crankshaft ω, which gives:

$$\mathbf{R}_A = r \cos \omega t \, \hat{\mathbf{i}} + r \sin \omega t \, \hat{\mathbf{j}}$$

(14.13)

$$\mathbf{a}_A = -r\omega^2 \cos \omega t \, \hat{\mathbf{i}} - r\omega^2 \sin \omega t \, \hat{\mathbf{j}}$$

The total inertia force $\mathbf{F}_i$ is the sum of the centrifugal (inertia) force at point A and the inertia force at point B.

$$\mathbf{F}_i = -m_A \, \mathbf{a}_A - m_B \, \mathbf{a}_B$$
(14.14a)

Breaking it into x and y components:

$$F_{i_x} = -m_A \left(-r\omega^2 \cos \omega t \right) - m_B \, \ddot{x}$$
(14.14b)

$$F_{i_y} = -m_A \left(-r\omega^2 \sin \omega t \right)$$
(14.14c)

Note that only the x component is affected by the acceleration of the piston. Substituting equation 14.3e into equation 14.14b:

$$F_{i_x} \cong -m_A \left(-r\omega^2 \cos \omega t \right) - m_B \left[-r\omega^2 \left(\cos \omega t + \frac{r}{l} \cos 2\omega t \right) \right]$$

(14.14d)

$$F_{i_y} = -m_A \left(-r\omega^2 \sin \omega t \right)$$

Notice that the x directed inertia forces have primary components at crank frequency and secondary (second harmonic) forces at twice crank frequency. There are also small magnitude, higher, even harmonics which we truncated in the binomial expansion of the piston displacement function. The component due to the rotating mass at point A has only a primary component.

The **shaking force** was defined in Section 12.8 to be *the sum of all forces acting on the ground plane*. The shaking force $\mathbf{F}_s$ is the equal and opposite reaction force to the inertia force.

$$\mathbf{F}_s = -\mathbf{F}_i \qquad (14.14e)$$

Negating equation 14.14d, for constant shaft ω,

$$F_{s_x} \cong m_A\left(-r\omega^2\cos\omega t\right) + m_B\left[-r\omega^2\left(\cos\omega t + \frac{r}{l}\cos2\omega t\right)\right]$$

$$(14.14f)$$

$$F_{s_y} = m_A\left(-r\omega^2\sin\omega t\right)$$

Note that the gas force from equation 14.4 does not contribute to the shaking force. Only inertia forces and external forces are felt as shaking forces. The gas force is an internal force which is cancelled within the mechanism. It acts equally and oppositely on both the piston top and the cylinder head as shown in Figure 14.7.

Program ENGINE calculates the shaking force at constant ω for any combination of linkage parameters input to it. Figure 14-13 shows the shaking force plot for the same unbalanced built-in example engine #3 as shown in the acceleration plot (Figure 14-8c). The linkage orientation is the same as in Figure 14-12 with the x axis horizontal. The x component is larger than the y component due to the high acceleration of the piston. The forces are seen to be quite large despite this being a relatively

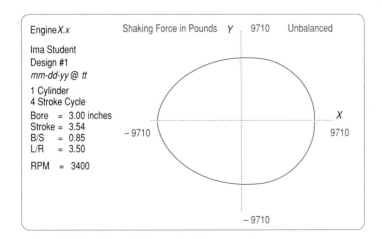

FIGURE 14-13

Shaking force in an unbalanced slider-crank linkage

small (0.4 liter per cylinder) engine running at moderate speed (3400 RPM). We will soon investigate techniques to reduce or eliminate this shaking force from the engine. It is an undesirable feature which creates noise and vibration.

14.6   INERTIA AND SHAKING TORQUES

The **inertia torque** results from the action of the inertia forces at a moment arm. The inertia force at point A in Figure 14-12 has two components, radial and tangential. The radial component has no moment arm. The tangential component has a moment arm of crank radius r. If the crank ω is constant, the mass at A will not contribute to inertia torque. The inertia force at B has a moment arm except when the piston is at TDC or BDC. As we did for the gas torque, we can express the inertia torque in terms of the forces $\mathbf{F}_{i14}$ or $\mathbf{F}_{i41}$ which act always perpendicular to the motion of the slider (neglecting friction), and the distance x, which is their instantaneous moment arm about O_2. The inertia torque $\mathbf{T}_{i21}$ acting on the shaft is:

$$\mathbf{T}_{i21} = -\left(F_{i14} \cdot x\right) \, \hat{\mathbf{k}} \tag{14.15a}$$

Substituting for F_{i14} (see Figure 14-12b) and for x, (see Eq. 14.3c) we get:

$$\mathbf{T}_{i21} \doteq -\left(-m_B \ddot{x} \tan\phi\right)\left[1 - \frac{r^2}{4l} + r\left(\cos\omega t + \frac{r}{4l}\cos 2\omega t\right)\right]\hat{\mathbf{k}} \tag{14.15b}$$

We previously developed expressions for x *double dot* (Eq. 14.3e) and $\tan\phi$ (Eq. 14.7d) which can now be substituted.

$$\begin{aligned}\mathbf{T}_{i21} \doteq m_B &\left[-r\omega^2\left(\cos\omega t + \frac{r}{l}\cos 2\omega t\right)\right] \\ &\cdot \left[\frac{r}{l}\sin\omega t\left(1 + \frac{r^2}{2l^2}\sin^2\omega t\right)\right] \\ &\cdot \left[1 - \frac{r^2}{4l} + r\left(\cos\omega t + \frac{r}{4l}\cos 2\omega t\right)\right]\hat{\mathbf{k}}\end{aligned} \tag{14.15c}$$

Expanding this and then dropping all terms with coefficients containing r/l to powers higher than one gives the following approximate equation for inertia torque with constant shaft ω:

$$\mathbf{T}_{i21} \doteq -m_B r^2 \omega^2 \sin\omega t\left(\frac{r}{2l} + \cos\omega t + \frac{3r}{2l}\cos 2\omega t\right)\hat{\mathbf{k}} \tag{14.15d}$$

This contains products of sine and cosine terms. Putting it entirely in terms of harmonics will be instructive, so substitute the identities:

$$2\sin\omega t\cos 2\omega t = \sin 3\omega t - \sin\omega t$$

$$2\sin\omega t\cos\omega t = \sin 2\omega t$$

to get: (14.15e)

$$\mathbf{T}_{i21} \cong \frac{1}{2} m_B r^2 \omega^2 \left(\frac{r}{2l} \sin \omega t - \sin 2\omega t - \frac{3r}{2l} \sin 3\omega t \right) \hat{\mathbf{k}}$$

This shows that the **inertia torque** has a third harmonic term as well as a first and second. The second harmonic is the dominant term as it has the largest coefficient since r/l is always less than $2/3$.

The **shaking torque** $\mathbf{T}_s$ is the negative of this inertia torque and acts to rock the engine block about the crankshaft.

$$\mathbf{T}_s = -\mathbf{T}_{i21} \qquad (14.15f)$$

Program ENGINE calculates the inertia torque and the shaking torque from equations 14.15e and 14.15f. Figure 14-14 shows a plot of the inertia torque for the built-in example engine #3. Note the dominance of the second harmonic. The ideal magnitude for the inertia torque is zero, as it is parasitic. Its average value is always zero, so *it contributes nothing to the net driving torque*. It merely creates large positive and negative oscillations in the total torque which increase vibration and roughness. We will soon investigate means to reduce or eliminate this inertia and shaking torque in our engine designs. It is possible to cancel their effects by proper arrangement of the cylinders in a multi-cylinder engine, as will be explored in the next chapter.

14.7 TOTAL ENGINE TORQUE

The total engine torque is the sum of the gas torque and the inertia torque.

$$\mathbf{T}_{total} = \mathbf{T}_g + \mathbf{T}_i \qquad (14.16)$$

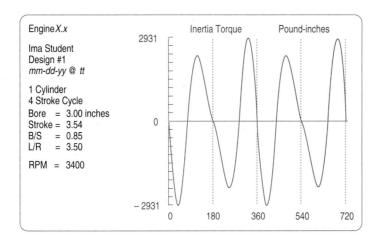

FIGURE 14-14

Inertia torque in the slider-crank linkage

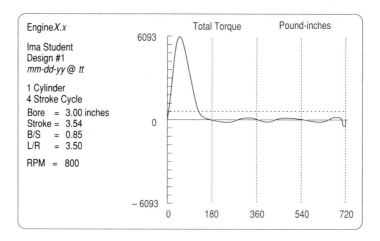

(*a*) Low speed

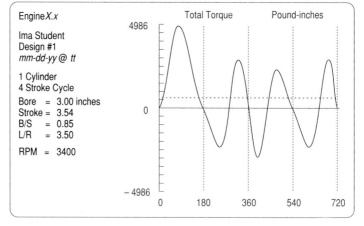

(*b*) Medium speed

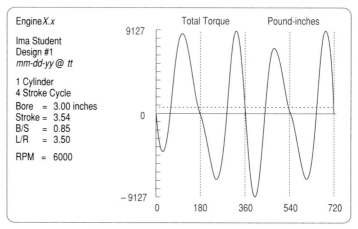

(*c*) High speed

FIGURE 14-15

The total torque function's shape and magnitude vary with crankshaft speed

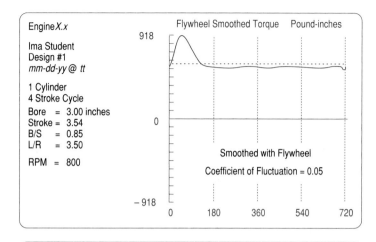

(a) Low speed

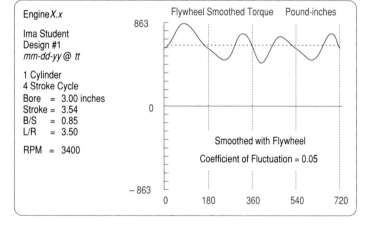

(b) Medium speed

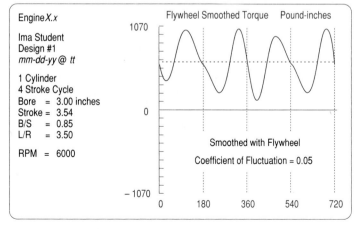

(c) High speed

FIGURE 14-16

The total torque function's variation is reduced by the addition of a flywheel to the system

The gas torque is less sensitive to engine speed than is the inertia torque, which is a function of w^2. So the relative contributions of these two components to the total torque will vary with engine speed. Figure 14-15a shows the total torque for example engine #3 plotted by program ENGINE for an idle speed of 800 RPM. Compare this to the gas torque plot of the same engine in Figure 14-9a. The inertia torque component is negligible at this slow speed compared to the gas torque component. Figure 14-15c shows the same engine run at 6000 RPM. Compare this to the plot of inertia torque in Figure 14-14. The inertia torque component is dominating at this high speed. At the midrange speed of 3400 RPM (Figure 14-15b), a mix of both components is seen.

14.8 FLYWHEELS

We saw in Chapter 12 that large oscillations in the torque-time function can be significantly reduced by the addition of a flywheel to the system. The single-cylinder engine is a prime candidate for the use of a flywheel. The intermittent nature of its power strokes makes one mandatory as it will store the kinetic energy needed to carry the piston through the Otto cycle's exhaust, intake, and compression strokes during which work must be done on the system. Even the two-stroke engine needs a flywheel to drive the piston up on the compression stroke.

The procedure for designing an engine flywheel is identical to that described in Section 12.11 for the fourbar linkage. The total torque function for one revolution of the crank is integrated, pulse by pulse, with respect to its average value. These integrals represent energy fluctuations in the system. The maximum change in energy under the torque curve during one cycle is the amount needed to be stored in the flywheel. Equation 12.18 expresses this relationship. Program ENGINE does the numerical integration of the total torque function and presents a table similar to the one shown in Figure 12-17 on p. 487. These data and the designer's choice of a coefficient of fluctuation k (see equation 12.19) are all that are needed to solve equations 12.20 and 12.21 for the required moment of inertia of the flywheel.

The calculation must be done at some average crank ω. Since the typical engine operates over a wide range of speeds, some thought needs to be given to the most appropriate speed to use in the flywheel calculation. The flywheel's stored kinetic energy is proportional to ω^2 (see equation 12.17). Thus at high speeds a flywheel can have a small moment of inertia and still be effective. The slowest operating speed will require the largest flywheel and should be the one used in the computation of required flywheel size.

Program ENGINE plots the flywheel-smoothed total torque for a user-supplied coefficient of fluctuation k. Figure 14-16 shows the smoothed torque functions for $k = 0.05$ corresponding to the unsmoothed ones in Figure 14-15. Note that the smoothed curves shown for each engine speed are what would result with the flywheel size necessary to obtain that coefficient of fluctuation at that speed. In other words, the flywheel applied to the 800-RPM engine is much larger than the one on the 6000-RPM engine, in these plots. Compare corresponding rows (speeds) between Figures 14-15 and 14-16 to see the effect of the addition of a flywheel. But do not directly compare parts a, b, and c within Figure 14-16 as to the amount of smoothing since the flywheel sizes used are different at each operating speed.

An engine flywheel is usually designed as a flat disk, bolted to one end of the crankshaft. One flywheel face is typically used for the clutch to run against. The clutch is a friction device which allows disconnection of the engine from the drive train (the wheels of a vehicle) when no output is desired. The engine can then remain running at idle speed with the vehicle or output device stopped. When the clutch is engaged, all engine torque is transmitted through it, by friction, to the output shaft.

14.9 PIN FORCES IN THE SINGLE-CYLINDER ENGINE

In addition to calculating the overall effects on the ground plane of the dynamic forces present in the engine, we also need to know the magnitudes of the forces at the pin joints. These forces will dictate the design of the pins and the bearings at the joints. Though we were able to lump the mass due to both conrod and piston, or conrod and crank, at points A and B for an overall analysis of the linkage's effects on the ground plane, we cannot do so for the pin force calculations. This is because the pins feel the effect of the conrod pulling on one "side" and the piston (or crank) pulling on the other "side" of the pin as shown in Figure 14-17. Thus we must separate the effects of the masses of the links joined by the pins.

We will calculate the effect of each component due to the various masses, and the gas force, and then superpose them to obtain the complete pin force at each joint. We need a bookkeeping system to keep track of all these components. We have already used some subscripts for these forces, so we will retain them and add others. The resultant bearing loads have the following components:

1 The gas force components, with the subscript g, as in F_g

2 The inertia force due to the piston mass, with subscript ip, as in F_{ip}

3 The inertia force due to the mass of the conrod at the wrist pin, with subscript iw, as in F_{iw}.

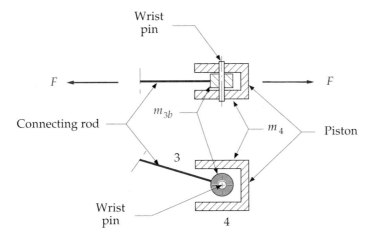

FIGURE 14-17

Forces on a pivot pin

4   The inertia force due to the mass of the conrod at the crank pin, with subscript *ic*, as in F_{ic}.

5   The inertia force due to the mass of the crank at the crank pin, with subscript *ir*, as in F_{ir}

The link number designations will be added to each subscript in the same manner as before, indicating the link from which the force is coming as the first number and the link being analyzed as the second. (See Section 12.2 for further discussion of this notation.)

Figure 14-18 shows the free-body diagrams for the forces due only to the acceleration of the mass of the piston, m_4. Those components are:

$$\mathbf{F}_{ip41} = -m_4 a_B \tan \phi \, \hat{\mathbf{j}}$$ (14.17a)

$$\mathbf{F}_{ip34} = m_4 a_B \, \hat{\mathbf{i}} - m_4 a_B \tan \phi \, \hat{\mathbf{j}}$$ (14.17b)

$$\mathbf{F}_{ip32} = -\mathbf{F}_{ip34}$$ (14.17c)

$$\mathbf{F}_{ip21} = \mathbf{F}_{ip32}$$ (14.17d)

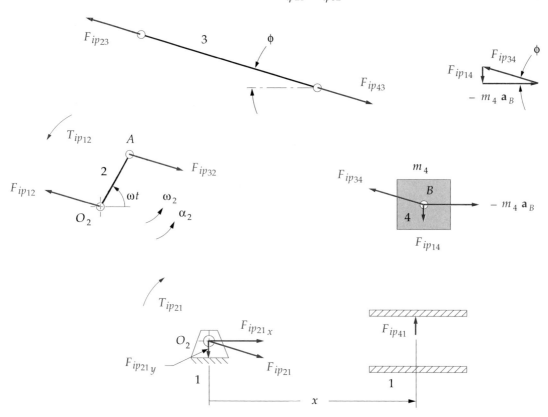

FIGURE 14-18

Free-body diagrams for forces due to the piston mass

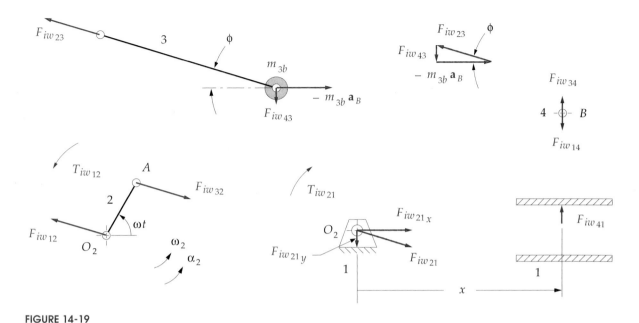

FIGURE 14-19

Free-body diagrams for forces due to the conrod mass concentrated at the wristpin

Figure 14-19 shows the free-body diagrams for the forces due only to the acceleration of the mass of the conrod located at the wristpin, m_{3b}. Those components are:

$$\mathbf{F}_{iw41} = -m_{3b}a_B \tan\phi \;\hat{\mathbf{j}} \tag{14.18a}$$

$$\mathbf{F}_{iw34} = \mathbf{F}_{iw41} \tag{14.18b}$$

$$\mathbf{F}_{iw32} = -m_{3b}a_B \;\hat{\mathbf{i}} + m_{3b}a_B \tan\phi \;\hat{\mathbf{j}} \tag{14.18c}$$

$$\mathbf{F}_{iw21} = \mathbf{F}_{iw32} \tag{14.18d}$$

Figure 14-20a shows the free-body diagrams for the forces due only to the acceleration of the mass of the conrod located at the crankpin, m_{3a}. That component is:

$$\mathbf{F}_{ic32} = -m_{3a}\mathbf{a}_A \tag{14.19a}$$

Substitute equation 14.13 for $\mathbf{a}_A$:

$$\mathbf{F}_{ic32} = m_{3a}r\omega^2\left(\cos\omega t \;\hat{\mathbf{i}} + \sin\omega t \;\hat{\mathbf{j}}\right) \tag{14.19b}$$

Figure 14-20b shows the free-body diagrams for the forces due only to the acceleration of the lumped mass of the crank at the crankpin, m_{2a}. These affect only the mainpin at O_2. That component is:

$$\mathbf{F}_{ir21} = m_{2a}r\omega^2\left(\cos\omega t \;\hat{\mathbf{i}} + \sin\omega t \;\hat{\mathbf{j}}\right) \tag{14.19c}$$

The gas force components were shown in Figure 14-7 and defined in equations 14.5.

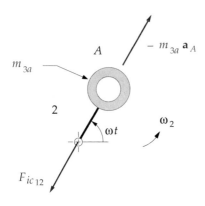

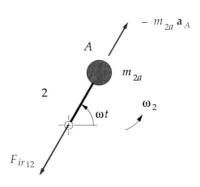

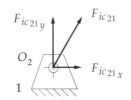

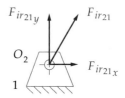

(a) Conrod mass at crankpin

(b) Crank mass at crankpin

FIGURE 14-20

Free-body diagrams for forces due to masses at the crankpin

We can now sum the components of the forces at each pin joint. For the sidewall force $\mathbf{F}_{41}$ of the piston against the cylinder wall:

$$\mathbf{F}_{41} = \mathbf{F}_{g41} + \mathbf{F}_{ip41} + \mathbf{F}_{iw41}$$
$$= -F_g \tan\phi\,\hat{\mathbf{j}} - m_4 a_B \tan\phi\,\hat{\mathbf{j}} - m_{3b} a_B \tan\phi\,\hat{\mathbf{j}} \qquad (14.20)$$
$$= -\big[(m_4 + m_{3b})a_B + F_g\big]\tan\phi\,\hat{\mathbf{j}}$$

The total force $\mathbf{F}_{34}$ on the wristpin is:

$$\mathbf{F}_{34} = \mathbf{F}_{g34} + \mathbf{F}_{ip34} + \mathbf{F}_{iw34}$$
$$= \big(F_g\,\hat{\mathbf{i}} - F_g \tan\phi\,\hat{\mathbf{j}}\big) + \big(m_4 a_B\,\hat{\mathbf{i}} - m_4 a_B \tan\phi\,\hat{\mathbf{j}}\big) + \big(-m_{3b} a_B \tan\phi\,\hat{\mathbf{j}}\big) \qquad (14.21)$$
$$= \big(F_g + m_4 a_B\big)\hat{\mathbf{i}} - \big[F_g + (m_4 + m_{3b})a_B\big]\tan\phi\,\hat{\mathbf{j}}$$

The total force $\mathbf{F}_{32}$ on the crankpin is:

$$\mathbf{F}_{32} = \mathbf{F}_{g\,32} + \mathbf{F}_{ip\,32} + \mathbf{F}_{iw\,32} + \mathbf{F}_{ic\,32}$$

$$= \left(-F_g\,\hat{\mathbf{i}} + F_g\tan\phi\,\hat{\mathbf{j}}\right) + \left(-m_4 a_B\,\hat{\mathbf{i}} + m_4 a_B\tan\phi\,\hat{\mathbf{j}}\right)$$

$$+ \left(-m_{3b} a_B\,\hat{\mathbf{i}} + m_{3b} a_B\tan\phi\,\hat{\mathbf{j}}\right) + \left[m_{3a} r\omega^2\left(\cos\omega t\,\hat{\mathbf{i}} + \sin\omega t\,\hat{\mathbf{j}}\right)\right] \qquad (14.22)$$

$$= \left[m_{3a} r\omega^2\cos\omega t - \left(m_{3b} + m_4\right)a_B - F_g\right]\hat{\mathbf{i}}$$

$$+ \left\{m_{3a} r\omega^2\sin\omega t + \left[\left(m_{3b} + m_4\right)a_B + F_g\right]\tan\phi\right\}\hat{\mathbf{j}}$$

The total force $\mathbf{F}_{21}$ on the mainpin is:

$$\mathbf{F}_{21} = \mathbf{F}_{32} + \mathbf{F}_{ir\,21} \qquad (14.23)$$

Note that, unlike the inertia force in equation 14.14 which was unaffected by the gas force, these pin forces **are** a function of the gas force as well as of the $-m\mathbf{a}$ forces. Engines with larger piston diameters will experience greater pin forces as a result of the explosion pressure acting on their larger piston area.

Program ENGINE calculates the pin forces on all joints using equations 14.20 to 14.23. Figure 14-21 shows the wristpin force on the same unbalanced engine example #3 as shown in previous figures, for three engine speeds. The "bow tie" loop is the inertia force and the "teardrop" loop is the gas force portion of the force curve. An interesting tradeoff occurs between the gas force components and the inertia force components of the pin forces. At a low speed of 800 RPM (Figure 14-21a), the gas force dominates as the inertia forces are negligible at small ω The peak wristpin force is then about 4200 lb. At high speed (6000 RPM), the inertia components dominate and the peak force is about 4500 lb (Figure 14-21c). But at a midrange speed (3400 RPM), the inertia force cancels some of the gas force and the peak force is only about 3200 lb (Figure 14-21b). These plots show that the pin forces can be quite large even in a moderately sized (0.4 liter/cylinder) engine. The pins, links, and bearings all have to be designed to withstand hundreds of millions of cycles of these reversing forces without failure.

14.10 BALANCING THE SINGLE-CYLINDER ENGINE

The derivations and figures in the preceding sections have shown that significant forces are developed both on the pivot pins and on the ground plane due to the gas forces and the inertia/shaking forces. Balancing will not have any effect on the gas forces, which are internal, but it can have a dramatic effect on the inertia/shaking forces. The mainpin force can be reduced, but the crankpin and wristpin forces will be unaffected by any crankshaft balancing done. Figure 14-13 shows the unbalanced shaking force as felt on the ground plane of our 0.4 liter per cylinder example engine #3 from program ENGINE. It is about 9700 lb even at the moderate speed of 3400 RPM. At 6000 RPM it increases to over 30,000 lb! The methods of Chapter 13 can be applied to this mechanism to balance the members in pure rotation and reduce these large shaking forces.

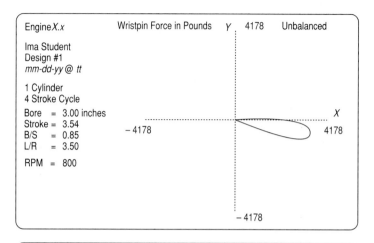

(a) Low speed

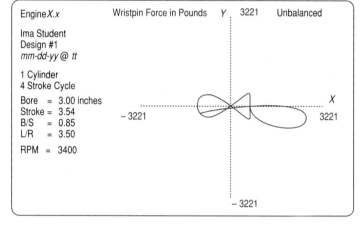

(b) Medium speed

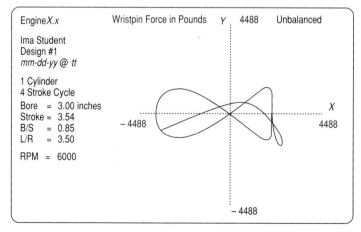

(c) High speed

FIGURE 14-21

Forces on the wristpin of the single-cylinder engine at various speeds

Figure 14-22a shows the dynamic model of our slider-crank with the conrod mass lumped at both crankpin A and wristpin B. We can consider this single-cylinder engine to be a single-plane device, thus suitable for static balancing (see Section 13.1). It is straightforward to statically balance the crank. We need a balance mass at some radius, 180° from the lumped mass at point A whose mr product is equal to the product of the mass at A and its radius r. Applying equation 13.2 to this simple problem we get:

$$m_{bal}r_{bal} = -m_A r_A \qquad (14.24)$$

Any combination of mass and radius which gives this product, placed at 180° from point A will balance the crank. For simplicity of example, we will use a balance radius equal to r. Then a mass equal to m_A placed at A' will exactly balance the rotating masses. The CG of the crank will then be at the fixed pivot O_2 as shown in the figure. In a real crankshaft, actually placing the counterweight at this large a radius would not work. The balance mass has to be kept close to the centerline to clear the piston at BDC. Figure 14-2c shows the shape of typical crankshaft counterweights.

Figure 14-23a shows the shaking force from the same engine as in Figure 14-13 after the crank has been exactly balanced in this manner. The Y component of the shaking force has been reduced to zero and the x component to about 3300 lb at 3400 RPM. This is a factor of three reduction over the unbalanced engine. Note that the only source of Y directed inertia force is the rotating mass at point A (see equations 14.14). What remains after balancing the rotating mass is the force due to the acceleration of the piston and conrod masses at point B which are in linear translation along the X axis, as shown by the inertia force at point B in the figure.

To completely eliminate this reciprocating unbalanced shaking force would require the introduction of another reciprocating mass, which oscillates 180 degrees out of phase with the piston. Adding a second piston and cylinder, properly arranged, can accomplish this. One of the principal advantages of multicylinder engines is their ability to reduce or eliminate the shaking forces. We will investigate this in the next chapter.

In the single-cylinder engine, there is no way to completely eliminate the reciprocating unbalance with a single, rotating counterweight, but we can reduce the shaking force still further. Figure 14-22b shows an additional amount of mass m_P added to the counterweight at point A'. (Note that the crank's CG has now moved away from the fixed pivot.) This extra balance mass creates an additional inertia force $(-m_P r \omega^2)$ as is shown, broken into x and y components, in the figure. The y component is not opposed by any other inertia forces present, but the x component will always be opposite to the reciprocating inertia force at point B. Thus this extra mass, m_P, which *overbalances the crank*, will reduce the x directed shaking force at the expense of adding back some y directed shaking force. This is a useful tradeoff as the direction of the shaking force is usually of less concern than is its magnitude. Shaking forces create vibrations in the supporting structure which are transmitted through it and modified by it. As an example, it is unlikely that you could define the direction of a motorcycle engine's shaking forces by feeling their resultant vibrations in the handlebars. But you **will** detect an increase in the magnitude of the shaking forces from the larger amplitude of vibration they cause in the cycle frame.

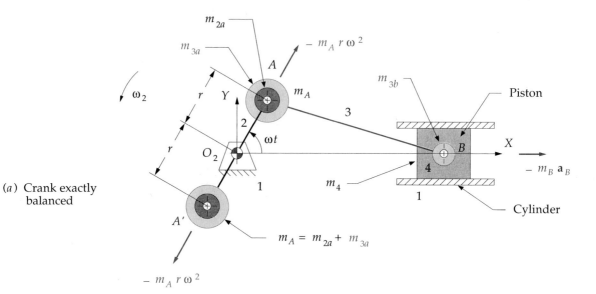

(a) Crank exactly
    balanced

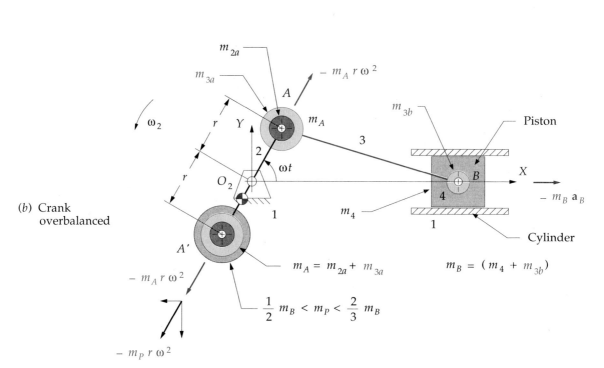

(b) Crank
    overbalanced

FIGURE 14-22

Balancing and overbalancing the single-cylinder engine

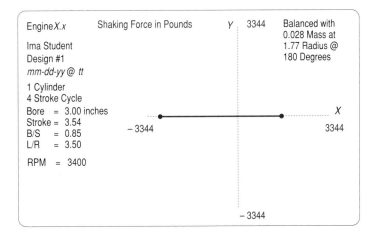

(a) Crank exactly balanced

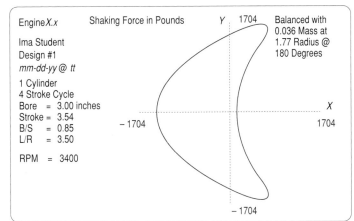

(b) Crank overbalanced

FIGURE 14-23

Effects of balancing and overbalancing on the shaking force in the slider-crank linkage

The correct amount of additional "overbalance" mass needed to minimize the peak shaking force, regardless of its direction, will vary with the particular engine design. It will usually be between one-half and two-thirds of the reciprocating mass at point B (piston plus conrod at wristpin), if placed at the crank radius r. Of course, once this mass-radius product is determined, it can be achieved with any combination of mass and radius. Figure 14-23b shows the minimum shaking force achieved for this engine with the addition of 65.5% of the mass at B acting at radius r. The shaking force has now been reduced to 1700 lb at 3400 RPM, which is **18% of its original unbalanced value**. The benefits of balancing, and of overbalancing in the case of the single-cylinder engine, should now be obvious.

14.11 DESIGN TRADEOFFS AND RATIOS

In the design of any system or device, no matter how simple, there will always be conflicting demands, requirements, or desires which must be traded off to achieve the best design compromise. This single-cylinder engine is no exception. There are two dimensionless design ratios which can be used to characterize an engine's dynamic behavior in a general way. The first is the crank/conrod ratio r/l, introduced in Section 14.2 above or its inverse, the **conrod/crank ratio** l/r. The second is the **bore/stroke ratio** B/S.

Conrod/Crank Ratio

The crank/conrod ratio r/l appears in all the equations for acceleration, forces, and torques. In general, the smaller the r/l ratio, the smoother will be the acceleration function and thus all other factors which it influences. Program ENGINE uses the inverse of this ratio as an input parameter. The **conrod/crank ratio** l/r must be greater than about two to obtain acceptable transmission angles in the slider-crank linkage. The ideal value for l/r from a kinematic standpoint would be infinity as that would result in the piston acceleration function being a pure harmonic. The second and all subsequent harmonic terms in equations 14.3 would be zero in this event, and the peak value of acceleration would be a minimum. However, an engine this tall would not package very well, and package considerations often dictate the maximum value of the l/r ratio. Most engines will have an l/r ratio between three and five which values give acceptable smoothness in a reasonably short engine.

Bore/Stroke Ratio

The bore B of the cylinder is essentially equal to the diameter of the piston. (There is a small clearance.) The stroke S is defined as the distance travelled by the piston from TDC to BDC and is twice the crank radius, $S = 2r$. The bore appears in the equation for gas force (Eq. 14.4) and thus also affects gas torque. The crank radius appears in every equation. An engine with a B/S ratio of 1 is referred to as a "square" engine. If B/S is larger than 1 it is "oversquare", if less, "undersquare." The designer's choice of this ratio can have a significant effect on the dynamic behavior of the engine. Assuming that the displacement, or stroke volume V, of the engine has been chosen and is to remain constant, this displacement can be achieved with an infinity of combinations of bore and stroke ranging from a "pancake" piston with tiny stroke to a "pencil" piston with very long stroke.

$$V = \frac{\pi B^2}{4} S \qquad\qquad (14.25)$$

There is a classic design tradeoff here between B and S for a constant stroke volume V. A large bore and small stroke will result in high gas forces which will affect pin forces adversely. A large stroke and small bore will result in high inertia forces which will affect pin forces (as well as other forces and torques) adversely. So there should be an optimum value for the B/S ratio in each case, which will minimize these adverse effects. Most production engines have B/S ratios in the range of about 0.75 to 1.5.

It is left as an exercise for the student to investigate the effects of variation in the B/S and l/r ratios on forces and torques in the system. Program EN-GINE will demonstrate the effects of changes made independently to each of these ratios, while all other design parameters are held constant. The student is encouraged to experiment with the program to gain insight into the role of these ratios in the dynamic performance of the engine.

Materials

There will always be a strength/weight tradeoff. The forces in this device can be quite high, due both to the explosion and to the inertia of the moving elements. We would like to keep the masses of the parts as low as possible as the accelerations are typically very high, as seen in Figure 14-8c. But the parts must be strong enough to withstand the forces, so materials with good strength to weight ratios are needed. Pistons are usually made of an aluminum alloy, either cast or forged. Conrods are most often cast iron or forged steel, except in very small engines (lawn mower, chain saw, motorcycle) where they may be aluminum alloy. High performance engines may have titanium connecting rods. Crankshafts are usually cast iron or forged steel, and wristpins are of hardened steel tubing or rod. Plain bearings of a special soft, nonferrous metal alloy called babbitt are usually used. In the four-stroke engine these are pressure lubricated with oil pumped through drilled passageways in the block, crankshaft, and connecting rods. In the two-stroke engine, the fuel carries the lubricant to these parts. Engine blocks are cast iron or cast aluminum alloy. The chrome-plated steel piston rings seal and wear well against cast iron cylinders. Most aluminum blocks are fitted with cast iron liners around the cylinder bores. Some are unlined and made of a high-silicon aluminum alloy which is specially cooled after casting to precipitate the hard silicon in the cylinder walls for wear resistance.

14.12 BIBLIOGRAPHY

Heywood, J. B., *Internal Combustion Engine Fundamentals*, McGraw-Hill Inc. New York, 1988.

Taylor, C. F., *The Internal Combustion Engine in Theory and Practice*, 2nd ed., MIT Press, Cambridge, Mass., 1966.

14.13 PROBLEMS

*14-1 A slider-crank linkage has $r = 3$ and $l = 12$. It has an angular velocity of 200 rad/sec at time $t = 0$. Its initial crank angle is zero. Calculate the piston acceleration at $t = 1$ sec. Use two methods, the exact solution, and the approximate Fourier series solution and compare the results.

14-2 Repeat problem 14-1 for $r = 4$ and $l = 15$ and $t = 0.9$ sec.

*14-3 A slider-crank linkage has $r = 3$ and $l = 12$, and a piston bore $B = 2$. The peak gas pressure in the cylinder occurs at a crank angle of 10° and is 1000 pressure units. Calculate the gas force and gas torque at this position.

14-4 A slider-crank linkage has $r = 4$ and $l = 15$, and a piston bore $B = 3$. The peak gas pressure in the cylinder occurs at a crank angle of 5° and is 600 pressure units. Calculate the gas force and gas torque at this position.

---

* Answers in Appendix E

*14-5　Repeat problem 14-3 using the exact method of calculation of gas torque and compare its result to that obtained by the approximate expression in equation 14.8b. What is the percent error?

14-6　Repeat problem 14-4 using the exact method of calculation of gas torque and compare its result to that obtained by the approximate expression in equation 14.8b. What is the percent error?

*14-7　A connecting rod of length $l = 12$ has a mass $m_3 = 0.020$. Its mass moment of inertia is 0.620. Its CG is located at 0.4 l from the crankpin, point A.

　　　a.　Calculate an exact dynamic model using two lumped masses, one at the wristpin, point B, and one at whatever other point is required. Define the lumped masses and their locations.
　　　b.　Calculate an approximate dynamic model using two lumped masses, one at the wristpin, point B, and one at the crankpin, point A. Define the lumped masses and their locations.
　　　c.　Calculate the error in the mass moment of inertia of the approximate model as a percentage of the original mass moment of inertia.

14-8　Repeat problem 14-7 for these data: $l = 15$, mass $m_3 = 0.025$, mass moment of inertia is 1.020. Its CG is located at 0.25 l from the crankpin, point A.

*14-9　A crank of length $r = 3.5$ has a mass $m_2 = 0.060$. Its mass moment of inertia about its pivot is 0.300. Its CG is located at 0.30 r from the mainpin, point O_2. Calculate a statically equivalent two-lumped mass dynamic model with the lumps placed at the mainpin and crankpin. What is the % error in the model's moment of inertia about the crank pivot.

14-10　Repeat problem 14-9 for a crank length $r = 4$, a mass $m_2 = 0.050$, a mass moment of inertia about its pivot of 0.400. Its CG is located at 0.40 r from the mainpin, point O_2.

*14-11　Combine the data from problems 14-7 and 14-9. Run the linkage at a constant 2000 RPM. Calculate the inertia force and inertia torque at $\omega t = 45°$. Piston mass = 0.012.

*14-12　Combine the data from problems 14-7 and 14-10. Run the linkage at a constant 3000 RPM. Calculate the inertia force and inertia torque at $\omega t = 30°$. Piston mass = 0.019.

*14-13　Combine the data from problems 14-8 and 14-9. Run the linkage at a constant 2500 RPM. Calculate the inertia force and inertia torque at $\omega t = 24°$. Piston mass = 0.023.

*14-14　Combine the data from problems 14-8 and 14-10. Run the linkage at a constant 4000 RPM. Calculate the inertia force and inertia torque at $\omega t = 18°$. Piston mass = 0.015.

14-15　Combine the data from problems 14-7 and 14-9. Run the linkage at a constant 2000 RPM. Calculate the pin forces at $\omega t = 45°$. Piston mass = 0.022. $F_g = 300$.

14-16　Combine the data from problems 14-7 and 14-10. Run the linkage at a constant 3000 RPM. Calculate the pin forces at $\omega t = 30°$. Piston mass = 0.019. $F_g = 600$.

14-17　Combine the data from problems 14-8 and 14-9. Run the linkage at a constant 2500 RPM. Calculate the pin forces at $\omega t = 24°$. Piston mass = 0.032. $F_g = 900$.

14-18 Combine the data from problems 14-8 and 14-10. Run the linkage at a constant 4000 RPM. Calculate the pin forces at $\omega t = 18°$. Piston mass = 0.014. $F_g = 1200$.

*14-19 Using the data from problem 14-11:

    a. Exactly balance the crank and recalculate the inertia force.

    b. Overbalance the crank with approximately two-thirds of the mass at the wristpin placed at radius $-r$ on the crank and recalculate the inertia force.

    c. Compare these results to those for the unbalanced crank.

*14-20 Repeat problem 14-19 using the data from problem 14-12.

*14-21 Repeat problem 14-19 using the data from problem 14-13.

*14-22 Repeat problem 14-19 using the data from problem 14-14.

14-23 Combine the necessary equations to develop expressions that show how each of these dynamic parameters varies as a function of the crank/conrod ratio alone:

    a. Piston acceleration

    b. Inertia force

    c. Inertia torque

    d. Pin forces

    Plot the functions. Check your conclusions with program ENGINE.

    Hint: *Consider all other parameters to be temporarily constant. Set the crank angle to some value such that the gas force is nonzero.*

14-24 Combine the necessary equations to develop expressions that show how each of these dynamic parameters varies as a function of the bore/stroke ratio alone:

    a. Gas force

    b. Gas torque

    c. Inertia force

    d. Inertia torque

    e. Pin forces

    Plot the functions. Check your conclusions with program ENGINE.

    Hint: *Consider all other parameters to be temporarily constant. Set the crank angle to some value such that the gas force is nonzero.*

14-25 Develop an expression to determine the optimum bore/stroke ratio to minimize the wristpin force. Plot the function.

14-26 Use program ENGINE, your own computer program, or an equation solver to calculate the maximum value and the polar-plot shape of the force on the mainpin $(I_{2,1})$ of a 1 cubic-inch displacement, single-cylinder engine with bore = 1.12838 inches for the following situations.

    a. Piston, conrod and crank masses = 0

    b. Piston mass = 1 blob, conrod and crank masses = 0

    c. Conrod mass = 1 blob, piston and crank masses = 0

    d. Crank mass = 1 blob, conrod and piston masses = 0

    Place the *CG* of the crank at 0.5 *r* and the conrod at 0.33 *l*. Compare and explain the differences in the mainpin force under these different conditions with reference to the governing equations.

14-27　　Repeat problem 14-26 for the crankpin.

14-28　　Repeat problem 14-26 for the wristpin.

14-29　　Use program ENGINE, your own computer program, or an equation solver to calculate the maximum value and the polar plot shape of the force on the mainpin $(I_{2,1})$ of a 1 cu. in. single-cylinder engine with bore = 1.12838 in. for the following situations.
　　　　a.　　Engine unbalanced
　　　　b.　　Crank exactly balanced against mass at crankpin
　　　　c.　　Crank optimally overbalanced against masses at crankpin and wristpin
　　　　Piston, conrod and crank masses = 1. Place the *CG* of the crank at 0.5 *r* and the conrod at 0.33 *l*. Compare and explain the differences in the mainpin force under these conditions with reference to the governing equations.

14-30　　Repeat problem 14-29 for the crankpin force.

14-31　　Repeat problem 14-29 for the wristpin force.

14-32　　Repeat problem 14-29 for the shaking force.

14.14　PROJECTS

These are loosely structured design problems intended for solution using program ENGINE. All involve the design of a single-cylinder engine and differ only in the specific data for the engine. The general problem statement is:

Design a single-cylinder engine for a specified displacement and stroke cycle. Optimize the conrod/crank ratio and bore/stroke ratio to minimize shaking forces, shaking torque, and pin forces, also considering package size. Design your link shapes and calculate realistic dynamic parameters (mass, CG location, moment of inertia) for those links using the methods shown in Chapter 11 and Section 12.12. Dynamically model the links as described in this chapter. Balance or overbalance the linkage as needed to achieve these results. Overall smoothness of total torque is desired. Design and size a minimum weight flywheel by the method of Chapter 12 to smooth the total torque. Write an engineering report on your design.

P14-1　　Two-stroke cycle with a displacement of 0.125 liters.

P14-2　　Four-stroke cycle with a displacement of 0.125 liters.

P14-3　　Two-stroke cycle with a displacement of 0.25 liters.

P14-4　　Four-stroke cycle with a displacement of 0.25 liters.

P14-5　　Two-stroke cycle with a displacement of 0.50 liters.

P14-6　　Four-stroke cycle with a displacement of 0.50 liters.

P14-5　　Two-stroke cycle with a displacement of 0.75 liters.

P14-6　　Four-stroke cycle with a displacement of 0.75 liters.

Chapter 15

MULTICYLINDER ENGINES

Look long on an engine,
it is sweet to the eyes
MACKNIGHT BLACK

15.0 INTRODUCTION

The previous chapter discussed the design of the slider-crank mechanism as used in the single-cylinder internal combustion engine and piston pumps. We will now extend the design to multicylinder configurations. Some of the problems with shaking forces and torques can be alleviated by proper combination of multiple slider-crank linkages on a common crankshaft. Program ENGINE, included with this text, will calculate the equations derived in this chapter and allow the student to exercise many variations of a design in a short time. Some examples are provided as diskfiles to be read into the program. These are noted in the text. The student is encouraged to investigate these examples with program ENGINE in order to develop an understanding of and insight to the subtleties of this topic. A user manual for program ENGINE is provided in the next chapter which can be read or referred to out of sequence, with no loss in continuity, in order to gain familiarity with the program's operation.

As with the single-cylinder case, we will not address the thermodynamic aspects of the internal combustion engine beyond the definition of the combustion forces necessary to drive the device presented in the previous chapter. We will concentrate on the engine's kinematic and mechanical dynamics aspects. It is not our intention to make an "engine designer" of the student so much as to apply dynamic principles to a realistic design problem of general interest and also to convey the complexity and fascination involved in the design of a more complicated dynamic device than the single-cylinder engine.

(a) Opposed four

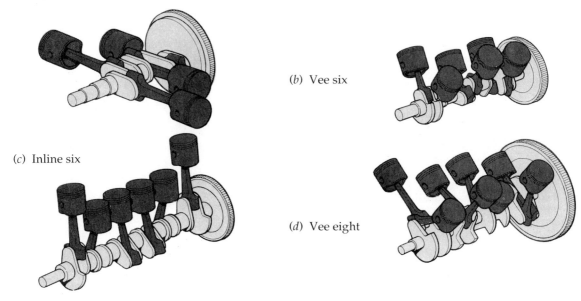

(b) Vee six

(c) Inline six

(d) Vee eight

FIGURE 15-1

Various multicylinder engine configurations
Illustrations Copyright Eaglemoss Publications/Car Care Magazine - Reprinted with permission

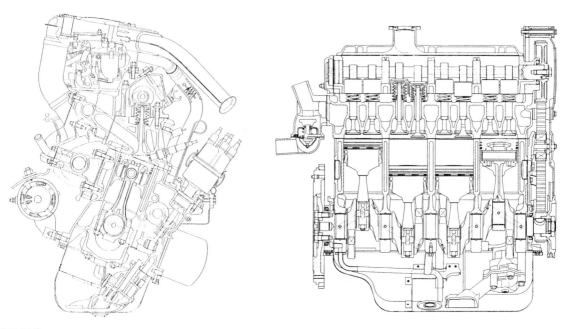

FIGURE 15-2

Cutaway views of a four-stroke, four-cylinder inline engine - *Courtesy of FIAT Corporation, Italy*

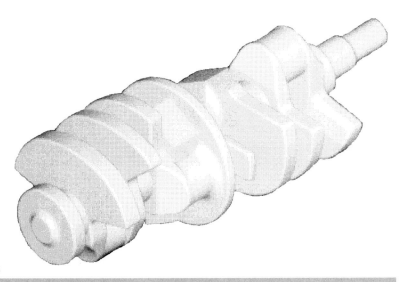

FIGURE 15-3

Crankshaft from a Corvette V8 engine - *Courtesy of Chevrolet Div.,General Motors Co.*

15.1 MULTICYLINDER ENGINE DESIGNS

Multicylinder engines are designed in a wide variety of configurations from the simple inline arrangement to vee, opposed and radial arrangements some of which are shown in Figure 15-1. These arrangements may use any of the stroke cycles discussed in Chapter 14, Clerk, Otto, or Diesel.

INLINE ENGINES The most common and simplest arrangement is an inline engine with its cylinders all in a common plane as shown in Figure 15-2. Two-[1], three-[1], four-, five-, and six-cylinder **inline engines** are in common use. Each cylinder will have its individual slider-crank mechanism consisting of a crank, conrod, and piston. The cranks are formed together in a common **crankshaft** as shown in Figure 15-3. Each cylinder's crank on the crankshaft is referred to as a **crank throw**. These crank throws will be arranged with some **phase angle** relationship one to the other, in order to stagger the motions of the pistons in time. It should be apparent from the discussion of shaking forces and balancing in the previous chapter that we would like to have pistons moving in opposite directions to one another at the same time in order to cancel the reciprocating inertial forces. The optimum phase angle relationships between the crank throws will differ depending on the number of cylinders and the stroke cycle of the engine. There will usually be one (or a small number of) viable crank throw arrangements for a given engine configuration to accomplish this goal. The engine in Figure 15-2 is a four-stroke cycle, four-cylinder, inline engine with its crank throws at 0, 180, 180, and 0° phase angles which we will soon see are optimum for this engine.

VEE ENGINES in two-[1], four-[1], six-, eight-, ten-[2] and twelve-cylinder[3] versions are produced, with vee six and vee eight being the most common configurations. Figure 15-4 shows a cross section and Figure 15-5 a cutaway of a 60° vee twelve engine. **Vee engines** can be thought of as *two inline engines grafted together*

---

[1] Mainly in motorcycles. [2] Honda, Chrysler. [3] BMW, Jaguar.

FIGURE 15-4

Cross section of a BMW 5 liter V12 engine - *Courtesy of BMW of North America Inc.*

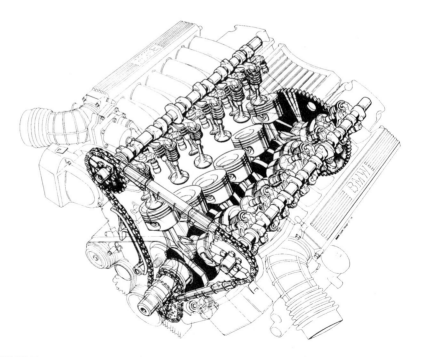

FIGURE 15-5

Cutaway view of a BMW 5 liter V12 engine - *Courtesy of BMW of North America Inc.*

onto a common crankshaft. The two "inline" portions, or **banks**, are arranged with some **vee angle** between them. The crankshaft in Figure 15-3 is from a vee eight. Its crank throws are at 0, 90, 270, and 180° respectively. A vee eight's vee angle is 90°. The geometric arrangements of the crankshaft (phase angles) and cylinders (vee angle) have a significant effect on the dynamic condition of the engine. We will soon explore these relationships in detail.

OPPOSED ENGINES are essentially vee engines with a vee angle of 180°. The pistons in each bank are on opposite sides of the crankshaft as shown in Figure 15-6. This arrangement promotes cancellation of inertial forces and is popular in aircraft engines.[4] It has also been used in some automotive applications.[5]

RADIAL ENGINES have their cylinders arranged radially around the crankshaft in nearly a common plane. These were common on World War II vintage aircraft as they allowed large displacements, thus high power, in a compact form whose shape was well suited to that of an airplane. Typically air cooled, the cylinder arrangement allowed good exposure of all cylinders to the airstream. Large versions had multiple rows of radial cylinders, rotationally staggered to allow cooling air to reach the back rows. The gas turbine jet engine has rendered these radial aircraft engines obsolete.

ROTARY ENGINES were an interesting variant on the aircraft radial engine. Similar in appearance and cylinder arrangement to the radial engine, the anomaly was that the crankshaft was the stationary ground plane. The propellor was attached to the crankcase (block) which rotated around the crankshaft! It is a kinematic inversion of the radial engine. At least it didn't need a flywheel!

Many other configurations of engines have been tried over the century of development of this ubiquitous device. The bibliography at the end of this chapter contains several references which describe other engine designs, the usual, unusual, and exotic. We will begin our detailed exploration of multicylinder engine design with the simplest configuration, the inline engine, then progress to the vee and opposed versions.

---

4   Continental six cylinder.
5   VW "Beetle" four cylinder, Subaru four, Honda motorcycle six, Ferrari twelve, Porsche six, the ill-fated Corvair six, and the short-lived Tucker (Continental) six.

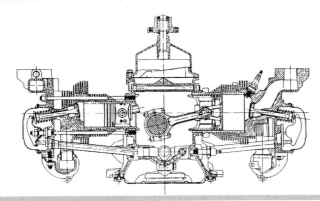

FIGURE 15-6

Chevrolet Corvair horizontally opposed six cylinder engine
Courtesy of Chevrolet Division, General Motors Co.

15.2   THE CRANK PHASE DIAGRAM

Fundamental to the design of any multicylinder engine (or piston pump) is the arrangement of crank throws on the crankshaft. We will use the four-cylinder inline engine as an example. Many choices are possible for the crank phase angles in the four-cylinder engine. We will start, for example, with the one that seems most obvious from a common sense standpoint. There are 360° in any crankshaft. We have four cylinders, so an arrangement of 0, 90, 180, and 270° seems appropriate. The **delta phase angle** between throws is then 90°. In general, for maximum cancellation of inertia forces, which have a period of one revolution, the optimum delta phase angle will be:

$$\Delta\phi_{inertia} = \frac{360°}{n} \tag{15.1}$$

where n is the number of cylinders.

We must establish some convention for the measurement of these phase angles which will be:

1   The first (front) cylinder will be number 1 and its phase angle will always be zero. It is the reference cylinder for all others.

2   The phase angles of all other cylinders will be measured with respect to the crank throw for cylinder 1.

3   Phase angles are measured internal to the crankshaft, that is, with respect to a rotating coordinate system embedded in the first crank throw.

4   Cylinders will be numbered consecutively from front to back of the engine.

The phase angles are defined in a **crank phase diagram** as shown in Figure 15-7 for a four-cylinder, inline engine. Figure 15-7a shows the crankshaft with the throws numbered clockwise around the axis. The shaft is rotating counterclockwise. The pistons are oscillating horizontally in this diagram, along the x axis. Cylinder 1 is shown with its piston at top dead center (TDC). Taking that position as the starting point for the abscissae (thus time zero) in Figure 15-7b, we plot the velocity of each piston for two revolutions of the crank (to accommodate one complete four-stroke cycle). Piston 2 arrives at TDC 90° after piston 1 has left. Thus we say that cylinder 2 lags cylinder 1 by 90 deg. By convention a *lagging event is defined as having a negative phase angle*, shown by the clockwise numbering of the crank throws. The velocity plots clearly show that each cylinder arrives at TDC (zero velocity) 90° later than the one before it. Negative velocity on the plots in Figure 15-7b indicates piston motion to the left (down stroke) in Figure 15-7a; positive velocity indicates motion to the right (up stroke).

For the discussion in this chapter we will assume counterclockwise rotation of all crankshafts and all phase angles will thus be negative. We will, however, omit the negative signs on the listings of phase angles with the understanding that they follow this convention.

Figure 15-7 shows the timing of events in the cycle and is a necessary and useful aid in defining our crankshaft design. However, it is not necessary to go to the trouble of drawing the correct sinusoidal shapes of the velocity plots to

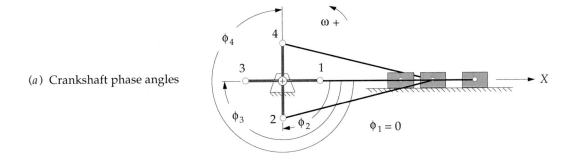

(a) Crankshaft phase angles

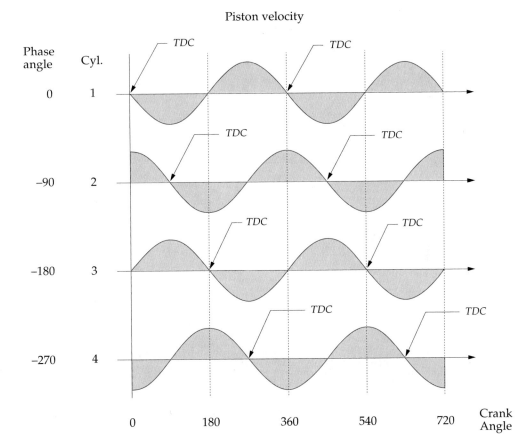

(b) The crank phase diagram

FIGURE 15-7

Crank phase angles and the phase diagram

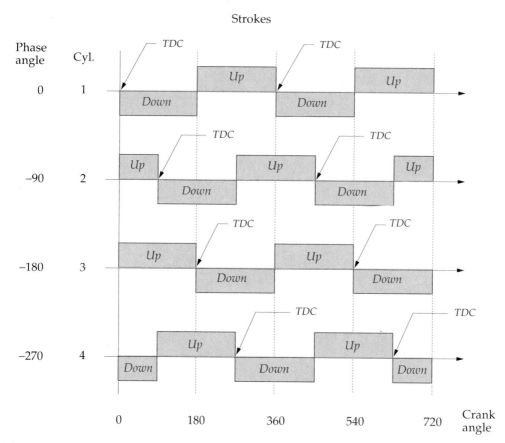

FIGURE 15-8

The schematic crank phase diagram

obtain the needed information. All that is needed is a schematic indication of the relative positions within the cycle of the ups and downs of the various cylinders. This same information is conveyed by the simplified crank phase diagram shown in Figure 15-8. Here the piston motions are represented by rectangular blocks with a negative block arbitrarily used to denote a piston downstroke and a positive one a piston upstroke. It is strictly schematic. The positive and negative values of the blocks imply nothing more than that stated. Such a schematic crank phase diagram can (and should) be drawn for any proposed arrangement of crankshaft phase angles. To draw it, simply shift each cylinder's blocks to the right by its phase angle with respect to the first cylinder.

15.3   SHAKING FORCES IN INLINE ENGINES

We want to determine the overall shaking force which results from our chosen crankshaft phase angle arrangement. The individual cylinders will each contribute to the total shaking force. We can superpose their effects, taking their phase shifts into account. Equation 14.14f defines the shaking force for one cylinder with the crankshaft rotating at constant ω.

for $\alpha = 0$:

$$\mathbf{F_s} \cong \left[-m_A \, r\omega^2 \cos\omega t - m_B r\omega^2 \left(\cos\omega t + \frac{r}{l}\cos 2\omega t\right)\right]\hat{\mathbf{i}} - \left[m_A r\omega^2 \sin\omega t\right]\hat{\mathbf{j}} \quad (15.2a)$$

This expression is for an unbalanced crank. It is standard practice in multi-cylinder engines to exactly balance each crank throw in order to eliminate the shaking force effects of the combined mass m_A of crank and conrod assumed concentrated at the crankpin. (See Section 14.10 and equation 14.24.) Note that there is normally *no need to overbalance* the crank throws in a multicylinder engine, as we did for the single cylinder, provided that we can arrange the crank phase angles to cancel the effects of the masses at the wrist pins. This will usually be possible except in some two-cylinder, four-stroke, inline engines.

If we provide balance masses with an mr product equal to $m_A r_A$ on each crank throw as shown in Figure 15-3, the terms in equation 15.2a which include m_A will be eliminated, reducing it to:

$$\mathbf{F_s} \cong -m_B r\omega^2 \left(\cos\omega t + \frac{r}{l}\cos 2\omega t\right)\hat{\mathbf{i}} \quad (15.2b)$$

Recall that these are approximate expressions which exclude all harmonics above the second.

We will assume that all cylinders in the engine are of equal displacement and that all pistons and all conrods are interchangeable. This is desirable both for dynamic balance and for lower production costs. If we let the crank angle ωt represent the instantaneous position of the reference crank throw for cylinder 1, the corresponding positions of the other cranks can be defined by their phase angles as shown in Figure 15-7. The total shaking force for a multicylinder inline engine is then:

$$\mathbf{F_s} \cong -m_B r\omega^2 \sum_{i=1}^{n}\left[\cos(\omega t - \phi_i) + \frac{r}{l}\cos 2(\omega t - \phi_i)\right]\hat{\mathbf{i}} \quad (15.2c)$$

where n = number of cylinders and $\phi_1 = 0$. Substitute the identity:

$$\cos(a - b) = \cos a \cos b + \sin a \sin b$$

and factor to:

$$\mathbf{F_s} \cong -m_B r\omega^2 \left[\begin{array}{l}\cos\omega t \sum_{i=1}^{n}\cos\phi_i + \sin\omega t \sum_{i=1}^{n}\sin\phi_i + \\[2mm] \frac{r}{l}\left(\cos 2\omega t \sum_{i=1}^{n}\cos 2\phi_i + \sin 2\omega t \sum_{i=1}^{n}\sin 2\phi_i\right)\end{array}\right]\hat{\mathbf{i}} \quad (15.2d)$$

The ideal value for the shaking force is zero. This expression can only be zero for all values of ωt if:

$$\sum_{i=1}^{n} \cos \phi_i = 0 \qquad\qquad \sum_{i=1}^{n} \sin \phi_i = 0 \qquad (15.3a)$$

$$\sum_{i=1}^{n} \cos 2\phi_i = 0 \qquad\qquad \sum_{i=1}^{n} \sin 2\phi_i = 0 \qquad (15.3b)$$

Thus, there are some combinations of phase angles ϕ_i which will cause cancellation of the shaking force through the second harmonic. If we wish to cancel higher harmonics, we could replace those harmonics' terms truncated from the Fourier series representation and find that the fourth and sixth harmonics will be cancelled if:

$$\sum_{i=1}^{n} \cos 4\phi_i = 0 \qquad\qquad \sum_{i=1}^{n} \sin 4\phi_i = 0 \qquad (15.3c)$$

$$\sum_{i=1}^{n} \cos 6\phi_i = 0 \qquad\qquad \sum_{i=1}^{n} \sin 6\phi_i = 0 \qquad (15.3d)$$

Equations 15.3 provide us with a convenient predictor of the shaking force behavior of any proposed inline engine design. Program ENGINE calculates equations 15.3 and displays a table of their values. Note that **both the sine and cosine summations of any multiple of the phase angles must be zero for that harmonic of the shaking force to be zero**. The calculation for our example of a four-cylinder engine with phase angles of $\phi_1 = 0$, $\phi_2 = 90$, $\phi_3 = 180$, $\phi_4 = 270°$ in Table 15-1 shows that the shaking forces are zero for the first, second, and sixth harmonics and nonzero for the fourth. So, our common sense choice in this instance has proven a good one as far as shaking forces are concerned. The coefficients of the fourth and sixth harmonic terms are miniscule, so their contributions, if any, can be ignored. The primary component is of most concern, because of its potential magnitude. The secondary (second harmonic) term is less critical than the primary as it is multiplied by r/l which is generally less than $1/3$. An unbalanced secondary harmonic of shaking force is undesirable but can be lived with, especially if the engine is of small displacement (less than about $1/2$ liter per cylinder).

TABLE 15-1   **Force Balance State of a 0, 90, 180, 270° Crankshaft Four-Cylinder Engine**

Primary forces:	$\displaystyle\sum_{i=1}^{n} \sin\phi_i = 0$	$\displaystyle\sum_{i=1}^{n} \cos\phi_i = 0$
Secondary forces:	$\displaystyle\sum_{i=1}^{n} \sin 2\phi_i = 0$	$\displaystyle\sum_{i=1}^{n} \cos 2\phi_i = 0$
Fourth harmonic forces:	$\displaystyle\sum_{i=1}^{n} \sin 4\phi_i = 0$	$\displaystyle\sum_{i=1}^{n} \cos 4\phi_i = 4$
Sixth harmonic forces:	$\displaystyle\sum_{i=1}^{n} \sin 6\phi_i = 0$	$\displaystyle\sum_{i=1}^{n} \cos 6\phi_i = 0$

To see the results of this engine configuration, run program ENGINE and select example #6 from the **Input** menu (main menu #1). **Calculate** (main menu #3) for any engine speed, **Balance** (main menu #6) the crank in any fashion you choose (including zero balance mass), and **Assemble** (main menu #7) the engine from the main menu, accepting all default values. Then **Plot** (main menu #5) **Shaking Force**. See Chapter 16 for more detailed instructions on the use of the program ENGINE.

15.4 INERTIA TORQUE IN INLINE ENGINES

The inertia torque for a single-cylinder engine was defined in Section 14.6 and equation 14.15e. We are concerned with reducing this inertia torque, preferably to zero, because it adds to the gas torque to form the total torque. (See Section 14.7.) Inertia torque adds nothing to the net driving torque as its average value is always zero, but it does create large oscillations in the total torque which detracts from its smoothness. Inertia torque oscillations can be masked to a degree with the addition of sufficient flywheel to the system, or their external, net effect can be cancelled by the proper choice of phase angles. However, the torque oscillations, even if hidden from the outside observer, or made to sum to zero, are still present within the crankshaft and can lead to torsional fatigue failure if the part is not properly designed. The one cylinder inertia torque equation, approximately, for three harmonics is:

$$\mathbf{T}_{i21} \cong \frac{1}{2} m_B r^2 \omega^2 \left(\frac{r}{2l} \sin \omega t - \sin 2\omega t - \frac{3r}{2l} \sin 3\omega t \right) \hat{\mathbf{k}} \qquad (15.4a)$$

Summing for all cylinders and including their phase angles:

$$\mathbf{T}_{i21} \cong \frac{1}{2} m_B r^2 \omega^2 \sum_{i=1}^{n} \left[\frac{r}{2l} \sin(\omega t - \phi_i) - \sin 2(\omega t - \phi_i) - \frac{3r}{2l} \sin 3(\omega t - \phi_i) \right] \hat{\mathbf{k}} \qquad (15.4b)$$

Substitute the identity:

$$\sin(a - b) = \sin a \cos b - \cos a \sin b$$

and factor to:

$$\mathbf{T}_{i21} \cong \frac{1}{2} m_B r^2 \omega^2 \begin{bmatrix} \dfrac{r}{2l} \left(\sin \omega t \displaystyle\sum_{i=1}^{n} \cos \phi_i - \cos \omega t \displaystyle\sum_{i=1}^{n} \sin \phi_i \right) \\[2mm] - \left(\sin 2\omega t \displaystyle\sum_{i=1}^{n} \cos 2\phi_i - \cos 2\omega t \displaystyle\sum_{i=1}^{n} \sin 2\phi_i \right) \\[2mm] - \dfrac{3r}{2l} \left(\sin 3\omega t \displaystyle\sum_{i=1}^{n} \cos 3\phi_i - \cos 3\omega t \displaystyle\sum_{i=1}^{n} \sin 3\phi_i \right) \end{bmatrix} \hat{\mathbf{k}} \qquad (15.4c)$$

This can only be zero for all values of ωt if:

$$\sum_{i=1}^{n} \sin \phi_i = 0 \qquad \sum_{i=1}^{n} \cos \phi_i = 0 \qquad (15.5a)$$

$$\sum_{i=1}^{n} \sin 2\phi_i = 0 \qquad \sum_{i=1}^{n} \cos 2\phi_i = 0 \qquad (15.5b)$$

$$\sum_{i=1}^{n} \sin 3\phi_i = 0 \qquad \sum_{i=1}^{n} \cos 3\phi_i = 0 \qquad (15.5c)$$

Equations 15.5 provide us with a convenient predictor of the inertia torque behavior of any proposed inline engine design. Calculation for our example of a four-cylinder engine with phase angles of $\phi_1 = 0$, $\phi_2 = 90$, $\phi_3 = 180$, $\phi_4 = 270°$ shows that the inertia torque components are zero for the first, second, and third harmonics. So, our current example is a good one for inertia torques as well.

15.5   SHAKING MOMENT IN INLINE ENGINES

We were able to consider the single-cylinder engine to be a single plane, or two dimensional, device and thus could statically balance it. The multicylinder engine is three dimensional. Its multiple cylinders are distributed along the axis of the crankshaft. Even though we may have cancellation of the shaking forces, there may still be unbalanced moments in the plane of the engine block. We need to apply the criteria for dynamic balance. (See Section 13.2 and equation 13.3.) Figure 15-9 shows a schematic of an inline four-cylinder engine with crank phase angles of $\phi_1 = 0$, $\phi_2 = 90$, $\phi_3 = 180$, $\phi_4 = 270°$. The spacing between the cylinders is normally uniform. We can sum moments in the plane of the cylinders about any convenient point such as one end of the crankshaft,

$$\sum M_L = \sum_{i=1}^{n} z_i \mathbf{F_{s_i}} \, \hat{\mathbf{j}} \qquad (15.6a)$$

where $\mathbf{F}_{si}$ is the shaking force and z_i the moment arm of the ith cylinder. Substituting equation 15.2d for $\mathbf{F}_{si}$:

$$\sum M_L \cong -m_B r \omega^2 \left[\begin{array}{c} \cos \omega t \sum_{i=1}^{n} z_i \cos \phi_i + \sin \omega t \sum_{i=1}^{n} z_i \sin \phi_i + \\[2mm] \dfrac{r}{l}\left(\cos 2\omega t \sum_{i=1}^{n} z_i \cos 2\phi_i + \sin 2\omega t \sum_{i=1}^{n} z_i \sin 2\phi_i \right) \end{array} \right] \hat{\mathbf{j}} \qquad (15.6b)$$

This expression can only be zero for all values of ωt if:

$$\sum_{i=1}^{n} z_i \cos \phi_i = 0 \qquad \sum_{i=1}^{n} z_i \sin \phi_i = 0 \qquad (15.7a)$$

$$\sum_{i=1}^{n} z_i \cos 2\phi_i = 0 \qquad \sum_{i=1}^{n} z_i \sin 2\phi_i = 0 \qquad (15.7b)$$

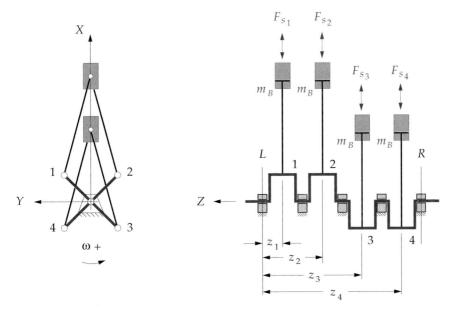

FIGURE 15-9

Moment arms of the shaking moment

These will guarantee no shaking moments through the second harmonic. We can extend this to higher harmonics as we did for the shaking force.

$$\sum_{i=1}^{n} z_i \cos 4\phi_i = 0 \qquad \sum_{i=1}^{n} z_i \sin 4\phi_i = 0 \qquad (15.7c)$$

$$\sum_{i=1}^{n} z_i \cos 6\phi_i = 0 \qquad \sum_{i=1}^{n} z_i \sin 6\phi_i = 0 \qquad (15.7d)$$

Note that both the sine and cosine summations of any multiple of the phase angles must be zero for that harmonic of the shaking moment to be zero. The calculation for our example of a four-cylinder engine with phase angles of $\phi_1 = 0$, $\phi_2 = 90$, $\phi_3 = 180$, $\phi_4 = 270°$, and an assumed cylinder spacing of one length unit ($z_1 = 1$) in Table 15-2 shows that the shaking moments are not zero for any of these harmonics. So, our choice of phase angles which is a good one for shaking forces and torques fails the test for zero shaking moments. The coefficients of the fourth and sixth harmonic terms in the moment equations are miniscule, so they can be ignored. The secondary (second harmonic) term is less critical than the primary as it is multiplied by r/l which is generally less than $1/3$. An unbalanced secondary harmonic of shaking moment is undesirable but can be tolerated, especially if the engine is of small displacement (less than about $1/2$ liter per cylinder). The primary component is of most concern, because of its magnitude. If we wish to use this crankshaft configuration we will need to apply a balancing technique to the engine as described in a later section to at least eliminate the primary moment. A large

TABLE 15-2   Moment Balance State of a 0, 90, 180, 270° Crankshaft Four-Cylinder
                 Engine with $z_1 = 1$, $z_2 = 2$, $z_3 = 3$, $z_4 = 4$

Primary moments:	$\displaystyle\sum_{i=1}^{n} z_i \sin\phi_i = -2$	$\displaystyle\sum_{i=1}^{n} z_i \cos\phi_i = -2$
Secondary moments:	$\displaystyle\sum_{i=1}^{n} z_i \sin 2\phi_i = 0$	$\displaystyle\sum_{i=1}^{n} z_i \cos 2\phi_i = -2$
Fourth harmonic moments:	$\displaystyle\sum_{i=1}^{n} z_i \sin 4\phi_i = 0$	$\displaystyle\sum_{i=1}^{n} z_i \cos 4\phi_i = 6$
Sixth harmonic moments:	$\displaystyle\sum_{i=1}^{n} z_i \sin 6\phi_i = 0$	$\displaystyle\sum_{i=1}^{n} z_i \cos 6\phi_i = -2$

shaking moment is undesirable as it will cause the engine to **pitch** forward and back (like a bucking bronco) as the moment oscillates from positive to negative in the plane of the cylinders. *Do not confuse this shaking moment with the shaking torque* which acts to **roll** the engine back and forth about the axis of the crankshaft.

Figure 15-10 shows the primary and secondary components of the shaking moment for this example engine for two revolutions of the crank. Each is a pure harmonic of zero average value. The total moment is the sum of these two components. Use example engine #6 in program ENGINE to view these functions. They are choices 14 and 16 on the Plot or Print menu. Follow the instructions in Section 15.3 above to generate the data. See also Chapter 16 for help with the program.

15.6   EVEN FIRING

The inertial forces, torques, and moments are only one set of criteria which need to be considered in the design of multicylinder engines. Gas force and gas torque considerations are equally important. In general, it is desirable to create a firing pattern among the cylinders that is evenly spaced in time. If the cylinders fire unevenly, vibrations will be created which may be unacceptable. Smoothness of the power pulses is desired. The power pulses depend on the stroke cycle. If the engine is a two-stroke, there will be one power pulse per revolution in each of its n cylinders. The optimum delta phase angle between the cylinders' crank throws for evenly spaced power pulses will then be:

$$\Delta\phi_{two\ stroke} = \frac{360°}{n} \tag{15.8a}$$

For a four-stroke engine there will be one power pulse in each cylinder every two revolutions. The optimum delta phase angle of the crank throws for evenly spaced power pulses will then be:

$$\Delta\phi_{four\ stroke} = \frac{720°}{n} \tag{15.8b}$$

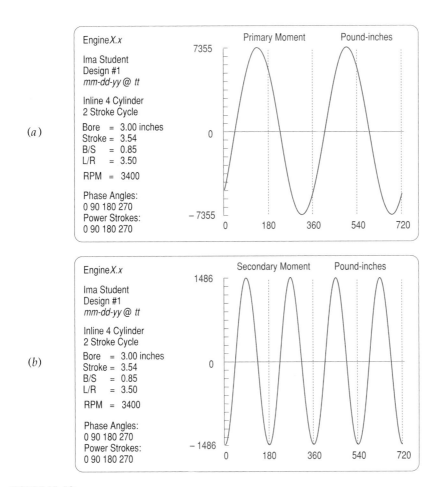

FIGURE 15-10

Primary and secondary moments in the 0 - 90 - 180 - 270° crankshaft four-cylinder engine

Compare equations 15.8a and b to equation 15.1 which defined the optimum delta phase angle for cancellation of inertia forces. A two-stroke engine can have both even firing and inertia balance but a four-stroke engine has a conflict between these two criteria. Thus some design tradeoffs will be necessary to obtain the best compromise between these factors in the four-stroke case.

Two-Stroke Cycle Engine

To determine the firing pattern of an engine design we must return to the crank phase diagram. Figure 15-11 reproduces Figure 15-8 and adds new information to it. It shows the power pulses for a **two-stroke cycle, four-cylinder engine** with the $\phi_i = 0, 90, 180, 270°$ phase angle crank configuration. Note that each cylinder's negative block in Figure 15-11 is shifted to the right by its phase angle with respect to the reference cylinder 1. In this schematic representation, only the negative blocks

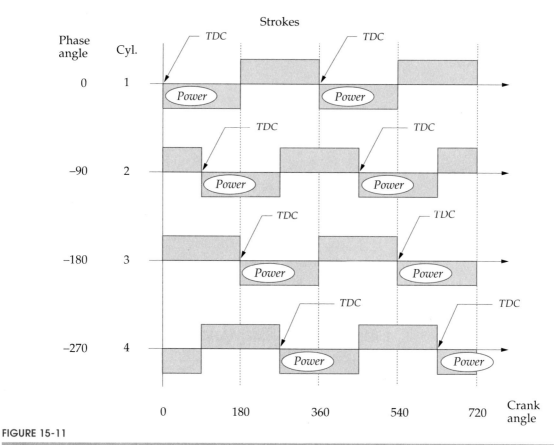

FIGURE 15-11

Two-stroke inline four-cylinder engine crank phase diagram with 0 - 90 - 180 - 270° crankshaft phase angles

on the diagram are available for power pulses as they represent the downstroke of the piston. By convention, cylinder one fires first, so its negative block at 0° is labelled *Power.* The other cylinders may be fired in any order, but their power pulses should be as evenly spaced as possible across the interval.

The available power pulse spacings are dictated by the crank phase angles. There may be more than one firing order which will give even firing, especially with large numbers of cylinders. In this simple example the firing order 1, 2, 3, 4 will work as it will provide successive power pulses every 90° across the interval. The **power stroke angles** ψ_i are the angles in the cycle at which the cylinders fire. They are defined by the crankshaft phase angles and the choice of firing order in combination and in this example are $\psi_i = 0, 90, 180$, and 270°. In general, ψ_i are not equal to ϕ_i. Their correspondence with the phase angles in this example results from choosing the consecutive firing order 1, 2, 3, 4.

For a **two-stroke engine**, the power stroke angles ψ_i must be *between* 0 *and* 360°. We always want them to be evenly spaced in that interval with a delta power stroke angle defined by equation 15.8c. For our four-cylinder, two-stroke

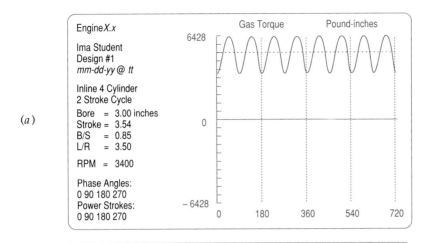

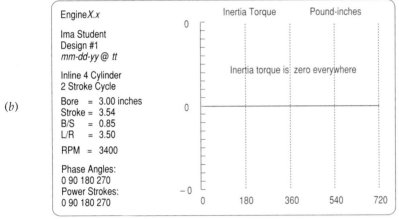

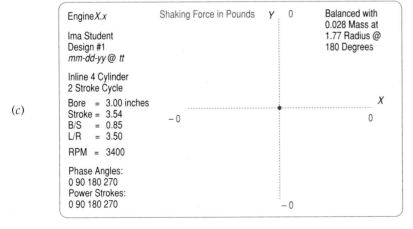

FIGURE 15-12

Torque and shaking force in the two-stroke four-cylinder inline engine

engine, ideal power stroke angles are then $\psi_i = 0, 90, 180, 270°$, which we have achieved in this example.

We define the **delta power stroke angle** differently for each stroke cycle. For the two-stroke engine:

$$\Delta\psi_{two\ stroke} = \frac{360°}{n} \qquad (15.8c)$$

For the four-stroke engine:

$$\Delta\psi_{four\ stroke} = \frac{720°}{n} \qquad (15.8d)$$

The gas torque for a one cylinder engine was defined in equation 14.8b. The combined gas torque for all cylinders must sum the contributions of n cylinders, each phase-shifted by its power stroke angle ψ_i:

$$\mathbf{T}_{g21} \cong F_g\, r \sum_{i=1}^{n} \left\{ \sin(\omega t - \psi_i)\left[1 + \frac{r}{l}\cos(\omega t - \psi_i)\right] \right\} \hat{\mathbf{k}} \qquad (15.9)$$

Figure 15-12 shows the gas torque, inertia torque and shaking force for this two-stroke four-cylinder engine plotted from program ENGINE. The shaking moment components are shown in Figure 15-10. Except for the unbalanced shaking moments, this design is otherwise acceptable. The inertia force and inertia torque are both zero which is ideal. The gas torque consists of uniformly shaped and spaced pulses across the interval, four per revolution. Note that program ENGINE plots two full revolutions to accommodate the four-stroke case thus eight power pulses are seen. Read the diskfile FIG15-12 into the program to exercise this example.

Four-Stroke Cycle Engine

Figure 15-13 shows a crank phase diagram for the *same crankshaft design* as in Figure 15-11 except that it is designed as a *four-stroke cycle engine*. There is now only one power stroke every 720° for each cylinder. The second negative block for each cylinder must be used for the intake stroke. Cylinder 1 is again fired first. An evenly spaced pattern of power pulses among the other cylinders is again desired, but is now not possible with this crankshaft. Whether the firing order is 1, 3, 4, 2 or 1, 2, 4, 3 or 1, 4, 2, 3, or any other chosen, there will be both gaps and overlaps in the power pulses. The first firing order listed, 1, 3, 4, 2, has been chosen for this example. This results in the set of power stroke angles $\psi_i = 0, 180, 270, 450°$. These **power stroke angles** define the points in the **720° cycle** where each cylinder fires. Thus for a four-stroke engine, the power stroke angles ψ_i must be between 0 and 720°. We would like them to be evenly spaced in that interval with a delta angle defined by equation 15.8d. For our four-cylinder, four-stroke engine, the ideal power stroke angles would then be $\psi_i = 0, 180, 360, 540°$. We clearly have not achieved them in this example. Figure 15-14 shows the resulting gas torque. Read the diskfile FIG15-14 into program ENGINE to exercise this example.

The uneven firing in Figure 15-14 is obvious. This uneven gas torque will be perceived by the operator of any vehicle containing this engine as rough running

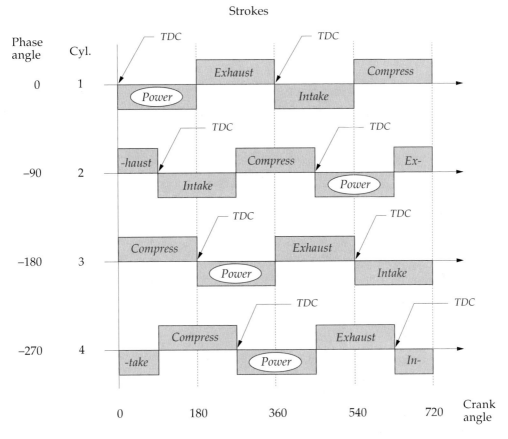

FIGURE 15-13

Four-stroke inline four-cylinder engine crank phase diagram with 0 - 90 - 180 - 270° crankshaft phase angles

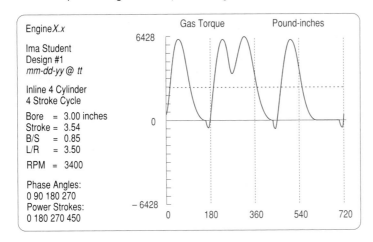

FIGURE 15-14

Uneven firing four-stroke, four-cylinder inline engine with 0 - 90 - 180 - 270° crankshaft

and vibration, especially at idle speed. At higher engine speeds the flywheel will tend to mask this roughness, but flywheels are ineffective at low speeds. It is this fact that causes most engine designers to *favor even firing over elimination of inertia effects* in their selection of crankshaft phase angles. The inertia force, torque, and moment are all functions of engine speed squared. But, as engine speed increases the magnitude of these factors, the same speed is also increasing the flywheel's ability to mask their effects. Not so with gas-torque roughness due to uneven firing. It is bad at all speeds and the flywheel won't hide it at low speed.

We therefore must reject this crankshaft design for our four-stroke, four-cylinder engine. Equation 15.8b indicates that we need a delta phase angle $\Delta\phi_i = 180°$ in our crankshaft to obtain even firing. We need four crank throws and all crank phase angles must be less than 360°. So, we must repeat some angles if we use a delta phase angle of 180°. One possibility is $\phi_i = 0, 180, 0, 180°$ for the four crank throws.[6] The crank phase diagram for this design is shown in Figure 15-15. The power strokes can now be evenly spaced over 720°. A firing order of 1, 4, 3, 2 has been chosen

---

[6]   Note that 0, 180, 360, 540, modulo 360 is the same as 0, 180, 0, 180.

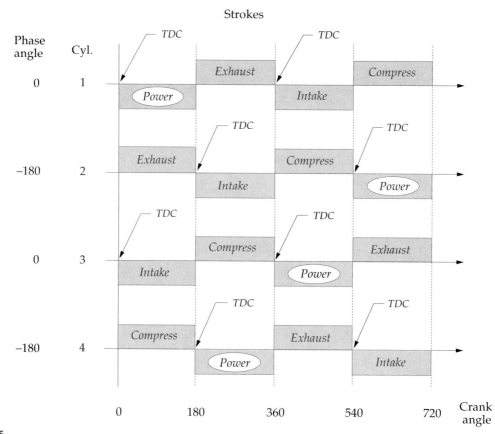

FIGURE 15-15

Even firing four-stroke, four-cylinder engine crank phase diagram with 0 - 180 - 0 - 180° crankshaft phase angles

which gives the desired sequence of power stroke angles, $\psi_i = 0, 180, 360, 540°$. (Note that a firing order of 1, 2, 3, 4 would also work with this engine.)[7]

The inertial balance condition of this design must now be checked with equations 15.3, 15.5 and 15.7. These show that the primary inertia force is zero, but the primary moment, secondary force, secondary moment, and inertia torque are all nonzero as shown in Table 15-3. So, this even-firing design has compromised the very good state of inertia balance of the previous design in order to achieve even firing. The inertia torque variations can be masked by a flywheel. The secondary forces and moments are relatively small in a small engine and can be tolerated. The nonzero primary moment is a problem which needs to be addressed. To see the results of this engine configuration, run Program ENGINE and select example #4 from the **Input** menu (main menu #1). **Calculate** (main menu #3) for any engine speed, **Balance** (main menu #6) the crank in any fashion you choose, and **Assemble** (main menu #7) the engine from the main menu, accepting all default values. Then **Plot** (main menu #5) the results. See Chapter 16 for more detailed instructions on the use of program ENGINE.

---

[7] Note the pattern of acceptable firing orders (FO). Write two revolutions worth of any acceptable FO, as in 1, 4, 3, 2, 1, 4, 3, 2. Any set of four successive numbers in this sequence, *either forward or backward*, is an acceptable FO. If we require the first to be cylinder 1 then the only other possibility here is the backward set 1, 2, 3, 4.

TABLE 15-3   Force and Moment Balance State of a 0, 180, 0, 180° Crankshaft Four- Cylinder Inline Engine with $z_1 = 1$, $z_2 = 2$, $z_3 = 3$, $z_4 = 4$

Primary forces:	$\sum_{i=1}^{n} \sin\phi_i = 0$	$\sum_{i=1}^{n} \cos\phi_i = 0$
Secondary forces:	$\sum_{i=1}^{n} \sin 2\phi_i = 0$	$\sum_{i=1}^{n} \cos 2\phi_i = 4$
Fourth harmonic forces:	$\sum_{i=1}^{n} \sin 4\phi_i = 0$	$\sum_{i=1}^{n} \cos 4\phi_i = 4$
Sixth harmonic forces:	$\sum_{i=1}^{n} \sin 6\phi_i = 0$	$\sum_{i=1}^{n} \cos 6\phi_i = 4$
Primary moments:	$\sum_{i=1}^{n} z_i \sin\phi_i = 0$	$\sum_{i=1}^{n} z_i \cos\phi_i = -2$
Secondary moments:	$\sum_{i=1}^{n} z_i \sin 2\phi_i = 0$	$\sum_{i=1}^{n} z_i \cos 2\phi_i = 6$
Fourth harmonic moments:	$\sum_{i=1}^{n} z_i \sin 4\phi_i = 0$	$\sum_{i=1}^{n} z_i \cos 4\phi_i = 6$
Sixth harmonic moments:	$\sum_{i=1}^{n} z_i \sin 6\phi_i = 0$	$\sum_{i=1}^{n} z_i \cos 6\phi_i = 6$

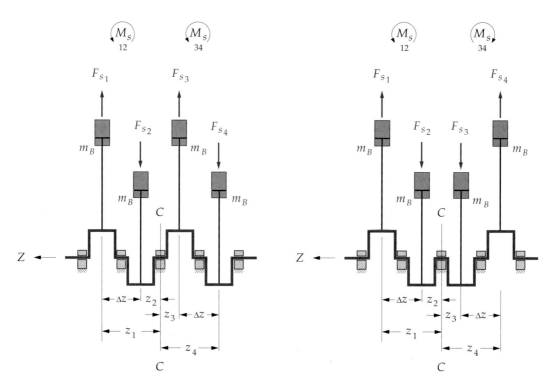

(a) Non-symmetric 0, 180, 0, 180° crankshaft   (b) Symmetric 0, 180, 180, 0° crankshaft

FIGURE 15-16

Mirror symmetric crankshafts cancel primary moments

We shall soon discuss ways to counter an unbalanced moment with the addition of balance shafts, but there is a more direct approach available in this example. Figure 15-16 shows that the shaking moment is due to the action of the individual cylinders' inertial forces acting at moment arms about some center. If we consider that center to be point C in the middle of the engine, it should be apparent that any primary force-balanced crankshaft design which is mirror symmetric about a transverse plane through point C would also have balanced primary moments as long as all cylinder spacings were uniform and all inertial forces equal. Figure 15-16a shows the 0, 180, 0, 180° crankshaft which is not mirror symmetric. The couple $\mathbf{F}_{s_1}\Delta z$ due to cylinder pairs 1, 2 has the same magnitude and sense as the couple $\mathbf{F}_{s_3}\Delta z$ due to cylinders 3 and 4, so they add. Figure 15-16b shows the 0, 180, 180, 0° crankshaft which is *mirror symmetric*. The couple $\mathbf{F}_{s_1}\Delta z$ due to cylinder pairs 1, 2 has the same magnitude but opposite sense to the couple $\mathbf{F}_{s_3}\Delta z$ due to cylinders 3, 4, so they cancel. We can then achieve both even firing and balanced primary moments by changing the sequence of crank throw phase angles to $\phi_i = 0, 180, 180, 0°$ which is *mirror symmetric*.

The crank phase diagram for this design is shown in Figure 15-17. The power strokes can still be evenly spaced over 720°. A firing order of 1, 3, 4, 2 has

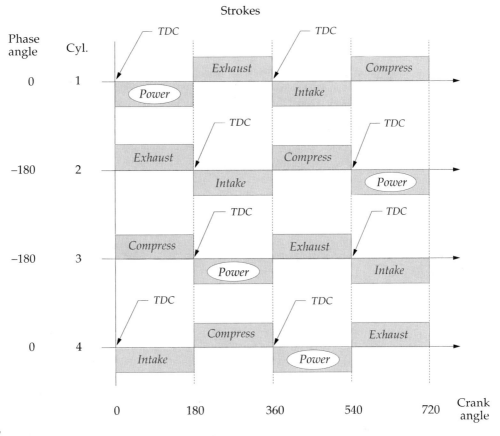

FIGURE 15-17

Even firing four-stroke, four-cylinder engine crank phase diagram with mirror symmetric 0 - 180 - 180 - 0° crankshaft

been chosen which gives the same desired sequence of power stroke angles, $\psi_i = 0$, 180, 360, 540°. (Note that a firing order of 1, 2, 4, 3 would also work with this engine.)[8] Equations 15.3, 15.5, and 15.7 and Table 15-4 show that the primary inertia force and primary moment are both now zero, but the secondary force, secondary moment, and inertia torque are still nonzero.

This $\phi_i = 0$, 180, 180, 0 crankshaft is considered the best design tradeoff and is the one universally used in these four-cylinder inline, four-stroke production engines. Figure 15-2 shows such a four-cylinder design. Inertia balance is sacrificed to gain even firing for the reasons cited above. Figure 15-18 shows the gas torque, inertia torque, and total torque for this design. Figure 15-19 shows the secondary shaking moment, secondary shaking force component, and a polar plot of the total shaking force for this design. Note that Figures 15-19b and c are just different views of the same parameter. The polar plot of the shaking force in Figure 15-19c

---

[8]   In carbureted, inline engines a nonconsecutive firing order (i.e., not 1, 2, 3, 4) is usually preferred so that adjacent cylinders near the ends of the engine do not fire sequentially. This allows the intake manifold more time to recharge locally between intake strokes.

(a)

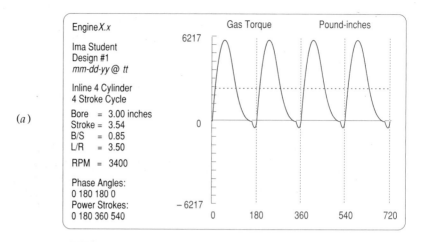

(b)

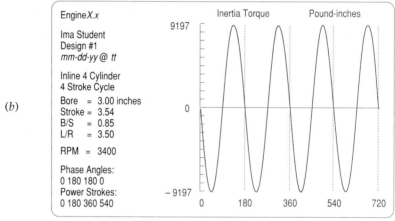

(c)

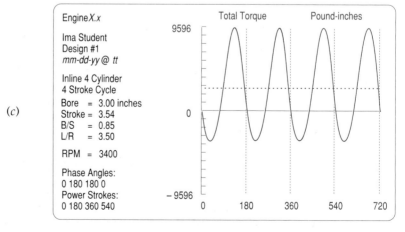

FIGURE 15-18

Torque in the four-stroke, four-cylinder inline engine with 0 - 180 - 180 - 0° crankshaft

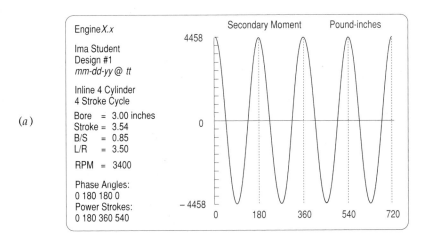

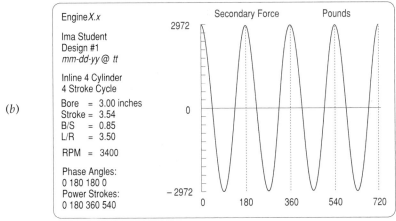

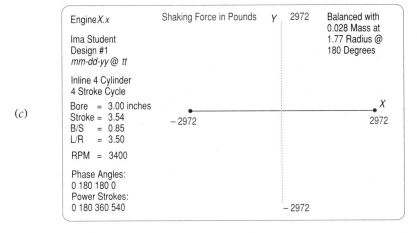

FIGURE 15-19

Shaking forces and moments in the four-stroke, four-cylinder 0 - 180 - 180 - 0° engine

TABLE 15-4   Force and Moment Balance State of a 0, 180, 180, 0° Crankshaft
                Four-Cylinder Inline Engine with $z_1 = 1$, $z_2 = 2$, $z_3 = 3$, $z_4 = 4$

Primary forces:	$\sum_{i=1}^{n} \sin\phi_i = 0$	$\sum_{i=1}^{n} \cos\phi_i = 0$
Secondary forces:	$\sum_{i=1}^{n} \sin2\phi_i = 0$	$\sum_{i=1}^{n} \cos2\phi_i = 4$
Fourth harmonic forces:	$\sum_{i=1}^{n} \sin4\phi_i = 0$	$\sum_{i=1}^{n} \cos4\phi_i = 4$
Sixth harmonic forces:	$\sum_{i=1}^{n} \sin6\phi_i = 0$	$\sum_{i-1}^{n} \cos6\phi_i = 4$
Primary moments:	$\sum_{i=1}^{n} z_i \sin\phi_i = 0$	$\sum_{i=1}^{n} z_i \cos\phi_i = 0$
Secondary moments:	$\sum_{i=1}^{n} z_i \sin2\phi_i = 0$	$\sum_{i=1}^{n} z_i \cos2\phi_i = 6$
Fourth harmonic moments:	$\sum_{i=1}^{n} z_i \sin4\phi_i = 0$	$\sum_{i=1}^{n} z_i \cos4\phi_i = 6$
Sixth harmonic moments:	$\sum_{i=1}^{n} z_i \sin6\phi_i = 0$	$\sum_{i=1}^{n} z_i \cos6\phi_i = 6$

is a view of the shaking force looking at the end of the crankshaft axis with the
piston motion horizontal. The cartesian plot in Figure 15-19b shows the same
force on a time axis. Since the primary component is zero, this total force is due
only to the secondary component. We will soon discuss ways to eliminate these
secondary forces and moments.

To see the results of this engine configuration, run program ENGINE and
select example #3 from the **Input** menu (main menu #1). **Calculate** (main menu #3)
for any engine speed, **Balance** (main menu #6) the crank in any fashion you choose
(including zero balance mass), and **Assemble** (main menu #7) the engine from the
main menu, accepting all default values. Then **Plot** or **Print** results (main menu #5).
See Chapter 16 for more detailed instructions on the use of the program.

15.7   VEE ENGINE CONFIGURATIONS

The same design principles which apply to inline engines also apply to vee and
opposed configurations. Even firing takes precedence over inertia balance and
mirror symmetry of the crankshaft balances primary moments. In general, a vee
engine will have similar inertia balance to that of the inline engines from which it

is constructed. A vee six is essentially two three-cylinder inlines on a common crankshaft, a vee eight is two four-cylinder inlines, etc. The larger number of cylinders allows more power pulses to be spaced out over the cycle for a smoother (and larger average) gas torque. The existence of a **vee angle** between the two inline engines introduces an additional phase shift of the inertial and gas events which is analogous to, but independent of, the phase angle effects. This vee angle is the designer's choice, but there are good and poor choices. The same criteria of even firing and inertia balance apply to its selection.

The **vee angle** 2γ is defined as shown in Figure 15-20. Each bank is offset by its **bank angle** γ referenced to the central X axis of the engine. The crank angle ωt is measured from the X axis. Cylinder one in the right bank is the reference cylinder. Events in each bank will now be phase shifted by its bank angle as well as by the crankshaft phase angles. These two phase shifts will superpose. Taking any one cylinder in either bank as an example, let its instantaneous crank angle be represented by:

$$\theta = (\omega t - \phi_i) \tag{15.10a}$$

Consider first a two-cylinder vee engine with one cylinder in each bank and with both sharing a common crank throw. The shaking force for a single cylinder in the direction of piston motion **u** *hat* with θ measured from the piston axis is:

$$\mathbf{F_s} \cong -m_B r \omega^2 \left(\cos\theta + \frac{r}{l}\cos 2\theta \right)\, \hat{\mathbf{u}} \tag{15.10b}$$

The total shaking force is the vector sum of the contributions from each bank.

$$\mathbf{F_s} = \mathbf{F_{s_L}} + \mathbf{F_{s_R}} \tag{15.10c}$$

We now want to refer the crank angle to the central X axis. The shaking forces for the left (L) and right (R) banks, in the planes of the respective cylinder banks can then be expressed as:

$$\mathbf{F_{s_L}} \cong -m_B r\omega^2 \left[\cos(\theta + \gamma) + \frac{r}{l}\cos 2(\theta + \gamma) \right]\, \hat{\mathbf{l}}$$
$$\tag{15.10d}$$
$$\mathbf{F_{s_R}} \cong -m_B r\omega^2 \left[\cos(\theta - \gamma) + \frac{r}{l}\cos 2(\theta - \gamma) \right]\, \hat{\mathbf{r}}$$

Note that the bank angle γ is added to or subtracted from the crank angle for each cylinder bank to reference it to the central X axis. The forces are still directed along the planes of the cylinder banks. Substitute the identities:

$$\cos(\theta + \gamma) = \cos\theta\cos\gamma - \sin\theta\sin\gamma$$
$$\tag{15.10e}$$
$$\cos(\theta - \gamma) = \cos\theta\cos\gamma + \sin\theta\sin\gamma$$

to get:

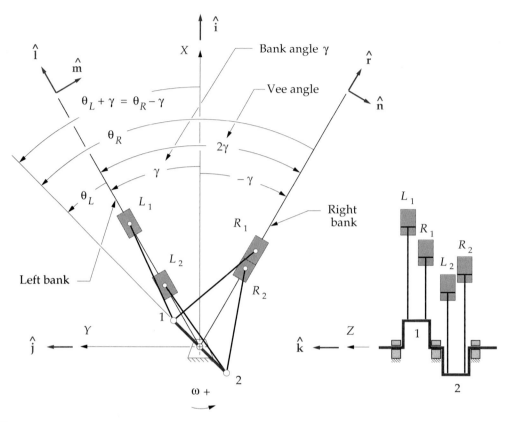

FIGURE 15-20

Vee-engine geometry

$$\mathbf{F}_{s_L} \cong -m_B r \omega^2 \left[\begin{array}{c} \cos\theta\cos\gamma - \sin\theta\sin\gamma \\ +\dfrac{r}{l}\left(\cos 2\theta\cos 2\gamma - \sin 2\theta\sin 2\gamma\right) \end{array} \right] \hat{\mathbf{i}}$$

(15.10f)

$$\mathbf{F}_{s_R} \cong -m_B r \omega^2 \left[\begin{array}{c} \cos\theta\cos\gamma + \sin\theta\sin\gamma \\ +\dfrac{r}{l}\left(\cos 2\theta\cos 2\gamma + \sin 2\theta\sin 2\gamma\right) \end{array} \right] \hat{\mathbf{r}}$$

Now, to account for the possibility of multiple cylinders, phase shifted within each bank, substitute equation 15.10a for θ and replace the sums of angle terms with products from the identities:

$$\cos(\omega t - \phi_i) = \left(\cos\omega t\cos\phi_i + \sin\omega t\sin\phi_i\right)$$

(15.10g)

$$\sin(\omega t - \phi_i) = \left(\sin\omega t\cos\phi_i - \cos\omega t\sin\phi_i\right)$$

After much manipulation, the expressions for the contributions from left and right banks reduce to:

for the left bank:

$$\mathbf{F}_{s_L} \cong -m_B r \omega^2 \begin{bmatrix} \left(\cos \omega t \cos \gamma - \sin \omega t \sin \gamma\right) \sum_{i=1}^{\frac{n}{2}} \cos \phi_i \\[2em] +\left(\cos \omega t \sin \gamma + \sin \omega t \cos \gamma\right) \sum_{i=1}^{\frac{n}{2}} \sin \phi_i \\[2em] +\frac{r}{l}\left(\cos 2\omega t \cos 2\gamma - \sin 2\omega t \sin 2\gamma\right) \sum_{i=1}^{\frac{n}{2}} \cos 2\phi_i \\[2em] +\frac{r}{l}\left(\cos 2\omega t \sin 2\gamma + \sin 2\omega t \cos 2\gamma\right) \sum_{i=1}^{\frac{n}{2}} \sin 2\phi_i \end{bmatrix} \hat{\mathbf{i}} \qquad (15.10h)$$

and for the right bank:

$$\mathbf{F}_{s_R} \cong -m_B r \omega^2 \begin{bmatrix} \left(\cos \omega t \cos \gamma + \sin \omega t \sin \gamma\right) \sum_{i=\frac{n}{2}+1}^{n} \cos \phi_i \\[2em] -\left(\cos \omega t \sin \gamma - \sin \omega t \cos \gamma\right) \sum_{i=\frac{n}{2}+1}^{n} \sin \phi_i \\[2em] +\frac{r}{l}\left(\cos 2\omega t \cos 2\gamma + \sin 2\omega t \sin 2\gamma\right) \sum_{i=\frac{n}{2}+1}^{n} \cos 2\phi_i \\[2em] -\frac{r}{l}\left(\cos 2\omega t \sin 2\gamma - \sin 2\omega t \cos 2\gamma\right) \sum_{i=\frac{n}{2}+1}^{n} \sin 2\phi_i \end{bmatrix} \hat{\mathbf{r}} \qquad (15.10k)$$

Note that the criteria for zero shaking force through the second harmonic for each bank are the same as those found for the inline engine in equation 15.3. We can now resolve the shaking forces for each bank into components along and normal to the central X axis of the vee engine:

$$F_{s_x} = \left(F_{s_L} + F_{s_R}\right) \cos \gamma \, \hat{\mathbf{i}}$$
$$F_{s_y} = \left(F_{s_L} - F_{s_R}\right) \sin \gamma \, \hat{\mathbf{j}} \qquad (15.10m)$$
$$\mathbf{F}_s = F_{s_x} \, \hat{\mathbf{i}} + F_{s_y} \, \hat{\mathbf{j}}$$

The **shaking moment** equations are easily formed from the shaking force equations by multiplying each term by the moment arm summations as was done in equations 15.6. The moments exist within each bank and their vectors will be orthogonal to the respective cylinder planes. For the left bank we define a moment unit vector **m** *hat* perpendicular to the **l** *hat*-Z plane in Figure 15-20. For the right bank we define a moment unit vector **n** *hat* perpendicular to the **r** *hat*-Z plane in Figure 15-20.

$$
\mathbf{M}_{s_L} \cong -m_B r \omega^2
\begin{vmatrix}
(\cos \omega t \cos \gamma - \sin \omega t \sin \gamma) \sum_{i=1}^{\frac{n}{2}} z_i \cos \phi_i \\[2ex]
+(\cos \omega t \sin \gamma + \sin \omega t \cos \gamma) \sum_{i=1}^{\frac{n}{2}} z_i \sin \phi_i \\[2ex]
+\frac{r}{l}(\cos 2\omega t \cos 2\gamma - \sin 2\omega t \sin 2\gamma) \sum_{i=1}^{\frac{n}{2}} z_i \cos 2\phi_i \\[2ex]
+\frac{r}{l}(\cos 2\omega t \sin 2\gamma + \sin 2\omega t \cos 2\gamma) \sum_{i=1}^{\frac{n}{2}} z_i \sin 2\phi_i
\end{vmatrix}
\hat{\mathbf{m}}
\qquad (15.11a)
$$

$$
\mathbf{M}_{s_R} \cong -m_B r \omega^2
\begin{bmatrix}
(\cos \omega t \cos \gamma + \sin \omega t \sin \gamma) \sum_{i=\frac{n}{2}+1}^{n} z_i \cos \phi_i \\[2ex]
-(\cos \omega t \sin \gamma - \sin \omega t \cos \gamma) \sum_{i=\frac{n}{2}+1}^{n} z_i \sin \phi_t \\[2ex]
+\frac{r}{l}(\cos 2\omega t \cos 2\gamma + \sin 2\omega t \sin 2\gamma) \sum_{i=\frac{n}{2}+1}^{n} z_i \cos 2\phi_i \\[2ex]
-\frac{r}{l}(\cos 2\omega t \sin 2\gamma - \sin 2\omega t \cos 2\gamma) \sum_{i=\frac{n}{2}+1}^{n} z_i \sin 2\phi_i
\end{bmatrix}
\hat{\mathbf{n}}
\qquad (15.11b)
$$

Note that the criteria for **zero shaking moment** through the second harmonic for each bank are the same as those found for the inline engine in equation 15.7. We can now resolve the shaking moments for each bank into components along and normal to the central X axis of the vee engine:

$$
M_{s_x} = \left(M_{s_L} - M_{s_R} \right) \sin \gamma \,\hat{\mathbf{i}}
$$

$$
M_{s_y} = \left(-M_{s_L} - M_{s_R} \right) \cos \gamma \,\hat{\mathbf{j}} \qquad (15.11c)
$$

$$
\mathbf{M}_s = M_{s_x} \,\hat{\mathbf{i}} + M_{s_y} \,\hat{\mathbf{j}}
$$

The **inertia torques** from the left and right banks of a vee engine are:

$$\mathbf{T}_{i21_L} \cong \frac{1}{2} m_B r^2 \omega^2 \left[\begin{array}{c} \dfrac{r}{2l}\left(\sin(\omega t + \gamma)\displaystyle\sum_{i=1}^{\frac{n}{2}}\cos\phi_i - \cos(\omega t + \gamma)\displaystyle\sum_{i=1}^{\frac{n}{2}}\sin\phi_i \right) \\[2em] -\left(\sin 2(\omega t + \gamma)\displaystyle\sum_{i=1}^{\frac{n}{2}}\cos 2\phi_i - \cos 2(\omega t + \gamma)\displaystyle\sum_{i=1}^{\frac{n}{2}}\sin 2\phi_i \right) \\[2em] -\dfrac{3r}{2l}\left(\sin 3(\omega t + \gamma)\displaystyle\sum_{i=1}^{\frac{n}{2}}\cos 3\phi_i - \cos 3(\omega t + \gamma)\displaystyle\sum_{i=1}^{\frac{n}{2}}\sin 3\phi_i \right) \end{array} \right] \hat{\mathbf{k}} \qquad (15.12a)$$

$$\mathbf{T}_{i21_R} \cong \frac{1}{2} m_B r^2 \omega^2 \left[\begin{array}{c} \dfrac{r}{2l}\left(\sin(\omega t - \gamma)\displaystyle\sum_{i=\frac{n}{2}+1}^{n}\cos\phi_i - \cos(\omega t - \gamma)\displaystyle\sum_{i=\frac{n}{2}+1}^{n}\sin\phi_i \right) \\[2em] -\left(\sin 2(\omega t - \gamma)\displaystyle\sum_{i=\frac{n}{2}+1}^{n}\cos 2\phi_i - \cos 2(\omega t - \gamma)\displaystyle\sum_{i=\frac{n}{2}+1}^{n}\sin 2\phi_i \right) \\[2em] -\dfrac{3r}{2l}\left(\sin 3(\omega t - \gamma)\displaystyle\sum_{i=\frac{n}{2}+1}^{n}\cos 3\phi_i - \cos 3(\omega t - \gamma)\displaystyle\sum_{i=\frac{n}{2}+1}^{n}\sin 3\phi_i \right) \end{array} \right] \hat{\mathbf{k}} \qquad (15.12b)$$

Add the contributions from each bank for the total. For **zero inertia torque** through the third harmonic it is sufficient (but not necessary) that:

$$\sum_{i=1}^{n}\sin\phi_i = 0 \qquad\qquad \sum_{i=1}^{n}\cos\phi_i = 0$$

$$\sum_{i=1}^{n}\sin 2\phi_i = 0 \qquad\qquad \sum_{i=1}^{n}\cos 2\phi_i = 0 \qquad (15.12c)$$

$$\sum_{i=1}^{n}\sin 3\phi_i = 0 \qquad\qquad \sum_{i=1}^{n}\cos 3\phi_i = 0$$

which are the same criteria as for the inline engine (Eq. 15.5). The **gas torque** is:

$$\mathbf{T}_{g21} \cong F_g r \sum_{i=1}^{n}\left(\sin[\omega t - (\psi_i + \gamma_k)]\left\{ 1 + \frac{r}{l}\cos[\omega t - (\psi_i + \gamma_k)] \right\} \right) \hat{\mathbf{k}} \qquad (15.13)$$

where the left bank has a *bank angle*, $\gamma_k = +\gamma$ and the right bank a *bank angle*, $\gamma_k = -\gamma$.

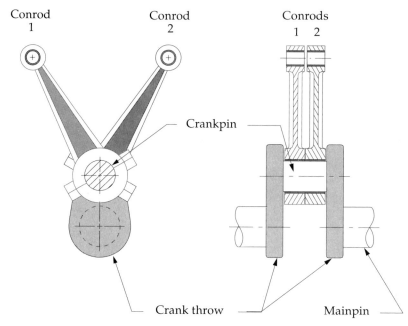

FIGURE 15-21

Two connecting rods on a common crank throw

　　　　It is possible to design a vee engine which has as many crank throws as cylinders, but this is seldom done for several reasons. The principal advantage of a vee engine over an inline of the same number of cylinders is its more compact size and greater stiffness. It can be about half the length of a comparable inline engine (at the expense of greater width), provided that the crankshaft is designed to accommodate two conrods per crank throw as shown in Figure 15-21. Cylinders in opposite banks then share a crank throw. One bank of cylinders is shifted along the crankshaft axis by the thickness of a conrod. The shorter, wider cylinder block and the shorter crankshaft are much stiffer in both torsion and bending than are those for an inline engine with the same number of cylinders. Figure 15-22 shows computer simulations of several bending and one torsional mode of vibration for a four throw crankshaft. The deflections are greatly exaggerated. The necessarily contorted shape of a crankshaft makes it difficult to control these deflections by design. If excessive in magnitude, they can lead to structural failure.

　　　　As an example we will now design the crankshaft for a four-stroke cycle, vee eight engine. We could put two $\phi_i = 0, 180, 180, 0°$ four-cylinder engines together on one such crankshaft and have the same balance conditions as the four-stroke, four-cylinder engine designed in the previous section (primaries balanced, secondaries unbalanced). However, the motivation for choosing that crankshaft for the four-cylinder was the need to space the four available power pulses evenly across the cycle. Equation 15.8b then dictated a 180° delta phase angle for that engine. Now we have eight cylinders available and equation 15.8b defines a delta phase angle of 90° for optimum power pulse spacing. This means we could use the

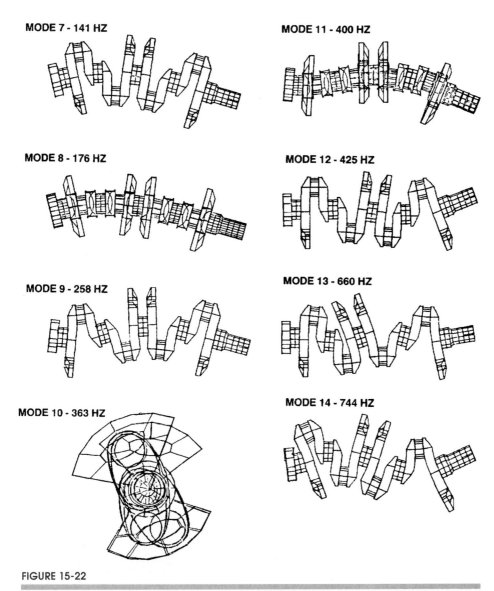

MODE 7 - 141 HZ

MODE 8 - 176 HZ

MODE 9 - 258 HZ

MODE 10 - 363 HZ

MODE 11 - 400 HZ

MODE 12 - 425 HZ

MODE 13 - 660 HZ

MODE 14 - 744 HZ

FIGURE 15-22

Bending and torsional modes of vibration in a four-throw crankshaft
Courtesy of Chevrolet Division, General Motors Co., Detroit, Michigan

$\phi_i = 0, 90, 180, 270°$ crankshaft designed for the two-stroke four-cylinder engine shown in Figure 15-11, and take advantage of its better inertia balance condition as well as achieve even firing in the four-stroke eight-cylinder vee engine.

The $\phi_i = 0, 90, 180, 270°$ four-cylinder crankshaft has all inertia factors equal to zero except for the primary and secondary moments. We learned that arranging the crank throws with mirror symmetry about the mid-plane would balance the primary moment. Some thought and/or sketches will reveal that it is not possible to obtain this mirror symmetry with any four throw, 90° delta phase angle crank-

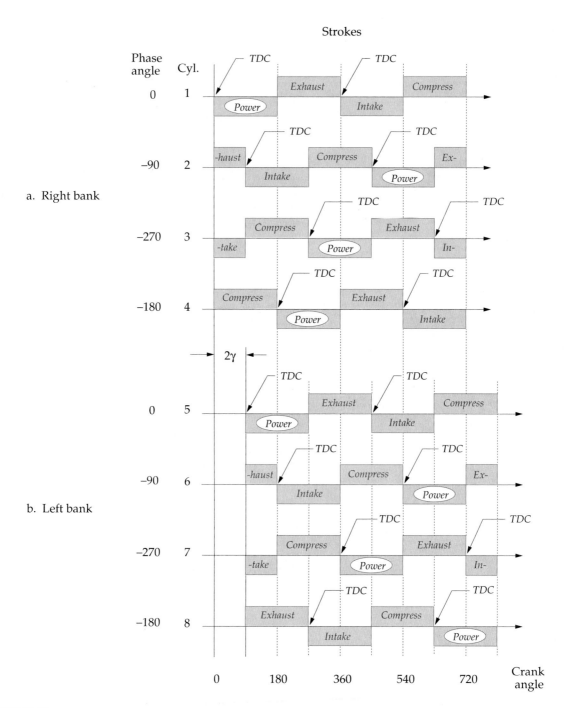

FIGURE 15-23

Four-stroke vee eight crank phase diagram with 0 - 90 - 270 - 180° crankshaft phase angles

shaft arrangement. However, just as rearranging the order of the crank throws from $\phi_i = 0, 180, 0, 180°$ to $\phi_i = 0, 180, 180, 0°$ had an effect on the shaking moments, rearranging this crankshaft's throw-order will as well. A crankshaft of $\phi_i = 0, 90, 270, 180°$ has all inertia factors equal to zero except for the primary moment. The secondary moment is now gone. This is an advantage worth taking. We will use this crankshaft for the vee eight and deal with the primary moment later.

Figure 15-23a shows the crank phase diagram for the right bank of a vee eight engine with a $\phi_i = 0, 90, 270, 180°$ crankshaft. Figure 15-23b shows the crank phase diagram for the second (left) bank which is identical to that of the right bank (as it must be since they share crank throws), but it is *shifted to the right by the vee angle* 2γ. Note that in Figure 15-20, the two pistons are driven by conrods on a common crank throw with positive ω, and the piston in the right bank will reach TDC before the one in the left bank. Thus as we show it, the left bank's piston motions lag those of the right bank. Lagging events occur later in time, so we must shift the second (left) bank rightward by the vee angle on the crank phase diagram.

We would like to shift the second bank of cylinders such that its power pulses are evenly spaced among those of the first bank. A little thought (and reference to equation 15.8b) should reveal that, in this example, each four-cylinder bank has potentially $720/4 = 180°$ between power pulses. Our chosen crank throws are spaced at $90°$ increments. A $90°$ bank angle will be optimum in this case as the phase angles and bank angles will add to create an effective spacing of $180°$. Every vee engine design will have one or more optimum vee angles that will give approximately even firing with any particular set of crank phase angles.

Several firing orders are possible with this many cylinders. Vee engines are often arranged to fire cylinders in opposite banks successively to balance the fluid flow demands in the intake manifold. Our cylinders are numbered from front to back, first down the right bank and then down the left bank. The firing order shown in Figure 15-23b is 1, 5, 4, 3, 7, 2, 6, 8 and results in power stroke angles $\psi_i = 0, 90, 180, 270, 360, 450, 540, 630°$. This will clearly give even firing with a power pulse every $90°$.

Figure 15-24 shows the total torque for this engine design. The total torque in this case is equal to the gas torque as the inertia torque is zero. Table 15-5 and Figure 15-25 show the only significant unbalanced inertial component in this engine to be the primary moment, which is quite large. The fourth harmonic terms have negligible coefficients in the Fourier series, and we have truncated them from the equations. We will address the balancing of this primary moment in a later section.

Any vee cylinder configuration may have one or more desirable vee angles which will give both even firing and acceptable inertia balance. However, vee engines of fewer than twelve cylinders will not be completely balanced by means of their crankshaft configuration. The desirable vee angles will typically be an integer multiple (including one) or submultiple of the optimum delta phase angle as defined in equations 15.8 for that engine. Ninety degrees is the optimum vee angle (2γ) for an eight-cylinder vee engine. To see the results for this vee eight engine configuration, run program ENGINE and select example #8 from the **Input** menu. See Chapter 16 for detailed instructions on the use of the program.

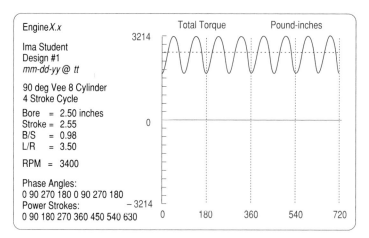

FIGURE 15-24

Total torque in the 90° V-8 engine with 0 - 90 - 270 -180° crankshaft phase angles

TABLE 15-5   **Force and moment balance state of a 0, 90, 270, 180° crankshaft vee eight engine with $z_1 = 1$, $z_2 = 2$, $z_3 = 3$, $z_4 = 4$**

Primary forces:	$\sum_{i=1}^{n} \sin\phi_i = 0$	$\sum_{i=1}^{n} \cos\phi_i = 0$
Secondary forces:	$\sum_{i=1}^{n} \sin 2\phi_i = 0$	$\sum_{i=1}^{n} \cos 2\phi_i = 0$
Fourth harmonic forces:	$\sum_{i=1}^{n} \sin 4\phi_i = 0$	$\sum_{i=1}^{n} \cos 4\phi_i = 8$
Sixth harmonic forces:	$\sum_{i=1}^{n} \sin 6\phi_i = 0$	$\sum_{i=1}^{n} \cos 6\phi_i = 0$
Primary moments:	$\sum_{i=1}^{n} z_i \sin\phi_i = -4$	$\sum_{i=1}^{n} z_i \cos\phi_i = -2$
Secondary moments:	$\sum_{i=1}^{n} z_i \sin 2\phi_i = 0$	$\sum_{i=1}^{n} z_i \cos 2\phi_i = 0$
Fourth harmonic moments:	$\sum_{i=1}^{n} z_i \sin 4\phi_i = 0$	$\sum_{i=1}^{n} z_i \cos 4\phi_i = 28$
Sixth harmonic moments:	$\sum_{i=1}^{n} z_i \sin 6\phi_i = 0$	$\sum_{i=1}^{n} z_i \cos 6\phi_i = 0$

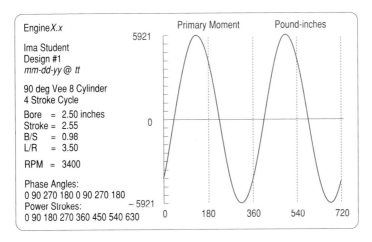

Engine X.x

Ima Student
Design #1
mm-dd-yy @ tt

90 deg Vee 8 Cylinder
4 Stroke Cycle

Bore = 2.50 inches
Stroke = 2.55
B/S = 0.98
L/R = 3.50

RPM = 3400

Phase Angles:
0 90 270 180 0 90 270 180
Power Strokes:
0 90 180 270 360 450 540 630

Primary Moment   Pound-inches

FIGURE 15-25

Unbalanced primary moment in the 90° V-8 engine with 0 - 90 - 270 -180° crankshaft

15.8   OPPOSED ENGINE CONFIGURATIONS

An opposed engine is essentially a vee engine with a 180° vee angle. The advantage, particularly with a small number of cylinders such as two or four, is the relatively good balance condition possible. A four-stroke opposed twin[9] with 0, 180° crank has even firing plus primary force balance, though the primary moment and all higher harmonics of force and moment are nonzero. A four-stroke, opposed four-cylinder engine (flat four) with a 0, 180, 180, 0°, four-throw crank has primary force balance but must fire its cylinders in pairs, so its firing pattern looks like a twin. A four-stroke flat four with a two-throw, 0, 180° crank will have even firing and the same balance condition as the inline four with 0, 180, 180, 0° crank. Program ENGINE will calculate the parameters for opposed as well as vee and inline configurations.

15.9   BALANCING MULTICYLINDER ENGINES

With a sufficient number (n) of cylinders, properly arranged, an engine can be inherently balanced. In a two-stroke engine with its crank throws arranged for even firing, all harmonics of shaking force will be balanced except those whose harmonic number is a multiple of n. In a four-stroke engine with its crank throws arranged for even firing, all harmonics of shaking force will be balanced except those whose harmonic number is a multiple of $n/2$. Corresponding shaking moment components will be balanced if the crankshaft is mirror symmetric about the central transverse plane. A four-stroke inline configuration then requires at least six cylinders to be inherently balanced through the fourth harmonic. We have seen that an inline four with 0, 180, 180, 0° crankshaft has nonzero secondary forces and moments as well as nonzero inertia torque. The inline six with a mirror symmetric crank of $\phi_i = 0, 240, 120, 120, 240, 0°$ will have zero shaking forces and moments through the fourth harmonic, though the inertia torque's third harmonic will still be present. To see the results of this six-cylinder inline engine configuration, run program ENGINE and select example #7 from the **Input** menu.

---

[9]   As in the BMW motorcycle.

A vee twelve is then the smallest vee engine with this state of near perfect balance, as it is two inline sixes on a common crankshaft. We have seen that vee engines take on the balance characteristics of the inline banks from which they are made. Equations 15.10 and 15.11 introduced no new criteria for balance in the vee engine over those already defined in equations 15.3 and 15.5 for shaking force and moment balance in the inline engine. Input the diskfile BMWV12 into program ENGINE to see the results for a vee twelve engine. The common V8 engine with crankshaft phase angles of $\phi_i = 0, 90, 270, 180$ has an unbalanced primary moment as does the inline four from which it is made. See example #8 in program ENGINE.

A vee six engine with 0, 240, 120° crankshaft has unbalanced primary and secondary moments as does the 3-cylinder inline from which it is made. Vee sixes should have a 60° or 120° vee angle for even firing and most do. But, some vee sixes are made with 90° vee angles to allow their assembly on the same tooling as 90° vee eights made by the same manufacturer. These 90° V6 engines are very rough running due to their uneven firing unless the crankshaft is redesigned to shift (or splay) the two conrods on each pin by 30° to correct the firing error. This results in a more complicated and expensive crankshaft but gives an even firing, smooth running engine.

Unbalanced inertia torques can be smoothed with a flywheel as was shown in Section 14.10 for the single-cylinder engine. Note that even an engine having zero inertial torque will require a flywheel to smooth its variations in gas torque. The total torque function should be used to determine the energy variations to be absorbed by a flywheel as it contains both gas torque and inertia torque (if any). The method of Section 12.11 also applies to calculation of the flywheel size needed in an engine, based on its variation in the total torque function. Program ENGINE will compute the areas under the total torque pulses needed for the calculation. See the referenced sections for the proper flywheel design procedure.

Unbalanced shaking forces and shaking moments can be cancelled by the addition of one or more rotating balance shafts within the engine. To cancel the primary components requires two balance shafts rotating at crank speed, one of which may be the crankshaft itself. To cancel the secondary components requires two balance shafts rotating at twice crank speed, gear or chain driven from the crankshaft. Figure 15-26a shows a pair of counter-rotating shafts which are fitted with eccentric masses arranged 180° out of phase. As shown, the unbalanced centrifugal forces from the equal, unbalanced masses will add to give a shaking force component in the vertical direction of twice the unbalanced force from each mass, while their horizontal components will exactly cancel. Pairs of counter-rotating eccentrics can be arranged to provide a harmonically varying force in any one plane. The harmonic frequency will be determined by the rotational speed of the shafts.

If we arrange two pairs of eccentrics, with one pair displaced some distance along the shaft from the other, and also rotated 180° around the shaft from the first, as shown in Figure 15-26b, we will get a harmonically varying couple in one plane. There will be cancellation of the forces in one direction and summation in an orthogonal direction.

Thus to cancel the shaking moment in any plane, we can arrange a pair of shafts, each containing two eccentric masses displaced along those shafts, 180° out of phase, and gear the shafts together to rotate in opposite directions at any

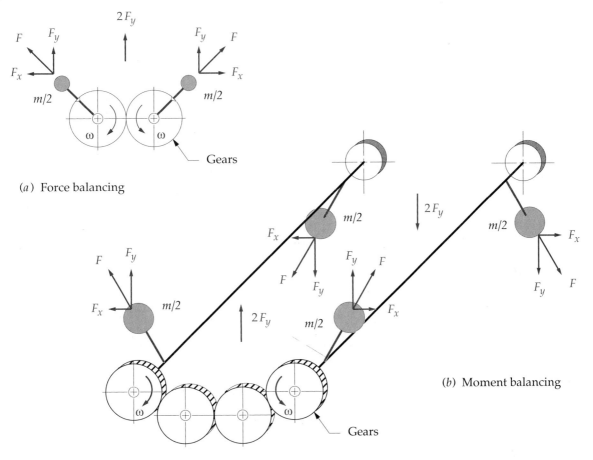

(a) Force balancing

(b) Moment balancing

FIGURE 15-26

Counterrotating eccentric masses can balance forces and moments

multiple of crankshaft speed. To cancel the shaking force as well, it is only necessary to provide sufficient additional unbalanced mass in one of the pairs of eccentric masses to generate a shaking force opposite to that of the engine, over and above that needed to generate the forces of the couple. Figure 15-27 shows such an arrangement used to cancel the secondary forces and moments in some production four-cylinder inline engines.[10]

Note that in an inline engine, the unbalanced forces and moments are all confined to the single plane of the cylinders as they are due entirely to the reciprocating masses assumed concentrated at the wristpin. (We are assuming that all crank throws are exactly balanced rotationally to cancel the effects of mass at the crankpin.) In a vee engine, however, the shaking forces and moments have both x and y components as shown in equations 15.10 and 15.11 and Figure 15-20. Each bank's shaking effects are acting within the plane of that bank's cylinders, and the bank angle γ is used to resolve them into x and y components. Figure 15-28 shows

10   Mitsubishi, also Chrysler and Porsche under license from Mitsubishi.

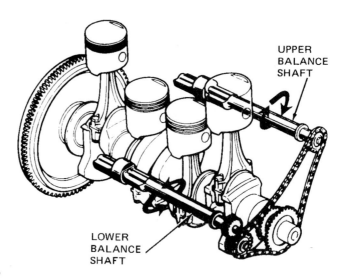

UPPER
BALANCE
SHAFT

LOWER
BALANCE
SHAFT

FIGURE 15-27

Balance shafts used to eliminate the secondary unbalance in the four-cylinder inline engine
Courtesy of the Chrysler Corporation

the two-dimensional shaking force present in a two-cylinder 90° vee engine. The two conrods share a common crank throw at a phase angle of zero. The force due to each piston is confined to the reciprocating plane (bank) of that piston, but the vee angle between the cylinder banks creates the pattern shown when the primary and secondary components of each piston force are added together in their proper phase relationship. To completely balance these forces with rotating counter-balances will require multiple shafts arranged to provide both x and y components at primary and secondary frequencies and in proper phase.

If the required multiple of crankshaft speed for balancing is 1 (for primary forces or moments), then it makes sense to actually use the crankshaft as one of the balance shafts and provide only one new shaft, geared to counter-rotate at crankshaft speed. This will save cost. Such an arrangement is sometimes used in the vee six engine with 60° bank angle, to cancel its primary moment component. The 90° vee eight engine, which has only an unbalanced primary moment, presents a special case. The 90° angle between the banks results in equal horizontal and vertical components of the primary shaking moment which reduces it to a couple of constant magnitude rotating about the crank axis at crankshaft speed as shown in Figure 15-29 which is looking at the end of the crankshaft axis. Thus with this engine, the primary moment can be balanced by merely adding two eccentric counterweights of proper size and orientation to the crankshaft alone. No independent, second balance shaft is needed in the 90° vee eight engine. The counterweights are typically placed near the ends of the crankshaft (180° apart) to obtain the largest moment arm possible and thus reduce their size. Figure 15-3 shows a vee eight crankshaft on which these extra counterweights can be seen at either end. The common 60° vee six engine has a nonconstant magnitude, rotating, unbalanced moment as shown in Figure 15-30. Its secondary component (Figure 15-30b) is of constant magnitude, but the primary component (Figure 15-30a) is an ellipse with its major axis lying along the midline between the cylinder banks.

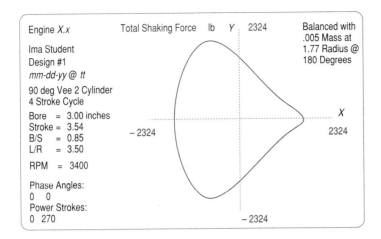

FIGURE 15-28

Shaking force in a 90 degree vee twin engine (looking end-on to the crankshaft axis)

The sum of these two components in Figure 15-30c is asymmetric due to the different frequencies of the two components.

The calculation of the magnitude and location of the eccentric balance masses needed to cancel any shaking forces or moments is a straightforward exercise in **static balancing** (for forces) and **two-plane dynamic balancing** (for moments) as discussed in Sections 13.1 and 13.2. The unbalanced forces and moments are calculated from the appropriate equations in this chapter. Two correction planes must be selected along the length of the balance shafts/crankshaft being designed. The magnitude and angular locations of the balance masses can then be calculated by the methods described in the referenced sections of Chapter 13.

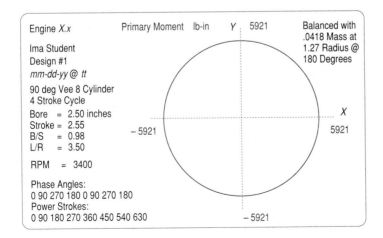

FIGURE 15-29

Primary moment in the 90 degree vee eight engine (looking end-on to the crankshaft axis)

(a)

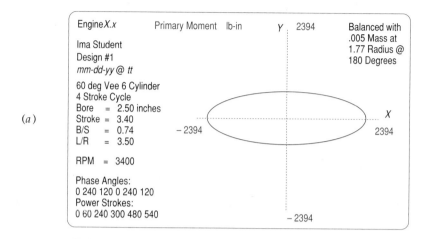

(b)

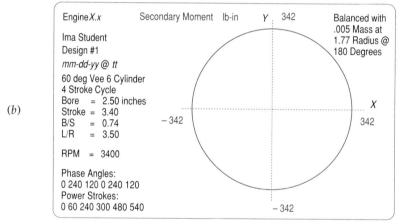

(c)

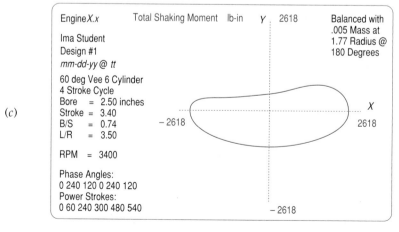

FIGURE 15-30

Shaking moments in the 60 degree vee six engine (looking end-on to the crankshaft axis)

15.10 BIBLIOGRAPHY

Crouse, W. H., *Automotive Engine Design*, McGraw Hill Inc, New York, 1970.

Jennings, G., "A Short History of Wonder Engines," *Cycle Magazine*, May 1979, p. 68ff.

Setright, L. J. K., *Some Unusual Engines*, Mechanical Engineering Publications Ltd., The Inst. of Mech. Engr., London, 1975.

Thomson, W., *Fundamentals of Automotive Balance*, Mechanical Engineering Publications Ltd., London, 1978.

15.11 PROBLEMS

15-1 Draw a crank phase diagram for a three-cylinder inline engine with a 0, 120, 240° crankshaft and determine all possible firing orders for:

a. Four-stroke cycle   b. Two-stroke cycle

Select the best arrangement to give even firing for each stroke cycle.

15-2 Repeat problem 15-1 for an inline four-cylinder engine with a 0, 90, 270, 180° crankshaft.

15-3 Repeat problem 15-1 for a 45° vee, four-cylinder engine with a 0, 90, 270, 180° crankshaft.

15-4 Repeat problem 15-1 for a 45° vee, two-cylinder engine with a 0, 90° crankshaft.

15-5 Repeat problem 15-1 for a 90° vee, two-cylinder engine with a 0, 180° crankshaft.

15-6 Repeat problem 15-1 for a 180° opposed, two-cylinder engine with a 0, 180° crankshaft.

15-7 Repeat problem 15-1 for a 180° opposed, four-cylinder engine with a 0, 180, 180, 0° crankshaft.

15-8 Calculate the shaking force, torque, and moment balance conditions through the second harmonic for the engine design in problem 15-1.

15-9 Calculate the shaking force, torque, and moment balance conditions through the second harmonic for the engine design in problem 15-2.

15-10 Calculate the shaking force, torque, and moment balance conditions through the second harmonic for the engine design in problem 15-3.

15-11 Calculate the shaking force, torque, and moment balance conditions through the second harmonic for the engine design in problem 15-4.

15-12 Calculate the shaking force, torque, and moment balance conditions through the second harmonic for the engine design in problem 15-5.

15-13 Calculate the shaking force, torque, and moment balance conditions through the second harmonic for the engine design in problem 15-6.

15-14 Calculate the shaking force, torque, and moment balance conditions through the second harmonic for the engine design in problem 15-7.

15.15 Derive expressions, in general terms, for the magnitude and angular location with respect to the first crank throw, of the mass-radius products needed on the crankshaft to balance the shaking moment in a 90° vee eight engine with a 0, 90, 270, 180° crankshaft.

15.16 Repeat problem 15.15 for a 90° vee six with 0, 240, 120° crankshaft.

15.17 Repeat problem 15.15 for a 90° vee four with 0, 180° crankshaft.

15.12 PROJECTS

These are loosely structured design problems intended for solution using program ENGINE. All involve the design of a multicylinder engine and differ only in the specific data for the engine. The general problem statement is:

Design a multicylinder engine for a specified displacement and stroke cycle. Optimize the conrod/crank ratio and bore/stroke ratio to minimize shaking forces, shaking torque, and pin forces, also considering package size. Design your link shapes and calculate realistic dynamic parameters (mass, CG location, moment of inertia) for those links using the methods shown in Chapter 11 and Section 12.12. Dynamically model the links as described in this chapter. Balance or overbalance the linkage as needed to achieve these results. Choose crankshaft phase angles to optimize the inertial balance of the engine. Choose a firing order and determine the power stroke angles to optimize even firing. Trade off inertia balance if necessary to achieve even firing. Choose an optimum vee angle if necessary. Overall smoothness of total torque is desired. Design and size a minimum weight flywheel by the method of Chapter 12 to smooth the total torque. Write an engineering report on your design.

P15-1	Two-stroke cycle inline twin with a displacement of 1 liter.
P15-2	Four-stroke cycle inline twin with a displacement of 1 liter.
P15-3	Two-stroke cycle vee twin with a displacement of 1 liter.
P15-4	Four-stroke cycle vee twin with a displacement of 1 liter.
P15-5	Two-stroke cycle opposed twin with a displacement of 1 liter.
P15-6	Four-stroke cycle opposed twin with a displacement of 1 liter.
P15-7	Two-stroke cycle vee four with a displacement of 2 liters.
P15-8	Four-stroke cycle vee four with a displacement of 2 liters.
P15-9	Two-stroke cycle opposed four with a displacement of 2 liters.
P15-10	Four-stroke cycle opposed four with a displacement of 2 liters.
P15-11	Two-stroke inline five cylinder with a displacement of 2.5 liters.
P15-12	Four-stroke inline five cylinder with a displacement of 2.5 liters.
P15-13	Two-stroke cycle vee six with a displacement of 3 liters.
P15-14	Four-stroke cycle vee six with a displacement of 3 liters.
P15-15	Two-stroke cycle opposed six with a displacement of 3 liters.
P15-16	Four-stroke cycle opposed six with a displacement of 3 liters.
P15-17	Two-stroke inline seven cylinder with a displacement of 3.5 liters.
P15-18	Four-stroke inline seven cylinder with a displacement of 3.5 liters.
P15-19	Two-stroke inline eight cylinder with a displacement of 4 liters.
P15-20	Four-stroke inline eight cylinder with a displacement of 4 liters.
P15-21	Two-stroke vee ten cylinder with a displacement of 5 liters.
P15-22	Four-stroke vee ten cylinder with a displacement of 5 liters.

Chapter 16

PROGRAM ENGINE

Try again, fail again, fail better
SAMUEL BECKETT

16.0   INTRODUCTION

The previous two chapters presented the mathematical theory behind the design of slider crank mechanisms for single- and multicylinder internal combustion engines. This chapter will serve as a manual for the use of the computer program ENGINE, supplied with the text. This program performs the calculations derived in the previous two chapters and provides a means to investigate the variation of parameters in the design of these devices. Program ENGINE is similar in many respects to the other programs provided and discussed in previous chapters. Chapter 8 should be read before this one to obtain basic information on the hardware and software requirements. See Section 8.2 for instruction in how to get the program running. (Essentially, type the word ENGINE when the computer is toggled to the disk drive containing it.) The caveats discussed in Section 8.0 about proper uses and possible misuses of the programs apply as well to this one and are repeated here in abbreviated form.

Program ENGINE is designed to be a learning tool to aid in the understanding of the relevant subject matter and *is not intended to be used for commercial purposes in the design of hardware* **and should not be so used**. It is quite possible to obtain inappropriate (but mathematically correct) results to any problem solved with these programs, due to incorrect or inappropriate input of data. In other words, the user is expected to understand the kinematic and dynamic theory underlying the program's structure, and to also understand the mathematics on which the program's algorithms are based. This information on the underlying theory and mathematics is derived and described in the noted chapters of this text. Most equations used in the program are presented or derived in this textbook.

```
Engine  Designer and Analyzer - Version X.x - mm/dd/yy

   1 – INPUT              the engine data

   2 – DISPLAY           the engine data

   3 – CALCULATE        the results (for ONE cylinder)

   4 – PRINT             the results

   5 – PLOT              the results

   6 – BALANCE          the crank (for ONE cylinder)

   7 – ASSEMBLE         the engine  (with all cylinders)

   8 – FLYWHEEL         calculations

   9 – HELP             screens

     <RET> - QUIT PROGRAM

                                  CHOICE ?
```

FIGURE 16-1

Main menu for program ENGINE

Commercial software for use in design or analysis needs to have built-in safeguards against the possibility of the user providing incorrect, inappropriate or ridiculous values for input variables, in order to guard against erroneous results due to user ignorance or inexperience. **The programs with this text are not commercial software and deliberately do not contain such safeguards against improper input data**, on the premise that to do so would "short circuit" the student's learning process. We learn most from our failures. These programs provide a consequence-free environment to explore failure of your designs and in the process come to a more thorough and complete understanding of the subject matter.

The very rapid computation speed of this program allows the student to explore a much larger number and variety of potential solutions to more realistic and comprehensive problems than could be accomplished using only hand calculator solutions of these complicated systems of equations. This is both an advantage and a danger. The advantage is that the student can use the programs like a "flight simulator" to "fly" his or her potential design solutions through their paces with no consequences from a "crash" of the design. If the student diligently attempts to interpret the program's results and relate them to the relevant theory, a more thorough understanding of the kinematics and dynamics can result. On the other hand, there is a great temptation to use these programs with *brute force and ignorance* (BFI) to somewhat randomly try solutions without regard to what the theory and equations are telling you and hope that somehow a usable solution will "pop out." The student who succumbs to this temptation will not obtain much benefit from the exercise and will probably have a poor design result.

```
╭─────────────────────────────────────────────────────────────╮
│                    SELECT INPUT METHOD                        │
│            ─────────────────────────────────────             │
│                                                               │
│         1 – TYPE data from KEYBOARD                           │
│                                                               │
│         2 – READ data from DISKFILE                           │
│                                                               │
│         3 – Create 4  Cyl Inline Engine w/ 0 - 180 - 180 - 0 crank │
│                                                               │
│         4 – Create 4  Cyl Inline Engine w/ 0 - 180 - 0 - 180 crank │
│                                                               │
│         5 – Create 4  Cyl Inline Engine w/ 0 - 90 - 270 - 180 crank │
│                                                               │
│         6 – Create 4  Cyl Inline Engine w/ 0 - 90 - 180 - 270 crank │
│                                                               │
│         7 – Create 6  Cyl Inline Engine                       │
│                                                               │
│         8 – Create 8  Cyl Vee-Type Engine                     │
│                                                               │
│         <Ret> – EXIT                                          │
│                                                               │
│                                        Choice  ?              │
│                                                               │
╰─────────────────────────────────────────────────────────────╯
```

FIGURE 16-2

Input data menu - program ENGINE

16.1   PROGRAM OPERATION

The main menu is shown in Figure 16-1. Its structure is similar to that of the other programs with some significant exceptions. The first five items, **Input**, **Display**, **Calculate**, **Print**, and **Plot**, perform the same functions as their counterparts in the other programs. These five functions plus item 6, the **Balance** selection, are used initially to design and compute the parameters for a single-cylinder mechanism which will be later **Assembled** into whatever multicylinder configuration is desired. Thus the program can be thought of as being used in two stages, which each roughly corresponds to the topics in Chapters 14 and 15, respectively. That is, items 1 to 6 on the main menu deal with single-cylinder engines, and item 7 with multicylinder engines. Item 8, **Flywheel** calculations, can be used with either a single- or multi-cylinder design. We will now discuss each main menu selection in turn, following the engine design process as developed in the previous two chapters.

1 - Input Data

The input menu is shown in Figure 16-2. The **Type data from keyboard** and **Read data from diskfile** selections operate in the same manner as in the other programs except that different data must be supplied. Six **built-in examples** of engine configurations are also available from this menu. Most are discussed in detail in the previous chapters where they were used as examples in the derivations. Any of these examples may be selected and then all default values accepted to calculate, balance, and assemble the engines in order to become familiar with program operation.

TABLE 16-1   **Display Data from Program ENGINE**

Total displacement = 100 cubic inches for 8 cylinders

Each cylinder has 12.5 cubic inches displacement

Piston mass = 0.005	Crank mass = 0.03 at 50 %
Conrod mass = 0.02	CR CG at 33 % from crankpin to wristpin
Idle RPM = 800	Redline RPM = 6000
Engine is 4-Stroke Cycle	8 cylinder vee 90 degrees
Bore = 2.5 inches	Stroke = 2.55 inches
Bore/Stroke ratio = .98	L/R ratio = 3.5
Piston area is 4.909 square inches	Bearing friction coefficient = 0.02
Main bearing journal dia. = 2	Crankpin journal dia = 1.5

Selecting *Type data from keyboard* brings up a sequence of prompts for the data needed to define a single-cylinder engine. In the order requested these are:

The stroke cycle, either two or four
The engine type, inline, vee, or opposed
The total displacement for all cylinders in the engine
The number of cylinders
The cylinder bore (or stroke or crank radius at your option)[1]
The conrod/crank ratio (l/r), which is the inverse of the r/l ratio in the equations
The mass of one piston
The mass of one conrod
The mass of one crank throw which consists of two webs (flanges), some portion of the mainpins and the crankpin as in Figure 16-3
The location of the CG of the unbalanced crank throw, expressed as a percent of the crank radius r from the mainpin centerline (see Figure 16-3)
The location of the CG of the conrod, expressed as a percent of the conrod length l from the crankpin centerline (see Figure 16-3)
The slowest (idle) speed of the engine
The fastest (redline) speed of the engine
The assumed coefficient of friction for the bearing surfaces
The diameter of the mainpin journals
The diameter of the crankpin journals.

2 - Display Data

From these data the program will immediately compute:

The displacement (stroke volume) of one cylinder (from total engine displacement and number of cylinders)
The stroke or bore (from cylinder displacement and bore, stroke or crank radius as supplied)
The bore/stroke ratio (B/S)
The piston area

---

[1] Note that choosing *crank radius* as the input geometry parameter allows this program to also be conveniently used as a slider-crank linkage analyzer without regard for the other engine parameters.

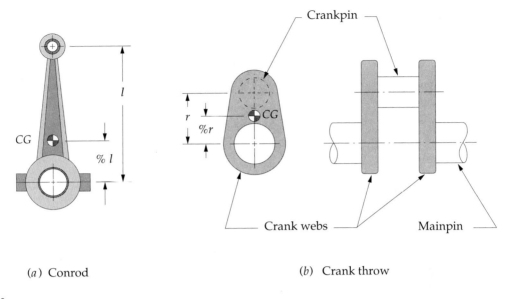

(a) Conrod

(b) Crank throw

FIGURE 16-3

Conrod and crank throw centers of gravity (CG) location data

These supplied and computed engine data will later be used in all calculations. The **Display data** selection prints these data on two screens and can be accessed at any time to determine the current engine configuration. Table 16-1 shows the data displayed. As with all other programs the display data option also allows the information to be sent to a printer or to a diskfile for later reinput to the program. Thus the input data can be saved for a later recalculation or editing.

3- Calculate

Once the data has been input from the keyboard or from a diskfile, you must **Calculate** the results. That choice will present a prompt to select the engine speed at which to calculate, among idle, redline, or midrange RPM (which is the average of the other two speeds). The piston position, velocity, and acceleration; the inertia forces, torques, and moments; and pin forces plus gas force and gas torque are then calculated for two revolutions of the crankshaft. These data are computed for one cylinder of the engine only, regardless of how many cylinders were specified. You may recalculate at any time for the same or a different engine speed. It will use the current input data as shown in the display data screens. The gas force and gas torque calculation is based on the built-in gas pressure curve similar to the one shown in Figure 14-4. This gas pressure function in the program is kept the same at all engine speeds as discussed in Section 14.1. Though this is not accurate in a thermodynamic sense, it is both necessary and appropriate for the purpose of comparing designs based solely on their kinematic and dynamic factors. The calculation is done for all parameters at 3° increments over two crankshaft revolutions, giving 241 data points per variable.

```
                              PLOT SELECT MENU

                    1 –  Piston Position

                    2 –  Piston Velocity

                    3 –  Piston Acceleration

                    4 –  Gas Force

                    5 –  Gas Torque

                    6 –  Inertia Torque

                    7 –  Total Torque

                    8 –  Friction Torque          18 –  Shaking Torque

                    9 –  Shaking Force

                   10 –  Pin Forces               20 –  Animate Engine

                     <Ret> –EXIT
                                          Choice ?
```

FIGURE 16-4

Plot menu for a single-cylinder engine

4 - Print and 5 - Plot

Figure 16-4 shows the **Plot** menu for the one-cylinder data. It is the same as the **Print** menu. More selections will later appear on these menus after assembly of a multi-cylinder configuration. Examples of plots of these variables can be seen in Figures 14-6, 14-8, 14-9, 14-10, 14-11, 14-16, 14-17, and 14-23. The equations used for the calculation of these variables are derived in Chapter 14 and are listed in Table 16-3. The **friction torque** (item 8) is calculated by multiplying the user-selected coefficient of friction by the forces calculated at the piston-cylinder interface and at the mainpin and crankpin journals. These last two friction forces are multiplied by the journal radii supplied by the user to obtain friction torque. The piston friction creates friction torque through the geometry described in the gas torque equation 14.8. The other torque values (gas torque, inertia torque, and total torque) are **not** reduced in the program by the amount of the calculated friction torque.

All plots except those of shaking force, shaking moment and pin forces are drawn on cartesian axes over 720° of crankshaft rotation in order to accommodate a four-stroke cycle. The force and moment plots are drawn as polar diagrams or hodographs as done in the other programs. These hodographs show the vectors plotted on an XY coordinate system with its origin at the crankshaft Z axis. One of three styles of hodograph may be chosen: the complete force vectors shown, the envelope of the tips of the vectors shown, or just the vector tips shown. The information presented is the same in each case. See Sections 8.3 and 16-3 for further information on polar plots. Item 20 **animates the engine** in either its single-cylinder form or shows two cylinders of any multicylinder inline, vee or opposed configuration. The bore, stroke and conrod dimensions are to scale in the animation.

This routine allows you to balance each crank

CRANKSHAFT MASS "concentrated" at Crankpin is   .0150 at 1.768 R.

You have 1 Connecting Rod on each Crankpin

CON-ROD MASS "concentrated" at Crankpin is   .0134 at 1.768 R.

COMBINED MASS "concentrated" @ Crankpin is   .0284 at 1.768 R.

CON-ROD MASS "concentrated" at Wristpin is   .0066

Present PISTON MASS is   .0050

COMBINED MASS "concentrated" at each Wristpin is .0116

Enter MASS of COUNTERWEIGHT you wish to place on CRANK ?

Enter RADIUS at which counterweight will be placed ?

Enter ANGLE vs. crankpin at which counterweight will be placed ?

FIGURE 16-5

Balancing information provided by program ENGINE

6 - Balance the Crank

This selection brings up the dialog shown in Figure 16-5 which reports the values of the equivalent masses assumed to be concentrated at the crankpin and at the wristpin. These correspond to the variables m_{2a}, m_{3a}, m_{3b}, and m_4 defined in equations 14.10 and 14.11 and in Figure 14-14. From these data for the current design, the user can determine the size of the balance mass needed at a chosen radius on the crank throw to either exactly balance the crank throw or overbalance it to counter some of the effect of the reciprocating mass at the wristpin. This procedure is detailed in Section 14.10 and Figures 14-24 and 14-25. Note that if the engine design has enough cylinders to allow a crankshaft arrangement that will cancel the inertial forces, then there is no advantage to overbalancing the crank. But, for a single-cylinder engine and some two-cylinder engines (twins), the procedure of overbalancing the crank can significantly reduce the shaking force.

7 - Assemble the Engine

Once the single-cylinder configuration is satisfactorily designed and balanced, the engine can be assembled. Two help screens are presented as shown in Figure 16-6. These remind of the fact that the crank phase diagram (called a *Power Chart* in program ENGINE) as defined in Section 15.2 and shown in Figures 15-8, 15-11, 15-13, 15-15, 15-17, and 15-23 should be defined and drawn manually before proceeding with the assembly of the engine. The assumptions and conventions used in the program to number the cylinders and banks are also stated on these screens. These are the same conventions as defined in Section 15.7. The upper limit stated in

+++++++++++++++ Warning +++++++++++++++++
Before proceeding you must work out, on paper, the crank's
Phase Angles, the engine's Firing Order and the Power Stroke angles
(the angles at which the cylinders fire).

This program uses the Phase Angles and Power Stroke Angles
and Vee Angle to calculate the results. Firing Order is used as a pointer.
If you change only the Firing Order, it will make the results incorrect.
It is your responsibility to ensure that all these factors agree!

This program requires that these rules be followed when designing
and numbering your crankshaft phase and firing angles, and vee angle:

1 - The engine's front cylinder in the right bank is always number 1
2 - Number one cylinder is the reference cylinder for all others
3 - Number one cylinder always has a Crank Phase Angle of zero degrees
4 - Number one cylinder always fires first in Firing Order .
5 - Thus the Power Stroke Angle for number one is also zero degrees
6 - Crank Phase Angles are always between 0 and 360 degrees
7 - Power Stroke Angles are between 0 and 720 deg. and in ascending order
8 - Phase Angles shift to the right versus cylinder 1 on the Crank Phase Diagram
9 - The Vee Angle shifts the second bank to the Right on the Crank Phase Diagram
10 - Cylinders are numbered first down right bank and then down left bank

You must draw your crank phase diagram to decide these values before proceeding!

FIGURE 16-6

Help screens for assembling the engine

item 7 of Figure 16-6 for the acceptable range of power stroke angles will be either
360° for a two-stroke engine or 720° (as shown) for a four-stroke engine. The program
will insert the correct value based on your choice of stroke cycle.

The program then requests the crankshaft phase angles in cylinder order,
the firing order, and the power stroke angles at which the cylinders fire. Note that
the program does not do any internal check on the compatibility of these data.
It will accept any combination of phase angles, firing orders, and power stroke
angles you provide. It is up to you to ensure that these data are compatible and
realistic. The program will also ask for the distance between cylinders and offer
a default choice of one. This information is used to define the z_i in equation 15.11.
If you want to compute the correct magnitude of the shaking moment, the actual

TABLE 16-2   Display Data for Vee Eight Multicylinder Engine

Crankpin Phase Angles, °							
0	90	270	180	0	90	270	180

Cylinder Firing Order							
1	5	4	3	7	2	6	8

Crankshaft Power Stroke Angles (in Firing Order), °							
0	90	180	270	360	450	540	630

```
                          PLOT SELECT MENU

          1 – Piston Position        11 – Power Chart

          2 – Piston Velocity        12 – Balance Condition

          3 – Piston Acceleration    13 – Primary Shaking Force

          4 – Gas Force              14 – Secondary Shaking Force

          5 – Gas Torque             15 – Primary Shaking Moment

          6 – Inertia Torque         16 – Secondary Shaking Moment

          7 – Total Torque           17 – Complete Shaking Moment

          8 – Friction Torque        18 – Shaking Torque

          9 – Shaking Force          19 – Flywheel Smoothed Total Torque

          10 – Pin Forces            20 – Animate Engine

                      <Ret> –EXIT
                                        Choice ?
```

FIGURE 16-7

Plot menu for multicylinder engines

cylinder spacing of your design needs to be supplied. If you only wish to compare two or more designs on the basis of relative shaking moment, then the default value of one will suffice. The computation sums moments about the first cylinder, so

```
          For this Vee Engine with 8 Cylinders and Vee Angle of 90 deg
              you have chosen the following firing arrangement:

   cyl. 1 in Bank 1  with crank at :    0  and Power Stroke at    0 deg.

   cyl. 5 in Bank 2  with crank at :    0  and Power Stroke at   90 deg.

   cyl. 4 in Bank 1  with crank at : 180  and Power Stroke at  180 deg.

   cyl. 3 in Bank 1  with crank at : 270  and Power Stroke at  270 deg.

   cyl. 7 in Bank 2  with crank at : 270  and Power Stroke at  360 deg.

   cyl. 2 in Bank 1  with crank at :   90  and Power Stroke at  450 deg.

   cyl. 6 in Bank 2  with crank at :   90  and Power Stroke at  540 deg.

   cyl. 8 in Bank 2  with crank at : 180  and Power Stroke at  630 deg.
```

FIGURE 16-8

Data on cylinder arrangements chosen

(a)

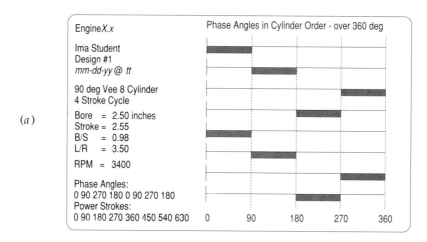

(b)

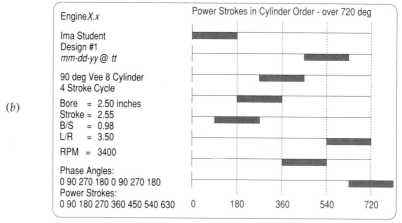

(c)

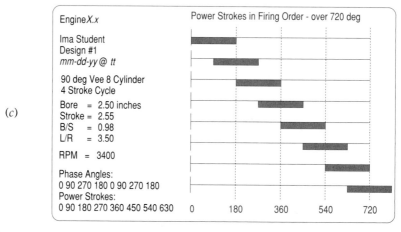

FIGURE 16-9

Crank phase angle and power chart diagrams

its moment arm is always zero. The small offset between a vee engine's banks, due to having two conrods per crank throw, is ignored in the moment calculation.

After assembling the engine, the **Display Data** option includes the multi-cylinder data as shown in Table 16-2, and the **Plot** and **Print** menus will have a number of additional items on them as seen in Figure 16-7. The **Power Chart** selection (#11) brings up a table of data on the angles chosen as shown in Figure 16-8 and then presents three plots in succession. Figure 16-9a is a plot of the crankshaft phase angles over 360° for the 90° vee eight engine designed in Section 15.7. Figure 16-9b shows the power stroke angles in cylinder order over 720° for this vee eight. This is the same as the crank phase diagram shown in Figure 15-23 and indicates when the power pulses will occur in each cylinder during the cycle. Figure 16-9c shows the power stroke angles in firing order over 720°. This plot rearranges the order of the cylinders on the power chart to match that of the selected firing order. If these three factors, *crankshaft phase angles, firing order, and power stroke angles* have all been correctly chosen for compatibility, this third power chart plot will appear *as a staircase*. The power pulses should occur in equispaced steps across the cycle if even firing has been achieved. Thus these power chart plots can serve as a check on the correct choices of these factors. Note that the program only draws the boxes on the crank phase diagrams (power charts) which represent the power strokes of each cylinder. The other piston strokes as shown in Figures 15-8, 15-11, 15-13, 15-15, 15-17, and 15-23 are not drawn.

Item 12 on the **Plot** or **Print** menu brings the **Balance Condition** charts to the screen which show the results of equations 15.3 and 15.7 for the first through sixth harmonic balance factors of shaking forces and shaking moments. These data were shown in Table 15-5. Items 13 to 16 plot the magnitudes of the primary and secondary components of the shaking forces and moments calculated from equations 15.10 and 15.11, as projected onto the central (X) axis of the engine. Item 9 plots the total shaking force and its primary and secondary components in XY polar form. Item 17 plots the total shaking moment and its components in XY polar form. These forces and moments for the XY polar plots are calculated by a different numerical method than that used for items 13-16. Examples of these plots are shown in Figures 15-19, 15-25, 15-28 ,15-29, and 15-30.

8 - Flywheel Calculations

Item 8 on the main menu does the same pulse by pulse integration of the total torque curve as is done in program DYNAFOUR. See Section 12.11 and Figure 12-16. The areas under the pulses of the total torque curve are a measure of the energy variations in the cycle. The flywheel size is based on these energy variations as defined in Section 12.11. The total torque curve for this example vee eight engine was shown in Figure 15-24. The table of areas under its pulses is printed to the screen as shown in Figure 16-10. Note that for a properly designed, even firing, multicylinder engine, the absolute values of the areas under each pulse will be equal. In that case the energy value needed for the flywheel calculation is simply the area under one pulse. These data are used in equation 12.22 to compute the needed flywheel moment of inertia based on a chosen coefficient of fluctuation k. Program ENGINE then calculates a smoothed total torque function based on that chosen k. Figure 16-11

Areas of Torque Pulses in Order During One Cycle

For Crankshaft Phase Angles: 0 90 270 180 0 90 270 180

Crankshaft Omega is 3400 RPM

Energy Units are Pounds - Inches - Radians

Order	Neg Area	Pos Area
1	−150.8115	355.9779
2	−355.9769	355.9779
3	−355.9769	355.9779
4	−355.9769	355.9779
5	−355.9769	355.9779
6	−355.9769	355.9779
7	−355.9769	355.9779
8	−355.9769	355.9779
9	−164.2168	0

Enter Desired Coefficient of Fluctuation for Flywheel ?

FIGURE 16-10

Areas under the total torque-time diagram in the 90° vee eight engine

shows the flywheel smoothed torque, from selection 19 on the plot menu, for the example vee eight engine with $k = 0.1$. Compare Figure 16-11 to Figure 15-24

TABLE 16-3   Equations Used in Program ENGINE

Variable	Equation
Piston position	14.3c
Piston velocity	14.3d
Piston acceleration	14.3e
Gas force	14.4
Gas torque	14.8b
Inertia torque	14.15e, 15.4c
Total torque	14.16
Shaking force	14.14f, 15.2d
Shaking torque	14.15f, negate 15.4c
Sidewall force	14.20
Wristpin force	14.21
Crankpin force	14.22
Mainpin force	14.23
Primary and secondary forces	15.2d, 15.10h, 15.10k
Primary and secondary moments	15.6b, 15.11a, 15.11b
Balance conditions	15.3, 15.7

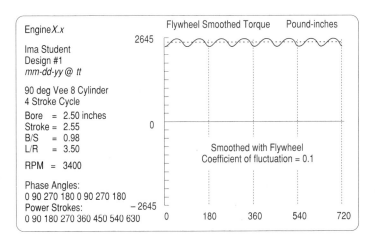

FIGURE 16-11

Flywheel smoothed total torque in the 90° vee eight engine with 0 - 90 - 270 -180° crankshaft

which shows the total torque for the same engine before the flywheel was added. The torque function with the flywheel added is nearly a constant value, which is desirable. The pulses due to the individual cylinder's explosions have been effectively masked by the flywheel.

9 - Help

This choice brings up the same help menu as in the other programs. See Section 8.3 and Figure 8-23 for further information.

16.2   EQUATIONS USED IN THE PROGRAM

Table 16-3 is a list of all the equations used in program ENGINE identified by their equation numbers in this text. The student should study these equations in concert with the use of program ENGINE in order to understand the kinematic and dynamic theory and the mathematics on which the programs are based.

16.3   SOME COMMON QUESTIONS

Some questions asked by students using this program have been:

How do I determine the masses and locations of the centers of gravity for the piston, conrod, and crank throw which the program asks for?

Draw some sketches of your proposed engine design, preferably to scale, and make a first approximation of the shapes and sizes of these parts. Then do some simple calculations of volumes of these parts and convert volume to mass using the **mass** density of the material chosen. Do summations of moments on the parts to locate their centers of gravity. See Sections 12.12 and 14.11 for some hints on how to do these tasks. An equation solver program such as *TKSolver* will be a big help in doing the calculations. Also, if you can obtain any sample pistons, conrods, and crankshafts of existing engines from your instructor or other source, it will help give you some ideas on how to design them.

What does the program mean by "crank throw"?

See Figure 16-3. For a multicylinder engine you will need to lump about half the axial length of each mainpin with each crank throw, as well as its two crank webs and the crankpin. The program will duplicate this amount of mass for each cylinder in the engine.

Should I add the vee angle to my crankshaft phase angles when putting two inline engines together into a vee engine?

No, the phase angles and the vee angle are independent. Both affect the power pulse distribution within the cycle. The program accounts for both effects independently. See Section 15.7 and Figure 15-23.

Why is the method of numbering a vee engine's cylinders different in this program than that used by some engine manufacturers?

Largely for convenience in programming. Such choices are arbitrary as far as the purpose of the program is concerned.

What is the optimum bore/stroke ratio and conrod/crank ratio to use?

See Section 14.11 and do problems 14.23 and 14.24 for clues to the answer.

How is the engine oriented with respect to the axes of the polar plots used to display shaking forces and moments ?

The polar plots are a view looking directly at the end of the crankshaft (the Z-axis), thus show an XY plane transverse to the engine block. An inline engine's pistons oscillate horizontally along the X-axis of the polar plot, consistent with the orientation of the slider crank in Figures 14-3, 14-7, 14-12 and 14-22. A vee engine has the midline axis between its banks aligned with the X-axis of the polar plot, consistent with the notation in Figure 15-20. (The X-axis is shown vertical in Figure 15-20 but is horizontal in the polar plots.) An opposed engine's pistons oscillate along the Y-axis in the polar plots since it is, in effect, a 180° vee engine. Thus all the engines in these plots are "on their sides" compared with their usual orientation in an automobile chassis. See *Animate* on the plot menu.

I want to design balance shafts to cancel shaking forces and moments. How do I get information from the program needed to do this?

The engine must first be assembled within the program. The plot and print menus (Figure 16-7) then provide data on primary and secondary forces and moments. The phase angles of the peak force and moment components (primary or secondary) are found from the printed tables of values. The crank angle at which the peak value occurs is the phase angle of that component in the XZ plane with respect to the TDC of cylinder 1. The phase of the peak balancing force or couple must be oriented 180° from this value in order to cancel it. The program also plots force and moment data in the XY plane as defined in Figure 15-20 . The envelopes of the rotating primary and secondary force and moment vectors can be seen in the polar plots from items 9 and 17 on the plot menu. (See Figures 15-28, 15-29 and 15-30). These plots show the XY plane orientation required for the counter forces and couples to be supplied from balance shafts. Also see Section 15.7 and equations 15.11 and 15.12.

Chapter 17

CAM DYNAMICS

The universe is full of magical things
patiently waiting for our wits to grow sharper
EDEN PHILLPOTS

17.0 INTRODUCTION

Chapter 9 presented the kinematics of cams and followers and methods for their design. Program DYNACAM was introduced and explained. A user manual for this program is included as the last section of Chapter 9. The reader can refer to that section at any time without loss of continuity in order to review the program's operation. The program's features not explained in that section will be described in this chapter. We will now extend the study of cam-follower systems to include considerations of the dynamic forces and torques developed. While the discussion in this chapter is limited to examples of cams and followers, the principles and approaches presented are universally applicable to all dynamic systems. The cam-follower system should be considered as a useful and convenient example for the presentation of topics such as creating lumped parameter dynamic models and defining equivalent systems. These techniques as well as the discussion of natural frequencies, effects of damping, and analogies between physical systems will be found useful in the analysis of all dynamic systems regardless of their type.

17.1 LUMPED PARAMETER DYNAMIC MODELS

Figure 17-1a shows a simple plate or disk cam driving a spring-loaded, roller follower. This is a force-closed system which depends on the spring force to keep the cam and follower in contact at all times. Figure 17-1b shows a lumped parameter model of this system in which all the **mass** which moves with the follower train is lumped together as m, all the springiness in the system is lumped within the **spring constant** k, and all the **damping** or resistance to movement is lumped together as a

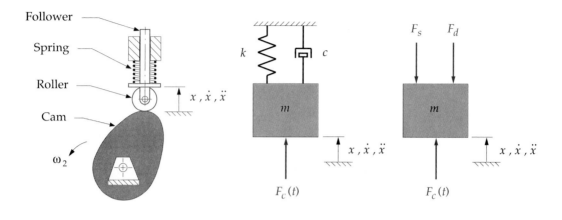

(a) Physical system   (b) Lumped model   (c) Free-body diagram

FIGURE 17-1

One *DOF* lumped parameter model of a cam-follower system

damper with coefficient c. The sources of mass which contribute to m are fairly obvious. The mass of the follower stem, the roller, its pivot pin, and any other hardware attached to the moving assembly all add together to create m. Figure 17-1c shows the free-body diagram of the system acted upon by the cam force F_c, the spring force F_s, and the damping force F_d. There will of course also be the effects of mass times acceleration on the system.

TABLE 17-1   Notation Used in this Chapter

c = damping coefficient
c_c = critical damping coefficient
k = spring constant
F_c = force of cam on follower
F_s = force of spring on follower
F_d = force of damper on follower
m = mass of moving elements
t = time in seconds
T_c = torque on camshaft
θ = camshaft angle, in degrees or radians
ω = camshaft angular velocity, rad / sec
ω_d = damped circular natural frequency, rad / sec
ω_f = forcing frequency, rad / sec
ω_n = undamped circular natural frequency, rad / sec
x = follower displacement, length units
$\dot{x} = v$ = follower velocity, length / sec
$\ddot{x} = a$ = follower acceleration, length / sec^2
ζ = damping ratio

Spring Constant

We have been assuming all links and parts to be rigid bodies in order to do the kinematic analyses, but to do a more accurate force analysis we need to recognize that these bodies are not truly rigid. The springiness in the system is assumed to be linear, thus describable by a spring constant k. A spring constant is defined as the force per unit deflection.

$$k = \frac{F_s}{x} \qquad (17.1)$$

The total spring constant k of the system is a combination of the spring constants of the actual coil spring, plus the spring constants of all other parts which are deflected by the forces. The roller, its pin, and the follower stem are all springs themselves as they are made of elastic materials. The spring constant for any part can be obtained from the equation for its deflection under the applied loading. Any deflection equation relates force to displacement and can be algebraically rearranged to express a spring constant. An individual part may have more than one k if it is loaded in several modes as, for example, a camshaft with a spring constant in bending and also one in torsion. We will discuss the procedures for combining these various spring constants in the system together into a combined, effective spring constant k in the next section. For now let us just assume that we can so combine them for our analysis and create an overall k for our lumped parameter model.

Damping

The friction, more generally called **damping**, is the most difficult parameter of the three to model. It needs to be a combination of all the damping effects in the system. These may be of many forms. **Coulomb friction** results from two dry or lubricated surfaces rubbing together. The contact surfaces between cam and follower and between the follower and its sliding joint can experience coulomb friction. It is generally considered to be independent of velocity magnitude but has a different, larger value when velocity is zero (static friction force F_{st} or *stiction*) than when there is relative motion between the parts (dynamic friction F_d). Figure 17-2a shows a plot of coulomb friction force versus relative velocity v at the contact surfaces. Note that friction always opposes motion, so the friction force abruptly changes sign at $v = 0$. The stiction F_{st} shows up as a larger spike at zero v than the dynamic friction value F_d. Thus, this is a **nonlinear** friction function. It is multivalued at zero. In fact, at zero velocity, the friction force can be any value between $-F_{st}$ and $+F_{st}$. It will be whatever force is needed to balance the system forces and create equilibrium. When the applied force exceeds F_{st}, motion begins and the friction force suddenly drops to F_d. This nonlinear damping creates difficulties in our simple model since we want to describe our system with linear differential equations having known solutions.

Other sources of damping may be present besides coulomb friction. **Viscous damping** results from the shearing of a fluid (lubricant) in the gap between the moving parts and is considered to be a linear function of relative velocity as shown in Figure 17-2b. **Quadratic damping** results from the movement of an object through a fluid medium as with an automobile pushing through the air or a boat through the water. This factor is a fairly negligible contributor to a cam-

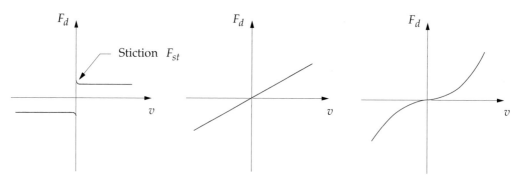

(a) Coulomb damping   (b) Viscous damping   (c) Quadratic damping

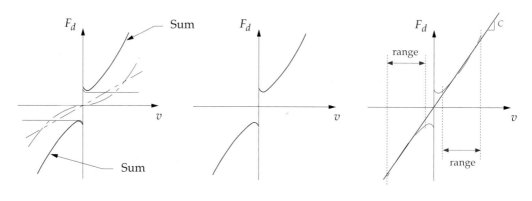

(d) Combined damping   (e) Sum of a, b, and c   (f) Linear approximation

FIGURE 17-2

Modelling damping

follower's overall damping unless the speeds are very high or the fluid medium very dense. Quadratic damping is a function of the square of the relative velocity as shown in Figure 17-2c. The relationship of the dynamic damping force F_d as a function of relative velocity for all these cases can be expressed as:

$$F_d = cv|v|^{r-1} \tag{17.2a}$$

where c is the constant damping coefficient, v is the relative velocity, and r is a constant which defines the type of damping.

For coulomb damping, $r = 0$ and:

$$F_d = \pm c \tag{17.2b}$$

For viscous damping, $r = 1$ and:

$$F_d = cv \tag{17.2c}$$

For quadratic damping, $r = 2$ and:

$$F_d = \pm cv^2 \qquad\qquad (17.2d)$$

If we combine these three forms of damping, their sum will look like Figure 17-2d and e. This is obviously a nonlinear function. But we can approximate it over a reasonably small range of velocity as a linear function with a slope c which is then a *pseudo-viscous damping coefficient*. This is shown in Figure 17-2f. While not an exact method to account for the true damping, this approach has been found to be acceptably accurate for a first approximation during the design process. The damping in these kinds of mechanical systems can vary quite widely from one design to the next due to different geometries, pressure or transmission angles, types of bearings, lubricants or their absence, etc. It is very difficult to accurately predict the level of damping (i.e., the value of c) in advance of the construction and testing of a prototype, which is the best way to determine the damping coefficient. If similar devices have been built and tested, their history can provide a good prediction. For the purpose of our dynamic modelling, we will assume *pseudo-viscous damping* and some value for c.

17.2   EQUIVALENT SYSTEMS

More complex systems than that shown in Figure 17-1 will have multiple masses, springs, and sources of damping connected together as shown in Figure 17-6. These models can be analyzed by writing dynamical equations for each subsystem and then solving the set of differential equations simultaneously. This allows a multi-degree of freedom analysis, with one *DOF* for each subsystem included in the analysis. Several publications which deal with multi*DOF* analysis are listed in the bibliography at the end of this chapter. Koster[1] found in his extensive study of vibrations in cam mechanisms that a five *DOF* model which included the effects of both torsional and bending deflection of the camshaft, backlash (see Section 10.2) in the driving gears, squeeze effects of the lubricant, nonlinear coulomb damping, and motor speed variation gave a very good prediction of the actual, measured follower response. But he also found that a single *DOF* model as shown in Figure 17-1 gave a reasonable simulation of the same system. We can then take the simpler approach and lump all the subsystems of Figure 17-6 together into a single *DOF* **equivalent system** as shown in Figure 17-1. The combining of the various springs, dampers, and masses must be carefully done to properly approximate their dynamic interactions with each other.

There are only two types of variables active in any dynamic system. These are given the general names of *through variable* and *across variable*. These names are descriptive of their actions within the system. A **through variable** *passes through the system.* An **across variable** *exists across the system.* The power in the system is the product of the through and across variables. Table 17-2 lists the through and across variables for various types of dynamic systems.

We commonly speak of the voltage across a circuit and the current flowing through it. We also can speak of the velocity across a mechanical "circuit" or system and the force which flows through it. Just as we can connect electrical elements

TABLE 17-2   Through and Across Variables in Dynamic Systems

System Type	Through Variable	Across Variable	Power Units
Electrical	Current (i)	Voltage (e)	ei = watts
Mechanical	Force (F)	Velocity (v)	Fv = (in-lb)/sec
Fluid	Flow (Q)	Pressure (P)	PQ = (in-lb)/sec

such as resistors, capacitors, and inductors together in series or parallel or a combination of both to make an electrical circuit, we can connect their mechanical analogs, dampers, springs, and masses together in series, parallel, or a combination thereof to make a mechanical system. Table 17-3 shows the analogs between three types of physical systems. The fundamental relationships between through and across variables in electrical, mechanical, and fluid systems are shown in Table 17-4.

Recognizing a series or parallel connection between elements in an electrical circuit is fairly straightforward, as their interconnections are easily seen. Determining how mechanical elements in a system are interconnected is more difficult as their interconnections are sometimes hard to see. The test for series or parallel connection is best done by examining the forces and velocities (or the integral of velocity, displacement) that exist in the particular elements. If two elements have the same force passing through them, they are in series. If two elements have the same velocity or displacement, they are in parallel.

TABLE 17-3   Physical Analogs in Dynamic Systems

System Type	Energy Dissipator	Energy Storage	Energy Storage
Electrical	Resistor (R)	Capacitor (C)	Inductor (Coil) (L)
Mechanical	Damper (c)	Mass (m)	Spring (k)
Fluid	Fluid Resistor (R_f)	Accumulator (C_f)	Fluid Inductor (L_f)

TABLE 17-4   Relationships Between Through and Across Variables in Dynamic Systems

System Type	Resistance	Capacitance	Inductance
Electrical	$i = \dfrac{1}{R} e$	$i = C \dfrac{de}{dt}$	$i = \dfrac{1}{L} \int e \, dt$
Mechanical	$F = c\,v$	$F = m \dfrac{dv}{dt}$	$F = k \int v \, dt$
Fluid	$Q = \dfrac{1}{R_f} P$	$Q = C_f \dfrac{dP}{dt}$	$Q = \dfrac{1}{L_f} \int P \, dt$

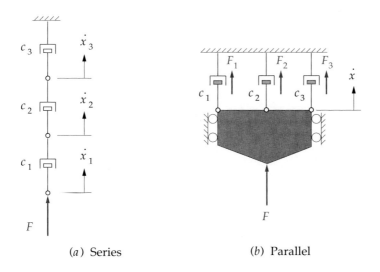

(a) Series   (b) Parallel

FIGURE 17-3

Dampers in series and in parallel

Combining Dampers

DAMPERS IN SERIES Figure 17-3a shows three dampers in series. The force passing through each damper is the same and their individual displacements and velocities are different.

$$F = c_1(\dot{x}_1 - \dot{x}_2) = c_2(\dot{x}_2 - \dot{x}_3) = c_3\dot{x}_3$$

or:

$$\frac{F}{c_1} = \dot{x}_1 - \dot{x}_2; \qquad \frac{F}{c_2} = \dot{x}_2 - \dot{x}_3; \qquad \frac{F}{c_3} = \dot{x}_3$$

combining:

$$\dot{x}_{total} = (\dot{x}_1 - \dot{x}_2) + (\dot{x}_2 - \dot{x}_3) + \dot{x}_3 = \frac{F}{c_1} + \frac{F}{c_2} + \frac{F}{c_3}$$

then:

$$\dot{x}_{total} = F\frac{1}{c_{eff}} = F\left(\frac{1}{c_1} + \frac{1}{c_2} + \frac{1}{c_3}\right)$$

$$\frac{1}{c_{eff}} = \frac{1}{c_1} + \frac{1}{c_2} + \frac{1}{c_3}$$

$$c_{eff} = \frac{1}{\dfrac{1}{c_1} + \dfrac{1}{c_2} + \dfrac{1}{c_3}} \qquad\qquad (17.3a)$$

The reciprocal of the effective damping of the dampers in series is the sum of the reciprocals of their individual damping coefficients.

DAMPERS IN PARALLEL   Figure 17-3b shows three dampers in parallel. The force passing through each damper is different, and their displacements and velocities are all the same.

$$F = F_1 + F_2 + F_3$$
$$F = c_1\dot{x} + c_2\dot{x} + c_3\dot{x}$$
$$F = (c_1 + c_2 + c_3)\dot{x}$$
$$F = c_{eff}\dot{x}$$

$$c_{eff} = c_1 + c_2 + c_3 \qquad (17.3b)$$

The effective damping of the three is the sum of their individual damping coefficients.

Combining Springs

Springs are the mechanical analog of electrical inductors. Figure 17-4a shows three springs in series. The force passing through each spring is the same, and their individual displacements are different. A force F applied to the system will create a total deflection which is the sum of the individual deflections. The spring force is defined from the relationship in equation 17.1:

$$F = k_{eff}x_{total}$$

where:

$$x_{total} = (x_1 - x_2) + (x_2 - x_3) + x_3 \qquad (17.4a)$$

$$\left(x_1 - x_2\right) = \frac{F}{k_1} \qquad\qquad \left(x_2 - x_3\right) = \frac{F}{k_2} \qquad\qquad x_3 = \frac{F}{k_3} \qquad (17.4b)$$

Substituting, we find that the effective k of **springs in series** is the sum of the reciprocals of their individual spring constants.

$$\frac{F}{k_{eff}} = \frac{F}{k_1} + \frac{F}{k_2} + \frac{F}{k_3}$$

$$(17.4c)$$

$$k_{eff} = \frac{1}{\dfrac{1}{k_1} + \dfrac{1}{k_2} + \dfrac{1}{k_3}}$$

Figure 17-4b shows three springs in parallel. The force passing through each spring is different, and their displacements are all the same. The total force is the sum of the individual forces.

$$F_{total} = F_1 + F_2 + F_3 \qquad (17.5a)$$

Substituting equation 17.4b we find that the effective k of **springs in parallel** is the sum of the individual spring constants.

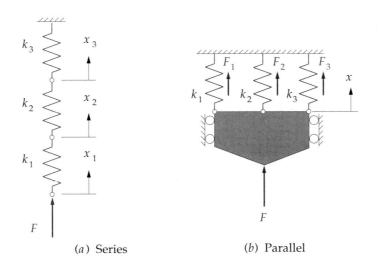

FIGURE 17-4

Springs in series and in parallel

$$k_{eff}x = k_1x + k_2x + k_3x$$

$$k_{eff} = k_1 + k_2 + k_3$$

(17.5b)

Combining Masses

Masses are the mechanical analog of electrical capacitors. The inertial forces associated with all moving masses are referenced to the ground plane of the system because the acceleration in $F = ma$ is absolute. Thus all masses are connected in parallel and combine in the same way as do capacitors in parallel with one terminal connected to a common ground.

$$m_{eff} = m_1 + m_2 + m_3 \qquad (17.6)$$

Lever and Gear Ratios

Whenever an element is separated from the point of application of a force or from another element by a **lever ratio** or **gear ratio**, its effective value will be modified by that ratio. Figure 17-5a shows a spring at one end (A) and a mass at the other end (B) of a lever. We wish to model this system as a single DOF lumped parameter system. There are two possibilities in this case. We can either transfer an equivalent mass m_{eff} to point A and attach it to the existing spring k, as shown in Figure 17-5b, or we can transfer an equivalent spring k_{eff} to point B and attach it to the existing mass m as shown in Figure 17-5c. In either case, for the lumped model to be equivalent to the original system, it must have the same energy in it.

Let's first find the effective mass that must be placed at point A to eliminate the lever. Equating the kinetic energies in the masses at points A and B:

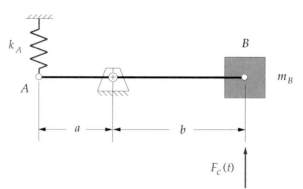

(a) Physical system

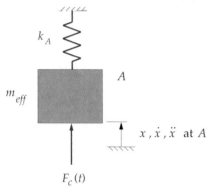

(b) Equivalent mass at point A

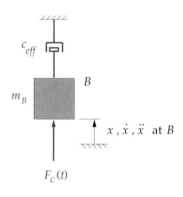

(c) Equivalent spring at point B

(d) Physical system

(e) Equivalent damper at point B

FIGURE 17-5

Lever or gear ratios affect the equivalent system

$$\frac{1}{2}m_B v_B^2 = \frac{1}{2}m_{eff} v_A^2 \qquad (17.7a)$$

The velocities at each end of the lever can be related by the lever ratio:

$$v_A = \left(\frac{a}{b}\right)v_B$$

substituting:

$$m_B v_B^2 = m_{eff}\left(\frac{a}{b}\right)^2 v_B^2$$

$$m_{eff} = \left(\frac{b}{a}\right)^2 m_B \qquad (17.7b)$$

The effective mass varies from the original mass by the square of the lever ratio. Note that if the lever were instead a pair of gears of radii a and b, the result would be the same.

Now find the effective spring that would have to be placed at B to eliminate the lever. Equating the potential energies in the springs at points A and B:

$$\frac{1}{2}k_A x_A^2 = \frac{1}{2}k_{eff} x_B^2 \qquad (17.8a)$$

The deflection at B is related to the deflection at A by the lever ratio:

$$x_B = \left(\frac{b}{a}\right)x_A$$

substituting:

$$k_A x_A^2 = k_{eff}\left(\frac{b}{a}\right)^2 x_A^2$$

$$k_{eff} = \left(\frac{a}{b}\right)^2 k_A \qquad (17.8b)$$

The effective k varies from the original k by the square of the lever ratio. If the lever were instead a pair of gears of radii a and b, the result would be the same. So, gear or lever ratios can have a large effect on the lumped parameters' values in the simplified model.

Damping coefficients are also affected by the lever ratio. Figure 17-5d shows a damper and a mass at opposite ends of a lever. If the damper at A is to be replaced by a damper at B then the two dampers must produce the same moment about the pivot, thus:

$$F_{d_A} a = F_{d_B} b \qquad (17.8c)$$

Substitute the product of the damping coefficient and velocity for force:

$$\left(c_A \dot{x}_A\right)a = \left(c_{B_{eff}} \dot{x}_B\right)b \qquad (17.8d)$$

The velocities at points A and B in Figure 17.5d can be related from kinematics:

$$\omega = \frac{\dot{x}_A}{a} = \frac{\dot{x}_B}{b}$$

$$\dot{x}_A = \dot{x}_B \frac{a}{b} \tag{17.8e}$$

Substituting in equation 17.8d we get an expression for the effective damping coefficient at B resulting from a damper at A.

$$\left(c_A \dot{x}_B \frac{a}{b} \right) a = \left(c_{B_{eff}} \dot{x}_B \right) b$$

$$c_{B_{eff}} = c_A \left(\frac{a}{b} \right)^2 \tag{17.8f}$$

Again, the square of the lever ratio determines the effective damping. The equivalent system is shown in Figure 17-5e.

✒️EXAMPLE 17-1

Creating a Single *DOF* Equivalent System Model of a Multielement Dynamic System

Given: An automotive valve cam with translating flat follower, long pushrod, rocker arm, valve, and valve spring is shown in Figure 17-6a.

Problem: Create a suitable, approximate, single *DOF*, lumped parameter model of the system. Define its effective mass, spring constant, and damping in terms of the individual elements' parameters.

Solution:

1 Break the system into individual elements as shown in Figure 17-6b. Each significant moving part is assigned a lumped mass element which has a connection to ground through a damper. There is also elasticity and damping within the individual elements, shown as connecting springs and dampers. The rocker arm is modelled as two lumped masses at its ends, connected with a rigid, massless rod as was done in Section 14.4 for the crank and conrod of the slider crank linkage. The breakdown shown represents a six *DOF* model as there are six independent displacement coordinates, x_1 through x_6.

2 Define the individual spring constants of each element which represents the elasticity of a lumped mass from the elastic deflection formula for the particular part. For example, the pushrod is loaded in compression, so its relevant deflection formula and its k are,

$$x = \frac{Fl}{AE} \qquad \text{and} \qquad k_{pr} = \frac{F}{x} = \frac{AE}{l} \tag{a}$$

where A is the cross-sectional area of the pushrod, l is its length, and E is Young's modulus for the material. The k of the tappet element will have the same expression. The expression for the k of a helical coil compression spring, as used for the valve spring, can be found in any spring design manual or machine design text and is:

$$k_{sp} = \frac{d^4 G}{8D^3 N} \tag{b}$$

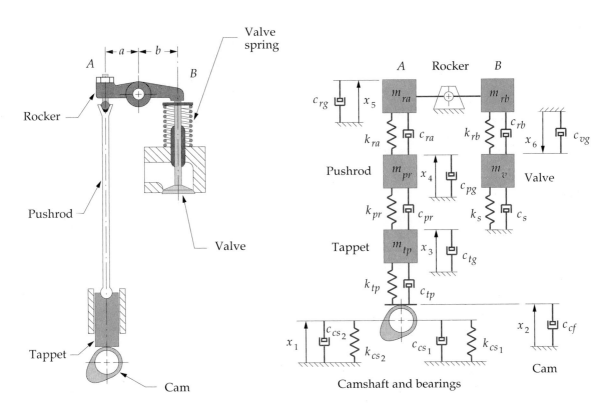

(a) Physical model

(b) Six DOF model

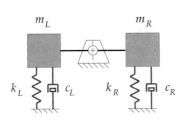

(c) One DOF model with lever arm

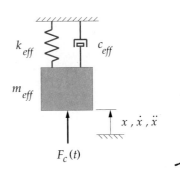

(d) One DOF lumped model

FIGURE 17-6

Lumped parameter models of an engine valve cam-follower system

where d is the wire diameter, D is the mean coil diameter, N is the number of coils and G is the modulus of rupture of the material.

The rocker arm also acts as a spring, as it is a beam in bending. It can be modelled as a double cantilever beam with its deflection on each side of the pivot considered separately. These spring effects are shown in the model as if they were compression springs, but that is just schematic. They really represent the bending deflection of the rocker arms. From the deflection formula for a cantilever beam with concentrated load:

$$x = \frac{Fl^3}{3EI} \qquad \text{and} \qquad k_{ra} = \frac{3EI}{l^3} \qquad\qquad (c)$$

where I is the cross-sectional second moment of area of the beam, l is its length, and E is Young's modulus for the material. The spring constants of any other elements in a system can be obtained in similar fashion from their deflection formulae.

3 The dampers shown connected to ground represent the friction or viscous damping at the interfaces between the elements and the ground plane. The dampers between the masses represent the internal damping in the parts, which typically is quite small. These values will either have to be estimated from experience or measured in prototype assemblies.

4 The rocker arm provides a lever ratio which must be taken into account. The strategy will be to combine all elements on each side of the lever separately into two lumped parameter models as shown in Figure 17-6c, and then transfer one of those across the lever pivot to create one, single *DOF* model as shown in Figure 17-6d.

5 The next step is to determine the types of connections, either series or parallel, between the elements. The masses are all in parallel as they each communicate their inertial force directly to ground and have independent displacements. On the left and right sides, respectively, the effective masses are:

$$m_L = m_{tp} + m_{pr} + m_{ra} \qquad\qquad m_R = m_{rb} + m_v \qquad\qquad (d)$$

The two springs shown representing the bending deflection of the camshaft split the force between them, so they are in parallel and thus add directly.

$$k_{cs} = k_{cs_1} + k_{cs_2} \qquad\qquad (e)$$

Note that, for completeness, the torsional deflection of the camshaft should also be included but is omitted in this example to reduce complexity. The combined camshaft spring constant and all the other springs shown on the left side are in series as they each have independent deflections and the same force passes through them all. The same is true of the springs on the right side. The effective spring constants for each side are then:

$$k_L = \frac{1}{\dfrac{1}{k_{cs}} + \dfrac{1}{k_{tp}} + \dfrac{1}{k_{pr}} + \dfrac{1}{k_{ra}}} \qquad\qquad k_R = \frac{1}{\dfrac{1}{k_{rb}} + \dfrac{1}{k_{sp}}} \qquad\qquad (f)$$

The dampers are in a combination of series and parallel. The pair of dampers c_{cs1} and c_{cs2} shown supporting the camshaft represent the friction in the two camshaft bearings and are in parallel.

$$c_{cs} = c_{cs_1} + c_{cs_2} \qquad\qquad (g)$$

The ones representing internal damping are in series with one another and with the combined shaft damping.

$$c_{in_L} = \frac{1}{\dfrac{1}{c_{tp}} + \dfrac{1}{c_{pr}} + \dfrac{1}{c_{ra}} + \dfrac{1}{c_{cs}}} \qquad\qquad c_{in_R} = \frac{1}{\dfrac{1}{c_{rb}} + \dfrac{1}{c_s}} \qquad (h)$$

where c_{in_L} is all internal damping on the left side and c_{in_R} is all internal damping on the right side of the rocker arm pivot. The combined internal damping c_{in_L} goes to ground through c_{rg} and the combined internal damping c_{in_R} goes to ground through the valve spring c_s. These two series combinations are then in parallel with all the other dampers that go to ground. The combined damping for each side of the system is then:

$$c_L = c_{tg} + c_{pg} + c_{rg} + c_{in_L} \qquad\qquad c_R = c_{vg} + c_{in_R} \qquad (k)$$

6 The system can now be reduced to a single *DOF* model with masses and springs lumped on either end of the rocker arm as shown in Figure 17-6c. We will bring the elements at point *B* across to point *A*. Note that we have reversed the sign convention across the pivot so that positive motion on one side results also in positive motion on the other. The damper, mass, and spring constant are affected by the square of the lever ratio as shown in equations 17.7 and 17.8. The result is:

$$m_{eff} = m_L + \left(\frac{b}{a}\right)^2 m_R$$

$$k_{eff} = k_L + \left(\frac{b}{a}\right)^2 k_R \qquad (m)$$

$$c_{eff} = c_L + \left(\frac{b}{a}\right)^2 c_R$$

These are shown in Figure 17-6d on the final, one *DOF* lumped model of the system.

Note that this one *DOF* model provides only a relatively crude approximation of this complex system's behavior. Even though it may be an oversimplification, it is nevertheless useful as a first approximation and serves in this context as an example of the general method involved in modelling dynamic systems. A more complex model with multiple degrees of freedom will provide a better approximation of the dynamic system's behavior.

17.3 DYNAMIC FORCE ANALYSIS OF THE FORCE-CLOSED CAM FOLLOWER

In Chapter 11 we discussed the two approaches to dynamic analysis, commonly called the forward and the inverse dynamics problems. The forward problem assumes that all the forces acting on the system are known and seeks to solve for the resulting displacements, velocities, and accelerations. The inverse problem is, as its name says, the inverse of the other. The displacements, velocities and accelerations are known and we solve for the dynamic forces that result. In this chapter we will explore the application of both methods to cam-follower dynamics. This section explores the forward solution. A later section will present the inverse solution. Both are instructive in this application of a force-closed (spring-loaded) cam-follower system.

Undamped Response

Figure 17-7 shows an even simpler lumped parameter model of the same system as in Figure 17-1 but which omits the damping altogether. This is referred to as a *conservative model* since it conserves energy with no losses. This is not a realistic or safe assumption in this case but will serve a purpose in the path to a better model which will include the damping. The free-body diagram for this mass-spring model is shown in Figure 17-7c. We can write Newton's equation for this one *DOF* system:

$$\sum F = ma = m\ddot{x}$$
$$F_c(t) - F_s = m\ddot{x}$$

From equation 17.1:

$$F_s = kx$$

then:

$$m\ddot{x} + kx = F_c(t) \qquad (17.9a)$$

This is a second order ordinary differential equation (DE) with constant coefficients. The complete solution will consist of the sum of two parts, the transient (homogeneous) and the steady state (particular). The homogeneous DE is,

$$m\ddot{x} + kx = 0$$

$$\ddot{x} = -\frac{k}{m}x \qquad (17.9b)$$

which has the well-known solution,

$$x = A\cos\omega t + B\sin\omega t \qquad (17.9c)$$

where A and B are the constants of integration to be determined by the initial conditions. To check the solution, differentiate it twice, assuming constant ω, and substitute in the homogeneous DE.

$$-\omega^2(A\cos\omega t + B\sin\omega t) = -\frac{k}{m}(A\cos\omega t + B\sin\omega t)$$

This is a solution provided that:

$$\omega^2 = \frac{k}{m} \qquad\qquad \omega_n = \sqrt{\frac{k}{m}} \qquad (17.9d)$$

The quantity ω_n (rad/sec) is called the *circular natural frequency* of the system and is the frequency at which it wants to vibrate if left to its own devices. This represents the *undamped natural frequency* since we ignored damping. The

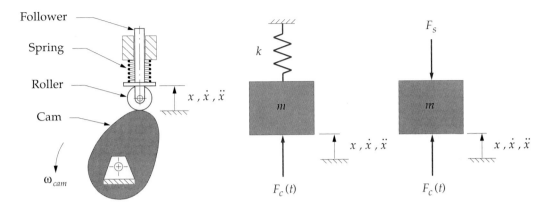

(a) Physical system   (b) Lumped model   (c) Free-body diagram

FIGURE 17-7

Conservative one *DOF* lumped parameter model of a cam-follower system

damped natural frequency will be slightly lower than this value. Note that ω_n is a function only of the physical parameters of the system m and k; thus it is completely determined and unchanging with time once the system is built. By creating a one *DOF* model of the system, we have limited ourselves to one natural frequency which is an "average" natural frequency usually close to the lowest, or fundamental, frequency.

Any real physical system will also have higher harmonics of its fundamental frequency which in general will not be integer multiples of the fundamental. In order to find them we need to create a multi-degree of freedom model of the system. The fundamental tone at which a bell rings when struck is its natural frequency defined by this expression. The bell also has harmonics or overtones which are its other, higher, natural frequencies. The fundamental frequency tends to dominate the transient response of the system.[1]

The circular natural frequency ω_n (rad/sec) can be converted to cycles per second (hertz) by noting that there are 2π radians per revolution and one revolution per cycle:

$$f_n = \frac{1}{2\pi}\omega_n \quad \text{hertz} \tag{17.9e}$$

The constants of integration, A and B in equation 17.9c, depend on the initial conditions. A general case can be stated as,

when $t = 0$;   let $x = x_0$ and $v = v_0$,   where x_0 and v_0 are constants

which gives a general solution to the homogeneous DE, equation 17.9b of:

$$x = x_0 \cos \omega_n t + \frac{v_0}{\omega_n} \sin \omega_n t \tag{17.9f}$$

Equation 17.9f can be put into polar form by computing the magnitude and phase angle:

$$X_0 = \sqrt{x_0^2 + \left(\frac{v_0}{\omega_n}\right)^2} \qquad\qquad \phi = \arctan\left(\frac{v_0}{x_0\omega_n}\right)$$

then:

$$x = X_0 \cos(\omega_n t - \phi) \qquad\qquad (17.9g)$$

Note that this is a pure harmonic function whose amplitude X_0 and phase angle ϕ are a function of the initial conditions and the natural frequency of the system. It will **oscillate forever** in response to a single, transitory input if there is truly no damping present.

Damped Response

If we now reintroduce the damping of the model in Figure 17-1b and draw the free-body diagram as shown in Figure 17-1c, the summation of forces becomes:

$$F_c(t) - F_d - F_s = m\ddot{x} \qquad\qquad (17.10a)$$

Substituting equations 17.1 and 17.2c:

$$m\ddot{x} + c\dot{x} + kx = F_c(t) \qquad\qquad (17.10b)$$

HOMOGENEOUS SOLUTION We again separate this differential equation into its homogeneous and particular components. The homogeneous part is:

$$\ddot{x} + \frac{c}{m}\dot{x} + \frac{k}{m}x = 0 \qquad\qquad (17.10c)$$

The solution to this DE is of the form:

$$x = Re^{st} \qquad\qquad (17.10d)$$

where R and s are constants. Differentiating versus time:

$$\dot{x} = Rse^{st}$$
$$\ddot{x} = Rs^2e^{st}$$

and substituting in equation 17.10c:

$$Rs^2e^{st} + \frac{c}{m}Rse^{st} + \frac{k}{m}Re^{st} = 0$$

$$\left(s^2 + \frac{c}{m}s + \frac{k}{m}\right)Re^{st} = 0 \qquad\qquad (17.10e)$$

For this solution to be valid either R or the expression in parentheses must be zero as e^{st} is never zero. If R were zero, then the assumed solution, in equation 17.10d, would also be zero and thus not be a solution. Therefore, the quadratic equation in parentheses must be zero.

$$\left(s^2 + \frac{c}{m}s + \frac{k}{m}\right) = 0 \qquad (17.10f)$$

This is called the characteristic equation of the DE and its solution is:

$$s = \frac{-\frac{c}{m} \pm \sqrt{\left(\frac{c}{m}\right)^2 - 4\frac{k}{m}}}{2}$$

which has the two roots:

$$s_1 = -\frac{c}{2m} + \sqrt{\left(\frac{c}{2m}\right)^2 - \frac{k}{m}}$$

$$\qquad (17.10g)$$

$$s_2 = -\frac{c}{2m} - \sqrt{\left(\frac{c}{2m}\right)^2 - \frac{k}{m}}$$

These two roots of the characteristic equation provide two independent terms of the homogeneous solution:

$$x = R_1 e^{s_1 t} + R_2 e^{s_2 t} \qquad \text{for} \quad s_1 \neq s_2 \qquad (17.10h)$$

If $s_1 = s_2$, then another form of solution is needed. The quantity s_1 will equal s_2 when:

$$\sqrt{\left(\frac{c}{2m}\right)^2 - \frac{k}{m}} = 0 \qquad \text{or:} \qquad \frac{c}{2m} = \sqrt{\frac{k}{m}}$$

and:

$$c = 2m\sqrt{\frac{k}{m}} = 2m\omega_n = c_c \qquad (17.10j)$$

This particular value of c is called the **critical damping** and is labelled c_c. The system will behave in a unique way when critically damped, and the solution must be of the form:

$$x = R_1 e^{s_1 t} + R_2 t e^{s_2 t} \qquad \text{for} \quad s_1 = s_2 = -\frac{c}{2m} \qquad (17.10k)$$

It will be useful to define a dimensionless ratio called the **damping ratio** ζ which is the actual damping divided by the critical damping.

$$\zeta = \frac{c}{c_c}$$

$$\qquad (17.11a)$$

$$\zeta = \frac{c}{2m\omega_n}$$

and then:

$$\zeta \omega_n = \frac{c}{2m} \qquad (17.11b)$$

The damped natural frequency w_d is slightly less than the undamped natural frequency w_n and is:

$$\omega_d = \sqrt{\frac{k}{m} - \left(\frac{c}{2m}\right)^2} \qquad (17.11c)$$

We can substitute equations 17.9d and 17.11b into equations 17.10g to get an expression for the characteristic equation in terms of dimensionless ratios:

$$s_{1,2} = -\omega_n \zeta \pm \sqrt{(\omega_n \zeta)^2 - \omega_n^2} \qquad (17.12a)$$

$$s_{1,2} = \omega_n \left(-\zeta \pm \sqrt{\zeta^2 - 1}\right)$$

This shows that the system response is determined by the damping ratio z which dictates the value of the discriminant. There are three possible cases:

CASE 1: $\zeta > 1$ Roots real and unequal
CASE 2: $\zeta = 1$ Roots real and equal   (17.12b)
CASE 3: $\zeta < 1$ Roots complex conjugate

Let's consider the response of each of these cases separately.

CASE 1: $\zeta > 1$ overdamped

The solution is of the form in equation 17.10h and is:

$$x = R_1 e^{\left(-\zeta + \sqrt{\zeta^2 - 1}\right)\omega_n t} + R_2 e^{\left(-\zeta - \sqrt{\zeta^2 - 1}\right)\omega_n t} \qquad (17.13a)$$

Note that since $\zeta > 1$, both exponents will be negative making x the sum of two decaying exponentials as shown in Figure 17-8. This is the transient response of the system to a disturbance and dies out after a time. There is no oscillation in the output motion. An example of an overdamped system which you have probably encountered is the tone arm on a good-quality record turntable with a "cueing" feature. The tone arm can be lifted up, then released, and it will slowly "float" down to the record. This is achieved by putting a large amount of damping in the system, at the arm pivot. The arm's motion follows an exponential decay curve such as in Figure 17-8.

CASE 2: $\zeta = 1$ critically damped

The solution is of the form in equation 17.10k and is:

$$x = R_1 e^{-\omega_n t} + R_2 t e^{-\omega_n t} = (R_1 + R_2 t) e^{-\omega_n t} \qquad (17.13b)$$

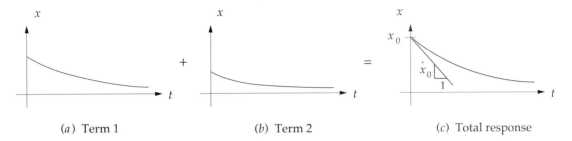

(a) Term 1   (b) Term 2   (c) Total response

FIGURE 17-8

Transient response of an overdamped system

This is the product of a linear function of time and a decaying exponential function and can take several forms depending on the values of the constants of integration, R_1 and R_2, which in turn depend on initial conditions. A typical transient response might look like Figure 17-9. This is the transient response of the system to a disturbance and dies out after a time. There is fast response but no oscillation in the output motion. An example of a critically damped system is the suspension system of a new sports car in which the damping is usually made close to critical in order to provide crisp handling response without either oscillating or being slow to respond. A critically damped system will, when disturbed, return to its original position within one bounce. It may overshoot but will not oscillate and will not be sluggish.

CASE 3: $\zeta < 1$ underdamped

The solution is of the form in equation 17.10h and s_1, s_2 are complex conjugate. Equation 17.6a can be rewritten in a more convenient form as:

$$s_{1,2} = \omega_n\left(-\zeta \pm j\sqrt{1-\zeta^2}\right); \qquad j = \sqrt{-1} \qquad\qquad (17.13c)$$

Substituting in equation 17.10a:

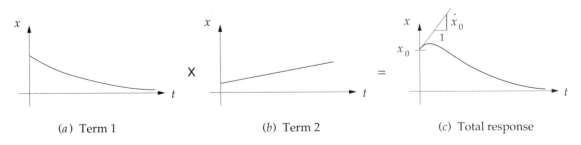

(a) Term 1   (b) Term 2   (c) Total response

FIGURE 17-9

Transient response of a critically damped system

$$x = R_1 e^{\left(-\zeta+j\sqrt{1-\zeta^2}\right)\omega_n t} + R_2 e^{\left(-\zeta-j\sqrt{1-\zeta^2}\right)\omega_n t}$$

and noting that:

$$y^{a+b} = y^a y^b$$

$$x = R_1\left[e^{-\zeta\omega_n t} e^{\left(j\sqrt{1-\zeta^2}\right)\omega_n t}\right] + R_2\left[e^{-\zeta\omega_n t} e^{\left(-j\sqrt{1-\zeta^2}\right)\omega_n t}\right]$$

factor:

$$x = e^{-\zeta\omega_n t}\left[R_1 e^{\left(j\sqrt{1-\zeta^2}\right)\omega_n t} + R_2 e^{\left(-j\sqrt{1-\zeta^2}\right)\omega_n t}\right] \qquad (17.13d)$$

Substitute the Euler identity from equation 4.4a:

$$x = e^{-\zeta\omega_n t}\left\{ \begin{array}{l} R_1\left[\cos\left(\sqrt{1-\zeta^2}\,\omega_n t\right) + j\sin\left(\sqrt{1-\zeta^2}\,\omega_n t\right)\right] \\[2mm] +R_2\left[\cos\left(\sqrt{1-\zeta^2}\,\omega_n t\right) - j\sin\left(\sqrt{1-\zeta^2}\,\omega_n t\right)\right] \end{array} \right\}$$

and simplify: (17.13e)

$$x = e^{-\zeta\omega_n t}\left\{ (R_1 + R_2)\left[\cos\left(\sqrt{1-\zeta^2}\,\omega_n t\right) + (R_1 - R_2)j\sin\left(\sqrt{1-\zeta^2}\,\omega_n t\right)\right]\right\}$$

Note that R_1 and R_2 are just constants yet to be determined from the initial conditions, so their sum and difference can be denoted as some other constants:

$$x = e^{-\zeta\omega_n t}\left\{ A\left[\cos\left(\sqrt{1-\zeta^2}\,\omega_n t\right) + jB\sin\left(\sqrt{1-\zeta^2}\,\omega_n t\right)\right]\right\} \qquad (17.13f)$$

We can put this in polar form by defining the magnitude and phase angle as:

$$X_0 = \sqrt{A^2 + B^2} \qquad\qquad \phi = \arctan\frac{B}{A}$$

then: (17.13g)

$$x = X_0 e^{-\zeta\omega_n t}\cos\left[\left(\sqrt{1-\zeta^2}\,\omega_n t\right) - \phi\right]$$

This is the product of a harmonic function of time and a decaying exponential function where X_0 and f are the constants of integration determined by the initial conditions.

Figure 17-10 shows the transient response for this **underdamped case**. The response overshoots and oscillates before finally settling down to its final position. Note that if the damping ratio ζ is zero, equation 17.13g reduces to equation 17.9g which is a pure harmonic.

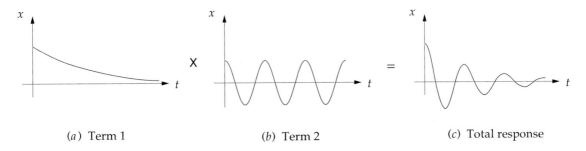

(a) Term 1　　　　　　　(b) Term 2　　　　　　　(c) Total response

FIGURE 17-10

Transient response of an underdamped system

An example of an **underdamped system** is a diving board which continues to oscillate after the diver has jumped off, finally settling back to zero position. *Many real systems in machinery are underdamped, including the typical cam-follower system.* This often leads to **vibration problems**. It is not usually a good solution simply to add damping to the system as this causes heating and is very energy inefficient. It is better to design the system to avoid the vibration problems.

PARTICULAR SOLUTION Unlike the homogeneous solution which is always the same regardless of the input, the particular solution to equation 17.10b will depend on the forcing function $F_c(t)$ which is applied to the cam-follower from the cam. In general the output displacement x of the follower will be a function of similar shape to the input function but will lag the input function by some phase angle. It is quite reasonable to use a sinusoidal function as an example since any periodic function can be represented as a Fourier series of sine and cosine terms of different frequencies (see equation 14.2).

Assume the forcing function to be:

$$F_c(t) = F_0 \sin \omega_f t \qquad (17.14a)$$

where F_0 is the amplitude of the force and ω_f is its circular frequency. Note that ω_f is unrelated to ω_n or ω_d and may be any value. The system equation then becomes:

$$m\ddot{x} + c\dot{x} + kx = F_0 \sin \omega_f t \qquad (17.14b)$$

The solution must be of harmonic form to match this forcing function and we can try the same form of solution as used for the homogeneous solution.

$$x_f(t) = X_f \sin(\omega_f t - \psi) \qquad (17.14c)$$

where:

X_f = amplitude

ψ = Phase angle between applied force and displacement

ω_f = angular velocity of forcing function

The factors X_f and ψ are not constants of integration here. They are constants determined by the physical characteristics of the system and the forcing function's frequency and magnitude. They have nothing to do with the initial conditions. To find their values, differentiate the assumed solution twice, substitute in the DE, and get:

$$X_f = \frac{F_0}{\sqrt{\left(k - m\omega_f^2\right)^2 + \left(c\omega_f\right)^2}}$$

(17.14d)

$$\psi = \arctan\left[\frac{c\omega_f}{\left(k - m\omega_f^2\right)^2}\right]$$

Substitute equations 17.9d, 17.10j, and 17.11a and put in dimensionless form:

$$\frac{X_f}{\left(\dfrac{F_0}{k}\right)} = \frac{1}{\sqrt{\left[1 - \left(\dfrac{\omega_f}{\omega_n}\right)^2\right]^2 + \left(2\zeta\dfrac{\omega_f}{\omega_n}\right)^2}}$$

(17.14e)

$$\psi = \arctan\left[\frac{2\zeta\dfrac{\omega_f}{\omega_n}}{1 - \left(\dfrac{\omega_f}{\omega_n}\right)^2}\right]$$

The ratio ω_f/ω_n is called the **frequency ratio**. Dividing X_f by the static deflection F_0/k creates the **amplitude ratio** which defines the relative dynamic displacement compared to the static.

COMPLETE RESPONSE The complete solution to our system differential equation for a sinusoidal forcing function is the sum of the homogeneous and particular solutions:

$$x = X_0 e^{-\zeta\omega_n t} \cos\left[\left(\sqrt{1 - \zeta^2}\,\omega_n t\right) - \phi\right] + X_f \sin\left(\omega_f t - \psi\right)$$

(17.15)

The homogeneous term represents the **transient response** of the system which will die out in time but is reintroduced any time the system is again disturbed. Thus with a cam containing long dwells, each subsequent rise or fall represents a new excitation which triggers the transient response. The **particular term** represents the **forced response** or **steady-state response** to a sinusoidal forcing function which will continue as long as the forcing function is present.

Note that the solution to this equation, shown in equations 17.13 and 17.14, depends only on two ratios, the damping ratio ζ which relates the actual damping relative to the critical damping, and the *frequency ratio* ω_f/ω_n which relates the forcing frequency to the natural frequency of the system. Koster[1] found that a typical value for the damping ratio in cam-follower systems is $\zeta = 0.06$, so they are underdamped and can **resonate** if operated at frequency ratios close to 1.

The initial conditions for the specific problem are applied to equation 17.15 to determine the values of X_0 and ϕ. Note that these constants of integration are contained within the homogeneous part of the solution.

17.4   RESONANCE

The natural frequency (and its harmonics) are of great interest to the designer as they define the frequencies at which the system will **resonate**. The single *DOF* lumped parameter systems shown in Figures 17-1 and 17-7 are the simplest possible to describe a dynamic system, yet they contain all the basic dynamic elements. Masses and springs are energy storage elements. A mass stores kinetic energy, and a spring stores potential energy. The damper is a dissipative element. It uses energy and converts it to heat. Thus all the losses in the model of Figure 17-1 or 17-6 occur through the damper.

These are "pure" idealized elements which possess only their own characteristics. That is, the spring has no damping and the damper no springiness, etc. Any system which contains more than one energy storage device (a mass and a spring, two or more masses, two or more springs, etc.) will possess at least one natural frequency. If we excite the system at its natural frequency, we will set up the condition called resonance in which the energy stored in the system's elements will oscillate from one element to the other at that frequency. The result can be violent oscillations in the displacements of the moveable elements in the system as the energy moves from potential to kinetic form and vice versa.

Figure 17-11 shows a plot of the amplitude and phase angle of the displacement response X of the system to a sinusoidal input forcing function at various frequencies ω_f. In our case, the forcing frequency is the angular velocity at which the cam is rotating. The plot normalizes the forcing frequency as the frequency ratio ω_f / ω_n. The response amplitude X is normalized by dividing the dynamic deflection x_d by the static deflection $x_s = F_0/k$ that the same force amplitude would create on the system. Thus at a frequency of zero, the output is one, equal to the static deflection of the spring at the input force amplitude. As the forcing frequency increases toward the natural frequency ω_n, the amplitude of the output motion, for zero damping, increases rapidly and becomes theoretically infinite when $\omega_f = \omega_n$. Beyond this point the amplitude decreases rapidly and asymptotically toward zero at high frequency ratios. It is obvious that we must avoid driving this system at or near its natural frequency. The addition of damping reduces the amplitude of vibration at the natural frequency, but very large damping ratios (> 0.6) are needed to keep the output amplitude less than or equal to the input amplitude. This is much more damping than is found in cam-follower systems and most machinery. One result of operation of an underdamped cam-follower system near ω_n can be **follower jump**. The system of follower mass and spring can oscillate violently at its natural frequency and leave contact with the cam. When it does reestablish contact, it may do so with severe impact loads that can quickly fail the materials.

The designer has a degree of control over resonance in that he can tailor the system's mass m and stiffness k to move its natural frequency away from any required operating frequencies. A common rule of thumb is to design the system to have a fundamental natural frequency ω_n at least ten times the highest forcing

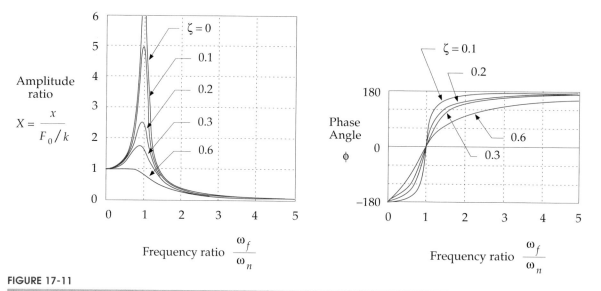

FIGURE 17-11

Amplitude ratio and phase angle of system response

frequency expected in service, thus keeping all operation well below the resonance point. This is often difficult to achieve in mechanical systems. One tries to achieve the largest ratio ω_n/ω_f possible nevertheless, even if it is less than 10.

For cams with finite jerk having a dwell after the rise of the same order of duration as the rise, Koster[1] defines a dimensionless ratio,

$$\tau = \frac{T_n}{t_r} \qquad\qquad T_n = \frac{1}{\omega_n} \qquad\qquad (17.16)$$

where T_n is the period of the follower system's natural frequency and t_r is the time to complete the rise portion of the follower motion. To minimize the effects of the transient vibrations on the system this ratio should be as small as possible and always less than 1. A small value of τ is equivalent to a large ratio of ω_n/ω_f. The error in acceleration of the follower due to vibrations during the dwell period will be proportional to the first power of τ and the error in follower position will be proportional to the third power of τ, for values of $\tau < 0.5$. If the dwell's duration is about as long or longer than that of the rise, these transient vibrations will tend to die out by the end of the dwell. The next fall or rise will again provide an input to the system and cause a new transient response.

Thus the response of a cam-follower system of this type will be dominated by the recurring transient responses rather than by the forced, or steady-state, response. It is very important to adhere to the fundamental law of cam design and use cam programs with finite jerk in order to minimize these residual vibrations in the follower system. Koster[1] reports that the cycloidal and 3-4-5 polynomial programs both gave low residual vibrations in the double-dwell cam. The modified sine acceleration program will also give good results. All these have finite jerk. (See Section 9.8 and Table 9-4.)

Some thought and observation of equation 17.9d will show that we would like our system members to be both light (low m) and stiff (high k) to get high values for ω_n and thus low values for τ. Unfortunately, the lightest materials are seldom also the stiffest. Aluminum is one-third the weight of steel but is also about a third as stiff. Titanium is about half the weight of steel but also about half as stiff. Some of the exotic composite materials such as carbon fiber/epoxy offer better stiffness-to-weight ratios but their cost is high and processing is difficult. Another job for *Unobtainium 208*!

Note in Figure 17-11 that the amplitude of vibration at large frequency ratios approaches zero. So, if the system can be brought up to speed through the resonance point without damage and then kept operating at a large frequency ratio, the vibration will be minimal. An example of systems designed to be run this way are large devices that must run at higher speed such as electrical power generators. Their large mass creates a lower natural frequency than their required operating speeds. They are "runup" as quickly as possible through the resonance region to avoid damage from their vibrations and "rundown" quickly through resonance when stopping them. They also have the advantage of long duty cycles of constant speed operation in the safe frequency region between infrequent starts and stops.

17.5 KINETOSTATIC FORCE ANALYSIS OF THE FORCE-CLOSED CAM FOLLOWER

The analysis in the previous sections is of the "**forward dynamic analysis**" type as defined in Section 11.1. That is, the applied force [$F_c(t)$ in equation 17.10b] is presumed to be known, and the equation is solved for the resulting displacement x from which its derivatives can also be determined. Cam-follower force analysis problems, like linkage analysis problems, lend themselves well to solution by the **inverse dynamics**, or **kinetostatics**, approach. Here the displacement and its derivatives are defined from the kinematic design of the cam based on an assumed angular velocity ω of the cam, and we want to solve for the force $F_c(t)$. Equation 17.10b can be solved directly for the force in a spring-loaded cam-follower system provided that values for mass m, spring constant k, and damping factor c are known in addition to the displacement, velocity, and acceleration functions.

The designer has a large degree of control over the system spring constant k_{eff} as it tends to be dominated by the k_s of the physical return spring. The elasticities of the follower parts also contribute to the overall system k_{eff} but are usually much stiffer than the physical spring. If the follower stiffness is in series with the return spring, as it often is, equation 17.4c shows that the softest spring in series will dominate the effective spring constant. Thus the return spring will virtually determine the overall k unless some parts of the follower train have similarly low stiffnesses.

The designer will choose or design the return spring and thus can specify both its k and the amount of preload to be introduced at assembly. Preload of a spring occurs when it is compressed (or extended if an extension spring) from its *free length* to its initial assembled length. This is a necessary and desirable situation as we want some residual force on the follower even when the cam is at its lowest displacement.

This will help maintain good contact between cam and follower at all times. This spring preload F_{pl} adds a constant term to equation 17.10b which becomes:

$$F_c(t) = m\ddot{x} + c\dot{x} + kx + F_{pl} \qquad\qquad (17.17a)$$

The value of m is determined from the effective mass of the system as lumped in the single DOF model of Figure 17-1. The value of c for most cam-follower systems can be estimated for a first approximation to be about 0.05 to 0.15 of the critical damping c_c as defined in equation 17.10j. Koster[1] found that a typical value for the damping ratio in cam-follower systems is $\zeta = 0.06$.

Calculating the damping c based on an assumed value of ζ requires specifying a value for the overall system k and for the value of preload. The choice of k will affect both the natural frequency of the system for a given mass and the available force to keep the joint closed. Some iteration will probably be needed to find a good compromise. A selection of data for commercially available helical coil springs is provided in Appendix C. Note in equation 17.17a that the terms involving acceleration and velocity can be either positive or negative. The terms involving the spring parameters k and F_{pl} are the only ones always positive. So, to keep the overall function always positive requires that the spring force terms be large enough to counteract any negative values in the other terms. Typically, the acceleration is larger numerically than the velocity, so the negative acceleration usually is the principal cause of a negative force F_c.

The principal concern in this analysis is to keep the cam force always positive in sign as its direction is defined in Figure 17-1. The cam force is shown as positive in that figure. In a force-closed system the cam can only push on the follower. It cannot pull. The follower spring is responsible for providing the force needed to keep the joint closed during the negative acceleration portions of the follower motion.   The damping force also can contribute, but the spring must supply the bulk of the force to maintain contact between cam and follower. If the force F_c goes negative at any time in the cycle, the follower and cam will part company, a condition called *follower jump*. When they meet again, it will be with large and potentially damaging impact forces. The follower jump, if any, will occur near the point of maximum negative acceleration as shown in Figure 17-12. Thus we must select the spring constant and preload to guarantee a positive force at all points in the cycle. In automotive engine valve cam applications follower jump is also called *valve float*, because the valve (follower) "floats" above the cam, also periodically impacting the cam surface. This will occur if the cam RPM is increased to the point that the larger negative acceleration makes the follower force negative. The "redline" maximum engine RPM often indicated on its tachometer is to warn of impending valve float above that speed which will damage the cam and follower.

Program DYNACAM allows the iteration of equation 17.17a to be done quickly and painlessly for any cam whose kinematics have been defined in that program. The main menu selection **Dynamic Forces and Torques** will solve equation 17.17a for all values of camshaft angle, using the displacement, velocity, and acceleration functions previously calculated for that cam design in the program. The program prompts for the desired values of effective system mass m, effective spring constant k, preload F_{pl}, and the assumed value of the damping ratio ζ. These values need to be determined for the model by the designer using the methods described in

Sections 17.1 and 17.2. The calculated force at the cam-follower interface can then be calculated and plotted or its values printed in tabular form. The system natural frequency is also reported when the tabular force data are printed.

✎ EXAMPLE 17-2

Kinetostatic Force Analysis of a Force-Closed (Spring-loaded) Cam Follower System

Given: A translating roller follower as shown in Figure 17-1 is driven by a force-closed radial plate cam which has the following program:

Segment 1:   Rise 1 inch in 50° with modified sine acceleration
Segment 2:   Dwell for 40°
Segment 3:   Fall 1 inch in 50° with cycloidal displacement
Segment 4:   Dwell for 40°
Segment 5:   Rise 1 inch in 50° with 3-4-5 polynomial displacement
Segment 6:   Dwell for 40°
Segment 7:   Fall 1 inch in 50° with 4-5-6-7 polynomial displacement
Segment 8:   Dwell for 40°
Camshaft angular velocity is 18.85 rad/sec.
Follower effective mass is 0.0738 mass units.
Damping is 15% of critical ($\zeta = 0.15$).

Problem: Compute the necessary spring constant and spring preload to maintain contact between cam and follower and calculate the dynamic force function for the cam. Calculate the system natural frequency with the selected spring. Keep the pressure angle under 30°.

Solution: Note that this is the same cam design as is shown in Figures 9-31, 9-50, 9-54, and 9-55 (pp. 336 - 364) and is example #5 on DYNACAM's input menu.

1   Calculate the kinematic data (follower displacement, velocity, acceleration, and jerk) for the specified cam functions. The acceleration for this cam is shown in Figure 9-31 on p. 336 and has a maximum value of 3504 in/sec². See Chapter 9 to review this procedure.

2   Calculate the radius of curvature and pressure angle for trial values of prime circle radius, and size the cam to control these values. Figure 9-50 (p. 358) shows the pressure angle function and Figure 9-54 the radii of curvature for this cam with a prime circle radius of 4 and zero eccentricity. The maximum pressure angle is 29.2° and the minimum radius of curvature is 1.7. Figure 9-55 shows the finished cam profile. See Chapter 9 to review these calculations.

3   With the kinematics of the cam defined, we can address its dynamics. To solve equation 17.17a for cam force, we must assume values for the spring constant k and the preload F_{pl}. The value of c can be calculated from equation 17.11a using the given mass m, the damping factor ζ, and assumed k and preload. The kinematic parameters are known.

4   Program DYNACAM does this computation for you. The dynamic force that results from an assumed k of 150 and a preload of 75 is shown in Figure 17-12a. The damping coefficient $c = 0.998$. Note that the force dips below the zero axis in two places during negative acceleration. These are locations of follower jump. The follower has left the cam during the fall because the spring does not have enough available force to keep the follower in contact with the rapidly falling cam. Run example #5 in the program and provide the specified k and F_{pl} to see this example. Another iteration is needed to improve the design.

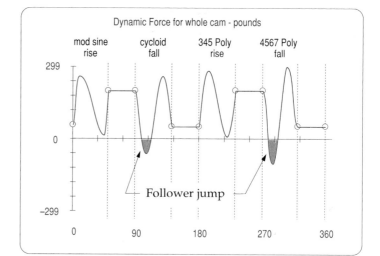

Insufficient spring
(a) force allows
follower jump

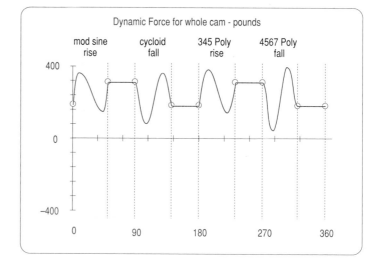

Sufficient spring
(b) force keeps the
dynamic force positive

FIGURE 17-12

Dynamic forces in a force-closed cam-follower system

5 Figure 17-12b shows the dynamic force for the same cam with a spring constant of
   $k = 200$ and a preload of 150. The damping coefficient $c = 1.153$. This additional force
   has lifted the function up sufficiently to keep it positive everywhere. There is no follower jump in this case. The maximum force during the cycle is 400.4. A margin of
   safety has been provided by keeping the minimum force comfortably above the zero
   line at 36.9. Run example #5 in the program and provide the specified spring constant
   and preload values to see this example.

6 The fundamental natural frequency, both undamped and damped, can be calculated for
   the system from equations 17.9d and 17.11c and are:

$$\omega_n = 52.06 \text{ rad/sec}; \qquad\qquad \omega_d = 51.98 \text{ rad/sec}$$

17.6 KINETOSTATIC FORCE ANALYSIS OF THE FORM-CLOSED CAM FOLLOWER

Section 9.1 described two types of joint closure used in cam-follower systems, **force closure** and **form closure**. Force closure uses an open joint and requires a spring or other force source to maintain contact between the elements. Form closure provides a geometric constraint at the joint such as the cam groove shown in Figure 17-13a or the conjugate cams of Figure 17-13b. No spring is needed to keep the follower in contact with these cams. The follower will run against one side or the other of the groove or conjugate pair as necessary to provide both positive and negative forces. Since there is no spring in this system, its dynamic force Eq. 17.17a simplifies to:

$$F_c(t) = m\ddot{x} + c\dot{x} \qquad (17.17b)$$

Note that there is now only one energy storage element in the system (the mass), so, theoretically, resonance is not possible. There is no natural frequency for it to resonate at. This is the chief advantage of a form-closed system over a force-closed one. Follower jump will not occur, short of complete failure of the parts, no matter how fast the system is run. This arrangement is sometimes used in high-performance or racing engine valve trains to allow higher redline engine speeds without valve float. In engine valve trains, a form-closed cam-follower valve train is called a *desmodromic* system.

As with any design, there are tradeoffs. While the form-closed system typically allows higher operating speeds than a comparable force-closed system, it is not free of all vibration problems. Even though there is no physical return spring in the system, the follower train, the camshaft, and all other parts still have their own spring constants which combine to provide an overall effective spring constant for the system as shown in Section 17-2. The positive side is that this spring constant will typically be quite large (stiff) since properly designed follower parts are designed to be stiff. The effective natural frequency will then be high (see equation 17.9d) and possibly well above the forcing frequency as desired.

Another problem with form-closed cams, especially the grooved or track type shown in Figure 17-13a, is the phenomenon of **crossover shock**. Every time the acceleration of the follower changes sign, the inertial force also does so. This causes the follower to abruptly shift from one side of the cam groove to the other. There cannot be zero clearance between the roller follower and the groove and still have it operate. Even if the clearance is very small, there will still be an opportunity for the follower to develop some velocity in its short trip across the groove, and it will impact the other side. Track cams of the type shown in Figure 17-13a typically fail at the points where the acceleration reverses sign, due to many cycles of crossover shock. Note also that the roller follower has to reverse direction every time it crosses over to the other side of the groove. This causes significant follower slip and high wear on the follower compared to an open, force-closed cam where the follower will have less than 1% slip.

Because there are two cam surfaces to machine and because the cam track, or groove, must be cut and ground to high precision to control the clearance, form-closed cams tend to be more expensive to manufacture than force-closed cams. Track cams usually must be ground after heat treatment to correct the distortion of the

groove resulting from the high temperatures. Grinding significantly increases cost.
Many force-closed cams are not ground after heat treatment and are used as-milled.
Though the conjugate cam approach avoids the groove tolerance and heat treat distortion problems, there are still two matched cam surfaces to be made per cam. Thus,
the desmodromic cam's dynamic advantages come at a significant cost premium.

We will now repeat the previous example cam design, modified for desmodromic operation. This is simple to do with program DYNACAM. We will merely
specify the spring constant and preload values to be zero, which then assumes that
the follower train is a rigid body. A more accurate result can be obtained by calculating and using the effective spring constant of the combination of parts in the follower train, once their geometries and materials are defined. The dynamic forces will
now be negative as well as positive, but a form-closed cam can both push and pull.

✍ EXAMPLE 17-3

Dynamic Force Analysis of a Form-Closed (Desmodromic) Cam-Follower System

Given: A translating roller follower as shown in Figure 17-13a is driven by a form-
                closed radial plate cam which has the following program:

                Segment 1: Rise 1 inch in 50° with modified sine acceleration
                Segment 2: Dwell for 40°
                Segment 3: Fall 1 inch in 50° with cycloidal displacement
                Segment 4: Dwell for 40°
                Segment 5: Rise 1 inch in 50° with 3-4-5 polynomial displacement
                Segment 6: Dwell for 40°
                Segment 7: Fall 1 inch in 50° with 4-5-6-7 polynomial displacement
                Segment 8: Dwell for 40°
                Camshaft angular velocity is 18.85 rad/sec.
                Follower effective mass is 0.0738 mass units.
                Damping is 15% of critical ($\zeta = 0.15$).

Problem: Compute the dynamic force function for the cam. Keep the pressure angle
                under 30° .

Solution: Note that this is the same kinematic cam design as is shown in Figures 9-31,
                9-50, 9-54 and 9-55 (pp. 336 - 364) and is example #5 on DYNACAM's input
                menu.

1   Calculate the kinematic data (follower displacement, velocity, acceleration and jerk) for
    the specified cam functions. The acceleration for this cam is shown in Figure 9-31 and
    has a maximum value of 3504 in/sec². See Chapter 9 to review this procedure.

2   Calculate radius of curvature and pressure angle for trial values of prime circle radius,
    and size the cam to control these values. Figure 9-50 shows the pressure angle function
    and Figure 9-54 the radii of curvature for this cam with a prime circle radius of 4 and
    zero eccentricity. The maximum pressure angle is 29.2° and the minimum radius of cur-
    vature is 1.7. Figure 9-55 shows the finished cam profile. See Chapter 9 to review these
    calculations.

3   With the kinematics of the cam defined, we can address its dynamics. To solve equation
    17.17b for the cam force, we assume zero values for the spring constant *k* and the pre-

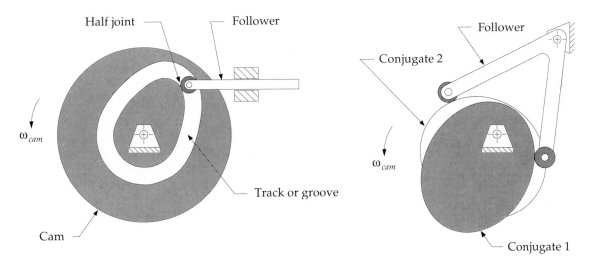

(a) Grooved or track cam (b) Conjugate cams on common shaft

FIGURE 17-13

Form-closed cam-follower systems

load F_{pl}. The value of c is assumed to be the same as in the previous example, 1.153. The kinematic parameters are known.

4 Program DYNACAM does this computation for you. The dynamic force that results is shown in Figure 17-14. Note that the force is now more nearly symmetric about the axis and its peak value is 287. Run example #5 in the program and provide the specified values to see this example. Crossover shock will occur each time this force changes sign.

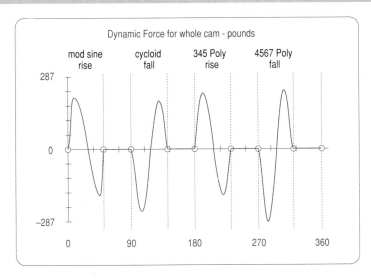

FIGURE 17-14

Dynamic force in a form-closed cam-follower system

Compare the dynamic force plots for the force-closed system (Figure 17-12b) and the form-closed system (Figure 17-14). The absolute peak force magnitude on either side of the track in the form-closed cam is less than that on the spring-loaded one. This shows the penalty that the spring imposes on the system in order to keep the joint closed. Thus, either side of the cam groove will experience lower stresses than will the open cam, except for the areas of crossover shock mentioned above.

17.7 CAMSHAFT TORQUE

The kinetostatic analysis assumes that the camshaft will operate at some constant speed ω. As we saw in the case of the fourbar linkage in Chapter 12 and with the slider-crank mechanism in Chapter 14, the input torque must vary over the cycle if the shaft velocity is to remain constant. The torque can be easily calculated from the power relationship, ignoring losses.

$$\text{Power in} = \text{Power out}$$
$$T_c\omega = F_c v \qquad\qquad (17.18)$$
$$T_c = \frac{F_c v}{\omega}$$

Once the cam force has been calculated from either equation 17.17a or 17.17b, the camshaft torque T_c is easily found since the follower velocity v and camshaft ω are both known. Figure 17-15a shows the camshaft input torque needed to drive the force-closed cam designed in Example 17-1. Figure 17-15b shows the camshaft input torque needed to drive the form-closed cam designed in Example 17-2. Note that the torque required to drive the force-closed (spring-loaded) system is significantly higher than that needed to drive the form-closed (track) cam. The spring force is also extracting a penalty here as energy must be stored in the spring during the rise portions which will tend to slow the camshaft. This stored energy is then returned to the camshaft during the fall portions, tending to speed it up. The spring loading causes larger oscillations in the torque.

A flywheel can be sized and fitted to the camshaft to smooth these variations in torque just as was done for the fourbar linkage in Section 12.11 and in Section 14.18 for the slider crank mechanism. See those sections for the design procedure. Program DYNACAM integrates the camshaft torque function pulse by pulse and prints those areas to the screen. These energy data can be used to calculate the required flywheel size for any selected coefficient of fluctuation.

One useful way to compare alternate cam designs is to look at the torque function as well as at the dynamic force. A smaller torque variation will require a smaller motor and/or flywheel and will run more smoothly. Three different designs for a single-dwell cam were explored in Chapter 9. (See Examples 9-5, 9-6, and 9-8.) All had the same lift and duration but used different cam functions. One was a double harmonic, one cycloidal and one a sixth-degree polynomial. On the basis of their kinematic results, principally acceleration magnitude, we found that the polynomial was superior. We will now revisit this cam as an example and compare its dynamic force and torque among the same three programs.

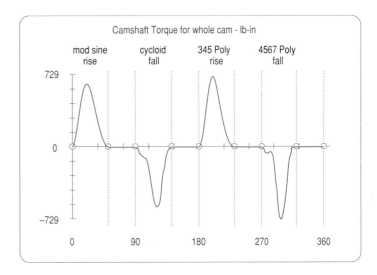

Force-closed
(a) (spring-loaded)
cam-follower

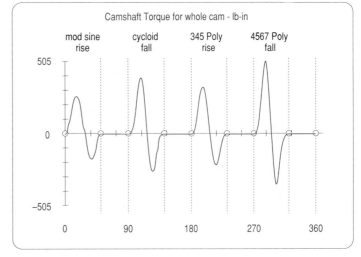

Form-closed
(b) (desmodromic)
cam-follower

FIGURE 17-15

Input torque in force- and form-closed cam-follower systems

EXAMPLE 17-4

Comparison of Dynamic Torques and Forces Among Three Alternate Designs of the
Same Cam

Given: A translating roller follower as shown in Figure 17-1 is driven by a force-
             closed radial plate cam which has the following program:

             Design 1
             Segment 1: Rise 1 inch in 90° double harmonic displacement
             Segment 2: Fall 1 inch in 90° double harmonic displacement
             Segment 3: Dwell for 180°

Design 2
Segment 1:   Rise 1 inch in 90° cycloidal displacement
Segment 2:   Fall 1 inch in 90° cycloidal displacement
Segment 3:   Dwell for 180°

Design 3
Segment 1:   Rise 1 inch in 90° and Fall 1 inch in 90° with polynomial
                    displacement
Segment 2:   Dwell for 180°

Camshaft angular velocity is 15 rad/sec. Follower effective mass is
0.0738 mass units. Damping is 15% of critical ($\zeta = 0.15$).

Find: The dynamic force and torque functions for the cam. Compare their peak
                  magnitudes for the same prime circle radius.

Solution: Note that these are the same kinematic cam designs as are shown in Figures
                  9-27, 9-25, and 9-35, respectively. (See pp. 328 - 340.)

1   Calculate the kinematic data (follower displacement, velocity, acceleration, and jerk)
    for each of the specified cam designs. See Chapter 9 to review this procedure.

2   Calculate the radius of curvature and pressure angle for trial values of prime circle
    radius, and size the cam to control these values. A prime circle radius of 3 inches
    gives acceptable pressure angles and radii of curvature. See Chapter 9 to review
    these calculations.

3   With the kinematics of the cam defined, we can address its dynamics. To solve equa-
    tion 17.17a for the cam force, we will assume a value of 50 lb/in for the spring constant
    k and adjust the preload F_{pl} for each design to obtain a minimum dynamic force of
    about 10 lb. For design 1 this requires a spring preload of 28 lb.; for design 2, 15 lb;
    and for design 3, 10 lb.

4   The value of damping c is calculated from equation 17.11d. The kinematic parame-
    ters x, v, and a are known from the prior analysis.

5   Program DYNACAM will do these computations for you. The dynamic forces that
    result from each design are shown in Figure 17-16 and the torques in Figure 17-17.
    Note that the force is largest for design 1 at 82 lb. peak and least for design 3 at 53 lb.
    peak. The same ranking holds for the torques which range from 96 lb-in for design
    1 to 52 lb-in for design 3. These represent reductions of 35% and 46% in the dynamic
    loading due to a change in the kinematic design. Not surprisingly, the sixth-degree
    polynomial design which had the lowest acceleration also has the lowest forces and
    torques and is the clear winner. Read the diskfiles EX9-5, EX9-6, and EX9-8 into pro-
    gram Dynacam to see these results.

17.8   POLYDYNE CAMS

The term *polydyne* is a contraction of the words *polynomial* and *dynamic*. It was coined
to describe a cam design approach which uses polynomial functions and takes the

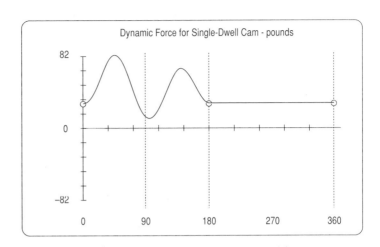

(a) Double harmonic rise - double harmonic fall

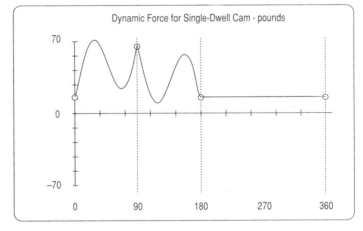

(b) Cycloidal rise - cycloidal fall

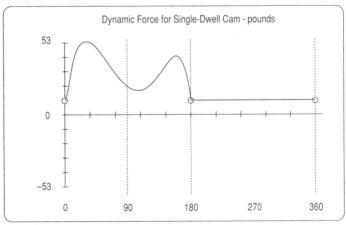

(c) Sixth-degree polynomial

FIGURE 17-16

Dynamic forces in three different designs of a single-dwell cam

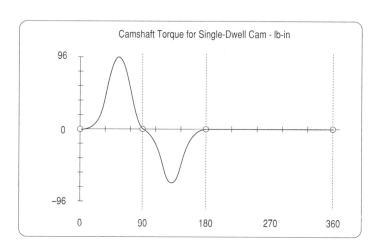

(a) Double harmonic rise - double harmonic fall

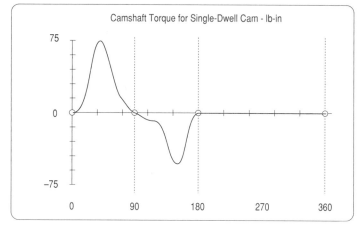

(b) Cycloidal rise - cycloidal fall

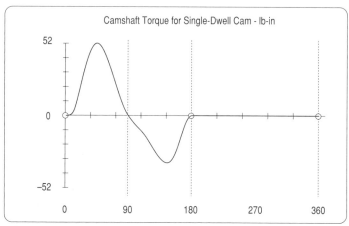

(c) Sixth-degree polynomial

FIGURE 17-17

Dynamic input torque in three different designs of a single-dwell cam

dynamics of the follower train into consideration when designing the cam. W. Dudley[2] first suggested and used the polydyne method to design automotive valve cams in 1948, though Stoddart[3] seems to have given it the name *polydyne* in 1952. Other work was done by Engemann[4], Thoren[4] and Barkan[5]. The problem that inspired this solution was severe follower jump at high speeds in internal combustion engine valve trains. It was later applied to other applications such as high speed typewriter and printer mechanisms[6]. Measurements and high-speed movies made of the motion of engine valves at the time[4] showed clearly that their actual dynamic motion at high speeds was only vaguely related to the theoretical displacement function designed into the cam. The reasons for the difference were the elastic deformations of the relatively flexible, overhead valve follower train as shown in Figure 17-6. The long pushrod and other elastic elements in the system were deflecting significantly under the high forces and causing the valve motion to lag the cam's input displacement at the start of the rise, oscillate during the motion, and overshoot on the fall.

The concept of a polydyne cam is simple, but its execution is complex. The motion desired for the output end of the follower train (the valve in Figure 17-6 for example) is defined based on the required timing in the cycle. Instead of just applying this motion program directly to the cam, as is done for a nonpolydyne design, the follower train is modelled as a lumped parameter dynamic system as described in Section 17.3 and its response to the defined input function is computed for one input speed. (A multi-degree of freedom lumped parameter model as shown in Figure 17-6b can also be used for greater accuracy if desired.) This dynamic response function (equation 17.15) is then used in combination with the desired valve output function to define the required cam displacement function which, when applied to the input end of the follower train (the tappet), will result in the desired (valve) output motion. In effect, the dynamic response of the follower train is "subtracted" from the desired output motion to account for the distortion which it will introduce at the design speed. The cam displacement function then becomes more complicated than it otherwise would have been. The mathematical flexibility of polynomial functions allows them to be tailored to these complicated demands.

The details of the polydyne cam design process are quite complicated and are beyond the scope and space of this introductory text. The reader is referred to the noted literature for more details. Additional references are provided in the bibliography at the end of this chapter. Some limitations of polydyne cams are that they tend to result in very high order polynomial cam functions which may be difficult to manufacture accurately, and the design can only be optimized for one camshaft speed. Despite these drawbacks, significant improvement in the high-speed behavior of cam-follower systems with flexible follower trains has been obtained with polydyne cams. They have been used quite successfully in various applications. Note that another, more direct approach to the problem is to design the follower train and camshaft to be short and stiff. This can be more easily done in *overhead camshaft* engine designs where the camshaft is very close to the valve. Then a nonpolydyne cam may be satisfactory.

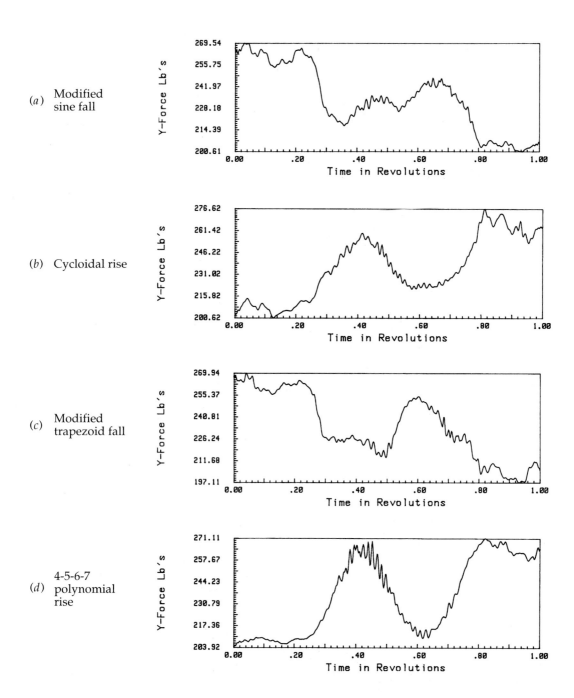

(a) Modified sine fall

(b) Cycloidal rise

(c) Modified trapezoid fall

(d) 4-5-6-7 polynomial rise

FIGURE 17-18

Experimentally measured follower radial forces on a force-closed (spring loaded) cam-follower system

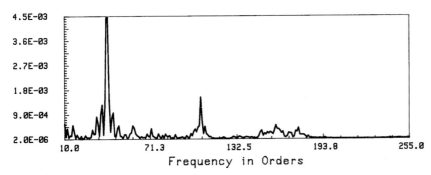

FIGURE 17-19

Experimentally measured frequency response of a cam-follower system

17.9 DYNAMIC FORCE MEASUREMENT

As described in previous sections, cam-follower systems tend to be under-damped. This allows significant oscillations and vibrations to occur in the follower train. Section 9.6 discussed the effects of manufacturing errors on the actual acceler-ations of cam-followers and showed dynamic measurements of acceleration. The dynamic forces, like the accelerations, can be measured fairly easily in operating cam systems. Compact, piezoelectric force transducers are readily available with frequency responses in the high thousands of hertz. Figure 17-18 shows follower forces measured experimentally on the cam dynamic test fixture which was de-scribed in Section 9.6. The cam is similar to the one whose theoretical follower force functions are shown in Figure 17-12b. A comparison of the measured forces to the theoretical ones shows considerable noise present. The noise is due to poorly damped vibrations in the follower train resulting from its dynamic re-sponse to small roughness and irregularities on a real, thus imperfect, though well-made cam.

Figure 17-19 shows the frequency response of this follower train to the excitation of the cam. The frequency units are orders or harmonics of the funda-mental frequency of the camshaft rotation which was 6 hertz. To convert orders to frequency in hertz for this experiment, multiply these order values by six. The follower train's fundamental frequency is at the 35th order (harmonic) of the forcing function, or 212 hertz. It is the largest spike, but many higher-frequency harmonics can also be seen. Frequencies lower than the follower train's fundamental are also present from sympathetic vibrations of the cam machine's structure. Everything in a machine tends to vibrate in sympathy at its own natural frequency when excited by any input forcing function. Sensitive transducers will pick up all vibra-tions as they are transmitted through the structure.

17.10 PRACTICAL CONSIDERATIONS

Koster[1] proposes some general rules for the design of cam-follower systems for high-speed operation based on his extensive dynamic modelling and experimentation.

To minimize the positional error and residual acceleration:

1 Keep the total lift of the follower to a minimum.

2 If possible, arrange the follower spring to preload all pivots in a consistent direction to control backlash in the joints.

3 Keep the duration of rises and falls as long as possible to minimize τ.

4 Keep follower train mass low and follower train stiffness high to increase natural frequency.

5 Any lever ratios present will change the effective stiffness of the system by the square of the ratio. Try to keep lever ratios close to 1.

6 Make the camshaft as stiff as possible **both in torsion and in bending**. This is perhaps the most important factor in controlling follower vibration.

7 Reduce pressure angle by increasing the cam pitch circle diameter.

8 Use low backlash or antibacklash gears in the camshaft drive train.

17.11 REFERENCES

1. Koster, M. P., *Vibrations of Cam Mechanisms*, Phillips Technical Library Series, Macmillan Press Ltd., London, 1974.

2. Dudley, W. H., "New Methods in Cam Design," *Trans. SAE*, vol 2, no. 1, January, 1948, pp. 1933.

3. Stoddart, D. A., "Polydyne Cam Design," *Machine Design*, January, 1953, pp. 121-135; February, 1953, pp. 146-154; March 1953, pp. 149-164.

4. Thoren, T. R., Engemann, H. H., and Stoddart, D. A., "Cam Design as Related to Valve Train Dynamics," *SAE Q. Trans*, Jan 1952, vol. 6, no. 1, pp. 1-14.

5. Barkan, P., "Calculation of High Speed Valve Motion with a Flexible Overhead Linkage," *SAE Trans.*, vol 61, 1953, pp. 687-700.

6. Kansaki, K., and Itao, K., "Polydyne Cam Mechanisms for Typehead Positioning," *ASME J. Eng. Industry*, vol 94, February, 1972, pp. 250-254.

17.12 BIBLIOGRAPHY

Barkan, P., McGarrity, R. V., "A Spring-Actuated Cam Follower System: Design Theory and Experimental Results," *ASME J. Eng. Industry*, August, 1965, pp. 279-286.

Chen, F.,Y., "A Survey of the State of the Art of Cam System Dynamics," *Mechanism and Machine Theory*, vol. 12, 1977, pp. 201-224.

Chen, F.,Y., *Mechanics and Design of Cam Mechanisms*, Pergamon Press, London, 1982.

Freudenstein, F., "On the Dynamics of High Speed Cam Profiles," *Int. J. Mech. Sci.* vol. 1, 1960, pp. 342-349.

Freudenstein, F., Vitagliano, V., Woo, L. S., and Hao, C., "Dynamic Response of Mechanical Systems," IBM Corp, New York Scientific Center, Report no. 320-2967, March 1969.

Hrones, J. A., "An Analysis of the Dynamic Forces in a Cam Driven System," *Trans. ASME* vol. 70, 1948, pp. 473-482.

Johnson, A. R., Motion Control for a Series System of "N" Degrees of Freedom Using Numerically Derived and Evaluated Equations," *ASME J. Eng. Industry*, May 1965, pp. 191-204.

Knight, B. A., and Johnson, H. L., "Motion Analysis of Flexible Cam-Follower Systems," ASME Paper no. 66-Mech-3, 1966.

Matthew, G., and Tesar, D., "Cam System Design: The Dynamic Synthesis and Analysis of the One Degree of Freedom Model," *Mechanism and Machine Theory*, vol. 11, 1976, pp. 247-257.

Matthew, G., and Tesar, D., "The Design of Modelled Cam Systems for the Two Degree of Freedom Model," *ASME J. Eng. Industry*, November 1975, pp. 1175-1180.

Mihda, A., and Turic, D. A., "On the Periodic Response of Cam Mechanisms with Flexible Follower and Camshaft," *J. Dyn. Sys. Meas. Control*, vol. 102, December 1980, pp. 255-264.

Rothbart, H., *Cams, Design, Dynamics and Accuracy*, John Wiley & Sons, New York, 1956.

17.13 PROBLEMS

Program DYNACAM *may be used to solve these problems where applicable. Where units are unspecified, work in any consistent units system you wish.*

Appendix C contains some pages from a catalog of commercially available helical coil springs to aid in designing realistic solutions to problems.

*17-1 Two springs are connected in series. One has a k of 34 and the other a k of 3.4. Calculate their effective spring constant. Which spring dominates? Repeat with the two springs in parallel. Which spring dominates?

17-2 Repeat problem 17-1 with $k_1 = 125$ and $k_2 = 25$.

17-3 Repeat problem 17-1 with $k_1 = 125$ and $k_2 = 115$.

*17-4 Two dampers are connected in series. One has a damping factor $c_1 = 12.5$ and the other, $c_2 = 1.2$. Calculate their effective damping constant. Which damper dominates? Repeat with the two dampers in parallel. Which damper dominates?

17-5 Repeat problem 17-4 with $c_1 = 12.5$ and $c_2 = 2.5$.

17-6 Repeat problem 17-4 with $c_1 = 12.5$ and $c_2 = 10$.

*17-7 A mass of $m = 2.5$ and a spring with $k = 42$ are attached to one end of a lever at a radius of 4. Calculate the effective mass and effective spring constant at a radius of 12 on the same lever.

17-8 A mass of $m = 1.5$ and a spring with $k = 24$ are attached to one end of a lever at a radius of 3. Calculate the effective mass and effective spring constant at a radius of 10 on the same lever.

17-9 A mass of $m = 4.5$ and a spring with $k = 15$ are attached to one end of a lever at a radius of 12. Calculate the effective mass and effective spring constant at a radius of 3 on the same lever.

17-10 Refer to Figure 17-6 and Example 17-1. The data for the valve train are:

Tappet is a solid cylinder 0.75 diameter by 1.25 long

* Answers in Appendix E

Pushrod is a hollow tube. 0.375 outside diameter by 0.25 inside diameter by 12 long.
Rocker arm has an average cross section of 1 wide by 1.5 high
Length $a = 2$, $b = 3$.
Camshaft is 1 diameter by 3 between bearing supports, cam in center.
Valve spring $k = 200$
All parts are steel

Calculate the effective spring constant and effective mass of a single *DOF* equivalent system placed on the cam side of the rocker arm.

*17-11   Design a double-dwell cam to move a follower of mass = 2.2 from 0 to 2.5 inches in 60° with modified sine acceleration, dwell for 120°, fall 2.5 inches in 30° with cycloidal motion, and dwell for the remainder. The total cycle must take 4 sec. Size a return spring and specify its preload to maintain contact between cam and follower. Calculate and plot the dynamic force and torque. Assume damping of 0.2 times critical. Repeat for a form-closed cam. Compare the dynamic force, torque and natural frequency for the form-closed design and the force-closed design.

*17-12   Design a double-dwell cam to move a follower of mass = 1.4 from 0 to 1.5 inches in 45° with 3-4-5 polynomial motion, dwell for 150°, fall 1.5 inches in 90° with 4-5-6-7 polynomial motion, and dwell for the remainder. The total cycle must take 6 sec. Size a return spring and specify its preload to maintain contact between cam and follower. Calculate and plot the dynamic force and torque. Assume damping of 0.1 times critical. Repeat for a form-closed cam. Compare the dynamic force, torque and natural frequency for the form-closed design and the force-closed design.

*17-13   Design a single dwell cam to move a follower of mass = 3.2 from 0 to 2 inches in 60°, fall 2 inches in 90°, and dwell for the remainder. The total cycle must take 5 sec. Use a seventh-degree polynomial. Size a return spring and specify its preload to maintain contact between cam and follower. Calculate and plot the dynamic force and torque. Assume damping of 0.15 times critical. Repeat for a form-closed cam. Compare the dynamic force, torque, and natural frequency for the form-closed design and the force-closed design.

*17-14   Design a three dwell cam to move a follower of mass = 0.4 from 0 to 2.5 inches in 40°, dwell for 100°, fall 1.5 inches in 90°, dwell for 20°, fall 1 inches in 30°, and dwell for the remainder. The total cycle must take 10 sec. Choose suitable programs for rise and fall to minimize dynamic forces and torques. Size a return spring and specify its preload to maintain contact between cam and follower. Calculate and plot the dynamic force and torque. Assume damping of 0.12 times critical. Repeat for a form-closed cam. Compare the dynamic force, torque, and natural frequency for the form-closed design and the force-closed design.

TABLE P17-1

Data for Prob.17-16

	m	k	c
a.	1.2	14	1.1
b.	2.1	46	2.4
c.	30.0	2	0.9
d.	4.5	25	3.0
e.	2.8	75	7.0
f.	12.0	50	14.0

*17-15   Design a four dwell cam to move a follower of mass = 1.25 from 0 to 2.5 inches in 40°, dwell for 100°, fall 1.5 inches in 90°, dwell for 20°, fall 0.5 inches in 30°, dwell for 40°, fall 0.5 inches in 30°, and dwell for the remainder. The total cycle must take 15 sec. Choose suitable programs for rise and fall to minimize dynamic forces and torques. Size a return spring and specify its preload to maintain contact between cam and follower. Calculate and plot the dynamic force and torque. Assume damping of 0.18 times critical. Repeat for a form-closed cam. Compare the dynamic force, torque, and natural frequency for the form-closed design and the force-closed design.

*17-16   A mass-spring damper system as shown in Figure 17-1b has the values shown in Table P17-1. Find the undamped and damped natural frequencies and the value of critical damping for the system(s) assigned.

Chapter 18

ENGINEERING DESIGN

It is a good answer that knows when to stop
ITALIAN PROVERB

18.0 INTRODUCTION

Of all the myriad activities that the practicing engineer engages in, the one that is at once the most challenging and potentially the most satisfying is that of design. Doing calculations to analyze a clearly defined and structured problem, no matter how complex, may be difficult, but the exercise of creating something from scratch, to solve a problem that is often poorly defined, is *very* difficult. The sheer pleasure and joy at conceiving a viable solution to such a design problem is one of life's great satisfactions for anyone, engineer or not. The preceding chapters have attempted to present the particular subject matter in a way and context that will not only enhance the student's fundamental understanding of the topic, but also encourage and promote his or her creative efforts toward the solution of design problems.

As a closing to a book that began with a discussion of a design process in Chapter 1, it is perhaps appropriate to present something of a "case study" of the design process to conclude the discussion. Some years ago, a very creative engineer of the author's acquaintance, George A. Wood Jr., heard a presentation by another creative engineer of the author's acquaintance, Keivan Towfigh, about one of his designs. Years later, Mr. Wood himself wrote a short paper about creative engineering design in which he reconstructed Mr. Towfigh's presumed creative process when designing the original invention. Both Mr. Wood and Mr. Towfigh have kindly consented to the reproduction of that paper here. It serves, in this author's opinion, as an excellent example and model for the student of engineering design to consider when pursuing his or her own design career.

18.1   A DESIGN CASE STUDY

Educating for Creativity in Engineering[1]

BY GEORGE A. WOOD JR.

One facet of engineering, as it is practiced in industry, is the creative process. Let us define creativity as Rollo May does in his book, The Courage to Create.[2] *It is* **"the process of bringing something new into being."** *Much of engineering has little to do with creativity in its fullest sense. Many engineers choose not to enter into creative enterprise, but prefer the realms of analysis, testing and product or process refinement. Many others find their satisfaction in management or business roles and are thus removed from engineering creativity as we shall discuss it here.*

From the outset, I wish to note that the less creative endeavors are no less important or satisfying to many engineers than is the creative experience to those of us with the will to create. It would be a false goal for all engineering schools to assume that their purpose was to make all would-be engineers creative and that their success should be measured by the "creative quotient" of their graduates.

On the other hand, for the student who has a creative nature, a life of high adventure awaits if he can find himself in an academic environment which recognizes his needs, enhances his abilities and prepares him for a place in industry where his potential can be realized.

In this talk I will review the creative process as I have known it personally and witnessed it in others. Then I shall attempt to indicate those aspects of my training that seemed to prepare me best for a creative role and how this knowledge and these attitudes toward a career in engineering might be reinforced in today's schools and colleges.

During a career of almost thirty years as a machine designer, I have seen and been a part of a number of creative moments. These stand as the high points of my working life. When I have been the creator I have felt great elation and immense satisfaction. When I have been with others at their creative moments I have felt and been buoyed up by their delight. To me, the creative moment is the greatest reward that the profession of engineering gives.

Let me recount an experience of eight years ago when I heard a paper given by a creative man about an immensely creative moment. At the First Applied Mechanisms Conference in Tulsa, Oklahoma, was a paper entitled The Four-Bar Linkage as an Adjustment Mechanism.[3] *It was nestled between two "how to do it" academic papers with graphs and equations of interest to engineers in the analysis of their mechanism problems. This paper contained only one very elementary equation and five simple illustrative figures; yet, I remember it now more clearly than any other paper I have ever heard at mechanism conferences. The author was Keivan Towfigh and he described the application of the geometric characteristics of the instant center of the coupler of a four bar mechanism.*

His problem had been to provide a simple rotational adjustment for the oscillating mirror of an optical galvanometer. To accomplish this, he was required to rotate the entire galvanometer assembly about an axis through the center of the mirror and perpendicular to the pivot axis of the mirror. High rigidity of the system after adjustment was essential with very limited space available and low cost required, since up to sixteen of these galvanometer units were used in the complete instrument.

His solution was to mount the galvanometer elements on the coupler link of a one-piece, flexure hinged, plastic four bar mechanism so designed that the mirror center was at the instant center of the linkage at the midpoint of its adjustment. (see Fig 4.) It is about this particular geometric point (see Fig 1.) that pure rotation occurs and with proper selection of linkage dimensions this condition of rotation without translation could be made to hold sufficiently accurately for the adjustment angles required.

Unfortunately, this paper was not given the top prize by the judges of the conference. Yet, it was, indirectly, a description of an outstandingly creative moment in the life of a creative man.

Let us look at this paper together and build the steps through which the author probably progressed in the achievement of his goal. I have never seen Mr. Towfigh since, and I shall therefore describe a generalized creative process which may be incorrect in some details but which, I am sure, is surprisingly close to the actual story he would tell.

The galvanometer problem was presented to Mr. Towfigh by his management. It was, no doubt, phrased something like this: "In our new model, we must improve the stability of the adjustment of the equipment but keep the cost down. Space is critical and low weight is too. The overall design must be cleaned up, since customers like modern, slim-styled equipment and we'll lose sales to others if we don't keep ahead of them on all points. Our industrial designer has this sketch that all of us in sales like and within which you should be able to make the mechanism fit."

Then followed a list of specifications the mechanism must meet, a time when the new model should be in production and, of course, the request for some new feature that would result in a strong competitive edge in the marketplace.

I wish to point out that the galvanometer adjustment was probably only one hoped-for improvement among many others. The budget and time allowed were little more than enough needed for conventional redesign, since this cost must be covered by the expected sales of the resulting instrument. For every thousand dollars spent in engineering, an equivalent increase in sales or reduction in manufacturing cost must be realized at a greater level than the money will bring if invested somewhere else.

In approaching this project, Mr. Towfigh had to have a complete knowledge of the equipment he was designing. He had to have run the earlier models himself. He must have adjusted the mirrors of existing machines many times. He had to be able to visualize the function of each element in the equipment in its most basic form.

(research)

Secondly, he had to ask himself (as if he were the customer) what operational and maintenance requirements would frustrate him most. He had to determine which of these

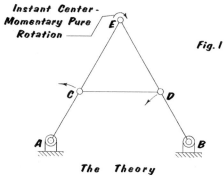

Instant Center-
Momentary Pure
Rotation

Fig. 1

The Theory

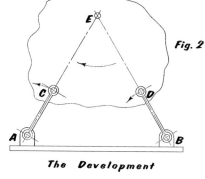

Fig. 2

The Development

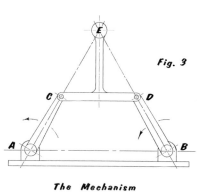

Fig. 3

The Mechanism

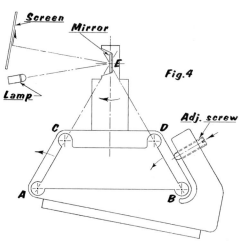

Screen

Mirror

Fig.4

Lamp

Adj. screw

The Final Product of Keivan Towfigh

*might be improved within the design time available. In this case he focused on the mirror
adjustment. He considered the requirement of rotation without translation. He determined
the maximum angles that would be necessary and the allowable translation that would
not affect the practical accuracy of the equipment. He recognized the desirability of a one
screw adjustment. He spent a few hours thinking of all the ways he had seen of rotating
an assembly about an arbitrary point. He kept rejecting each solution as it came to him as
he felt, in each case, that there was a better way. His ideas had too many parts, involved
slides, pivots, too many screws, were too vibration sensitive or too large.*

(ideation)

(frustration)

*He thought about the problem that evening and at other times while he proceed-
ed with the design of other aspects of the machine. He came back to the problem several
times during the next few days. His design time was running out. He was a mechanism
specialist and visualized a host of cranks and bars moving the mirrors. Then one day,
probably after a period when he had turned his attention elsewhere, on rethinking of the
adjustment device an image of the system based on one of the elementary characteristics
of a four bar mechanism came to him.*

(incubation)

(Eureka!)

I feel certain that this was a visual image, as clear as a drawing on paper. It was probably not complete but involved two inspirations. First was the characteristics of the instant center. (See Figs 1, 2, 3.) Second was the use of flexure hinge joints which led to a one-piece plastic molding. (See Fig 4.) I am sure that at this moment he had a feeling that this solution was right. He knew it with certainty. The whole of his engineering background told him. He was elated. He was filled with joy. His pleasure was not because of the knowledge that his superiors would be impressed or that his security in the company would be enhanced. It was the joy of personal victory, the awareness that he had conquered.

The creative process has been documented before by many others far more qualified to analyze the working of the human mind than I. Yet I would like to address, for the remaining minutes, how education can enhance this process and help more engineers, designers and draftsmen extend their creative potential.

*The key elements I see in creativity that have greatest bearing on the quality that results from the creative effort are **visualization** and **basic knowledge** that gives strength to the feeling that the right solution has been achieved. There is no doubt in my mind that the fundamental mechanical principles that apply in the area in which the creative effort is being made must be vivid in the mind of the creator. The words that he was given in school must describe real elements that have physical, visual significance. F = ma must bring a picture to his mind vivid enough to touch.*

If a person decides to be a designer, his training should instill in him a continuing curiosity to know how each machine he sees works. He must note its elements and mentally see them function together even when they are not moving. I feel that this kind of solid, basic knowledge couples with physical experience to build ever more critical levels at which one accepts a tentative solution as "right."

It should be noted that there have been times for all of us when the inspired "right" solution has proven wrong in the long run. That this happens does not detract from the process but indicates that creativity is based on learning and that failures build toward a firmer judgment base as the engineer matures. These failure periods are only negative, in the growth of a young engineer, when they result in the fear to accept a new challenge and beget excessive caution which then stifles the repetition of the creative process.

(analysis)

What would seem the most significant aspects of an engineering curriculum to help the potentially creative student develop into a truly creative engineer?

***First** is a **solid, basic knowledge** in **physics**, **mathematics**, **chemistry** and those subjects relating to his area of interest. These fundamentals should have physical meaning to the student and a vividness that permits him to explain his thoughts to the untrained layman. All too often technical words are used to cover cloudy concepts. They serve the ego of the user instead of the education of the listener.*

***Second** is the growth of the student's **ability to visualize**. The creative designer must be able to develop a mental image of that which he is inventing. The editor of the book* Seeing with the Mind's Eye,[4] *by Samuels and Samuels, says in the preface:*

"... visualization is the way we think. Before words, images were. Visualization is the heart of the bio-computer. The human brain programs and self-programs through its images. Riding a bicycle, driving a car, learning to read, baking a cake, playing golf - all skills are acquired through the image making process. Visualization is the ultimate consciousness tool."

Obviously, the creator of new machines or products must excel in this area.

*To me, a course in **Descriptive Geometry** is one part of an engineer's training that enhances one's ability to visualize theoretical concepts and graphically reproduce the result. This ability is essential when one sets out to design a piece of new equipment. First, he visualizes a series of complete machines with gaps where the problem or unknown areas are. During this time, a number of directions the development could take begin to form. The best of these images are recorded on paper and then are reviewed with those around him until, finally, a basic concept emerges.*

*The **third** element is the building of the student's **knowledge of what can be or has been done by others** with different specialized knowledge than he has. This is the area to which experience will add throughout his career as long as he maintains an enthusiastic curiosity. Creative engineering is a building process. No one can develop a new concept involving principles about which he has no knowledge. The creative engineer looks at problems in the light of what he has seen, learned and experienced and sees new ways for combining these to fill a new need.*

***Fourth** is the development of the **ability** of the student **to communicate** his knowledge to others. This communication must involve not only skills with the techniques used by technical people but must also include the ability to share engineering concepts with untrained shop workers, business people and the general public. The engineer will seldom gain the opportunity to develop a concept if he cannot pass on to those around him his enthusiasm and confidence in the idea. Frequently, truly ingenious ideas are lost because the creator cannot transfer his vivid image to those who might finance or market it.*

***Fifth** is the development of a student's **knowledge of the physical result of engineering**. The more he can see real machines doing real work, the more creative he can be as a designer. The engineering student should be required to run tools, make products, adjust machinery and visit factories. It is through this type of experience that judgement grows as to what makes a good machine, when approximation will suffice and where optimization should halt.*

It is often said that there has been so much theoretical development in engineering during the past few decades that the colleges and universities do not have time for the basics I have outlined above. It is suggested that industry should fill in the practice areas that colleges have no time for, so that the student can be exposed to the latest technology. To some degree I understand and sympathize with this approach, but I feel that there is a negative side that needs to be recognized. If a potentially creative engineer leaves college without the means to achieve some creative success as he enters his first job, his enthusiasm for creative effort is frustrated and his interest sapped long before the most enlightened company can fill in the basics. Therefore, a result of the "basics later" approach often is to remove from the gifted engineering student the means to express himself visually and physically. Machine design tasks therefore become the domain of the graduates of technical and trade schools and the creative contribution by many a brilliant university student to products that could make all our lives richer is lost.

As I said at the start, not all engineering students have the desire, drive and enthusiasm that are essential to creative effort. Yet I feel deeply the need for the enhancement of the potential of those who do. That expanding technology makes course decisions difficult for both student and professor is certainly true. The forefront of academic thought has a compelling attraction for both the teacher and the learner. Yet I feel that the development of strong basic knowledge, the abilities to visualize, to communicate, to respect what has been done, to see and feel real machinery, need not exclude or be excluded by the excitement of the new. I believe that there is a curriculum balance that can be achieved which will enhance the latent creativity in all engineering and science students. It can give a firm basis for those who look towards a career of mechanical invention and still include the excitement of new technology.

I hope that this discussion may help in generating thought and providing some constructive suggestions that may lead more engineering students to find the immense satisfaction of the creative moment in the industrial environment. In writing this paper I have spent considerable time reflecting on my years in engineering and I would close with the following thought. For those of us who have known such times during our careers, the successful culminations of creative efforts stand among our most joyous hours.

18.2 CLOSURE

Mr. Wood's description of his creative experiences in engineering design and the educational factors which influenced them closely parallel this author's experience as well. The student is well advised to follow his prescription for a thorough grounding in the fundamentals of engineering and communication skills. A most satisfying career in the design of machinery can result.

18.3 REFERENCES

1. Wood, G. A. Jr., "Educating for Creativity in Engineering," Presented at the 85th Annual Conf. ASEE, Univ. of No. Dakota, 1977.

2. May, R., *The Courage to Create*, Bantam Books, New York, 1976.

3. Towfigh, K., "The Four-Bar Linkage as an Adjustment Mechanism," 1st Applied Mechanism Conf., Oklahoma State Univ. , Tulsa Okla., 1969, pp. 27-1– 27-4.

4. Samuels and Samuels, *Seeing with the Mind's Eye: the History, Techniques and Uses of Visualization*, Random House, New York, 1975.

THE
END

Appendix A

MATERIAL PROPERTIES

For selected engineering materials - many other alloys are available.

Material	Weight Density lb/in^3	Young's Modulus psi x 10^6	Tensile Yield Strength psi x 10^3	Ultimate Tensile Strength psi x 10^3	Compres-sive Strength psi x 10^3	Endur-ance Limit psi x 10^3
Aluminum						
2024-T4	.10	10	47	68	47	20[1]
6061-T6	.10	10	40	45	40	14[1]
7075-T6	.10	10	40	45	40	14[1]
Cast -195	.10	10	16	32	17	
Gray Cast Iron						
Class 20	.26	12		20	95	10
Class 40	.26	17		40	143	19
Class 60	.26	20		60	170	24
Nodular or Ductile Cast Iron						
60-40-18	.26	23.5	52	70	140	35
80-55-06	.26	23.5	67	100	200	50
120-90-02	.26	23.5	108	135	270	63

Continued ...

Material	Weight Density lb/in^3	Young's Modulus psi x 10^6	Tensile Yield Strength psi x 10^3	Ultimate Tensile Strength psi x 10^3	Compres-sive Strength psi x 10^3	Endur-ance Limit psi x 10^3
			Engineering Plastics			
ABS	.04	0.3[4]		5	5	
Acetal	.04	0.5[4]		9	5	
Acrylic	.04	0.4[4]		9	14	
Nylon	.04	0.1[4]		7	3	
Polycarbo nate	.04	0.2[4]		18	18	
			Magnesium			
AZ61A-F	.06	6.5	26	37	13	18[1]
			Steel			
1020 HR[3]	.28	30	43	65	43	32
1020 CD[3]	.28	30	66	78	66	33
1045 HR[3]	.28	30	59	98	59	49
1045 CD[3]	.28	30	90	103	90	51
4340 HR[3]	.28	30	69	101	69	50
4340 CD[3]	.28	30	99	110	99	55
			Stainless Steel			
304 CD[3]	.28	29	75	110	75	34
410 CD[3]	.28	29	85	100	85	40
			Titanium			
	.16	17	100[2]	150[2]		

[1] Measured at 5x10^8 cycles

[2] Approximate - varies with alloy

[3] CD = cold drawn, CDA = Cold drawn annealed, HR = hot rolled, HRA = hot rolled annealed

[4] Approximate values - Most plastics do not obey Hookes Law - These apparent moduli of elasticity vary with time and temperature.

GEOMETRIC PROPERTIES

DIAGRAMS AND FORMULAS TO CALCULATE THE FOLLOWING PARAMETERS FOR SEVERAL COMMON GEOMETRIC SOLIDS

Volume *(V)*

Mass *(m)*

Center of Gravity *(Cg)*

Mass Moment of Inertia *(I)*

Radius of Gyration *(k)*

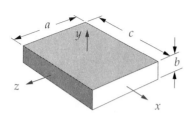

$$V = abc \qquad m = V \cdot \text{mass density}$$

$$x_{Cg} @ \frac{c}{2} \qquad y_{Cg} @ \frac{b}{2} \qquad z_{Cg} @ \frac{a}{2}$$

$$I_x = \frac{m(a^2+b^2)}{12} \qquad I_y = \frac{m(a^2+c^2)}{12} \qquad I_z = \frac{m(b^2+c^2)}{12}$$

$$k_x = \sqrt{\frac{I_x}{m}} \qquad k_y = \sqrt{\frac{I_y}{m}} \qquad k_z = \sqrt{\frac{I_z}{m}}$$

(*a*) Rectangular prism

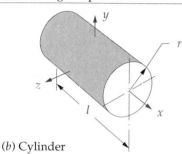

$$V = \pi r^2 l \qquad m = V \cdot \text{mass density}$$

$$x_{Cg} @ \frac{l}{2} \qquad y_{Cg} \text{ on axis} \qquad z_{Cg} \text{ on axis}$$

$$I_x = \frac{mr^2}{2} \qquad I_y = I_z = \frac{m(3r^2+l^2)}{12}$$

$$k_x = \sqrt{\frac{I_x}{m}} \qquad k_y = k_z = \sqrt{\frac{I_y}{m}}$$

(*b*) Cylinder

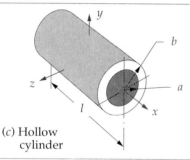

$$V = \pi(b^2-a^2)l \qquad m = V \cdot \text{mass density}$$

$$x_{Cg} @ \frac{l}{2} \qquad y_{Cg} \text{ on axis} \qquad z_{Cg} \text{ on axis}$$

$$I_x = \frac{m(a^2+b^2)}{2} \qquad I_y = I_z = \frac{m(3a^2+3b^2+l^2)}{12}$$

$$k_x = \sqrt{\frac{I_x}{m}} \qquad k_y = k_z = \sqrt{\frac{I_y}{m}}$$

(*c*) Hollow
cylinder

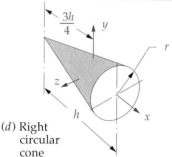

$$V = \pi \frac{r^2 h}{3} \qquad m = V \cdot \text{mass density}$$

$$x_{Cg} @ \frac{3h}{4} \qquad y_{Cg} \text{ on axis} \qquad z_{Cg} \text{ on axis}$$

$$I_x = \frac{3}{10}mr^2 \qquad I_y = I_z = \frac{m(12r^2+3h^2)}{80}$$

$$k_x = \sqrt{\frac{I_x}{m}} \qquad k_y = k_z = \sqrt{\frac{I_y}{m}}$$

(*d*) Right
circular
cone

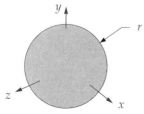

$$V = \frac{4}{3}\pi r^3 \qquad m = V \cdot \text{mass density}$$

$$x_{Cg} \text{ at center} \qquad y_{Cg} \text{ at center} \qquad z_{Cg} \text{ at center}$$

$$I_x = I_y = I_z = \frac{2}{5}mr^2$$

$$k_x = k_y = k_z = \sqrt{\frac{I_y}{m}}$$

(*e*) Sphere

Appendix C

SPRING DATA

The following catalog pages of helical compression and extension spring data provided courtesy of *Hardware Products Co., Chelsea, Massachusetts.*

COMPRESSION SPRINGS

Will go in hole In.		7/16			1/2				5/8				3/4				7/8			
Wire Dia. In.		.031	.047	.062	.047	.062	.078	.094	.047	.062	.078	.094	.062	.078	.094	.125	.062	.078	.094	.125
FREE LENGTHS 7/16	Catalog No.	247	248	249																
	Price Code	HB	HB	HC																
	lbs./in.	12	55	180																
	Max. Defl.	.32	.23	.16																
1/2	Catalog No.	250	251	252	283	284	285	286												
	Price Code	HB	HB	HC	HB	HB	HE	HE												
	lbs./in.	10	47	150	37	110	320	840												
	Max. Defl.	.37	.27	.19	.29	.22	.15	.10												
5/8	Catalog No.	253	254	255	287	288	289	290	331	332	333	334								
	Price Code	HB	HB	HC	HB	HB	HE	HE	HB	HB	HD	HE								
	lbs./in.	7.9	36	175	29	85	240	610	20	54	140	320								
	Max. Defl.	.47	.35	.25	.38	.29	.20	.14	.43	.35	.27	.20								
3/4	Catalog No.	256	257	258	291	292	293	294	335	336	337	338	375	376	377	378				
	Price Code	HB	HB	HC	HB	HB	HE	HE	HB	HB	HD	HE	HD	HD	HF	HG				
	lbs./in.	6.4	29	90	23	68	185	470	16	43	105	250	32	78	170	650				
	Max. Defl.	.58	.44	.32	.48	.37	.26	.18	.53	.44	.34	.26	.59	.40	.32	.19				
7/8	Catalog No.	259	260	261	295	296	297	298	339	340	341	342	379	380	381	382	419	420	421	422
	Price Code	HB	HB	HC	HB	HB	HE	HE	HB	HB	HD	HE	HD	HD	HF	HG	HF	HF	HJ	HK
	lbs./in.	5.4	24	75	19	56	155	384	13	36	90	204	27	65	140	520	21	49	100	350
	Max. Defl.	.68	.52	.39	.57	.44	.32	.23	.64	.53	.42	.32	.58	.48	.39	.24	.62	.53	.44	.30
1	Catalog No.	262	263	264	299	300	301	302	343	344	345	346	383	384	385	386	423	424	425	426
	Price Code	HB	HB	HC	HB	HC	HE	HE	HB	HC	HD	HE	HD	HD	HF	HG	HF	HF	HJ	HK
	lbs./in.	4.7	21	65	17	48	130	320	11	31	77	170	23	55	115	430	18	42	86	290
	Max. Defl.	.79	.60	.45	.66	.51	.37	.27	.74	.62	.49	.38	.68	.57	.46	.29	.73	.63	.53	.36
1¼	Catalog No.	265	266	267	303	304	305	306	347	348	349	350	387	388	389	390	427	428	429	430
	Price Code	HB	HC	HC	HC	HC	HF	HF	HC	HC	HD	HE	HE	HE	HF	HG	HG	HG	HJ	HK
	lbs./in.	3.7	16	50	13	38	100	245	9.0	24	59	130	18	42	89	320	14	32	66	220
	Max. Defl.	1.0	.77	.58	.84	.66	.49	.35	.94	.80	.64	.49	.88	.74	.60	.39	.94	.81	.69	.47
1½	Catalog No.	268	269	270	307	308	309	310	351	352	353	354	391	392	393	394	431	432	433	434
	Price Code	HB	HC	HC	HC	HC	HF	HF	HD	HD	HE	HF	HE	HE	HF	HG	HG	HG	HJ	HK
	lbs./in.	3.1	14	41	11	31	83	200	7.4	20	48	105	15	34	72	260	12	26	53	175
	Max. Defl.	1.2	.94	.70	1.0	.81	.60	.43	1.1	.98	.78	.61	1.1	.91	.74	.48	1.1	1.0	.85	.59
1¾	Catalog No.	271	272	273	311	312	313	314	355	356	357	358	395	396	397	398	435	436	437	438
	Price Code	HC	HD	HE	HC	HD	HF	HF	HD	HD	HC	HF	HE	HE	HF	HG	HG	HG	HJ	HK
	lbs./in.	2.6	11	35	9.1	26	70	170	6.2	17	41	90	12.4	29	61	216	9.9	22	45	147
	Max. Defl.	1.4	1.1	.84	1.2	.96	.71	.52	1.3	1.1	.93	.73	1.3	1.1	.89	.58	1.35	1.2	1.0	.71
2	Catalog No.	274	275	276	315	316	317	318	359	360	361	362	399	400	401	402	439	440	441	442
	Price Code	HD	HD	HE	HC	HD	HF	HF	HD	HD	HE	HF	HE	HE	HF	HG	HG	HG	HJ	HL
	lbs./in.	2.3	10	30	7.9	23	60	145	5.4	14	35	77	11	25	52	185	8.6	19	38	125
	Max. Defl.	1.6	1.3	.96	1.4	1.1	.82	.60	1.5	1.3	1.1	.85	1.4	1.2	1.0	.68	1.5	1.4	1.2	.83
3	Catalog No.	277	278	279	319	320	321	322	363	364	365	366	403	404	405	406	443	444	445	446
	Price Code	HD	HE	HE	HF	HF	HG	HG	HD	HD	HF	HG	HF	HF	HG	HK	HJ	HJ	HK	HL
	lbs./in.	1.5	6.6	20	5.2	15	39	94	3.6	9.4	23	50	7	16	34	115	5.6	12	25	80
	Max. Defl.	2.4	1.9	1.4	2.1	1.7	1.2	.93	2.4	2.0	1.6	1.3	2.2	1.9	1.6	1.0	2.4	2.1	1.8	1.3
4	Catalog No.	280	281	282	323	324	325	326	367	368	369	370	407	408	409	410	447	448	449	450
	Price Code	HE	HE	HE	HE	HF	HG	HG	HE	HE	HF	HG	HF	HF	HG	HL	HJ	HJ	HL	HM
	lbs./in.	1.1	4.9	15	3.9	11	29	69	2.6	6.9	17	37	5.2	12	25	86	4.2	9.2	18	59
	Max. Defl.	3.3	2.6	2.0	2.8	2.3	1.7	1.2	3.2	2.7	2.2	1.8	3.0	2.6	2.1	1.4	3.2	2.8	2.5	1.8
6	Catalog No.				327	328	329	330	371	372	373	374	411	412	413	414	451	452	453	454
	Price Code				HF	HF	HG	HG	HF	HF	HG	HG	HG	HG	HJ	HN	HK	HK	HL	HO
	lbs./in.				2.5	7.0	17	45	1.8	4.6	11	24	3.4	7.9	16	56	2.7	6.0	12	38
	Max. Defl.				4.4	3.5	2.5	2.	4.8	4.2	3.4	2.7	4.6	3.9	3.3	2.2	4.9	4.3	3.8	2.7
8	Catalog No.												415	416	417	418	455	456	457	458
	Price Code.												HJ	HJ	HK	HO	HL	HL	HM	HP
	lbs./in.												2.6	6	11	40	2.0	4.5	8.9	28
	Max. Defl.												6.1	5.2	4.5	3.0	6.5	5.8	5.1	3.7
Maximum Load		3.7	12.7	29	11	25	45	88	8.3	19	38	66	15.8	31.2	54	125	13.4	26.3	45	105
Will work free over		.347	.315	.285	.375	.345	.313	.281	.505	.475	.443	.411	.585	.554	.522	.460	.700	.670	.638	.576
Pitch		.195	.141	.128	.173	.151	.141	.141	.259	.214	.188	.177	.284	.240	.217	.204	.371	.306	.268	.239
Solid Stress (000 omitted)		125	118	113	118	113	109	105	118	113	109	105	113	109	105	99	113	109	105	99

Pounds per inch figure is a constant for each spring, and represents the number of pounds required to compress the spring 1″. To compress the spring ½″ or ¼″ requires ½ or ¼ of this value.

Maximum Deflection is the amount spring deflects to give the maximum load. This value subtracted from the free length gives the solid or compressed length.

NOTE: Stock springs can be ordered in stainless steel or plated. Prices quoted upon request.

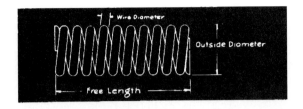

Will go in hole In		1				1 1/4			1 1/2			2			3			4			6	
Wire Dia. In		.078	.094	.125	.187	.094	.125	.187	.125	.187	.250	.187	.250	.375	.250	.375	.500	.375	.500	.750	.750	1.000
1	Catalog No.	459	460	461	462																	
	Price Code	HK	HL	HL	HR																	
	lbs./in.	34	67	210	1500																	
	Max. Defl.	.67	.58	.41	.19																	
1¼	Catalog No.	463	464	465	466	499	500	501														
	Price Code	HL	HM	HM	HS	HN	HN	HM														
	lbs./in.	26	52	160	1100	35	100	600														
	Max. Defl.	.87	.76	.55	.26	.85	.67	.37														
1½	Catalog No.	467	468	469	470	502	503	504	526	527	528											
	Price Code	HL	HM	HN	HS	HN	HO	HT	HR	HX	HAC											
	lbs./in.	21	42	130	870	29	82	460	60	300	1200											
	Max. Defl.	1.0	.93	.69	.34	1.1	.84	.48	.95	.60	.35											
1¾	Catalog No.	471	472	473	474	505	506	507	529	530	531											
	Price Code	HL	HM	HN	HS	HN	HO	HT	HR	HX	HAC											
	lbs./in.	18	35	108	712	24	68	379	50	244	960											
	Max. Defl.	1.3	1.1	.83	.41	1.3	1.0	.59	1.1	.74	.44											
2	Catalog No.	475	476	477	478	508	509	510	532	533	534	553	554	555								
	Price Code	HM	HN	HO	HT	HO	HP	HU	HS	HZ	HAE	HAA	HAG	HZZ								
	lbs./in.	16	30	93	600	21	59	320	43	200	800	115	390	3000								
	Max. Defl.	1.4	1.3	.97	.49	1.4	1.2	.70	1.3	.87	.53	1.1	.77	.34								
3	Catalog No.	479	480	481	482	511	512	513	535	536	537	556	557	558	577	578	579					
	Price Code	HN	HO	HP	HZ	HP	HR	HAA	HT	HAD	HAL	HAE	HAN	HZZ	HAR	HZZ	HZZ					
	lbs./in.	10	19	59	370	13	37	200	27	130	480	73	230	1650	105	560	2300					
	Max. Defl.	2.2	2.0	1.5	.79	2.2	1.8	1.1	2.1	1.4	.89	1.8	1.3	.61	1.8	1.1	.64					
4	Catalog No.	483	484	485	486	514	515	516	538	539	540	559	560	561	580	581	582	598	599	610		
	Price Code	HP	HR	HS	HAC	HS	HT	HAD	HW	HAG	HAQ	HAJ	HAR	HZZ	HAT	HZZ	HZZ	HZZ	HZZ	HZZ		
	lbs./in.	7.4	14	43	270	9.9	27	144	20	93	340	53	170	1150	76	390	1500	210	720	4600		
	Max. Defl.	3.0	2.7	2.1	1.1	3.0	2.5	1.5	2.8	1.9	1.2	2.5	1.8	.96	2.1	1.4	.8					
6	Catalog No.	487	488	489	490	517	518	519	541	542	543	562	563	564	583	584	585	600	601	611	616	621
	Price Code	HR	HT	HU	HAD	HU	HW	HAE	HX	HAJ	HAT	HAM	HAW	HZZ	HAZ	HZZ	HZZ	HZZ	HZZ	HZZ	HZZ	HZZ
	lbs./in.	4.9	9.4	28	175	6.5	18	93	13	60	220	34	105	710	49	240	920	130	430	2840	850	3500
	Max. Defl.	4.6	4.1	3.2	1.7	4.7	3.9	2.4	4.3	3.0	1.9	3.8	2.8	1.4	4.0	2.6	1.6	3.4	2.4	1.3	1.9	1.4
8	Catalog No.	491	492	493	494	520	521	522	544	545	546	565	566	567	586	587	588	602	603	612	617	622
	Price Code	HS	HU	HW	HAL	HW	HX	HAM	HAA	HAP	HAW	HAR	HAZ	HZZ	HBD	HZZ	HZZ	HZZ	HZZ	HZZ	HZZ	HZZ
	lbs./in.	3.6	7.0	21	125	4.8	13	68	9.6	44	160	25	79	510	36	175	660	95	310	2050	630	2500
	Max. Defl.	6.2	5.6	4.3	2.3	5.2	5.2	3.2	5.9	4.1	2.7	5.2	3.9	1.9	5.4	3.6	2.2	4.5	3.4	1.8	2.7	2.0
12	Catalog No.	495	496	497	498	523	524	525	547	548	549	568	569	570	589	590	591	604	605	613	618	623
	Price Code	HT	HW	HZ	HAP	HX	HAA	HAR	HAC	HAU	HBA	HAZ	HBE	HZZ	HBK	HZZ	HZZ	HZZ	HZZ	HZZ	HZZ	HZZ
	lbs./in.	2.4	4.6	14	84	3.2	8.7	45	6.3	29	105	16	52	330	23	110	420	61	195	1325	400	1580
	Max. Defl.	9.4	8.4	6.5	3.5	9.5	7.9	5.0	8.9	6.2	4.1	8.0	5.9	3.0	8.3	5.5	3.5	7.3	5.3	2.6	4.3	3.1
16	Catalog No.								550	551	552	571	572	573	592	593	594	606	607	614	619	624
	Price Code								HAE	HAW	HBD	HAZ	HBG	HZZ	HBL	HZZ	HZZ	HZZ	HZZ	HZZ	HZZ	HZZ
	lbs./in.								4.7	21	74	12	38	240	17	83	310	45	145	975	300	1170
	Max. Defl.								11.9	8.5	5.6	10.7	8.0	4.1	11.3	7.5	4.0	10	7.3	3.8	6.1	4.3
24	Catalog No.											574	575	576	595	596	597	608	609	615	620	625
	Price Code											HBA	HBL	HZZ	HBP	HZZ	HZZ	HZZ	HZZ	HZZ	HZZ	HZZ
	lbs./in.											7.8	23.4	150	11.4	54	200	29	94	640	175	760
	Max. Defl.											16.3	12.1	7.0	11.4	7.3	3.6	15.2	11.1	5.8	9.4	6.5
Maximum Load		23	39	90	295	30	69	224	57	180	428	131	307	1000	195	624	1470	449	1040	3700	2000	4800
Will work free over		.784	.752	.690	.565	1.00	.940	.815	1.19	1.06	.940	1.52	1.39	1.14	2.33	2.08	1.83	3.08	2.83	2.25	4.25	3.75
Pitch		.382	.328	.279	.268	.481	.384	.327	.516	.403	.388	.596	.518	.514	.917	.741	.736	1.08	.969	1.00	1.3	1.4
Solid Stress (000 omitted)		109	105	99	90	105	99	90	99	90	85	90	85	77	85	77	73	77	73	70	70	65

Pounds per inch figure is a constant for each spring, and represents the number of pounds required to compress the spring 1". To compress the spring ½" or ¼" requires ½ or ¼ of this value.

Maximum Deflection is the amount spring deflects to give the maximum load. This value subtracted from the free length gives the solid or compressed length.

NOTE: Stock springs can be ordered in stainless steel or plated. Prices quoted upon request.

Hardware Products Company, Inc.

EXTENSION SPRINGS

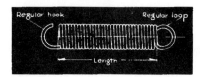

Regular hook   Regular loop

Length

ORDER BY:
SE LENGTH x O.D. x WIRE DIA.

SPECIFY HOOKS OR LOOPS

The figures given for "Maximum Extension" and "lbs. per inch" are for a spring 1" long. For other lengths multiply the "Maximum Extension" and divide the "lbs. per inch" by the length in inches. The "Maximum Load" and "Initial Tension" remain constant for any length.

Example: A spring ½" diam. .062" wire and 4" long will have a safe maximum extension of 3.2" and it will require 4 lbs. to deflect it 1 in. The spring will hold approximately 3.3 lbs. before it starts to extend, and will hold a maximum of 16.1 lbs. without permanent stretch. If 8.5 lbs. is hung on the spring it will deflect 1.3". 8.5 lbs. minus 3.3 lbs. divided by 4 lbs. per inch equals 1.3".

FOR QUICK DELIVERY
Call 617-884-9410

NOTE: Stock springs can be ordered in stainless steel or plated. Prices quoted upon request.

Outside dia.	Wire dia.	Catalog No.	Price code	Safe maximum load in pounds	Safe maximum extension - in.	Approx. initial tension in pounds	Pound per inch extension	Stress at max. load (000 omitted)	Weight per foot (lbs.)
1/8	.012	01	EHE	.6	1.9	.07	.27	100	.012
	.016	02	EHD	1.3	.9	.2	1.2	93	.015
	.023	03	EHD	4.2	.35	.9	9.0	90	.02
5/32	.012	04	EHE	.47	3.5	.01	.12	100	.015
	.016	05	EHD	1.1	1.7	.15	.55	93	.019
	.023	06	EHD	3.2	.7	.5	3.9	90	.027
3/16	.016	07	EHD	.87	2.5	.1	.3	93	.024
	.023	08	EHD	2.6	1.0	.4	2.2	90	.032
	.031	09	EHD	6.5	.45	1.5	10.7	88	.04
7/32	.016	10	EHD	.75	4.0	.01	.18	93	.028
	.023	11	EHD	2.3	1.6	.32	1.2	90	.039
	.031	12	EHD	5.5	.7	1.0	6.5	88	.048
1/4	.023	13	EHD	1.8	1.9	26	.8	90	.044
	.031	14	EHD	4.7	1.0	.75	3.8	88	.055
	.047	15	EHE	16.0	.3	3.5	40.0	83	.082
5/16	.023	16	EHE	1.5	3.5	.16	.38	90	.058
	.031	17	EHE	3.6	1.6	.55	1.9	88	.072
	.047	18	EHE	12.5	.9	2.2	10.8	83	.108
3/8	.031	19	EHE	2.9	2.5	.37	1.0	88	.084
	.047	20	EHE	10.5	.9	1.7	9.5	83	.13
	.062	21	EHF	23.0	.39	5.3	45.0	79	.16
7/16	.031	22	EHF	2.5	3.5	.26	.63	88	.105
	.047	23	EHF	8.5	1.2	1.4	5.7	83	.163
	.062	24	EHF	20.0	.6	4.3	26.0	79	.2
1/2	.047	25	EHF	7.3	1.6	1.1	3.7	83	.18
	.062	26	EHF	17.0	.8	3.3	16.0	79	.23
	.078	27	EHG	34.0	.45	8.0	57.0	77	.28
	.094	28	EHJ	57.0	.25	16.0	160.0	74	.32
5/8	.047	29	EHG	6.0	3.0	.7	1.7	83	.24
	.062	30	EHG	13.3	1.4	2.1	7.6	79	.3
	.078	31	EHG	27.0	.9	5.2	23.0	77	.37
	.094	32	EHJ	45.0	.4	11.0	73.0	74	.44
3/4	.062	33	EHG	10.5	2.2	1.5	4.1	79	.36
	.078	34	EHJ	22.0	1.3	3.5	14.0	77	.46
	.094	35	EHJ	36.0	.7	8.0	38.0	74	.51
	.125	36	EHK	85.0	.3	22.0	180.0	69	.64
7/8	.062	37	EHK	9.2	3.3	1.1	2.4	79	.4
	.078	38	EHK	18.0	1.7	2.6	8.7	77	.59
	.094	39	EHL	31.0	1.0	6.0	25.0	74	.64
	.125	40	EHM	72.0	.5	17.0	107.0	69	.8
1	.078	41	EHL	16.0	2.5	2.0	5.5	77	.67
	.094	42	EHL	26.0	1.5	4.5	13.7	74	.70
	.125	43	EHN	65.0	.75	14.0	68.0	69	.90
	.187	44	EHW	200.0	.23	60.0	600.0	63	1.4
1 1/4	.094	45	EHM	21.0	2.6	2.8	6.8	74	.94
	.125	46	EHO	47.0	1.2	9.0	31.0	69	1.3
	.187	47	EHZ	148.0	.3	40.0	290.0	63	1.8
1 1/2	.125	48	EHS	39.0	1.9	6.0	17.0	69	1.4
	.187	49	EHAA	122.0	.6	33.0	150.0	63	2.2
	.250	50	EHAC	290.0	.27	90.0	720.0	60	2.6
2	.187	51	EHAD	90.0	1.3	20.0	54.0	63	3.1
	.250	52	EHAG	210.0	.6	55.0	260.0	60	3.7

Carried in stock in 3-foot lengths - cut to length and looped to order.

Hardware Products Company, Inc.

191 WILLIAMS STREET • CHELSEA, MA 02150

ATLAS OF GEARED FIVEBAR LINKAGE COUPLER CURVES

C. Zhang, R. L. Norton, T. Hammond

The following pages of coupler curve data are excerpted from the complete work. See sections 3.6, 4.8, 6.8 and 7.4 for more information on the geared fivebar linkage. Use program FIVEBAR to investigate other linkage geometries.

> *Alpha* = Coupler Link 3/Link 2
>
> *Beta* = Ground Link 1/Link 2
>
> *Lambda* = Gear Ratio = Gear 5/Gear 2.*
>
> *Phase angle* is noted on each plot of a coupler curve.
>
> The dots along curves are at every 10 degrees of Link 2's rotation.
>
> Linkage is symmetrical: Link 2 = Link 5 and Link 3 = Link 4

---

* Note that this lambda is the **inverse** of the λ which is defined in Sections 4.8, 6.8, and 7.4. See also Figures P4-4 on p. 139, P6-4 on p. 229, and P7-4 on p. 263. For example, a gear ratio of 2 in this atlas corresponds to a λ of 0.5 in the text and in program FIVEBAR. (The difference merely corresponds to a mirroring of the linkage from left to right.)

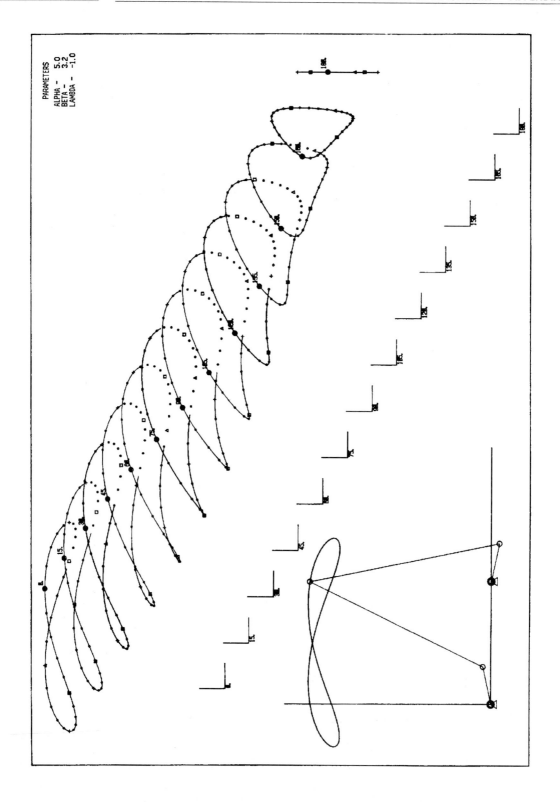

PARAMETERS
ALPHA – 5.0
BETA – 3.2
LAMBDA – –1.0

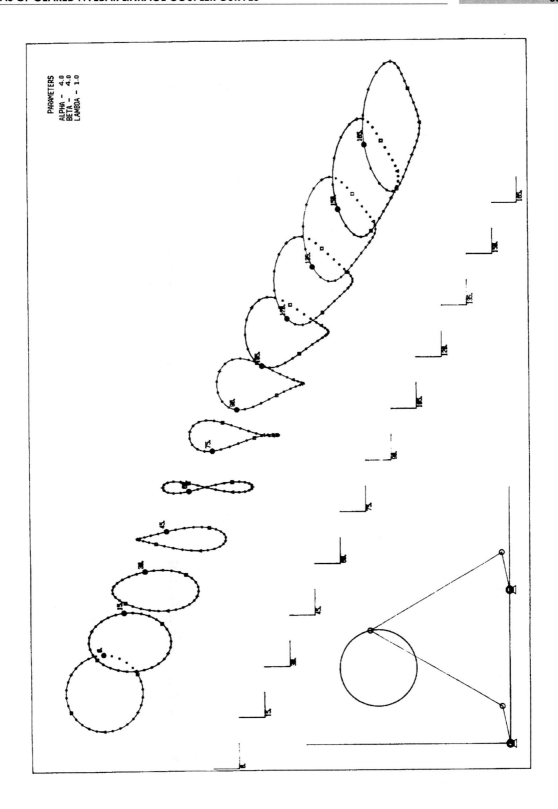

PARAMETERS
ALPHA - 4.0
BETA - 4.0
LAMBDA - 1.0

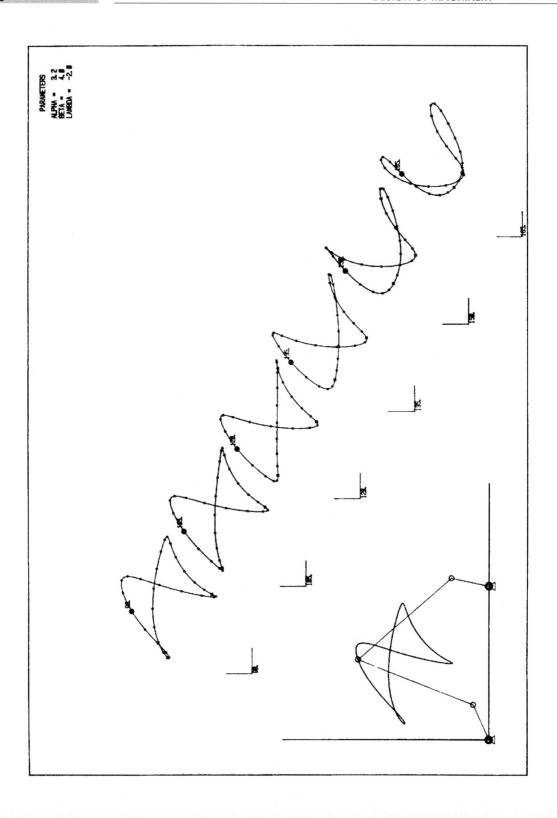

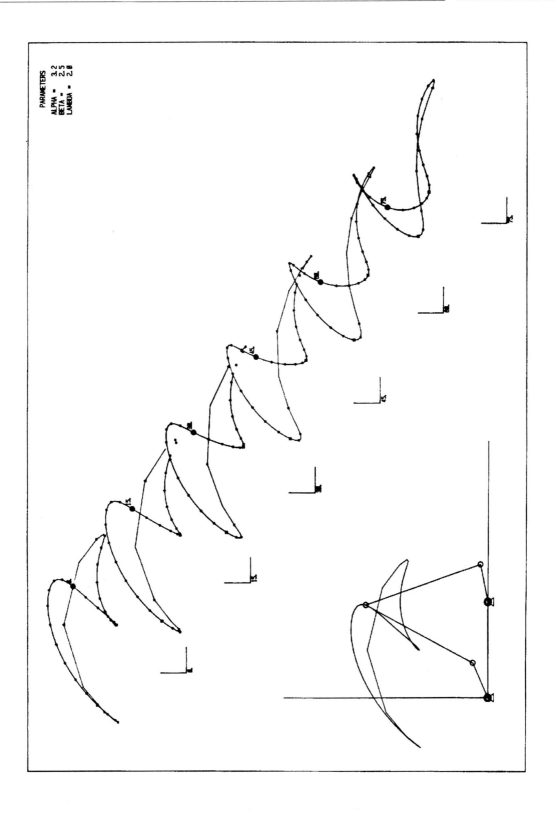

PARAMETERS
ALPHA = 3.2
BETA = 2.5
LAMBDA = 2.8

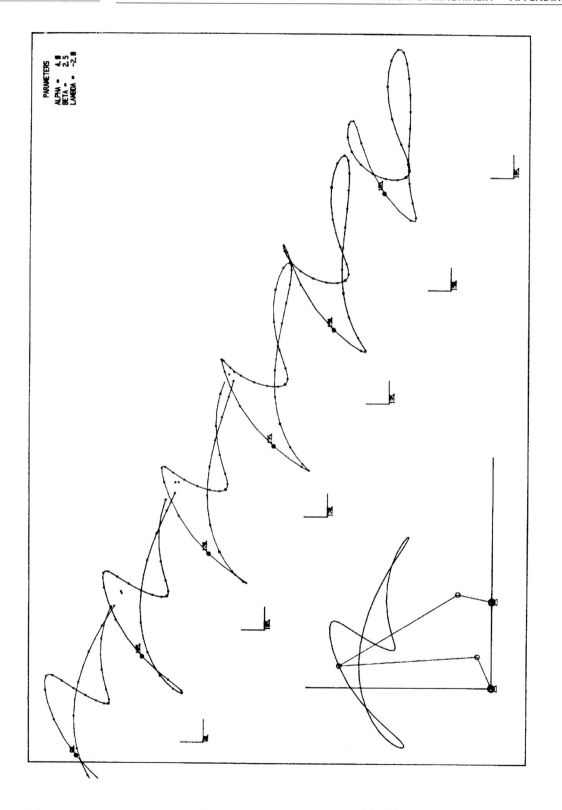

PARAMETERS
ALPHA = 4.0
BETA = 2.5
LAMBDA = -2.0

ANSWERS TO SELECTED PROBLEMS

CHAPTER 2

2-1

a. 1, b. 1, c. 2, d. 4, e. 7, f. 4, g. 4, h. 4, i. 4, j. 2, k. 1, l. 1, m. 2, n. 2, o. 4, p. as many as it has sections less one, q. 3.

2-5

a. 6, b. 6, c. 3, d. 3, e. 10, f. 3

2-7

a. pure rotation
b. complex planar motion
c. pure translation
d. pure translation
e. pure rotation
f. complex planar motion
g. pure translation
h. pure translation
i. complex planar motion

2-8

a. 1, b. 3, c. 3, d. 3, e. 1

2-9 force closed

2-10

a. structure - $DOF = 0$.
b. mechanism - $DOF = 1$
c. mechanism - $DOF = 1$
d. mechanism - $DOF = 3$

2-15

a. Grashof   b. non - Grashof   c. special case Grashof

CHAPTER 3

3-1

     a.   Path generation
     b.   Motion generation
     c.   Function generation
     d.   Path generation
     e.   Path generation

Note that synthesis problems have many valid solutions. We cannot provide a "right answer" to all of these design problems. Check your solution with a cardboard model and/or by putting it into one of the programs supplied with the text.

3-3   One possible solution:

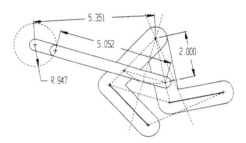

3-5   One possible solution:

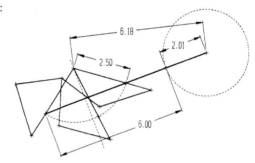

3-6   A unique solution:

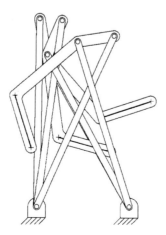
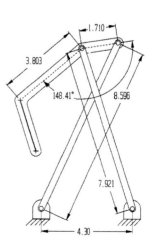

3-8 One possible solution:

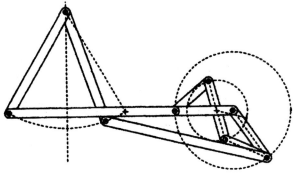

3-10 The solution using Fig 3-17: (Use program FOURBAR to check your solution.)

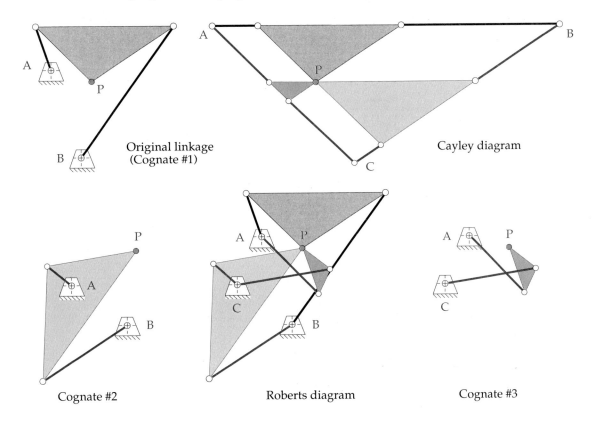

Original linkage
(Cognate #1)

Cayley diagram

Cognate #2

Roberts diagram

Cognate #3

SOLUTION TO PROBLEM 3-10

Finding the cognates of a fourbar linkage

3-11 Use program FOURBAR to check your solution.

TABLE S4-1   Solution for Problems 4-6 and 4-7

Row	θ_3 open	θ_4 open	Trans Ang	θ_3 crossed	θ_4 crossed	Trans Ang
a	88.8	117.3	28.4	−115.2	−143.6	28.4
c	−53.1	16.5	69.6	173.3	103.6	69.6
e	7.5	78.2	70.7	− 79.0	− 149.7	70.7
g	−16.3	7.2	23.5	155.7	132.2	23.5
i	−1.5	103.1	75.4	−113.5	141.8	75.4
k	−13.2	31.9	45.2	−102.1	−147.3	45.2
m	−3.5	35.9	39.4	−96.5	−135.9	39.4

TABLE S4-2   Solution for Problems 4-9 and 4-10

Row	θ_3 open	slider open	θ_3 crossed	slider crossed
a	180.1	4.99	−0.43	−3.0
c	205.9	9.8	−25.9	−4.6
e	175	16.4	4.2	−23.5
g	212.7	27.1	−32.7	−14.9

TABLE S4-3   Solution for Problems 4-11 and 4-12

Row	θ_3 open	θ_4 open	R_B open	θ_3 crossed	θ_4 crossed	R_B crossed
a	232.7	142.7	1.79	−79	−169	−1.79
c	91.4	46.4	2.72	208.7	163.7	−11.2
e	158.2	128.2	6.17	-36.2	-66.2	-9.63

TABLE S4-4   Solution for Problems 4-16 and 4-17

Row	θ_3 open	θ_4 open	θ_3 crossed	θ_4 crossed
a	173.6	-177.7	−115.2	-124
c	17.6	64	-133.7	180
e	-164	-94.4	111.2	41.6
g	44.2	124.4	-69.1	-149.3
i	37.1	120.2	-67.4	-150.5

CHAPTER 4

CHAPTER 5

5-8 Given: $\alpha_2 = -62.5°$, $P_{21} = 2.47$, $\delta_2 = 120°$

For left dyad: Assume: $z = 1.075$, $\phi = 204°$ $\beta_2 = -27°$

Calculate: $\mathbf{W} = 3.67 \,@\, -113.5°$

For right dyad: Assume: $s = 1.24$, $\psi = 74°$ $\gamma_2 = -40°$

Calculate: $\mathbf{U} = 5.46 \,@\, -125.6°$

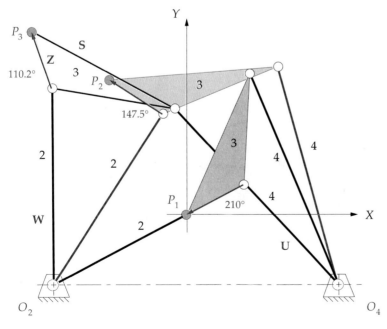

SOLUTION TO PROBLEM 5-11

Read diskfile P5-11 into program FOURBAR for linkage dimensions and animation

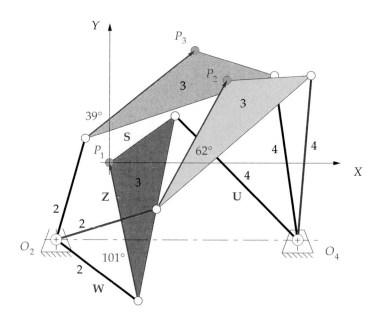

SOLUTION TO PROBLEM 5-15

Read diskfile P5-15 into program FOURBAR for dimensions and animation

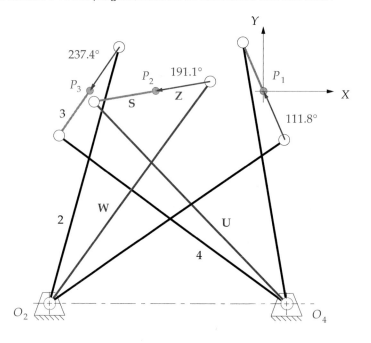

SOLUTION TO PROBLEM 5-19

Read diskfile P5-19 into program FOURBAR for dimensions and animation

5-11 Link 1 = 4.35,   Link 2 = 3.39,   Link 3 = 1.94,   Link 4 = 3.87

5-15 Link 1 = 3.95,   Link 2 = 1.68,   Link 3 = 3.05,   Link 4 = 0.89

5-19 Link 1 = 2,   Link 2 = 2.5,   Link 3 = 1,   Link 4 = 2.5

CHAPTER 6

6-4 and **6-5**   See Table S 6-1 and program FOURBAR

6-6 and **6-7**   See Table S 6-2

6-8 and **6-9**   See Table S 6-3

6-10 and **6-11**   See Table S 6-4 and program FIVEBAR

CHAPTER 7

7-3 and **7-4**   See Table S 7-1 and program FOURBAR

7-5 and **7-6**   See Table S 7-2

7-7 and **7-8**   See Table S 7-3

7-9   See Table S 7-4 and program FIVEBAR

CHAPTER 9

All the problems in this cam chapter are design problems with more that one correct solution. Use program DYNACAM to explore various solutions and compare them to find the best one for the constraints given in each problem.

CHAPTER 10

10-1 pitch diameter = 5.5, circular pitch = 0.785, addendum = 0.25, dedendum = 0.313, tooth thickness = 0.393, and clearance = 0.063.

10-5 a. $p_d = 4$,   b. $p_d = 2.67$

10-6 Assume a minimum no. of teeth = 14 then: pinion = 14t and 1.84″ pitch dia. Gear = 126t and 16.58″ pitch dia.

10-7 Assume a minimum no. of teeth = 14 then: pinion = 14t and 2.33″ pitch dia. Gear = 112t and 18.67″ pitch dia. An idler gear of any dia is needed to get the positive ratio.

10-10 Two stages of 8:1 and 8.75:1 give 70:1. Using 14 teeth minimum gives stage 1 = 14t (1.4 dia.) to 112t (dia. 11.2 dia.). Stage 2 = 16t (1.6 dia.) to 140t (14 dia)

10-12 The square root of 150 is > 10 so need three stages. 5 x 5 x 6 = 150. Using a min no. of teeth = 15 gives 15:75, 15:75 and 15:90 teeth. Pitch dias. are 2.5, 12.5 and 15″.

10-14 The factors 5 x 6 = 30. The ratios 14:70 and 12:72 revert to same center distance of 4.2. Pitch dias. are 1.4, 1.2, 7 and 7.2.

TABLE S6-1   Solution for Problems 6-4 and 6-5

Row	ω_3 open	ω_4 open	V_P Mag	V_P Ang	ω_3 crossed	ω_4 crossed	V_P Mag	V_P Ang
a	−6	−4	40.8	58.2	−.66	−2.66	22	129.4
c	−12.7	−19.8	273.8	−53.3	−22.7	−15.7	119.1	199.9
e	1.85	−40.8	260.5	−12.1	−23.3	19.3	139.9	42
g	76.4	146.8	798.4	92.9	239	168.6	1435.3	153.9
i	−25.3	25.6	103.1	−13.4	56.9	6.0	476.5	70.4
k	−56.2	−94.8	436.0	−77.4	−55.6	−16.9	362.7	79.3
m	18.3	83	680.8	149.2	7.73	−57	571.3	133.5

TABLE S6-2   Solution for Problems 6-6 and 6-7

Row	V_A mag	V_A ang	ω_3 open	V_B mag open	ω_3 crossed	V_B mag crossed
a	14	135	−2.47	−9.87	2.47	−9.92
c	45	−120	5.42	−41.5	−5.42	−3.54
e	250	135	−8.86	−189.7	8.86	−163.8
g	700	60	−28.8	738.9	28.8	−38.9

TABLE S6-3   Solution for Problems 6-8 and 6-9

Row	ω_3 open	ω_4 open	V_{slip} open	ω_3 crossed	ω_4 crossed	V_{slip} crossed
a	−10.3	−10.3	33.46	3.64	3.64	−33.46
c	23.75	23.75	73	33	33	−73
e	−2.7	−2.7	−176	−8.24	−8.24	176

TABLE S6-4   Solution for Problems 6-10 and 6-11

Row	ω_3 open	ω_4 open	ω_3 crossed	ω_4 crossed
a	32.6	16.9	−75.2	−59.6
c	10.7	−2.6	−8.2	5.1
e	−158.3	−81.3	−116.8	−193.9
g	−8.9	−40.9	−48.5	−16.5
i	−40.1	47.9	59.6	−28.4

TABLE S7-1 Solution for Problems 7-3 and 7-4

Row	α_3 open	α_4 open	A_P Mag	A_P Ang	α_3 crossed	α_4 crossed	A_P Mag	A_P Ang
a	26.1	53.3	419	240.4	77.9	50.7	298	−11.3
c	−154.4	−71.6	4400	238.9	−65.2	−148.0	3554	100.6
e	331.9	275.6	10260	264.8	1287.7	1344.1	19340	−65.5
g	−23510	−19783	172688	191	−43709	−47436	273634	−63
i	−344.6	505.3	9492	−81.1	121.9	−728.0	27871	150
k	−2693	−4054	56271	220.2	311.0	1672.1	27759	−39.1
m	680.8	149.2	35149	261.5	9266.1	10303.0	63831	103.9

TABLE S7-2 Solution for Problems 7-5 and 7-6

Row	A_A mag	A_A ang	α_3 open	A_B mag open	A_B ang open	α_3 cross	A_B mag cross	A_B ang cross
a	140	−135	24.8	123.7	180	−24.8	74.3	180
c	676	153	−29.0	709.1	180	29.0	490.0	180
e	12500	45	−447.1	6652.7	0	447.1	11095.7	0
g	70000	150	−1136.0	62687.9	180	1136.0	58429.6	180

TABLE S7-3 Solution for Problems 7-7 and 7-8

Row	α_3 open	α_4 open	A_{slip} open	α_3 crossed	α_4 crossed	A_{slip} crossed
a	130.5	130.5	−3111.7	−9.9	−9.9	1125.2
c	−212.9	−212.9	3327.9	−217.8	−217.8	6776.4
e	896.3	896.3	2020.1	595.6	595.6	−1044.4

TABLE S7-4 Solution for Problem 7-9

Row	α_3 open	α_4 open	α_3 crossed	α_4 crossed
a	3191	2492	−6648	−5949
c	314	228	87.2	147
e	2171	−6524	7781	5414
g	−22064	−23717	−5529	−29133
i	−5697	−3380	−2593	−7184

10-16   The factors 7.5 x 10 = 75. The ratios 22:165 and 17:170 revert to same center distance of 6.234.   Pitch dias. are 1.467, 1.133, 11 and 11.33.

10-19   The factors 2 x 1.5 = 3. The ratios 15:30 and 18:27 revert to the same center distance of 3.75.   Pitch dias. are 2.5, 5, 3 and 4.5. The reverse train uses the same 1:2 first stage as the forward train, so needs a second stage of 1:2.25 which is obtained with a 12:27 gearset. The center distance of the 12:27 reverse stage is 3.25 which is less than that of the forward stage. This allows the reverse gears to engage through an idler of any suitable diameter to reverse output direction.

10-21   For the low speed of 6:1, the factors 2.333 x 2.571 = 6. The ratios 15:35 and 14:36 revert to the same center distance of 3.125. Pitch dias. are 1.875, 4.375, 1.75 and 4.5. The second speed train uses the same 1:2.333 first stage as the low speed train, so needs a second stage of 1:1.5 which is obtained with a 20:30 gearset which reverts to the same center distance of 3.125.   The additional pitch dias. are 2.5 and 3.75. The reverse train also uses the same 1:2.333 first stage as both forward trains, so needs a second stage of 1:1.714 which is obtained with a 14:24 gearset. The center distance of the 14:24 reverse stage is 2.375 which is less than that of the forward stages. This allows the reverse gears to engage through an idler of any suitable diameter to reverse output direction.

10-25   a. $\omega_2 = 790$,   c. $\omega_{arm} = -4.544$,   e. $\omega_6 = -61.98$

10-26   a. $\omega_2 = -59$,   c. $\omega_{arm} = 61.54$,   e. $\omega_6 = -63.33$

10-27   a. 560.2 rpm and 3.57 to 1,   b. $x = 560.2 * 2 - 800 = 320.4$ rpm

CHAPTER 11

11-1   CG @ 7.89 in. from handle end, $I_{zz} = 0.1873$ in-lb-sec^2, $k = 8.39$ in.

11-2   CG @ 6.74 in. from handle end, $I_{zz} = 0.0918$ in-lb-sec^2, $k = 7.38$ in.

11-4

a. $x =$ 3.547, $y = 4.8835$, $z = 1.4308$, $w = -1.3341$
b. $x = -62.029$, $y = 0.2353$, $z = 17.897$, $w = 24.397$

CHAPTER 12

12-3   **12-4** and **12-7**   Input diskfile P12-3x (where x is the row letter) into program DYNAFOUR to check your solution.

12-5   and **12-6** Input diskfile P12-5x (where x is the row letter) into program DYNAFOUR to check your solution.

CHAPTER 13

13-1

a. $m_b r_b = 0.934$, $\theta_b = -75.5°$
c. $m_b r_b = 5.932$, $\theta_b = 152.3°$
e. $m_b r_b = 7.448$, $\theta_b = -80.76°$

13-5

a.	$m_a r_a = 0.814,$	$\theta_a = -175.2°,$	$m_b r_b = 5.50,$	$\theta_b = 152.1°$
c.	$m_a r_a = 7.482,$	$\theta_a = -154.4°,$	$m_b r_b = 7.993,$	$\theta_b = 176.3°$
e.	$m_a r_a = 6.254,$	$\theta_a = -84.5°,$	$m_b r_b = 3.671,$	$\theta_b = -73.9°$

13-6 $m_a = 0.0926,$ $\quad \theta_a = -138.6°,$ $\quad m_b = 0.0801,$ $\quad \theta_b = 43.5°$

13-7 $m_a = 0.0947,$ $\quad \theta_a = 97.21°,$ $\quad m_b = 0.0942,$ $\quad \theta_b = -67.99°$

13-8 Input diskfile P12-5x (where x is the row letter) into program DYNAFOUR to check your solution. Use the program to calculate the flywheel data.

CHAPTER 14

14-1 Exact solution = – 45,013.39 in/sec @ 300° and 200 rad/sec

Fourier series approximation = – 45,000

Error = 0.03% (0.0003)

14-3 Gas torque = 2039.5 (approx.), $\qquad$ Gas force = 3141.6

14-5 Gas torque = 2039.5 (approx.), $\qquad$ Gas torque = 2025.7 (exact)

$$\text{Error} = 0.68\% \quad (0.0068)$$

14-7

a. $m_b = 0.00748$ at $l_b = 7.2,$ $\qquad$ $m_p = 0.01251$ at $l_p = 4.31$
b. $m_b = 0.00800$ at $l_b = 7.2,$ $\qquad$ $m_a = 0.01200$ at $l_a = 4.80$
c. $I_{model} = 0.6912$ $\qquad\qquad$ Error = 11.48% (0.1148)

14-9 $m_{2a} = 0.018$ at $r_a = 3.5,$ $\qquad I_{model} = 0.2205,$ $\quad$ Error = –26.5% (–0.265)

14-11 to **14-14** and **14-19** to **14-22** Use program ENGINE to check your solutions.

CHAPTER 15

Use program ENGINE to check your solutions.

CHAPTER 17

17-1

a.	In series:	$k_{eff} = 3.09,$	Softer spring dominates
b.	In parallel:	$k_{eff} = 37.4,$	Stiffer spring dominates

17-4

a.	In series:	$c_{eff} = 1.09,$	Softer damper dominates
b.	In parallel:	$c_{eff} = 13.7,$	Stiffer damper dominates

17-7 $k_{eff} = 4.667,$ $\qquad m_{eff} = 0.278$

17-11 to **17-15** $\qquad$ Use program DYNACAM to solve these problems.

17-16 See Table S 17-1

TABLE S 17-1

Solutions to Prob. 17-16

	ω_n	ω_d	c_c
a.	3.42	3.38	8.2
b.	4.68	4.65	19.7
c.	0.26	0.26	15.5
d.	2.36	2.33	21.2
e.	5.18	5.02	29.0
f.	2.04	1.96	49.0

INDEX

ALSO AVAILABLE FROM MCGRAW-HILL

Schaum's Outline Series in Mechanical Engineering

Most outlines include basic theory, definitions, and hundreds of solved problems and supplementary problems with answers.

TITLES ON THE CURRENT LIST INCLUDE:

Acoustics
Basic Equations of Engineering
Continuum Mechanics
Engineering Economics
Engineering Mechanics, 4th edition
Fluid Dynamics, 2nd edition
Fluid Dynamics & Hydraulics, 2nd edition
Heat Transfer
Introduction to Engineering Calculations
Lagrangian Dynamics
Machine Design
Mathematical Handbook of Formulas and Tables
Mechanical Vibrations
Operations Research
Statics & Mechanics of Materials
Strength of Materials, 2nd edition
Theoretical Mechanics
Thermodynamics, 2nd edition

Schaum's Solved Problems Books

Each title in this series is a complete and expert source of solved problems containing thousands of problems with worked out solutions.

RELATED TITLES ON THE CURRENT LIST INCLUDE:

3000 *Solved Problems in Calculus*
2500 *Solved Problems in Differential Equations*
2500 *Solved Problems in Fluid Mechanics and Hydraulics*
1000 *Solved Problems in Heat Transfer*
3000 *Solved Problems in Linear Algebra*
2000 *Solved Problems in Calculus*
2000 *Solved Problems in Mechanical Engineering Thermodynamics*
700 *Solved Problems in Vector Mechanics for Engineers: Dynamics*
800 *Solved Problems in Vector Mechanics for Engineers: Statics*

Available at your College Bookstore. A complete list of Schaum titles may be obtained by writing to: Schaum Division, McGraw-Hill Inc., Princeton Road, S-1, Hightstown, NJ 08520

PROGRAM DISK INSTRUCTIONS

TO RUN ANY PROGRAM FROM THIS DISK - JUST TYPE ITS NAME - ONE OF:

FOURBAR, FIVEBAR, SIXBAR, DYNAFOUR, DYNACAM, ENGINE or **MATRIX.**

These programs require an IBM™ compatible PC with 640k of memory and CGA, EGA, or VGA graphics card emulation. A math coprocessor will greatly speed the computations but is not required. Limited help is available within the programs and complete explanations of, and manuals for these programs can be found in Chapters 8, 9, 12, and 16 of this text. The programs are provided on a 1.4 megabyte MS DOS disk and are compressed. They can be run directly from the original floppy or from any copy of it made with the DOS **DISKCOPY** command. At least 256k of free disk space is needed on the disk to expand and run each program from the original floppy or a copy of it. (You should store any data files which you create with these programs on a different disk in order to preserve free space on the program disk.) Several batch files are provided to transfer the files to other media:

HARD DISK INSTALLATION To expand and install these compressed programs from drive **A:** or from drive **B:** onto the hard disk **C:**, type **AINSTALL** or **BINSTALL** (depending on which drive the 3.5" original disk is in) The programs collectively require 1.9 megabytes of hard-disk space after expansion.

EXPAND AND COPY TO (8) 5.25" DISKS To copy the programs from the supplied disk onto (8), pre-formatted 360k or 1.2 meg, 5.25" floppy disks, type **COPYA2B** or **COPYB2A** (depending on which drive the 3.5" original disk is in) and follow the prompts.

COPY EXPANDED FILES FROM HARD DISK TO FLOPPIES Once either INSTALL.BAT program has been used, more floppies can be made from the hard disk with:

- Type **COPYC2A** to download the expanded programs from **drive C:** to (8) 5.25" floppies in **drive A:**
- Type **COPYC2B** to download the expanded programs from **drive C:** to (8) 5.25" floppies in **drive B:**
- You may also copy any of the programs to high-density floppy disks with the DOS **COPY** command (after they have been expanded with one of the installer routines listed above).

DUPLICATE THE ORIGINAL DISK Use DISKCOPY to make a duplicate backup disk of the compressed files on one 3.5" 1.4 megabyte floppy disk.

The batch files also create, for each program, a subdirectory in which the program and its data files are stored. These subdirectories are put at the root on a floppy disk. Either INSTALL.BAT routine creates a subdirectory called C:\DOM\ on the hard disk in which the subdirectories are placed and various files used by all the programs are also put.

Note:

- You can run any batch file by typing its name. The .BAT extension is not needed, i.e., type AINSTALL to run AINSTALL.BAT.
- If the programs will not print your plots with SHIFT PRTSCRN, you probably do not have the MS DOS GRAPHICS program loaded. Return to DOS and type **GRAPHICS**. Then rerun the program as instructed above.
- If your own personal computer does not have a high-density 3.5" (1.4 megabyte) disk drive, you will need to take this disk to your college computer center or to a computer store and have them transfer the contents to other media as described above.